Scientific Basis for Nuclear
Waste Management V

Scientific Basis for Nuclear Waste Management V

Proceedings of the Materials Research Society Fifth International Symposium on the Scientific Basis for Nuclear Waste Management, held June 7-10, 1982, in Berlin, Germany

EDITOR:

Werner Lutze

Hahn-Meitner-Institut, Berlin, Germany

NORTH-HOLLAND

NEW YORK · AMSTERDAM · OXFORD

Published by:

Elsevier Science Publishing Company, Inc.
52 Vanderbilt Avenue, New York, New York, 10017

Sole distributors outside the USA and Canada:

Elsevier Science Publishers B.V.
P.O. Box 211, 1000 AE Amsterdam, The Netherlands

Library of Congress Cataloging in Publication Data

International Symposium on the Scientific Basis for Nuclear Waste Management
(5th: 1982: Berlin, Germany)

Scientific basis for nuclear waste management V.
(Materials Research Society symposia proceedings; v. 11)

Includes indexes.
1. Radioactive waste disposal—Congresses. I. Lutze, Werner. II. Materials
Research Society. III. Title. IV. Series.
TD898.I57 1982   621.48'38   82-18386
ISBN 0-444-00725-3

Manufactured in the United States of America

Contents

PROPERTIES OF HIGH-LEVEL RADIOACTIVE WASTE FORMS

GLASS

Reaction with Aqueous Media

Mechanical Properties, Crystallization

Composition, Structure, Quality

CERAMICS

Igneous and Metamorphic Rocks

*invited papers

Preface

The fifth International Symposium on the Scientific Basis for Nuclear Waste
Management was held in Berlin (West), Germany, on June 7-10, 1982. It was the
first symposium in this series held by the Materials Research Society in Europe.
The purpose of the symposium was to provide an interdisciplinary forum for the
discussion of scientific work dealing with radioactive waste.

The scope of the symposium was guided by the program and steering
committees. The members of these committees included:

Program Committee

W. Lutze (Chairman), Hahn-Meitner-Institut, Berlin, Germany
R. C. Ewing, University of New Mexico, Albuquerque, USA
J. Hamstra, Energieonderzoek Centrum Nederland, Petten, The Netherlands
L. L. Hench, University of Florida, Gainesville, USA
J. A. C. Marples, Atomic Energy Research Establishment, Harwell, England
L. B. Nilsson, Svensk Kärnbränsleförsörjning, Stockholm, Sweden
C. Sombret, Commissariat à l'Energie Atomique, Marcoule, France
R. P. Turcotte, Pacific Northwest Laboratories, Richland, USA

Steering Committee

R. Bonniaud, CEA Marcoule, France
D. Bünemann, GKSS Geesthacht, Germany
H. R. Franzen, CNEN, Brazil
A. Gandellini, Agip Nucleare, Italy
J. A. Goedkoop, E.C.N., Petten, The Netherlands
O. Heinonen, Technical Research Centre, Finland
H. D. Holland, Harvard University, USA
H. W. Levi, GSF Neuherberg, Germany
P. Macedo, Catholic University, Washington D.C., USA
E. Merz, KFA Jülich, Germany
G. Oertel, Department of Energy, USA
N. S. S. Rajan, Bhabha Atomic Research Centre, India
K. D. Reeve, Australian Atomic Energy Commission
A. E. Ringwood, Australian National University
E. L. J. Rosinger, Atomic Energy of Canada Limited

R. Roy, Pennsylvania State University, USA
R. Simon, CEC Brussels, Belgium
T. Westermark, KTH Stockholm, Sweden
P. Zühlke, WAK, Germany

Contributed papers on the science underlying the following subjects were especially solicited: Waste forms (vitreous, ceramic, spent fuel, cement, development and processing); waste isolation (container materials, backfill, repository); nuclide migration, modeling and safety assessment. High-level radioactive waste was emphasized in this symposium.

The program committee selected 98 contributions. There were 129 abstracts submitted (48 on vitreous, 19 on crystalline materials, 40 on waste isolation, 22 on nuclide migration). The symposium featured two special sessions: "In-situ Solidification of Low- and Medium-Level Waste" and a panel discussion on "Research Priorities for Nuclear Waste Isolation".

In addition to members of the program and steering committee the following scientists served as session chairmen:

P. Offermann, J. A. Goedkoop, T. Westermark, J. E. Mendel,
G. Malow, H. C. Burkholder, H. W. Levi, J. Neretnieks,
J. D. Mather, B. Grundfelt, D. M. Roy.

The papers presented in these proceedings represent work from 56 institutions from 14 countries. This volume contains all 92 papers presented at the symposium. All papers have been reviewed prior to their inclusion in this volume.

The summary of the panel discussion on research priorities was sent to the panel members and their comments included by the editor.

The preparation of these proceedings was made possible by support of the Hahn-Meitner-Institut für Kernforschung Berlin GmbH and the Department of Energy, Washington, USA.

Werner Lutze
Hahn-Meitner-Institut für
Kernforschung Berlin GmbH
Berlin 39
Germany
July, 1982

THE SCIENTIFIC BASIS FOR NUCLEAR WASTE MANAGEMENT V

MRS SYMPOSIA PROCEEDINGS VOLUME 11

EDITED BY:

Werner Lutze
Hahn-Meitner-Institut,
Berlin, Germany

ASSOCIATE EDITORS:

Rodney C. Ewing
University of New Mexico,
Albuquerque, New Mexico,
U.S.A.

J. Angwin C. Marples
Atomic Energy Research Establishment,
Harwell, England

Jan Hamstra
Energieonderzoek Centrum
Nederland, Petten,
The Netherlands

Lars B. Nilsson
Svensk Kärnbränsleförsörjning,
Stockholm, Sweden

Larry L. Hench
University of Florida,
Gainesville, Florida,
U.S.A.

Claude Sombret
Commissariat à l'Energie Atomique,
Marcoule, France

Rheal P. Turcotte
Pacific Northwest Laboratories,
Richland, Washington,
U.S.A.

EDITORIAL ASSISTANT:

Helga Fuchs
Hahn-Meitner-Institut,
Berlin, Germany

PROPERTIES OF HIGH-LEVEL RADIOACTIVE WASTE FORMS

GLASS

Reaction with Aqueous Media

SCIENTIFIC BASIS FOR RADIOACTIVE WASTE MANAGEMENT - V
Werner.Lutze, editor

EFFECTS OF WASTE COMPOSITION AND LOADING ON THE CHEMICAL DURABILITY OF A BOROSILICATE GLASS

D.E. CLARK, C.A. MAURER, A.R. JURGENSEN AND L. URWONGSE
Department of Materials Science and Engineering, College of Engineering,
University of Florida, Gainesville, Florida 32611, USA.

ABSTRACT

The effects of waste composition and percent loading in a borosilicate glass designed for US defense high level wastes (HLW) have been evaluated. Three types of simulated wastes were investigated; high alumina, high iron and a composite representative of an average waste composition from Savannah River Plant (SRP) waste tanks. Corrosion resistance of the borosilicate glass is significantly enhanced by the presence of any of the three types of wastes. Additionally, corrosion resistance is improved as the % waste loading is increased in the glass. The best corrosion performance was obtained with the high alumina waste in deionized water.

INTRODUCTION

During the last several years a large number of glass compositions has been evaluated for encapsulating high level wastes (HLW) from both commercial nuclear power plants and defense applications. Two important variables in the vitrification process are the composition and the percentage of the waste in the base glass. Both of these may affect the processing and will alter physical and chemical properties of the waste form.

The single most important property of the waste form will be its ability to minimize the release of hazardous radionuclides during the period when the radiation level is siginificant (ie, several thousand years after burial). The most likely form of release will occur by contact of the waste form with groundwater resulting in a leaching of the radionuclides. Previous investigators[1,2] have shown that leaching of simple (ie, two and three component glasses) glasses in contact with aqueous solution can occur via two mechanisms; ion exchange and network dissolution. In ion exchange protons (or hydronium ions) from the solution exchange with mobile species from the glass resulting in the development of a surface film.

Network dissolution may also occur if the solution pH is sufficiently high[3], and this process results in the removal of the species in the same proportions as they are present in the glass. Even in simple systems additional processes may become important if the solution in contact with the glass contains species other than H^+ and OH^- or becomes concentrated (or saturated) with respect to any of the species from the glass. As the solution approaches saturation the leach rate due to both ion exchange and network dissolution will decrease, and precipitation from solution will occur when solubility limits are exceeded[4]. Thus, the ratio of the surface area of the glass to volume of solution (SA/V) is an important leaching variable. High SA/V's will result in a rapid approach towards solution saturation and equilibrium while small SA/V's will require longer times.

The nuclear waste glasses exhibit more complicated behavior than do the simple glasses and consequently their leaching mechanisms and kinetics are more difficult to characterize[5]. In addition to ion exchange and network dissolution, complex surface films may form during leaching. The formation, stability and protectiveness of these films depend on the composition of the waste form, solution pH and chemistry and SA/V.

EXPERIMENTAL

The glass frit and simulated wastes used in this study were supplied by Savannah River Laboratory (SRL). Their compositions are shown in Table 1.

TABLE 1
COMPOSITION OF GLASS FRIT AND SIMULATED WASTES (WT%)

Glass Frit Composition		Simulated Waste Compositions			
			Waste Type		
Component		Component	High Al_2O_3	TDS-3A	High Fe
SiO_2	57.9	Fe_2O_3	13.8	47.3	59.1
B_2O_3	14.7	MnO_2	11.3	13.6	4.0
Na_2O	17.7	Zeolite*	10.2	10.2	9.7
Li_2O	5.7	Al_2O_3	49.3	9.5	1.4
TiO_2	1.0	NiO	2.0	5.8	10.1
MgO	2.0	SiO_2	4.5	4.1	2.9
La_2O_3	0.5	CaO	0.9	3.5	4.0
ZrO_2	0.5	Na_2O	5.0	3.1	5.9
		Coal	2.3	2.3	2.1
		Na_2SO_4	0.7	0.6	0.1

*Linde Ion-Siv IE-95

The three types of simulated wastes investigated were high alumina, high iron and a composite (TDS-3A) representative of an average waste composition from Savannah River Plant (SRP) waste tanks. These wastes in the form of dried powders were mixed with the alkali borosilicate glass frit shown in Table I and the mixtures were preheated to $900^{\circ}C$ for 1 hour and then homogenized at $1150^{\circ}C$ for 3.5 hours in a platinum crucible. Glasses containing up to approximately 30% wastes were prepared from all three wastes. Bars with dimensions of 1 cm^2 x 10 cm were cast into a graphite mold, annealed at $450^{\circ}C$ for 5 hours and furnace cooled. Corrosion specimens 0.2 cm thick were sliced from the bars using a diamond impregnated slow speed wafering blade, polished through 600 grit on all six sides with SiC paper and cleaned with acetone. This procedure provided an equivalent surface finish on all samples prior to corrosion. A modified MCC-1 static leach test[6] was used to evaluate the corrosion resistance of the glasses. Polyethylene containers were used instead of the PFA Teflon containers recommended by the MCC. Two samples were run for each matrix point. The majority of the samples were corroded in deionized (D.I) water (initial pH=5.6) but pH buffered solutions, simulated ground water and brine solution were also evaluated. Both solution and surface analyses were performed after corrosion. A detailed discussion of glass and sample preparation, corrosion procedures, solution and surface analyses is given elsewhere[7].

RESULTS AND DISCUSSION

Leach data are shown in Table 2 for the unloaded base glass and for the same glass with three different types of wastes and two levels of loading. The normalized leach rates for the individual elements were calculated using the methods specified by MCC[6]. The zeros in the table indicate that the leach rate was less than 0.05 g/m^2-d for that species, and all other values correspond to averages of at least two samples run under identical conditions. Although not shown in the table, Al, Fe, Mn, Ni, Ti and Ca were searched for but in nearly all solutions their concentrations were below the detection limits of our instrument. These elements either remained on the glass surface during leaching or were precipitated as fast as they were leached.

4

When the normalized leach rates are equal for all elements the glass
is dissolving congruently. The absence of many species in solution,
particularly the major ones such as Al and Fe, indicates that incongruent
dissolution (ie, selective leaching) was the major mechanism of leaching.
This type of leaching usually results in the development of a surface
enriched in those elements not found in solution.

In general the leaching rate of all elements decreased as the temperature
was decreased. Figure 1 summarizes the normalized leaching rates for Si
at 40°C, 70°C and 90°C after 4 weeks for the glasses listed in Table 1.
Si could not be detected in the solutions in contact with the glass
containing 28.7% high Al_2O_3 waste after 4 weeks at either 40°C or 70°C.

With the exception of the unloaded base glass, the leach rate for all
glasses decreased with corrosion time in D.I. water. Additionally the pH
of the solutions, with the exception of the ones in contact with the SRL
131 + 29.8% TDS glass, also increased with time. At equivalent temperatures
and times the rate of leaching decreased as the percentage waste was increased in the glass (up to 30%), and the glasses containing high Al_2O_3
waste were more resistant to leaching in D.I. water than were the ones
with the composite and high iron wastes. Similar results have been
obtained by G. Wicks and coworkers at SRL[8] with TDS-waste. Observe that
the unloaded glass experienced much larger weight losses than did the
ones containing the wastes. Of the three types of wastes investigated the
glasses containing the high iron waste exhibited the highest leach rates.

Table 2 also gives the results of corroding the SRL base glass in
simulated silicate (initial pH=7.6) and brine solutions (inital pH=6.0)[6].
The leach rates and WT losses were significantly decreased in the brine
solution but were not appreciably altered in the silicate solution
compared to deionized water. Corrosion studies involving simple alkali
silicate glasses in the same solutions gave similar results. The effectiveness of the brine solution in reducing the leach rate is thought to be
due to the formation of a $Mg(OH)_2$ precipitate on the glass surface[4]. The
formation of this precipitate provides a physical barrier against contact
of the glass surface and also prevents large increase in the solution pH,
both of which are beneficial in reducing the leach rate. The table
presents two sets of 28 day data for D.I. water corresponding to experiments conducted several months apart. The D.I. water data, silicate and
brine solution data that are grouped together in the table were obtained

at the same time. Na and Mg leach rates are not given for the brine solution because large concentrations ($>10^4$ ppm) were present in this solution before corrosion.

Severe surface flaking was observed on the SRL 131 base glass upon removal from solution and air drying. Other glasses, with the exception of the one containing 28.7% high Al_2O_3 waste, exhibited various extents of surface flaking upon drying at 110°C (MCC procedure). Surface film stability (during drying) appeared to improve on the glasses containing TDS-3A waste as the % waste in the glass was increased up to 30%.

TABLE 2
NORMALIZED LEACH RATES (g/m^2-d) AND OTHER RELEVANT DATA FOR GLASSES CORRODED IN DEIONIZED WATER, SIMULATED GROUND WATER[6] AND BRINE SOLUTION[6], 90°C, SA/V=0.1 cm^{-1}

Glass	Corr time(d)	Final pH	%WT loss	Si	B	$(g/m^2$-d) Na	Li	Mg
SRL 131-un-loaded	1	8.8	0.1	9.7	9.6	5.7	8.7	4.6
" " "	7	10.4	10.3	19.1	32.1	18.3	27.1	0.7
" " "	28	11.1	41.8	20.1	45.2	15.0	35.3	0.3
In silicate " water	28	11.2	27.0	14.2	28.3	32.6	30.5	0.7
In brine " solution	28	7.5	0.3	0.1	0.8	–	1.1	–
In D.I.water	28	11.0	30.0	18.3	30.3	25.9	23.6	0.3
SRL 131 + 15% High Al_2O_3 Waste	1	6.3	0.0	0.0	0.9	0.8	0.7	0.1
	7	7.8	0.1	0.1	0.5	0.4	0.5	0.1
	28	8.0	0.2	0.1	0.2	0.3	0.2	0.0
SRL 131 + 15% TDS-3A Waste	1	8.8	0.2	3.5	3.0	1.9	3.4	1.3
	7	8.6	0.5	1.8	1.3	0.9	1.6	0.4
	28	10.0	1.8	2.1	2.4	1.0	1.3	0.0
SRL 131 + 15% High Fe Waste	1	9.6	0.2	4.3	5.7	6.3	5.3	3.4
	7	9.7	1.1	3.5	5.0	4.6	4.9	0.1
	28	10.0	2.0	1.6	4.1	3.6	3.6	0.0
SRL 131 + 28.7% High Al_2O_3 Waste	1	6.9	0.0	0.0	0.3	0.3	0.2	0.0
	7	7.7	0.1	0.2	0.4	0.2	0.3	0.3
	28	7.2	0.2	0.1	0.1	0.1	0.1	0.1
SRL 131 + 29.8% TDS-3A Waste	1	6.8	0.1	1.6	0.0	1.3	1.3	0.6
	7	6.4	0.2	0.6	0.8	0.7	0.7	0.3
	28	6.3	0.2	0.2	0.2	0.2	0.2	0.1
SRL 131 + 29.8% High Fe Waste	1	9.4	0.2	4.6	4.1	4.2	4.2	2.4
	7	9.0	0.3	1.7	2.4	2.0	2.0	0.3
	28	8.6	0.6	0.7	1.0	0.9	0.9	0.2

The effect of glass surface area to solution volume ratio (SA/V) on leach rate is shown in Table 3.

TABLE 3
NORMALIZED LEACH RATES (g/m^2-d) AND OTHER RELEVANT DATA FOR SRL 131
CORRODED IN DEIONIZED WATER, $70^{\circ}C$, 1 WEEK

SA/V (cm^{-1})	Final pH	%WT loss	Si	B	(g/m^2-d) Na	Li	Ti	Mg
0.01	9.4	1.8	6.5	8.3	3.4	6.0	0	5.4
0.10	9.5	0.4	1.7	1.6	1.3	1.5	0	1.1
0.40	9.8	1.0	2.3	2.3	1.9	1.1	0.9	1.6
0.70	10.4	1.3	4.1	5.6	3.6	5.4	1.2	1.5
1.00	10.6	1.6	4.7	8.2	3.3	7.1	1.1	1.4

A minimum in the leach rates and %WT loss occurs at an SA/V=0.1 cm^{-1} for the SRL 131 base glass. However, two other unloaded alkali borosilicate glasses, SRL 21 and SRL 211, showed a gradual decrease in their leaching rates as the SA/V was increased from 0.01 to 1.0 cm^{-1}. The %WT loss for these glasses versus SA/V is illustrated in Figure 2. The decrease in leach rate with increased SA/V is thought to be due to solution solubility effects. However, the behavior exhibited by SRL 131 suggests that the influence of solution chemistry on leaching is complicated and apparently very sensivitve to pH. Most certainly the solubility of SiO_2 is altered as the pH is increased by ion exchange between the glass and solution. The effect of solution pH on leach rates is shown in Table 4 for SRL 131 + 30 WT% TDS.

TABLE 4
NORMALIZED LEACH RATES (g/m^2-d) AND OTHER RELEVANT DATA FOR SRL + 30 WT%
TDS-3A CORRODED IN VARIOUS SOLUTIONS, $90^{\circ}C$, 4 WEEKS

Solution	Initial pH	Final pH	%WT loss	Si	B	(g/m^2-d) Na	Li	Fe	Al
0.05 M THAM	10.5	9.2	0.4	0.3	0.3	0.5	0.5	0	0.2
10^{-4} M NaOH	10.0	9.6	0.4	0.3	0.3	–	0.5	0	0.2
10^{-3} M NaOH	11.0	10.1	0.5	0.3	0.4	–	0.7	0	0.2
10^{-3} M LiOH	11.0	10.5	0.9	0.6	1.0	0.7	–	0	0.4
10^{-2} M NaOH	12.0	11.6	2.3	1.6	1.1	–	0.7	0	1.1
0.02M Glycine sulfate	2.4	2.8	4.2	2.6	2.8	7.4	5.4		3.6
D.I. Water	5.6	6.3	0.2	0.2	0.2	0.2	0.2	0	0

Buffered solutions with pH's of 2.4, 5.9 and 10.4 were prepared from glycine sulfate, glycine and THAM, respectively[9]. High pH solutions were also prepared using NaOH and LiOH. In general, the pH's after corrosion were only slightly different from the initial pH's.The largest leach rates and WT losses occurred in the low pH solutions (ie, 2.4), and the smallest occurred in neutral solutions. The differences in leaching behavior were small between the THAM (pH=10.5) and the NaOH solution (pH=10.0 and pH=11.0). However, when the pH was increased to 12.0 using NaOH the leach rate increased significantly compared to the rate at pH=11.0 presumably due to increased solubility of both SiO_2 and Al_2O_3 in this pH range [10]. There were also some small differences in leach rates between the solution containing LiOH and the one containing NaOH although the pH was 11.0 for both solutions suggesting that the solution chemistry as well as pH is important. Table 5 summarizes the leach rates VS pH for several glasses.

TABLE 5
LEACH RATES BASED ON WT LOSS VS. pH FOR GLASSES CORRODED AT $90^\circ C$, 4 WEEKS, SA/V=0.1 cm^{-1}

Glass	% Al_2O_3	% Fe_2O_3	% SiO_2	pH=2.4	Leach rate (g/m^2-d) pH=5.9	pH=10.4	pH=12	D.I. water
SRL 131-unloaded	0	0	57.9	1.3	2.1	16.3	–	27.8
" "+15%Fe-waste	0.2	8.9	49.2	–	–		–	1.5
" "+29.8% "	0.4	17.8	40.6	–	–		–	0.4
" "+15%TDS-3A	1.4	7.1	49.2	2.9	0.9	0.6	–	1.3
" "+29.8% "	2.8	14.2	40.6	3.2	0.1	0.3	1.7	0.1
" "+15%Al_2O_3-waste	7.4	2.1	49.2	3.2	0.2	0.2	–	0.1
" "+28.7% "	13.7	4.0	41.3	3.4	0.1	0.2	–	0.1

For glasses containing wastes, the leach rates were the largest at pH=2.4 and in general were higher at 10.4 and 12.0 than at 5.9. In contrast, the unloaded base glass had the lowest leach rate at pH=2.4 and the highest in the alkaline solutions. Although other elements probably have an effect, the important parameter in controlling the leach rate appears to be the Al_2O_3 content of the glass. As the % Al_2O_3 is increased the resistance to leaching is decreased in acidic and increased in neutral and basic solutions. The importance of Al_2O_3 in decreasing the leach rate in neutral and basic solutions has been previously demonstrated by Dilmore et al[11]. Fe_2O_3 is also known to form stable species in the pH range of 3-12[12] but it is less effective at reducing the leach rate than is Al_2O_3.

In low pH solutions SiO_2 has a lower solubility (ie, few ppm; dissolves via $SiO_2 + H_2O \rightleftharpoons H_2SiO_3$) than most other oxides. The SRL 131 unloaded glass contains a sufficiently high % of SiO_2 to form a coherent framework even when most of the other species have been leached. The SiO_2 content of the waste glasses has been sufficiently reduced (Table 5), primarily with Al_2O_3 and Fe_2O_3, to prevent a continuous SiO_2 framework. Both Al_2O_3 and Fe_2O_3 are unstable in low pH solutions (dissolving as Al^{+3} and Fe^{+3} at pH<3)[12] and, when they are leached from the glass the SiO_2 concentration is insufficient to hold the remaining framework together. Consequently large concentrations of all species including Si are released into solution. In basic solutions (pH>9) SiO_2 is unstable and the unloaded glass readily dissolves. Even in D.I. water the rate of alkali extraction from the unloaded glass is high and the pH increases very quickly to values greater than 11. In this pH range the solubility of SiO_2 by the dissociation of silicic acid ($H_2SiO_3 \rightleftharpoons 2H^+ + SiO_3^=$) is appreciable, and the dissolution of SiO_2 from any glass will be high unless it is capable of forming a stable-protective film. Although some SiO_2 may dissolve from the glasses containing the wastes the remaining species, particularly Al_2O_3, can stabilize the surface against extensive deterioration up to a pH $\leq$ 10.7 above which dissolution via AlO_2^- becomes appreciable[12]. These data are consistent with the thermodynamic aqueous stability diagrams for oxides presented by Paul[10] and with other studies involving Al_2O_3 and Fe_2O_3 silicate glasses[12].

Infrared reflection spectra (IRRS) are presented in Figure 3 for the various glasses both before and after corroding at $90^{\circ}C$ for 4 weeks in deionized water with SA/V=0.1 cm^{-1}. Significant decreases in spectral intensity (reflectance %) occurred on all the samples after corrosion with the exception of the glass containing 28.7% high alumina waste. These changes in the intensity suggest compositional and/or morphological alteration of the glass surface; and the magnitude of the changes are proportional to the extent of surface alteration. The least surface alteration has occured on the glass containing 28.7% high Al_2O_3 waste and the greatest on the unloaded base glass and with the glass containing 29.8% high Fe waste.

Scanning electron micrographs (SEMs) are shown in Figure 4 of the samples used to obtain the infrared spectra in Figure 3. The SEM in Figure 4a corresponds to the surface of an uncorroded glass with no waste, but it is representative of all the glass surfaces prior to corrosion. Polishing lines (due to the 600 grit final polish with SiC) are faintly visible on the uncorroded surface; otherwise, the surface is nearly featureless. Two types of surface alteration are apparent in these micrographs; cracking and precipitation. Surface cracks are usually an indication of extensive leaching by ion exchange. There were no cracks observed on the glass containing the 28.7% high alumina waste and polishing lines are still visible, both of which indicate only minor surface alterations. In contrast the surfaces of the unloaded glass, 29.8% high iron waste, and 29.8% TDS-3A waste contain various extents of cracking. All of the glasses containing wastes have a surface precipitate, the morphology of which is clearly visible in Figure 4f. Surface precipitate is not seen on the unloaded glass.

Additional SEMs together with the ones in Figure 4 indicate that the type of surface alterations that occur are independent of temperature in the range of $40^{\circ}C$-$90^{\circ}C$. The effect of increasing the temperature is to reduce the time required for surface precipitate formation. The SEMs also suggest that the time required for precipitate formation (visible at 100X) decreases as the percentage of waste in the glass is decreased. The quantity of surface precipitate increases with time at all temperatures and percentages of loading. The SEMs, as do the solution and IRRS data, indicate that the most significant surface alteration occurred on the unloaded glass and the ones containing high Fe waste. The least alteration occurred on the glasses with the high Al_2O_3 waste.

Depth compositional profiles are shown in Figures 5 and 6 for the SRL 131 + 29.8% TDS waste. These profiles are for a surface equivalent to those in Figure 4d, e. The ESCA profile in Figure 5 shows that the major elements in the glass surface are Si, Al, B, Fe and 0 (not shown). The surface is enriched with respect to Si, Al and Fe, and is depleted of Na. SIMs analysis, Figure 6, gives a similar result and also illustrates the destribution of the minor elements. Ba, Zr and Ti are enriched while Ce, La and Li are depleted in the surface. Depth compositional profiles were obtained for other glasses under a variety of exposure conditions and although variations occurred with most elements from sample to sample, Fe and usually Al were enriched at the surfaces of the glasses containing wastes. The profile for an unloaded SRL 131 glass after corrosion is shown in Figure 7. Although the surface is enriched in the minor elements La, Ti and Ca (impurity), their presence and/or concentrations are insufficient to provide a protective film.

A Type IV surface[2] is formed on the unloaded glass corresponding to an extensively leached and non protective surface film. The glasses containing waste form dual layer Type III[2] surfaces. This type of surface consists of a leached layer over which a second layer has been deposited (see SEMs). The combination of depth compositional profiles and SEMs show that the morphology of the outer surface consists of a non-continuous precipitate comprised primarily of O, Al and Fe (probably as hydrated oxides). The degree of protection provided by this outer layer is probably small as illustrated by the surface cracks in Figure 4d. However, if the glass contains sufficient Al_2O_3 the underlying layer (ie, leached layer) may be stabilized against extensive leaching and cracking.

SUMMARY

Two types of surfaces may form on nuclear waste glass during leaching; non protective-non stable (Type IV), non protective-stable (Type III) and protective-stable (Type III). The leach resistance of the glasses is dependent on their ability to form stable and protective surface films which control the kinetics of leaching and thermodynamic stability. Increasing the % waste (all three types) in the base glass reduces the leach rate in neutral and basic solutions but increases it in acidic solutions. The important variable appears to be the Al_2O_3 content. In neutral and basic solutions (up to pH ∼11) the Al_2O_3 can produce stable and protective surfaces on the glasses if present in sufficient concentrations (0.4 to 2.8%). In acidic solutions (pH<3) the Al_2O_3 cannot form stable surfaces and the leach rate appears to be indirectly proportional to the SiO_2 concentration in the glass. The higher the SiO_2 the lower the leach rate. The formation of stable-protective surface films helps to control the pH in neutral solutions. However, when the solution pH and chemistry are altered by external sources, the protective films may become unstable when 3>pH>11.

The extent of leaching is decreased as SA/V is increased unless a change in solution chemistry (ie, pH) alters the solubilities of species in the glass. High SA/V ratios may cause the pH to increase very quickly to values where stable films do not form. Simulated silicate ground water had little effect on leach rate while the brine solution significantly reduced the leach rate probably due to the formation of a stable-protective $Mg(OH)_2$ surface film.

Thermodynamic stability diagrams are available for various oxides in aqueous solutions and may be useful in predicting oxide glass stability. The solubilities of the glasses reported in this study are consistent with the stability diagrams of Al_2O_3, Fe_2O_3 and SiO_2. However, when the oxides are part of the glass structure extensive leaching of the glass may be required for the establishment of equilibria phases at the surface

and these phases although stable may not be protective. This leaching may
result in significant alteration of the glass leading to the release of
many other species into solution. Although Fe_2O_3 is thermodynamically
more stable than Al_2O_3 in aqueous solutions[12] the rate of surface alteration
on the glasses with Al_2O_3 is less indicating that the latter is kinetically
more stable. Thus, an understanding of the kinetics of leaching, surface
film morphology and the thermodynamics of stable phase formation is
desirable for predicting glass corrosion.

ACKNOWLEDGEMENTS

This work was supported primarily by Savannah River Laboratory,
contract No. Ax-511740R. It was also supported in part by the U.S.
Nuclear Regulatory Commission, contract No. NRC-04-78-252. Partial support
was provided by SKBF/Div. KBS, Sweden.

REFERENCES

1. El-Shamy, T.M. and Douglas, R.W., (1972), Glass Technol., 13(3), 77-80.
2. Clark, D.E., Pantano, C.G., Jr and Hench, L.L., (1979), <u>Corrosion of Glass</u>,
   Magazines for Industry, Inc, New York.
3. El-Shamy, T.M., Lewins, J. and Douglas, R.W., (1972), Glass Technol., 13(3),
   81-87.
4. Chao, Y. and Clark, D.E., in preparation.
5. Clark, D.E. and Lue Yen-Bower, E., (1980), Surf. Science, 100(1), 53-70.
6. Materials Characterization Center, 1982, DOE/TIC-11400, Pacific Northwest
   Laboratory, Richland, Washington.
7. Hench, L.L. and Clark, D.E., October 1981, Annual Report to the U.S. Nuclear
   Regulatory Commission, Contract No. NRC-04-78-252.
8. Private communication.
9. American Nuclear Society, American Nuclear Society Standards Subcommittee
   16.1 (April, 1981).
10. Paul, A., Mater, J., (1977), Sci., 12, 2246-2268.
11. Dilmore, M.F., Clark, D.E. and Hench, L.L., (1978), Am. Ceram. Soc. Bull.,
   57(11), 1040-1045.
12. Paul, A. and Zaman, M.S., Mater, J., (1978), Sci., 13, 1499-1502.

12

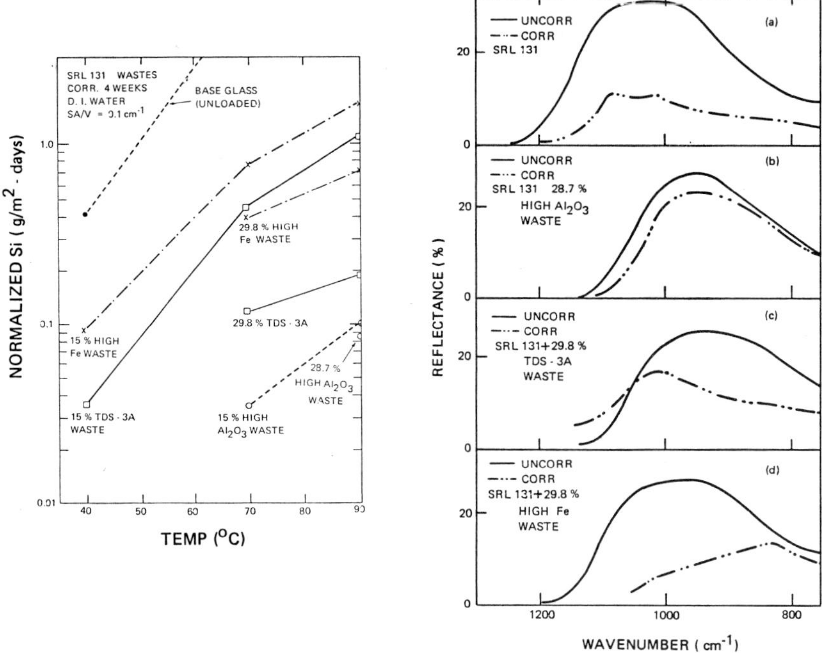

Fig.1. Normalized leach rates for Si VS temperature and glass composition.

Fig.3. IRRS of the SRL 131 unloaded and loaded glasses corroded at 90°C, 4 weeks, SA/V=0.1 cm⁻¹.

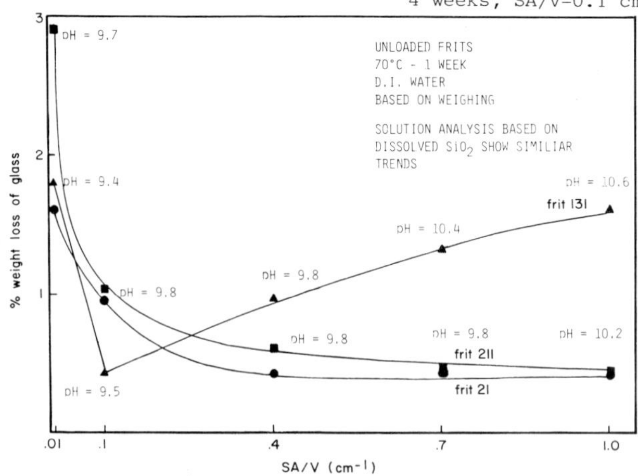

Fig.2. % WT loss VS. SA/V for three SRL alkali borosilicate glasses without waste.

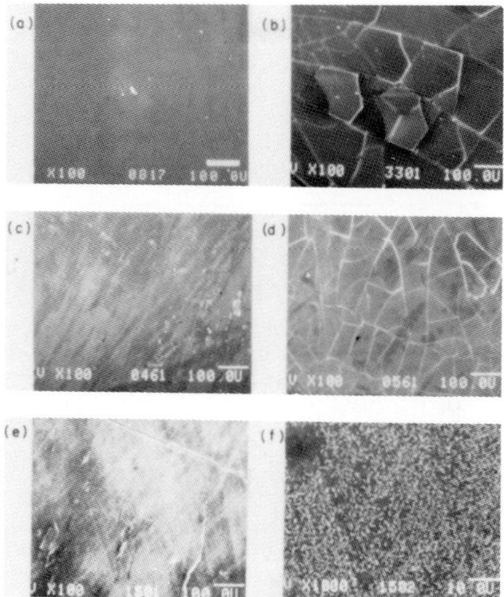

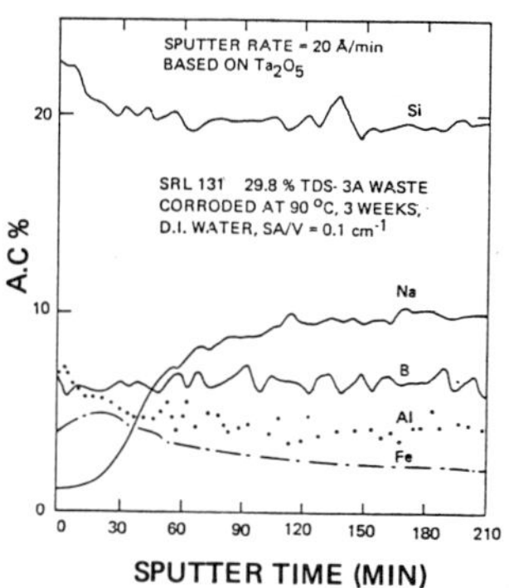

Fig.5. ESCA depth compositional profile of surface similar to that in Figs 4e, f.

Fig.4. SEMs of the SRL 131 unloaded and loaded glasses corroded at 90°C, SA/V=0.1 cm⁻¹: a)unloaded and uncorroded, 100X, b) unloaded and corroded 4 weeks, 100X, c) 28.7% high Al₂O₃ waste corroded 4 weeks, 100X, d) 29.8% high Fe waste corroded 4 weeks, 100X, e) 29.8% TDS-3A waste corroded 3 weeks, 100X, f) same as e but 1000X.

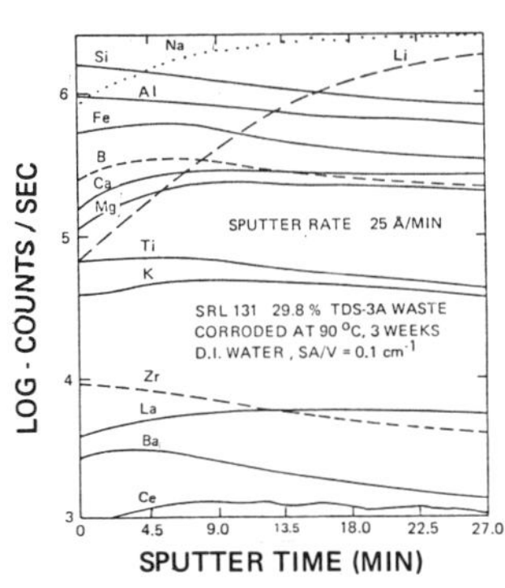

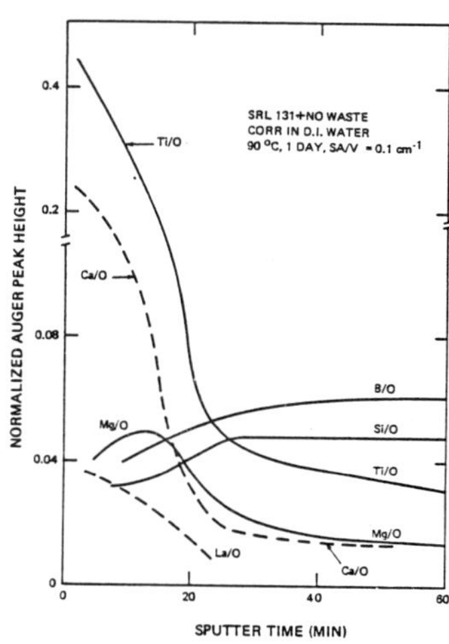

Fig.6. SIMs depth compositional profile of surface similar to that in Figs 4e,f.

Fig.7. AES depth compositional profile of the SRL 131 unloaded glass after corrosion.

Published 1982 by Elsevier Science Publishing Co
SCIENTIFIC BASIS FOR RADIOACTIVE WASTE MANAGEMENT - V
Werner. Lutze, editor

CHEMICAL DURABILITY OF GLASS CONTAINING SRP WASTE - LEACHABILITY CHARACTERISTICS, PROTECTIVE LAYER FORMATION, AND REPOSITORY SYSTEM INTERACTIONS

GEORGE G. WICKS, BARBARA M. ROBNETT, AND W. DUNCAN RANKIN
E. I. du Pont de Nemours & Co.
Savannah River Laboratory
Aiken, South Carolina 29808

LEACHABILITY AND FORMATION OF SURFACE LAYERS

Leachability is one of the most important properties of solidified nuclear waste forms because it provides information on the performance and the subsequent safety and reliability that the waste products will possess. One of the most important experimental findings in the leachability field has been the discovery and subsequent detailed characterization of protective surface layers that form on waste glass during leaching.[1,2] These layers can have a beneficial effect on product performance while in storage by improving product durability with time. As a result of surface layer formation and the effects on subsequent product leaching characteristics, new qualitative and quantitative leaching models have recently been proposed.[2]

The leaching of the waste glass can be described as a 3-stage corrosion process[2]: (1) interdiffusion, (2) matrix dissolution, and (3) surface layer formation and stabilization. Layers are formed primarily from the nonradioactive components of the waste. As these species are leached from the glass, they form insoluble compounds at the gel layer which then form the precipitated surface layer. For Savannah River composition 131/TDS waste glass (Table 1), the layer is generally enriched in major waste constituents such as Fe and Mn and depleted in major frit components such as Si and Na.[1] The surface layers as shown in Figure 1 adhere relatively well to the glass underneath and are usually several microns thick (Figure 1A). The surface layer thickness decreases and appears more adherent to the glass as the waste content of the glass increases. The thinner surface layers of higher waste-loaded glasses correlate with improved product durability or lower leach rates[3] (Figure 1B).

REPOSITORY SYSTEM TESTS

After burial in a repository, a new environment is introduced to the product, and the waste form becomes only one element of a multibarrier isolation

16

system. The reference waste isolation system for Savannah River includes the waste glass form as nucleus surrounded by a casting canister (304L SS), a potential over-pack (possibly a titanium-based alloy), potential backfill material (clay), and finally the host repository rock. While there have been experimental studies directed at sorption of radionuclides of interest with the various rock candidates, there have been very few systematic waste form/ system interaction experiments performed thus far. As a result, a series of scouting tests were performed to evaluate the separate and synergistic effects that elements of the multibarrier isolation system may have on waste glass durability. These data will be used in conjunction with other studies to define future system tests which can best simulate the most important conditions of the repository and for more detailed analyses. The Repository Systems Program at Savannah River consists of the following elements: (a) static repository tests, (b) slow flow groundwater tests, (c) dynamic leaching experiments, (d) long term leaching tests, and (e) in situ experiments. The scouting tests performed in the static repository program will be summarized and a new dynamic leaching concept will be introduced.

STATIC REPOSITORY TESTS

Abbreviated MCC-1 static leaching tests were used in the static repository program. The tests were conducted at 90°C for 7 days with deionized water as well as repository groundwaters as leachants. The static repository program consisted of three phases. Phase 1 of the study evaluated the leachability of SRP waste glass leached in the presence of salt, basalt, shale, granite, and tuff at three different ratios of surface areas of sample to volume of leachant (SA/V). These data were then compared to tests on similar glasses leached in the absence of rock. The second phase of this program evaluated the effects of five simulated groundwaters on the leachability of waste glass both with and without host rock present and compared these results to glass leached by deionized water. The groundwaters included many minor as well as major constituents. The third and final phase of the program evaluated waste glass leachability as a function of tests conducted in the presence of candidate canister and overpack metals, backfill mixtures, and five complete package systems.

Phase 1 - SA/V and Rock Series

The concentrations of six different elements of interest found in the deionized water leachate after leaching the waste glass (with and without five

different rock types) are given in Table 2 as a function of three different
SA/V ratios. The corresponding "effective leachabilities" (i.e., leachability
based on species in solution and not including rock adsorbed species) were
then calculated. Following are conclusions drawn from these data:

- Leach rates calculated from solution data (Table 2) show that leach rates
  decrease as the SA/V ratios increase for all samples based on mass loss
  and extraction of nine different elements. This effect has been observed
  before[4] and can be beneficial for long term repository storage because of
  the relatively small quantities of water that are anticipated.

- "Effective leachability" generally decreases but only slightly for leaching
  by deionized water in the presence of rocks (salt, basalt, shale, granite
  and tuff).

- Leach rates can vary significantly for different elements.

- At a given SA/V ratio, leachate concentration and subsequent leachability
  generally correlate well with final pH of the solution. This suggests a
  non-destructive means for monitoring leachability.

- Leaching in the presence of salt reduces leachability most appreciably,
  especially at high SA/V ratios. The salt dissolves more readily than any
  other rock and forms brine which is a less aggressive leachant than deionized water.

- There is no significant difference in the leachability of waste glass when
  leached in the presence of Carlsbad salt compared to Avery Island salt.

- Key elements such as Fe, Mn, and Mg show very low leach rates for all systems and SA/V ratios studied. Previous studies[1] have shown that the low
  leach rates correspond to retention of these species within the leached
  surface layers and generally result in a reduction of glass leachability.
  Therefore, glass leachability in a repository environment would also be
  expected to decrease with time.

Phase 2 - Groundwater Tests

Most of the leachability data that now exist are a result of short term
leaching tests using deionized water as the leachant. In a repository setting, the most realistic potential leachants are the inherent groundwaters
found in the geology. The primary objective of phase 2 of this study was
to compare glass leached in the various groundwater compositions with glass
leached in deionized water. Furthermore, these data were then compared to
leachability results using a rock-equilibrated shale groundwater, "standard"
brine and basalt groundwaters used in MCC tests, and an actual shale

groundwater extracted from the SRP site. The silicon leachate concentrations
are given in Table 3. Following are conclusions from this study:

- Glass leached in simulated groundwater generally results in a lower leachability than glass leached in deionized water. (This observation is based
  on silicon extraction after correcting for the initial silicon concentration in the leachant.) Hence, product performance involving groundwaters
  in a repository should be better than the results from most laboratory
  tests using deionized water as leachant.

- For leaching of glass using simulated groundwaters in the presence of host
  rock, the leachability is generally reduced further, although only slightly
  in most cases.

- The largest reductions in leachability were for brine solutions. The
  largest source of experimental error was also for salt systems due to the
  dissolving host rock.

- The simulated shale groundwater was only slightly more aggressive than
  the actual groundwater and the rock-equilibrated groundwater was the
  most aggressive leachate used except for deionized water.

- Leachabilities of glass using MCC groundwaters, which incorporate only
  major elements of the groundwaters, are similar to more complete simulated
  groundwater compositions. However, this comparison becomes less clear for
  leaching in the presence of rocks, perhaps due to inherent variations in
  rock compositions.

Phase 3 - Canister Metals, Backfills, and Complete System Tests

In phase 3 of the study, a salt repository was selected as an arbitrary
reference repository. Each of the potential elements of the multibarrier
isolation system was evaluated in the simulated salt environment, including
several canister or overpack metal candidates, three backfill mixtures, and
three additional "backfill" compositions. A reference "complete system" involving waste glass, a 304L stainless steel primary canister metal, a Ticode 12
overpack metal, and a backfill mixture of sand, bentonite and charcoal was then
selected for further tests. Each of these complete systems was then leached in
the presence of the five different rock geologies of salt, basalt, shale, granite, and tuff. As a result of recent studies performed by C. O. Buckwalter at
Battelle Pacific Northwest Laboratory, a third metal candidate, lead, was also
evaluated in a long term leachability experiment. These data are summarized in
Table 4, and following are conclusions from this study:

- In system tests involving waste glass, canister material, overpack metal, backfill, and five different rock types, the backfill constituent most strongly influenced the release rate of species of interest.
- Backfill can be beneficial, marginal, or perhaps detrimental to the release rate of species of interest depending on the choice of material.
- Waste glass release rates are significantly lowered in the presence of lead due to the formation of a lead enriched surface layer.

DYNAMIC REPOSITORY TESTS

An important consideration for burial of waste glass forms in a repository environment is the potential effect of groundwater flowrates on glass durability. In order to investigate this effect, a dynamic leaching test station has recently been developed and placed into operation at Savannah River. The station consists of a series of Teflon® (trademark of Du Pont) modular units which can contain the various elements of the multibarrier isolation system. All units are located in a large oven in which the atmosphere is controlled and the temperature held constant to $\pm 2°C$.

One objective of this station is to provide information on formation and stabilization of protective surface layers that form during leaching in an accelerated testing mode. This type of information will be valuable to an ongoing mechanism study and to leaching models now under development. The station will also provide a new accelerated testing approach for rapid evaluation of multibarrier elements now under consideration and what effects they may have on waste glass durability and release of mobile species to solution. Finally, a practical application of this system includes evaluation of product performance under low probability accident scenarios. These scenarios would include conditions such as flooding of repositories or breached canisters falling into a river during transportation events. Also, the system may simulate, to a limited degree, possible circulation cells that might exist in boreholes containing water in contact with the product.

Preliminary analysis of this system showed that only small increases in leachability of waste glass forms occurred even for very high flow rates of 100 to 200 mL/min. The leached glass surface layer was then examined by electron microprobe analysis. As shown in Figure 2A, iron-rich precipitates were observed on the leached glass surfaces. A more detailed analysis using Auger spectroscopy with argon ion milling showed that the surface layer was enriched in iron with a maximum concentration at 200 to 400 Å deep (Figure 2B). These analyses and corresponding leachability data suggest that surface layers are

just beginning to form and leaching has progressed to Stage 3 corrosion even
in a very dynamic system. Leachant flow rates for these tests were about
100,000 times greater than groundwater flow rates expected in some repository
environments now under consideration.

ACKNOWLEDGEMENT

The authors wish to thank J. L. Ehrhardt, P. E. O'Rourke, and P. G. Whitkop
for their assistance in this study.

The information contained in this article was developed during the course
of work under Contract No. DE-AC09-76SR00001 with the U.S. Department of
Energy.

REFERENCES

1. Wicks, G. G., Mosley, W. C., Whitkop, P. G. and Saturday, K. A., "Dura-
    bility of Simulated Waste Glass - Effects of Pressure and Formation of
    Surface Layers," Journal of Non-Crystalline Solids (in press).

2. Wicks, G. G. and Wallace, R. M., "Leachability of Waste Glass Systems
    - Physical and Mathematical Models," Journal of Nuclear Materials (in
    press).

3. Rankin, W. D. and Wicks, G. G., "The Chemical Durability of Savannah
    River Plant Waste Glass as a Function of Waste Loading," DP-MS-81-105,
    presented at the 84th Annual Meeting and Exposition of the American
    Ceramic Society, Cincinnati, Ohio, May 1982.

4. Altenhein, F. K., Lutze, W., and Malow, G., "The Mechanisms for Hydro-
    thermal Leaching of Glass and Glass-Ceramic Nuclear Waste Forms," Scien-
    tific Basis for Nuclear Waste Management, Vol. 3, Plenum Press, New York,
    New York, 1981.

TABLE 1. SIMULATED 131/TDS WASTE GLASS COMPOSITION

COMPONENT	131 FRIT (wt %)	COMPONENT	TDS WASTE (wt %)
SiO_2	39.5	SiO_2	1.2
Na_2O	12.1	Na_2O	0.9
B_2O_3	10.0	$CsCl$	1.0
TiO_2	0.7	$SrCl_2$	1.0
Li_2O	3.9	Fe_2O_3	14.1
MgO	1.4	MnO_2	4.1
ZrO_2	0.3	Al_2O_3	2.8
La_2O_3	0.3	NiO	1.7
		CaO	1.1
		Na_2SO_4	0.2
		COAL	0.7
		AW-500*	3.0

* Linde AW-500 zeolite (calcium-aluminum-silicate).

TABLE 2.   PHASE 1 - LEACHATE CONCENTRATIONS (PPM)

ELEMENT	SA/V	GLASS*	GLASS + SALT**	GLASS + BASALT**	GLASS + GRANITE**	GLASS + SHALE**	GLASS + TUFF**	FINAL pH*
Si	.5 cm^{-1}	57.34	20.78	45.93	56.50	42.46	45.08	10.02
	.1	27.18	12.55	21.11	26.56	25.30	24.59	9.59
	.03	11.68	10.82	8.55	11.30	11.76	10.03	9.09
Cs	.5	2.17	1.50	.44	1.69	.28	.38	
	.1	.87	.69	ND	.51	ND	ND	
	.03	ND	.51	ND	.21	ND	ND	
Sr	.5	.02	.80	.02	.01	ND	.03	
	.1	.05	ND	.02	.05	ND	.02	
	.03	.20	ND	.01	.18	.04	.02	
Fe	.5	.08	.11	.06	.09	.15	.21	
	.1	ND	.04	.24	ND	ND	ND	
	.03	ND	ND	.12	ND	.01	.01	
Mn	.5	.13	.22	.17	.11	.14	.15	
	.1	.02	.11	.05	.02	.04	.03	
	.03	.02	.01	.01	ND	.01	.01	

* Concentrations measured in solution after leaching of glass alone in deionized water.

** Concentrations represent corrected data obtained by leaching glass and rock and then subtracting the results for leaching rock alone.

TABLE 3.   PHASE 2 - Si LEACHATE CONCENTRATIONS (PPM)

LEACHANT	NO ROCKS			ROCKS			
	GLASS AND LEACHATE	LEACHATE	GLASS*	GLASS** AND LEACHATE AND ROCK	ROCK AND LEACHATE	GLASS*	FINAL pH*
DI H$_2$O	22.5	<0.5	**22.5**	-	-	**(22.5)**	-
SIM BASALT	25.7	5.8	**19.9**	23.79	8.7	**15.1**	9.18
GRANITE	16.9	1.98	**14.9**	15.05	3.3	**11.8**	9.27
SHALE	14.6	.31	**14.3**	18.45	2.8	**15.7**	8.13
TUFF	21.1	3.26	**17.8**	24.27	9.3	**15.0**	9.24
SALT	23.9	19.1	**4.8**	23.81	21.7	**2.1**	6.92
MCC SALT	27.1	22.0	**5.1**	30.58	24.6	**6.0**	7.10
MCC BASALT	39.4	28.2	**11.2**	37.34	30.7	**6.7**	9.39
REAL SHALE	11.2	1.61	**9.6**	14.56	2.8	**11.8**	8.36
ROCK EQUIL. SHALE				24.29	2.8	**21.5**	9.42

* Glass concentrations represent leachate concentrations corrected for initial Si content of leachant and Si extracted from rocks.

** In general, less than a total of 5 PPM each of Cs, Sr, Fe, Mg, Mn and Al were found in solution after leaching. Exceptions were higher Mg concentrations in test involving salt, basalt and shale and higher Sr content in the leachate for tests involving only shale. These total concentrations reflect elements from the initial groundwater leachant and species extracted from both the glass as well as host rock.

TABLE 4. PHASE 3 - MEASURED CONCENTRATIONS IN PPM (UNCORR.)*

ROCK	CANISTER MATERIAL	BACKFILL	Si	Cs	Sr	FINAL pH
GLASS ALONE	–	–	27.31	.969	.05	9.59
GLASS ALONE**	Pb	–	0.64		.007	
SALT	–	–	15.15	.918	1.35	9.25
SALT	304L	–	20.00	.822	.68	9.21
	Ticode	–	18.50	.756	.52	9.35
SALT	–	BF 1***	39.5	.264	1.47	8.44
	–	BF 2	52.5	.094	3.33	8.43
	–	BF 3	24.3	.390	2.76	8.24
SALT	–	IE 95	36.6	.043	4.17	7.43
	–	CLINO	53.2	.109	.95	6.55
	–	A-51	24.3	.361	.04	7.80
SALT	304L Ticode 12	BF 1	30.00	.333	1.46	8.33
AVERY ISLAND SALT	304L Ticode 12	BF 1	31.00	.456	1.30	8.23
BASALT	304L Ticode 12	BF 1	62.3	.037	.18	9.22
GRANITE	304L Ticode 12	BF 1	72.7	.053	.17	9.31
SHALE	304L Ticode 12	BF 1	112.0	.045	.40	9.23
TUFF	304L Ticode 12	BF 1	86.0	.069	.23	9.18

 * All tests conducted with 131/TDS-3A waste glass. Concentrations
   represent total concentrations measured in solution, not corrected for
   leaching from multibarrier elements such as backfill materials.

 ** All leach tests were conducted using MCC-1 static leach procedures. Except
    for experiments with lead, the tests were conducted at 90°C, SA/V = 0.1 cm^{-1}
    and for 7 days in duration. Tests with lead present were conducted under
    similar conditions except for the duration, 1-1/2 months.

*** BF 1 (30% Sandia bentonite, 20% coconut charcoal, 50% sand), BF 2 (100%
    Sandia bentonite), BF 3 (100% SRP bentonite), IE 95 (primarily chavazite),
    CLINO (clinoptilolite, a zeolite), and A-51 (synthetic A-type zeolite).

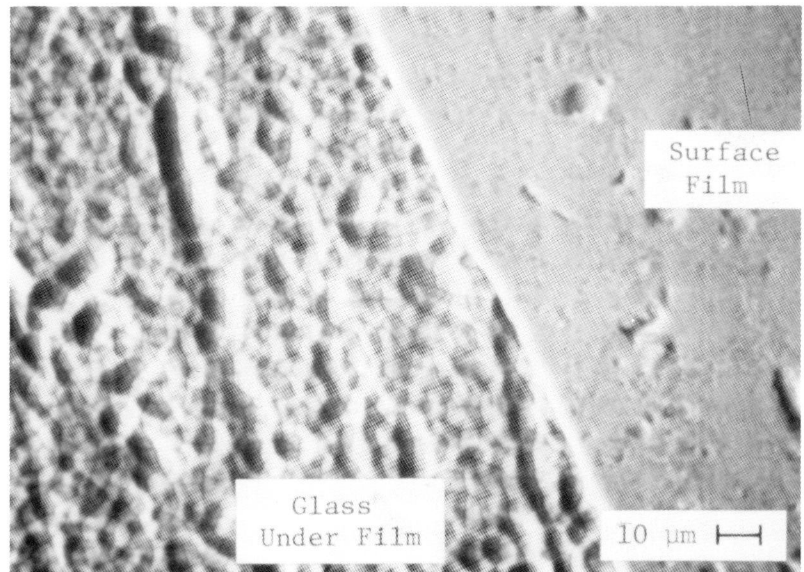

A. Top view of surface layer and glass underneath

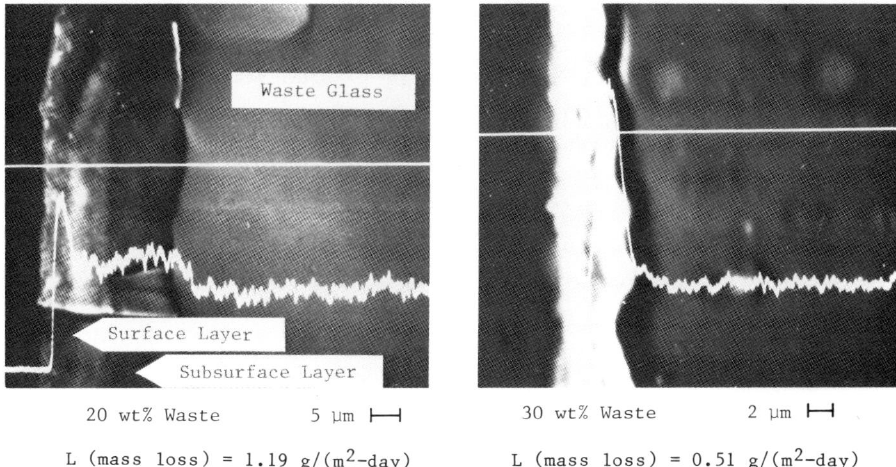

L (mass loss) = 1.19 g/(m^2-day)   L (mass loss) = 0.51 g/(m^2-day)

B. Cross sections of leached layers for 20% and 30% waste loaded
    glasses with x-ray line profiles of waste constituent Fe.

Figure 1. Leached Glass Surface Layers.

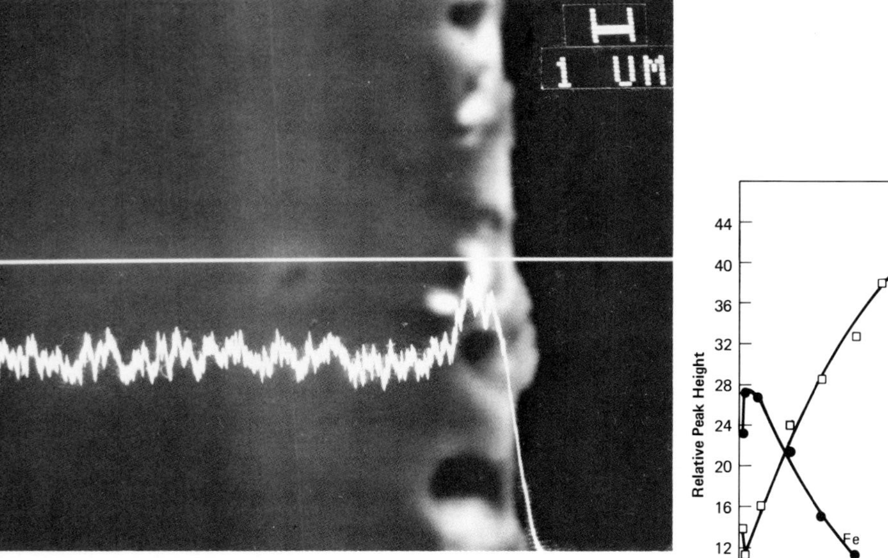

A. Electron microprobe analysis with Fe
x-ray line profile.

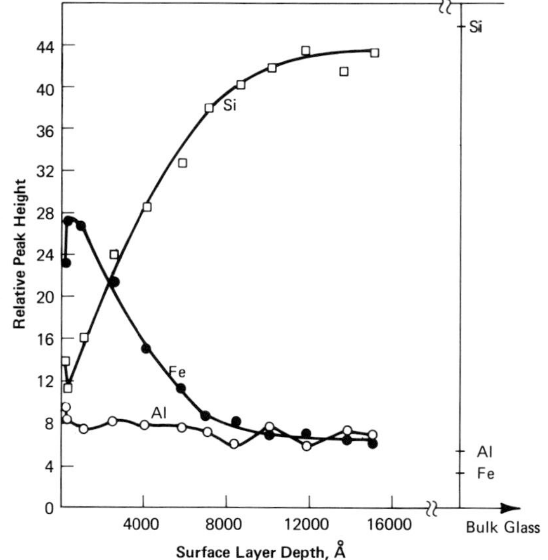

B. Auger analysis with argon ion milling and
depth profiles of Fe, Si, and Al.

Figure 2.   Leached Glass Surface Layers
(Flow rate = 250 mL/min, SA/V = 57.1 cm^{-1}, 90°C, 18 hours)

THE MECHANISMS FOR HYDROTHERMAL LEACHING OF NUCLEAR WASTE GLASSES: PROPERTIES AND EVALUATION OF SURFACE LAYERS

GÜNTER MALOW
Hahn-Meitner-Institut für Kernforschung Berlin GmbH.,
Glienicker Straße 100, D 1000 Berlin 39

INTRODUCTION

Remobilization of solidified high-level waste by corrosion and transport of radionuclides by groundwater are the only likely events for radioactivity to find its way back to the environment. The interaction of the waste form with aqueous solutions is the most important mechanism and must be known for safety and risk analysis. It has been shown that the activity release depends on a number of experimental and environmental parameters.[1] The corrosion mechanisms of the waste forms control the leach rate of individual components. Of particular importance to the resistance against water attack of various glass compositions is the formation of layers on the glass surface.[2] The results for leaching in deionized water revealed that the mechanism was strongly influenced by the formation of a surface layer which determined the further attack on the glass.[3] In this work the specific weight losses of the glass after leaching in rock salt and $MgCl_2$-$MgSO_4$-NaCl-KCl solutions were measured and surface layers investigated.

EXPERIMENTAL

The borosilicate glass C-31 3EC containing 20 w/o LWR-type (30.000 MWd/t_{HM}) simulated waste oxides was investigated.[4,5]

Glass beads were used for hydrothermal leaching experiments having size and shape comparable to beads which will be produced in the German prototype vitrification plant PAMELA.[6]

The German final repository concept foresees the waste disposal in salt formations. Present model calculations of heat release from HLW glass blocks comprize temperatures up to 200°C at the canister/salt interface.[7]

The compositions of the leachants (brines) can be derived from mining experience and experiments.[5]

Compositions of the brines for 55°C used in this work are given in table 1. Glass beads were leached in Teflon lined autoclaves at 200°C at equilibrium pressure, i.e. $\sim$ 15 bar in saturated salt solutions.

Hydrothermal leaching experiments in pure water revealed a volume and time dependency of the specific weight loss which seems to disappear at about 300 ml solution and 10 d resp.[3] Therefore, 300 ml stagnate salt solutions and leaching periods up to 30 d were applied.

After leaching, specific weight losses were determined and the leached surfaces were investigated by X-ray diffraction, Scanning Electron Microscopy (SEM) and Electron Probe Microanalysis (EPMA).

TABLE 1

COMPOSITIONS OF SALT SOLUTIONS (AFTER J.D'ANS, "DIE LÖSUNGSGLEICHGEWICHTE DER SYSTEME DER SALZE OZEANISCHER SALZABLAGERUNGEN", BERLIN 1933)

System	Point	Mole per 1000 Mole H_2O (T = 55°C)			
		NaCl	KCl	$MgCl_2$	$MgSO_4$
$NaCl-H_2O$		113.4	–	–	–
Quinary System	Ẑ	1.0	2.6	111.2	2.1
$NaCl-KCl-MgCl_2-Na_2SO_4-H_2O$	Q	6.8	17.4	77.3	3.2

RESULTS AND DISCUSSION

Specific weight loss measurements

The results of some preliminary leaching experiments in various salt solutions showed no obvious dependency upon chemical composition of the leachants and values of the specific weight loss are within one order of magnitude.[5] NaCl-, Q- and Z-solutions were selected for a more detailed investigation of the leaching process. The results of the specific weight loss measurements and the leach rates for various times and solutions are given in table 2.

The specific weight loss increased about a factor of 3 and the leaching rates decreased about a factor of 1/4 with time, indicating a retardation in the leaching process except for Z-solution.

After 3 d leaching the highest specific weight loss was measured in the Q- and the lowest in the Z-solution.*) The difference is about a factor of 3. After 30 d leaching the highest specific weight loss was measured in Z- and the lowest in N.

---

*) In the following text NaCl-, Q- and Z-solution will be abbreviated by N, Q and Z.

TABLE 2

SPECIFIC WEIGHT LOSSES AND LEACH RATES OF GLASS BEADS AFTER LEACHING AT 200°C
IN SOLUTIONS SATURATED AT 55°C (SEE TABLE 1)

Leachant		Specific weight loss m(t)/mg·cm^{-2}; leaching rate R(t)/mg·cm^{-2}d^{-1}		
		3 d	10 d	30 d
NaCl-	$\overline{m}$(t)	3.8 ± 0.3	5.1 ± 0.8	9.4 ± 0.5
solution	R(t)	1.3 ± 0.1	0.5 ± 0.1	0.31 ± 0.02
Q-	m(t)	5.4 ± 1.6	10 ± 3	16 ± 2
solution*)	R(t)	1.8 ± 0.5	1.0 ± 0.3	0.55 ± 0.08
Z-	m(t)	1.7 ± 0.4	9.8 ± 0.7	18 ± 3
solution	R(t)	0.6 ± 0.1	1.0 ± 0.1	0.6 ± 0.1

*) at 55°C saturated with NaCl for 200°C after Conradt et al.[8]

SEM- and EPMA-investigation of surface layers

Figure 1a exhibits an SEM photomicrograph of a sectioned glass bead after 10 d
leaching in N. The leached surface consist of two different layers. The inner
layer is about 15 μm thick and looks like a typical gel layer with drying
cracks, whereas the outer one appears more dense and is relatively thin with a
thickness of about 1.5 μm.

Figures 1b to r show EPMA concentration profiles in the form of X-ray line
scans obtained at the white line marked in figure 1a. As seen, elements such as
Ca, Ba, Cs are nearly completely leached from the layers, whereas Mo and Zn are
strongly depleted in the inner layer, but their concentration increases again in
the outer layer. Si, Al, Mg are also depleted in the inner layer but again
have a higher concentration in the outer layer, whereas U, Ni, Fe are enriched
in the outer layer when compared with the pristine glass. The elements Ti, Zr,
La, Ce, Nd are enriched in both layers.

After 30 d leaching in N the appearance of the surface changes (figure 2a).
The outer layer appeared to be thicker whereas the inner layer seems to become
denser and thinner. X-ray line scans of the elements Si, Al, Mg, Zr, U indicated two concentration maxima, figures 2b - f, which implies a periodically
alternating composition in the scale. This becomes also evident from figures
3a,b, showing surfaces leached 10 d in Q and Z, resp. Figure 3c shows the surface of a glass bead leached for one year in quinary brine from the ASSE II salt
mine in Germany.

28

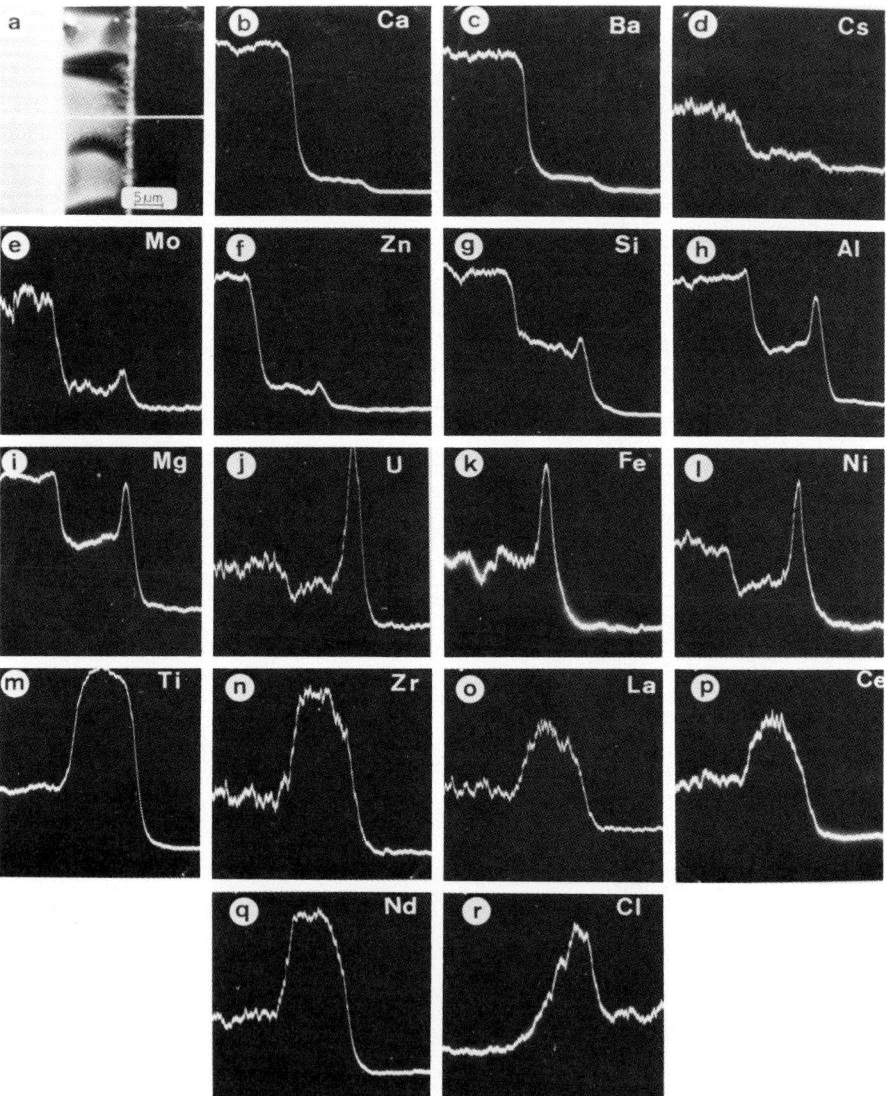

Fig. 1. SEM photomicrograph and X-ray line scans of a glass bead leached 10 d
at 200°C in saturated NaCl-solution.
    (a) SE-micrograph with line scan position
    (b) - (r) line scans of elements as indicated

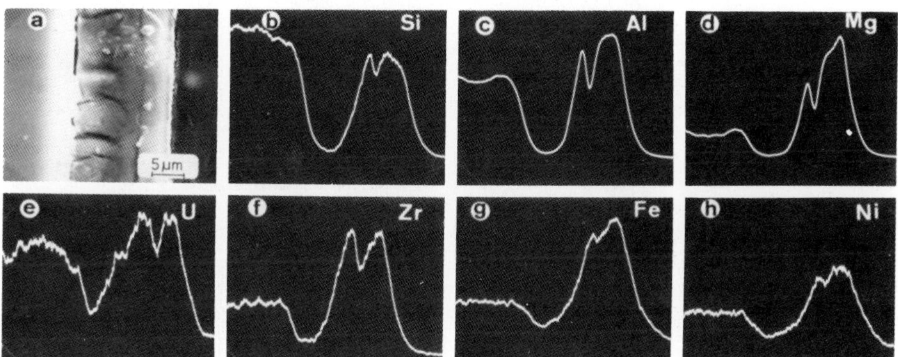

Fig. 2. SEM photomicrograph and X-ray line scans of a glass bead leached 30 d
at 200°C in saturated NaCl-solution.
(a) SE-micrograph with line scan position
(b) - (f) line scans of elements as indicated

The line scans of the 10 d leached surface in Q for Ca, Ba, Cs and Mo are the
same as shown in figure 1 for N. However, Zn has disappeared and Si is not increased in the outer layer. Again Mg and Al have a concentration profile with
steps as shown in figures 4b and c resp. As seen in figures 4d and e, the U and
Fe profiles look significantly different when compared with the corresponding
surfaces leached in N, since there is no U and little Fe left. Ni, Ti, Zr, Cr
are strongly enriched but have varying concentrations in the surface layer
(figures 4f - h).

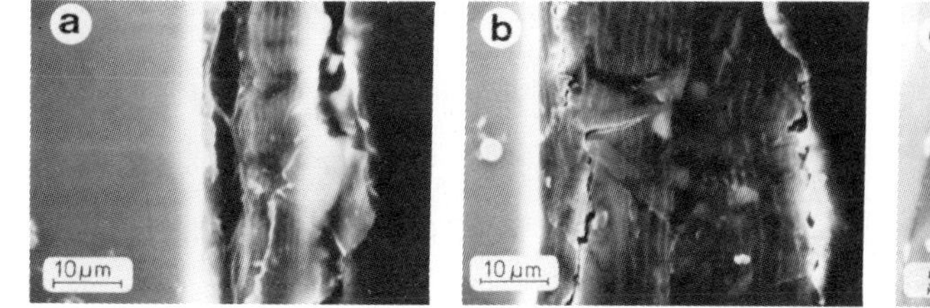
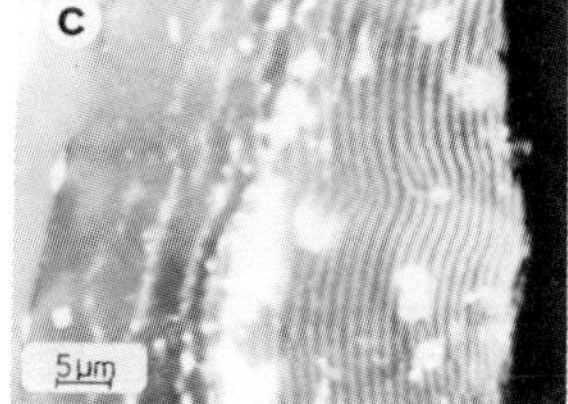

Fig. 3. SEM photomicrographs of glass beads leached at 200°C.
(a) 10 d in Q-solution, (b) 10 d in Z-solution,
(c) one year in quinary brine

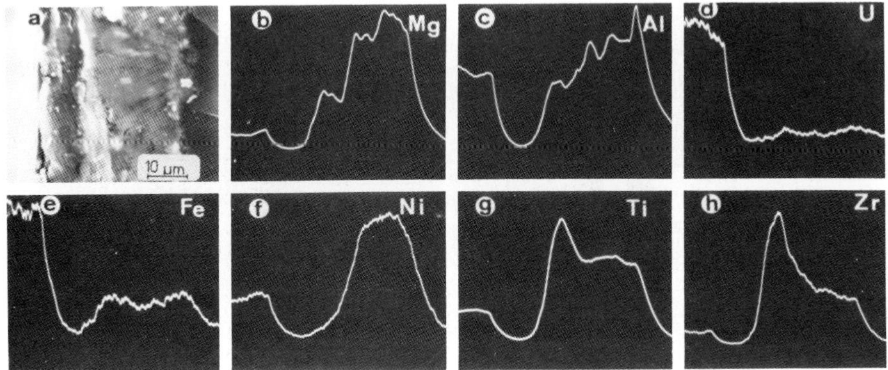

Fig. 4. SEM-photomicrograph and X-ray line scans of a glass bead leached
        10 d at 200°C in Q-solution
        (a) SE-micrograph with line scan position
        (b) - (h) line scans of elements as indicated

Figure 5 shows the results for glass leaching in Z. The lamellar structure of
the scale and an outer thin scale can be seen. Remarkable differences in line
scans from surfaces leached in N and Q become evident when comparing figures 5b,
c with figures 1 and 4. The Z-leached layer showed strong enrichment of Ti and
Zr only in the outermost scale, whereas enrichment of these elements is apparent in both layers of the N-leached and in the Q-leached samples in the inner
layer, figures 1 and 4.

The Q- and Z-leached surfaces showed an enrichment of Cl, S and K (figures
5d - f) which are components of the leachants. Obviously, they contributed to
deposits in the layers. These deposits probably precipitate at higher temperatures, i.e. $\sim 200°C$ but not during the cooling period of the autoclave. This was
inferred from the observation that the phases, containing brine constituents,
formed near the boundary of the pristine glass rather than on the outer surface
of the scale. Figure 6a shows an SEM-photomicrograph of the cross-sectioned surface of a bead leached for one year in quinary brine (from the ASSE II salt
mine.) Figure 6b shows an energy-dispersive X-ray spectrum. The electron beam
position is indicated by the black point in figure 6a. As seen the layer contains S and Ba, which is probably $BaSO_4$.

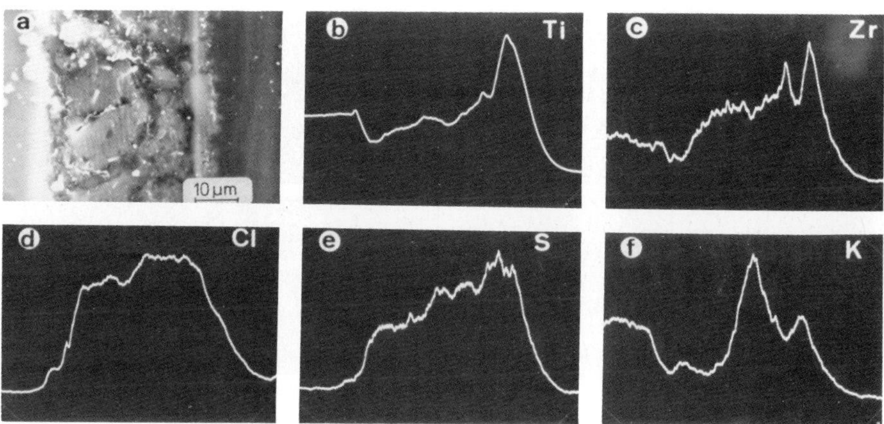

Fig. 5. SEM photomicrograph and X-ray line scans of a glass bead leached
        10 d at 200°C in Z-solution.
        (a) SE-micrograph with line scan position
        (b) - (f) line scans of elements as indicated

The various structures of the surface layer can be seen in the figures 7

and 8 which show SEM photomicrographs and EDX spectra of a surface layer re-

moved from the glass bead after 10 d leaching in Q. In top view an upper layer

was found, figure 7a, which was removed leaving an under layer shown in

figure 8a. The upper layer appears amorphous, figure 7a, whereas the one beneath

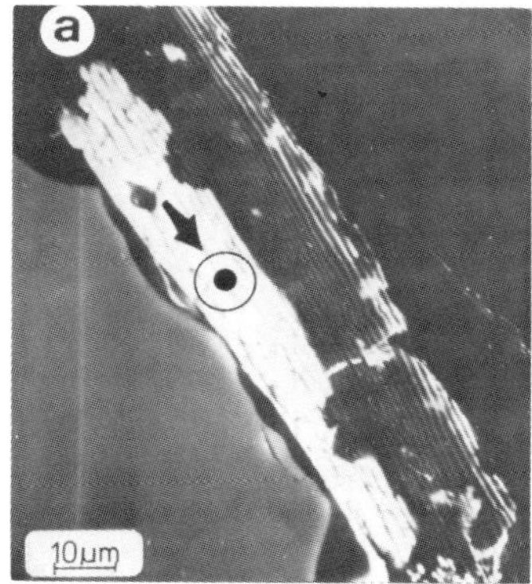
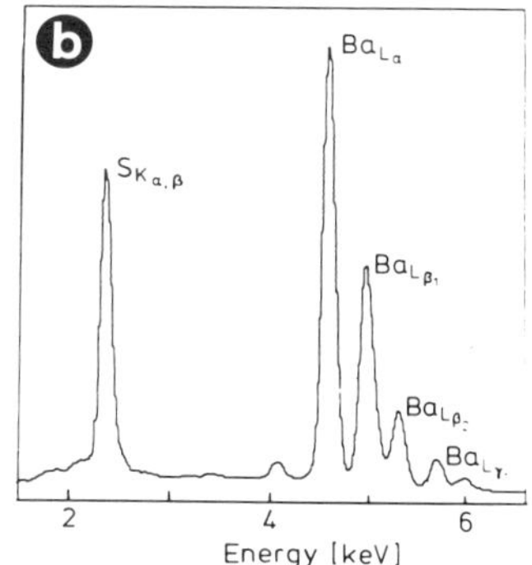

Fig. 6. (a) SEM photomicrograph of a cross-sectioned glass surface leached
        one year at 200°C in quinary brine
        (b) EDX spectrum of the phase indicated by the black point in
        figure (a)

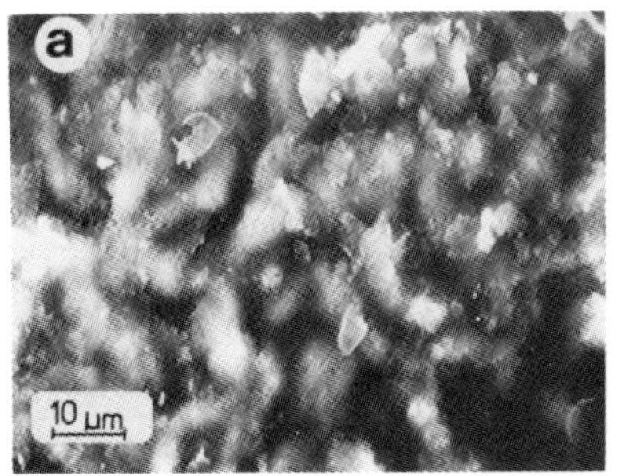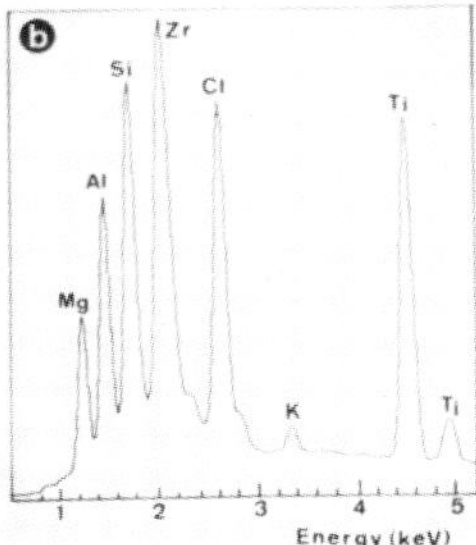

Fig. 7. (a) SEM photomicrograph of the surface layer of a glass bead
leached 10 d at 200°C in Q-solution
(b) EDX spectrum of the surface layer in figure (a)

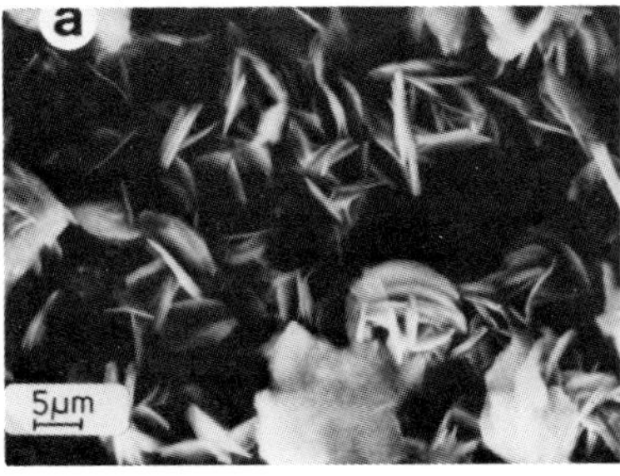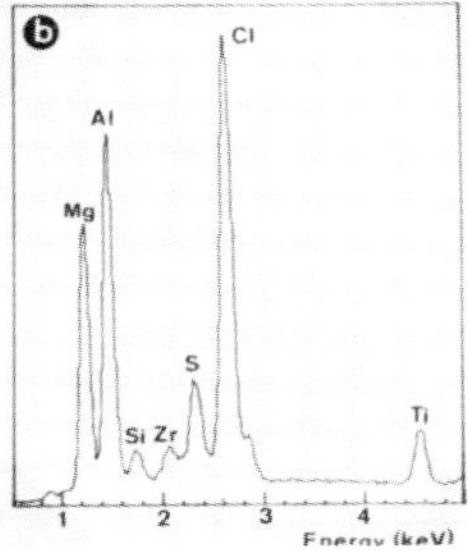

Fig. 8. (a) SEM photomicrograph of a surface layer of a glass bead after
10 d leaching at 200°C in Q-solution
(b) EDX spectrum of the surface layer in figure (a)

seems to contain crystals (figure 8a). Microprobe analyses revealed remarkable
differences of the layers' composition, figures 7b and 8b. The spectrum of the
crystalline phase, figure 8b, appears to belong to a Mg-Al-Cl containing com-
pound.The upper layer shows high contents of Si, Zr and Ti in addition to Mg,
Al and Cl (figure 7b).

X-ray diffraction patterns of the surfaces leached in N, Q and Z provided
evidence for the presence of various crystalline phases in all layers. The
phase identification is still underway. Until now only $BaSO_4$ could be identified.

The results from semiquantitative wavelength dispersive EPMA of the surface
layers are given in table 3. For the purpose of comparison the last column
contains the values of the surface layer formed during leaching in deionized
water. These results are based on our earlier work.[3] As seen the elements Ca,
Ba, Cs, Mo and Zn are strongly depleted in the layers leached in salt solutions,

TABLE 3

RELATIVE COMPOSITION OF SURFACE LAYERS ON GLASS, FORMED DURING LEACHING
AT 200°C IN VARIOUS SOLUTIONS

Element	NaCl-solution/10d		NaCl-solution/30d		Q-solution/10d		Z-solution/10d		H₂O/3d	
	inner	outer	inner	outer	inner	outer	inner	outer	inner	outer
		scale		scale		scale		scale		scale
Ca	− −	− −	− −	− −	− −	− −	− −	− −	−	−
Ba	− −	− −	− −	− −	− −	− −	− −	− −	−	−
Cs	− −	− −	− −	− −	− −	− −	− −	− −	− −	− −
Mo	− −	− −	− −	− −	− −	− −	− −	− −	− −	− −
Zn	− −	−	− −	− −	− −	− −	− −	− −	−	+ +
Si	− −	−	− −	o	−	− −	− −	− −	− −	−
Al	− −	−	−	+	+	+ +	o	+ +	−	+ +
Mg	−	o	−	+ +	+ +	+ +	+ +	+ +	−	+ +
U	− −	+ +	−	+	− −	− −	− −	− −	o	+
Fe	o	+ +	−	+ +	−	−	− −	− −	o	+ +
Ni	−	+ +	−	+ +	+ +	+ +	−	o	−	+ +
Ti	+ +	+ +	+ +	+ +	+ +	+ +	o	+ +	+ +	+ +
Zr	+ +	+ +	−	+ +	+ +	+ +	+	+ +	−	+ +
La	+ +	+ +	+ +	+ +	− −	− −	− −	− −	+	+
Ce	+ +	+ +	−	−	− −	− −	− −	− −	−	−
Nd	+ +	+ +	+ +	+ +	− −	− −	− −	− −	+ +	+ +
Pr			+ +	+ +	− −	− −	− −	− −	+	+
Cr					+ +	+ +	+ +	+ +	o	o
Cl	+	+ +	o	+ +	+ +	+ +	+ +	+ +		
S					+	+ +	+ +	+ +		
K					+	o	+ +	+		

− −	strongly depleted:		$c_l/c_g < 0.5$
−	depleted:		$0.5 < c_l/c_g < 1$
o	unchanged:		$c_l/c_g \approx 1$
+	enriched:		$1.5 > c_l/c_g > 1$
+ +	strongly enriched:		$c_l/c_g > 1.5$

C_g: concentration of element in the unleached glass
c_l: concentration of element in the surface layer

whereas Zn is strongly enriched in the outer scale of the water leached surface.
Al and Mg are enriched in the Q- and Z-leached surfaces and also in samples
leached 30 d in N, whereas the samples leached in water are higher in Al and Mg
after 3 d. Ti and Zr concentrations were high in layers formed from all solutions. La, Ce, Nd and Pr are enriched only in the layers leached in N and are
strongly depleted in Q- and Z-leached surfaces. Therefore, most of the elements
are enriched in the H_2O-leached surface and least of all in Q- and Z-leached
surfaces. In other words, in Q- and Z-brines most elements of the glass were
leached. This means that the tendency toward selective leaching decreases and
congruent dissolution increases in the following order: H_2O, NaCl, Q, Z.

Nevertheless, as mentioned previously, the specific weight losses do not show differences beyond one order of magnitude. Thus, all types of surface layers protect the underlying glass. This may mainly be due to the ions Ti, Zr and Ni, which are present in the outer scales of all layers. Additionally the rare earth elements Ce, Nd and La are constituents of layers leached in N.

In all types of surface layers, independent of the salt brine, chlorine was present, and S and K in layers leached in Q and Z. When comparing the concentrations of Si and Mg in the layers it was found that the Q leached layer contained only 1.6 w/o Si and 13 w/o Mg in contrast to $\sim$ 16 w/o Si and 0.8 w/o Mg in the unleached glass resp. This means that only 10 % of the original silicon amount is in the layer, but the Mg-content was increased about a factor of 10. In terms of the glass structure then only every 10th SiO_4-tetraeder, the main network former, is still present but certainly unable to preserve the glass structure. Therefore, when leaching the glass in Q the glass network is nearly completely destroyed and the layer formed on the surface is assumed to have no longer structural similiarities with the original glass structure. This presumption is further supported by the fact that the surface layers are partly crystalline and do not look homogeneous. There are enrichments of some elements, as seen in figures 9b - i, showing steps in X-ray line scans.

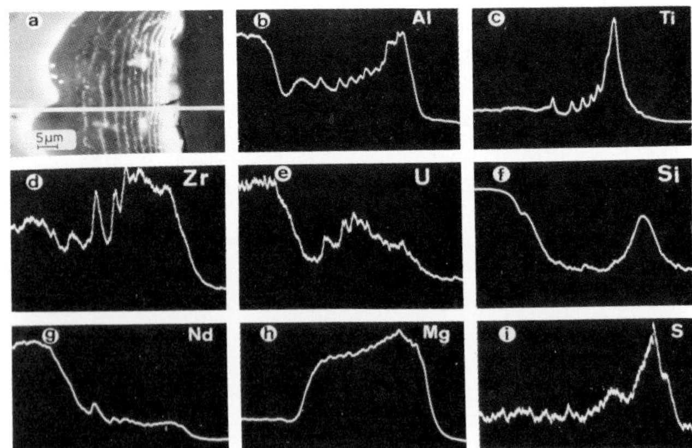

Fig. 9. SEM photomicrograph and X-ray line scans of a glass bead leached one year in quinary brine at 200°C.
(a) SE micrograph with line scan position
(b) - (i) line scans of elements as indicated

The surface layer of the one year leached sample adhered very strongly to the surface and could not be removed while on the contrary layers peeled off

more readily after short leaching time. Longer leaching periods yielded denser
surface layers. Obviously, with increasing leaching time the mechanical stability of the surface layer increases but the layer does not become thicker after
30 days. The specific weight loss after one year was about 5 to 6 mg·cm^{-2}, the
same as after 3 - 10 d leaching, table 2. The leaching rate slowed down close to
zero. Hence, the release of matter was discontinued but the surface layer became increasingly impervious for both the leachants and leachates.

CONCLUSIONS

Leaching with various leachants leads to the formation of surface layers.
Their composition and structures are complex. From the relative concentrations
of the elements in the layers it must be assumed that the layers do not represent a partially leached glass phase and are formed by instantaneous precipitation of insoluble compounds, formed upon leaching.

The drop in the leach rates suggests that the surface layer protects the
underlying glass against further attack. Therefore, the chemical and mechanical
long-term stability of the layers are the most important properties which must
be known to understand the corrosion kinetics and leaching mechanisms of the
glass. The mass **release seems to** be best fitted by the equation $m_{(t)} = c \cdot t^n$,
but data are lacking for a precise calculation of the parameters n and c for all
relevant conditions; n was found to be well below 0.5. Hence, congruent dissolution and diffusion controlled ion exchange, i.e. selective leaching, can be
excluded as rate determining leaching mechanisms.

The significance of the stability of the layers and the value of n for pre-
diciton of activity release from nuclear waste glasses over longer periods of
time was demonstrated by Altenhein and Lutze.[1]

Although the surface layers formed in the various leachants are different in
nature and composition, the protective property for the underlying glass can be
assumed to be a common feature.

ACKNOWLEDGEMENT

The author thanks W. Lutze for discussing the manuscript.

This work was in part supported by the Commission of the European Communities,
under Contract WAS-122-53D..

REFERENCES

1. Altenhein, F.K., Lutze, W. (1982) in Scientific Basis for Nuclear Waste
   Management, Vol.11, Lutze, W. ed., Elsevier North-Holland, New York, to
   be published.
2. Hench, L.L., Clark, D.E., Yen-Bower, E.L. (1980),
   Nucl. Chem. Waste Management 1, 59-79.
3. Altenhein, F.K., Lutze, W., Malow, G. (1981) in Scientific Basis for
   Nuclear Waste Management, Vol. 3, Moore, J.G. ed.,
   Plenum Press, New York, pp. 363-370.
4. Marples, J.A.C. et al. (1981) European Appl. Res. Rept.-Nucl. Sci.Technol.3,
   pp. 395-484, EUR 7138 EN.
5. Malow, G. et al. (1982) European Appl. Res. Rept-Nucl. Sci. Technol.,
   to be published.
6. Heimerl, W. (1977) Annals of Nuclear Energy, 4, 237-77.
7. Röthemeyer, H. (1980) Int. Atomic Energy Agency, Vienna, 1, pp. 297-310.
8. Conradt, R., Engelke, H., Kaiser, A. (1982) in Scientific Basis for
   Nuclear Waste Management, Vol.11, Lutze, W. ed., Elsevier North-Holland,
   New York, to be published.

STABILITY AND ALTERATION OF NATURALLY OCCURRING LOW-SILICA GLASSES: IMPLICATIONS FOR THE LONG TERM STABILITY OF WASTE FORM GLASSES

CARLTON C. ALLEN
Department of Geology and Institute of Meteoritics, University of New Mexico, Albuquerque, NM   87131   USA

ABSTRACT

Volcanic glass of basaltic composition (approximately 50 wt.% SiO_2) was studied as a natural analogue to proposed nuclear waste form glasses. Basaltic glass can resist devitrification and remain glassy for periods in excess of 10 m.y. Surface alteration of the glass occurs rapidly upon exposure to water or hydrothermal fluids, but is essentially absent if the glass remains dry. Alteration is restricted to a rind of clay-like, cryptocrystalline palagonite, generally less than 100 μm thick. Formation of this rind protects the underlying glass from further alteration. Similar rinds, or "gel layers", have been produced in laboratory and field tests with borosilicate glasses. Major cations within the basaltic glass are selectively depleted or enriched relative to the glass in the alteration rind. Qualitatively-similar depletion and enrichment patterns characterize the short-term leaching behavior of candidate waste form glasses.

INTRODUCTION

If low-silica glasses are to be a medium for the disposal of nuclear waste, the stability and leaching behavior of such glasses over periods of $10^4 - 10^7$ years must be understood. This problem has been extensively studied in laboratory experiments, and the short-term behavior of proposed waste form glasses in a variety of environments is well known. However, laboratory experiments are limited in duration to at most a few years, so that the long-term behavior of waste form glass must be predicted based on large extrapolations. An alternative procedure has been to study natural glass samples in the geologic record. In this case the appropriate time scales can sometimes be approximated. Previous studies of natural glasses have focused on high-silica compositions subjected to alteration conditions which were not always well understood. The present investigation concerns low-silica glass of a variety of ages which was altered under conditions similar to those predicted in a nuclear waste repository.

SAMPLES AND METHODS

The samples consist of millimeter-sized shards of the orange basaltic glass sideromelane. The glass was formed by water quenching of volcanic magma in subglacial and submarine eruptions. Samples were collected from sites in the United States, Iceland and Antarctica[1]. Representative chemical compositions for these samples and several proposed nuclear waste form glasses are listed in Table 1. These proposed waste form glass compositions have roughly the same SiO_2 and alkali ($CaO + Na_2O$) contents as basaltic glass. Engineering considerations, however, dictate that the synthetic glasses contain considerable B_2O_3 and little or no Al_2O_3.

TABLE 1

CHEMICAL COMPOSITIONS (WT. %) OF NATURAL AND WASTE FORM GLASSES

| | Sideromelane* | Simulated Waste Form Glasses (excluding radionuclides) | |
	Iceland[1]	Battelle 76-101[2]	Battelle 3008[2]
SiO_2	48.1	59.7	44.3
B_2O_3	----	14.2	11.4
TiO_2	2.8	4.5	4.8
Al_2O_3	12.9	----	----
Fe_2O_3	2.3	----	12.3
FeO	11.4	----	----
ZnO	----	7.5	5.7
NiO_2	----	----	2.4
Cr_2O_3	----	----	0.5
MgO	6.5	----	----
CaO	12.0	3.0	3.0
Na_2O	2.5	11.2	15.7
K_2O	0.4	----	----
P_2O_5	0.4	----	----
TOTAL	99.3	100.1	100.1

*mean of 6 microprobe point analyses; empirical Fe_2O_3/FeO partitioning = 0.2

Samples were studied using optical and scanning electron microscopes (Hitachi S-450 SEM) and an electron microprobe (ARL EMX-SM). Polished sections were analyzed in the microprobe using 15 KeV accelerating potential, 18 nA sample current and a 3-5 μm beam spot size. Analyses were calibrated with mineral standards and a Bence-Albee data reduction technique, and the procedure was checked against a well-characterized synthetic basalt glass.

RESULTS

Stability against devitrification. Freshly-quenched sideromelane is usually clear and isotropic. This condition implies quenching from magmatic temperatures to essentially 0° C in a few seconds[3]. Several grains in the present samples contain abundant plagioclase microcrystals as long as 100 μm. These are primary crystallization products probably formed in the supercooled melt during quenching.

The microcrystals incorporated in otherwise pristine glass are easily distinguished from the masses of fine crystals formed by later devitrification in some samples. A 600,000 year old sample from Alaska appears completely glassy and isotropic, while a 22 m.y. old sample from Antarctica is extensively devitrified, with only a few glassy areas remaining. Glass produced by an eruption 13 m.y. ago and then isolated from contact with water for most of its existence remains strikingly fresh and transparent[4]. Thus devitrification of basaltic glass at temperatures near 0° C requires on the order of 10 m.y.

Sideromelane is the best natural analogue to waste form glass in terms of devitrification rates, due to the similar abundance of network-forming oxides $SiO_2 + Al_2O_3$ (60 - 65 wt.% in sideromelane, approximately 45 - 55 wt.% in waste form glass)[5]. Other common natural glasses, such as obsidian, have $SiO_2 + Al_2O_3$ greater than 80 wt.%. Basaltic glass may be expected to devitrify more easily than those glasses with higher contents of network-forming oxides.

Surface alteration. Basaltic glass preserved in dry environments can remain unaltered for exceedingly long periods of time. Several of the present samples which are tens to hundreds of thousands of years old show no sign of alteration. Moreover, the 13 m.y. old glass discussed above was reported to be essentially unaltered[4].

Many sideromelane samples, however, do show the effects of alteration in the form of a surface rind (Fig. 1). This rind is composed of a complex mixture of iron-rich, cryptocrystalline clays which are grouped under the general term "palagonite"[1]. Alteration rinds on basaltic glass are strikingly similar in morphology to "gel layers" observed in laboratory[6] and field[7,8] studies of the hydrothermal alteration of borosilicate glasses (Fig. 2).

The rinds on our natural glass samples range in thickness from near zero to 100 μm, though the characteristic rind thickness in any given hand specimen remains relatively constant from grain to grain. The rinds are, in many cases, separated from the underlying glass by a gap of a few microns and are

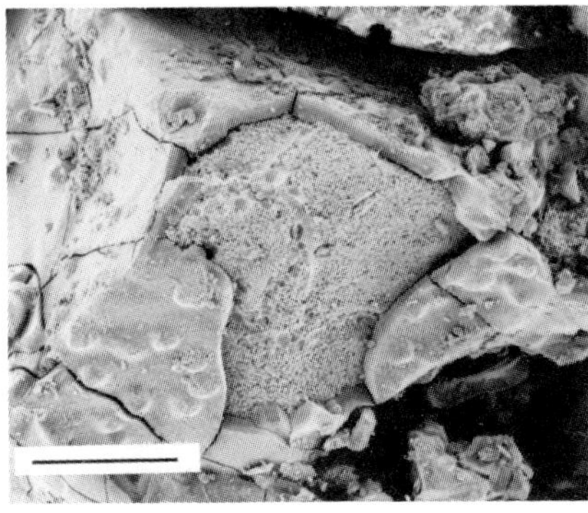

Fig. 1a   Sideromelane grain with 15 μm rind, formed by hydrothermal alteration in Iceland. Note dehydration cracks and separation of the rind from underlying glass. Secondary electron SEM photo (Fig. 3 of [1]). Scale bar equals 100 μm.

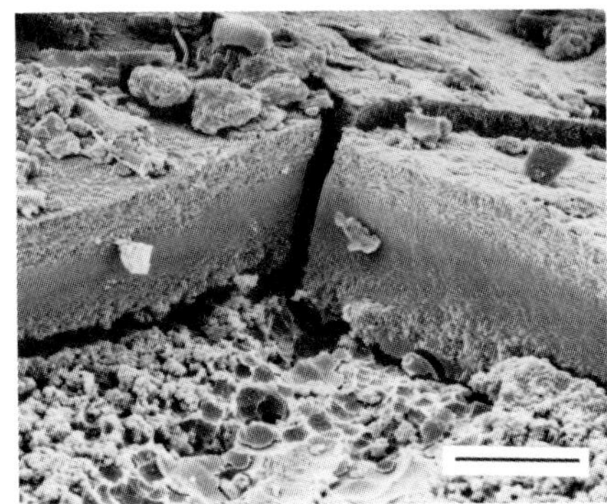

Fig. 1b   Palagonite rind overlying sideromelane (detail of Fig. 1a). The rind exhibits distinct layering but only a single generation of dehydration cracks. Secondary electron SEM photo. Scale bar equals 10 μm.

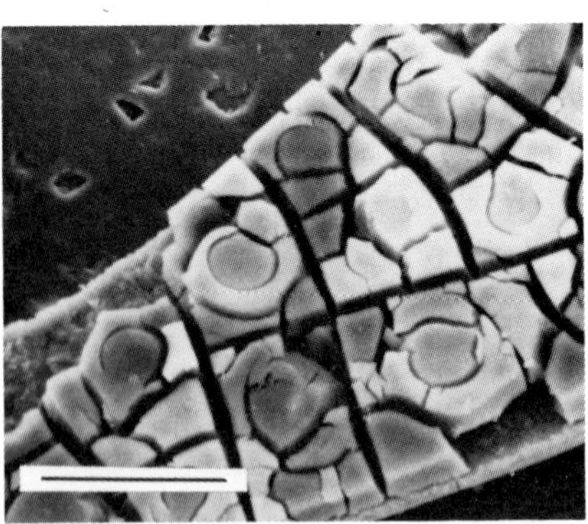

Fig. 2a   Interior surface of gel layer on borosilicate (PNL 75-25) glass developed in 243-day buried heater experiment in Conasauga shale. Note extensive dehydration cracking. Secondary electron SEM photo (Fig. 49 of [8]). Scale bar equals 50 μm.

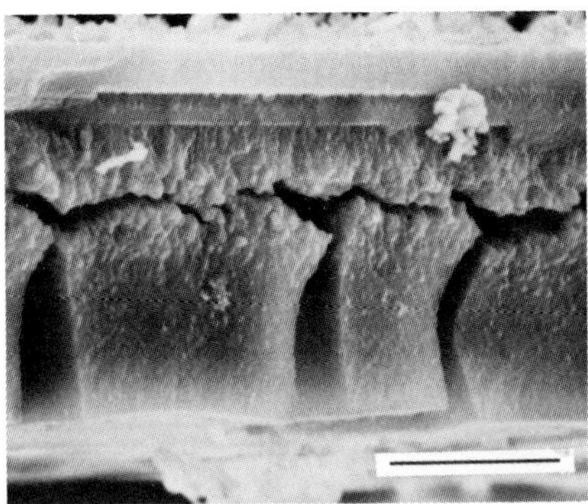

Fig. 2b   Cross section of gel layer on borosilicate glass shown in Fig. 2a. The layer consists of four distinct portions, reflecting several episodes of alteration, but only the layer nearest the glass (bottom) is extensively cracked. Secondary electron SEM photo (Fig. 46 of [8]). Scale bar equals 10 μm.

characteristically broken into a number of discrete curving plates by cracks oriented normal to the glass interface (Fig. 1). This cracking and separation, attributed to post-hydrothermal dehydration[1], is closely approximated in the gel layers of borosilicate glass (Fig. 2).

Palagonite rinds on sideromelane form in response to interaction with ground water or hydrothermal fluids. Rinds which are indistinguishable in morphology and bulk chemical composition coat basaltic glass shards exposed to infiltrating rain water in Hawaii and steam in Iceland[1]. Alteration to depths of tens of microns has been observed to occur within a period of a few years in a 100° C hydrothermal field on the Icelandic island of Surtsey[9]. In laboratory leaching experiments at 200° C in water, gel layers 5 µm thick were formed on borosilicate glass in three days[6].

Alteration of natural basaltic glass is a surface phenomenon which is effectively halted after the formation of the initial palagonite rind. Below the palagonite interface the glass remains pristine, with the exception of localized alteration within large vesicles. In the natural environment a palagonite coating tens of microns thick acts as an effective barrier to alteration of the underlying glass. A similar result has been reported for leaching experiments on time scales of weeks using borosilicate glass[6].

The alteration rinds in the present natural samples are characterized by distinct internal layering (Fig. 1b) but only a single generation of dehydration cracks. This indicates that they were subjected to only a single cycle of alteration/wetting followed by dehydration. More complex morphology is seen in waste form glass from a 243-day buried heater experiment[7,8] in which the glass was subjected to several distinct episodes of steam interaction followed by drying. The resulting gel layer consists of four overlying portions, only one of which displays pervasive cracks (Fig. 2b). Laboratory studies of waste form glass leaching has produced similar results. After leaching and dehydration, portions of the gel layer spall off the glass, and renewed leaching can lead to the formation of subsequent gel layers as the newly-exposed glass is attacked (W. Lutze, personal communication, 1981). It should be noted that no cases of multiple alteration rinds, and only limited spalling of dehydrated rinds, were observed in the present natural samples.

<u>Selective leaching of major elements</u>. The alteration of basaltic glass involves hydration of the glass and selective leaching of the various elements. Palagonite rinds in geological samples contain approximately 10 - 20 wt.% water, based on deficits in microprobe totals[1]. Since these samples generally exhibit

42

cracking due to partial dehydration, the water content of freshly altered
material may be even higher.

The oxides of eight major cations (Si, Ti, Al, Fe, Mg, Ca, Na, K) comprise at
least 98 wt.% of most basaltic glass samples. Each cation shows a unique geochemical behavior in the alteration process. Na is leached almost totally from
the alteration rind and the loss of Ca is considerable. Al and Si remain relatively constant in concentration. Fe is concentrated throughout the rind, while
Mg, K and Ti are concentrated in distinct layers parallel to the glass interface. Fig. 3 shows X-ray maps of the distribution of five of these cations.

Studies of gel layers formed by hydrothermal alteration of borosilicate glass
have shown patterns of cation depletion and concentration generally similar to
those described for sideromelane. X-ray line scans for glass leached at 200° C
in water show strong enhancement of Mg, Ti, Fe and Al in the gel layer, with Ca
strongly depleted[6]. In contrast to the present results, Si is also depleted
in the waste form samples, while Na is enriched. A more limited water leaching
study, conducted at 75° C, showed concentration of Ti and Fe and depletion of Si
and Na in the gel layers of waste form glasses[2].

These results suggest that, for waste form glasses, leaching behavior is a
complex function of glass composition and alteration conditions. In the natural
samples, however, depletion and enhancement patterns of glass altered under
different conditions are strikingly similar. Basaltic glass altered by rain
water in Hawaii and by hydrothermal fluids in Iceland, Alaska and Antarctica
displays nearly identical patterns of major element concentration changes[1].

Nuclear waste form studies are less concerned with the loss of the major
elements than with potentially hazardous radionuclides. The two reports discussed above contain data on the leaching behavior of U and the lanthanides.
Leaching at 75° C[2] produced enhanced surface layer concentrations of these
elements. The 200° C experiments[6] also showed concentration of U in the gel
layer, along with La and Pr. However, the Ce content was somewhat reduced in
the gel layer, relative to the glass. In both studies the leaching behavior of
U imitated that of Ti. Thus, our natural analogue studies predict that at least
some of the radionuclides of interest in a repository, while being mobilized
when glass is leached, should remain trapped within the alteration rind.

Radionuclides which are removed from waste form glass by leaching may still
not be lost from the system. Common byproducts of natural palagonitization are
zeolites formed between glass grains. Chabazite and analcite are the most
common zeolites in altered sideromelane[1]. Zeolites have excellent properties as
molecular sieves for the removal of radioactive wastes, and have been so used in

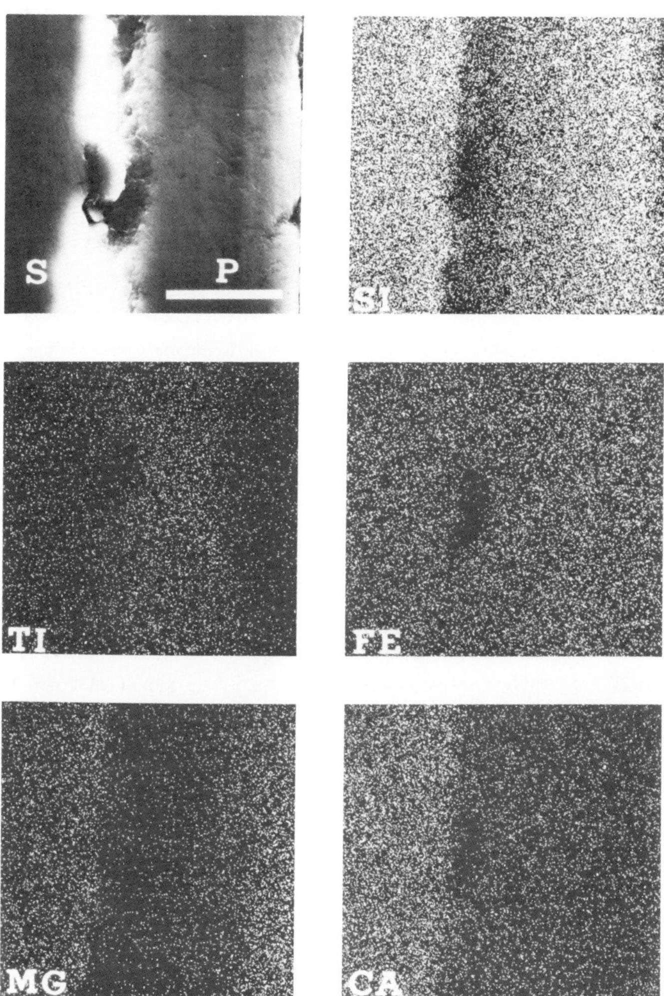

Fig. 3   Cross section through sideromelane (S) and palagonite (P) rind in a sample from Iceland.   X-ray maps of the same area show the relative concentrations of Si, Ti, Fe, Mg and Ca in the glass and rind.   Secondary electron SEM photo of a polished section.   Scale bar equals 10 μm.

a number of experimental and operational processes[10]. Thus, if zeolites are produced during the alteration of waste form glass they could effectively trap part of the radioactive material lost from the glass during leaching.

CONCLUSIONS

Low-silica basaltic glass is a useful natural analogue to borosilicate glass in studies of stability and leaching behavior. Basaltic glasses can resist devitrification for up to 10^7 years. If the glass is isolated from contact with water, little or no alteration occurs. If water and/or steam are introduced into the system, surface alteration will occur rapidly, penetrating to depths of tens of microns in less than a year. With the exception of glass subjected to multiple wetting/drying cycles, however, the alteration rind will protect the underlying glass from further leaching for millions of years. U and some lanthanides which are mobilized by leaching may remain trapped within the rinds. Zeolites, common byproducts of glass alteration, should trap a portion of the remaining volume of released radionuclides.

ACKNOWLEDGMENTS

The author thanks J. Allen, J. Calhoun, M. Fillmon and J. Salas for technical support and Drs. R. Waterman and J. Krumhansl for several of the SEM photos. Dr. R. Ewing made the author aware of the value of basaltic glass as a waste form analogue and provided considerable support and assistance. Figs. 1a,b are used with the permission of Academic Press, Inc. This research was conducted under NASA Grant NSG-7579 to Dr. K. Keil.

REFERENCES

1. Allen, C.C. et al. (1981) Icarus, 45, 347.
2. Karim, D.P. et al. (1980) in Scientific Basis for Nuclear Waste Management, Moore, J.G. ed., Plenum, N.Y., 3, 397.
3. Moore, J.G. (1976) American Scientist, 63, 269.
4. Schmincke, H.-U. et al. (1978) in Initial Reports of Deep Sea Drilling Project, Dimitriev, L. et al. eds, XLVI, 341.
5. Dran, J.C. et al. (1980) in Scientific Basis for Nuclear Waste Management, Moore, J.G. ed., Plenum, N.Y., 3, 449.
6. Altenhein, F.K. et al. (1980) in Scientific Basis for Nuclear Waste Management, Moore, J.G. ed., Plenum, N.Y., 3, 363.
7. Ewing, R.C. and Krumhansl, J.L. (1979) Amer. Nuclear Soc. Trans., 33, 280.
8. Krumhansl, J.L. (1979) Sandia Laboratories SAND79-1855.
9. Jakobsson, S.P. (1978) Bull. Geol. Soc. Denmark, 27, 91.
10. Mumpton, F.A. (1977) in Mineralogy and Geology of Natural Zeolites, Mumpton, F.A. ed., Mineralogical Society of America Short Course, Southern Printing Co., Blacksburg, Va., 4, 177.

Published 1982 by Elsevier Science Publishing Co
SCIENTIFIC BASIS FOR RADIOACTIVE WASTE MANAGEMENT - V
Werner.Lutze, editor

LONG-TERM RADIOACTIVITY RELEASE FROM SOLIDIFIED HIGH-LEVEL WASTE -
PART I: AN APPROACH TO EVALUATING EXPERIMENTAL DATA

Friedrich K. Altenhein and Werner Lutze
Hahn-Meitner-Institut für Kernforschung Berlin, Glienicker Strasse 100,
D 1000 Berlin 39, Germany
Rodney C. Ewing
The University of New Mexico, Albuquerque, New Mexico 87131, USA

INTRODUCTION

Safety and risk analyses for the isolation of radioactive waste in a repository must begin with a source term to quantify the amount of radioactivity released from the waste form under a specific set of conditions. The interaction of the waste form with aqueous solutions is the most important mechanism to consider, as any material released may be dissolved and reach the biosphere. In this regard the behaviour of heat generating high-level waste is of particular importance, because reaction rates are higher at elevated temperatures. A long-term leach rate was derived from previous and continuing experimental work. The purpose of this paper is not to describe the "real case" release but rather to provide guidelines for the design of leaching experiments and determine the required precision for the data. This can be derived from the relative sensitivity of extrapolated leach rates for various parameters measured in laboratory experiments.

EXPERIMENTAL DATA

Experiments have been completed and are in progress in order to investigate the leaching mechanism when pure water, concentrated rock salt (NaCl) and mixtures of NaCl, KCl, $MgCl_2$ and Na_2SO_4 which come into contact with the glass waste form (C-31-3EC) at temperatures $\leq 200°C$. The purpose was to understand the long-term performance of borosilicate glass, in the (unlikely) event, that parts of the repository are flooded before sealing. It was inferred from leaching data, SEM and EPMA investigations that a layer formed on the glass surface upon leaching. The scale was found to be adhesive in all cases. Its composition was different in the three leachants. The scale's composition and structure changed as a function of time, the strongest changes occurring during the first three days. Simultaneously the leach rate dropped sharply. The leaching data for NaCl indicated a $t^{0.5}$ dependency of the integrated mass loss up to three days, at $200°C$. Solid state diffusion is probably rate determining.

46

The growing film assumed a layered structure and the exponent of t dropped to
0.3. It appears that the film becomes increasingly impervious and prevents
further reactions with the underlying glass. A discussion of these findings is
contained in the paper by Malow presented at this symposium [2]. Additionally,
the role of such layers in accounting for the long-term stability of silicates
in aqueous solutions has been of considerable interest to geochemists [3].

SOURCE TERM

In this section some parameters will be derived from experimental work and
the source term for both mass and radioactivity release formulated. A plot of
log m(t), the integrated mass loss of the glass sample, vs. log t, yields a
straight line, in Figure 1. From this figure a slope, n, was calculated. Values
for n may depend on temperature [4]. There is no simple explanation for the temperature dependency of n. Values of n are effected by the leaching mechanism,
solubility limits (the ratio of sample surface area (SA) to solution volume
(V)), and the pH of the solution. There is experimental and geologic evidence
that n will change as a function of temperature and time [5]. Equilibria (saturation of the solution in a closed system) are attained earlier and impervious
layers may form faster at higher temperatures. If SA/V is high or solubilities
are low, n can approach zero and the reaction ceases. If, however, selective
leaching of single glass constituents lasts very long before a steady state is
reached and the process is not solubility limited, the dissolution rate of the
surface film finally determines the total leach rate [6]. In this case n approaches one. Until there is a more detailed understanding of the leaching
mechanism, n must be treated as an empirical parameter which is varied within
certain limits. Long-term measurements of n at temperatures lower than 200°C
are underway; however, the temperature dependency of the leach rate, dm/dt, has
been measured in solutions N and C, and the leach rate may be given as:

$$L(t) = \frac{dm(t)}{dt} = n \cdot A \cdot \exp(-b/T) \cdot t^{n-1} \tag{1}$$

A and b are determined from an Arrhenius-type plot, with b containing an activation energy for the rate determining process and A being a constant. A
physical – chemical interpretation can only be obtained when more data are
available on leaching mechanism. Westsik et al.[4] have found the same relationship when measuring B, Mo, Na, Si concentrations up to 341 days in glass/water
leaching experiments between 25° and 250°C with a value of n $\approx$ 0.67 between
50°C and 150°C. A similar equation has been used to describe the dissolution

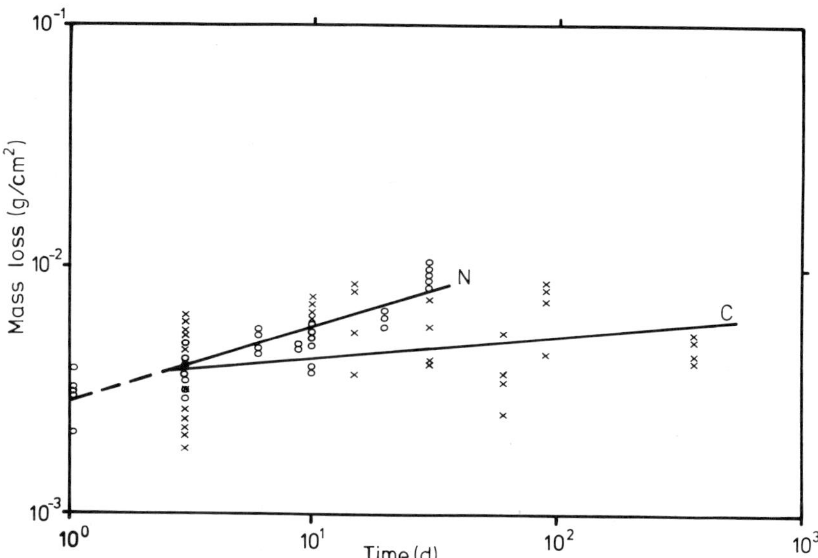

Fig. 1.Integrated mass loss of glass at 200°C as a function of time
(N = NaCl solution, C = NaCl, KCl, MgCl$_2$, Na$_2$SO$_4$ solution).

of naturally occurring silicates over long periods of time [7].

RELEASE SCENARIO FOR A ROCK SALT REPOSITORY

It is assumed that large quantities of brine (SA/V $\approx$ 0.01 cm^{-1}) have access
to the waste package. Once in place, the brine is heated to a temperature of
200°C and becomes saturated with NaCl. This temperature is presently used as an
upper limit for the temperature at the salt-canister interface [8]. The corrosion
of the canister is not considered. Instead a series of distinct life-times have
been attributed to the canister, after which it disappears at that instant. The
thermal history of the repository was taken from model calculations for a northern German repository in rock salt [9]. An empirical function was fitted to the
data for the period of temperature decrease:

$$\frac{1}{T} = a_o - a_1 \exp(-a_2 t) = f(t) \tag{2}$$

48

with a_o = 3.165 $\cdot$ 10^{-3} K^{-1}, a_1 = 1.051 $\cdot$ 10^{-3} K^{-1} and a_2 = 2.3 $\cdot$ $10^{-3} y^{-1}$.

(1) then becomes

$$L(t) = \frac{dm(t)}{dt} = n \cdot A \cdot \exp(-b \cdot f(t)) \cdot t^{n-1}. \tag{3}$$

With L(t) the decrease of the glass block´s cylinder radius can be calculated

$$r(t) = \frac{1}{\varrho} \int_o^t L(t)dt \tag{4}$$

ϱ is the density of the glass. r(t) was calculated in discrete time steps:

$$r(t_i) = r(t_{i-1}) + \frac{1}{\varrho} \int_{t_{i-1}}^{t_i} L(t)dt \tag{5}$$

The time intervals $\Delta t = t_i - t_{i-1}$ must be small in terms of the function f(t), equation (2). In this case, the exponential function with f(t) can be brought before the integral as an approximation

$$r(t_i) = r(t_{i-1}) + \frac{1}{\varrho} A \exp(-b \cdot f(t_i)) [t_i^n - t_{i-1}^n]. \tag{6}$$

$r(t_i)$ is then used to calculate $m(t_i)$ values for the glass block taking the respective formula from Ewest´s work [10].

RESULTS

All results were calculated with the computer code QTERM.

Case 1: Contact of brines N and C immediately after emplacement. The canister is absent. One block is leached, and its total geometric surface area is exposed to the leachant. Block dimensions and activity are: height = 1.0 m, radius = 0.1493 m, density = 3.27 g/cm^3, mass = 229 kg, waste load = 11.9 weight percent, total activity = 3.6 $\cdot$ 10^5 Ci, six years after discharge from the reactor. The results are shown in Figure 2 for brine C and in Figure 3 for brine N. The figures show the total activity in solution and the integrated mass loss of the glass block in (%) as a function of time.

Case 2: Sames as case 1, solution C, but it was assumed that n increases from n = 0.09 to n = 1 at a time beyond the duration of the leach experiment. The leaching is incongruent in the beginning [1,2]. It was, therefore, assumed but not yet observed that the process approaches a steady state with a constant dissolution rate (n = 1), two examples are given. In Figure 4 it was assumed that a steady state is reached after 3 years, whereas in Figure 5 this happens between 10^3 and $1.1 \cdot 10^3$ years.

Case 3: Same as case one, but alternating detachment and reformation of the surface layer was assumed. Figure 6 shows the release of radioactivity if successive detachment of the layer is assumed after it has reached a critical thickness of 10 μm. The first layer was arbitrarily detached after 1 year, as experiments in progress for 1 year have shown no layer detachment. Subsequent to this initial layer detachment, there are only six detachment events based on the calculated time required to form layers of 10 μm. The figure shows the extrapolated released activity and mass loss for the leaching process in solution C.

Case 4: Same as case one, except the canister was given life-times of 50, 100, 200, 400 and 1000 years, thus preventing contact of water with the waste form. Figure 7 shows the results. The curve for the life-time 1000 years is off scale. There is only the starting point of the curve to be seen on the abscissa. The shaded area in the figure will be explained in the discussion section.

DISCUSSION

It should be noted that in the model $m(t)$ values comprise all constituents released from the glass (the sum of dissolved ionic species, $m_{D+}(t)$ and $m_{D-}(t)$, colloids, $m_C(t)$, and precipitates, $m_P(t)$). The released radioactivity is distributed between the fractions of $m(t)$. Theoretically not included in $m(t)$ is material in the surface layer as long as it adhers to the glass surface. In practice some flakes may have been detached when preparing the leached samples for mass loss measurements. Mass balance calculations showed that m_c and m_p are $\leq$ 10 % in the isothermal experiment. This may explain the scatter of the data points in Figure 1. Solution analysis and EPMA analysis have revealed that certain elements are retained in the surface layer upon leaching (U, Am and Zr [1, 11]). Due to the morphology of the surface layer and the relative enrichment of elements in different layers, the mechanism of formation may be precipitation. The equilibrium concentration and hence the mass of these ions in solution would then be determined by their solubilities. The mass loss and related radioactivity may be lower than that calculated after (1). The $m(t)$ values in Figures 2-6 have been interpreted as mass loss of a congruently dissolving glass, where all constituents in solution are found in the same stoichiometric ratios as in the solid. This assumption was made since there are not yet enough data available to justify a specification of the mass loss values. The present approach over estimates the quantity of specific radionuclides released to the liquid phase, when these are retained in the layer. It under estimates their concentration in solution when they are depleted in the layer.

The results presented in Figures 2-7 are based on assumptions which should be discussed in more detail. In Figure 2 (case 1) the leaching process is assumed to continue with the slope $t^{0.09}$ for all time. This yields the lowest activity released when compared with case 2 and 3. The maximum fraction of activity in solution leached from one block is $3 \cdot 10^{-2}$%. This peak value is attained after one year. The result was obtained by extrapolating experimental data for solution C, Figure 1. The mass loss rate drops to low values after 100 years and the total loss does not exceed $4 \cdot 10^{-2}$%. This corresponds to an alteration depth of about 40 μm into the glass, including the layer. In solution N, it is 240 μm. The layer thickness was found to be 35 % of the total penetration depth when the glass was leached in DI water. This value was assumed to be the same in solutions N and C where respective measurements of the layer thickness are less accurate. The above value of 40 μm total penetration depth in 10^7 years in solution C yields a calculated maximum layer thickness of 13 μm. Since this thickness is reached only once in 10^7 years, the assumption of a detachment of layers is reasonable only if the "critical" thickness for detachment is less than 13 μm. It was assumed in the calculations (case 3, Figure 6) that the layer is detached whenever it reaches a thickness of 10 μm in solution C. Whenever the layer is released $m(t)$ and $A(t)$ increase stepwise. The mass of the layer is known from experiments. As said before, there are only six consecutive detachments, and these take place in the first 60 years. After 60 years the effect of the lowered temperature on the leach rate is already so large that the next 10 μm layer and the respective corrosion depth of about 40 μm is not reached within 10^7 years. There are surprisingly few detachments, but there may be even less if one takes into account that the attack of a pristine glass surface may be slower after layer detachment than before, in a solution that already contains leached glass constituents.

Below the layer the glass remains pristine as seen from the very sharp and narrow concentration profiles taken from that part of the samples [12]. The underlying pristine glass, the layer thickness, its protective character as a reaction barrier, and its formation time are common features of the borosilicate glass under investigation and of natural basaltic glasses [13]. These glasses have low SiO_2 contents, comparable to waste glasses. Upon leaching they show alteration effects in the form of a surface layer named "palagonite". The leaching process ceases after a few years for unknown reasons, eventhough these volcanic glasses have been exposed to repeated periods of wetting and drying

over long periods of time. There is no correlation of the alteration depths
with the age of the glasses. This provides strong evidence that layers can form
on glass surfaces and protect the underlying glass against further leaching.
Basaltic glasses serve as natural analogues to glass waste forms and may provide
an understanding of long-term leaching mechanisms and verification of the
results of modelling studies.

A measure of the influence of n on A(t) and m(t) can be obtained when comparing
Figures 2 and 3 though the values for A and b are different. A parametric study
showed that the influence of b on A(t) is smaller than that of n when increasing b from $2 \cdot 10^3$ K to $9 \cdot 10^3$ K and n from 0.1 to 0.5. In Figure 3, n is 0.30
instead of 0.09, and the maximum fraction of activity in solution increases by
a factor of 4 and the mass loss by a factor of 6. The value of n influences the
slope of the long-term mass loss curve in the first one or two hundred
years; therefore, more long-lived radionuclides are set free. If n is arbitrarily increased to one after a short time, i.e. 1 to 3 years, Figure 4, more
material is released at shorter times and the activity level is increased for
all time. An increase of n after 1,000 years is almost negligible in terms of
activity level in solution as seen from Figure 5 compared to Figure 2. Whenever n was increased in the calculations, A was decreased to the extent that
the resulting absolute value of the total leach rate L(t) was not higher than
before the increase of n.

One can also use this model to analyze the effect of variable canister life-
times. In this analysis the only role of the canister is to prevent solutions
from reaching the waste form. Canister failure is considered to be instantaneous, after which the entire geometric surface of the waste form is exposed
to solutions present in the repository. Figure 7 illustrates the effect of
canister life-times of 0, 50, 100, 200, 400 and 1,000 years on the total activity release. Protecting the glass for 400 years, during the thermal period
of the repository, decreases the maximum released activity by a factor of
$3 \cdot 10^4$.

One can also analyze the interdependence of repository variables (e.g. canister
life-time) with waste form properties (e.g. b, A and n). The lightly shaded
area of Figure 7 may be considered a "release envelope" for the borosilicate
glass, but for variable canister life-times. The upper boundary of the envelope is the release curve for the waste form without a canister. The lower
boundary is a locus of points of maximum release of activity for canisters with
life-times up to 1,000 years. For the borosilicate glass, in which the release

is very temperature dependent, the release envelope should be broad, indicating the quite important effect of the canister. In comparison, one can consider a waste form (e.g. SYNROC) which gives minimum release at high temperatures, but whose behaviour is not as temperature dependent as that of the borosilicate glass. Based on the limited data available for some crystalline waste forms [14], the apparent activation energy in b, might decrease by a factor of two or three; the pre-exponential factor, A, might be decreased by many orders of magnitude; and the mechanism, n, may be in the range of the borosilicate glass. The release envelope for a waste form with such improved high-temperature performance is shown in dark-grey. Note, that for no canister, there is a two order of magnitude decrease in release from the "high temperature" waste form during the first 200 years; however, because the performance of such a waste form is not very temperature dependent, the release envelope is narrow for canisters of variable life-times. The combination of borosilicate glass and a 400 year canister results in released activity that is less than that of the "high-temperature" waste form. Two related points can be made: (1) a repository´s ability to exclude water from the waste form for as little as 10,000 years can have a tremendous effect on the total released activity and has the effect of minimizing the difference in behaviour between waste forms and (2) the final shape of the curve is a complex result of the values of A, b and n. These are rarely determined in present experiments.

The following general points can be made by considering the curves generated by this model:

1.) The long-term behaviour of a waste form cannot be judged independent of a specific set of conditions. One must consider not only the intrinsic properties of the waste form (leach rate as expressed by A, n and b) but also specific canister and repository conditions. The model clearly shows that the properties of the waste form can be weighed against such variables as canister life-time.

2.) The long-term activity is most dependent on short-term events (e.g. less than 200 years). Even with increased mass loss in the range of 10^4 to 10^6 years, there is only a minor increase in activity (Figure 4). The total activity depends most heavily on how many radionuclides are placed into solution during the early history of the repository. Because these events occur at elevated temperatures and over relatively short periods of time, long-term predictions based on laboratory experiments (designed for specific repository conditions) are justified. The agreement between the results of the model and those presented by Allen for natural glasses further substaniates this approach [13].

The importance of short term events also suggest that the repository may be considered as a short-term barrier - in contrast to the long-term geologic behaviour that is now commonly required.

3.) Such models can clearly indicate research-priorities. The "best" borosilicate glass and a 400-year canister may provide for a lower amount of released activity than a new more complicated or costly "high temperature" waste form. The multi-barrier system as classically presented (e.g. each barrier independent of the other) must consider possible positive and negative interactions between barriers.

REFERENCES

1. Altenhein, F.K., Lutze, W. and Malow, G. (1981) Scientific Basis for Nuclear Waste Management, Vol. 3, Moore, J.G. ed., Plenum Press, New York, pp. 363-370.
2. Malow, G. (1982) Scientific Basis for Nuclear Waste Management, Vol. 11, Lutze, W. ed.
3. Petrović, R. (1976), Geochimica et Cosmochimica Acta, 40, pp. 1509-1521
4. Westsik, J.H. and Peters, R.D. (1981) Scientific Basis for Nuclear Waste Management, Vol. 3, Moore, J.G. ed., Plenum Press, New York, pp. 355-362.
5. Fisher, G.W. (1978) Geochimica et Cosmochimica Acta, 42, pp. 1035-1050.
6. Macedo, P.B. et al. (1979) Proc. of the Conference on High-Level Solid Waste Forms, Denver 1978, Casey, L.A. ed., US-NRC, Washington, pp. 81-89.
7. Helgeson, H.C. (1971) Geochimica et Cosmochimica Acta, 35, pp. 421-469 see errata: Helgeson, H.C. (1972) Geochimica et Cosmochimica Acta, 36, pp. 1067-1070.
8. Röthemeyer, H. (1980) Underground Disposal of Radioactive Wastes, Vol. 1, Int. Atomic Energy Agency, Vienna, pp. 297-310.
9. Delisle, G. (1980) Zeitschrift der Deutschen Geologischen Gesellschaft, 131, pp. 461-484.
10. Ewest, E. (1979) Scientific Basis for Nuclear Waste Management, Vol. 1, McCarthy, G.J. ed., Plenum Press, New York, pp. 161-168.
11. Malow, G., Offermann, P., Behrend, U., Matiske, H. and Müller, R. (1982) European Appl. Res. Rept. in press
12. Scholze, H., Conradt, R., Engelke, H., Roggendorf, H. (1982) Scientific Basis for Nuclear Waste Management, Vol. 11, Lutze, W. ed.
13. Allen, C.C. (1982) Scientific Basis for Nuclear Waste Management, Vol. 11, Lutze, W. ed.
14. Ringwood, A.E., Oversby, V.M., Kesson, S.E., Sinclair, W., Ware, N., Hibberson, W., Major, A. (1981) Nuclear and Chemical Waste Management, Vol. 2, pp. 281-305

Part II will be presented at the MRS 1982 Annual Meeting in Boston, Mass, November 1-4.

54

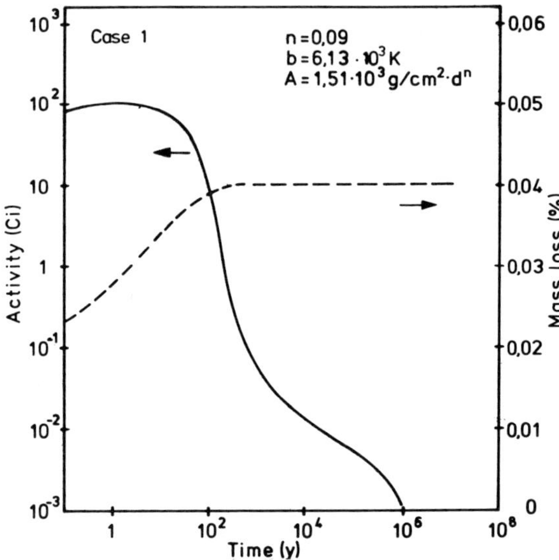

Fig. 2. Activity and mass loss from borosilicate glass in carnallite brine

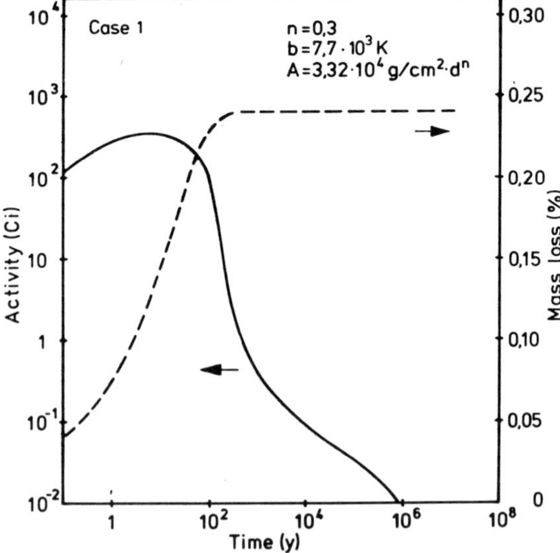

Fig. 3. Activity and mass loss from borosilicate glass in sodium chloride brine .

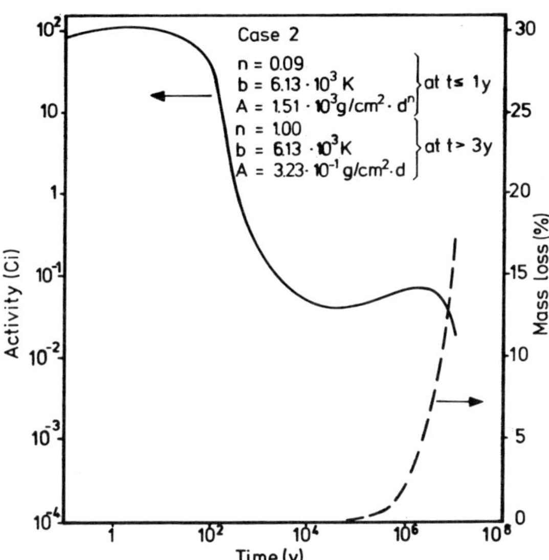

Fig. 4. Activity and mass loss from borosilicate glass in carnallite brine.

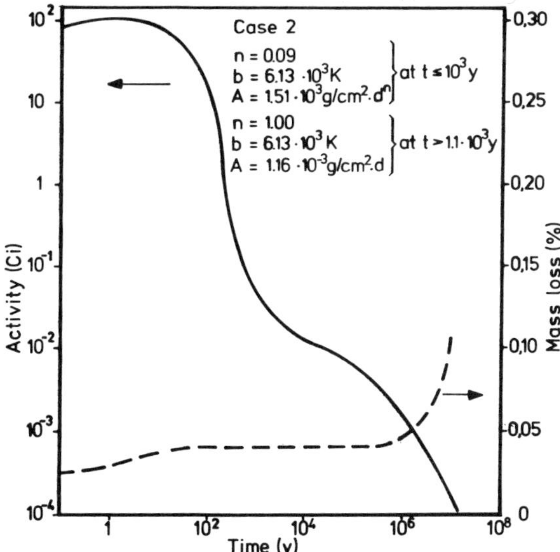

Fig 5. Activity and mass loss from borosilicate glass in carnallite brine.

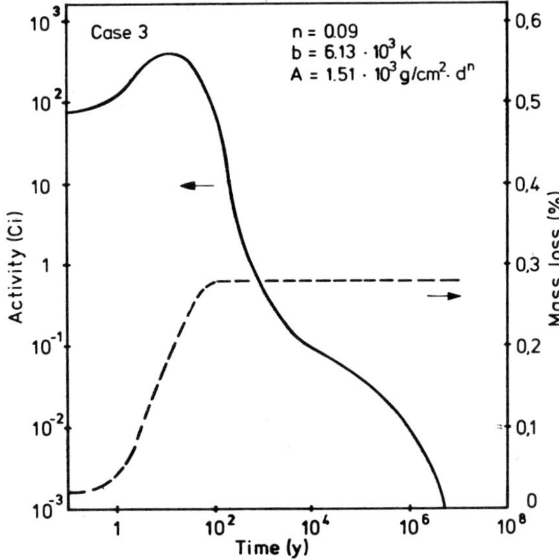

Fig. 6. Activity and mass loss from borosilicate glass with successive layer detachment. Critical layer thickness = 10 μm.

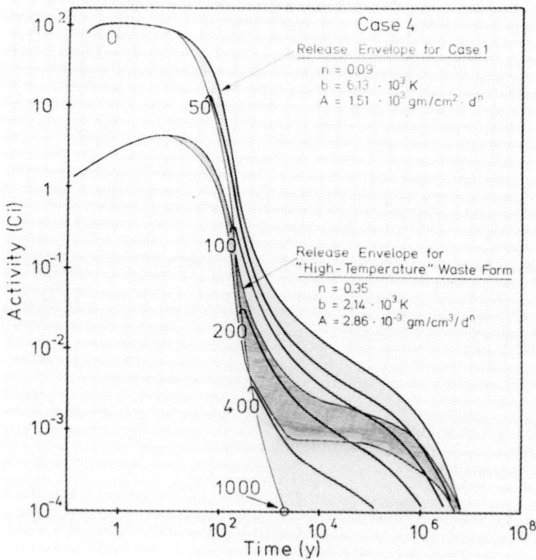

Fig. 7. Activity and mass loss from a borosilicate glass and a high temperature waste form. Canister life-times 0, 50, 100, 200, 400, 1000 years

A FLOW MODEL FOR THE KINETICS OF DISSOLUTION OF NUCLEAR WASTE
FORMS; A COMPARISON OF BOROSILICATE GLASS, SYNROC AND HIGH-SILICA
GLASS

PEDRO B. MACEDO, AARON BARKATT, AND JOSEOPH H. SIMMONS
Vitreous State Laboratory, The Catholic University of America,
Washington, D.C. 20064, USA

INTRODUCTION

A model has been developed to predict the long-term leach
or release rates of various waste-form materials under repository conditions. The basic elements of the model are as follows:

(i) Due to the effects of interaction with the solid (pH
increase, approach to saturation with respect to SiO_2, Al_2O_3,
etc.) the reactivity of the medium and the resulting leach rates
strongly depend on N_V, the rate at which the volume of water in
contact with the solid is replaced by fresh leachant.

(ii) In the long term, the concentration of each leached
species i in the surrounding medium reaches a constant level C_i

$$C_i = L_i \sigma_i \cdot \frac{S}{V} \cdot \frac{1}{N_V} = L_i \sigma_i \cdot \frac{R}{N_V} \tag{1}$$

where L_i is the normalized leach rate of the species i in units
of $g \cdot m^{-2} \cdot d^{-1}$, σ_i is the fraction of the component i in the solid, and $R = S/V$ is the ratio between the surface area of the
solid and the volume of water in effective contact with it in
units of m^{-1}. L_i itself is a sensitive function of C_i due to
the effects mentioned above (pH excursion, saturation).

(iii) Under repository conditions, N_V is related to the
flow rate f through the relationship

$$N_V = g \cdot \frac{P_R}{D} \cdot f \tag{2}$$

where P_R is the rock porosity, f is the interstitial flow rate,
D is the minimum distance required for leachant exchange, typically the smallest dimension of the canister, and g is a
geometrical factor, e.g. $g = \frac{4}{\pi(1-\alpha)}$ where the canister is

cylindrical and a fraction α of its volume is occupied by the waste-form. N_v is expressed in units of d^{-1}.

(iv) The annual fractional loss, θ_i, which is the critical figure of merit characterizing the release of component i from a given waste-form, is given by

$$\theta_i = 10^{-6} \frac{C_i V_L N_v}{\rho \sigma_i V_G} \tag{3}$$

where ρ is the density of the waste-form and V_G and V_L are the volumes of the waste-form and of the leachant in contact with it, respectively. θ_i is dimensionless.

(v) Since C_i is a function of R/N_v (see Eq. 1), it is possible to cover the range of leachant exchange rates expected in a repository, including very low rates, by carrying out an experimental flow test at a surface-to-volume ratio R_{exp} and a rate of leachant exchange $(N_v)_{exp}$ until the concentration of the component i approaches a constant value C_i. Under repository conditions ($R=R_{rep}$; typically, $R_{rep} = 60$ m^{-1}) the same concentration will occur when the rate of leachant exchange $(N_v)_{rep}$ is

$$(N_v)_{rep} = \frac{R_{rep}}{R_{exp}} (N_v)exp \tag{4}$$

(vi) since both f and θ_i are related to N_v (see Eqs. 2-3), a series of flow tests giving C_i as a function of $(N_v)_{exp}$ and the use of Eq. (4) permit complete mapping of the dependence of θ_i on f.

EXPERIMENTAL

The complete details of the flow test are given elsewhere.[1,2] A fully interactive dynamic test has been developed to generate the data-base required for the use of the model.[3] The test method is based on exposing a sample with a known surface area to contact with a standard volume of water in a tightly sealed, inert vessel, further protected from the penetration of ambient air through the use of an outer container. A constant fraction of the water is removed at fixed intervals by means of a syringe and replaced with fresh water. The test continues until the concentration of the leachate fractions removed and replaced by water at the end of each interval becomes constant. This generally requires the total exchanged volume to reach 3-4 times the leach volume, and the total time to be sufficient for completion of any rapid process of surface layer build-up. Ionic balance calculations are carried out on the leachate composition to ensure the complete identification of all major species which control the pH and the reactivity of the medium, and to distinguish between the respective contributions of the leach products and of extraneous factors (CO_2, leach container components) to the ionic composition.

RESULTS

The flow model has been applied to results of dynamic tests carried out over the range of flow rates between 0.5-300 m/yr at a temperature of 70°C on three glassy and two crystalline waste-form materials. These materials include Battelle PNL 76-68 borosilicate glass[4], Savannah River Laboratory TDS-131 borosilicate glass[5], CU PGM (Porous Glass Matrix) high-silica glass (see below), Rockwell International polyphase ceramics (with nepheline as the major cesium host phase) supplied as representative of Tailored Ceramics[6], and Lawrence Livermore Laboratory Synroc-D.[7] The results obtained for the dependence of the annual fractional loss θ on the flow rate f in the cases of Si and Cs are shown in Figs. 1a and 1b, respectively.

Quantitatively, the results shown in Fig. 1 indicate that the two borosilicate glasses, PNL 76-68 and SRL TDS-131, and the two ceramic waste-forms, Tailored Ceramics and Synroc-D, all exhibit approximately the same magnitudes of θ (within a factor of 2) over the entire range of flow rates. In the case of CU PGM (Porous Glass Matrix) high-silica glass, however, θ is lower by at least one order of magnitude, and at the higher flow rate region by two orders of magnitude, than the annual fractional loss exhibited by all the other waste-forms. Most significantly, the results show that high-silica glass is the only material to fulfill the durability criterion required of high-level waste solids[8], i.e. a long-term value of θ smaller than $1 \times 10^{-5} \dfrac{g}{g \cdot year}$ under repository conditions. Results obtained with other components[2] show that the loss rates based on B and Na show less decline than those based on Si or Al as the flow rates decrease, but in all cases the loss rates measured for CU PGM glass are lower by 1-2 orders of magnitude than those obtained with the other waste-form materials under the same flow conditions.

At the high flow rate limit, the higher durability of high-silica glass (by 1-2 orders of magnitude) is confirmed by frequent-exchange tests[9,10] carried out under high-dilution conditons (see Fig. 2). These tests consist of suspending diamond-cut (600 grit) blocks of the specimens, each with a surface area of 4.0 cm^2, near the center of a volume of 40 cm^3 of de-ionized water in a 60-cm^3 PFA Teflon vessel at 90°C. The leachate is completely removed for analysis and replaced by 40 cm^3 of water every day during the first week, twice during the second week, once a week for the next six weeks, and monthly thereafter. The results of static MCC-1 tests[11], although inapplicable to flow situations, also lead to a similar conclusion. The results of these MCC-1 tests on the various waste-form materials mentioned above, based on tests carried out at the authors' laboratory as well as on results reported elsewhere,[12,13] are detailed in Table I.

The great difference in durability between the high-silica glass can be ascribed to two major reasons. First, this glass, unlike the other materials included in this study, has a silica content higher than two-thirds on a mole basis (PGM glasses corresponding to various immobilized waste compositions generally have a silica content greater than 70% by mole; the composite waste glass described in the present study contains 71.235 mol% SiO_2). Based on studies of both experimental and medieval glasses[14], silicate glasses show a sudden drop in durability as the silica content decreases below 66.67 mol%, since the integrity of the silica network is only maintained above this level. Studies of natural glasses (tektites, obsidians)[15,16] show that glasses with a silica content higher than 67% can last tens of millions of years. Most glasses with a lower silica content, on the other hand, survive for periods no longer than a few thousand years.[14,15] A recent study of rhyolite glass[17] showed that high-silica glasses retain their structural integrity and display low solubility rates even when subject to prolonged high-temperature hydrothermal leaching, unlike borosilicate glasses, which disintegrate and release the radioactive components uncontrollably under similar conditions. The study concludes that high-silica glass is much more suitable for long-term disposal of radioactive wastes than borosilicate glass.

The second cause of the superiority of the high-silica glass over the other waste-form materials is the fact that due to the presence of a sufficiently high ratio of acidic components (SiO_2, B_2O_3 and, in particular, P_2O_5) to alkaline components the pH of the leachate in contact with the. glass never rises beyond 7.8, even under conditions of very high surface-to-volume ratios and very long residence times. On the other hand, the other four waste-form materials give rise under similar conditions to pH values between 9.5 and 12 (see also Ref. 5). The corrosion rates of silicate-based glasses (and, probably, other solids) show a steep rise when the pH reaches values beyond 8.5-9 due to the onset of rapid matrix corrosion[18] and conversion of silica to a soluble ionic form[19] (for silicic acid, pKa = 9.7; for boric acid, pKa = 9.1). Accordingly, high-silica glass provides the only case where the corrosion rates will remain in the low range which corresponds to near-neutral media under all the flow/dilution conditions possible in a repository environment.

PREPARATION OF CU PGM HIGH-SILICA GLASS SPECIMENS

The Porous Glass Matrix process[20] leads to the formation of glasses with a high silica content, while specific oxides can be added to the glass composition in order to improve the overall chemical durability of the fixation medium, as well as to assist in the immobilization of specific waste components. The glass waste-form is formed by sintering, avoiding the very high melting temperatures (on the order of 1700-1800°C) typical of high-silica glasses and making it possible to minimize the volatilization of components with high vapor pressure (e.g. Cs and Ru oxides). The process is relatively insensitive to the specific waste composition and to the relative liquid/solid proportion in the waste stream.

There are three major stages in the fixation of radioactive waste material in the PGM medium. These stages include:

i. <u>Preparation of the PGM glass powder:</u> A borosilicate-type base glass is melted to produce a composition required to achieve the necessary phase separation (immiscibility) characteristics. The homogeneous base glass is then subjected to a suitable heat treatment schedule to induce phase segregation with the phases possessing the desired interpenetrating microstructural morphology. The resulting two-phase composite is crushed and the powder sieved to obtain well-defined grain sizes. The powder is leached in 3M HCl to dissolve the low-silica phase and then washed in clean water to remove the leaching products. The material which remains is a fine powder (grain sizes in the range 10 μm to 100 μm, typically about 50 μm) of porous high-silica glass. The final preparation step is the addition of the appropriate dopants to the porous glass.

ii. <u>Porous glass - waste stream mixing:</u> This waste loading stage consists of first mixing the liquid fraction of the waste stream with porous high-silica glass powder. The dissolved waste components such as Cs^+ are absorbed or ion-exchanged in the pores of the glass grains. (The pore size is between about 0.01 μm and 0.02 μm.) The loaded glass is dried by gradual heating under vacuum to 850°C.

The solid fraction of the waste stream is dried under vacuum (18 hours at 350°C, 1 hour at 850°C), ground or ball-milled to reduce the grain size to approximately 25 μm, and mixed with the dried glass previously loaded with the dissolved wastes (see above) and with fresh dry porous high-silica glass powder to achieve the desired waste loading of 30%. The solid powders are mixed together in a blender. (An alternative procedure, which was not used for preparing the samples characterized in the present study, consists of simultaneous mixing of the glass, the dissolved wastes and the solid wastes, following by spray-calcining.) This causes the solid waste grains to become dispersed among the grains of the glass. This stage, which results in the formation of a mixture of solid waste grains with high-silica glass grains which have incorporated the dissolved waste species, is illustrated in Fig. 3a.

iii. <u>Sintering:</u> Drying and heating the mixture then effects the waste fixation. For this stage the mixture is placed either in a metallic can or in an expansion-matched glass tube which is held vertically in a zone heating furnace. As each horizontal layer, beginning at the bottom (in order to minimize volatilization) is heated, the calcined waste-porous glass powder mixture undergoes several transformations:

(a) Between 600°C and 900°C the reactive liquid - dissolved waste, components which are within the pores of the glass grains on the grain surface, react with the glass.

(b) At the same time, the pores themselves collapse, thus incorporating most of the dissolved waste into the glass structure.

(c) Between 900°C and about 1200°C, the glass grains sinter together to form solid glass in which the remainder of any dissolved wastes and the waste solids become trapped.

These transformations are shown schematically in Figure 3b. If a glass tube is used to house the porous glass - radwaste mixture, continued heat treatment at 1200 C causes this tube to collapse on and fuse to the core to form a solid glass object. In this case, the end product consists of a glass core, fixating the waste elements, surrounded by a almost pure silica glass protective layer which is free of radioactive elements, and which has extremely high durability.[20] If the sintering is carried out in a metal can, this outer glass envelope is, of course, not present as a part of the waste form. All the results described in the previous sections were obtained with the loaded glass core without an outer glass envelope.

ACKNOWLEDGEMENT

The authors are indebted to Dr. H. G. Sutter and Dr. C. J. Montrose for over-all co-operation, to W. Sousanpour, Al. Barkatt, M. A. Boroomand, C. Fisher, M. Penafiel and C. Smeal for technical assistance, and to P. Szoke and V. L. Rogers for their analytical work. This study has been supported in part by the Electric Power Research Institute under Contract No. RP1579-6.

TABLE I

Results of MCC-1 Leach Tests on Various Waste-form
Materials, Deionized Water, 90°C, 28 days

Material	Normalized Leach Rates, $g \cdot m^{-2} \cdot d^{-1}$					
	Si	B	Na	Al	Sr	Cs
PNL 76-68[a]	1.53	2.25	2.41			
MCC 76-68[a,d]	0.68	1.00	1.11		0.13	0.74
MCC 76-68[b,d]	0.85	1.29			0.064	0.99
SRL TDS-131[a]	0.35	0.64	0.59	0.23	0.27	0.36
Comp. Synroc-D[c,e]	0.67		0.62	0.18	0.38	0.79
Hi-Fe Synroc-D[c,f]	0.89		0.84	0.33	0.56	0.80
CU PGM[a,e]	0.007	0.013	0.075	0.003	0.031	<0.1
FEU PGM[a,f]	0.008	0.009	0.080	0.002	<0.01	<0.1

a Results obtained at the authors' laboratory
b Results quoted from Ref. 12
c Results quoted from Ref. 13
d Glass of 76-68 composition electro-melted for the Materials Characterization Center[12]
e Waste-form material loaded with composite defense waste
f Waste-form material loaded with high-iron defense waste

REFERENCES

1. P. B. Macedo, A. Barkatt and J. H. Simmons, "A Flow Model for the Kinetics of Dissolution of Nuclear Waste Glasses", Nucl. Chem. Waste Manage., $\underline{3}$, 13 (1982).

2. A. Barkatt and P. B. Macedo, to be published.

3. A. Barkatt, Al. Barkatt, P. E. Pehrsson, P. Szoke and P. B. Macedo, "Static and Dynamic Tests for the Chemical Durability of Nuclear Waste Glass", Nucl. Chem. Waste Manage., $\underline{2}$, 151 (1981).

4. W. A. Ross, "Annual report on the characterization of high-level waste glasses", Pacific Northwest Laboratory Rept. PNL-2625 (1978).

5. G. G. Wicks, W. C. Mosley, P. G. Whitkop and K. A. Saturday, "Durability of Simulated Waste Glasses - Effects of Pressure and Formation of Surface Layers", J. Non-cryst. Solids, 1982.

6. P. E. D. Morgan et al., "High-alumina Tailored Nuclear Waste Ceramics", J. Am. Ceram. Soc., $\underline{64}$, 249 (1981).

7. J. Campbell et al., "Properties of SYNROC-D Nuclear Waste Form: A State-of-the-Art Review", Lawrence Livermore Laboratory, Rept. no. UCRL-53240, January 1982.

8. United States Nuclear Regulatory Commission Rules and Regulation, Title 10, Chapter 1, Code of Federal Regulations - Energy, Part 60 - Disposal of High-level Radioactive Wastes in Geologic Repositories; Proposed Rule. Federal Register 46 FR 35280, July 8 (1982).

9. E. D. Hespe, "Leach Testing of Immobilized Radioactive Waste Solids", At. Energy Rev., $\underline{9}$, 195 (1971).

10. International Standards Organization, Long-term Leach Testing of Radioactive Waste Solidifiation Products, Draft ISO Standard ISO/TC85/SC5/WG5/N22, April 1980.

11. MCC-1P Static Leach Test Method, in Nuclear Waste Materials Handbook- Waste Form Test Methods, Materials Characterization Center, J. E. Mendel, mgr., DOE/TIC-11400 (1981).

12. J. W. Johnston and J. L. Daniel, Summary Report for the Interlaboratory Round Robin on the MCC-1 Static Leach Test Method, Materials Characterization Center Rept. PNL-4249, March 1982.

13. J. Campbell, C. Hoenig, F. Bazan, F. Ryerson, M. Guinan, R. Van Konynenburg and R. Rozsa, Properties of Synroc-D Nuclear Waste Form: A State-of-the-Art Review, Lawrence Livermore Lab. Rept. UCRL-52340, January 1982.

14. T. M. El-Shamy, "The Chemical Durability of K_2O-CaO-MgO-SiO_2 Glasses", Phys. Chem. Glasses, __14__, 1 (1973).

15. G. W. Morey, "The Properties of Glass", Reinhold Publ. Corp., New York, N.Y., 1938, Chap. III-The Composition of Glass.

16. J. A. O'Keefe, "Tektites and their Origin", Elsevier, Amsterdam, 1976.

17. A. P. Dickin, "Hydrothermal Leaching of Rhyolite Glass in the Environment Has Implications for Nuclear Waste Disposal", Nature, __294__, 342 (1981).

18. E. C. Ethridge, D. E. Clark, and L. L. Hench, "Effects of glass surface area to solution volume ratio on glass corrosion", Phys. Chem. Glasses __20__, 35 (1979).

19. T. M. El-Shamy, J. Lewins and R. W. Douglas, "The Dependence on the pH of the Decomposition of Glasses by Aqueous Solutions", Glass Tech., __13__, 81 (1972).

20. J. H. Simmons, P. B. Macedo, A. Barkatt and T. A. Litovitz, "Fixation of Radioactive Waste in High-silica Glasses", Nature, __278__, 729 (1979).

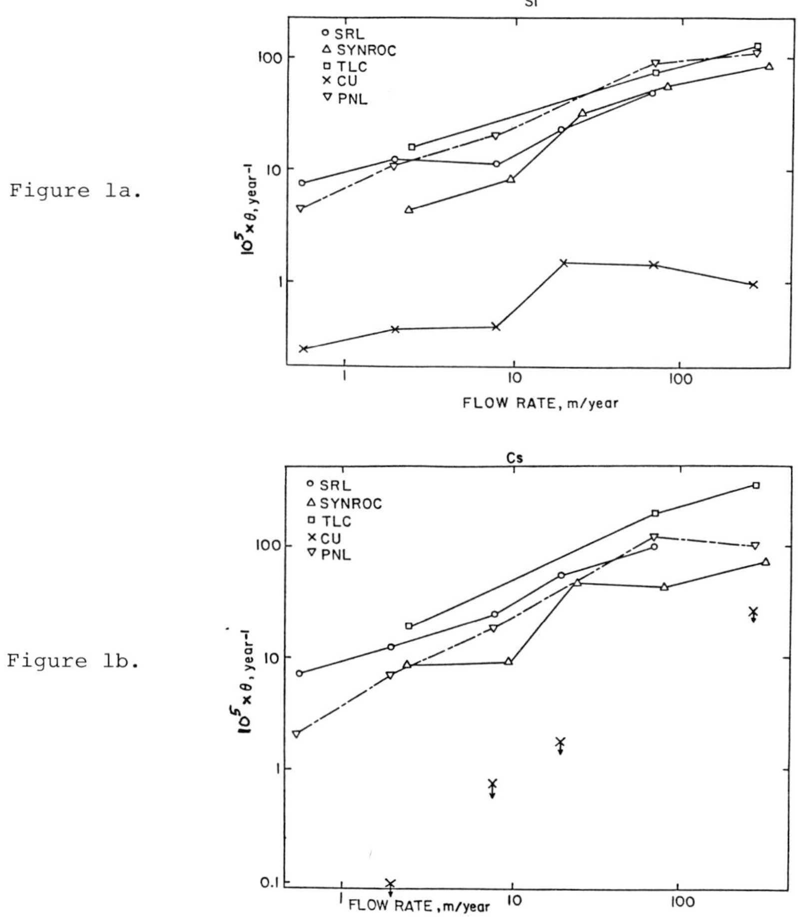

Figure 1a.

Figure 1b.

Figure 1. Results of flow tests on various waste-forms; DI
water leachant; 70°C: ▽ PNL 76-68 glass; O SRL
TDS-131 glass; △ Synroc; ◻ Tailored Ceramics;
x PGM high-silica composite defense waste glass.
la. Annual fractional loss based on Si as a
function of equivalent flow rate.
lb. Annual fractional loss based on Cs as a
function of equivalent flow rate.

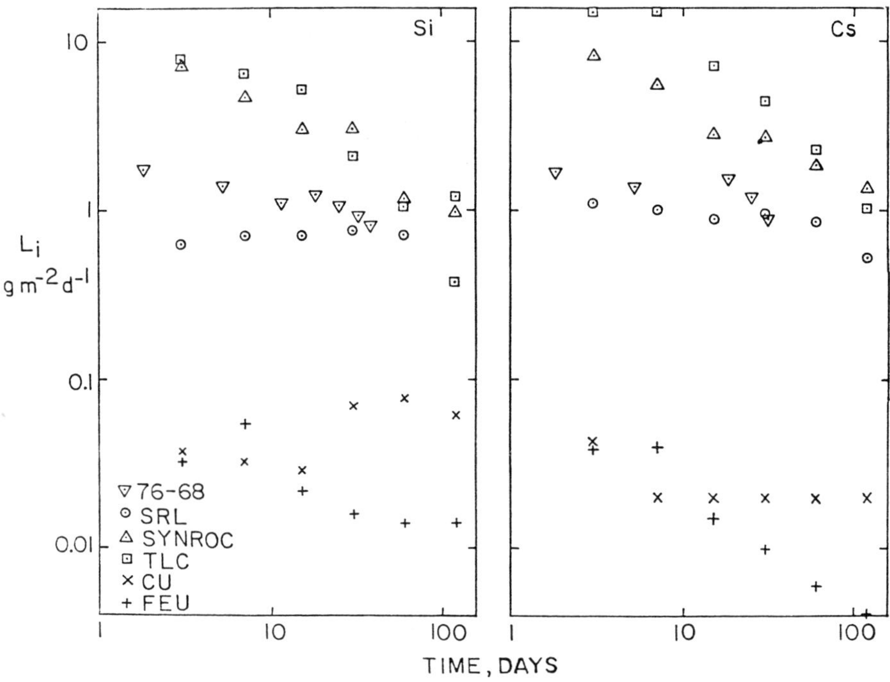

Figure 2a.   Figure 2b.

Figure 2.   Results of high-dilution (S/V = 10 m^{-1}), frequent-
             exchange IAEA-type tests on various waste-forms;
             DI water leachant; 90 C: ⊽ Electrically melted
             76-68 glass; ⊙ SRL TDS-131 glass; △ Synroc; ▫
             Tailored Ceramics; x PGM high-silica composite
             defense waste glass; + PGM high-silica high-iron
             defense waste glass.

             2a. Normalized leach rates based on Si.
             2b. Normalized leach rates based on Cs.

Figure 3a.

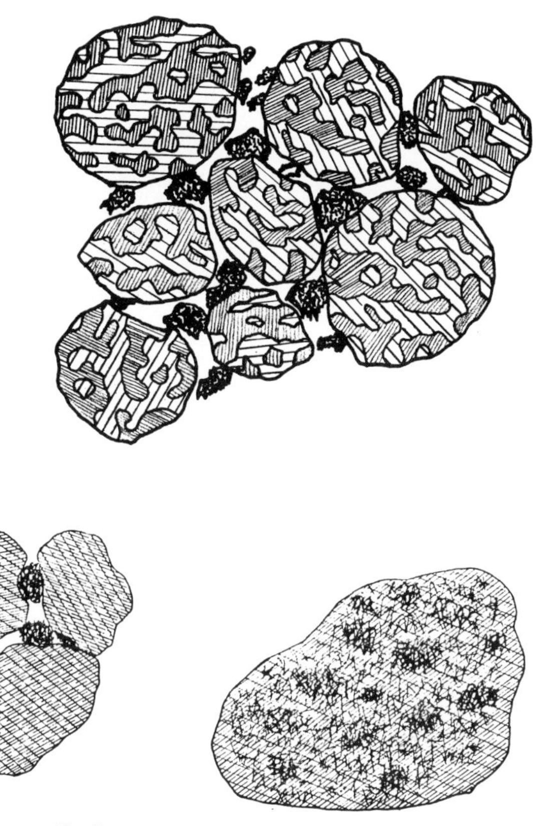

Figure 3b.i.

Figure 3b.ii.

Figure 3.   (a)   Schematic diagram of the porous glass powder -
radioactive waste mixture. Typical sizes; pores,
0.01 µm; glass grains, 50 µm; solid waste particles,
25 µm.
(b)   The transformations which occur during the
heat treatment of the waste-glass mixture: (i)
Dissolved wastes are incorporated into the glass
structure as the pores collapse; (ii) solid
precipitate wastes are immobilized in the glass
as the grains sinter together.

WASTE GLASS/REPOSITORY INTERACTIONS

D.E. CLARK AND C.A. MAURER
Department of Materials Science and Engineering, College of Engineering
University of Florida, Gainesville, Florida 32611, USA.

ABSTRACT

The effects of repository material and a tailored backfill or overpack
on the leaching behavior of glass have been studied. Two types of glasses
were used in this investigation: 1) Model glasses comprised of 33 mol%
alkali oxide-67 mol% SiO_2, and 2) alkali borosilicate glasses with and
without simulated wastes. Several types of repository material were
placed in the same containers as the glasses to determine if their
presence would alter the extent of leaching. A backfill material consisting of phosphate slime/sand mixtures was evaluated with the same procedure. The results indicate that the leaching behavior of the glass may be
influenced by the presence of some materials. Preliminary results of a
$16\frac{1}{2}$ month burial experiment in Florida are also discussed.

INTRODUCTION

Water chemistry is probably the single most important variable in
glass leaching. Orders of magnitude variations in both the leach rate and
extent of degradation can be achieved by modifying the composition and pH
of the water in contact with the glass. Investigations have shown that
the presence of certain species in solution, such as Al, Zn, Ca and Mg
are effective in improving the leach resistance of both alkali silicate
and alkali borosilicate glasses under certain conditions[1-4]. In a repository the water chemistry will inevitably be altered and possibly controlled
by the materials in close proximity to the waste form (ie, repository and
overpack materials). Additionally, the chemistry of the water in contact
with the waste form will be complex and probably vary with time, temperature, flow rate, geology, radiation field and the ratio of surface area
of material to volume of water in the repository. Since the waste form is
expected to provide the first barrier to release of radionuclides it is
important to understand its leaching behavior over the widest possible
range of potential repository conditions.

Other barriers such as tailored overpacks and backfills may provide
additional protection against radionuclide release from the repository.
Some of these barriers, including bentonite, clay/charcoal mixtures and

metals, have been discussed by other investigators at this conference[5-6].
These tailored materials are designed specifically to perform one or all
of the following functions:

. minimize water flow throught the repository

. adsorb hazardous species released from the first barrier (ie, the
waste form)

. reduce the rate and/or extent of leaching from the waste form by
controlling the water chemistry.

The last function can be effective in immobilizing the radionuclides
by producing protective surface films on the waste form and by controlling saturation and buffering capacity of the repository water. In the
present investigation we examined the effects of several types of repository and overpack materials on the leaching performance of glass. Preliminary results from a burial experiment are also presented to better
illustrate the importance of water chemistry and environment on leaching
behavior.

EXPERIMENTAL

The compositions of the glasses used in this investigation are shown
in Table 1.

TABLE 1
GLASS COMPOSITION USED IN THIS INVESTIGATION

| | | | | Glass | |
	33N (mol%)	33L (mol%)	SRL 21 (WT%)	ABS 41 (WT%)	PNL 140 (mol%)
SiO_2	67	67	52.5	52.0	42.0
B_2O_3			10.0	15.9	6.0
Na_2O	33		18.5	9.4	16.0
Li_2O		33	4.0	3.0	
CaO			5.0		14.0
TiO_2			10.0		
Fe_2O_3				3.6	
ZnO				3.0	6.0
Al_2O_3				2.5	15.0
MgO					1.0
UO_2				1.6	
Simulated fission product				9.0	

Corrosion specimens 0.2 cm thick were sliced from glass bars using a
diamond wafering saw, polished through 600 grit with SiC paper on all six
sides, and cleaned with acetone. This procedure produced an equivalent
surface finish on all samples prior to corrosion. Samples of basalt, granite and other geological materials were cut, polished and cleaned in the

same manner. Tailored overpack materials were prepared using mixtures of
Florida phosphate slime and course sand. These mixtures were extruded,
dried and sintered to produce consolidated bodies with sufficient strength
for handling. Samples were cut and polished with the same procedure used
for the glass and rock specimens. The glass samples were corroded using a
modified MCC-1 static leach test[7]. This test involves suspending the
sample in the solution inside the corrosion cell. In order to evaluate
the interactive effects, rock specimens or tailored phosphate slime
specimens were suspended in the corrosion cells containing the glass as
shown in Figure 1. In all cases the glass and either rock or tailored
backfill material were immersed simultaneously into the D.I. water. The
dimensions of the rocks and phosphate slime samples were the same as the
dimensions of the glass specimens. The surface area to solution volume
ratio (SA/V) was based on the geometric area of the glass specimens and
was maintained at 0.1 cm^{-1} for all tests. A more detailed description of
sample preparation, corrosion testing and post-corrosion analyses is
given elsewhere[8]. Petrographic analysis was used to determine the mineral
content of the geological minerals.

RESULTS AND DICUSSION

<u>Glass/Geological Material Interactions.</u> Solution and WT loss data are
presented in Tables 2-4 for the SRL 21 glass (contained no waste) leached
in deionized (D.I.) water containing one of five types of geological
materials.

TABLE 2
SOLUTION AND WT LOSS DATA OF SRL 21 LEACHED IN COMBINATION WITH VARIOUS
GEOLOGICAL MATERIALS AT 90°C, 216 HOURS, SA/V=0.1 cm^{-1} (BASED ON GLASS)

Sample	Final	%WT	Solution concentration (mg/l)			
pair	pH	loss	Si	B	Na	Ca
SRL 21 alone	8.5	2.2	113	26	84	1.6
21/Cal.Gran.	8.0	1.9	94	18	68	4.4
21/Swed.Gran.	8.6	1.8	100	23	76	3.8
21/G.B.Obsid.	7.6	–	87	29	61	5.0
21/F.G.Bas.	7.6	1.7	83	23	58	5.2
21/FL.Chert	8.5	2.6	125	24	94	0.9

TABLE 3
SAME AS TABLE 2 BUT LEACHED FOR 4000 HOURS (167 DAYS)

Sample	Final	%WT	Solution concentration (mg/l)			
pair	pH	loss	Si	B	Na	Ca
SRL 21 alone	9.1	5.2	215	79	340	1.5
21/Cal.Gran.	9.7	5.9	310	89	360	0.6
21/Swed.Gran.	8.2	3.8	200	55	240	5.7
21/G.B.Obsid.	9.4	4.7	245	80	320	1.0
21/F.G.Bas.	9.2	4.5	320	86	380	1.3
21/FL.Chert	8.0	4.6	180	52	220	5.9

TABLE 4
SAME AS TABLE 2 BUT LEACHED AT $200^\circ C$, 185 psi, 1 HOUR, PFA Teflon
Container

Sample pair	Final pH	%WT loss	Solution concentration (mg/l)			
			Si	B	Na	Ca
SRL 21 alone	9.9	1.3	316	27	58	0.6
21/Cal.Gran.	9.8	1.2	265	21	48	0.9
21/Swed.Gran.	9.8	1.5	295	17	54	0.7
21/G.B.Obsid.	9.9	0.9	288	18	44	1.0
21/F.G.Bas.	9.9	1.1	312	27	46	0.8
21/FL.Chert	9.6	1.4	380	15	51	0.3

Cal.Gran. = California granite, primary mineral phases are Plagioclase,
Hypersthene, Angite, Biotite and Quartz; Swed.Gran.= Swedish Granite;
G.B.Obsid. = Glass Buttes Obsidian, primarly glass with <1% Diopsidic
Angite and plagioclase; F.G.Bas.= Fine Grain Basalt, primarly mineral
phases are plagioclase feldspar, hornblende and illmenite; FL.Chert =
Florida Chert, primarly mineral phase is microcrystalline quartz.

All solution concentrations and pH values were measured at room
temperature. The pH of the solutions containing both glass and geological
material increased during exposure as did the pH of the solution containing
glass alone. Longer exposure times produced higher pH values for all
solutions indicating that the geological materials (rocks) do not provide
much if any solution buffering capacity. The pHs of the solutions containing
only the rocks did not change appreciably under identical exposure conditions. Therefore, the pH increases observed in the solutions containing
both glass and rocks were due primarily to the leaching of the glass.
Although not shown, the WT losses and solution concentrations were
obtained for the rocks exposed to solutions without glasses. These were
small compared to those obtained for the glass leached alone or in
combination with the rocks. Normalized leach rates of the glass in g/m^2-d
were not calculated because the rocks also contributed species to the
solution. Under any given exposure condition it is not possible to
accurately evaluate the effects the geological materials on leaching from
these data. However, the data do show that if there is an effect, either
positive or negative, its magnitude is at most a factor of 2 or less.
Previous studies on 33N leached in combination with the same materials
indicated that the granites and chert slightly reduced the extent of
leaching while the basalt had no appreciable effect[9].

Glass/Tailored Overpack Interactions. One of the major by-products
from mining Florida phophates is a slime material containing nearly equal
quantities of fine sand, clay and fluroapatite. At present this by-product
has no commercial value, is available in large quantities, and is inexpensive. Furthermore, it is known that small quantities of uranium

present in the ore become concentrated[10] in the slimes during benefication
suggesting that they might provide a good barrier against leaching of
this element from a repository. The nominal composition and mineral
content of these slimes are shown in Table 5.

TABLE 5
CHEMICAL AND MINERALOGICAL COMPOSITION OF FLORIDA PHOSPHATE SLIMES[11]

Analysis	Range(WT%)	Mineral	Range(WT%)
P_2O_5	9-17	Carbonate-Fluorapatite	20-25
SiO_2	31-46	Quartz	30-35
Fe_2O_3	3-7	Montmorillonite	20-25
Al_2O_3	6-18	Attapulgite	5-10
CaO	14-23	Wavellite	4-6
MgO	1-2	Feldspar	2-3
CO_2	Trace-1%	Heavy minerals	2-3
F	Trace-1%	Dolomite	1-2
LoI(1000°C)	9-16%	Miscellaneous	0-1
BPL	19-37%		

The theoretical composition of the Fluorapatite is $Ca_{10}(PO_4, CO_3)_6 F_{2-3}$
and is thought to be the mineral phase that immobilizes the uranium.

Various mixtures of phosphate slime/course sand mixtures were extruded
and sintered at 800°C to produce bars with controlled porosity and
slaking resistance. Samples were prepared from these bars and placed into
D.I. water together with samples of 33N glass as shown in Figure 1. Leach
data are shown in Table 6.

TABLE 6
SOLUTION AND WT LOSS DATA FOR 33N GLASS LEACHED IN COMBINATION WITH
VARIOUS FLORIDA PHOSPHATE SLIME/SAND MIXTURES AT 25°C, 1 WEEK, SA/V=
0.1 cm^{-1} (BASED ON GLASS)

Glass	Slime/sand	pH	%WT loss	Si	Concentration(mg/l) Na	Ca	Al	P
33N	100/0	11.45	7.13	525	485	0	12	12
33N	90/10	11.44	6.42	595	475	0	17	32
33N	80/20	11.52	6.06	420	390	0	22	31
33N	70/30	11.59	5.50	303	445	0	35	41
33N	60/40	11.47	5.18	466	375	0	36	50
33N alone	–	11.44	13.25	1044	650	0	0	0

Each value in Table 5 corresponds to an average of data from two
samples run under identical conditions. The pH of all solutions increased
from 5.6 to approximately 11.5 during corrosion indicating that the
phosphate slime/sand samples provide little if any buffering capacity.
Samples of the tailored mixtures were also leached alone in solution and
the pH increased only slightly. Thus, the pH increase in the solutions
containing both glass and the tailored materials is due primarily to the
reaction of the glass with the water.The Al and P concentrations in the
solutions are due entirely to the partial dissolutions of the tailored
materials; Na is entirely from the glass; and the Si may come from both

the glass and the tailored materials.

The %WT loss and solution data (ie, Si and Na concentrations) show that the presence of the tailored materials did reduce the extent of leaching by a factor of about 2. Based on %WT loss and Na concentration the 60/40 mixture appears to be the most effective in reducing the leaching. The solution containing this mixture did have a higher Si concentration but this is thought to be due to a higher Si contribution from the tailored mixture and not the glass. Higher concentrations of Al and P were also found in the solution containing the 60/40 mixture. This is due to the more open structure (ie, higher porosity) and hence more surface area of material available for dissolution from this mixture.

Electron microprobe analyses (EMP) are shown in Table 7 of the glass corroded alone and the glasses corroded in the solution with two of the tailored materials.

TABLE 7
EMP ANALYSIS OF 33N GLASS CORRODED ALONE, IN THE SAME SOLUTION WITH A 100% PHOSPHATE SLIME PELLET, AND IN THE SAME SOLUTION WITH A 60/40 PELLET, SAME CORROSION CONDITIONS AS IN TABLE 6

Glass	Slime/sand	Intensity(cps)		
		Ca	Al	P
33N	100/0	64	56	6
33N	60/40	47	84	6
33N(alone)	–	12	4	4

The intensities (cps) in this table correspond to averages of 15 spots on each sample. These varied slightly over the sample surface and in areas where surface flaking had occurred the intensities were close to those of the 33N corroded alone in D.I. water. These data show that the glass surfaces corroded in solutions containing the tailored materials are enriched in Ca, Al and P (to a small extent) compared to the glass corroded alone in D.I. water.

Scanning electron micrographs (SEMs) are shown in Figure 2 for the samples listed in Table 7. Extensive surface degradation can be seen on the 33N glass leached alone (Figure 6a). The mechanisms responsible for these alterations are ion exchange ($Na^+_{(g)} + H^+_{(s)} \rightleftharpoons Na_{(s)} + H^+_{(g)}$) and network dissolution. A type IV^{12} non protective surface film is formed on the 33N glass leached alone. In contrast, the SEMs in Figure 2 b-d show the formation of a much different type of film on the 33N leached in combination with the phosphate slime/sand mixtures.

Infrared reflection spectra (IRRS) of the 33N surfaces are presented in Figure 3. These spectra indicate that significant alterations occurred

on all glass surfaces during corrosion but that the final surface films
were much different. The spectral changes in Figure 3a are typical of
glass that has undergone both ion exchange and network dissolution[9]. The
spectra in Figures 3 b-c can be interpreted in two ways; extensive
network dissolution leading to severe surface roughening and general
decrease in spectral intensity, or formation of a dual layer type III[12]
surface film. The EMP and SEM analyses suggest that the latter alternative
is most probable. It is thought that better protection of 33N could be
achieved by equilibrating the D.I. water with the phosphate slime/sand
mixtures prior to immersion of the glass. In this manner the films might
form earlier on the glass preventing large pH increases that are
detrimental to stable film development.

A similar experiment was conducted using an alkali borosilicate glass
(ABS 41) containing depleted uranium and simulated fission products
(F.P.). This glass was leached in combination with both bentonite and
tailored phosphate slime/sand mixtures (ie, three different samples in
the same solution). Bentonite is under consideration as a potential
overpack in the Swedish nuclear waste program. It is expected to swell
and help seal the repository once it comes into contact with water.
Additionally, it may adsorb potentially hazardous species that are
leached from the glass. Previous work by Hench et al[5] has shown that the
presence of bentonite accelerates the rate of leaching on ABS 41. It was
thought that the presence of a phosphate slime/sand mixture together with
the bentonite might provide better resistance to leaching as well as
serve as a sealant and radionuclide adsorber. IRRS data shown in Figure 4
indicate that the presence of phosphate slime can indeed reduce the
extent of leaching (Figure 4b) compared to that in either D.I. water or
in the presence of bentonite alone (Figure 4c). However, these data also
suggest that the position of the phosphate slime/sand is important. It is
not effective when mixed together with bentonite in the same pellet
(Figure 4a). Furthermore, the closer to the glass, the more effective is
the phosphate slime/sand mixture in reducing the extent of leaching (ie,
side 1 in Figure 4b was closer to the phosphate slime/sand mixture than
was side 2).

Burial experiment. Thirty three samples of ten types of glasses and
ceramic materials were buried in direct contact with the soil in the
vicinity of a well characterized archaelogical site near Gainesville,
Florida. The pit, which was 1 meter in depth, consisted of two types of

soil. Up to ∿0.5 meter the soil was primarily sand and below this depth was a
sand/clay mixture, approximately 50% smectite, 30% Kaolin, 20% quartz. The water
table was also ∿0.5 meter below the surface and samples buried below this depth
were continuously wet (static leaching) while those above this level were cyclic
wet and semi-dried (flow leaching) depending on the frequency of rainfall. The
pH of the soil at all depths was 3.5-4.0. The average temperature in the pit for
the first $16\frac{1}{2}$ months ranged from $11^{\circ}C$ to $30^{\circ}C$ and the total rainfall was
63 inches. A more complete description of the experiment and analyses
will be presented in a future report[13].

After $16\frac{1}{2}$ months the first set of samples was retrieved and placed
inside plastic bags containing soil samples to prevent drying prior to
analysis. Optical examination revealed that all samples were covered with
a thin gel-like surface film which firmly bonded soil grains to their
surfaces during drying. Surface cracking was observed on the majority of
the glass samples with the exceptions being SRL 21 and the natural
glasses (obsidians). The cracking occurred during optical examination on
some samples, specifically the PNL 140. Samples of 33N glass had almost
completed disintegrated with only small fragments remaining after $16\frac{1}{2}$
months. In contrast, similar 33N samples buried by Hench[14] and coworkers
in England in an alkaline soil (pH>9) showed minimal deterioration. These
results were attributed to the development of a protective Ca-rich
surface that developed during burial in the alkaline limestone soils. If
such surfaces did form on the Florida samples they were apparently non
protective in the low pH environment.

Optical micrographs are shown in Figure 5a-d for 33L and PNL-140
glasses after ultrasonically cleaning in acetone. Surface alterations can
be seen on all samples. Figure 5a-b illustrate the effect of sample
orientation on leaching. The sample in Figure 5a was oriented with the 1
cm^2 sides parallel to the ground surface (horizontally oriented) and the
one in Figure 5b was oriented with its 1 cm^2 sides perpendicular to the
ground surface (vertically oriented). The morphology of surface cracking
on the former (5a) sample suggests more extensive leaching. Figure 5c-d
show the influence of soil composition on leaching of PNL-140. Surface
alterations are less severe on the sample buried in the sand (Figure 5c)
than on the one buried in the sand/clay mixture (Figure 5d).

Infrared reflection spectra (IRRS) are shown in Figure 6 for the 33L
and PNL-140 surfaces buried and/or exposed to various environments. In
general these spectra show that leaching is more extensive on the buried

samples than on the control samples stored in the laboratory. Additionally, leaching of 33L was more extensive in the sand than in the sand/clay mixture. The reverse is true for the PNL-140. Additional analyses of these buried samples along with ones leached in the laboratory is now in progress.

SUMMARY

The influence of repository materials on glass leaching appears to be minimal under the conditions tested in our laboratory. In general the leaching of the geological materials was much less than for the glass, and the presence of these materials did not provide any significant buffering capacity to the solution. Certain species such as Al, Mg and Ca, which are known to retard leaching, are present in most of the geological materials, but these were apparently not leached in sufficient quantities to form protective surfaces on the glasses. Higher surface areas of geological materials such as crushed rocks or equilibration of the D.I. water with the rocks prior to immersion of the glass might provide higher concentrations of passivating species and better protection.

A tailored backfill/overpack material consisting of Florida phosphate slime and sand in the form of extruded pellets was effective in reducing the extent of leaching from a very corrodible glass (33N). Analysis showed that the tailored materials provided sufficient concentrations of Al, Ca and P to produce protective surface films. The pellets containing the highest surface areas gave the best protection. Similar effects were found for an alkali borosilicate glass containing simulated fission products and depleted uranium. However, the placement of the tailored material appears to be critical when bentonite is present. The tailored material must be sufficiently close to the glass to overcome accelerated leaching due to bentonite.

Preliminary data from a shallow pit burial experiment show that the soil composition and sample orientation are both important in leaching. Surface films form during burial and the protectiveness of these films depend on the soil composition.

80

ACKNOWLEDGMENTS

This work was supported by Savannah River Laboratory (SRL) under contract no AX-487975R. Samples of PNL-140 were supplied by G. McVay at Battelle Pacific Northwest Laboratory, and SRL 21 frit was supplied by John Wiley. Partial support was provided by SKBF/Div. KBS, Sweden.

REFERENCES

1. Dilmore, M.F., Clark, D.E. and Hench, L.L., (1979), Am. Ceram. Soc. Bull. 58(11), 1111-1124.
2. Iler, R.K., (1973), I. Colloid Interface Sci., 43(2), 399-408.
3. Materials Characterization Center (MCC), Results of Round Robin Leach Test, Battelle Pacific Northwest Laboratory, Richland, Washington, 1980.
4. Tait, J.C. and Jensen, C.D., 1981, in Proceedings of 6th University Conference on Glass Science, Penn. State University.
5. Werme, L., et al., This Proceedings, p. 135.
6. Wicks, G., et al., This Proceedings, p. 15.
7. Materials Characterization Center, DOE/TIC-11400, Pacific Northwest Laboratory, Richland, Washington, 1982.
8. Hench, L.L. and Clark, D.E., October 1981, Annual Report to the U.S. Nuclear Regulatory Commission, Contract No. NRC-04-78-252.
9. Clark, D.E., Urwongse, L. and Maurer, C., (1982), Nuclear Technology, 56, 212-225.
10. Moudgil, B.M., Private Communication.
11. Whitney, E.D., Moudgil, B.M. and Onoda, Jr., G.Y., 1977, Report to the Center for Research in Mining and Mineral Resources, College of Engineering, University of Florida, Gainesville, Florida.
12. Clark, D.E., Pantano, Jr. C.G. and Hench, L.L., 1979, Corrosion of Glass, Magazines for Industry, Inc., New York.
13. Maurer, C., Purdy, B. and Clark, D.E., in preparation.
14. Hench, L.L., Private communication.

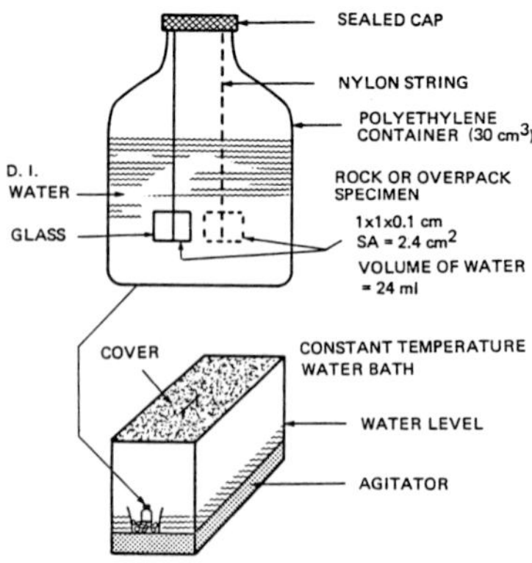

Fig. 1.
Corrosion cell assembly.

Fig. 2. SEMs of 33N glass after corrosion at 25°C, 1 week, SA/V=0.1cm⁻¹ (based on glass) in combination with a)33N alone, 1000X,b)100% Phosphate slime 100X,c)60/40 mixture of Phosphate slime/sand, 1000X,d)same as c, 1000X.

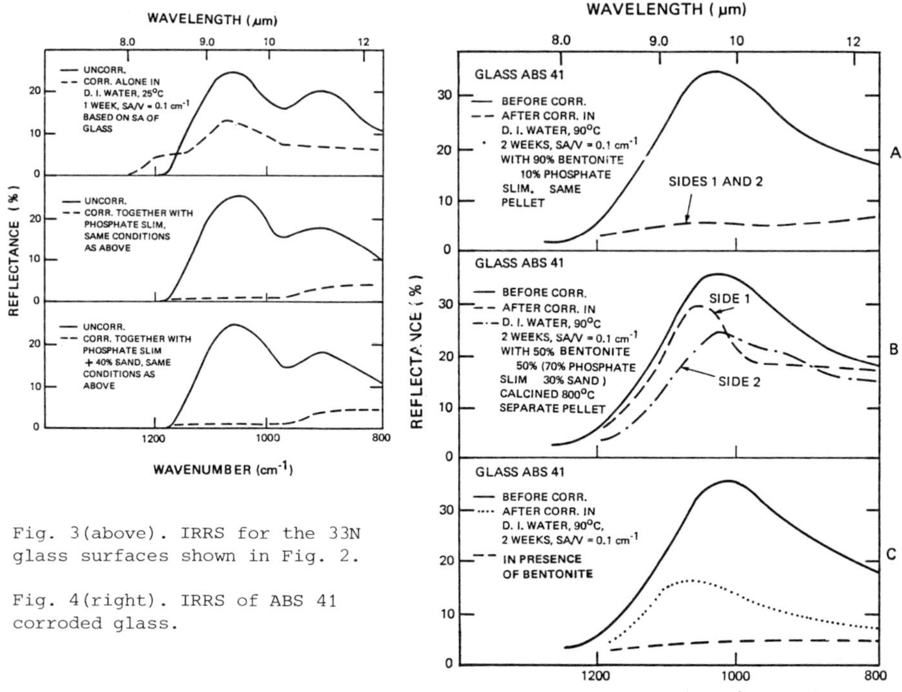

Fig. 3(above). IRRS for the 33N glass surfaces shown in Fig. 2.

Fig. 4(right). IRRS of ABS 41 corroded glass.

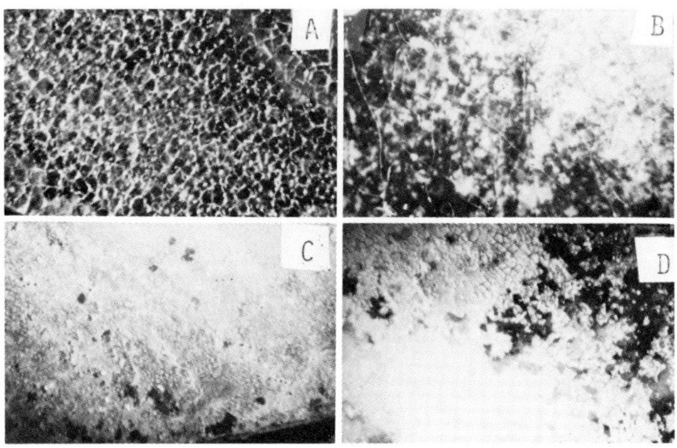

Fig.5. Optical micrographs of 33L and PNL glasses buried for 16½ months. a)33L-Horizontal Orientation, 80 cm below surface. b)33L-Vertical Orientation, same depth. c)PNL-140, 38 cm below surface, sand soil. d) PNL-140, 80 cm below surface, sand/clay soil.

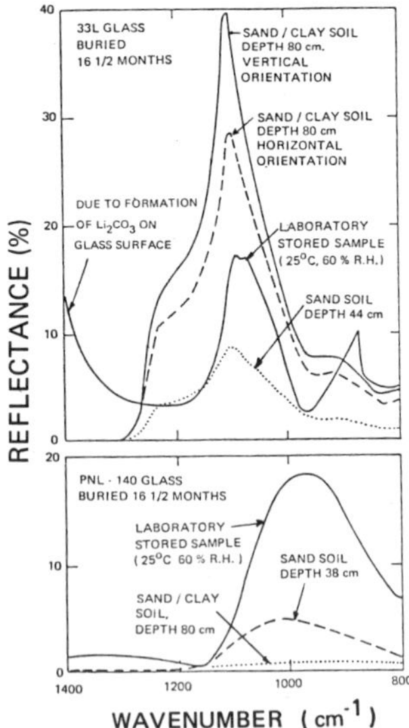

Fig.6. IRRS of 33L and PNL-140 glasses buried for 16½ months and for identical samples weathered in the laboratory at 25°C, 60% R.H.

Published 1982 by Elsevier Science Publishing Co
SCIENTIFIC BASIS FOR RADIOACTIVE WASTE MANAGEMENT - V
Werner.Lutze, editor

LEACHING OF VITRIFIED HIGH-LEVEL RADIOACTIVE WASTE *

A.R.HALL, A.HOUGH and J.A.C.MARPLES
Chemistry Division, AERE Harwell, Didcot, Oxon OX11 0RA, U.K.

INTRODUCTION

The work reported here is part of a study of the leaching of individual elements from waste glasses. The first part of the paper reports on leaching of the major constituents at $30^{\circ}C$, and leaching of the alkalis at $20^{\circ}C$ from glass UK189; the second part describes leaching of Sr, Cs, Tc and various actinides from glasses UK189 and UK209. Compositions are given in Table 1.

TABLE 1

COMPOSITIONS OF GLASSES UK189 AND UK209 (wt%)

Oxide	UK 189	UK 209
Glass formers		
SiO_2	41.51	50.88
B_2O_3	21.87	11.12
Na_2O	7.68	8.30
Li_2O	3.69	3.99
Waste from reprocessing Magnox Reactor Fuel		
MgO	6.23	6.34
Al_2O_3	5.03	5.11
Fe_2O_3	2.68	2.73
Cr_2O_3	0.55	0.56
NiO	0.36	0.36
ZnO	0.44	0.44
U_3O_4	0.06	0.06
FPox (full spectrum of inactive isotopes)	9.90	10.08

The constituents of a glass may be classified according to the way in which they are leached: (i) Major glass network elements which are leached at a constant rate, suggesting removal from the surface only (e.g. silicon): (ii) those elements which are leached by ion exchange with protons (or H_3O^+ groups) in the leachant with $\sqrt{time}$ dependence, suggesting removal from beneath the surface by a diffusion mechanism. The alkalis and other network modifiers typify this class. The process results in a hydrated surface layer depleted in elements of this class, often called the gel layer: (iii) elements which are leached at a rate slower than the main glass network and thus become concentrated in the gel layer. Fe, Al and some actinides appear to fall into this class.

DOE/RW/82.037.   *This work has been commissioned jointly by the UK Department of the Environment and the Commission of the European Communities as part of their Radioactive Waste Management Research Programmes. The results may be used in the formulation of UK Government policy but at this stage they do not necessarily represent such policy.

84

LEACHING OF THE MAJOR CONSTITUENTS

Experiments have been carried out in an effort to understand the effect of
(a) pH and (b) the onset of saturation effects due to accumulated dissolution
products in the leachate, on the leaching of the main glass constituents. In
addition, leach rates have been measured for certain of the fission products.
This gives a guide as to which elements fall into which of the three classes
listed above.

Experimental. Samples of $12cm^2$ surface area were cut from annealed castings,
ground and polished.   Final polishing was carried out dry, using $2\mu m$ diamond
powder on paper laps so as to provide a fresh surface untouched by water.
Leachate analysis was made by emission spectroscopy. For investigation of the
initial stages of alkali leaching the sample was placed into 75ml of stirred
leachant, which was analysed continuously for the first few minutes and then
intermittently for up to 1 hour. Longer tests of up to 3 months duration were
made by sealing samples in Teflon vessels containing the leachant placed in a
temperature-controlled shaking bath. Leachants were either distilled water,
very dilute HCl or NH_4OH, or buffer solutions.

Results. Figure 1 shows the leaching of sodium and lithium at $20^{\circ}C$ in
distilled water. The glass contains equimolar amounts of the two alkalis, and
they leached at the same molar rate. The quantity leached was proportional to
$\sqrt{time}$, except for a small non-linear region at the beginning. When the
experiment was repeated using a fused sample, the same result was obtained,
thus demonstrating that the initial non-linearity was not caused by contamination from polishing debris or by surface micro-roughness. The initial
region is shown in greater detail for a range of pH values in Figure 2.

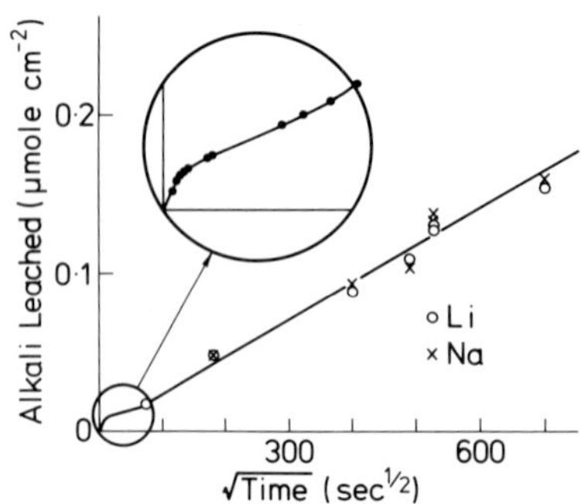

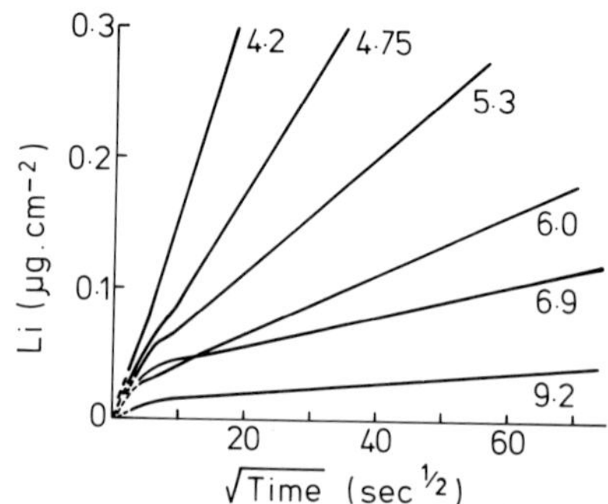

Figure 2. Effect of pH on lithium
leaching

Figure 1. Alkali Leaching at 20°C

Diffusion of an ionic species results in an electrical potential across the diffusion layer (the diffusion potential, Φ_D)   caused by the ion gradient. When two species are interdiffusing, both contribute to Φ_D, which has the effect of accelerating the slower-moving ion and decelerating the faster-moving ion so that the two ionic fluxes are in balance and Φ_D is constant. Electrical potentials ($\Delta\Phi$) were measured in thin glass bulbs which were filled with 1 molar KOH electrolyte and immersed in leachants of various pH. Figure 3 shows some results. $\Delta\Phi$ is of course the sum of Φ_D and the two boundary potentials caused by the difference in pH between the leachant and the electrolyte. It is evident from Figures 2 and 3 that the initial disturbance in $\Delta\Phi$ is matched by the anomalous initial leach rate, and represents a period when the ingoing protons and outgoing alkali ions are not yet in balance.

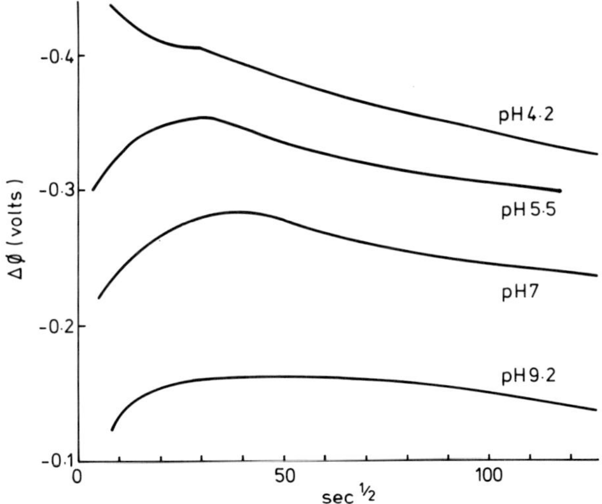

Figure 3. Electrical potentials. 20°C, various pH values

Effective diffusion coefficients (D_e) were calculated from the linear parts of graphs such as Figure 1, using the equation

$$D_e = \frac{\pi}{4t}\left(\frac{Q}{C_O}\right)^2$$

where Q is the amount of alkali leached in time t and C_O is the original alkali concentration in the glass. For any value of pH, D_e was identical for sodium, lithium and magnesium. When values of $\Delta\Phi$, taken from the same steady-state time region, were combined with values of D_e, the results gave close

86

agreement with $D_e = 7.2 \times 10^{-20}$ exp $(-F\Delta\Phi/RT)$, illustrating again the important inter-relationship between diffusion potential and alkali leaching.

Figure 4 shows leaching results for times up to 30 days, expressed as quantity removed $(g.m^{-2})$ ÷ original concentration $(g.m^{-3})$. This method of presentation clarifies the relationships between the leach rates of the various elements. Silicon is leached at a constant rate, i.e. $Q_{Si}/C_{Si} = At$. For other elements (B, Mg, Na, Li) $Q_m/C_m = At + Bt^{\frac{1}{2}}$ where A is the same as for silicon and B is the same for each element. This relationship applies at all pH, although of course the values of A and B vary with conditions. The curve at top left in Figure 4 represents the expression

$$Q_m/C_m = 2.21 \times 10^{-13}t + 2.64 \times 10^{-10} \, t^{\frac{1}{2}} \text{ (metres)}$$

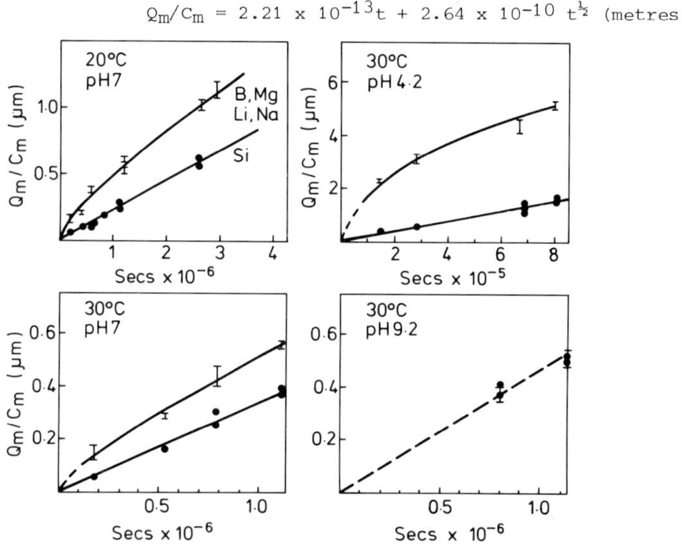

Figure 4. Leaching of major elements

and similar expressions were used to generate the other curves. The effect of pH is readily seen.

It is interesting that boron should leach diffusively and congruently with the alkalis at all values of pH. Two possible explanations are (i) Boron atoms cannot be attacked by water until an associated alkali has first been exchanged with an H^+ ion. Against this however is the fact that there is insufficient alkali to allow one alkali atom to each boron atom: (ii) the boron is removed by water which can penetrate only the gel layer readily and not the glass. Thus the glass must first be reduced to gel by removal of the alkalis before the boron can be removed. The proportions of 3- and 4-coordinated boron may also be relevant.

Leachates were also analysed for Al, Fe, Mo, Zr, Sr and Nd. In many cases the concentration was close to the detection limit and results were less accurate than for the major constituents. Aluminium appears to be unique in that it was leached most rapidly in alkali. At pH9.2 the rate was the same as that of silicon; in distilled water the rate was ca half that of silicon, and at pH4.2 aluminium could scarcely be detected. Iron could not be detected in any sample, the detection limit being 0.05ppm ≡ 0.03μm of glass. Molybdenum leached at a rate similar to boron and the alkalis within the rather poor limits of accuracy obtained. Zirconium and strontium appear to be partly retained in the gel layer. Neodymium was partly retained in the gel layer and leached more rapidly in acid than in distilled water.

Two experiments were made to examine the onset of saturation effects. The surface area to volume (SA/V) range covered was from 0.03 to 0.3. Figure 5 shows results of leaching for 10 days in leachant buffered at pH4.2. Fairly predictable trends are seen, with saturation effects beginning at ca SA/V = 0.1. By contrast the results for a 73 day leach in distilled water (Figure 6) show a totally different trend with very rapid saturation at the smallest SA/V. It must be remembered that pH would increase with time at rates which would differ with SA/V, so that the effect of pH change would reinforce any saturation effect. It is obvious from this one experiment that many similar tests will be needed to complete the picture.

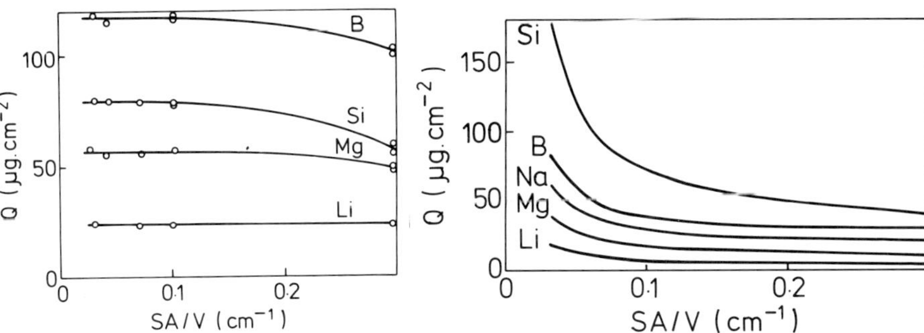

Figure 5. SA/V effect, pH4.2   Figure 6. SA/V effect, distilled water

LEACHING OF GLASSES SPIKED WITH RADIOISOTOPES

The radioisotopes selected for study were Cs-137 and Sr-90, which will be important during the first few hundred years after vitrification, and Tc-99, Np-237, Pu-239 and Am-241, which will be important in the longer term. Separate samples of glasses 189 and 209 containing a full spectrum of non-active fission products were doped with each of these isotopes. For Cs-137 and Sr-90 the level was arbitrarily set at 0.05mCi/g, so that they could readily be detected in the leachates. The levels of the other isotopes were Tc-99: 0.05mCi/g, Np-237: 0.003mCi/g, Pu: 0.05mCi/g and Am-241: 2mCi/g. These levels are approximately those which will occur in the real glasses except for Np, where the level is $\sim$ 10X higher so that it could be detected in the leachate. Two series of experiments have been carried out with these samples (a) in pure water at low flow rate, and (b) under almost static conditions over a column of crushed granite.

Flowing leachant experiments. Experiments were carried out in a temperature-controlled Teflon sample chamber of $\sim$ 2ml capacity. The glass sample, a disc of surface area $\sim$ 3cm^2, was mounted upright so that almost all the surface was freely exposed to the leachant. Water from a reservoir was pumped through the chamber and out to a collecting vessel through Teflon tubing. Initially, for the UK189 specimens, the flow rate was about 9ml/day, but this was later reduced to 1ml/day and then to 1ml/week, the latter reduction being achieved by running the pump intermittently for about 2 hours per day. Glass UK209 was studied only at a flow rate of 1ml per week.

At intervals the leachate that had collected was acidified and an aliquot evaporated to dryness on a counting tray. This was then counted in an α- or β- counter as appropriate. The count rates thus obtained were converted to leach rates by dividing by the surface area of the specimen, the time over which the sample had been collected and the specific activity of the glass.

The leach rates obtained at 60°C for the two glasses reached constant values after a few days and these values are listed in Table 2, where they are compared with previously published values measured at higher flow rates.

The leach rate obtained for UK189 by counting for Tc-99 did not change with flow-rate and was similar to that obtained previously by weight loss measurements after experiments at a high flow rate. This suggests that the release of Tc is limited by its rate of escape from the glass rather than by its solubility in the leachant. The leach rates for Cs, Sr and Np decreased by X3, X10 and X6 for a 60-fold decrease in flow-rate, suggesting that the solubility of the species in the leachant may have some effect.

TABLE 2

LEACH RATES IN DISTILLED WATER AT 60°C IN $g.cm^{-2}day^{-1}$

	Glass 189			Glass 209	
Flow rates Isotope	9ml/day	1ml/day	1ml/week	1ml/week	Ref(1)
Tc-99	7×10^{-5}	7×10^{-5}	6×10^{-5}	1.3×10^{-6}	
Sr-90	5×10^{-5}	2×10^{-5}	5×10^{-6}	4×10^{-7}	3.4×10^{-6}
Cs-137	6×10^{-5}	4×10^{-5}	2×10^{-5}	9×10^{-7}	1.1×10^{-5}
Np-237	1.8×10^{-5}	1×10^{-5}	3×10^{-6}	2×10^{-6}	Ce-144:
Pu	(5×10^{-7})	(1.3×10^{-7})	(1.5×10^{-8})	(1.5×10^{-7})	2×10^{-8}
Am-241	(4×10^{-9})	(6×10^{-10})	(6×10^{-10})	(8×10^{-9})	
Weight Loss (1)	At a high flow rate (1ml/min) 6.5×10^{-5}			1.8×10^{-5}	

Values in parentheses are uncertain – see text. The values quoted in the final
column were obtained by the Marcoule method where the water is circulated with
a 90-second cycle time and replaced daily.

The leach rates found for glass 209 are considerably less than those
previously found using high flow rates [1].   The final column of Table 2 gives
the average of values determined for Sr and Cs at 50°C and 70°C by the Marcoule
method, under conditions which were effectively equivalent to a very high flow
rate. These values are about 10X those obtained on the present work. The
leach rate determined at Marcoule for Ce-144 was very low, between those found
in the present work for Pu and Am-241.

The leach rate found for Np-237 was similar for the two glasses. This
would be the case if the leach rate is governed by the solubility of Np in
the leachate rather than by its rate of escape from the glass. However this
is not the case for Cs and Sr, despite the reduction in their leach rates with
flow rate.

After a total leaching time of 270 days the tests on glass 189 were stopped.
When the samples had been removed from the containers the apparatus was rinsed
with acid and this was then evaporated and counted in the same way as the
leachate to determine the amount of the various isotopes held on the walls of
the apparatus: the amounts in the specimen chamber, the outflow pipe and the
collecting container were measured separately. For comparison purposes the
values obtained were divided by the concentration of that isotope in the glass

to give a mass balance, although the various components of the glass would in fact be adsorbed to different extents. To complete the mass balance, the gel layer was removed, dissolved in acid and counted and the amounts that had previously been found in the leachates were summed. Finally the specimens were weighed to determine the total amounts lost. These data are given for all six samples in Table 3.

TABLE 3

APPARENT MASS BALANCES FOR SAMPLES OF GLASS 189 (grams)

Location	Tc-99	Sr-90	Cs-137	Np-237	Pu	Am-241
Sample container	2.2 E-4	5.8 E-4	3.0 E-4	3.0 E-4	2.0 E-4	5.8 E-5
Outflow pipe	3.8 E-5	1.8 E-5	4.9 E-5	1.7 E-5	3.7 E-5	5.1 E-6
Catch pot	9.3 E-5	1.3 E-4	1.5 E-4	3.9 E-5	8.4 E-5	4.6 E-6
Leachate total	5.5 E-2	1.4 E-2	2.6 E-2	6.8 E-3	5.7 E-5	1.9 E-6
Gel layer	9.5 E-3	2.3 E-2	1.2 E-2	7.5 E-2	6.8 E-2	3.2 E-2
Sum	6.5 E-2	3.7 E-2	3.9 E-2	8.2 E-2	6.8 E-2	3.2 E-2
Weight loss	5.0 E-2	3.5 E-2	3.5 E-2	4.8 E-2	4.7 E-2	2.4 E-2
Retained in gel layer (%)	15	62	31	91	99.4	99.8

The quantity of each isotope for each location, as determined by counting techniques, was divided by the specific activity of the glass to obtain a notional equivalent weight of glass.

In general the total weights lost by the samples as determined by counting agreed reasonably well with those measured directly, except for the Np-237 sample, where the error was almost a factor of two.

The average leach rate over the whole 270 days leaching was 4.6×10^{-5} g.cm^{-2}day^{-1} after removal of the gel layer. This compares with 6.5×10^{-5} g.cm^{-2}day^{-1} for the experiment in rapidly flowing water with pH7 at the same temperature, carried out previously [1] .

The quantities of Tc, Sr, Cs and Np retained in the apparatus compared to the totals found in the leachates were negligible but more plutonium and americium were retained in the apparatus than were found in the leachates. If we take the sum of the amounts found in the apparatus and the leachates, the true average leach rates over the 270 days would be as given in Table 4. The true values, although still very low, are 7X and 36X higher than the apparent values obtained from the leachates alone. Most of the adsorbed material was found in the specimen chamber in close proximity to the sample.

TABLE 4

APPARENT AND TRUE AVERAGE LEACH RATES FOR GLASS 189 AT 60°C $(g.cm^{-2}day^{-1})$

	Apparent	True
Plutonium	6.6×10^{-8}	4.4×10^{-7}
Am-241	2.2×10^{-9}	8.1×10^{-8}

In the last line of Table 3 we give the fraction of each isotope retained in the gel layer compared to that leached out. Of particular interest is the very high retention of the actinides Pu and Am. Also of interest is the 30% retention of Cs-137 since the alkali constituents of the glass Na and Li are readily released. This is probably a size effect since the ionic radii (for 6-fold coordination) are: Li^+ 0.78Å, Na^+ 0.98Å, Cs^+ 1.65Å.

Granite repository simulation. Here the specimens were from the same batches of doped glasses used in the flowing leachant tests. Each specimen (surface area $3cm^2$) was held in a Teflon support in a 1.5ml chamber at the top of a water-filled column of granite powder at 60°C. The only flow past the specimen occurred when 1ml samples of leachate were taken at intervals from the specimen chamber with a hypodermic syringe. Water to replace these samples was drawn up from the column where it had been exposed to the granite for several weeks before it contacted the specimen; fresh water entered at the bottom of the column. Initially samples were taken at 3-day intervals but later this interval was extended to weekly and then to monthly, giving a "flow rate" of 1ml per month.

The initial results for glass 189 are plotted in Figure 7. The results for Sr, Cs and Np are not entirely comparable with those for the other isotopes because they were started later and the sampling rate - and hence the water flow rate - was much lower. However, even taking this into account, the apparent leach rates obtained are very low. A difference between this experiment and the previous one is that the leachate has access to the granite and ion exchange might lower the apparent leach rates of Cs and Sr. Surprisingly, the apparent leach rates for Pu and Am are higher in this experiment than in the low-flow one described above, although still lower than the true leach rates after allowing for adsorption.

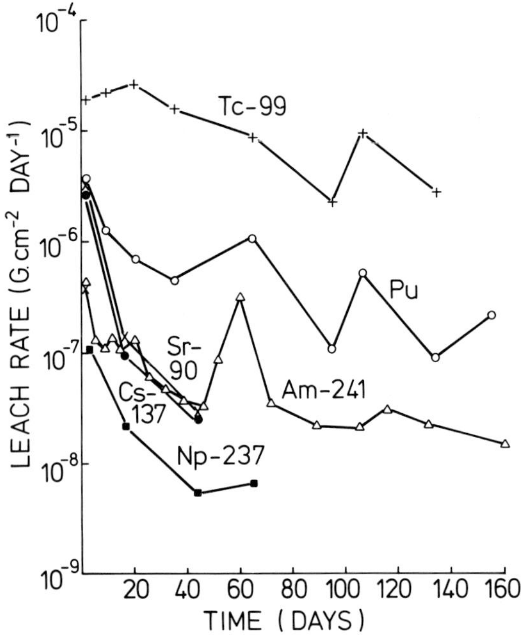

Figure 7. Leach rates of UK189 at 60°C in repository simulation.

REFERENCES

1. J.A.C. Marples et al. (1981) European Applied Research Reports $\underline{3}$ (3) 395.

Published 1982 by Elsevier Science Publishing Co
SCIENTIFIC BASIS FOR RADIOACTIVE WASTE MANAGEMENT - V
Werner.Lutze, editor

THE ROLE OF METAL ION SOLUBILITY IN LEACHING OF NUCLEAR WASTE GLASSES*

B. GRAMBOW
Pacific Northwest Laboratory, P. O. Box 999, Richland, Washington 99352, USA and
Hahn-Meitner-Institut für Kernforschung Berlin GmbH, 1000 Berlin 39 Glienicker
Str. 100, Berlin, Germany

INTRODUCTION

The leachability of solid nuclear waste forms has been studied by many investigators to evaluate the short-term kinetics of elemental release or to determine the effects of leachant composition or other system parameters. Some
general observations from these studies have included: incongruent leaching;
formation of reaction layers that contain rare-earth, alkaline-earth, or transition metal elements; and apparent saturation of some elements in solution while
others continue to be leached.[1,2,3] Before these observations can be used to
predict long-term performance of waste glasses under repository conditions, the
mechanisms controlling the release of elements from the solid must be understood.

Solubility-related phenomena have become increasingly important in interpreting many aspects of leaching behavior. Several previous studies have, for
example, qualitatively considered solubility limits to be important with respect
to reaction layer formation and the prediction of long-term behavior under
accident conditions in a repository.[4,5] Rai[6] showed that concentrations of Np
in solutions contacting crushed Np-doped PNL 76-68 glass were controlled by a Np
solid phase that is similar to that controlling the solubility of pure, crystalline NpO_2.

Since multicomponent waste glasses are unstable with respect to an assemblage
of alteration products, thermodynamic equilibrium between the glass phase and
the leaching solution can never be achieved. This paper shows that the alteration phases resulting from the leaching of the waste glass ultimately regulate the solution concentrations as if equilibria with metal hydroxides or carbonates are established.

EXPERIMENTAL METHODS

The composition of PNL 76-68 glass that was used in this study has been
reported by Strachan.[3] The results to be discussed come from the MCC-1P Static
Leach Test Method;[3,7,8] from a solubility test with glass powder;[9] or from

---

* This work was supported by the U.S. Department of Energy under Contract
DE-AC06-76RLO 1830.

experiments using the MCC-5S Soxhlet Leach Test Method.[10] The test conditions are summarized in Table 1. Details can be found in the references indicated.

TABLE 1

EXPERIMENTAL CONDITIONS OF LEACH TESTS CONSIDERED IN THIS PAPER

Test Method	Leachant	Time d	Temp. °C	SA/V m^{-1}	Final pH	Ref.
MCC 1 (Static)	HNO_3, NaOH or DI-H_2O	≤28	90	10	1 to 12	8
"	DI-H_2O	≤365	90	10	8.8 to 9.6	3
"	Silicate water	≤28	90	10	9.5	3
"	EDTA (0.1m)	≤28	90	10	8.3 & 9.0	b
Solubility (Static)	H_2O	79	95	1200, 5000	8.8 9.6	9 9
Soxhlet[a] (Dynamic)	H_2O	≤28	90	(0.01)	6.8	b

[a] Flow rate: 1 ml/min; pH measured in leach cup. SA/V = sample surface area divided by total volume of water that had contacted the sample after 28 d. The geometric SA/V was 30 m^{-1}.

[b] Data collected in this study

CALCULATION OF EQUILIBRIUM CONCENTRATIONS

Numerous solid phases can control solution concentrations by their solubility.[11-14] For this study, the set of possible solids was reduced by considering solution conditions and the glass composition. For instance, phosphate solids could not play a major role since there was insufficient phosphorous in PNL 76-68 glass to precipitate any of the elements.

Assuming a certain solid phase, the equilibrium concentration of an element in solution was calculated at the final pH (25°C), P_{CO_2} of the air, and test temperature by the use of equilibrium constants and reaction enthalpies. Hydrolyzed and other complex species were considered in the calculations. The experimental data were compared with these equilibrium concentrations. For dilute solutions, concentrations were compared rather than activities, with an expected maximum error of 20%. For solutions with moderately high ionic strengths, such as the water and silicate water leachates after a 365 d test, calculated activities were used for the comparisons. Colloidal particles are assumed to be negligible for this discussion.

Two derived units were used to compare all of the solution data for a particular element over a broad spectrum of tests and pH conditions. By normalizing all of the solution concentrations to the original composition of the glass, data over the entire pH range could be compared in terms of elemental losses and solubility limits. The normalized element mass loss, NL_i $(g \cdot m^{-2})$, is given by the equation:[7]

$$NL_i = \frac{C_i}{f_i \, SA/V}$$ 1)

where C_i = concentration of element i in solution in, $g \cdot \ell^{-1}$
   f_i = mass fraction of element i in the glass, dimensionless
   SA = specimen surface area, m^2
   V = volume of the leachate, ℓ

The solubility term, also normalized, is defined as the amount of glass that must dissolve to raise the concentration of element i in the leachate to its solubility limit. Hence, the normalized solubility, $NS_i = (g \cdot m^{-2})$, is given by:

$$NS_i = \frac{C_{i,sat}}{f_i \, SA/V}$$ 2)

where $C_{i,sat}$ = the solubility limited concentration in the leachate at the specified conditions, $g \cdot \ell^{-1}$.

A convenient parameter for comparison and an indicator of the ion selectivity of the leach process is the ratio of the experimentally determined values for the normalized loss for some element to that for a major structural element, e.g., silicon. In this paper, the ratios NL_i/NL_{Si} and NS_i/NL_{Si} will be used to provide important insights into the leaching mechanisms.

RESULTS AND DISCUSSION

The tests involved the solubility behavior of representative elements, Ca, Fe, Si, Sr, Zn, and rare earths. For a variety of test conditions, solution concentrations of these elements were studied at various pH values. As an example of raw data used, Fig. 1 shows the solution concentration of calcium. Figure 2 shows the ratio NL_i/NL_{Si} for each data point after normalizing those raw data and Si data. In this diagram, the solubility limit is expressed as NS_i/NL_{Si}. Since this ratio is dependent upon the silicon loss NL_{Si}, an average value for NL_{Si} $(20 \, g \cdot m^{-2})$ was used in order to get straight lines and to make the solubility behavior evident. This average line can be adjusted easily to the correct position by using the actual silicon leach data. Except for some data collected under highly acid conditions, the observed NL_{Si} from tests used in this study never deviated by more than a factor of 3 from the average value,

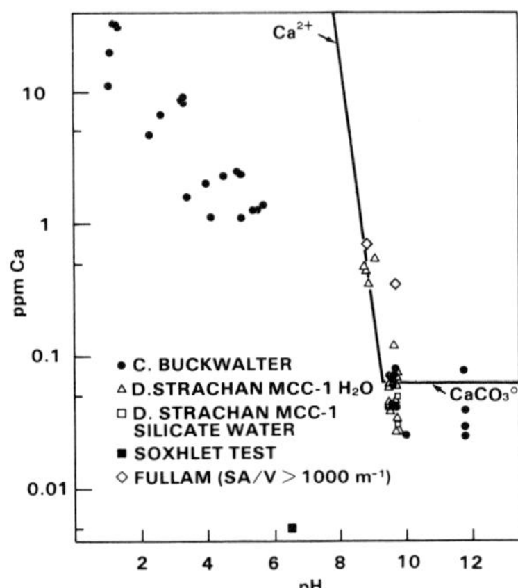

Fig. 1. Concentration of Ca in leachates from PNL 76-68 glass at various conditions. Total $[Ca]_T = [Ca^{2+}] + [CaCO_3^{\circ}]$ at 90°C in equilibrium with calcite at $\log[P_{CO_2}] = -3.52$

and therefore, to a first approximation, the solubility lines in Fig. 2 may be used directly to compare the experimental data with the equilibrium concentrations. The use of Fig. 2 for studying the leach mechanism will be discussed in detail in the following section on Ca, since the data available for this element are the most accurate at all pH values.

Interpretation of Ca leach data

Calcite ($CaCO_3$) frequently governs the calcium concentration in alkaline soils.[11] Chloride, nitrate, molybdate, phosphate, carbonate, sulfate, and boron complexes of calcium as well as the hydrolysis species were also considered in the calculations. At the low concentrations observed in the leachates, none of these complexes are important at pH < 9, and at pH > 9 only the $CaCO_3^{\circ}$ complex contributes significantly to the amount of soluble calcium.

The equilibrium concentration of Ca^{2+} governed by calcite is given by the equation:

$$\log[Ca^{2+}] = \log K_c - \log[P_{CO_2}] - 2\,pH \tag{3}$$

and the concentration of $CaCO_3^{\circ}$ by the equation:

$$\log CaCO_3^{\circ} = \log K_{cc} + \log[Ca^{2+}] + \log[P_{CO_2}] + 2\,pH \tag{4}$$

The total calcium concentration in equilibrium with calcite is:

$$[Ca]_T = [Ca^{2+}] + [CaCO_3^{\,o}] \tag{5}$$

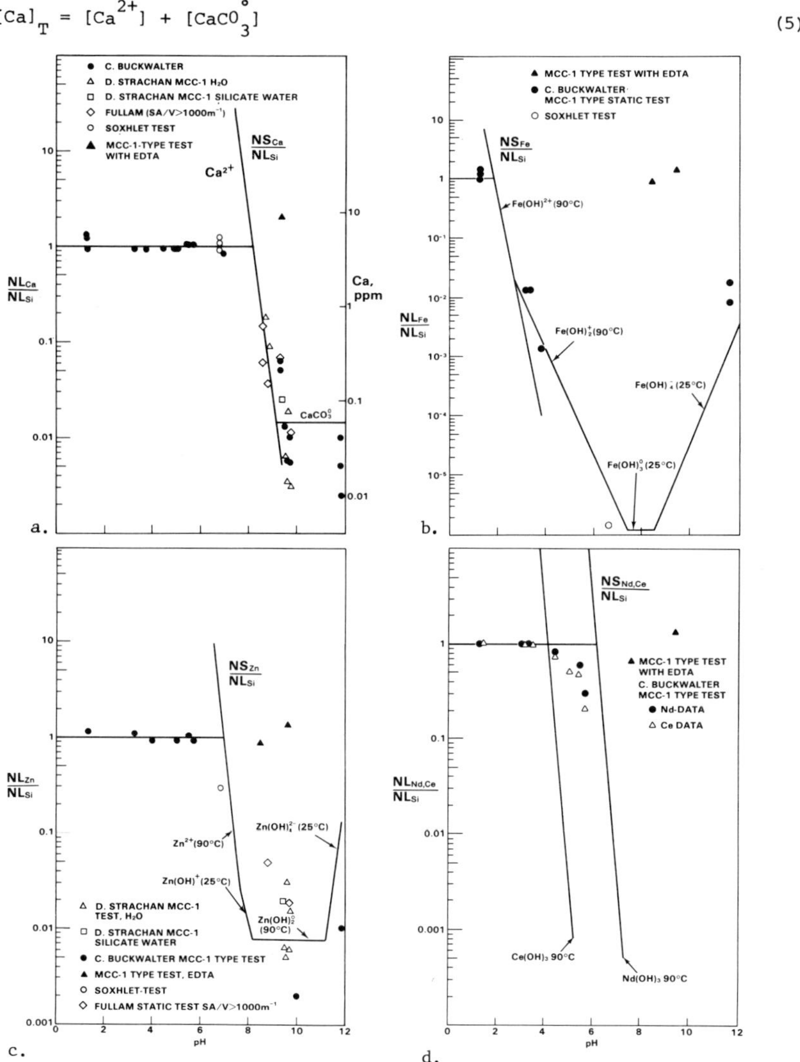

Fig. 2. Comparison of leach data expressed as the ratio of normalized elemental mass loss to the silicon loss with the ratio of normalized solubility to NL_{Si} = 20 g·m⁻² for (a) $CaCO_3$, (b) $Fe(OH)_3$, (c) $Zn(OH)_2$, (d) $Nd(OH)_3$ or $Ce(OH)_3$

98

Figure 2a shows that for pH > 8 there is a good fit of the observed data NL_{Ca}/NL_{Si} to the theoretical curve NS_{Ca}/NL_{Si}. This fit indicates that a solid phase regulates the solution concentration as if pure $CaCO_3$ is in equilibrium with the leachates from PNL 76-68 glass with pH > 8. In this pH range, the ratio of the normalized calcium loss to the silicon loss (NL_{Ca}/NL_{Si}) is < 1, indicating that silicon continues to leach, whereas Ca is controlled by solubility. Solid state analysis showed that calcium is concentrated in the leached layer.

At pH < 8 the above ratio is 1, indicating control of Ca release by matrix dissolution. In this pH range, the solubility is high and the leaching of the glass matrix does not raise the solution concentration of calcium to the equilibrium value. Hence, the ratio of the normalized solubility to the silicon loss (NS_{Ca}/NL_{Si}) is > 1. The introduction of complexing agents such as EDTA showed that even under alkaline conditions calcium is controlled by matrix dissolution ($NL_{Ca}/NL_{Si} \sim 1$), if the solubility restrictions are removed.

The same calculations were made for the strontium data. The experimental data agreed with the equilibrium concentration with respect to $SrCO_3$ at log P_{CO_2} = -3.52. The results will be published in a later document.

<u>Interpretation of Fe, Zn, and rare-earth data</u>

The influence of solubility limits on the mechanism controlling PNL 76-68 glass leaching can be studied in the same way for other elements. For example, amorphous $Fe(OH)_3$ is expected to be formed as a reaction product because it is commonly found in soils as a fresh precipitate that limits iron concentrations when Fe-salt solutions are added to soils. The alteration to more stable, i.e., less soluble, crystalline phases takes many months or even years. For $Fe(OH)_4^-$, the enthalpy of formation is not known. Therefore, the normalized solubility of $Fe(OH)_3$ for pH > 8 was calculated only for 25°C. The redox potential necessary to calculate the solubility was measured at 25°C by Strachan[3] and is expressed as pe + pH = 13 in deionized water and silicate water. At these conditions, at 90°C in contrast to 25°C, none of the Fe(II) species are significant.

Figure 2b shows the ratio of the experimentally observed normalized iron concentrations to the silicon loss and a comparison of the ratio of the normalized solubility to an average silicon loss of 20 $g \cdot m^{-2}$. Good agreement between experimental data and theoretical values was obtained, although the experimental data were somewhat limited because the concentrations frequently were below detection limits.

At a pH < 2, iron is leached similarly to silicon because the solubility limit of $Fe(OH)_3$ (amorphous) is not exceeded. At pH > 2, the normalized solubility is lower than the silicon loss (NS_{Fe}/NL_{Si} <1). Thus, $Fe(OH)_3$ (amorphous) precipitates on the glass surface by forming an insoluble layer. Because of the low solubility, Fe ions released by matrix dissolution probably never leave the surface layer but react immediately to form $Fe(OH)_3$ (amorphous). If, on the other hand, the solubility restrictions are removed by stabilizing high iron concentrations by complex formation with EDTA (▲ in 2b) iron leaches congruently with silicon even at pH 9.

For Zn, various hydroxides, willmenite (Zn_2SiO_4), and franklinite ($ZnFe_2O_4$) were considered, with $Zn(OH)_2$ (amorphous) best fitting the data, as illustrated in Figure 2c. For $Zn(OH)^+$, the enthalpy of formation, and therefore the contribution of this species to the total soluble Zn at 90°C, is not known. Even though the Zn data deviate by up to a factor of 5 from the solubility limit, it can be seen that at pH > 7 Zn is leached less than silicon because it is controlled by a phase with solubility similar to that of $Zn(OH)_2$. At pH < 7 or by leaching with EDTA, Zn is leached congruently with silicon. The deviation of data from the solubility limit might be due to adsorbed material at the container walls, and/or the inability to accurately calculate the contribution of $Zn(OH)^+$. Willmenite and franklinite are far too insoluble to account for the observed Zn concentrations, but alteration to these phases after longer times or higher temperatures is possible. It should be noted that the formation of willmenite was observed in hydrothermal tests with PNL 76-68 glass.[15]

For Nd and Ce, the oxides and hydroxides were considered for comparison with experimental data. As shown in Figure 2d, the measured pH dependence appears compatible with the steep drop expected for control by $Nd(OH)_3$ or $Ce(OH)_3$, considering that the thermodynamic data[14] are not certain. At pH >5.5, the formation of a rare-earth rich layer on the glass surface is to be expected. However, as shown in Figure 2d, even at pH 9 rare-earth element leaching is similar to silicon if solubility restrictions are removed by using EDTA as a complexing agent.

Boundary of control by solubility or matrix dissolution

Figure 3 summarizes the behavior of the various elements considered in this paper. The diagram makes possible a better understanding of the mechanism of leaching and layer formation under various test conditions. In particular, the composition of the leached layer can be predicted for various pH values, and the sequence of precipitation reactions can be predicted in experiments where the pH changes with time. In static leach tests of glasses, Fe will be the first element to precipitate near pH 6, followed by the rare-earth elements. As the

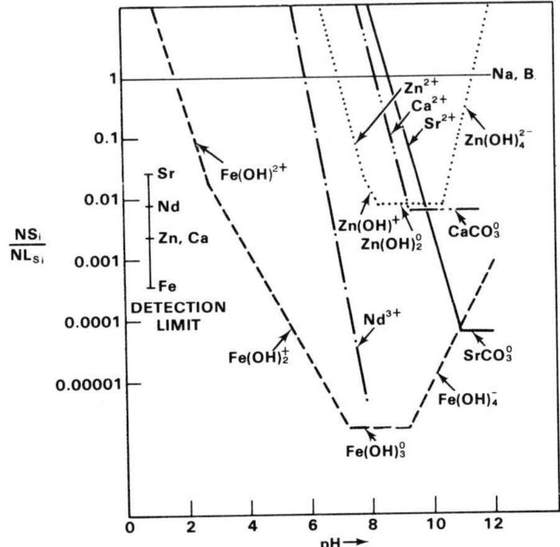

Fig. 3. Ratio of normalized solubility to NL_{Si} (20 g·m^{-2}) for $CaCO_3$, $SrCO_3$, $Nd(OH)_3$, $Fe(OH)_3$ and $Zn(OH)_2$ in MCC-1 test at 90°C, 28 d versus pH

glass makes the pH more basic due to alkali release, Zn will precipitate with the alkaline-earth elements following. The decrease in solution concentrations of Ca and Sr with time in the MCC-1 test[3] can be explained by this process. At very high pH values, some elements can be expected to show increased solubilities and leach rates. In the case of Fe, for example, a minimum in its dissolution rate should occur between pH 7 and pH 9.

At all pH values, the general observation is that there is a control of leach data by matrix dissolution as long as solubility limits are not exceeded. Figure 3 was calculated for PNL 76-68 glass for a fixed set of reference conditions: NL_{Si} = 20 g·m^{-2}, SA/V = 10 m^{-1}, 90°C and $\log[P_{CO_2}]$ = -3.52. To compare this calculation with the actual condition of an individual test, it is necessary to adjust the position of the curves according to the deviation of the test conditions from the point of reference. Figure 4 shows the influence of various parameters on the position of the lines, as calculated by Eq. 2 for $Zn(OH)_2$, for example. This diagram can be used to show that the alteration of $Zn(OH)_2$ to the less soluble willmenite ($ZnSiO_4$) increases the pH range at which solubility limits are predicted to be important. On the other hand, for the decrease in temperature from 90°C to 25°C it can be predicted that the solubility limit for Zn is less important.

The application to other glasses may require different conditions in regard to the formation of sparingly soluble solids or the formation of aqueous solution complexes. Phosphate-containing glasses, for example, may form Al, Fe, Ca,

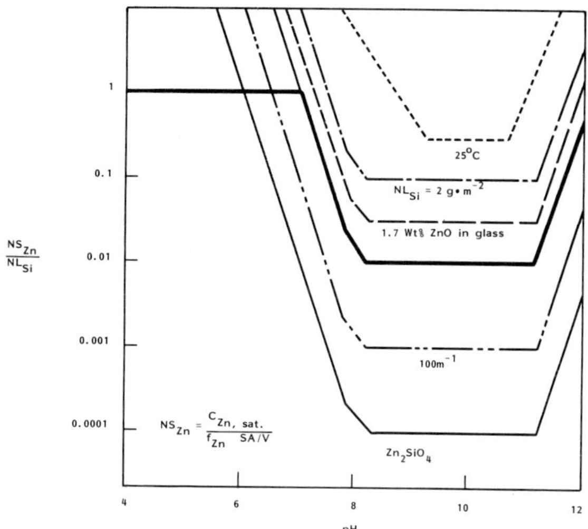

Fig. 4. Boundary between kinetic and solubility control. Reference: $Zn(OH)_2$, 90°C, NL_{Si} = 20 g·m-2, SA/V = 10 m-1, PNL 76-68 (5.1 wt% ZnO)

or rare-earth phosphate solids with solubilities lower than those considered here. Magnesium-containing glasses may form insoluble Mg-Ca-silicates, etc. Leachants containing phosphates or Mg may have the same effect.

CONCLUSIONS

From the results of a variety of experiments it can be concluded that reaction of the matrix is the fundamental process that occurs in the leaching of PNL 76-68 glass. This reaction has two aspects. Without solubility restrictions, congruent leaching behavior occurs at all pH values and leachant compositions. When this reaction raises solution concentrations of certain elements to the level at which new solid phases form, these phases will regulate the solution concentration. These solid phases are dominant constituents of the leached layer. For the leaching of PNL 76-68 glass, the solubilities of these reaction products regulate the solution concentration as if the solution is in equilibrium with pure $Fe(OH)_3$ (amorphous), $Zn(OH)_2$ (amorphous), $Nd(OH)_3$, $SrCO_3$ or $CaCO_3$. The experimental conditions, in particular the pH value, that govern the formation of solid reaction products and control of the solution concentrations can be identified.

ACKNOWLEDGMENT

Fruitful discussions and advice by D.M. Strachan made this study possible. Also, discussions with C.Q. Buckwalter and G.L. McVay helped in the understanding of the leaching behavior of PNL 76-68 glass.

REFERENCES

1. McVay, G.L. and Buckwalter, C.Q. (1980) Nuc. Tech. 51, 123-129.
2. Clark, D. and Hench, L. (1981) Nuc. and Chem. Waste Management, 2, 93-101.
3. Strachan, D.M. (1982) Results from a 1-Year Leach Test: Long-Term Use of MCC-1. PNL-SA-10036, Pacific Northwest Laboratory, Richland, Washington 99352, USA.
4. Barkatt, A., Simmons, J. and Macedo, P. (1981) Nuc. and Chem. Waste Management, 2:1, 3-23.
5. Strachan, D.M., Turcotte, R.P. and Barnes, B.O. (1981) PNL-SA-8737, Pacific Northwest Laboratory, Richland, Washington 99352, USA.
6. Rai, D. (1982) Neptunium Concentration in Solutions Contacting Actinide Doped Glass, Nuc. Tech. PNL-SA-9699, Pacific Northwest Laboratory, Richland, Washington 99352, USA.
7. Materials Characterization Center. (1981) Nuclear Waste Materials Handbook, DOE/TIC-11400, Pacific Northwest Laboratory, Richland, Washington 99352, USA.
8. Buckwalter, C.Q., McVay, G.L., Kuhn, W. and Peters, P. (1982) The Effect of pH on the Leaching of a Nuclear Waste Glass, PNL-SA-10321, Pacific Northwest Laboratory, Richland, Washington 99352, USA.
9. Fullam, H.T. (1981) PNL-3614, Pacific Northwest Laboratory, Richland, Washington 99352, USA.
10. MCC-Materials Characterization Center (1981), PNL-3990, Pacific Northwest Laboratory, Richland, Washington 99352, USA.
11. Lindsay, W. (1979) Chemical Equilibria in Soil, John Wiley & Sons, New York, USA.
12. Ames, L.L. and Rai, D. (1978) Radionuclide Interactions with Soil and Rock Media, 1, EPA 520/6-78-007-a, PB 292460.
13. Schumm, R., Wagman, D., Bailey, S., Evans, W. and Parker, V. (1973) NBS Technical Note 270-7.
14. Smyshlyaer, S.I., et al. (1977) Sevcro-Karkazskii Nauchnyi Tsentia Vysshei Shkoly Izvestia, 60-63.
15. Chick, L.A., McVay, G.L., Mellinger, G.B., and Roberts, F.P. (1980), PNL-3465, Pacific Northwest Laboratory, Richland, Washington 99352, USA.

RELATIONSHIP BETWEEN GLASS LEACHING MECHANISM
AND GEOCHEMICAL TRANSPORT OF RADIONUCLIDES

A. AVOGADRO and F. LANZA
Commission of the European Communities
Joint Research Centre - Ispra Establishment
I-21020 Ispra (Va), Italy

When the leaching of glasses is taken into consideration in risk analysis models, it is necessary to make a distinction between accidental and normal conditions in the repository. In accidental conditions, it is assumed that the glass will be leached by ground-water which can also be renewed. The composition of this water will be related to the aquifer existing around the repository. Under normal repository conditions the leaching process has to be analyzed separately for each geological formation considered.

In a salt repository the brine migrates, due to the thermal gradient, towards the conditioned waste. The leaching of the glass will correspond then to a leaching in a closed system evolving towards saturation.

In a hard rock system water flows slowly through the fracture network. The water around the glass will then be slowly renewed and only near saturation conditions will be reached.

In a porous system, like clay or marine sediments, water is practically stagnant and transport could only be due to diffusion of the soluble species. It is possible that, due to Brownian motion, even microcolloids could diffuse.

It is known that in a closed system the leaching of glass tends to decrease as the solution approaches saturation reaching eventually a zero or near zero value[1]. In the two other formations considered, a reduced transport exists due to diffusion in the interstitial water or to water flow in fractures which does not allow to attain complete saturation. The leaching rate will then depend upon the types and rate of transport of the dissolved species. Moreover, the transport will depend upon the physico-

104

chemical characteristics of the different species as well as on their interaction with the geological material; it is obvious therefore, that each element has to be considered separately.

In this paper an attempt has been made to subdivide the different elements present in the vitrified waste into classes in order to simplify the description of the release of the radioelements in the various repository conditions.

In the release of the different radionuclides from borosilicate glasses, a prominent role is played by the leaching rate of silicon. Silicon dioxide represents the skeleton of the glass matrix and its leaching rate establishes the rate of degradation of the glass. The leaching of silicon will be limited in a closed system by the saturation concentration of dissolved silica. Amorphous silica has a relatively high solubility in pure water, reaching at 80°C a value around 300 ppm[2,3]. Silica dissolves in pure water in a non-ionic form. However, when alkalies are present in the solution, even silicate ions will increase the solubility of silica; such an effect is reported to be effective already at pH 8-9[3]. It is usually assumed that an increase of the concentration of the silicon species in water decreases the leaching rate of the glass.

Some tests have been performed in order to follow the leaching rate of I-117 glass, for glass composition see Table 1, in distilled water in a closed system at 80°C. The obtained values were compared with those reported previously[4] in renewed distilled water (see Fig. 1). Weight loss of the glass was taken as an index of the leaching and saturation was considered attained when no more weight variationa were encountered.

The analysis of Si content in the leachate confirms that its saturation value was reached. However, a maximum value of Si concentration of 67 ppm was found which is well below the saturation concentration of amorphous silica but more comparable to that of quartz which is 33.5 ppm[2]. A similar effect was reported by Chapman et al.[5]. In order to explain this discrepancy either reprecipitation of hydrated crystals or a very slow increase of the silica content in the solution has to be postulated. This uncertainty on the practical maximum solubility of silica is an obstacle in determining an explicit mathematical expression relating the silicon leaching rate to the silicon concentration. It is evident, however, that only if silicon is transported out or reprecipitated, the leaching will continue.

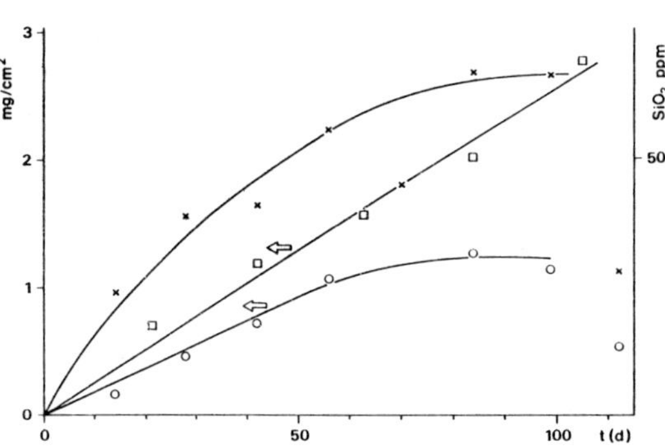

Fig. 1. Weight loss of I-117 glass at 80°C; ⊙ static conditions; ⊡ slowly renewed water; x silicon content in the leachate in static conditions.

The other elements composing the glass enter into solution as simple ions or complexed with legands present in the leaching solution. The maximum concentration (C_i) of an element in the leachate will be the sum of the saturation concentrations of all the possible species in solution. Comparing the leaching rate (K_i) and the saturation concentration (C_i) of the different elements with those of silicon $(K_{si}$ and $C_{si})$, three cases can be defined:

1. elements for which $K_i > K_{si}$
2. elements for which $K_i \leq K_{si}$ and $C_i > C_{si}$
3. elements for which $K_i \leq K_{si}$ and $C_i \leq C_{si}$.

These three classes are discussed separately.

Typical elements of the first class are the alkalies which have a leaching rate higher than that of silicon. The alkalies are not released due to the dissolution of silica but diffuse out of the glasses by ion exchange with H^+ ions present in the solution[6]. Consequently, the model of alkali release

is a model controlled by the Na^+, H^+ interdiffusion in the glass.
The leaching process will be influenced by the concentrations of alkali and
hydrogen ions. The transport of the leachate will have an influence particularly in semiclosed systems having originally a solution of high saline
content.

The diffusion of alkalies and in particular of Na which is generally the
most abundant element of the glass can influence the pH in a closed or
semiclosed system. Consequently, the leaching rate of the glass and the
solubility limit of the silica will increase. In the borosilicate glass boron
is leached at about the same rate as silicon. The boron will buffer the solution pH so that a complex equilibrium would be set up, which depends on
the glass composition.

The second class comprises soluble elements which cannot diffuse in
the glass; they are released to the ground-water when the SiO_2 matrix is
dissolved. The technetium is quite soluble in its higher oxidation state
whereas it has a significantly lower solubility in the reduced form; then if
technetium is leached in the heptavalent state it will belong to the second
class and its release will follow that of silicon.
Technetium is not homogeneously distributed in the borosilicate glass; at
the Pacific Northwest Laboratory[7], using autoradiographic techniques, the
technetium has been found to be concentrated in the gaseous inclusions
present in the glass matrix. This feature gives rise to an intermittent
release, characterized by sudden increase of activity when the water,
during the leaching process, comes into contact with the bubble inclusions.
In our leaching experiments ground glass has been used (size 1.5-2 mm) so
that the intermittent release was masked by the random distribution of the
gas inclusions. The leaching of the glass has been carried out at $22^{\circ}C$
with ground-water at 1.5 ppm dissolved oxygen. The water flow rate of
1.5 ml/h was such that saturation phenomena were avoided. Solvent extraction and coprecipitation methods have been employed to measure Tc (VII)
and its reduced species in leachate. Ion exchange separations have been
adopted to characterize cationic and anionic species of Tc in leachate.

Under these experimental conditions the leached technetium was always found as soluble pertechnetate ion.

Dialysis methods are being used to distinguish the eventual colloidal fraction from that solubilized under different conditions[8]. Under the experimental condition used, the technetium goes always in solution as pertechnetate ion and Tc (IV) is not stabilized by complexation with the constituents of ground-water, humic acid or NaCl solution.

The technetium leaching rate as a function of time is shown in Fig. 2. It can be noticed that after an initial period of decrease, the release of technetium is stabilized at a constant value which corresponds to a leaching rate of the glass of 2.2×10^{-7} g cm^{-2} day^{-1}. This value compares favourably with the long-term leaching rate using a large amount of water and obtained through weight loss measurements[4]; in this case the asymptotic value results to be 1.8×10^{-7} g cm^{-2} day^{-1}, but is approached in a longer time period, viz more than one year.

The release of technetium to the geosphere is therefore dependent upon the dissolution of the glass matrix and therefore on the parameters influen-

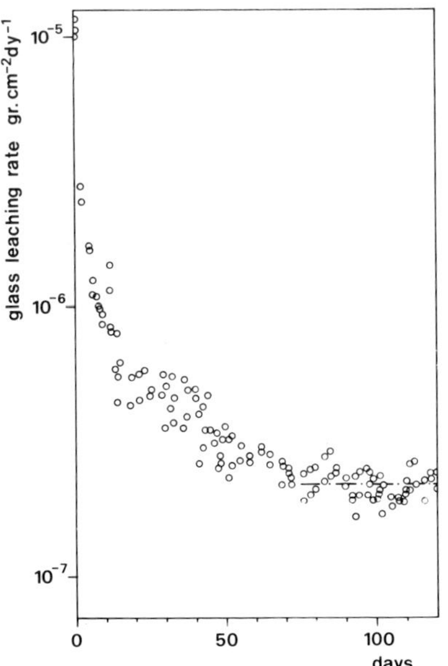

Fig. 2. Leaching rate of Tc at 22°C in flowing water.

cing the silicon leaching rate. The redox conditions of the geochemical environment also play an important role. In the case of the leaching solutions studied, the anionic pertechnetate formed is very mobile. It should be noticed, however, that a different situation may exist under considerably reduced dissolved oxygen levels as those found in interstitial water of clay formations and of certain types of sea sediments. Preliminary results from samples taken of the Oregon coast indicate that it may be possible to retain technetium on sediments which are mildly reducing[9]. At very low Eh values, the technetium will be present in reduced form and the sparingly soluble TcO_2 will be enriched on the surface layer of the vitrified waste. Under these conditions, the release will take place mainly by microcolloid formation which belongs to the last class considered below.

This last category comprises important elements such as plutonium and americium. Due to their very low solubility they tend to be enriched on the glass surface layer which releases these nuclides mainly by microcolloid formation.

A test similar to that described previously for technetium was performed on glass granules spiked with americium, neptunium and plutonium. The flowing water was filtered on membrane filters of different porosity in order to classify the dimensions of the released colloids. Fig. 3 shows the colloid distribution for each actinide. If their release was only related to the loss of the silica gel from the glass surface, all the elements would have shown the same retention patterns on the different filters. However, it can be supposed that a hydrated layer of silica is formed on the surface of the glass which will become slowly enriched in lower solubility elements. The subsequent loss of this layer to the water vector followed by the rapid dissolution of the siliceous material, releases the low soluble radionuclides partially by colloidal formation and partially by solubilization. It has been shown that under the present experimental conditions, the water is highly undersaturated with respect to silica.

In the case of americium it has been found[10] that in a flowing system more than 90% of this element is released in colloidal form. Fig. 4 shows the release rate of americium as a function of time obtained under the same

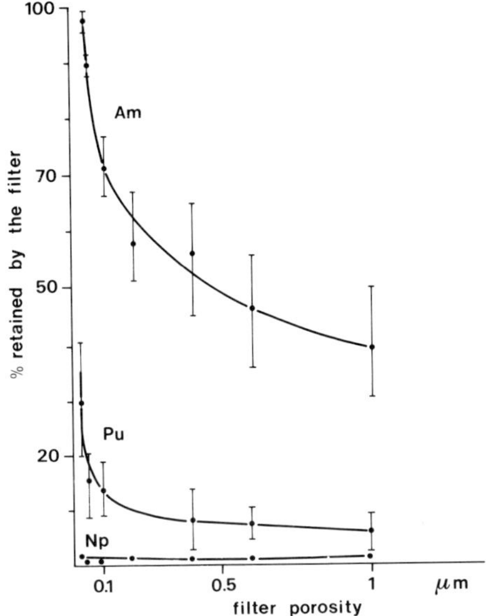

Fig. 3. Retention of actinide colloids on filters of different porosities.

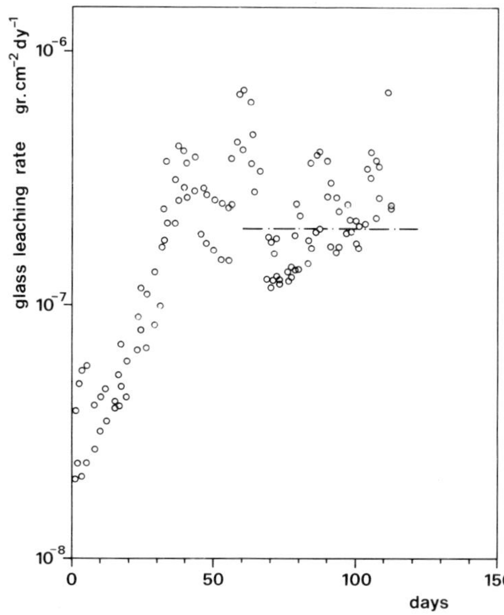

Fig. 4. Leaching rate of Am at 22°C in flowing water.

conditions as those adopted for technetium. The released activity reaches an equilibrium value dependent upon the steady state concentration of americium at the surface as well as on the rate of disaggregation of the gel layer.

A glass leaching rate of 2×10^{-7} g cm^{-2} day^{-1} has been calculated from americium data which corresponds also in this case to the dissolution rate of the glass matrix obtained from weight loss measurements. These results are apparently in contradiction with literature data giving actinide release rates lower than the leaching rate of the glass matrix[7,11]. The above observation can be explained taking into account the mechanical stability of the gel layer which, in turn, is a function of the geometry of the glass surface. In a flowing system the diminution of the glass grain size enhances the release of colloids from the surface.

In the case of a leaching process with a nearly stagnant water, the mechanical removal of colloids from the surface is reduced; the actinides will be mainly released by dissolution and their release depends on their solubility which can be enhanced by the presence of complexing agents.

From the discussion of the different classes it appears clearly that the evaluation of the release of the different radionuclides depends strongly on the transport in the geological medium. In addition, to evaluate the transport it is necessary to know the physico-chemical species under which the various nuclides are leached. Consequently, there is a strong relationship between transport and leaching; any separation of the two phenomena risks to be artificial.

Looking at the different repository options, the following considerations can be outlined:

- In a salt repository the reprecipitation of hydrothermal minerals seems the most important phenomenon which could lead to a continuous leaching.
- In a porous system, the formation of soluble species and their diffusion is predominant. The leaching rate and diffusion of silicon could dominate the release of radionuclides.
- In the hard rock system the small water flow defines the near saturation conditions; the transport of colloids could eventually play a role.

- The transport of colloids assumes the highest importance in the accidental scenarios involving the contact of the glass with a large flow of
water.

REFERENCES

1. Van Iseghem, P., Timmermans, W. and De Batist, R. (1981) Interaction of vitrified HLW with clay environment. Presented at the International Seminar on Chemistry and Process Engineering for HLW
   Solidification, Jülich, June 1981.
2. Hughes, A.E., Marples, J.A.C. and Stoneham, A.M. (1981) The significance of leach rates in determining the release of radioactivity
   from vitrified nuclear waste. AERE - R - 10190.
3. Iler, R.K. (1955) The colloid chemistry of silica and silicates. Cor-
   nell University Press, New York.
4. Lanza, F. and Parnisari, E. (1981) Influence of film formation and
   its composition on the leaching of borosilicate glasses. Nucl. and
   Chem. Waste Management, 2, 131.
5. Chapman, N.A., McKinley, I.G. and Savage, D. (1980) The effect of
   groundwater availability on the release source term in a low hydraulic
   conductivity environment. Proc. of the NEA Workshop on Radionuclide
   Release Scenario for Geological Repositories, Paris, September 1980.
6. Engelmann, C., Neus, B. and Trocellier, P. (1981) Mesure des pro-
   fils de concentration de l'hydrogène et du sodium à la surface des
   verres par des méthodes nucléaires. Verres Refract., 35, 3, 486.
7. Bradley, D.J., Harvey, C.O. and Turcotte, R.P. (1979) Leaching of
   actinides and technetium from simulated high-level waste glass, PNL-
   3152.
8. Avogadro, A., Bidoglio, G. and Chatt, A. (1982) Speciation of radionuclides in natural waters. Presented at the 7th Annual Meeting of the
   Seabed Working Group of the Nuclear Energy Agency, OECD, La Jolla
   (CA) USA, March 15-18, 1982.
9. Schreiner, F., Fried, S. and Friedman, A.M. (1980) Measurement
   of radionuclide mobility in ocean floor sediment and clay. Presented
   at the Symposium on the Backfill for Radioactive Waste Management,
   Las Vegas, August 1980.
10. Avogadro, A., Murray, C.N. and De Plano, A. (1980) Transport
    through deep aquifers of transuranic nuclides leached from vitrified
    high level wastes. Scientific Basis of Nuclear Waste Management,
    vol. 2, Clyde J.M. Northrup Jr., Ed. Plenum Press, p. 665.
11. Bonniaud, R.A., Jacquet-Francillon, and Sombret, C.G. (1980) The
    behaviour of actinides in alpha-doped glasses with regard to the long-
    term disposal of high-level radioactive materials. Ibid., p. 117.

T A B L E 1

Glass Composition

Component	Wt. %	Component	Wt. %
SiO_2	48	CeO_2	0.76
B_2O_3	15	SnO_2	0.03
Al_2O_3	5	ZrO	1.06
Na_2O	18.49	MoO_3	0.16
SrO	0.22	MnO_2	1.35
BaO	0.33	Fe_2O_3	3.25
Y_2O_3	1.28	NiO	0.32
La_2O_3	0.63	CuO	0.04
Pr_2O_3	0.31	ZnO	0.03
Nd_2O_3	0.99	U_3O_8	1.38
Sm_2O_3	1.12		

LEACHING OF ACTINIDES FROM SIMULATED NUCLEAR WASTE GLASS

S. PICKERING AND C.T. WALKER

Commission of the European Communities, Joint Research Centre,

Karlsruhe Establishment, European Institute for Transuranium Elements,

Postfach 2266, D-7500 Karlsruhe, Federal Republic of Germany

P. OFFERMANN

Hahn-Meitner-Institut für Kernforschung Berlin

Glienicker Strasse 100, D-1000 Berlin 39

INTRODUCTION

The aim of the work was to measure the resistance of simulated nuclear waste glasses to leaching by water and salt solution under the conditions of elevated temperature and pressure. Of particular interest was the behaviour of Am and Np.

EXPERIMENTAL

Glass preparation

The compositions of the glasses investigated, GP 98/12 and C31-3-EC, are listed in table I. The GP 98/12 glass[*] was doped with an additional 4.95 ^{W}/o UO_2 and 2.7 ^{W}/o PuO_2. Of the PuO_2, 8.6 ^{W}/o was initially Pu-241 which produced Am-241 by β-decay. The C31-3-EC glass was doped with 1 ^{W}/o Np-237 and 0.05 ^{W}/o Am-241 as oxides. The glasses were melted in a platinum crucible and cast in cylindrical graphite moulds. After annealing at 600°C for 3h the rods of glass were sectioned to provide specimens 10 mm ∅ and ∼ 5 mm high.

Leaching conditions

Leaching of the glass was carried out in stagnant conditions in PTFE lined autoclaves at 200°C under air at a pressure of $25-50 \times 10^5$ Pa. Distilled water and NaCl solution saturated at R.T. were the leachants used at leachant volume to specimen surface area ratios (V/S ratio) of 5 and 10 ml/cm^2 (10 ml/cm^2 corresponds to the MCC 1 recommendations[1]). A fresh specimen was used for each measurement to avoid complications due to the repeated drying of the surface layer that could arise from cumulative measurements on one specimen.

---

[*] Provided by the Institut für Nukleare Entsorgungstechnik, KfK, Karlsruhe

114

Post-leaching examination

A surface layer formed on both types of glass during leaching. This layer
contained a significant amount of loosely bound water. The layer could be dried
either by heating or by allowing the specimens to equilibrate with dry air at
R.T. The dried layer, at least on GP 98/12, however, was capable of reabsorbing
up to 25 ^{W}/o of water in only 20 mins from air with 50 % RH. After drying the
glass at 200°C the surface layer reabsorbed only 80 % of the water that was
reabsorbed after drying at $\leq$ 150°C. The procedure that was adopted to measure
the weight of material lost to the leachant therefore was weigh the specimens
in dry air after equilibration with dry air at R.T.

In the case of glass C31-3-EC doped with Am-241, the α-spectrum of the
specimens was then measured to record the increased Am concentration in the
surface layer after leaching. The weight of the surface layer was determined
by reweighing the glass after removing the layer by scraping.

The Np content of the leachant was determined by neutron activation analysis[*]
(NAA) after extraction of the Np from the leachant with thenoyltrifluoroacetone
(TTA). The Am in the leachant was determined by γ-spectrometry using Cs-134
as a spike. It was found that Am tended to precipitate out on the autoclave
walls so the autoclave was washed out with 4N HNO_3 to remove such deposits and
the acid was added to the leachant before analysis. The same procedure of
washing with HNO_3 was also adopted for Np analysis.

Sectioned glass specimens were also analysed with the microprobe for information on the composition of the surface layer.

RESULTS

Americium

The results for weight loss and leaching of Am from glass GP 98/12 leached
in distilled water at 200°C for times up to 96 h are given in fig.1. The leached
weight loss refers to the weight loss measured on the dried specimens before
scraping off the surface layer. The total weight loss is the weight loss measured after removal of the layer. The data are characterized by a relatively
rapid weight loss for the first 10 h after which there was no further measurable
weight change. The detection limit for "measurable weight change" was rather
poor due to the relatively large scatter in the data. This type of behaviour,
however, was also confirmed by more numerous experiments on inactive GP 98/12
and C31-3-EC glasses in water and salt brines [2,3]. The amount of Am in the
leachant was proportional to the leached and total weight losses of the glass.
The leaching ratio for Am as defined by:

[*] Performed by the Institut für Radiochemie, KfK, Karlsruhe

$$LR_{Am} = \frac{\text{Wt. Am in leachant}}{\text{leached wt. loss}} \times \frac{100 \ \%}{^{w}/o \ \text{Am in glass}}$$

was in the range $1.5 - 2.5 \times 10^{-2}$ indicating preferential retention of Am in the surface layer. Leaching ratios above 1 imply preferential leaching of an element from the glass relative to the sum of the weight losses of all elements. Leaching ratios below 1 imply preferential rentention of the element. Leaching ratio is a useful concept for comparing the behaviour of the same element in different glasses. The leaching ratios for Am in GP 98/12 and in glass C31-3-EC leached in distilled water at V/S ratios of 5 and 10 ml/cm^2 are given in fig.2. Although there was considerable scatter in the results, the leaching ratio for V/S = 10 ml/cm^2 was always smaller than for 5 ml/cm^2.

For leaching in distilled water at 200°C the glasses GP 98/12 and C31-3-EC both show the same trend of a rapid attainment of quasi-static weight loss conditions. For glass C31-3-EC leached in saturated NaCl (V/S = 10 ml/cm^2) at 200°C however the quasi-static conditions were not reached until after 1o days as shown by weight loss curves of fig.5b.

The conditions where no further weight loss is apparent are called quasi-static because there is evidence that changes in the surface layer continue to occur even though there is no further change in specimen weight. The α-energy spectra of glass C31-3-EC were measured after leaching as shown in figs 3 and 4 The measurements were made at 4.48 keV/channel so that 100 channels correspond to 2.65 μm of glass. Because the composition and density of the surface layer are unknown, the depth scale can only be defined as an equivalent depth in a material of known composition e.g. the glass. The spectra all indicate an enrichment of Am in the surface layer but in fig.4 the reduction in peak height for times between 3 and 30 days indicates a reduction in Am concentration in the layer even though, from weight loss measurements, the layer stopped growing after 10 days.

Neptunium

The results for the leaching of Np from glass C31-3-EC at 200°C in distilled water are given in figs 5a and 6. The weight losses were greater under V/S = 10 ml/cm^2 conditions than at V/S = 5 ml/cm^2 because whereas no further weight change occurred after 1o h at V/S = 5 ml/cm^2, the weight losses continued up to 30 h at V/S = 10 ml/cm^2. These weight losses were about 1/3 of those measured for glass GP 98/12 under similar conditions.

Leaching ratios for Np were calculated in the same way as those for Am. The results in fig. 6 indicate final values in the range 0.1 - 0.2 after initially higher values for V/S = 10 ml/cm^2 conditions. Thus, although Np was preferentially retained by the surface layer, the leaching ratios were about an order of magnitude greater than for Am.

Microprobe analysis of the surface layer by line-scans across the layer indicated that the maximum enrichment of Np in the layer was twice that of the concentration in the glass. Most enrichment occurred at the glass/layer interface.

Plutonium and Uranium

The concentrations of U in the surface layer on glass C31-3-EC and of U and Pu in the layer on glass GP 98/12 were measured by microprobe analysis. The typical results in fig. 7 show that U and Pu levels were somewhat higher in the layer than in the glass but that the distribution was not always homogeneous. Apparently there were U and Pu rich inclusions both in the layer and in the glass in the specimens of GP 98/12. Possibly solubility limits were exceeded in this glass[4]. No differences were apparent in the U and Pu profiles for glasses leached in water or 1 % NaCl solution. The specimens for microprobe analysis were all leached for 64 h at 200 °C and the surface layers were 30 to 60 μm thick according to glass and leachant composition.

Additional Observations

The surface layer was observed to consist of two distinct layers similar to those previously reported on inactive glass[5,6]. The outer layer was thin and continuous over a much thicker layer which was broken up as if a result of shrinkage. In glass GP 98/12, microprobe analysis showed the outer layer to be significantly enriched in Ca relative to the inner layer whereas in glass C31-3-EC both layers were depleted in Ca and in the outer layer Al was the only element significantly enriched relative to the inner layer (see fig. 7). Apart from the changes in Ca and Al content, no systematic differences in chemical composition between the two layers were detected.

The differences in chemical composition of the layers relative to the glass were: enrichment in Ce, Nd, Ti, Zn, Zr (particularly strong in glass C31-3-EC) U, Pu and Np and depletion in Na, Cs, Si, Ba and Ca (only in C31-3-EC). The composition of the layer was broadly similar to that described in detail in references 3 and 6.

The pH of the leachant from experiments with glass GP 98/12 was consistently higher than from experiments with glass C31-3-EC: Tests with inactive GP 98/12 showed that the pH rose rapidly ($\leq$ 10 mins at 200 °C) to about pH 9 and did not change thereafter up to 100 h which was the longest time investigated. In contrast, the pH values obtained with Am and Np doped C31-3-EC glass were usually in the range 6.5 - 7. The pH was measured at RT after the experiment in all cases.

DISCUSSION

The results, which may only be valid for stagnant conditions and short times were characterized by two main features. First, loss of specimen weight due to leaching of glass constituents, including actinides, occurred only for a relatively short period at the beginning of the tests. Second, the actinides investigated were preferentially retained within the surface layer that developed as a result of leaching. This is consistent with results on inactive C31-3-EC[3] and Pu doped GP98/12[7].

The Surface Layer

The surface layer was observed to consist of two layers. The outer layer was a mechanically intact layer enriched in Al on the C31-3-EC glass and enriched in Ca in the GP 98/12 glass. These differences in enrichment correlate with the glass composition: glass C31-3-EC contains 10.2 ^{w}/o Al_2O_3 whereas glass GP 98/12 contains only 2.2 ^{w}/o Al_2O_3. Both glasses have similar Ca content. It is believed that when leaching starts, Na goes into solution tending to increase the pH. Al in the glass in converted to $Al(OH)_3$ when exposed to the leachant. In the case of glasses with sufficiently high Al content, reaction of Al to form a hydroxide at the interface with the leachant is sufficient to buffer the pH of the leachant at its original value. In glasses such as GP 98/12 the Al content is insufficient to buffer the pH which rises due to continued Na dissolution until $Ca(OH)_2$ starts to precipitate at pH 9-10. Thus high Al_2O_3 glass will have an $Al(OH)_3$ rich surface layer, the leachant pH will remain near its original value and Ca will be soluble in the leachant. Low Al_2O_3 glass however will have a $Ca(OH)_2$ rich surface layer and the leachant will have a pH corresponding to that for $Ca(OH)_2$ precipitation, e.g. pH 9-10. In the experiments described here, very little CO_2 was available, otherwise the formation of $CaCO_3$ could limit Ca availability as described by Grambow[8].

It follows from the rapid rise to equilibrium pH measured on glass GP 98/12 that the Al or Ca rich layer must start to form very early in the leaching process and that, despite the mechanically sound appearance of the layer in SEM pictures, the layer does not prevent the subsequent formation of the thicker inner layer.

The trend to higher weight losses in glass GP 98/12 than in glass C31-4-EC is consistent with their respective equilibrium pH values of 9 and 6.5 and with the fact that break down of the silicate network in glass and hence the rate of weight loss increases with increasing pH.

The fact that specimen weight loss and the leaching of actinides cease after times dependent on the leachant volume to glass surface area implies that the cessation of further weight loss results from a build up of a glass constituent in the leachant not on the thickness of the surface layers. The cessation of further weight loss also occurs long after the pH has reached its equilibrium value. The thickness of the surface layer was not measured, but it is assumed that the onset of constant layer weight conditions and constant specimen weight imply that layer growth had stopped. Nevertheless some changes may still occur within the layer as evidenced by the α-energy spectra of fig. 4.

Unfortunately there was too much scatter in the weight loss data to enable conclusions to be drawn about the leaching mechanism, i.e. as to whether it was diffusion controlled.

The Leaching of Actinides

All data indicated that the actinides were preferentially retained within the surface layer but the α-spectra recorded for Am indicated that Am continued to migrate after the layer had apparently stopped growing so that the calculated leaching ratios should be viewed with some caution although the values for Am and Np measured by Marples et al [10] on glass UK 189 are in good agreement. The U and Pu additions to glass GP 98/12 may have been in excess of the solubility limits. The presence of U and Pu rich inclusions in the surface layer means that the leaching results are valid for a 2 phase system of U and Pu in solution +U,Pu oxide. For glass C31-3-EC the U was all in solution and the U distribution in the layer resembled that of the lanthanides Ce and Nd.

From the α-spectra of the leached glass, Am appears to be concentrated in the outer part of the surface layer. It was not possible to calculate the thickness of the Am enriched zone from the α-spectra because the necessary calibration of α-energy loss within the layer was not made. However, as stated earlier, the thickness of the Am rich zone was equivalent to $\sim$ 2.5 μm of glass. The true thickness of this zone would be greater than 2.5 μm by a factor depending on the ration of the density of the glass to that of the layer. Even allowing for a large correction factor, the true thickness of the Am rich zone must remain a small fraction of the $\sim$ 40 μm layer thickness.

SUMMARY

Two types of simulated nuclear waste glass doped with actinides were leached

at 200 °C in distilled water and salt solutions. Am, Np, Pu and U were all preferentially retained in the surface layer on the glass. Leaching ratios of
0.1 to 0.2 for Np and $\sim$ 0.02 for Am were measured. The losses of Am and Np to
the leachant were proportional to the total weight loss of the glass and were
larger at 10 ml leachant/cm^2 glass than at 5 ml/cm^2. Weight loss from the glass
occurred only at the start of the experiments for periods ranging from 10 h to
10 days according to leachant composition and volume. Wt losses from the
C31-3-EC glass were much greater in saturated NaCl solution than in distilled
water. Enrichments in the outer surface layer of Al or Ca according to glass
type could be correlated with leachant pH, glass composition and weight loss
measurements.

REFERENCES

1. Materials Characterization Center (1981), Nuclear Waste Materials Handbook,
   DOE/TIC-11400, Pacific Northwest Laboratory, Richland, Wa., USA

2. Pickering, S. (1980) J. Am. Cer. Soc., Vol. 63, pp.

3. G. Malow, (1982) Scientific Basis for Nuclear Waste Management, Vol. 11,
   Lutze, W., ed., MRS Symposia Proceedings.

4. Scheffler, K., Riege, U. and Walker, C.T. (1978) EUR 5750e

5. Altenhein, F.K., Lutze, W. and Malow, G. (1981) Scientific Basis for Nuclear
   Waste Management, Vol. 3, Moore, J.G., ed., pp. 363-370

6. Kahl, L. (1981) Nuclear and Chemical Waste Management, Vol. 2, pp. 143-146

7. Scheffler, K., Riege, U., Louwrier, K., Matzke, Hj., Ray, I. and Thiele, H.
   (1977) EUR 5509e

8. Grambow, B. (1982) Scientific Basis for Nuclear Waste Management, Vol. 11,
   Lutze, W., ed., MRS Symposia Proceedings.

9. Hench,L.L., Clark, D.E.and Yen-Bower, E.L. (1980) Nuclear and Chemical Waste
   Management, Vol. 1

10. Hall, A.R., Hough, A. and Marples, J.A.C. (1982) Scientific Basis for Nuclear
    Waste Management, Vol. 11. Lutzw, W., ed., MRS Symposia Proceedings.

120

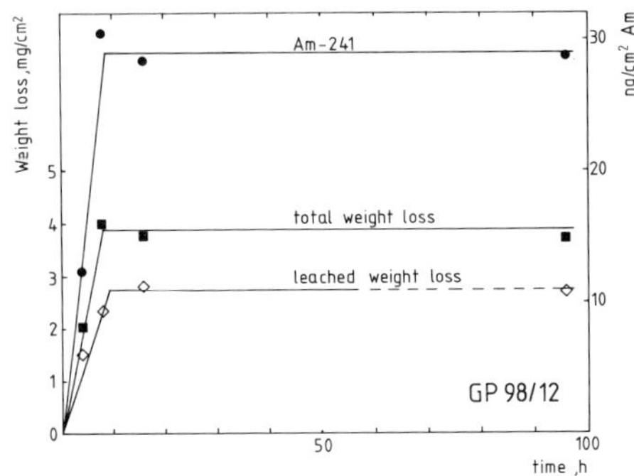

Fig. 1. Leaching of Am-doped GP 98/12 glass in distilled water (5ml/cm^2) at 200°C

Fig. 2. Leaching ratios for Am in glasses leached at 200°C in distilled water

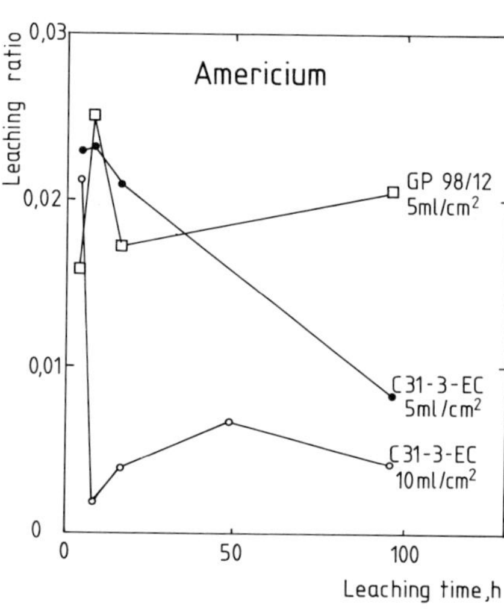

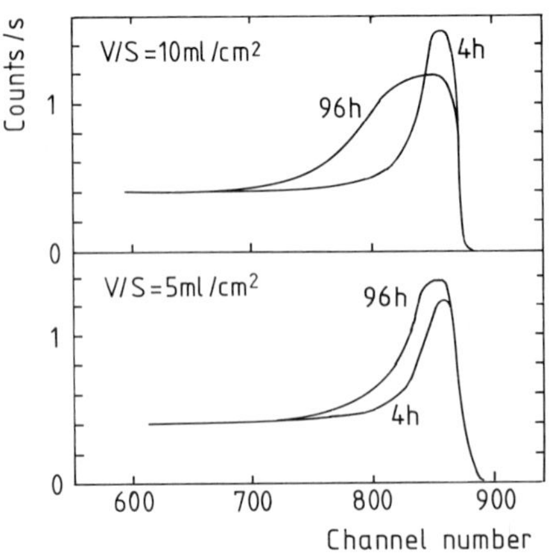

Fig. 3. α-energy spectra from the surface layer on glass C31-3-EC leached in distilled water at 200°C

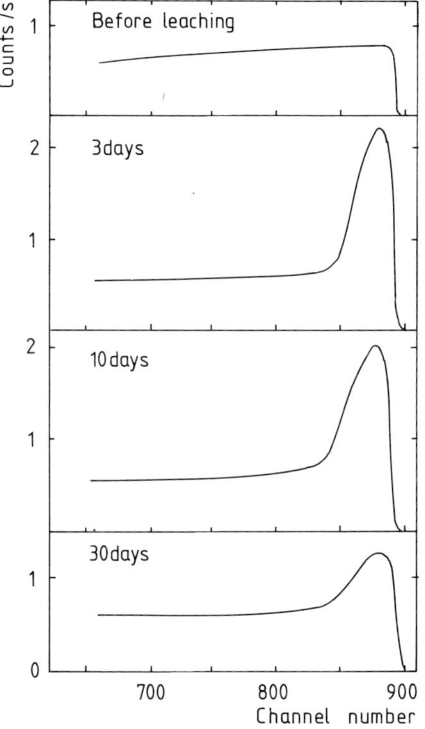

Fig. 4. α-energy spectra from the surface layer on glass C31-3-EC leached in sat. NaCl (10 ml/cm^2) at 200°C.

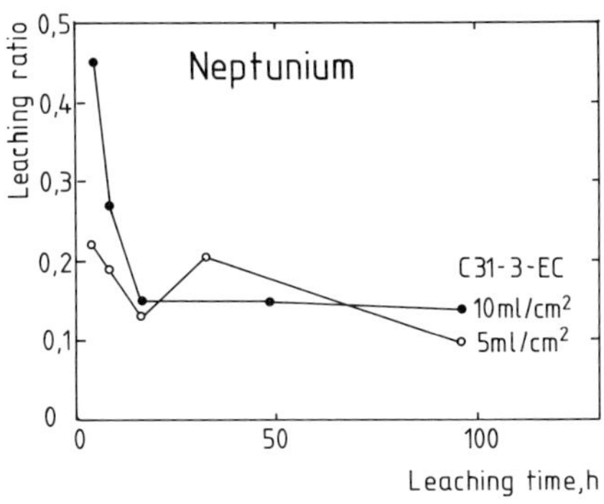

Fig. 6. Leaching ratios for Np in glass C31-3-EC leached at 200°C in distilled water.

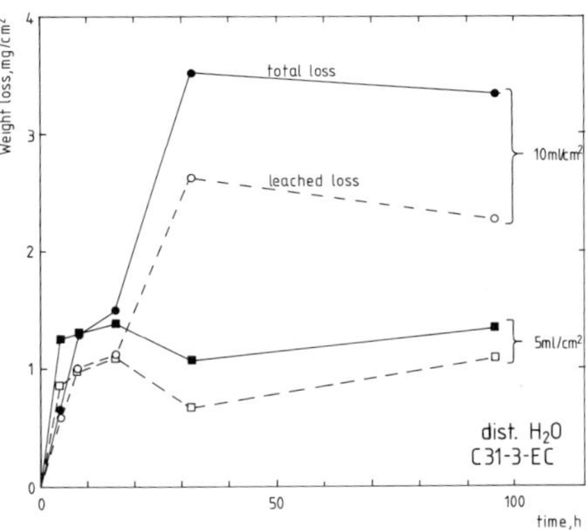

Fig. 5a. Weight loss from glass C-31-3-EC leached in distilled water at 200°C.

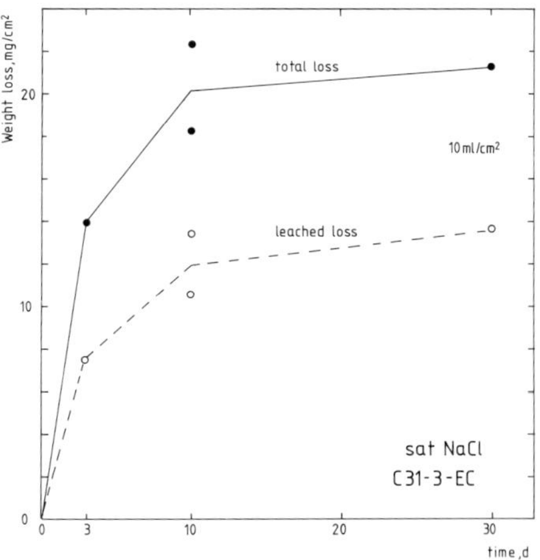

Fig. 5b. Weight loss from glass C31-3-EC leached in saturated NaCl at 200°C.

122

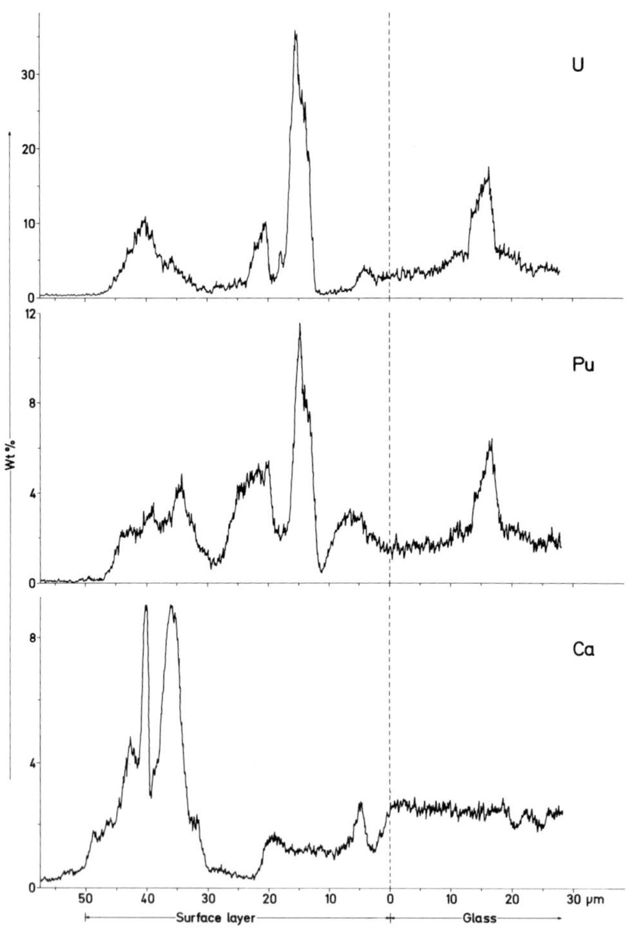

Fig. 7a. Microprobe line scans across the surface layer on glass GP 98/12 leached for 64 h at 200°C in distilled water.

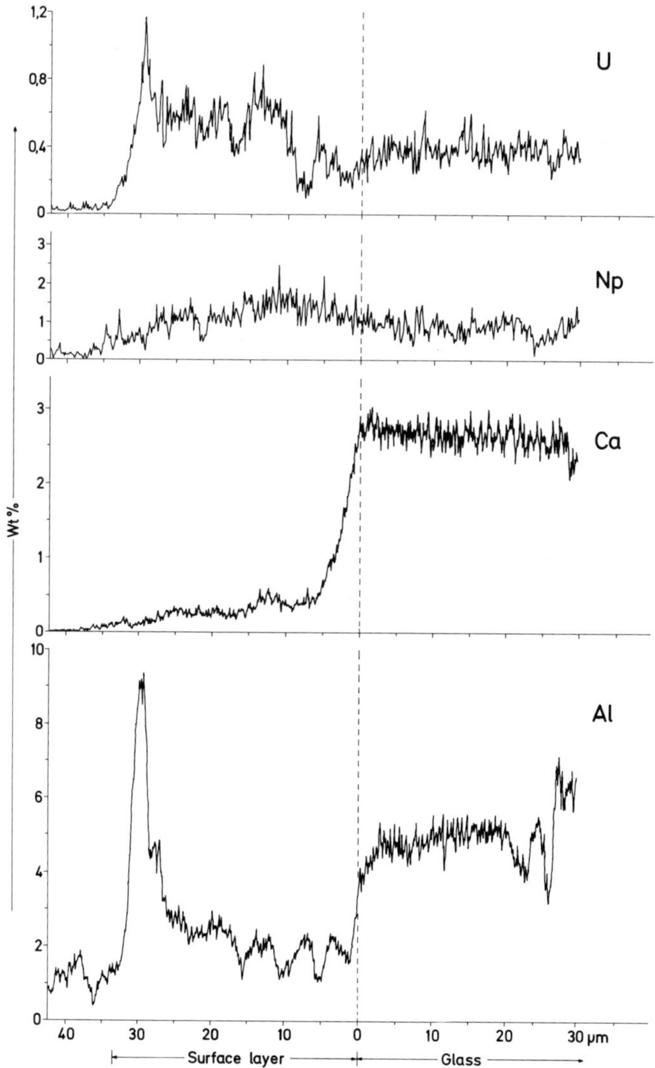

Fig. 7b. Microprobe line scans across the surface layer on glass C31-3-EC leached for 64 h at 200°C in distilled water.

TABLE 1
COMPOSITIONS OF SIMULATED GLASSES AND WASTES

A. Glass composition

Glass	Glass component (wt %)												
	SiO_2	B_2O_3	Na_2O	BaO	TiO_2	CaO	Al_2O_3	MgO	Li_2O	ZrO_2	ZnO	As_2O_3	Waste
GP 98/12	48.20	10.54	14.88	–	3.91	3.48	2.21	1.80	–	–	–	–	15.0
C31-3-EC	34.75	4.14	1.14	14.45	2.80	3.84	10.28	1.44	0.98	0.80	4.90	0.48	20.0

B. Waste composition

Glass	Principal waste components**(wt % of waste)											
	Rb_2O	Cs_2O	SrO	BaO	Y_2O_3	La_2O_3	CeO_2	Pr_2O_3	Nd_2O_3	Sm_2O_3	Eu_2O_3	MoO_3
GP 98/12	0.10	6.19	2.38	4.42	1.42	3.57	7.32	3.46	11.65	2.44	0.49	15.70
C31-3-EC	0.80	8.51	2.96	3.46	1.70	3.31	6.56	3.01	12.02	–	–	11.82

Glass	Principal waste components**(wt % of waste)											
	MnO	TeO_2	ZrO_2	Ru*	Rh*	Pd*	Na_2O	P_2O_5	Fe_2O_3	NiO	Cr_2O_3	U_3O_8
GP 98/12	–	1.54	12.07	5.18	0.91	3.41	–	6.09	1.75	0.69	0.65	7.46
C31-3-EC	1.70	1.30	12.62	4.61	1.20	0.55	10.87	–	7.36	1.15	2.11	2.30

*Added as oxides to C31-3-EC

**Only components added in quantities in excess of 0.5 wt % are given

INFLUENCE OF A BACKFILLING MATERIAL ON BOROSILICATE GLASS LEACHING

F. LANZA and C. RONSECCO
Commission of the European Communities, Joint Research Centre - Ispra
Establishment, I-21020 Ispra (Va), Italy

INTRODUCTION

When a canister containing a borosilicate glass incorporating highly radioactive waste is put into an underground repository, care must be taken in
favouring the dispersion of the generated decay heat by conduction in the surrounding rock. For this purpose the space existing between the glass container
and the repository walls is filled with a material having an acceptable thermal
conductivity. If the backfilling material is highly compacted, it can form a zone
of very low permeability[1] reducing in this way the flow of water around the
canister. In general, a mixture of clay and sand is considered as backfilling
material, sand being added in order to improve the heat conductivity of the
system. Clays present the advantage of having a good ion exchange capacity
(around 100 meq/100 g)[2], of being plastic and of increasing in volume when in
contact with water. It has to be noted, however, that when a dry pressed clay
is put into contact with water, the affinity of water with clay produced potential gradients which facilitate water diffusion and, in a relatively short time,
create homogeneous distribution of water in the clay[3]. Consequently, clay or
clay mixtures, though they prevent an elevated flow of water due to their very
low permeability, do not prevent water from coming into contact with the canister and, eventually, after the canister´s disintegration due to corrosion, to
come into contact with the glass conditioning the high level waste.

It seemed worthwhile to analyze the influence of the backfilling on the glass
leaching in order to better evaluate backfilling properties and to be able to
calculate the release and diffusion of radioactive elements in repository conditions. In particular it seems interesting to compare the values obtained in
clay with those obtained in closed systems to verify whether even in a porous
system it is possible to reach near-saturation conditions.

DESCRIPTION OF THE EXPERIMENT

Since the scope of this work is the characterization of the phenomena of leaching in a porous, absorbing medium, it was decided to choose a simple testing procedure. Mixing a clay with a limited amount of water is possible only under an elevated pressure due to the swelling properties of the mixture. On the contrary, using a dilute paste it is possible to obtain a fluid paste which can be contained in an ampoule at atmospheric pressure. For this reason pastes containing around 55% of water were used. To emphasize the influence of clay, a dry mixture composed of 80% of clay and 20% of sand was chosen.

The choice of the type of clay presented another difficulty. Bentonite was first proposed by KBS[1]. However, a dilute suspension of bentonite in distilled water tends to give a paste with a high pH value, which, as is well known[4], gives rise to elevated leaching rates. Montmorillonite, on the contrary, shows such an effect in a more limited way (Fig. 1) and anyhow within values not significantly affecting the leaching rate. Since some non-homogeneous zones may exist around the glass, montmorillonite was chosen. It also offers the advantage of giving a free-flowing paste with lower contents of water.

To prepare the paste two types of water have been used: distilled water and synthetic interstitial clay-water (s. i. c.)[5].

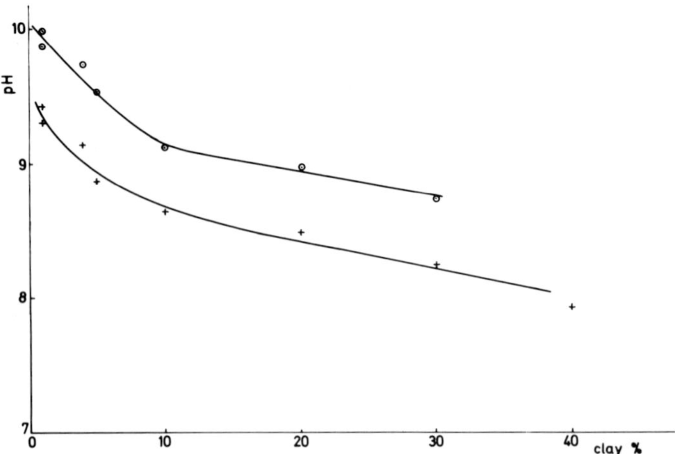

Fig. 1. Variation in pH of clay pastes with distilled water as a function of clay content. o bentonite; x montmorillonite.

With distilled water an acceptable fluidity is obtained with 55% of water, while with s.i.c. a fluid paste is obtained with 60% of water. In order to investigate the effect of the presence of corrosion products, two other types of paste were used, mixing the two solutions mentioned above with a clay mixture of 20% sand, 70% montmorillonite and 10% Fe_2O_3.

The glass investigated was a borosilicate glass containing 20% in weight of simulated waste oxides. Its silica contents is 48%; the detailed composition can be found in ref. 6. Samples were in the form of a cylinder, 10 mm in diameter, 5 mm high. The lateral surface was in the as-cast condition, while the end surfaces were obtained by diamond saw cutting.

Experiments were conducted in glass ampoules in which 30 cc of paste was introduced. After cleaning the glass samples in alcohol in an ultrasonic bath, they were put in the paste. This was a delicate operation due to the viscosity of the paste. In some cases small air bubbles may remain between the glass and the paste. Over the paste an equal volume of water was placed and then the ampoules were closed. Experiments were performed at 50 and 80°C. Every three weeks an ampoule was removed and the sample extracted. After careful washing the sample was dried for 4 h in a desiccator at room temperature and the sample weighed. Tests continued up to a maximum of 60 weeks. On some ampoules, before extracting the glass sample, the pH of the paste in contact with the glass was measured. As a comparison static tests with distilled water and with interstitial clay water were conducted at 80°C using stainless steel capsules. An analysis of the external part of the surface layer was performed by the ESCA method on those samples with the longest exposure time.

EXPERIMENTAL RESULTS AND DISCUSSION

From the examination of the samples extracted from the paste it was observed that on some samples large black areas appeared as if in this zone no leaching had occurred. It is possible that, irrespective of the care used in placing the sample in the paste, some air bubbles were trapped between the sample and the paste, thus protecting the sample from the paste's attack. As a consequence the scattering of the results is higher than that obtained in pure water.

128

In Figure 2 are collected the data obtained using pastes with and without Fe_2O_3 at $80^\circ C$ in distilled water, while in Figure 3 are presented the data obtained at $50^\circ C$ in distilled water. In the same figure, for the sake of comparison, are also presented results obtained from static tests in distilled water

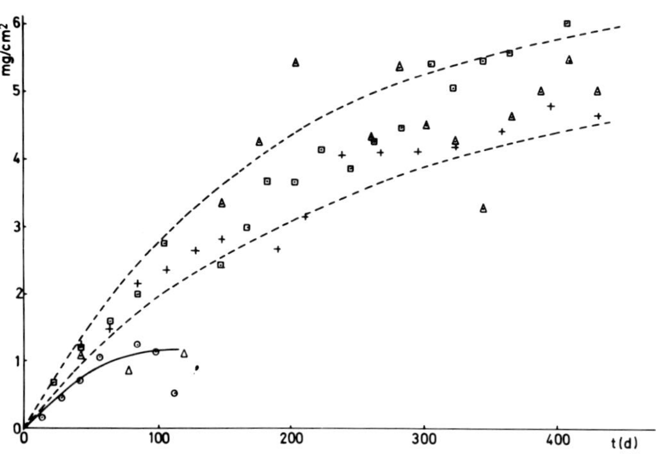

Fig. 2. Tests in distilled water at $80^\circ C$

o   distilled water   Δ montmorillonite and Fe_2O_3 paste

x   montmorillonite paste □ slowly renewed water

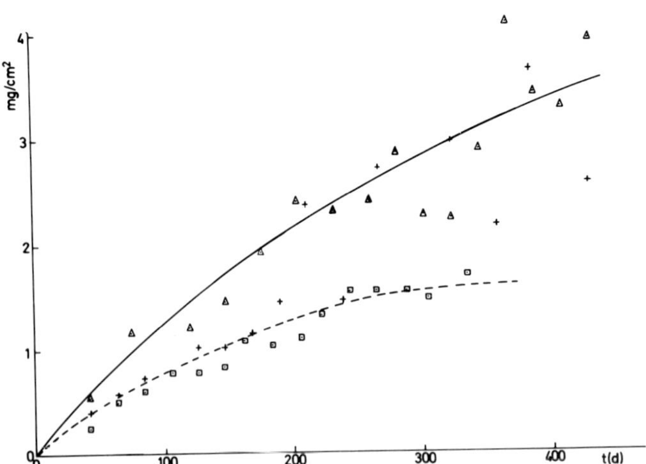

Fig. 3. Tests in distilled water at $50^\circ C$

+ montmorillonite paste; Δ montmorill. + Fe_2O_3 paste; □ slowly renewed water

and in slowly renewed water; these results are described in a previous work[6].
A marked difference exists between tests at 80°C and at 50°C.

At 80°C the values of weight loss with the two pastes and the comparison
values obtained in slowly renewed distilled water are essentially grouped together. Data obtained in pure static water evolve toward a rapid stabilization
to lower values, showing that under these conditions saturation effects are
encountered.

At 50°C the values obtained with the different pastes are of the same magnitude but higher than the values obtained with slowly renewed water, the difference reaching a factor of 2 after 300 days.

A similar picture is obtained with s. i. c. The data at 80°C are presented
in Figure 4 and the data at 50°C in Figure 5.

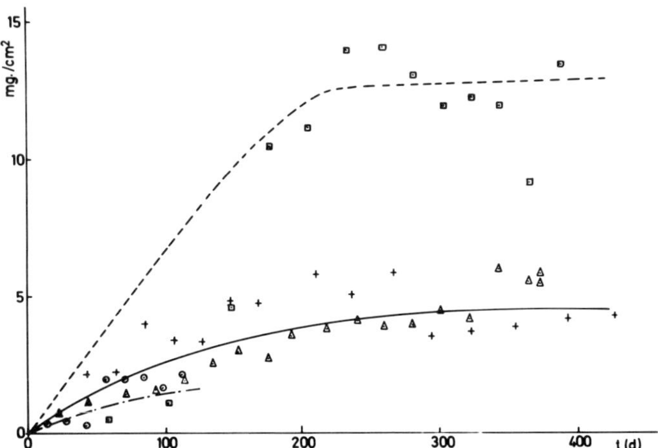

Fig. 4. Tests in synthetic interstitial clay water at 80°C
o synthetic interstitial clay water ◻ montmorill. +Fe$_2$O$_3$ paste
+ montmorillonite paste   Δ slowly renewed clay-cond. water

We note that in this case the reference data in slowly renewed water refer not
to s. i. c. , but to water conditioned with clay. At 80°C the weight loss values
obtained with a montmorillonite paste are similar to those obtained with slowly
renewed water. The addition of Fe$_2$O$_3$, on the contrary, increases sharply
the weight loss. A static test in the solution, as in pure water, encounters a

130

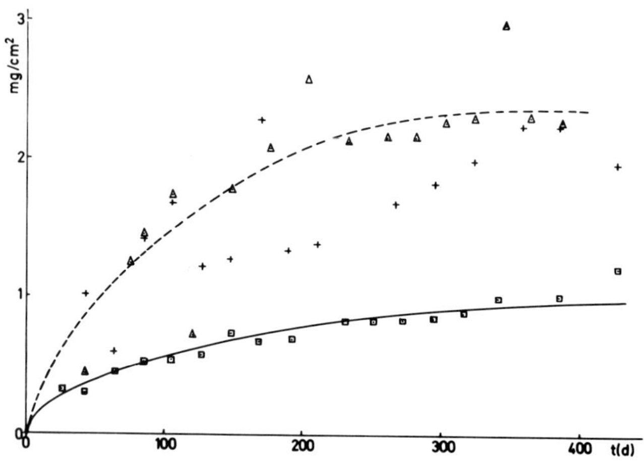

Fig. 5. Tests in synthetic interstitial clay-water at 50°C
+ montmorillonite paste; Δ montmorillonite + Fe_2O_3 paste;
☐ slowly renewed clay-conditioned water

rapid saturation. At 50°C, as with distilled water, values obtained with the pastes are about the same, but higher than those obtained with slowly renewed water.

The measurements of pH on the paste near to the leached sample show moderate variations generally lower than 0. 3 units.

An explanation of the similarity existing in specific weight losses at 80°C between renewed water and the pastes must take into account the elevated ion exchange capacity of the montmorillonite. It is possible that the elements escaping from the glass due to leaching, instead of remaining in the solution contributing to the saturation effect, tend to be absorbed by the clay. Thus, the glass continues to be leached as if in contact with the original solution.

The very high value obtained at 80°C with the s. i. c. paste containing Fe_2O_3 cannot easily be explained. We note however, that after an initial steep increase the specific weight loss remains practically constant. It may be that the large initial release has saturated the absorptive capacity of the clay surrounding the glass pellet. Subsequently, near-saturation conditions must exist which are controlled by diffusion of the leached species to more distant clay.

At 50°C the specific weight losses are even greater than those obtained with slowly renewed water. In a previous study[6] it was noted that at low temperature leaching became sensitive to water impurities. The relative low weight loss prevents to observe any saturation phenomena.

Some surface analyses were performed on the samples corresponding to the longest exposure time, using the ESCA method. The results obtained on samples leached in the various pastes are shown in Table 1. It can be seen that the main difference between the samples immersed in the pastes and those leached in water is the silicon content.

TABLE 1

Conditions	T °C	largely depleted	depleted	enriched
mont. paste with	80	Na	Fe, Ce, U	Si, Al, Zr
dist. water	50	Na	Fe, Ce, U	Si, Al, Zr
mont. paste with	80	Na, Ce	Zr, U	Si, Al, Fe
s. i. c.	50	Na, Ce	Zr, U	Si, Al, Fe
mont. + Fe_2O_3 paste with	80	Na, Fe	Ce, U	Si, Al, Zr
dist. water	50	Na	Fe, Ce, U	Si, Al, Zr
mont. + Fe_2O_3 paste with	80	Na, Fe, Ce	Zr, U	Si, Al
s. i. c.	50	Na, Fe, Ce	Zr, U	Si, Al
distilled water	80	Na	Si	Al, Fe, Zr, U, Ce
s. i. c.	80	Na	Si, Zr, U	Al, Fe, Ce

In the samples immersed in the paste the Si content increases while in those leached in water it decreases sharply. A similar effect was noted[6] in samples leached in large amounts of water; leaching with distilled water giving a lower Si content, while water conditioned with clay gives a higher Si content. Sodium is always largely depleted. Zirconium shows an enrichment on the surface in the samples leached with distilled water but is depleted in tests with s. i. c.

Iron is largely enriched in tests with only water. The presence of the paste reduces its concentration on the surface and in some case gives rise to depletion. The concentration of iron on the surface is possibly influenced by the redox conditions. Some tests are being conducted to clarify this point.

We note that at 80°C weight losses are independent of the leaching conditions and therefore of the surface layer composition. At 50°C (Figs. 2 and 4) the higher values obtained correspond to films which have a lower iron content.

CONCLUSIONS

The results of the leaching tests in different types of paste have shown that the presence of an absorbing medium around the glass sample tends to eliminate the saturation conditions that are present in water or water solution. It is conceivable that such an effect is transitory. When the clay surrounding the glass is saturated, the composition of the leachate will begin to evolve toward saturation and will probably be diffusion-controlled. However, at least for the first period, a model of dissolution in large amounts of water fits the experimental data better than a model based on saturation effects.

Tests performed with addition of Fe_2O_3 show that the influence of corrosion products is not negligible. Their presence partially saturates the absorptive centre of the clay, altering the leaching conditions. The use of Fe_2O_3 to simulate corrosion products is probably a crude approximation. A more realistic approach is needed.

If a similarity exists between leaching data tests in paste an in large amounts of water, the surface composition in the two cases is largely different. In the first case, in the external part of the film, silicon predominates while in the second case iron is prevalent.

At 80°C the surface composition does not seem to affect the capacity of the surface layer of reducing the leaching rate with time. At 50°C tests with the highest iron content present lower values of weight losses.

REFERENCES

1. Push, R. (1979), Proc. of the Workshop on the Use of Argillaceous Materials
   for the Isolation of Radioactive Waste. NEA, Paris, September, 1979.
2. Grim, R.E. (1962) Applied Clay Mineralogy, McGraw Hill Co.
3. Push, R. and Bergström, A. (1981) Scientific Basis for Nuclear Waste Management, Vol. 3, Moore, J.G., ed., Plenum Press.
4. Marples, J.A.C., Lutze, W. and Sombret, C. (1980) in Radioactive Waste
   Management and Disposal, Proc. of the First European Community Conference,
   pp. 305-323, Harwood Academic Publ., Brussels.
5. Van Iseghem, P.,Timmermans, W. and De Batist, R. (1981) International Seminar
   on Chemistry and Process Engineering for HLW Solidification, Jülich,
   June 1981.
6. Lanza, F. and Parnisari, E. (1981) Nucl. and Chem. Waste Man., Vol. 2, 131.

EFFECT OF OVERPACK MATERIALS ON GLASS LEACHING IN GEOLOGICAL BURIAL

LARS WERME*, L. L. HENCH** AND ALEXANDER LODDING***
*SKBF/Div. KBS, Stockholm (Sweden); **University of Florida, Gainesville,
Florida (USA); ***Chalmers University of Technology, Gothenburg (Sweden)

INTRODUCTION

This is one of two papers discussing the findings in an in situ-burial
experiment, presently being performed in the Stripa mine in Sweden. The purpose of the experiment is to evaluate the effects of various components in the
SKBF/KBS waste storage system on the leaching of the vitreous waste form. Two
configurations of glass, canister, overpack and buffer/backfill materials were
designed (Figs. 1 and 2). Both configurations were inserted into 56 mm
diameter boreholes in the Stripa mine and maintained at 90°C. One of the configurations (Fig. 2) was also kept at ambient temperature, 8°C. In the experiments two glass types were used, ABS 39 and ABS 41 (Table 1). These glasses,
developed by Dr. T. Lakatos of the Swedish Glass Research Institute, contain
9% simulated fission products by weight and are compatible with the French AVM
process.

TABLE 1

GLASS COMPOSITION (WEIGHT %)

Oxide	ABS 39	ABS 41
SiO_2	48.5	52.0
B_2O_3	19.1	15.9
Al_2O_3	3.1	2.5
Na_2O	12.9	9.9
Fe_2O_3	5.7	3.0
ZnO	0	3.0
Li_2O	0	3.0
UO_2	1.7	1.7
Simulated fission products	9	9

One configuration, Fig. 1, is designed to simulate closely a waste
package in a disposal hole, as described
in the KBS concept 2. The small "minicanisters" contained ABS 39 or ABS 41
glass cast into chromium-nickel steel
cylinders. Upper and lower surfaces
of the minicanisters were polished to
a 600 grit surface finish after a concentric center hole was bored for a
center heater rod. The minicanisters
were then arranged to provide glass-
glass or glass-betonite interfaces
(Fig. 2). Sleeves of lead and titanium
or copper overpacks were placed around
the steel wall of the minicanister,
and a bentonite buffer sleeve separated
the waste package from the walls of
the borehole in the granitic rock.

In the second configuration, described in detail in ref. 1, circular discs of glass, granite, bentonite and various overpack metals are stacked around a central tube containing a heater rod, to maximize the number of interfaces between glass and other materials (Fig. 3). This configuration is also referred to as the "pineapple slice" configuration, due to the shape of the discs.

Both configurations provide interfaces between the glass and the other components in the SKBF/KBS waste storage system. However, the interfaces are maximized in the "pineapple slice" configuration (28 vs 8) and consequently most experiments were performed using that configuration. In this paper only interactions between the glass and the metallic overpacks are discussed. Other burial effects are described elsewhere.[1]

EXPERIMENTAL METHODS

For the characterization of the glass surfaces, three different methods have been used: Infrared reflection spectroscopy (IRRS), optical microscopy, and secondary ion mass spectrometry (SIMS). The interpretation methods of IRRS have been described elsewhere[3,4,5] as well as application to these experiments.[1] Use of optical microscopy in both laboratory and burial glass corrosion studies is discussed in ref. 1. These two techniques were applied at the University of Florida for this study.

SIMS is a well established technique for surface analysis and diffusion studies.[6] It has been applied previously to glass corrosion studies but not as a routine method.[7-10]

In this study a SIMS instrument, a double focusssing modification of a CAMECA IMS 300 ion analyzer has been used. This instrument, available at Chalmers University of Technology, has proven a rapid and useful tool for routine SIMS analyses of corroded waste glass surfaces.

EXPERIMENTAL RESULTS

After burial the bentonite quickly absorbs water; a preliminary experiment showed that after 14 days the bentonite was water-saturated and leaching of the glass was clearly detected. The experience within the SKBF/KBS project is that for the experimental arrangements used here, water uptake will only take a few days. As a result of the swelling capacity of the bentonite, the bentonite flows out and fills the space between the stack of samples and the borehole walls. The force to remove the arrangement can therefore be considerable.

<u>Infrared Reflection Spectra (IRRS)</u>

Infrared reflection spectra were recorded from all glass-metal interfaces. At present, data from 28 days 90°C, 3 months 90°C and 3 months 8°C exposures are available. Examples of typical IRRS data are shown in Figs. 4-6. These spectra are recordings from the glasses with 3 months 90°C exposures. The over all corrosion attack of the glasses is low. However, glass ABS 39 tends to show some small dealkalization in contact with Cu and Ti (Figs. 4B and 5B), whereas ABS 41 shows very little evidence of attack at the interface with these metals. The glass-Pb interfaces show almost no change for either glass (Fig. 6). Compared with the glass-glass and glass-granite interfaces (see ref. 1) the effect of the metal overpack interface is a positive one. Using IRRS, both glasses ABS 39 and 41 with a Pb interface show no measurable attack after 3 months exposure. However, glass-glass interfaces for both ABS 39 and 41 show some damage at several spots on the surface.[1] The fact that some regions of these samples show no attack at all, whereas others do, may mean that some bentonite has penetrated into the interfaces at the extremity of the stack. All the metal overpacks seem to protect the glass from such damage. It is important to note that although Pb in particular is strongly adherant to the glass, the presence of Pb corrosion products between the metal and glass indicates that a water film had been present between the glass and the overpack during burial (see below).

The 28 days 90°C and 3 months 8°C exposures show negligible attack, by IRRS analysis, for all overpack metals.

<u>Optical Microscopy</u>

Optical micrographs of the glass-metal interfaces are shown in Figs. 7-9. Both ABS 39 and 41 glasses show very little change after 3 months 90°C contact with Cu and Ti. However, the Pb-glass interfaces for both glasses have a very thin film adherant to the glass. This film most probably consists of Pb corrosion products. The presence of the film results in no change in IRRS as shown above.

<u>Secondary Ion Mass Spectrometry (SIMS)</u>

Selected glass samples have been subjected to SIMS analyses. The most important operational conditions of the SIMS study are summarized in Table 2. For quantitative applications, the instrument has been calibrated using NBS glasses 611 and 613. The resulting sensitivity factors and detection limits are given in Table 3.

TABLE 2

IN-DEPTH PROFILING OF ELEMENTS IN GLASS TYPICAL PARAMETERS IN
ROUTINE SIMS ANALYSIS

Primary Beam:	0^-, 14.5 eV energy; 1 μA
Secondary Ions:	positive
Surface Layer:	10 nm gold on top of glass surface
Bombarded Area:	$10^2 - 10^3$ μm diameter
Analyzed Area:	$10 - 10^2$ μm diameter
Speed of Erosion:	0.1 - 10 μm/h (depends on degree of corrosion)
Effective Depth Resolution:	2 - 20 nm

TABLE 3

SENSITIVITY FACTORS AND DETECTION LIMITS FOR ELEMENTS IN A GLASS MATRIX

	Sensitivity Factors (Ion Yield/Atom-Conc.) Normalized to Si	Detection Limits (Atom-ppm, Approx.) At Profiling Rate 3μm/h
Li	6.5 ± 1.5	0.2 - 1.0
Na	13.5 ± 1.0	0.1
K	46.5 ± 2.5	0.02 - 1.0
Rb	36.0 ± 2.5	0.02 - 1.0
Cs	18.0 ± 5.0	0.05 - 1.0
Ca	9.6 ± 1.0	0.1 - 1.0
Sr	6.1 ± 1.5	0.2 - 1.0
Ba	7.6 ± 1.5	0.2 - 1.0
Y	1.0 ± 0.4	1.0 - 10
La	1.7 ± 0.5	0.5 - 10
Ti	6.3 ± 0.8	0.2 - 1.0
Zr	1.2 ± 0.4	1.0 - 10
Mn	3.5 ± 0.9	0.5 - 10
Fe	2.1 ± 0.7	$0.5 - 10^2$
Co	2.5 ± 0.5	0.5 - 10
Cu	2.0 ± 0.8	$0.5 - 10^2$
Ag	1.1 ± 0.5	0.5 - 10
Zn	0.36 ± 0.15	$5.0 - 10^2$
B	0.42 ± 0.05	2.0
Al	4.3 ± 0.3	0.2
Tl	2.2 ± 0.6	0.5 - 10
Pb	1.15 ± 0.45	1.0 - 10
Th	0.70 ± 0.40	$2.0 - 10^2$
U	0.75 ± 0.40	$2.0 - 10^2$

(Higher values in estimated detection limit dependent on spectral background.)

The sensitivity factors are deduced from the NBS glasses 611 and 613 except
for Li, Cs, Y, Zr and Fe, which are inferred from ABS 39 and 41. All
sensitivity factors are isotope corrected.

SIMS profiles from ABS 39 after 28 days and 3 months interface with Pb at 90°C are shown in Figs. 10 and 11. For comparison a 3 month glass-glass interface is shown in Fig. 12. The corrosion depth for the glass-Pb interface has not increased appreciably from 28 days to 3 months. The surface of the glass gives rise to a strong Pb signal that drops off rapidly with depth in the glass. At depths beyond approx. 0.15 μm, the Pb signal roughly follows the Ca (and K) signal, indicating that some diffusion of Pb into the glass matrix is taking place. Closer to the surface the Pb profiles are very steep and SIMS calibration indicates there are Pb concentrations as high as 10 atom %. This probably corresponds to surface deposits of lead salts or lead oxides, as mentioned **above**. A complementary study by means of a surface level monitor (Rank Talysurf) reveals a particular roughness of the glass after contact with Pb, manifested as "islands" some 1 μm in height, 1-5 μm in lateral size, and 5-100μm in spacing. The corrosion layer is very thin as also indicated by IRRS. For a glass-glass interface (Fig. 12) the corrosion layer is thicker (∿1μm), but as shown for the glass-Pb interface, the layer has not grown appreciably from 28 days to 3 months. The reason for this is most probably that the very thin water layers between the solids rapidly become saturated. A different situation is encountered in the glass-bentonite interface, where the corrosion layers have grown from approximately 5μm in 28 days to over 15μm in 3 months (see ref. 1). This is probably due to the ion-exchange capacity of bentonite, which is expected to also be pH dependent, i.e., an increased pH results in an increased ion-exchange capacity.[11]

CONCLUSIONS

The metallic overpacks (Pb, Cu, Ti) have little effect on the corrosion of borosilicate nuclear waste glasses under either 90°C or 8°C burial conditions in deep granite. Corrosion layers for metal-glass-interfaces are generally thinner than for glass-glass interfaces. One possible reason for this is that the ductile metallic materials allow only a very thin water film between the glass and the metal. This thin water film becomes rapidly saturated with corrosion products and further corrosion is consequently slower.

ACKNOWLEDGMENTS

The authors acknowledge support of SKBF/Project KBS and one of them (LLH) acknowledges assistance of the U.S. Dept. of Energy during the period of this research. They also acknowledge assistance of Dr. T. Lakatos for the glass samples, J. Wilson-Hench for optical microscopy, and S. D. Hench for IRRS.

REFERENCES

1. Hench,L. L., Werme, L. O. and Lodding,A., paper to this Conference.
2. KBS Report; Handling of Spent Fuel and Final Storage of Vitrified High Level Reprocessing Waste, Stockholm (1978).
3. Clark,D. E.,Pantano,C. G. and Hench,L. L. (1979) in Glass Corrosion, Books for Industry, New York.
4. Sanders,D. M., Person,W. D. and Hench, L. L. (1972) Appl. Spectroscopy, 26,530.
5. Sanders,D. M., Person,W. D. and Hench, L. L. (1974) Appl. Spectroscopy, 28, 247.
6. Lodding, A. (1982) Review of Analytical Chemistery.
7. Scherrer, S. and Nandin, F. (1977) Proc. XI Intern. Congr. Glass, Prague, vol. III, p. 301.
8. Hench, L. L. and Clark D. E. (1978) J. Non-Cryst. Solids, 28, 83.
9. Tournay, J-C., Thomassin, J-H., Baillif, P. and Scherrer, S. (1980) J. Non-Cryst. Solids, 38 and 39, 643.
10. McIntyre, N. S. and Strathdee, G. G. (1980) Suface Science 100, 71.
11. Marinsky, J., private communication.

Fig. 1. Minicanister experiment, Photograph of: granite disc, bentonite sleeve, holder, glass canister, and bentonite disc.

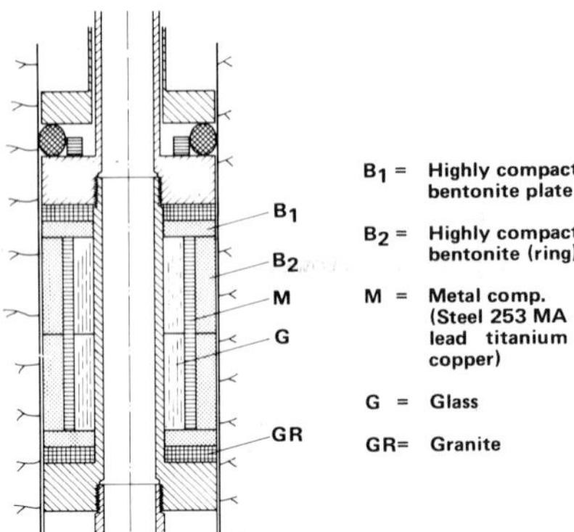

Fig.2. Schematic view of minicanister in borehole.

Fig. 3. Parts of a "pineapple slice" stack.

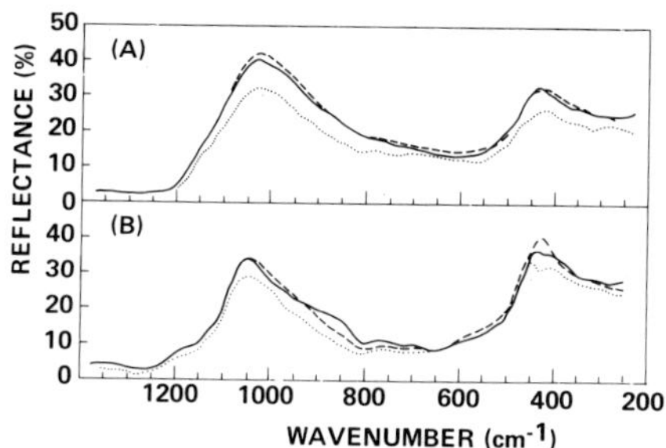

Fig. 4. IRRS recordings of glass surfaces interfaced with copper. (A): Glass ABS 41. (B): Glass ABS 39. Solid curve corresponds to one sample. Dashed and dotted curves correspond to two positions on another sample.

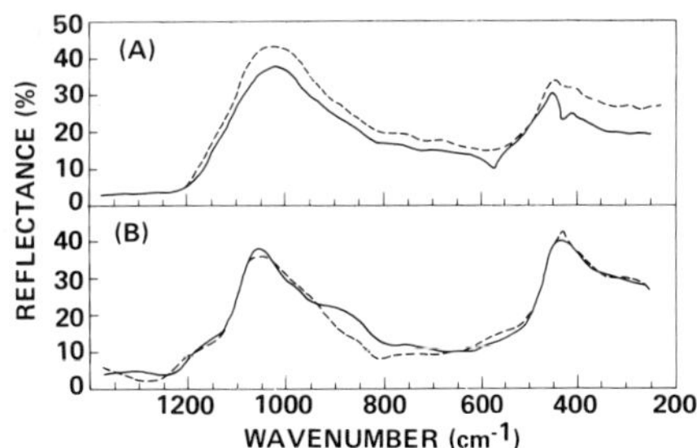

Fig. 5. IRRS recordings of glass surfaces interfaced with titanium. (A): Glass ABS 41. (B):Glass ABS 39. Solid and dashed curves correspond to different samples.

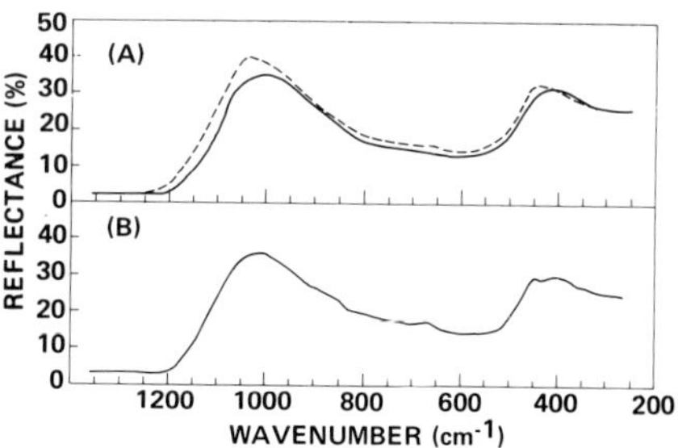

Fig. 6. IRRS recordings of glass surfaces interfaced with lead. (A): Glass ABS 41. (B): Glass ABS 39. Solid and dashed curves in (A) correspond to different positions on the sample.

EFFECTS OF OVERPACK METALS ON NUCLEAR WASTE
GLASS SURFACES (120X) AFTER 3 MO, 90°C STRIPA BURIAL

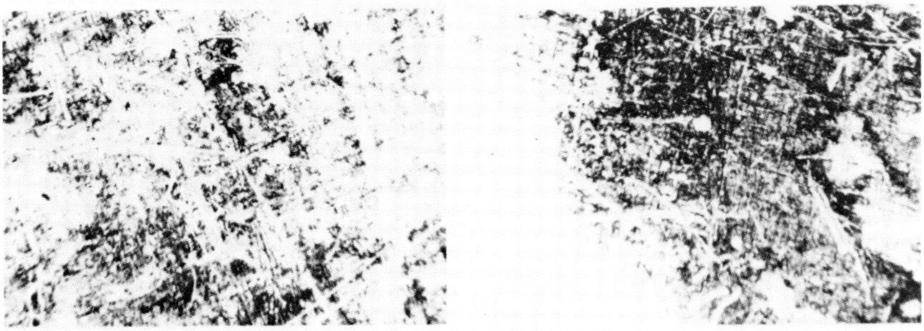

ABS 39 - Pb   Fig. 7   ABS 41 - Pb

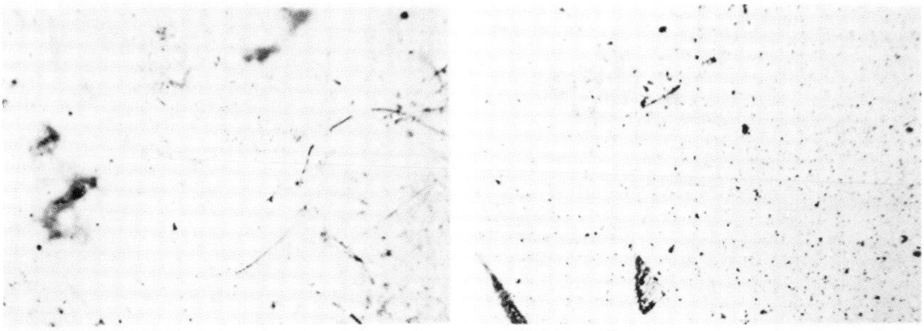

ABS 39 - Ti   Fig. 8   ABS 41 - Ti

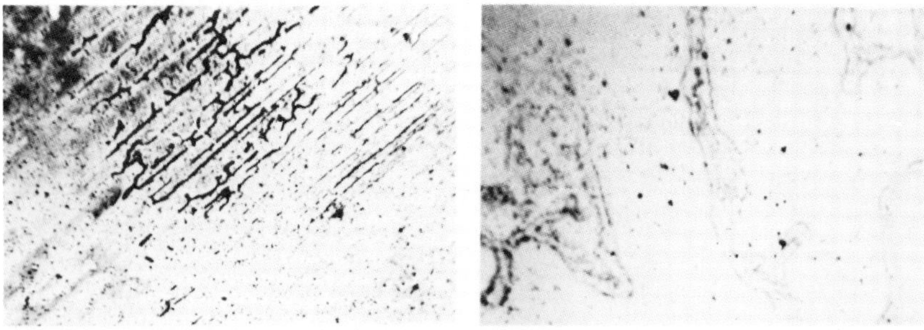

ABS 39 - Cu   Fig. 9   ABS 41 - Cu

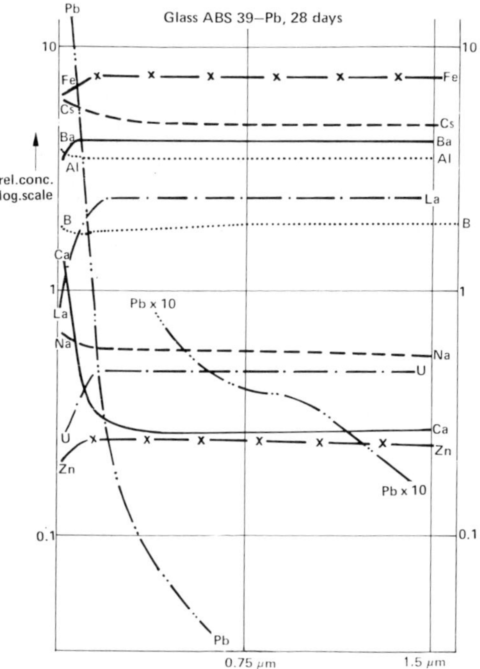

Fig.10. SIMS in-depth profile of a glass surface interfaced with lead for 28 days.

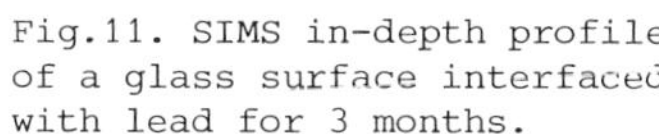

Fig.11. SIMS in-depth profile of a glass surface interfaced with lead for 3 months.

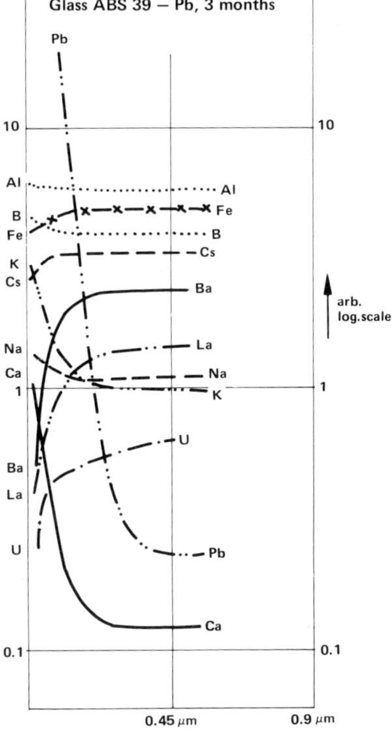

144

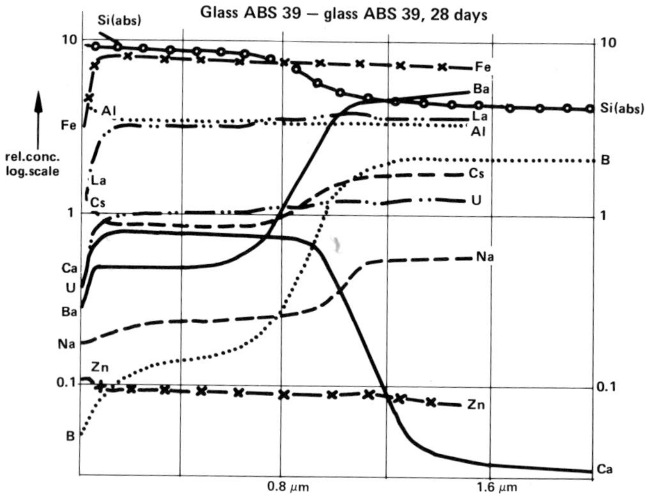

Fig.12. SIMS in-depth profile of a glass-glass interface
after 28 days.

SCIENTIFIC BASIS FOR RADIOACTIVE WASTE MANAGEMENT - V
Werner.Lutze, editor

THE INTERACTION OF BOROSILICATE GLASS AND GRANODIORITE AT 100°C, 50 MPa: IMPLICATIONS FOR MODELS OF RADIONUCLIDE RELEASE

DAVID SAVAGE AND JANE E. ROBBINS
Environmental Protection Unit, Institute of Geological Sciences, Building 151, Harwell Laboratory, Oxfordshire, OX11 ORA, United Kingdom

INTRODUCTION

An essential component of any assessment of HLRW geological disposal options is the quantitative prediction of radionuclide release rates from the near-field over time spans of the order of 10^3-10^6 years. Fundamental to this assessment is the investigation of the interaction of potential wasteforms with groundwater under repository conditions of temperature, pressure, and groundwater flow-rate. Consequently, many studies world-wide have been initiated to examine the kinetics of wasteform dissolution over a wide range of physical and chemical conditions. Although these studies have provided a considerable amount of invaluable data on wasteform-fluid interactions, they have tended to focus on breakdown of the wasteform itself, and not on the fate of released waste components in the near-field. For example, effects of saturation of species in solution, precipitation of secondary minerals or amorphous gels, and the effect of host rock chemistry on the products (solid and fluid) of waste-fluid interaction have largely been ignored or even specifically excluded in laboratory experiments. This is despite growing evidence from source term modelling studies which suggest that the above processes may well be the chief factors in governing rates of radionuclide release from the near-field, bearing in mind the limited availability of ground-water in typical crystalline rock environments[1,2].

Therefore, it was the intention of this study to examine the interaction of a typical wasteform (borosilicate glass) and a potential repository host rock (granodiorite) under near-field conditions of temperature, pressure and water flow-rate appropriate to the early (0-600 years), high temperature life of a repository, and to study the effects of saturation of waste components in the fluid phase on rates of radionuclide release.

EXPERIMENTAL

The experiment was carried out in a large volume pressure-vessel, the reactants being contained in a deformable, inert, gold cell. A gold and titanium lined sampling tube and valve block enabled fluids to be extracted without disturbing conditions of temperature and pressure[3]. The use of such zero-flow experimental apparatus is appropriate in modelling near-field geo-chemical behaviour given the likely long groundwater residence times in crystalline rocks[2] (in excess of two hundred days in 1 m^3 of rock).

Starting materials were a U.K. reference borosilicate waste glass (209/M22), a granodiorite, and de-ionised water. The glass contained approximately 25 wt.% of simulated fission product oxides and 0.005 wt.% uranium oxide. Major oxides in the glass were (wt.%) SiO_2: 50.9; B_2O_3: 11.1; Na_2O: 8.3; MgO: 6.3; Al_2O_3: 5.1; Li_2O: 4.0. Detailed chemical compositions of both waste glass and granodiorite were presented by Savage and Chapman[4]. Solid reactants were in powdered form, such that the surface area of the glass and granodiorite were approximately 0.5 and 1.9 m^2g^{-1}, respectively (as measured by nitrogen BET). The initial fluid : glass : granodiorite mass ratio was 20:1:1, falling to 12:1:1 at run completion. The initial fluid volume/glass surface area ratio was 3.9 x 10^{-3} cm, falling to 2.4 x 10^{-3} cm at run completion. Six fluid samples of approximately 13 cm^3 volume each were extracted during the course of the experiment. Temperature and pressure conditions were 100°C and 50 MPa, respectively. The total experiment duration was 200 days. Measurement of pH and analysis of 41 species in solution were performed at 25°C. Analytical techniques utilised included atomic absorption, colorimetry, ICP and neutron activation analysis.

RESULTS

The chemical components analysed in solution from the six samples are presented graphically as concentration versus time diagrams in Figs. 1 and 2. Data for Ca, Mg, K, F, Cu, Bi and Cd which were automatically included in the ICP spectra have been omitted for the sake of clarity. Data for the following components were at or below detection limits (ppm, in paraenthesis):- Ni (0.15); Ti (0.2); Co (0.06); Nd (0.3-3); Ru (0.2); Be (0.006); Ga (0.06); Ge (0.08); In (0.3); V (0.1); W (0.4); U (0.05-0.19). Precision of analyses lie between ± 5-10%.

Although the concentrations of a number of components in Figs. 1 and 2 will have been contributed to by both the glass and granodiorite e.g. SiO_2, Na, Al, Fe, Mg, Ba, and Sr (this is borne out by a parallel experiment conducted on the granodiorite-water system under the same conditions), certain components will almost exclusively have been leached from the waste glass, e.g. B, Mo, Li, Cs, Cr, Zr, La, Rb. A large number of these components show rapid (10-30 days or less) saturation in the fluid phase, apparently reaching equilibrium conditions thereafter (within the precision of the chemical analyses). These are (with

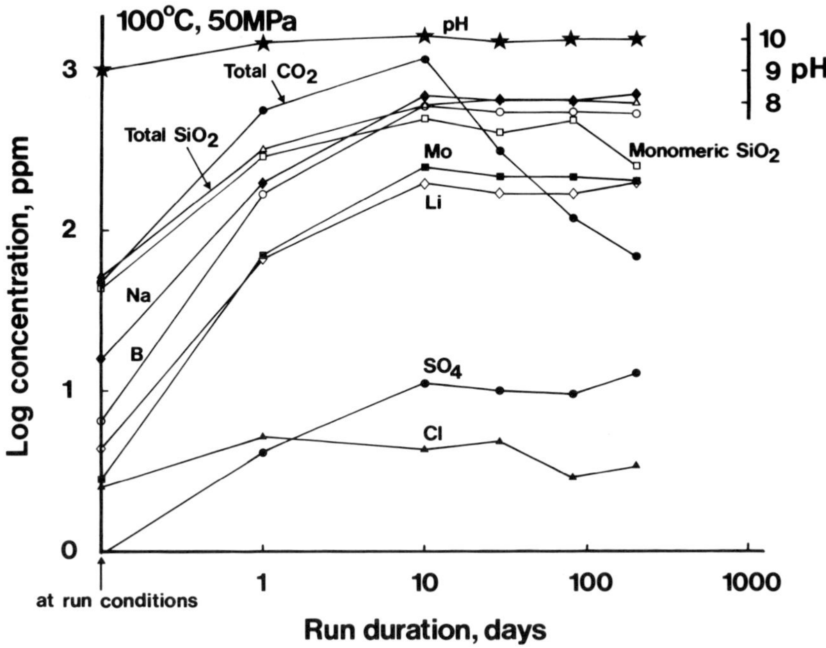

FIG. 1. Log. concentration (ppm) of total SiO_2, monomeric SiO_2, total CO_2, Na, B, Mo, Li, SO_4, and Cl versus run duration.

final concentrations in ppm in parenthesis):- total SiO_2 (635); monomeric SiO_2 (250); Na (710); B (535); Mo (205); Li (200); SO_4 (13); Cl (3.5); Sn (6); Al (0.95); Fe (0.065); and Cr (9.35). However, a number of components show consistently increasing concentrations throughout the experiment duration, without reaching any equilibrium value. These are:- Ba, Cs, Rb, Zn, Zr and Sr.

In order to be able to describe the observed solution chemistry fully, it is necessary to have complete information concerning the dissociation and speciation of all the chemical components present in the starting materials or formed during reaction at the physical conditions of the experiment. Unfortunately, because of the paucity of data available concerning the high-temperature behaviour of components such as boron and molybdenum, this is not possible, However, a number of observations may be made.

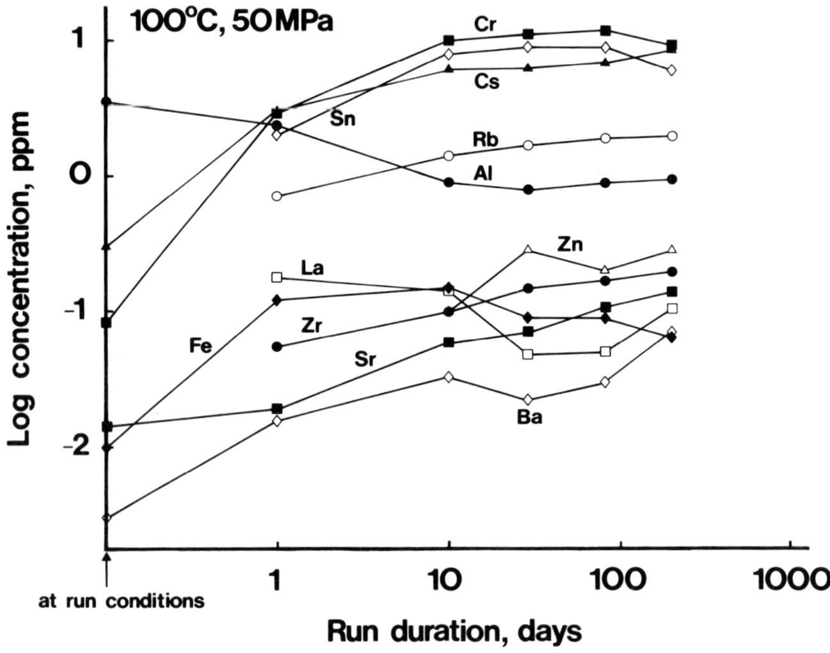

FIG. 2. Log. concentration (ppm) of Cr, Cs, Sn, Rb, Al, Zn, La, Zr, Fe, Sr and Ba versus run duration.

Solution pH shows an immediate rise to pH = 9, eventually levelling off at approximately pH = 10. Such alkaline conditions reflect the rapid extraction of alkalis from the glass and generally result in dissolution of the glass network, rather than selective element leaching.[5] Total silica solubility (i.e. silica as measured by ICP or atomic absorption) is considerably greater than that measured for the amorphous silica-water system at $100°C$, 1 atm (372 ppm).[6] This may be explained by increased silica solubility through partial ionisation of the solvated silica monomer (H_4SiO_4) to produce $H_3SiO_4^-$, which occurs in solutions with pH >8.[7] Carbonate and bicarbonate dominate the anionic species and are derived from solution of carbon dioxide from the air void at the top of the reaction cell. The decreasing total CO_2 concentration seen in Fig. 1 after 10 days reaction is due to outgassing of CO_2 from the reaction cell during sampling. Boron and molybdenum also probably contribute significantly to the anionic species, whereas sulphate and chloride (both dominantly derived from the granodiorite) are both less than 10 ppm each. Chromium is believed to be in solution as the chromate ion (CrO_4^{2-}) up to 80 days reaction, because of the pale yellow coloration of the extracted samples. The pale greenish colour of the solution extracted after 200 days suggests a combination of both Cr^{3+} (blue /green aqueous complexes) and chromate ions, and probably reflects decreasing oxygen activity in the fluid phase with increasing run duration. Such reactions may be used to define redox conditions within the reaction cell.[8]

In order to be able to compare relative elemental dissolution behaviour, the normalised molar release rates of various glass components have been calculated and plotted versus run duration in Figs, 3 and 4. These were obtained by

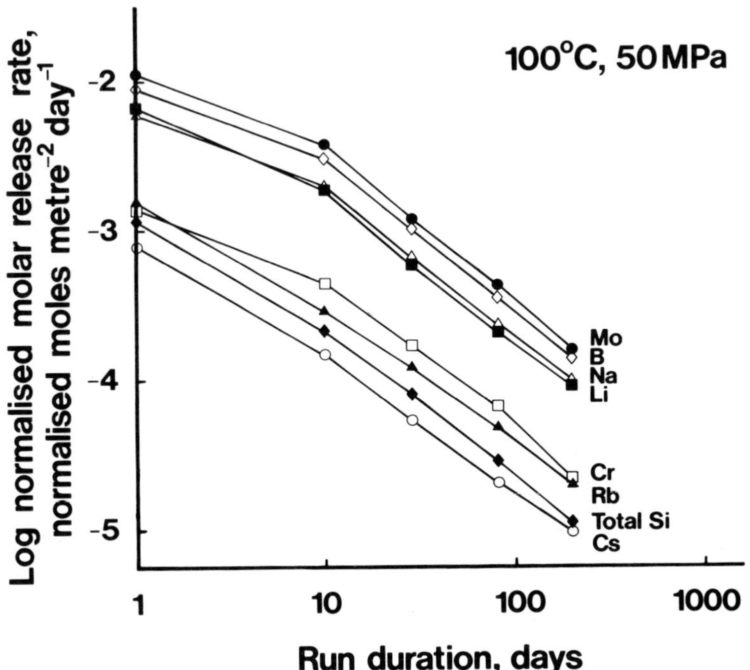

FIG. 3. Log normalised molar release rate (normalised moles metre^{-2} day^{-1}) for Mo, B, Na, Li, Cr, Rb, total Si and Cs versus run duration.

dividing the molar leach rates of each glass component by the proportion of the component in the original glass. For congruently dissolving glasses there-

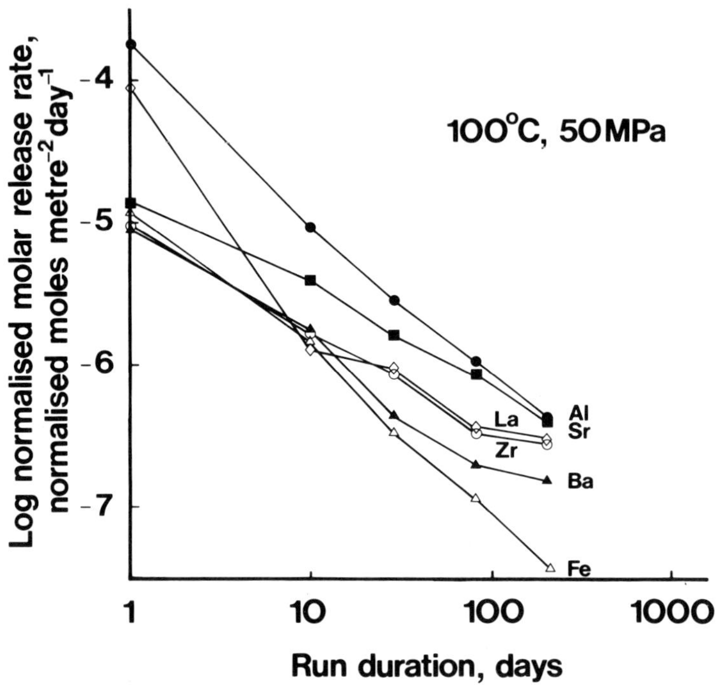

FIG. 4. Log. normalised molar release rate (normalised moles metre^{-2} day^{-1}) for Al, Sr, La, Zr, Ba, and Fe versus run duration.

fore, each normalised molar leach rate should be identical. However, in the experiment considered here, one would expect the components contributed to by both the dissolution of the glass and granodiorite to exhibit the greatest normalised molar leach rates.

The normalised molar release rates displayed in Figs. 3 and 4 cover a range of approximately five orders of magnitude from the most readily leached element after one day (Mo) to the most insoluble element after 200 days (Fe). Three or four orders of magnitude encompass the range of release rates at any one time interval. Typically, the leach rate for any one component decreases by approximately two orders of magnitude between one and two hundred days reaction time. It may be seen from Figs. 3 and 4 that there is no tendency for species extracted from both glass and granodiorite (SiO_2, Na, Al, Sr, Ba, Fe) to be vastly greater than species extracted from the glass alone. Indeed, boron and molybdenum are the most readily leached elements and can be considered to have been derived solely from the glass.

IMPLICATIONS FOR MODELS OF RADIONUCLIDE RELEASE

The results of this experiment demonstrate that if groundwater should breech canister and overpack materials during the early high-temperature life of a repository then chemical components of borosilicate waste glass will saturate in solution as a result of rock-waste-water interaction. This implies that the rate of release of the relevant radio-nuclides from the near-field during this period may be adequately modelled by a function of equilibrium solubility and groundwater flow-rate. However, absolute solubility values for specific components will depend upon a number of factors, including temperature, glass composition, groundwater composition, and fluid/glass surface area ratio[9], and further research is necessary to investigate the effects of these parameters on solubility variations.

The rate at which equilibrium is achieved is also of fundamental importance to this method of source term modelling, and has been shown to be critically dependent upon temperature and the fluid volume/glass surface area ratio.[6] For example, glass-water equilibration times may exceed nine months at $25^{\circ}C$,[10] and may thus exceed near-field groundwater residence times. In order to be able to predict component solubilities over 10^3-10^6 years, it is necessary to characterise the solid phases governing solubility equilibria. No XRD data on the solid products are yet available for the experiment described here, although previous work investigating glass 209-granodiorite-water interactions at $100^{\circ}C$[4] indicate largely amorphous solid products (other than the granodiorite mineralogy) with traces of poorly crystalline smectite. It may be that solubilities observed in this experiment were buffered by amorphous oxides/gels, although in the long term, crystalline reaction products may stabilise and become solubility controlling. More research is therefore needed to investigate these secondary alteration products.

Those waste constituents which show no evidence of equilibration with the fluid phase on the timescale of the experiment considered here (Ba, Cs, Rb, Zn, Zr and Sr) could not be approximated by a simple solubility-flow rate relationship as a source term model and, for these components, it is necessary to place more emphasis on the kinetics of release from the glass to be able to model their release from the near-field. Quite clearly, this work illustrates that near-field geochemical interactions will produce not only a wide range (possibly five orders of magnitude or more) of relative rates of radionuclide release under a given set of physical conditions, but also that each nuclide must be considered separately in terms of the mechanism of release to be defined by the source term model.

CONCLUSIONS

The interaction of a simulated borosilicate waste glass, granodiorite and de-ionised water at $100^{\circ}C$, 50 MPa under closed system experimental conditions has revealed the rapid achievement of steady-state fluid concentrations for many chemical components of interest, (e.g. SiO_2, La) and their rates of release from the near-field would be most appropriately modelled by a function of solubility and groundwater flow-rate. The conversion of these solubilities into conventional leach-rates has shown over five orders of magnitude range in relative release rates and emphasises the need for source-term models to consider each radionuclide separately in terms of mechanisms of release.

ACKNOWLEDGEMENTS

The authors would like to thank the following: Dr N A Chapman of I.G.S. for critically reviewing the manuscript; Mr J T Dalton of UKAEA for provision of the glass sample; Messrs J R Stevens, A R Lister, W Temple, T Sanders and Dr C Pickford of the UKAEA for chemical analyses of the fluid samples. The work was carried out by N.E.R.C./I.G.S. under contract to the Department of the Environment and the Commission of the European Communities and is published by kind permission of the Director, Institute of Geological Sciences (N.E.R.C.).

REFERENCES

1. American Physical Society (1978) Rev. Mod. Phys., 50, S1-S186.
2. Chapman, N.A., McKinley, I.G., and Savage, D. (1980) Proc. NEA workshop on 'Radionuclide release scenarios for geologic repositories', NEA/OECD, Paris, 91-103.
3. Seyfried, W.E., Gordon, P.C., and Dickson, F.W. (1979) Amer. Mineral., 64, 646-649.
4. Savage, D., and Chapman, N.A. (in press) Chem. Geol.
5. El-Shamy, T.M., Lewins, J., and Douglas, R.W. (1972) Glass Technol., 13, 81-87.
6. Rimstidt, J.D., and Barnes, H.L. (1980) Geochim. Cosmochim. Acta, 44, 1683-1699.
7. Helgeson, H.C. (1969) Amer. J. Sci., 267, 729-804.
8. Holloway, J.R., Jenkins, D.M., Kakoyannakis, J.F., and Apted, M.J. (1981) Rep. RHO-BWI-C-105 (Rockwell Hanford Operations).
9. Fullam, H.T. (1981) PNL Rep. 3614 (Pacific Northwest Laboratory).
10. Barkatt, A., Simmons, J.H., and Macedo, P.B. (1981) Nucl. Chem. Waste Management, 2, 3-23.

SCIENTIFIC BASIS FOR RADIOACTIVE WASTE MANAGEMENT - V
Werner.Lutze, editor

BURIAL EFFECTS ON NUCLEAR WASTE GLASS

LARRY L. HENCH[*], LARS WERME[**] AND ALEXANDER LODDING[***]
[*]Department of Materials Science and Engineering, University of Florida, Gainesville, Florida, USA; [**]SKBF/KBS, Stockholm, Sweden; [***]Chalmers University of Technology, Gothenburg, Sweden.

INTRODUCTION

The purpose of this experiment was to evaluate the effects of various components of the SKBF/KBS nuclear waste storage system on the leaching of the vitreous waste form. Two configurations of nuclear waste glasses, canisters, overpacks, and backfill material were inserted into 5.6 cm x 3 m deep boreholes located at the 350 m level in the STRIPA mine.[1] Some were maintained at 90°C. The others were allowed to equilibrate at the ambient temperature of the mine, approximately 8°C. Two borosilicate nuclear waste glass compositions (termed ABS 39 and ABS 41) compatible with the French AVM process containing 9 percent by weight of simulated fission products were compared. The two compositions (Table 1) bracket the range of $SiO_2/Na_2O/B_2O_3$ ratios likely to be selected for commercial vitrification operations at La Hague.

Table 1

Glass composition (weight %)

oxide \ glass	SiO_2	B_2O_3	Al_2O_3	Na_2O	Fe_2O_3	ZnO	Li_2O	UO_2	simulated fission products
ABS 39	48.5	19.1	3.1	12.9	5.7	0	0	1.7	9
ABS 41	52.0	15.9	2.5	9.9	3.0	3.0	3.0	1.7	9

One configuration tested (called "pineapple slices") involved thin, right circular discs of glass with a central hole for a heater rod. The sequence of interfaces are shown in Fig. 1. One side of each sample was polished to 600 grit SiC, the standard used for correlative lab tests. The alternate side was left rough in order to avoid mixing interfaces during disassembly. The second configuration, called minicans, is described in another publication.[2]

Static leaching laboratory experiments were also conducted at 90°C to compare with the in-situ burial results and to isolate the interfacial variables. A procedure similar to MCCI was used.[3] The variables investigated included: glass composition ratio of glass surface area to volume of leachant solution ($SA/V=cm^{-1}$); presence or absence of compacted bentonite exposed to the leachant solution; ratio of cm^3 bentonite to cm^3 water; presence of granite together with compacted bentonite, glass, and water; ratio of cm^3 granite to cm^3 water. Detailed results from this extensive series of tests are presented in another publication.[4]

EXPERIMENTAL METHODS

Surface analyses of the samples were made using both infrared reflection spectroscopy (IRRS)[5,6] and secondary ion mass spectroscopy (SIMS) with ion beam profiling.[7,8] Areas of samples used for IRRS were photographed using reflection optical microscopy at 120X magnification. From 4-8 spots were analyzed per sample with IRRS and the spectra reported represent the range of changes observed. All spectra were normalized to a value of 80 for the 1120 cm^{-1} Si-O-Si molecular stretching vibration of vitreous SiO_2 by use of a shutter in the reference beam of a Perkin-Elmer 467 IR Spectrometer. Details of the SIMS technique are presented in another publication.[2]

Results

IRRS Analysis

The assemblies removed from the mine were damp with water oozing from the end of the cylinder. Bentonite had expanded to produce considerable pressure within the borehole and led to difficulty in removing the cylinder.

Fig. 2 illustrates the finding for several of the burial and laboratory simulations investigated. Figure 2A shows that the IRRS spectra of ABS 41 nuclear waste glass after 28 days burial at 90°C in STRIPA granite is nearly equivalent to a freshly polished sample; i.e. surface reactions are less than 0.5 µm, the depth of the IRRS beam. In contrast, laboratory simulations involving only ABS 41 glass and water, $SA/V=0.1$ cm^{-1} do show attack and deterioration of the IRRS spectrum (Fig. 1B, dashed curve) after 28 days at 90°C. Addition of bentonite and STRIPA granite to the static leach test, maintaining $SA/V=0.1cm^{-1}$, 90°C, shows an even greater extent of glass surface attack (Fig.2B, full curve). If the quantity of water is increased in the laboratory simulation ($SA/V=0.03$ cm^{-1}) and (bentonite volume/water volume=0.03) there is even greater surface attack (Fig. 2B, dotted curve).

The effects of bentonite on nuclear waste glass-water interactions is also seen when post in-situ burial interfaces are examined. Fig. 2B, cross hatched curve, shows a reduction in IRRS intensity for the bentonite - glass interface after 28 days 90°C burial. However, the extent of surface - ion exchange is considerably less than observed for the laboratory experiments with bentonite and granite. A comparison of the effect of variable ratios of bentonite - water in the laboratory experiments to the post burial spectrum suggests that the "effective burial SA/V" ratio for this experiment is about 0.2 cm^{-1}.

Results for glass composition ABS 39, Fig. 3, are similar but more extensive suface attack occured. The 90°C burial condition is less severe than 28 days static lab leaching, compare 3A and 3B. Bentonite accelerates the leaching for ABS 39 also and the bentonite effect increases as the bentonite/H_2O ratio decreases, i.e., it is worse in larger volumes of leachant solution. The"effective burial SA/V" appears to be approximately 0.05 cm^{-1}, much less than observed for glass ABS 41.

A duplicate 28 day , 90°C STRIPA burial of a pineapple slice assembly was conducted to confirm these findings. IRRS analyses of all 28 sets of interfaces were nearly identical to the first experiment.

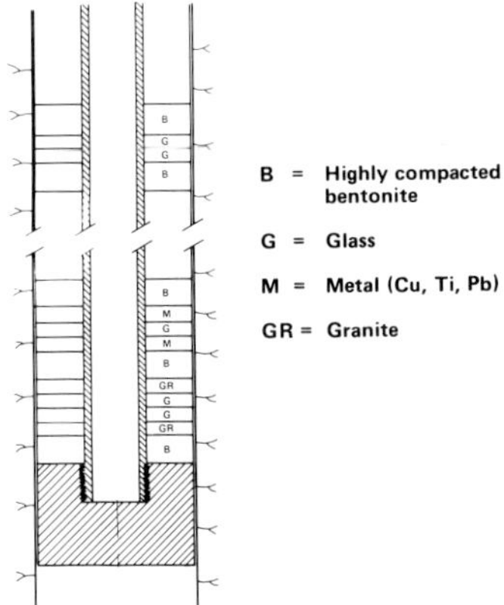

Fig. 1. Sequence of glass interfaces in the "pineapple slice" STRIPA burial configuration.

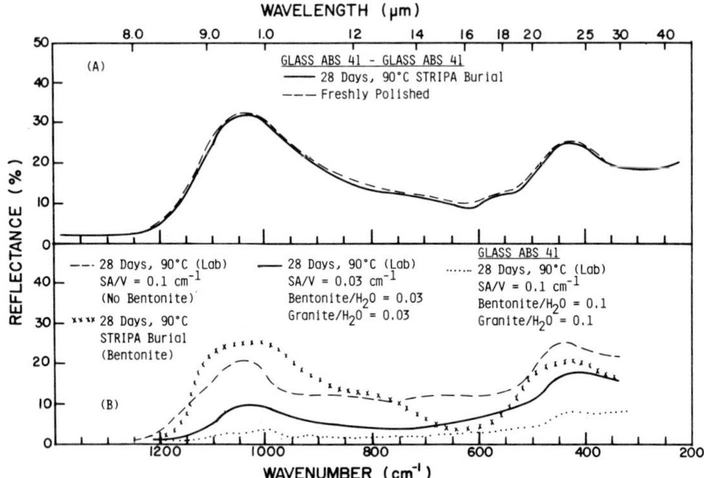

Fig. 2. Infrared reflection spectra of glass ABS 41 before and after 28 days, 90°C STRIPA burial and various 90°C laboratory simulations.

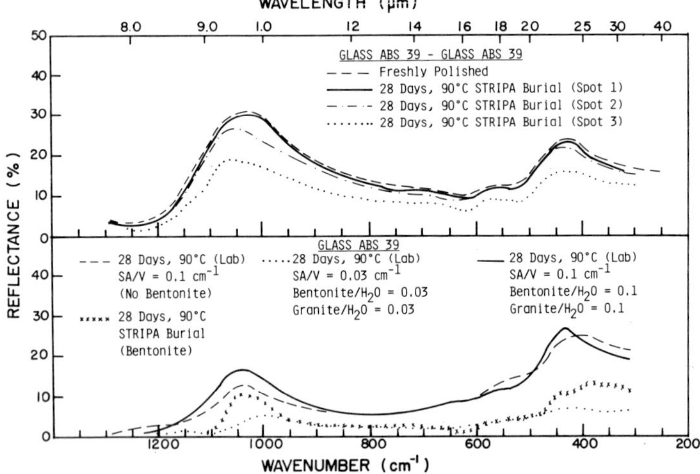

Fig. 3. IRRS spectra of glass ABS 39 before and after 28 days, 90°C STRIPA burial and various 90° C laboratory simulations.

IRRS analysis of the glass-glass interfaces and glass-bentonite interfaces of the 28 day, 90°C burial of minicans were also the same: glass ABS 39 showed more attack than ABS 41, both glasses had much less damage after burial than observed for 28 day, 90°C static laboratory

leaching in D.I. water, wet bentonite accelerated ion depletion of the
glass surface, the effect of bentonite was more severe for glass ABS 39
than glass ABS 41.

Two additional experiments were conducted using the pineapple slice configuration (Fig. 1) to determine: 1) whether the minimal glass-glass interface attack would progress or stabilize with time, 2) whether the bentonite
effect would increase or decrease with longer exposure, and 3) establish
the relative effect of storage temperature on the glass-glass and glass-
bentonite interactions. Figure 4 shows that glass-glass interfaces ABS 41
continue to show very little surface attack even after 3 mo. of 90°C exposure, Fig. 4A. No observable damage is present after 3 mo. at 8°C,ambient,
mine temperature either. In contrast the glass with the higher alkali
content, ABS 39 does show evidence of dealkalization after 3 mo. at 90°C,
Fig. 4B. It is also possible that some bentonite infiltration may have
been responsible by the appearance of the IRRS spectrum in Fig. 4B. Glass
ABS 39 however shows little indication of dealkalization during 3 mo.at 8°C.

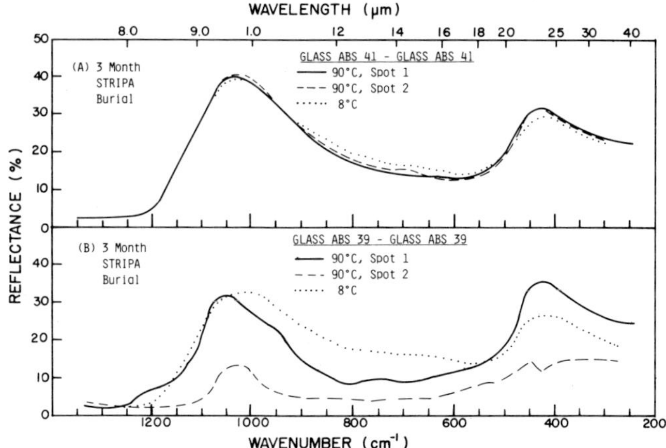

Fig. 4.   IRRS spectra of glass-glass interfaces after 3 mo.STRIPA
burial at 90°C and 8°C.

The presence of bentonite results in more surface attack in the 3 mo. experiment than occurs with glass-glass interfades. However, the extent of
bentonite enhanced ion depletion from the glass is not much more severe
than observed after 28 days at 90°C burial (Figs.2 and 3) for both glasses.

At the lower burial temperature of 8°C some bentonite effect is present for both glasses, compare Figs. 4 and 5, but it is minor.

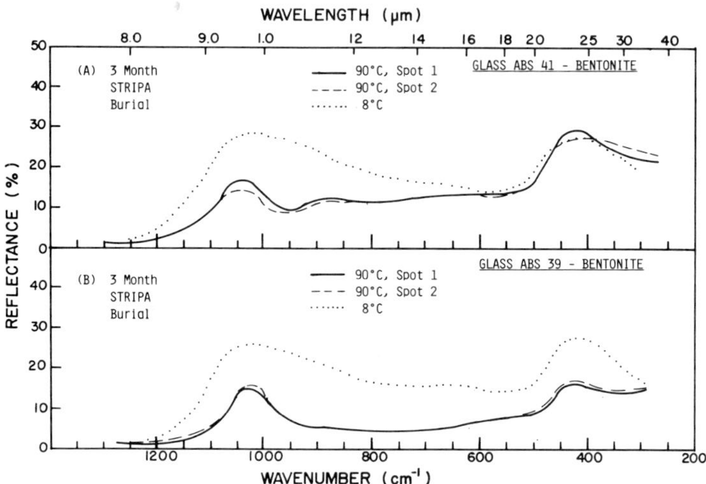

Fig. 5. IRRS spectra of glass-bentonite interfaces after 3 mo. STRIPA burial at 90°C and 8°C.

Optical Microscopy

The optical micrographs from this series of experiments shown in Fig. 6 represent the several hundred taken. After 3 mo.burial in STRIPA at 90°C the glass-glass interfaces show few features at 120X magnification (6A,B). Some evidence of glass surface attack exists for the glass-bentonite interfaces (6C,D), The extent of surface reaction for the 3 mo.90°C burial is considerably less than observed for just 28 days, 90°C laboratory experiments with water (6E) and very much less than exists when bentonite is present along with water in the lab tests (6F). The 3 mo., 8°C burial experiments have featureless glass-glass interfaces and even the glass-bentonite interfaces show almost no surface features after the 3 mo.,8°C burial.

SIMS Analysis

Selected glass samples from the burial experiments have been subjected to SIMS analysis. The instrument and its application to glass studies are briefly discussed in another publication[2].

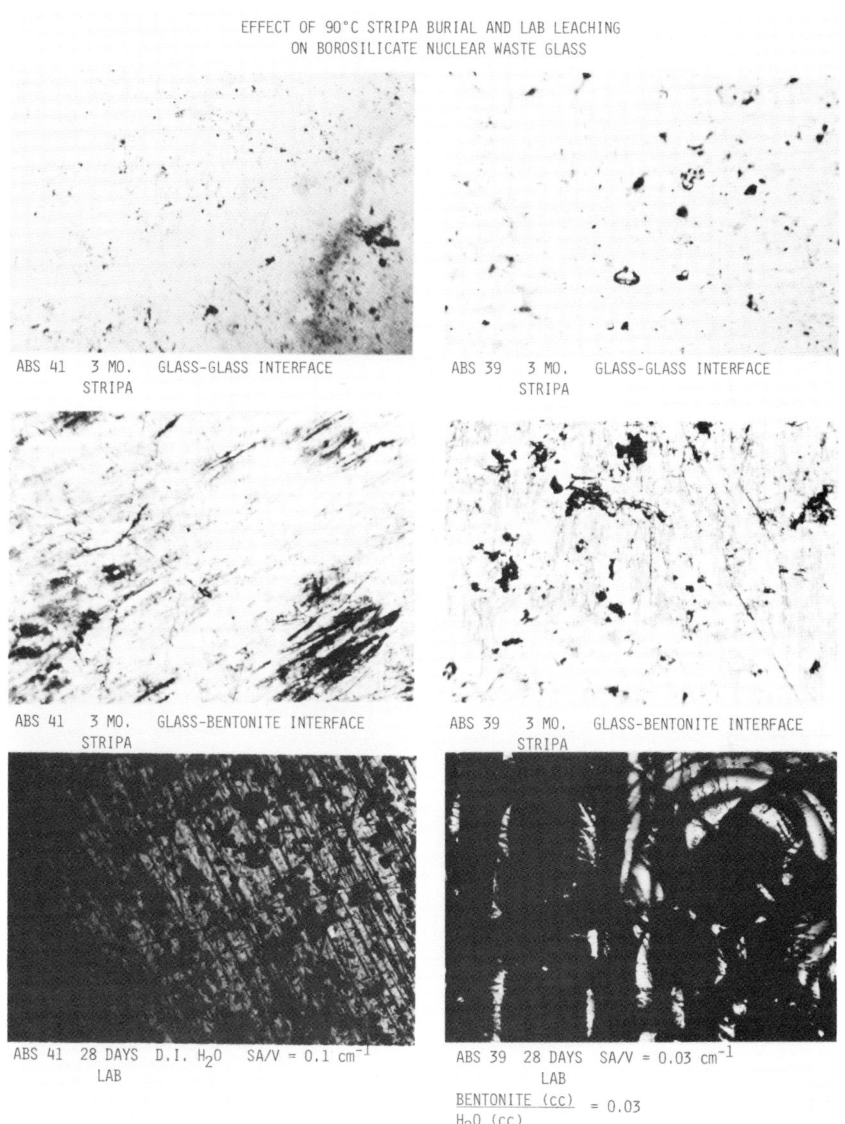

Fig. 6. Comparison of optical micrographs (120X) after 3 mo., 90°C STRIPA burial and 28 days, 90°C lab leaching tests.

SIMS in-depth profiles of ABS 39 glass in contact with bentonite are shown
in Figs. 7,8. As can be seen, in contact with bentonite the glass very rapidly
develops deep corrosion zones. These zones consisted of two layers; one thin
surface layer rich in Fe, Al and Si and with a Na concentration not very
different from bulk concentration is followed by a plateau region, extending
down to more than 10 µm depth in some cases. This plateau is generally also
enriched in Si and Al, although the concentrations are different. This plateau
also shows an enhanched Ca concentration, compared both with surface and bulk.
The leached species show a two step depletion, with a concentration drop down
to plateau concentrations followed by a nearly constant level and a further
drop in the surface layer (see e.g. B, La, Sr etc.).Surprising effect is a
superficial enrichment of the alkali Li, Cs, possibly due to a local formation
of compound corrosion products. The glass ABS 41 shows analogous tendencies.
Typical concentrations in the corrosion layers are given in Table 2.

Table 2

Stripa 3 months burial

Concentrations (atom-% of cations) in different parts of leached profiles
of ABS 39. Symbols: gl:glass against glass;bt: against bentonite.

	Surface		Plateau		Bulk
	gl	bt	gl	bt	
Na	5.3	20.3	5.3	7.9	19
Li	0.5	1.5	0.05	0.07	0.1
Cs	0.4	0.3	0.4	0.05	0.3
Ca	0.3	0.3	1.2	1.7	0.01
Sr	0.06	0.01	0.1	0.06	0.1
La	0.09	0.005	0.2	0.3	0.1
U	0.1	0.006	0.02	0.1	0.1
Fe	3.9	12.5	2.5	8.6	3.6
B	1.7	0.2	1.9	0.8	31
Al	6.5	21.2	6.7	10.8	3.2
Si	78	65	78	67	40

CONCLUSIONS

*   There are only minor differences in the ion depletion levels at surfaces
     of glass ABS 39 and 41 during 28 days and 3 mo., $90^{\circ}C$ or $8^{\circ}C$ burial in
     Stripa at 350m.
*   In-depth compositional profiles show growth rates of layers with selective
     depletion of Na,B, at up to ca $1.5 \cdot 10^{-7}$ m/day for glass 39 and up to
     ca $2.5 \cdot 10^{-8}$ m/day for glass abs 41.

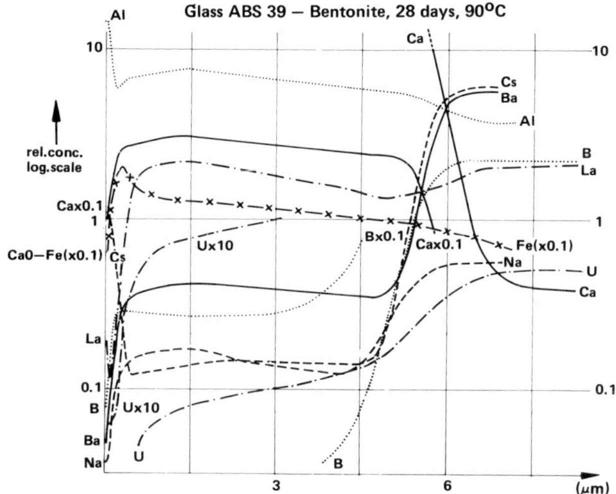

Fig. 7. SIMS in-depth profile of ABS 39 glass after contact with bentonite for 28 days at 90°C.

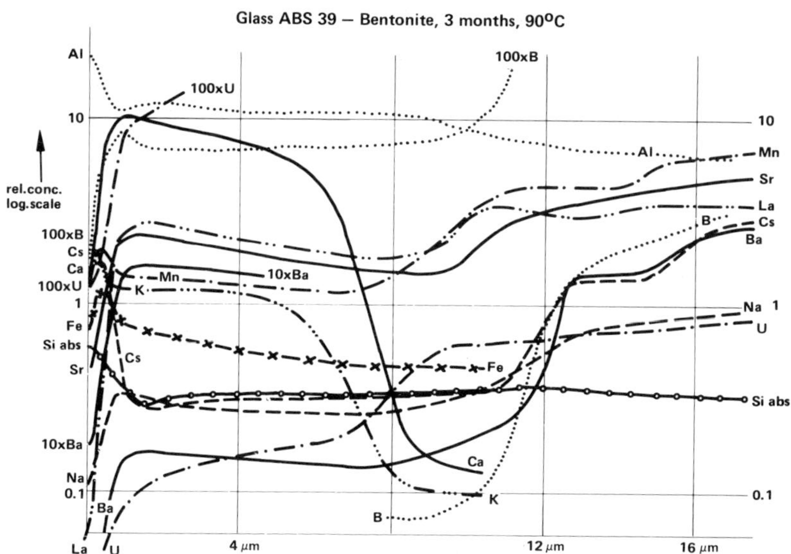

Fig. 8 . SIMS in-depth profile of ABS 39 glass after contact with bentonite for 3 months at 90°C.

162

*   A thin film rich in <u>Si</u>, <u>Fe</u> and <u>Al</u> forms on both glasses during burial.

*   Moist or wet bentonite accelerates the rate of ion depletion for both glasses, with glass ABS 39 the worst, in both laboratory and STRIPA burial experiments.

*   The bentonite effect is much less in burial environments than in lab simulations.

*   There is little difference in burial surface reactions between the "pineapple slice" configuration and "minicans".

ACKNOWLEDGEMENTS

The authors acknowledge the laboratory assistance of M.R. Wilson, S.A. Wilson, and S.D. Hench and optical microscopy of J.Wilson-Hench and Dr.T. Lakatos for the glass samples. We acknowledge financial support of SKBF/KBS and one of us (LLH) also acknowledges financial assistance of the U.S. Department of Energy during part of the course of these studies.

REFERENCES

1.   Witherspoon, P.A. and Degerman O., SKBF and Lawrence Berkeley Laboratory joint project, Report No LBL-7049, SAC-01.

2.   Werme, Lars, Lodding, Alexander and Hench, L.L. (1982) "Effect of Overpack Materials on Glass Leaching in Geologic Burial", to be published in the Proceedings of the Symposium on the Scientific Basis for Radioactive Waste Management, June 7-10 in Berlin.

3.   MCC-1 Static Leach Test (1982) Materials Characterization Center, Battelle Pacific Northwest Laboratory, Richland, Washington.

4.   Hench, L.L., "Effect of Bentonite on Nuclear Waste Glass Leaching" (to be published).

5.   Sanders, D.M., Person, W.B. and Hench, L.L. (1972) Appl. Spectroscopy, 26, 530-536.

6.   Sanders, D.M. and Hench, L.L. (1973) Am.Ceram. Soc. Bull.,52,666-669.

7.   Lodding, A., Reviews on Analyt. Chem. (1982), in press.

8.   McIntyre, N.S. and Strathdee, G.G., (1980) Surf. Science <u>100</u>, 71.

9.   Clark, D.E., Pantano, C.G. and Hench, L.L. (1972) in Glass Corrosion, Books for Industry, New York, New York.

Published 1982 by Elsevier Science Publishing Co
SCIENTIFIC BASIS FOR RADIOACTIVE WASTE MANAGEMENT - V
Werner.Lutze, editor

THE CHEMICAL DURABILITY OF SOME HLW GLASSES: EFFECTS OF HYDROTHERMAL CONDITIONS AND IONISING RADIATION

D.R. COUSENS, R.A. LEWIS, S. MYHRA, R.L. SEGALL, R.St.C. SMART AND
P.S. TURNER
School of Science, Griffith University, Nathan, Queensland 4111. Australia.

ABSTRACT

The time dependence of leaching and dissolution under hydrothermal
conditions at 250°C has been investigated in an autoclave designed for
continuous sampling; it is found that the rates are in accord with those
obtained by the commonly used procedures. The effect of pressure by *itself*
has been studied in an ultracentrifuge; ambient rates for Si and Na loss
remain unchanged up to about 40 MPa followed by a monotonic increase to
about 1.5 times the ambient rates at 160 MPa. The surface layers formed as
a result of chemical attack have been examined (XRD, IR, XPS and PIXE). It
is found that Zn present in the glass or in the solution is retained in the
siliceous layer which in some circumstances contains a crystalline zinc-
hydroxy-silicate phase. A method for enhancing rates of displacement damage
by a factor of ~10^6 has been developed. Powdered borosilicate glass and
glass doped with UO_2 have been neutron-irradiated in a high-flux reactor.
Damage is thereby introduced in the bulk as a result of $^{10}B(n,\alpha)^7Li$ and
(n,fission) events and subsequent slowing-down of energetic charged ions.
Irradiation doses equivalent to 10^6 years HLW storage have been achieved.
The resultant stored energy of ~10^2 kJ kg^{-1} is released over the range 300
to 800 K. The initial leaching/dissolution rates are found to be increased
by a factor of about three by irradiation; the changes in longer-term rates
are difficult to determine due to the effects of back-reactions and changing
particle size geometry and size distribution, but significant enhancement is
found. Rates for fission and activation products are similar to those for
bulk species.

INTRODUCTION

It is likely that high-level nuclear wastes from the commercial operation
of power reactors will be immobilized by incorporation in a glassy matrix.
The present pre-eminence of glasses over other waste forms is largely due
to the considerable research and development effort during the last twenty
years which has sought to ensure that no major unsuspected short-comings

will be found. The work discussed below has been undertaken in order to test
the validity of some specific techniques which are in widespread use for
studying the effects of hydrothermal conditions on glasses. Also, experiments
have been designed and carried out to evaluate the role of surface layers
in the dissolution of HLW glasses. Durability tests have been performed on
glasses subjected to extreme pressure and high doses of ionising radiation.
The object of these separate experiments is to establish that the chemical
durability of glasses under extreme conditions can be evaluated and that the
results exhibit no unusual features. The experiments have been optimized
in each case so as to potentially yield an effect which can be interpreted
in terms of its dependence on a minimal set of variables. The work therefore may not necessarily be directly related to an actual disposal situation.

HYDROTHERMAL DISSOLUTION

   Hydrothermal dissolution tests are now considered as essential for a
thorough evaluation of a particular waste form. The MCC and ISO tests[1,2],
for example, describe sets of preferred procedures. However, time dependent
measurements on a single sample involve certain difficulties. Periodic
withdrawal of leachant for analysis can clearly not be made without interrupting the hydrothermal process. New variables may therefore be introduced
as a result of thermal cycling, the manner in which leachant is replenished,
the possible effects on the specimen of handling and perhaps drying. A
number of procedural permutations are possible and it is not clear which of
these is most representative of repository conditions. We have employed a
continuous sampling autoclave with facilities for bleeding off 5 ml aliquots
from a 700 ml reservoir while leaving all other conditions constant. A
preliminary run has compared time dependent rates for a Harwell 209 glass
in deionised water at 250°C. The leachant (15 ml) from a 23 ml Bernas bomb
was exchanged in toto at specific intervals while the disc specimen was
dried and retained. Preliminary results (in scaled ppm) for Si and Na
from such an interrupted run are compared in Table 1 to corresponding data
using the continuous sampling autoclave. The agreement is surprisingly
good. However, more work is required before definitive conclusions can be
drawn.

TABLE 1

Integrated time (hrs)		2	4	8	12
Bernas bomb	Na(ppm)	0.3	1.1	1.3	1.6
	Si(ppm)	1.2	2.8	4.4	6.0
Autoclave	Na(ppm)	0.4	0.6	1.5	2.2
	Si(ppm)	1.4	3.0	6.2	6.7

ZINC-SILICATE SURFACE LAYERS

The various attack mechanisms which affect HLW glasses in aqueous solutions have been identified and discussed by Hench et al.[3,4] Broadly speaking, one may distinguish network breakdown, alkali leaching and the formation of modified surface layers. While the former two mechanisms have been studied extensively, less is known about the structure, composition, evolution and role of surface layers. The present work has investigated the effects of Zn in the glass or in solution. Two glasses were considered, the 72-68 formulation with 10% simulated waste and a "Marcoule" formulation. The compositions are given below in Table 2. The essential difference between the two glasses is that they contain 21% and 0% by weight Zn, respectively. A sequence of four experiments was performed under hydro-thermal conditions in a Parr bomb containing 15 ml solution. The disc specimens were attacked at 200°C for 18 hrs. The essential features of each type of run are listed below in Table 3 together with the corresponding characteristics of the resultant surface layer.

TABLE 2

	72-68 (10%)	"Marcoule"
B	4.22	4.92
O	39.24	47.40
Na	2.64	7.28
Mg	1.09	
Al		3.14
Si	15.57	27.31
K	4.11	
Ca	1.29	
Fe	0.77	0.71
Ni	0.23	0.21
Zn	20.90	
Sr	2.06	0.50
Others	7.89	6.53

TABLE 3

Run		Leachant	Surface Modification
1	72-68	Dist. Water	X-talline $ZnSiO_4$ in amorphous siliceous layer
2	72-68	0.1M EDTA	Amorphous siliceous layer
3	Marcoule	Dist. Water	No surface modification detected by XRD
4	Marcoule	0.1M $Zn(NO_3)_2$	X-talline Zn-phase in amorphous siliceous layer

The role of the EDTA (Run 2) is to act as a Zn-scavenger in solution so that this species cannot be readily redeposited. Run 4 was included in order to investigate the effects of Zn in solution. The $ZnSiO_4$ (Run 1) phase was identified by XRD. TEM investigations of scrapings from the surface revealed micron-sized crystallites in a largely amorphous matrix. The Zn-phase observed in Run 4 is as yet unidentified. The compositional changes following attack (Run 1) were investigated with Proton Induced X-ray Emission (PIXE) and X-ray Photoelectron Spectroscopy (XPS) coupled with Ar-ion beam milling. Both techniques show that the surface layer is enriched in Zn. Moreover, XPS has shown that Zn surface modifications are confined to the first 20 nm. The IR absorption spectra confirm that silicate structures are present and, furthermore, that molecular water is bound in the siliceous layer. These results taken together show that a species such as Zn in the glass or in solution plays an important role in the formation and characteristics of surface layers.

PRESSURE DEPENDENCE

Experiments which seek to determine the temperature dependence of glass dissolution rates above 100°C, performed in closed reaction vessels, contain pressure as an inseparable variable. Even though pressure *by itself* may conceivably affect dissolution rates it has been subject to little explicit experimental investigation. The present experiment examines this variable. An ultra-centrifuge has been used to achieve pressures of up to 160 MPa in a cellulose nitrate tube at a temperature of 20°C during runs of 45 hrs duration. The Harwell 189 precursor glass was selected as a specimen material with suitable dissolution rates at ambient temperatures so that the concentration of species in solution could readily be measured with standard flame atomic absorption techniques. Beads of 2.5 mm diameter were centrifuged at 28 to 40 Krpm (80 to 160 MPa) in 12.5 ml of deionised distilled water in parallel with blank water specimens; control experiments were also performed under identical conditions at ambient pressure. Loss of Na and Si was monitored. The results for Na are shown in Fig. 1 (Si exhibits a similar trend). It is seen that pressure has insignificant effect up to 80 MPa and enhances dissolution and leaching by less than 50% even at 160 MPa. Hence pressure by itself is an insignificant variable even under extreme hydrothermal conditions.

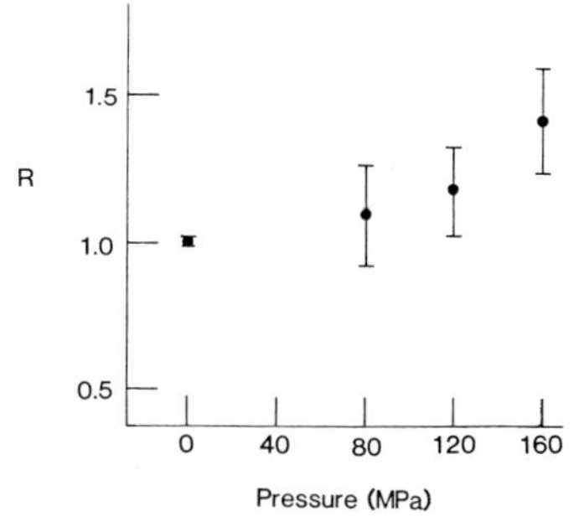

Fig. 1. The ratio, R, of the leach rate of Na at pressure to ambient leach rate.

RADIATION EFFECTS

The potential effects of long-term self-irradiation on fully active waste forms are many and complicated. The complexity of the problem precludes meaningful mathematical modelling. Furthermore, it is difficult to design realistic experiments that will enable simulation of the effects on a laboratory time-scale. The present state of affairs for HLW glasses has been discussed recently.[5,6] The work reported below has been undertaken in order to develop a new simulation technique and to investigate the effects of radiation damage on the chemical durability. The technique essentially amounts to doping a glass composition with enriched or natural abundance UO_2. The resultant glass is then irradiated in a high flux reactor so as to produce energetic fission fragments which produce primary knock-on defects as well as displacement cascades. Additional damage will be produced in borosilicate or lithiasilicate glasses as a result of (n,α) reactions with ^{10}B and ^7Li. The enhancement of the dose rate is dependent on the ^{235}U isotopic enrichment, the level of UO_2 doping, the concentration of ^{10}B and ^7Li and the thermal neutron flux level. It is well established that the main long-term contribution to displacement damage in actual waste forms is the result of actinide α-decay. The energetic α-particle (~5MeV) and the associated recoil nucleus (~100 KeV) will create ~150 and ~1000 displacements along track lengths of ~20μm and ~30nm, respectively. These characteristics are comparable to those for fission events (~4 x 10^5 displacements/event and ~10μm track length). The (n,α) mechanism of, for instance, $^{10}B(n,\alpha)^7Li$ produces ~100 displacements per energetic species per event along track lengths of 3-5μm. Both methods have the added advantage of producing uniform and isotropic bulk damage. Also, (n,f) reactions result in uniform deposition of radioactive waste species in the

168

bulk while a by-product of neutron irradiation is the production of activated nuclei. Subsequent leaching studies can therefore encompass the monitoring of HLW species in solution by highly sensitive radiometric methods. A technique similar to the one described above has been developed at Ispra by Lanza.[7] However, most studies of radiation effects have relied on the well-established but "slow" method of doping with a short half-life α-emitter.[6]

The chosen glass composition was based on the Harwell 189 precursor but with 9.5% by weight UO_2 and 4.8% SrO so that the leaching rates could be monitored for an alkali (Li), a network (Si) and an intermediate (Sr) species. Powder specimens with mean particle size of ~1μm were used in order to increase the specific surface area so that it could be determined accurately with the BET technique. Irradiated and control specimens were studied by leaching in a 0.02M glycine buffered solution at 60°C for up to ~45 hours. Species in solution were determined by FAA analysis and validated by ICP. The irradiations were carried out over a 24 day period in the AAEC HIFAR reactor at an average thermal neutron flux of 5.8×10^{17} $m^{-2}sec^{-1}$. The total number of (n,α) and (n,f) events was 2.8×10^{22} and 9.2×10^{25} m^{-3}, the estimated number of displacements was 1.6×10^{29} and the displacements per atom were ~2.3. This corresponds to an equivalent canister age of ~10^6 years (assuming typical LWR reprocessed waste loadings). Several irradiated thin rod specimens were annealed in a differential scanning calorimeter; the total stored energy was ~100 kJ kg^{-1} and it was released at a near uniform rate between 320 and 800K. The agreement with other workers is good.[8,9] Typical results for one of the dissolution experiments are shown in Fig. 2 and 3 while some of the features of these and other results are summarized in Tables 4 and 5 below. The results for (n,α) damage in 189 precursor glass powder will be presented in full elsewhere.

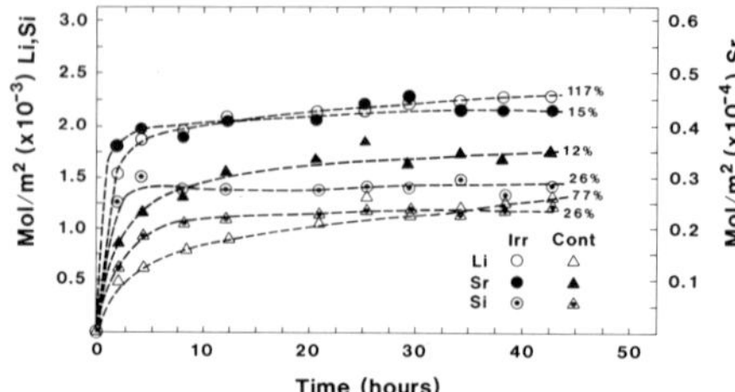

Fig. 2. Concentration in solution normalized to initial specimen surface area of three base glass species. Note that the Sr results refer to the ordinate on the right. Irradiated and control specimens are compared.

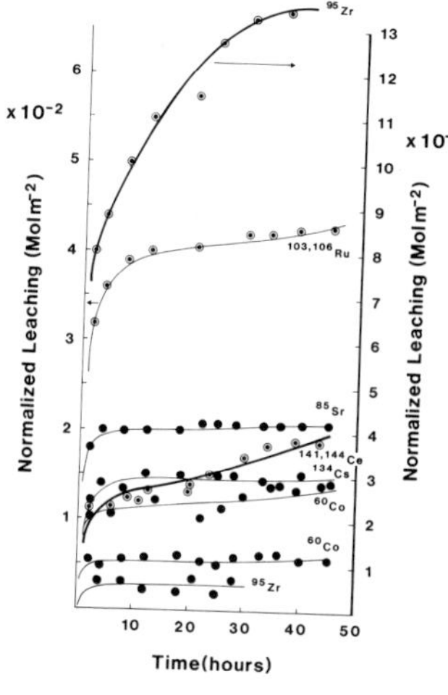

Fig. 3. Concentration of fission fragments and activation products in solution. Heavily drawn curves refer to the right-hand ordinate. The species in solution is normalized to initial concentration of species in the glass.

TABLE 4

Parameter	Species	Irr.	Cont.
t_k (hrs)	Li	5.6	10.9
	Sr	5.5	11.7
	Si	–	7.1
R $(t \to t_{max})$	Li x 10^{-4}	1.1 ± 0.3	2.1 ± 0.3
	Sr x 10^{-5}	0.8 ± 1.2	0.7 ± 1.2
mol m^{-2} h^{-1}	Si x 10^{-5}	0.14± 2.7	2.7 ± 2.7
D $(t = t_k)$ %	Li	99	54
	Sr	13	11
	Si	26	22

The time t_k corresponds to the position of the knee; this quantity is used purely as an event marker. R is the standard normalized measure of the rate at which species arrive into solution. D is the extent of dissolution.

TABLE 5

Isotope	Product	$D(t = t_{max})$%	$R(10^{-5}$ mol m^{-2}h^{-1})
^{60}Co	AP	38	0.34 ± 0.05
^{85}Sr	AP	47	5.1 ± 0.9
^{95}Zr	AP	4.1	-0.11 ± 1.1
^{95}Zr	FF	6	2.2 ± 0.1
^{103}Ru	FF	144	7.1 ± 1.6
^{106}Ru	FF	135	9.6 ± 2.6
^{134}Cs	AP	29	1.6 ± 3.4
^{141}Ce	FF	1.5	1.6 ± 0.9
^{144}Ce	FF	1.3	0.58 ± 0.1

Notation AP = activation product and FF = fission fragment. The extent of
dissolution for Ru is listed as exceeding 100%; this anomaly is a result
of uncertainties associated with calculating the initial mol fractions of
these isotopes present in the glass. ^{95}Zr occurs both as an AP and FF.

The most important conclusions about the effects of irradiation which
can be drawn from this experiment are: i) the position of the knee is
brought forward by a factor of ~3. The initial dissolution rate is therefore enhanced by that factor. ii) Rates above the knee are affected by
depletion of Li and the rapid approach to the solubility limit for Si in
solution. Data in Table 4 suggest little enhancement in the rates. However, three other runs exhibit average rates which are enhanced by ~50%.
Further analysis, considering the effect of back-reactions, suggests that
increases by a factor of three are typical. iii) The extent of dissolution
ranges from ~1% for Ce to total depletion for Ru at $t = t_{max}$. This suggests
that the formation of a retentive surface layer retards some species. The
total depletion of Ru remains a mystery. iv) There is a slight trend to
more rapid depletion when (n,f) rather than (n,α) reactions dominate
damage production.

ACKNOWLEDGEMENTS

This work was supported by the Australian Institute of Nuclear Science
and Engineering and the Australian Atomic Energy Commission.

REFERENCES

1. Materials Characterization Center MCC-1 Static Leach Test: Draft, PNL Report (1981).
2. Long-Term Leach Testing of Radioactive Waste Solidification Products, Draft Intl. Standard, ISO/DlS 6961.
3. Hench, L.L. and Clark, D.E. (1978) J. Noncryst. Sol. 28, 83.
4. Hench, L.L., Clark, D.E. and Lue Yen-Bower, E. (1980) Nucl. Chem. Waste Management, 1, 59.
5. "Characteristics of Solidified High-Level Waste Products" (1979) Tech. Rept. Series No. 187, IAEA.
6. Burns, W.C., Hughes, A.E., Marples, J.A.C., Nelson, R.S. and Stoneham, A.M. (1982), Nature, 295, 130.
7. Lanza, F. Proc. HFR Users Meeting Oct. 1977, p. 287.
8. Roberts, F.P., Jenks, G.H. and Bopp, C.D. (1976), "Radiation Effects in Solidified High-Level Wastes: I-Stored Energy", PNL Rep. BNWL-1944.
9. Scheffler, K., Riege, V. and Hild, W. (1976), "Stored in Borosilicate Glass", Rep. KFK-2333.

Published 1982 by Elsevier Science Publishing Co
SCIENTIFIC BASIS FOR RADIOACTIVE WASTE MANAGEMENT - V
Werner.Lutze, editor

DETERMINATION OF THE CORROSION MECHANISMS OF HIGH-LEVEL WASTE CONTAINING GLASS

HORST SCHOLZE, REINHARD CONRADT, HEINRICH ENGELKE AND HANS ROGGENDORF
Fraunhofer-Institut für Silicatforschung, Würzburg, Germany

INTRODUCTION

The German concept of high level waste final storage provides the use of
certain glasses containing radioelement oxides as glass components. These
waste forms are to be stored in rock salt formations in order to isolate the
waste from the biosphere. The efficiency of this isolation is a most important
question. The aim is to achieve a high safety standard that remains valid
under extreme conditions such as the uncontrolled water entrance to the
deposit.

Investigations were carried out with two model glasses doped with inactive
waste. Glass 1 relates to a waste glass developed from the glass ceramics
C31 3EC[1], glass 2 is a borosilicate glass. The theoretical compositions of the
two glasses are shown in table 1.

TABLE 1

COMPOSITIONS OF MODEL WASTE GLASSES IN WT.%

	glass 1	glass 2	
	total	frit	simulated waste
SiO_2	38	57	–
B_2O_3	10	12.5	–
Al_2O_3	11	1	1
TiO_2	3.5	4.5	–
Li_2O	1.5	3.5	–
Na_2O	6	4.5	3.5
MgO	–	2	–
CaO	7	4	–
BaO	15	–	–
rest	8	–	6.5
sum	100	89	11

The simulated waste in glass 2 contains 15 other components. The rest of glass 1 consists of 3.5 Fe_2O_3, 2 MoO_3, 1 ZrO_2 and 1.5 Nd_2O_3 (in wt.%). Glass 2, available as a finished product, was analysed by microprobe. Only 53 % SiO_2 were found, the remaining difference is explained by an excess of Al_2O_3 and K_2O.

The glasses were corroded in specific salt brines (see table 2) under static conditions under a pressure of 130 bar at 80, 120, 160 and 200°C, respectively. The Q"-solution was preliminarily used. The Q-solution according to D'Ans[2] is one of the reference brines proposed by the Physikalisch-Technische Bundesanstalt, Braunschweig, Germany, for the performance of corrosion tests.

TABLE 2

COMPOSITIONS OF THE CORRODING AGENTS IN WT.%

	Q"-solution	Q-solution
NaCl	1.9	1.4
KCl	3.3	4.7
$MgCl_2$	24.7	26.8
$MgSO_4$	2.4	1.4
H_2O	67.9	65.6
sum	100.2	99.9

The Q-solution was saturated at the test temperatures with additional amounts of NaCl.

Investigations under comparable conditions concerning the glasses and the experimental parameters were reported by Altenhein et al.[3] and by Malow[4]. As one of their most striking results they observed that during the corrosion process reaction layers were formed on the glass surfaces with a sharp boundary between the layers and the underlying glass matrices. Concentration profiles of single elements in the layers were presented. The intent of the present study was to investigate the effect of corrosion on the remaining glass matrix, i.e. the rest glass under the reaction layer.

EXPERIMENTAL PROCEDURE

First some properties of the glasses were determined:
for glass 1
- the transformation temperature T_g/°C ≈ 525
- the density at 20°C $\rho/g\cdot cm^{-3}$ $= 2.90\pm0.01$

for glass 2
- the transformation temperatur T_g/°C $= 470\pm3$
- the density at 20°C $\rho/g\cdot cm^{-3}$ $= 2.611\pm0.001.$

These data are necessary for appropriate tempering of the glass and for further evaluation of weight loss data, respectively.

The glass 1 samples were cut into rectangular pieces of different thickness.
Thus the corrosion tests were carried out with surface area to solution volume
ratios SA/V between 0.035 cm^{-1} and 0.05 cm^{-1} depending on the sample thickness.
The samples were polished to optical quality, pretreated in H_2O (1 h, 60°C)
and corroded in the Q"-solution for 4, 9, 19, 27 and 30 d.

The glass 2 samples were cut by a low-speed saw to a size of 1.2x0.9x0.2
cm^3. This size corresponds to a surface area to solution volume ratio SA/V =
0.033 cm^{-1}. Before the corrosion tests the samples were tempered (1 h, 490°C),
etched in $HF+HNO_3$ to improve the sample surface and pretreated with H_2O (1 h,
60°C). The corrosion tests were carried out for 8, 14, 30 and 60 d in Q-solution.

For the corrosion tests special reaction containers were used, one of which
is shown in figure 1. It is made of teflon and is sealed by a teflon foil
($\approx$ 1 mm thick) that allows the inside pressure to adjust to the environment.
The foil and the top edge of the container are pressed together by two steel
rings, the force being applied by steel springs. So the corrosion media (90 ml)
were in direct contact with teflon only. Five of these containers were placed
in one steel autoclave, respectively, where the pressure was applied. The autoclaves were kept at the appropriate temperatures in heating chambers.

After the corrosion runs the samples were found to be covered with reaction
layers. The layers were distinctly separate from the underlying glass. They
had a very poor mechanical stability and were of a structure and colour quite
different from the glass matrix. In some cases they had already partly spalled
off the samples. Before the samples were dried to constant weight and the
weight loss due to the corrosion process was recorded the layers were removed
by distilled water and a rubber scraper. Some glass 1 samples were checked by
microscope. An uniform glass surface was observed indicating that the reaction

Fig. 1. Reaction container

layers were completely removed. The complete removal of the glass 2 layers was
evidenced by the reproducibility of ESCA measurements.

ESCA (electron spectroscopy for chemical analysis) measurements were carried
out in combination with Ar ion milling and HF etching techniques in order to
determine the corrosion profiles of single elements in the rest glass.

RESULTS

Weight loss data of the rest glass

In figure 2A-B weight loss data due to the corrosion process are presented
as a function of corrosion time and temperature, for glass 1 and 2, respectively. Anticipating that any concentration profiles in the rest glass are very
thin, a matrix dissolution depth can be calculated (see right ordinate of diagrams). Obviously the time dependence is nonlinear in the initial stage.

Overall activation energies were calculated from the Arrhenius equation.
Their values at 30 d are approx. 46 kJ/mol for glass 1 and approx. 60 kJ/mol
for glass 2. This temperature dependence of weight loss is very strong. The
data of glass 1 at 200°C and 80°C, respectively, differ by almost two orders

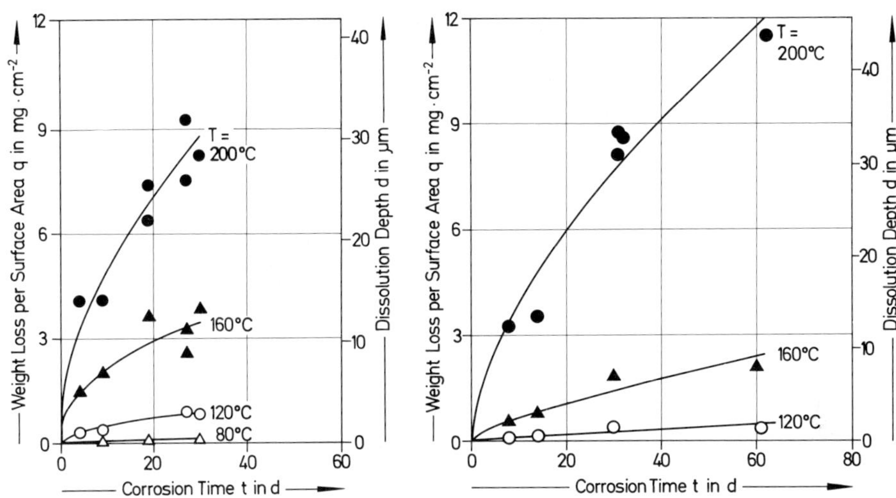

A   B

Fig. 2A-B. Weight loss per surface area q and matrix dissolution depth d of
glass 1 in Q"-solution (A) and glass 2 in Q-solution (B) at 130 bar as a
function of corrosion time t and temperature T.

of magnitude. The two glasses behave very similarly at 200°C. But glass 2 is
even more sensitive to temperature changes, which can also be concluded from
a comparison of the activation energies.

Corrosion profiles in the rest glass

In order to determine the corrosion profiles of the single elements in the
glass, ESCA measurements in combination with etching techniques were carried
out with single samples. Figures 3A to C present the concentration profiles of
Si, Na und Ba, respectively, after a 30 d corrosion test with glass 1 in terms
of relative intensities of the ESCA signals. The correlation between Ar ion
milling time and profile depth was taken from experiments with soda lime glass.
Therefore some systematic error might be involved which, however, does not
affect the following conclusions. As to the temperature dependence: for temperatures $\leq$ 160°C the profiles can hardly be distinguished from those in

178

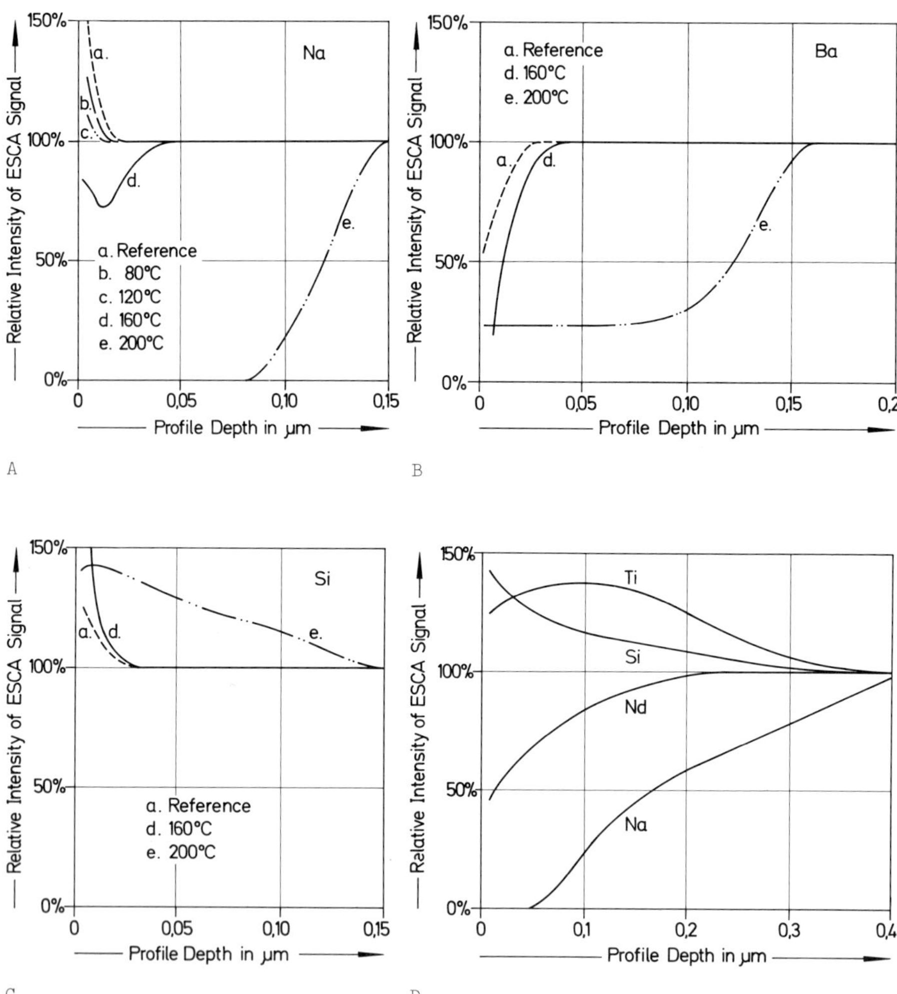

Fig. 3A-D. Concentration profiles of Na (A), Ba (B) and Si (C) after an approx. 30 d corrosion of glass 1 in Q"-solution at different temperatures and of Si, Ti, Na and Nd (D) after a 60 d corrosion of glass 2 in Q-solution at 200°C; at 130 bar.

uncorroded samples (marked as "reference"), which had passed through the same pretreatment. Only for 200°C notable profile changes occur. The temporal development of the leaching profiles (not presented) is not understood in detail. The experiments show that at 200°C after short periods (i.e. 4 d) the profiles reach almost half the depth of 30 d samples.

Further ESCA measurements in combination with a HF etching technique were carried out to examine the possibility of far-reaching profiles with very small slopes. At least beyond 0.7 μm no concentration change concerning Si, Na, Ca, Ba and Ti was found. Reproduction of the ESCA results concerning glass 1 is still required. Nevertheless there is a sufficient basis for the conclusions given below.

Figure 3D shows the relative intensities of the ESCA signals of several elements in a glass 2 sample corroded for 60 d as a function of the etching depth. Although the borosilicate glass (glass 2) was expected to show a stronger selective leaching and a less strong dissolution than the celsian type glass (glass 1), only very thin leaching profiles were found. Additionally a micro-probe analysis with the electron beam perpendicular to the sample surface and with varying acceleration voltages was carried out. The depth of analysis was ~ 6 μm. The elements under investigation were Si, Al, Ti, Na, K, Mg, Ca, Fe and Zr. In samples corroded for 8 d and 60 d at 200°C and 130 bar in Q-solution, no significant concentration changes were found.

Pressure dependence of a waste glass corrosion process

With glass 2 samples corrosion experiments were carried out during ≈ 30 d at 130 bar and at equilibrium pressures (i.e. ≈ 16 bar at 200°C and ≈ 6 bar at 160°C). The other conditions were the same as described above. The results may be summarized as follows: a pressure dependence cannot yet be proved although it seems that an increase of pressure from 16 to 130 bar slightly increases the weight loss. If there is a pressure dependence the effect is smaller than 20 %. This order of magnitude corresponds to results obtained by Ito et al.[5] from experiments with silica glass in water.

180

CONCLUSIONS

The observations and investigations of the corrosion behaviour of two rather
different model waste glasses in salt brines revealed the simultaneous action
of several corrosion mechanisms. The differences between the two glasses refer
to different extents of the same mechanisms. The glass matrix itself is
involved in matrix dissolution (disintegration) and corrosion profile formation
processes. The latter comprise silicon enrichment and alkali and alkaline earth
depletion. A comparison between the matrix dissolution depths and the profile
depths shows that the matrix is submitted to an almost congruent dissolution.
Nevertheless the overall time law has an exponent between 0.5 and 1. Taking
into account the results by Malow[4] one may assume that the reaction layers
which are formed on the glass surfaces during corrosion enforce the time law.
This assumption has to be modified by the fact that the profiles in the rest
glass pass through a temporal development, too. I.e. the conditions for the
reaction layer formation are continuously changed because of the surface concentration change of the underlying rest glass. The meaning of either processes
for the long term behaviour of waste glass still needs clarification.

ACKNOWLEDGEMENTS

This work was performed in the frame of contracts with the BMFT (Bundesmini-
sterium für Forschung und Technologie), Federal Republic of Germany, and with
the Commission of the European Communities under the contract numbers WAK
1628L2 and WAS-232-81-53 D (B), respectively.

REFERENCES

1. Marples,J.A.C. et al. (1981) Europ.Appl.Res.Rept.-Nucl.Sci.Technol. 3,
   p. 466

2. D'Ans,J. (1933). Die Löslichkeitsgleichgewichte der Systeme der Salze
   ozeanischer Salzablagerungen, Verlagsgesellschaft für Ackerbau mbH, Berlin

3. Altenheim,F.K., Lutze,W., Malow,G. (1981) in Sci.BasisNucl.WasteManage.
   Vol. 3, Moore,J.G.ed., Plenum Press, New York, pp. 363-370

4. Malow, G. (1982) Sci. Basis Nucl. Waste Manage., Vol. 11, Lutze, W. ed.,
   MRS Symposia Proceedings

5. Ito,S., Tomozowa,M. (1981) J.Amer.Ceram.Soc., 64, p.C-160

RESULTS FROM A ONE-YEAR LEACH TEST:
LONG-TERM USE OF MCC-1*

D. M. STRACHAN

Pacific Northwest Laboratory, P. O. Box 999, Richland, Washington, 99352, USA

INTRODUCTION

The Nuclear Waste Materials Characterization Center at Pacific Northwest
Laboratory is developing standard tests to obtain data on nuclear waste forms,
barriers, and backfills. These tests include performance measurements of thermal, radiation, mechanical, and chemical properties.[1] Five tests have been developed to determine the chemical durability of waste forms[1,2] under either
static (MCC-1P and MCC-2P) or flowing (MCC-4S and MCC-5S) leaching environments.
Maximum credible release by waste forms is determined using powders and stirred
solutions (MCC-3S).

This paper presents results from the long-term use of the MCC-1P Static Leach
Test Method which has already been used for short-term testing[3,4,5] and for
comparing waste forms.[3,6] Data were obtained on the leaching behavior of PNL
76-68 glass over a 1-year period. Complete solution analyses coupled with selected solid state analyses gave further insights into the mechanism of leaching
for this glass and show that the MCC-1P test method can be used for periods up
to at least 1 year.

EXPERIMENTAL METHODS

The MCC-1P Static Leach Test Method has been well documented elsewhere.[1]
Briefly, the test requires the use of monolithic specimens leached in three
leachants in Teflon® PFA leach vessels under static conditions. The three leachants are pure water; a silicic acid/sodium bicarbonate solution; and a K, Mg, and
Na chloride brine. Three temperatures may be used: 40°C, 70°C, and 90°C. Results at 40°C and 90°C are reported here. Triplicate specimens are required
periodically for error analysis. The test performed in the work for this paper
deviated slightly from the now-approved procedure but without any significant
impact on the results (for example, specimens were cut in an organic fluid and
washed in acetone, whereas the approved procedure calls for cutting in water).
The glass studied was the PNL 76-68 composition (Table 1).

---

* This work was supported by the U.S. Department of Energy under contract
DE-AC06-76RLO 1830.

TABLE 1.
COMPOSITION OF PNL 76-68 GLASS

Constituent	Mass %	Constituent	Mass %
Al_2O_3	0.46	MoO_3	1.89
B_2O_3	8.52	Na_2O	15.00
BaO	0.52	Nd_2O_3	4.12
CaO	2.15	NiO	0.25
CeO_2	1.03	P_2O_5	0.70
Cr_2O_3	0.49	RuO_2	0.78
Cs_2O	0.84	SiO_2	42.80
Fe_2O_3	10.80	SrO	0.36
Gd_2O_3	0.07	TeO_2	0.21
La_2O_3	0.49	TiO_2	3.05
MnO_2	0.04	ZnO	4.76
MgO	---	ZrO_2	1.66

Although the Teflon® vessels used in this test were thick walled (0.26 mm) and had tight fitting lids with an internal valve, some mass loss was noted after 182 d and after 365 d in the oven. Since the assembled leach vessels were not weighed at the start of the experiment, the mass loss was estimated from the concentration of the brine and silicate water blanks and from the known initial mass of leachant and specimen. These estimates yielded correction factors for the water and silicate water of 0.95 at 182 d and 0.80 at 365 d. For brine, the correction factors were 0.89 at 182 d and 0.80 at 365 d. The uncertainty in estimating these correction factors is included in error bars for the 182 d and 365 d data points shown in Fig. 1. No corrections were made to the 40°C data since the reduced temperature was not expected to yield significant water losses, and water blanks were not useful in determining concentration effects.

EQUIPMENT

All Teflon® PFA leach vessels were from Savillex (0102-53 MOD). Specimens were suspended from the lid by a 0.125 mm diameter Teflon® FEP monofilament. Blue M ovens (Model OV-490A-2) were used to maintain the test temperatures of 40°C and 90°C. All solutions were analyzed for 30+ elements with an inductively coupled argon plasma spectrometer (ICP; Jarrel-Ash, Model 975 ICP ATOM COMP); an atomic absorption spectrometer (Perkin-Elmer, Model AA-305B) was used for cesium analyses. An Orion 901 or Markson Model 88 pH meter was used to determine the pH of the solutions. The pH electrodes were Markson 8430B and Corning 476223. An Orion 701 pH meter with an Orion Combination Redox electrode (96-78-00) was used to measure the Eh of the 1-year leachates. A saturated solution of quin-

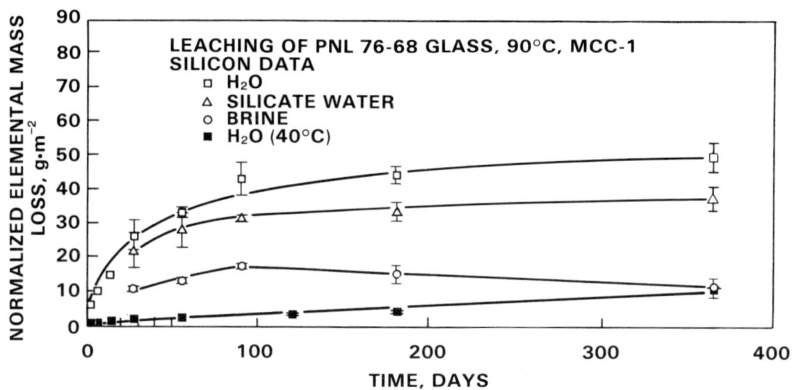

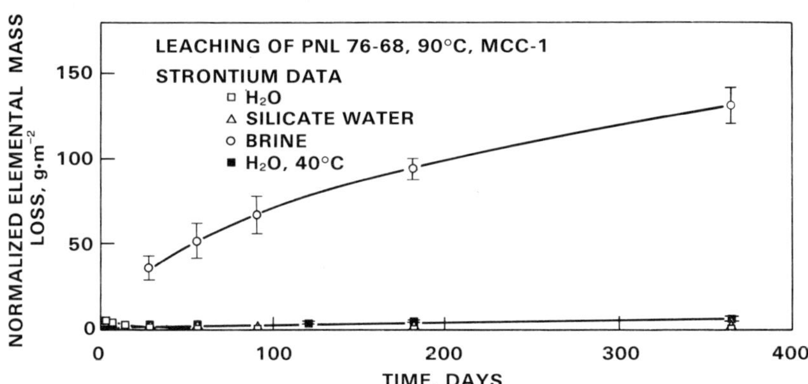

Fig. 1. Results for silicon, strontium, cesium, and pH from the leaching of PNL 76-68 glass by the MCC-1P static leach test method at 40°C and 90°C.

184

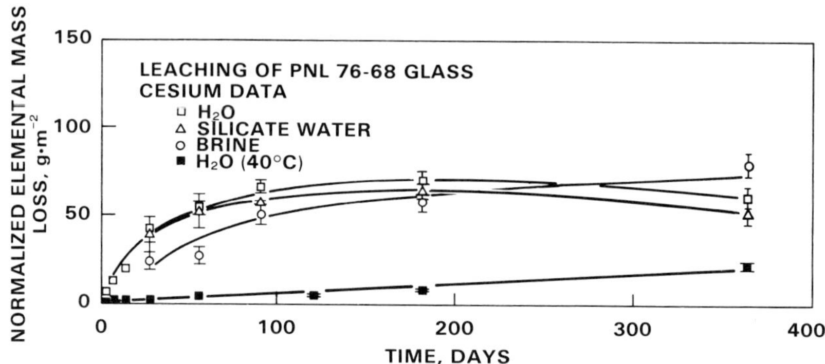

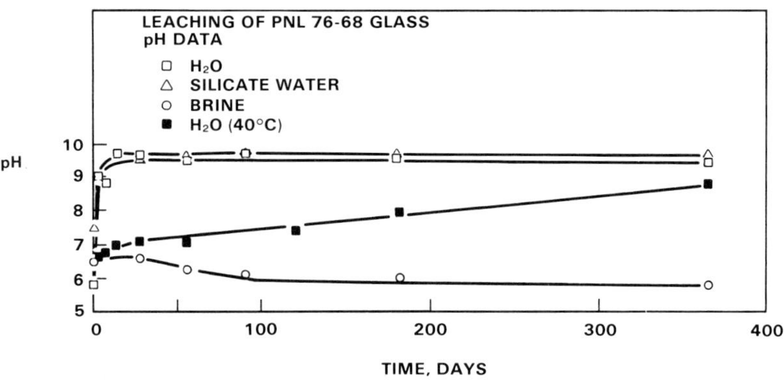

Fig. 1 (cont.). Results for silicon, strontium, cesium, and pH from the leaching of PNL 76-68 glass by the MCC-1P static leach test method at 40°C and 90°C.

hydrone was used to calibrate the redox electrode. Solid state analyses were
performed using a scanning electron microscope (JEOL, Model JSM-U3) with an
energy dispersive x-ray analyzer (Tracor Northern) (SEM/EDX). Depth profiles
were obtained using secondary ion mass spectroscopy (SIMS) and x-ray photoelectron spectroscopy (XPS); both were Physical Electronics, Model 550.

RESULTS AND DISCUSSION

Figure 1 summarizes the results from the leachate analyses. Silicon, cesium,
and strontium results are shown as representative elements for the three different leachants at 90°C, and for deionized water at 40°C. In general, smooth
curves were obtained for all of the data that were quantitative (defined as 10
times the detection limit).[7] The apparent standard deviation (1σ) from the triplicate data was observed to vary with time and with leachant type, though in no
systematic manner. The average standard deviation was less than ± 10%. Also
shown in Fig. 1 are the pH data from the leachates. The average standard deviation was ± 0.05 pH unit.

Deionized water and silicate water

At 90°C, the silicate and deionized water leachants yielded similar leach
results within 15 to 20% for most elements. In general, the normalized elemental
mass losses for the silicate water leachant were lower. The pH of these leachates rapidly rose from 5.8 to 9.6 ± 0.1 and remained constant for the duration
of the test (Fig. 1). At 1 year, the Eh of the leachates was found to be 0.20 V,
indicating an air-saturated condition. A small decrease in the silicon loss
in the silicate water relative to the deionized water leachant was due to the
higher initial SiO_4^{2-} concentration in the silicate water and the concomitant
blank correction. In both leachates the concentration of Si was 95 mg/ℓ
(204 mg SiO_2/ℓ) after 182 d and 123 mg/ℓ (264 mg SiO_2/ℓ) after 365 d. The
latter value agreed well with the range of reported equilibrium solubilities
for amorphous silica when other ions are present in solution[8,9] and for the
concentration found in a 1-year test with PNL 76-68 glass powder.[10]

Elements such as Sr, Cs, and Ca appeared to reach saturation relative to some
solid phase(s); for instance, the normalized Cs loss decreased from 182 d to
365 d (Fig. 1). Strontium and calcium losses decreased to near detection limits
within the first 28 d of the experiment and remained at these levels for the
duration of the experiment. The observed behavior for Ca and Sr indicated that
a solid with a solubility similar to $SrCO_3$ and $CaCO_3$ was controlling the concentrations of these elements.[11] From solution analyses it appeared that elements

such as B. Na, and Mo, which are usually soluble, also approached saturation in
the water leachant and saturated in the silicate water leachant. These results
were contrary to the solid state analyses (see below) and may have been a manifestation of the inability to assign accurate correction factors.

Solid state analyses of the leached specimens indicated a steady growth of
two layers. The outer layer grew by precipitation reactions on the original
surface of the glass (Fig. 2). This growth was detectable at 28 d but was only
examined in detail after the 182 d and 365 d time periods. The layer consisted
predominantly of zinc and silicon, thus indicating a zinc silicate phase(s).
Trace elements of Cs, Ca, and Ti also were found. Of note was that at 182 d the
Cs was at about the same concentration in the zinc silicate layer as in the bulk
glass (0.84 mass %), whereas at 365 d the Cs concentration in the zinc silicate
layer was 2.5 mass %. Hence, the zinc silicate phase(s) either accepts Cs structurally or causes some other phase that contains Cs to precipitate, which
accounts for the observed decrease in the solution concentration of Cs. Phase
identification was not possible due to broad peaks in the x-ray diffraction pattern, but two weak peaks at 1.12 and 1.15 nm suggest a clay-like phase is present
in one or both layers. The zinc silicate was 1 to 1.5 µm thick after 182 d of
leaching in deionized water and 2 to 2.5 µm thick after 365 d. After leaching
for 365 d in silicate water, this same layer was 1 to 1.5 µm thick and appeared
to be more dense. It is also possible that this material partially controlled
the silicon concentration in the leachates.

As reported in the past,[3] a gel layer or altered layer remained behind as the
aqueous solutions leached soluble and moderately soluble material from the glass
matrix. Thus, this layer was rich in Fe, Nd, La, Ti, etc. and depleted in B,
Cs, Na, and Mo. The growth of this layer with time is shown in Fig. 3 and was
determined by three methods. Two solid state methods, SEM and SIMS/XPS, were
used to obtain both depth and elemental profiles. The third method involved calculating the depth from the solution concentrations. Since the solid state
methods showed the gel layer to be depleted in B, Na, and Mo, the concentration
of these elements in solution was expected to be an accurate method of calculating the depth of the gel layer. In Fig. 3, the depths up to 182 d were calculated from the solution data, and all three methods were used for the 182 d and
365 d data.

Good agreement was found between the three methods for the specimens leached
in water and brine for 365 d but not for the specimens leached in silicate water.
However, the curve for the silicate water was drawn as if the solution data were
more accurate than the single SEM photomicrograph because the agreement between

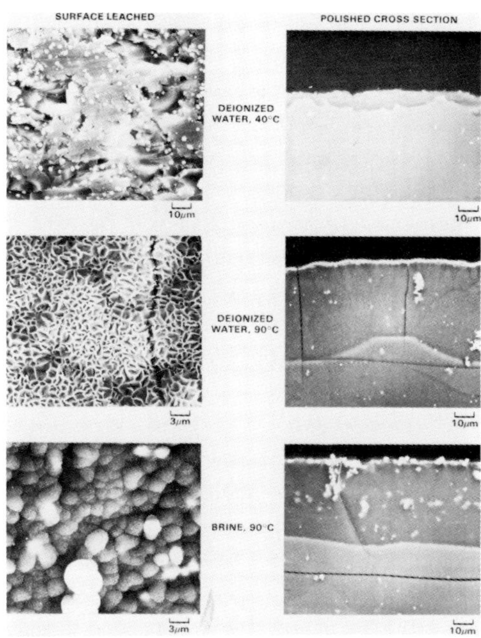

Fig. 2. Photomicrographs of leached specimens (364 d).

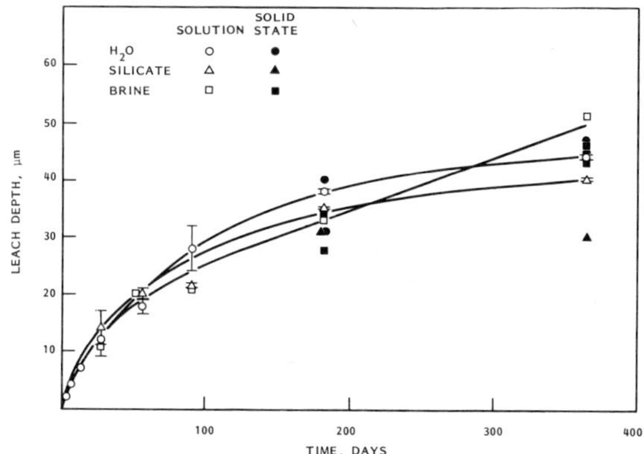

Fig. 3. Gel layer thicknesses for specimens leached in deionized water, silicate water and brine at 90°C for up to 1 year. The data are from solid state
analyses and calculated from solution analyses.

188

the three methods was quite good for the other two leachants. Also, as pointed
out above, the agreement between the two leachates in total Si concentration
would indicate parallel behavior. Some ambiguity exists in the data and confirmation data are being collected. Nevertheless, Fig. 3 indicates a dramatic decrease in the rate of leaching after approximately 91 d. Yet, even though many
elements became saturated with respect to alteration phases, PNL 76-68 glass
appears to continue to alter and/or leach. Also, it is evident from these data
that more than a 1-year period is necessary to confirm this conclusion.

The results in this section suggest at least one change in the MCC-1P procedure and reinforce a requirement in the MCC-1P test method. Specifically, the
data suggest better specimen preparation may be needed (see the cross section
photomicrograph in Fig. 2). A study to evaluate specimen preparation will begin
in the near future. Currently, the MCC-1P test method requires leach vessels
with tight fitting lids and accurate measurements of the leachant mass and/or
volume. Leach vessels with tighter fitting lids than those used in this study
are available, but water vapor through the Teflon® walls still occurs.

Deionized water at 40°C

As expected, at 40°C, losses for Cs and Si, as well as for most other elements, were lower than at 90°C. Strontium losses were lower during the early
periods than at 90°C, but by 365 d the normalized strontium loss at 40°C was 10
times greater than at 90°C (Fig. 1). This trend was consistent with the lower
pH (Fig. 1). At pH 8.8, $SrCO_3$ could not control the Sr concentration, an indication that the Sr concentration was controlled by release from the glass.[11]
Photomicrographs of a specimen that was leached for 364 d show that it had a
leach depth of 2 to 10 μm (Fig. 2). Small particles rich in Nd and Zr or P were
found on the surface. The gel layer was found to be enriched in Ti, Zr, Nd, Ce,
Zn, and Fe. Qualitative SEM/EDX examination of the leached specimens in cross
section revealed that the elemental profiles were similar to those found at 90°C,
except that the zinc-rich precipitation layer was not developed and the calcium
was not as enriched in the upper part of the gel layer. The difference in the
composition of the gel layer that developed at 40°C relative to the layers that
developed on the specimens leached at 90°C resulted from the lower temperature
and pH.[11] Also, the gradual increase in pH led to increases in Cs and Si in the
leachates. While increases in Cs and Si might be expected to continue until
above pH 9, other elements, such as Sr and Ca, should decrease. Comparing the
Si and Sr curves in Fig. 1 shows that the Sr is already being regulated by some

solid phase and not by release from the glass. If Sr was regulated by release from the glass, the normalized Sr loss would be nearly the same as that for Si. Longer times are needed to observe the decrease of Sr and Ca at 40°C.

Brine leachant at 90°C

Earlier published results from this test[3] suggested that the Cs losses were suppressed in the brine leachant as compared with deionized or silicate waters. However, the present results show that Cs continued to leach in brine, while Cs solution concentrations decreased in the other two leachants. The difference was due largely to the growth of a magnesium silicate layer on the original glass surface, whereas a zinc silicate layer developed on specimens leached in the other two leachants. The magnesium silicate layer and the gel layer contained no Cs. Thus, examination of the surface of the leached specimens explained the behavior of Cs in solution.

Strontium behavior can be explained by the pH of the brine solutions (Fig. 1). At 28 d the brine leachates had a pH of 6.6. This pH decreased uniformly with time to 5.78. At this pH, $SrCO_3$ could not control the Sr^{2+} solution concentration.[11] Depth profiles showed that Sr was depleted throughout the altered layers and, hence, must have gone into solution. The decrease in pH was probably caused by the formation of the aqueous complexes $Mg(OH)^+$, $Mg_4(OH)_4^{4+}$, and $MgHCO_3^+$, and/or small quantities of $MgCO_3$.

Microscopic analysis showed that the gel layer thickness was 34±2 μm at 182 d and 43±1 μm at 365 d (Figs. 2 and 3). These depths were in generally good agreement with SIMS, XPS, SEM, and solution data from B, Mo, and Sr (Fig. 3). The gel layer was depleted in B, Ca, Cs, Mo, Na, and Sr and enriched in Fe, Nd, and Ti. Silicon was only moderately depleted. Only Mg and Si were enriched in the magnesium silicate layer. This layer had a ratio of MgO to SiO_2 of nearly 1:1 at 182 d and 0.67:1.0 at 365 d. Surface profiles showed that the Mg remained constant and the Si was enriched, and that at 182 d the layer was 1 to 2 μm thick and 2 to 3 μm at 365 d. Both sepiolite $[Mg_2Si_3O_6(OH)_2]$ and talc $[Mg_3Si_4O_{10}(OH)_2]$ were identified, using x-ray diffraction, as possible phases in this surface layer. A rare-earth phosphate phase was observed as spherical shapes or as irregular particles at the interface between the magnesium silicate layer and the gel layer (Fig. 2). Identification of this phase by x-ray diffraction was not possible.

The Eh of the 1-year brine leachate was measured and found to be approximately 0.3 V. This measurement was likely subject to some error since it was uncertain how the probe performed in the brine solution. If the value was correct,

190

the redox and pH conditions indicated that Fe^{2+}, $FeCl^{+}$, and $FeCl_2^{\circ}$ were the dominant solution species,[12] although still below detectable quantities.

Since Cs was not found in any of the layers that formed on the brine-leached specimens, the behavior of Cs in brine solutions can be understood in terms of the solids that formed. Solid state analyses showed that Si was further enriched in the magnesium silicate layer between 182 d and 365 d, thus explaining the decrease in the Si concentration. Continued glass leaching, even though silicon had supersaturated with respect to the magnesium silicate phase, is clearly pointed out by the increasing normalized Sr losses.

CONCLUSIONS

The results of the 1-year leach test led to the following conclusions:

- The MCC-1P Static Leach Test Method has been demonstrated to be useful for studies up to 1 year at temperatures up to 90°C.

- Accurate measurements of leachant volume at the start and the end of the test are important.

- PNL 76-68 glass appears to continue to alter, albeit at a significantly reduced rate, even though the solution concentrations of many elements are saturated or supersaturated with respect to alteration phases.

- The original surface of the glass was still present after 1 year, evidence that the gel layer will slowly get thicker but dissolution will not occur, or will occur at extremely slow rates.

- At 40°C, more than 1 year is required to reach saturation of some important elements. Even at 90°C, testing beyond 1 year is needed to define alteration rates and details of the leaching mechanisms.

REFERENCES

1. Materials Characterization Center (1981) Nuclear Waste Materials Handbook DOE/TIC-11400, Pacific Northwest Laboratory, Richland, WA, USA
2. Materials Characterization Center (1981) Materials Characterization Center Test Methods - Preliminary Version, PNL-3990, Pacific Northwest Laboratory, Richland, WA, USA.
3. Strachan, D.M. and Barnes, B.O. (1981) The Leachability of Some Nuclear Waste Forms. PNL-SA-9118, Pacific Northwest Laboratory, Richland, WA, USA
4. Jantzen, C.M., Clarke, D.R., Morgan, P.E.D. and Harker, A.B. (1982) The Leaching of Tailored Polyphase Nuclear Waste Ceramics: Microstructural and Phase Characterization. J. Amer. Cer. Soc., Vol. 65 pp. 292-300
5. Campbell, J.H., Hoenig, C.L., Bazan, F., Ryerson, F.J. and Rozsa, R.B. (1981). Immobilization of Savannah River High-Level Wastes in SYNROC: Results from Performance Tests. UCRL-86753, Lawrence Livermore Laboratory, Livermore, CA, USA.

6. Stone, J.A. (1981) An Experimental Comparison of Alternative Solid Forms for Savannah River High-Level Wastes. DP-MS-81-102, Savannah River Laboratory, Aiken, SC, USA.
7. ACS (1980) Guidelines for Data Acquisition and Data Quality Evaluation in Environmental Chemistry. ACS Committee on Environmental Improvement. Analytical Chem. 52, 2242-2249.
8. Marshall, W.L. (1980) Amorphous Silica Solubilities-I. Behavior in Aqueous Sodium Nitrate Solutions; 25-300°C, 0-6 Molal. Geochemica et Cosmochemica Acta, 44, 907-913.
9. Iler, R.K. (1979) The Chemistry of Silica: Solubility, Polymerization, Colloid, and Surface Properties, and Biochemistry. John Wiley and Sons, New York, NY, USA.
10. Fullam, H.T. (1981) Solubility Effects in Waste-Glass/Demineralized Water Systems. PNL-3614. Pacific Northwest Laboratory, Richland, WA, USA.
11. Grambow, B.E.E. (1982) The Role of Metal Ion Solubility in the Leaching of Nuclear Waste Glasses. PNL-SA-10079. Pacific Northwest Laboratory, Richland, WA, USA (also this publication).
12. Kester, D.R., O'Conner, T.P., and Byrne, R.H. Jr. (1975) Solution Chemistry, Solubility, and Adsorption Equilibria of Iron, Cobalt, and Copper in Marine Systems. Thalassia Jugoslavia, 11(2), 121-134.

ACKNOWLEDGMENTS

This work would not have been possible without the capable assistance of talented people. Valerie Coburn performed a number of the experiments. Chemical analyses were performed by Frank Hara and Allen Lautensleger and the solid state analyses by Jim Coleman (SEM) and Dr. Larry Pederson (XPS and SIMS). Dr. Rheal Turcotte, John Mendel, Bernd Grambow and Susan Gano provided helpful discussions and assisted in editing this paper. Of course most papers would not be possible without the able and patient assistance of a good secretary - Sharon Sloot.

Published 1982 by Elsevier Science Publishing Co
SCIENTIFIC BASIS FOR RADIOACTIVE WASTE MANAGEMENT - V
Werner.Lutze, editor

UTILIZATION OF CHARGED PARTICLE BACKSCATTERING TO STUDY THE NEAR SURFACE REGION OF GLASSES. APPLICATION TO DEPTH PROFILING OF LANTHANIUM, CERIUM, THORIUM AND URANIUM INDUCED BY AQUEOUS LEACHING.*

Patrick Trocellier, Bernard Nens, Charles Engelmann
Département de Chimie Appliquée et d'Etudes Analytiques
Service d'Etudes Analytiques - Section d'Etudes et d'Analyse Isotopique
et Nucléaire.
CEN Saclay, 91191 Gif-Sur-Yvette Cédex, France

INTRODUCTION

The Rutherford backscattering technique is useful for the determination of the concentration profiles of some heavy elements in the near surface region of glasses, but is not able to provide chemical information on the elements detected.

Nevertheless, it permits a non destructive exploration of the surfaces of solids, and enables the actinides: Np, Pu, Am, ... to be detected.

Relatively simple glasses containing either lanthanum and uranium or cerium and thorium were made, leached and finally examined to specify on one hand the analytical possibilities of the process, on the other hand the experimental conditions required to prevent secondary effects during sample bombardment.

BASIS PRINCIPLE OF THE METHOD

It is essentially based on elastic collisions between projectiles (mass : m, energy : E_o) and target atoms (mass : M)

After interaction, the energy of the backscattered particles is fully determined by m, M, E_o and scattering angle θ according to the following relation:

$$E = K \cdot E_o$$

---

* Work performed under contract WAS 268-81-55 F with European Community Commission.

194

with the kinematic factor :

$$K = \left[\frac{m \cos \theta + (M^2 - m^2 \sin^2 \theta)^{1/2}}{m + M}\right]^2$$

For given incident ions and observation direction θ (Fig.1) the value of
energy E is characteristic of the target atom M.

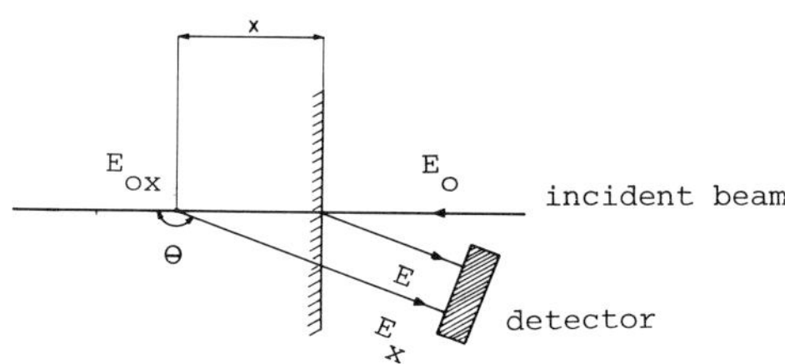

Fig.1. Basic principle of the analysis of the near surface region
of glass by charged particle elastic backscattering.

The capability of the process to separate   neighbouring elements increases
when the energy E_o is higher,the incident ions are heavier,the detection
angle θ is nearer 180° and the target atoms are lighter.
The analysis sensitivity depends on the differential scattering cross section:

$$\left(-\frac{d\sigma}{d\Omega}\right)_{E_o,\theta} = \left(-\frac{z}{2}\frac{Ze^2}{E_o}\right)^2 \cdot \left(\frac{1}{\sin^4\theta}\right)\cdot\left[1 + \frac{\cos\theta}{\left(1 - \left(-\frac{m}{M}\sin\theta\right)^2\right)^{1/2}}\right]$$

which represents the probability of elastic scattering in a direction θ for a
projectile (mass : m , energy: E_o, atomic number : z) colliding with a
target atom (mass : M , atomic number : Z).

The method is obviously more sensitive when the incident ions and the target
atoms are heavier,the energy E_o,and the detection angle are lower.
The number of backscattered ions detected is given by the following formula :

$$N_E = Q\left(-\frac{d\sigma}{d\Omega}\right)_{E_o,\theta} \cdot \varepsilon \cdot \Omega \cdot n$$

Q : number of incident particles ;

$\left(-\dfrac{d\sigma}{d\Omega}\right)_{E_o,\theta}$: scattering cross section ;

ε: detector efficiency (around 100 %);
Ω: subtended solid angle of detection;
n: superficial concentration (at/cm^2) of target atoms.

This relation implies that the irradiated sample is sufficiently thin
to avoid incident and backscattered ion energy alteration.So,
differential scattering cross section and detected ion energy are the same
for all the target atoms. Thus, the backscattering spectrum is reduced to an
unique ray centred on energy E.

However, if the above condition is not satisfied, incident particles
penetrating below the sample surface, loose a part of their energy before
interaction.At a depth x, their energy becomes :

$$E_{o,x} = E_o - \int_o^x \left(-\frac{dE}{dx}\right)dx$$

$-\dfrac{dE}{dx}$: stopping power of the target material with respect to the
projectiles used.

The energy of the ions leaving the sample is in agreement with the following
expression:

$$E_x = KE_{o,x} - \int_o^{x/\cos(\pi-\theta)} \left(-\frac{dE}{dx}\right) dx$$

That involves a continuous spectrum whose upper limit corresponds to the
energy of incident particles scattered by target atoms located
on the sample surface.

The rules which govern the stopping power of ions in matter allow the establishment of the wellknown correspondance between the detected ion energy E_x and
the interaction depth x.

A fortran computer program entitled VERDI leads directly from the
backscattering spectra to the concentration profiles of heavy elements, like
rare earths and actinides in the near surface region of glasses.

This program builds simulated spectra which are iteratively compared with the
experimental one by successive adjustments of parameters, until a good fit
is reached.

EXPERIMENTAL

The set up includes the main following parts :
- a 2 MV VAN DE GRAAFF accelerator ;
- the beam line with a deflecting magnet ;
- an irradiation chamber ;
- a multichannel analyser .

The principle scheme of the cylindrical irradiation chamber (diameter : 300 mm, height : 200 mm) is shown in figure 2.

It is fitted out with two observation windows (diameter : 30 and 110 mm respectively) in pyrex glass.The eight-target sample holder is attached to the lid.   The chamber is equipped with a surface barrier detector (active area : 25 mm^2, sensitive depth : 100 μm, intrinsic resolution : 13 KeV) whose orientation with respect to the incident beam direction can vary between 130 and 170°. The distance between the target and the surface barrier detector is in the range 50 - 120 mm. The detection area is limited by a tantalum collimator (diameter : 1 mm) to reduce pulse pile up.

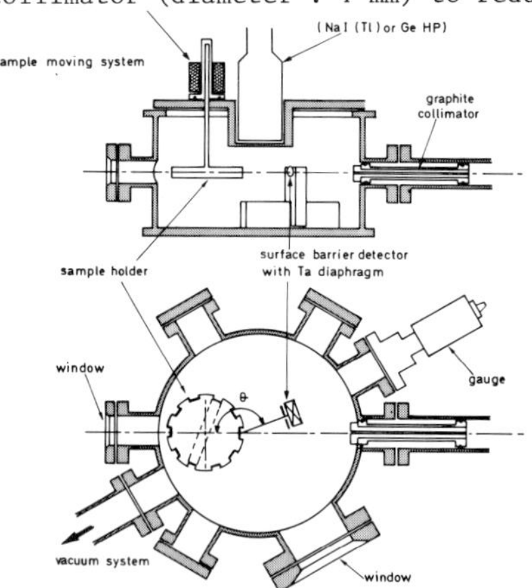

Fig.2. Scheme of the irradiation chamber.

Moreover, the chamber lid can carry a gamma detector (Na I (Tl) or Ge HP) to record prompt gamma rays emitted during nuclear resonant reactions and which can be used to determine concentration profiles of sodium and aluminum [1,2]. The irradiated area is defined by a graphite diaphragm (diameter : 5 mm) inserted in the beam line, near the chamber.

EXAMPLES

All the experiments described below were performed with a 1.5 MeV ^{4}He beam,
choosing a detection angle of 160°.
The chemical compositions of glasses   especially prepared for testing the
analytical method are given in table 1.

Table 1.
CHEMICAL COMPOSITIONS OF GLASSES STUDIED

Constituant		Weighted concentration (%)		
	Glass 1	Glass 2	Glass 3	Glass 4
Li	2.49	2.49	2.49	2.49
B	7.75	7.75	7.75	7.75
O	52.65	50.73	52.55	50.82
Na	8.89	8.89	8.89	8.89
Si	25.68	23.34	25.68	23.34
La	–	4.26	–	–
Ce	–	–	–	4.07
Th	–	–	2.64	2.64
U	2.54	2.54	–	–

The beam intensity and the examination duration are limited to
$I \leqslant 2.10^{-8}$ A   and $t \leqslant 5000$ s to prevent secondary effects occuring during
sample bombardment.

If this is not done,particularly if the beam intensity is higher than the
above mentionned value,the concentration profiles are distorted;in particular the heavy elements like La,Ce,Th and U move into the sample surface
during analysis.On the other hand if the beam intensity is restricted so that
the current density is less than $0.1uA/cm^{2}$ over the irradiated area,changes
in the concentration profiles are undetectable.

Figures 3 to 7 show backscattering spectra for the glasses listed in table 1.
These samples were leached in static deionized water except for the 100°C
test,for which a Soxlhet was used.The sample area/leachant volume ratio is
maintained at 0.6 cm^{-1}.

All the examinations are performed immediately after aqueous treatment
because storing the samples,notably in an ambient atmosphere produces slight
changes in the profiles.

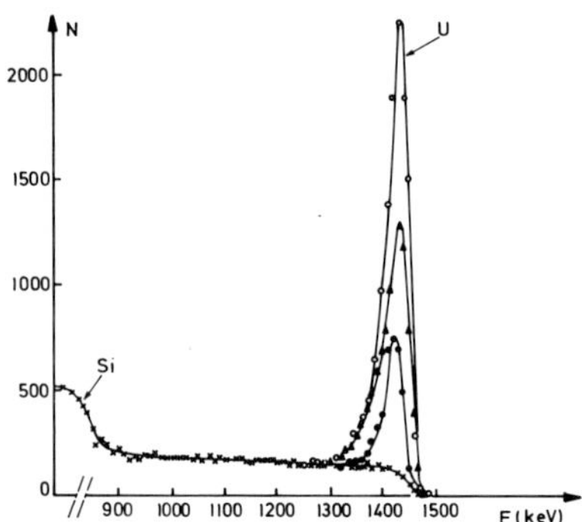

Fig.3. ^{4}He backscattering spectra (E_o = 1.5 MeV) determined before (x) and after leaching of glass samples 1 during 2(●), 9(▲) and 40 (o) days at 20°C.

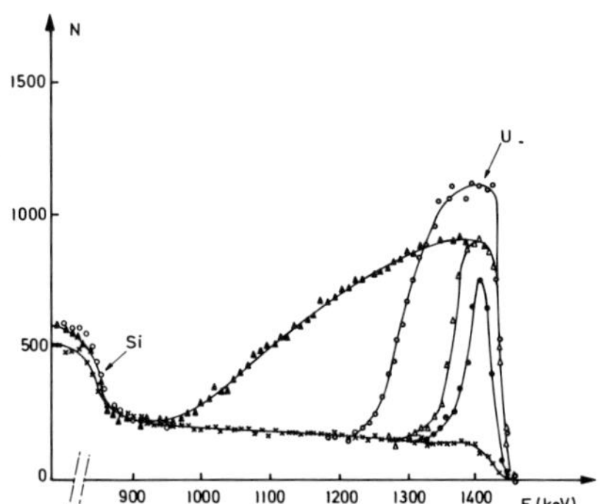

Fig.4. ^{4}He backscattering spectra (E_o = 1.5 MeV) obtained before (x) and after aqueous treatment of glass samples 1 during 2 days at 20 (●) , 40 (Δ) , 60 (o) and 100°C (▲).

Concentration profiles of lanthanum, cerium, thorium and uranium established using VERDI are given in figures 8 to 11.

It is obviously clear that aqueous leaching induces a strong concentration enhancement of these elements in the hydrated layer developed on the glass surface.This effect increases with the duration and the temperature of leaching.

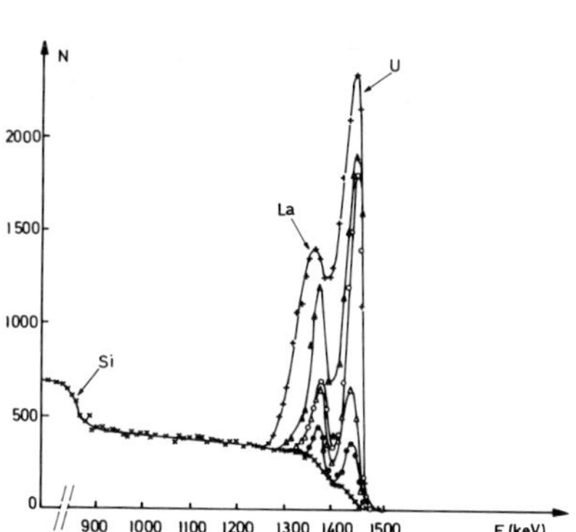

Fig.5. ^{4}He backscattering spectra (E_o = 1.5 MeV) obtained before (x) and after leaching of glass samples 2 during 1 ($\bullet$), 2 (Δ), 7 (o) , 20 ($\blacktriangle$) and 40 days (+) at 20°C.

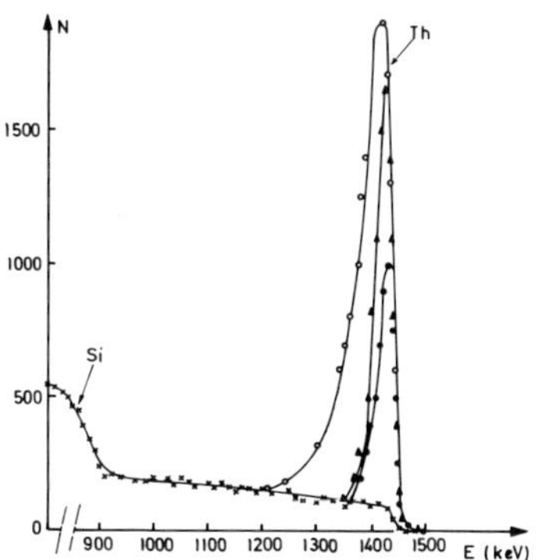

Fig.6. ^{4}He backscattering spectra (E_o = 1.5 MeV) obtained before (x) and after aqueous treatement of glass samples 3 during 1 ($\bullet$), 2 ($\blacktriangle$) , and 4 days (o) at 20°C.

200

Extending the leaching time makes the hydrated layer very brittle.

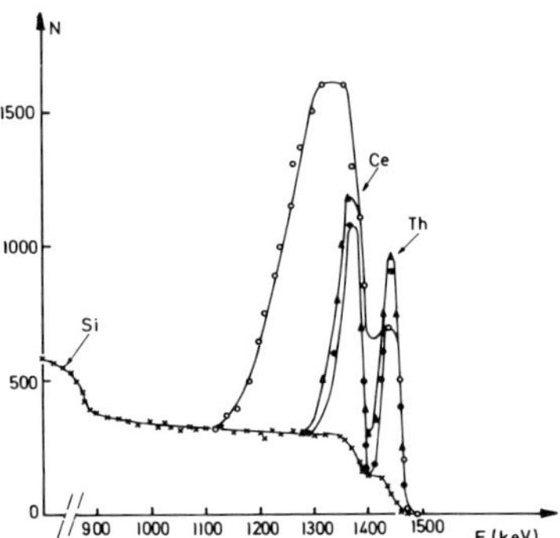

Fig.7. ^{4}He backscattering spectra (E_o = 1.5 MeV) determined before
(x) and after leaching of glass samples 4 during 1 ($\bullet$), 3 ($\blacktriangle$),
and 6 days (o) at 20°C.

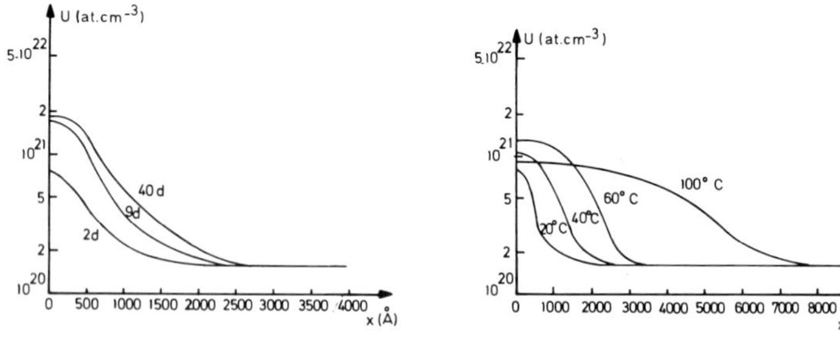

Fig.8. Concentration profiles of uranium induced by aqueous leaching
for glass samples 1.

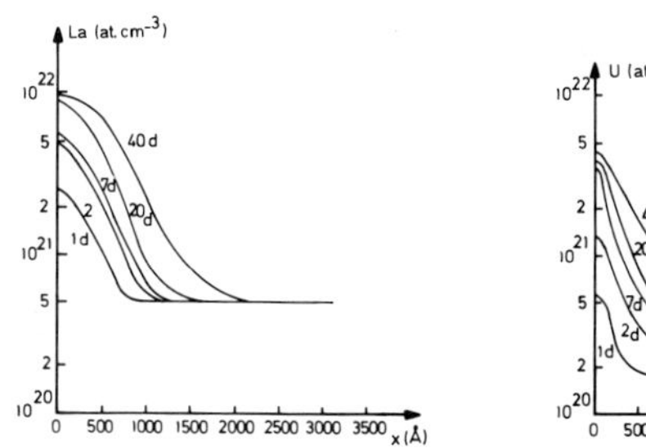

Fig.9. Concentration profiles of lanthanum and uranium induced by aqueous leaching for glass samples 2.

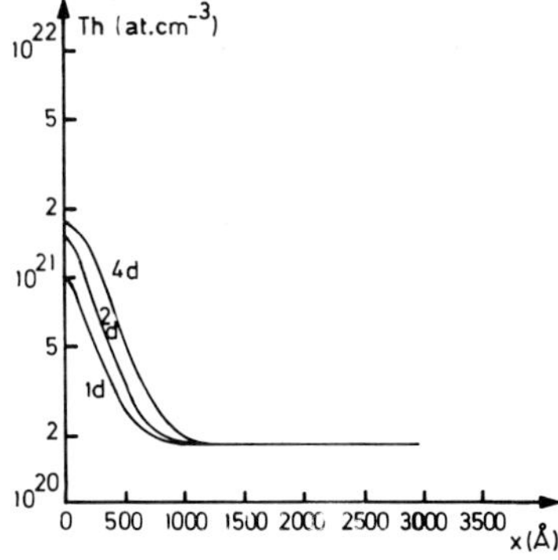

Fig.10. Concentration profiles of thorium induced by aqueous leaching for glass samples 3.

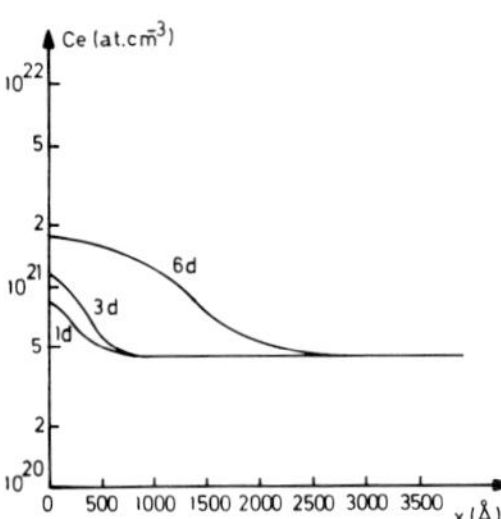
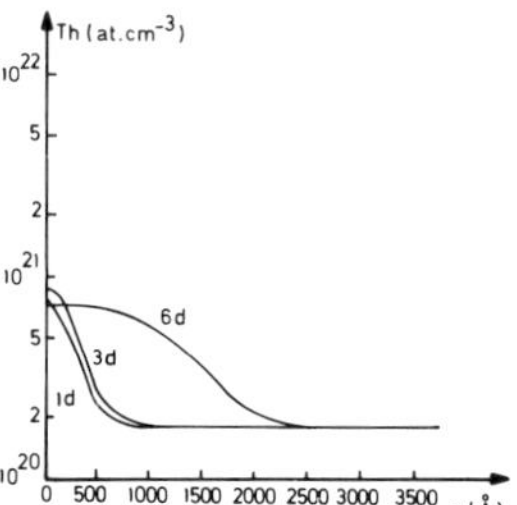

Fig.11. Concentration profiles of cerium and thorium induced by
aqueous leaching for glass samples 4.

CONCLUSION

The main advantage of this analytical method which enables the first
micrometer of the glass surface to be studied is its non destructive
character.
The results obtained clearly show that aqueous leaching of glass samples
containing either lanthanum and uranium or cerium and thorium induces
accumulation of these heavy elements in the hydrated layer developed on the
sample surface : the effect is apparently less pronounced for the "cerium
thorium" glass.

ACKNOWLEDGEMENTS

The authors are indebted to J.GOSSET and P.HSIUNG for their friendly and
constructive help during this study.

REFERENCES

1. ENGELMANN, CH., LOEUILLET, M.,NENS, B.,TROCELLIER, P.Silicates Industriels
   2 (1981) 47
2. ENGELMANN, CH., NENS , B., TROCELLIER,P.
   Verres Refract. 35 (1981) 486.

ON THE LEACHING BEHAVIOUR OF A SIMPLE BOROSILICATE GLASS IN A CONFINED
ENVIRONMENT

J.C. PETIT, Y. LANGEVIN, J.M. LAMEILLE, J.C. DRAN, Laboratoire René Bernas,
F. 91406 Orsay.

ABSTRACT

Cm-sized blocks of a simple borosilicate glass have been leached in hydrothermal conditions with various rates of water renewal. Weight loss measurements
reveal that restricted water access does not slow down the dissolution rate of
this particular glass as would be expected from solubility limits, but instead
accelerates its destruction and its transformation into a secondary solid phase
still ill-defined.

INTRODUCTION

In the course of our studies of radiation effects on the aqueous dissolution
of simulated HLW glasses, based on ion implantation (1), we have identified by
short term leach tests, involving very thin (~ 1000 Å) layers of material, several processes which could be radiation sensitive. When attempting to infer from
these observations, the long term durability of these glasses, we realized that
any valid extrapolation should need first to identify the process which rules the
kinetics of glass corrosion and which strongly depends on the chemical and thermodynamic conditions prevailing in the glass-leachant system. In the case of a
confined environment which is now considered by experts as the most likely for
water intrusion in a geological repository (2), such identification is particularly difficult, because the leaching conditions could considerably evolve with
time.

For several years, the effect of the SA/V ratio on the kinetics of glass
dissolution has been investigated (2,3) and two opposite consequences of restricted water access have been inferred : i) <u>pH increase</u> as a result of alkali diffusion in solution which would change the corrosion mechanism from a diffusion
controlled one ($\sqrt{t}$ - kinetics) to a process governed by the direct dissolution
of the network (t-kinetics)(2). The net effect of this pH-increase would then be
an enhanced dissolution rate. However, for borosilicate glasses, it has been
argued that the release of boron in solution would buffer the pH to moderate
values ~ 9, which could prevent self-accelerated dissolution (4). ii) <u>Saturation
effect</u> due to the accumulation in solution of dissolved species which would
progressively slow down the dissolution rate of the glass. According to
Hughes et al. (5) the rate of release of glass constituents would then be controlled by the effective saturation solubility of the glass in water and the rate of

204

mass loss in stagnant conditions could even be $\sim 10^6$ times lower than that measured in a Soxhlet test. Actually, such a saturation effect has been experimentally observed by Braithwaite (6) in the case of a simulated waste borosilicate glass containing copper and titanium and leached in various aqueous solutions at high temperature (250°C).

We have undertaken similar experiments with other glass compositions, and in particular those on which we have previously investigated radiations effects, in order to check whether or not this saturation effect is a common feature of silicate glass dissolution. However in the first stage of this program we have chosen to study a borosilicate glass of very simple composition, expecting that the results would be much more simple to interpret. The choice of this glass does not intend to simulate any HLW glass but rather to check the ability of classical glass corrosion models to predict the leaching behaviour of a particular quite simple borosilicate glass in a confined environment. Thus the results presented in this paper are preliminary.

EXPERIMENTAL PROCEDURE

Cm-sized cubes of glass cut with a diamond saw were subjected to leaching experiments in teflon-walled autoclaves with a high SA/V ratio (0.4 to 2 cm^{-1}) at temperatures ranging from 100 to 160°C. This relatively high temperature range was used in order to accelerate the chemical reactions and to obtain an observable effect in a reasonable time of experiment. The single parameter which was measured in this preliminary approach was the weight loss. This was done by interrupting the leaching experiment at regular intervals (usually 1 to 4 days), withdrawing the remaining glass block and filtering the leachant with a milli-pore filter of 0.45µm pore diameter, after cooling at room temperature. At each leach step, both the glass block and the filtered insoluble material (released fragments and colloïds) were carefully dried at 105°C during one hour and subsequently weighed. After weighing, the block and the filtered material were reintroduced in the leaching solution which could be renewed or not. Note that the separation of the insoluble material is only intended to avoid overestimation of the weight of the residual block but the weight of the material collected from the leachant after cooling at room temperature is certainly not representative of material which was insoluble at higher temperature. Moreover the separation of reaction products is not possible when the leach test is performed with "presaturated" solution (see below) or concentrated NaCl-brine, as such products are mixed with excess powdered glass or NaCl crystals. At each leach step, the pH of the solution is measured.

The leaching conditions investigated include :

- pure deionized water not renewed or renewed at each step of weight loss measurement :

- water previously "saturated" with the powdered glass ($\sim 5\mu$m grain size) of the same composition. This presaturated solution was obtained by mixing ~ 2.5 g of powdered glass (corresponding to the same mass of the glass block used on the subsequent leaching experiment) to 15 cm^3 of deionized water during one day at the same temperature as the leaching experiment. Solutions were either used as prepared i.e. with the residual undissolved glass powder, or filtered, but in both cases the leachant was not renewed.

- NaCl brine (400g.1^{-1}) not renewed.

This investigation has been performed with a borosilicate glass of the following composition : SiO_2 54.5 wgt% ; B_2O_3 29.2 ; Na_2O 10.3 ; Li_2O 4.9.

RESULTS AND DISCUSSION

In figure 1 we report for several typical cases the progress of glass dissolution in terms of the relative residual weight versus time, P_i/P_o (where P_i is the mass of the residual block of glass and P_o its initial value).

1) Let us first consider the case where the leaching solution is renewed, curve 1. We note that the pH of the solution initially of about 6 rapidly increases to a value between 10 and 12 after about 1 hr of leaching and remains constant until the end of each leach step. Despite this high pH, we observe at 70x the formation of a layer of hydrated silica gel, the thickness of which increases with time. This layer is likely formed by the superficial dealkalinisation of the glass. However the simple classical model of interdiffusion of alkali and hydronium ions where the gel acts as a diffusion barrier with constant diffusivities (7) cannot explain the observed <u>very sharp phase transition between this layer and the unaltered glass</u>. In contrast, such a sharp discontinuity can be conveniently described by assuming a drastic increase in the diffusion coefficients in the gel when compared to their values in the unaltered glass. In such a case, the hydrated silica gel can be considered as "transparent" to the diffusing ionic species. Similar conclusions have been inferred by Bunker et al (8) for (1-x) Na x K_2O $3SiO_2$ glasses leached in pure deionized water at lower temperature. In addition, optical micrographs of fragments of the hydrated silica layer itself formed in this particular glass reveal a multi-layered structure.

In our experiment the residual weight first decreases at a rate of $\sim 4.10^{-2}$ g.cm^{-2}.d^{-1} for the first 5 days while we only note the build up of the gel layer

206

without any sizable suspended material in solution. When reaching a critical thickness (θ_c) the gel layer flakes off in solution likely because of the mechanical stresses induced by swelling. Then a new layer builds up until its thickness reaches again θ_c, and this cycle is repeated until the glass block is completely disaggregated.The periodicity of this desquamation process is ~ 3 days at 160°C. However, at lower temperatures (100 and 140°C) this process seems to stop when the size of the residual glass block has been reduced to ~ 0.3 to 0.5 times its initial value, respectively. Moreover, the flakes and the remaining hydrated block exhibit a multi-layered structure when observed with an optical microscope. We note that the residual hydrated block has not been split although θ_c has been reached several times. This feature seems to indicate that the desquamation process only appears if sufficient mechanical stresses are applied and thus probably depends on geometric factors etc... Then the hydration proceeds inwards until the block is completely transformed into an hydrated silica gel.

Therefore, the thickness of the hydrated silica layer does not seem to reach a constant value on a 0.5 cm range, which indicates that a steady state for inter-diffusion through this layer has not been established. Moreover, θ_c and the texture of the layer depend on the temperature. In fact, when comparing leach tests conducted at two temperatures (100 and 140°C), one observes that the total mass of dissolved material does increase with T, but the mass of the residual block of glass at 100°C is also surprisingly lower than at 140°C and this seems to be due to an increase in θ_c with T for this particular glass.

Finally, after a period of 20 days at 160°C the pH does not increase anymore when renewing the solution. The remaining insoluble material consists of flakes and fragments of the residual hydrated block, the weight of which slowly decreases linearly with time. The rate of weight loss ($\sim 3.10^{-4}$ g/hr) likely reflects the dissolution rate of the hydrated silica gel formed from this parti-cular glass in deionized water.

2. If one considers now the shape of curve 2 obtained when the solution, constituted initially of deionized water, is not renewed, one notices that the mass of the glass block first rapidly decreases in the first two days, while showing the build up of an hydrated silica layer similar to that observed in the preceding case. This layer disappears after the first two days. Then the mass of the glass block decreases more slowly for about eight days and finally steeply decreases until complete disappearance after ~ 15 days. Furthermore, in this last part of the kinetics, the glass block does not show any evidence for an hydrated layer, remains perfectly transparent and probably dissolves nearly congruently. Therefore as regards to the destruction of the glass block itself, there is no

<u>apparent</u> saturation effect. In fact the advocated solubility limits which are reflected by an important reprecipitation when cooling the solution do not imply that the disagregation of the glass block stops. Again, the pH of the solution rapidly increases to a value between 10 and 12 and remains constant during all the dissolution process. The shape of curve 2 could be tentatively interpreted as follows : in the first stage of the leaching process, the interdiffusion is the dominent mechanism of glass dissolution as the leachant is deionized water. The hydrated silica layer builds up rapidly as it forms more rapidly than it dissolves, as in the case of renewed solution. This process progressively enriches the leachant in dissolved specis which in turn slow down and finally hamper further interdiffusion. Thereafter the layer of hydrated silica gel is attacked and eventually eliminated. Then dissolution proceeds congruently with an enhanced rate. This dissolution is accompanied by a precipitation of reaction products. This precipitate can be amorphous or crystalline depending on the glass and solution chemical compositions. For example, in the case of the aqueous dissolution of a phosphate glass, Bunker (9) has identified the precipitation of apatite. Moreover, the amorphous material can transform on the long term in an assemblage of crystalline minerals (10).

3. In a third experiment, we used as a leachant in unrenewed conditions a "presaturated" solution prepared with powdered glass as described in the experimental section. Curves 3 and 4 refer to the kinetics of the glass block dissolution obtained when the leachant is used as prepared, i.e. with the remaining glass powder, or filtered. For curve 3, after a very short initial stage (2 days) where a very thin hydrated layer is visible, the glass block rapidly dissolves while remaining completely transparent. Such a curve is well fitted by a variation law of the type : $P_i/P_o = \frac{(tf-t)^3}{tf^3}$, where tf is the time necessary for complete dissolution, P_o is the initial weight, and P_i the weight measured at a time t. Such variation corresponds to dissolution at a constant rate V if the surface reduction is taken into account. Moreover, the value of t_f = a/2V when a is the site of the block, allows the direct measurement of the dissolution rate V, which is $\sim 10^{-1}$ g.cm^{-2}d^{-1} at 160°C. When compared to curve 2, curve 3 shows that the initial stage of hydrated layer build-up has been considerably shortened as a result of "presaturating" the solution. Moreover, its shape is very similar to that of the second stage (congruent dissolution) of curve 2. Finally, we carefully examined the solid phase formed during the test, and noticed that it has been compacted and is more like a cement than a glass powder. X-ray diffraction does not show any crystalline structure for these corrosion products and complete characterization is in progress.

In contrast, when the "presaturated" solution has been filtered, the degradation of the glass block is drastically slower and the kinetic curve of its residual mass looks more like that of curve 2 than that of curve 3. This unexpected result demonstrates the importance of the residual undissolved glass powder and/or possible colloïds on the corrosion process. The nature and the chemistry of these species retained on the $0.45 \mu m$ filter is still to be achieved. Such information is necessary to fully understand the precise mechanism of gel layer dissolution.

4. Finally, in a brine of NaCl $400g.l^{-1}$ at 160°C (curve 5), one observes a very slow decrease of the weight of the glass block, which exhibits a very thin and fragile hydrated layer. Such a feature can likely be ascribed to the high sodium concentration which inhibits the interdiffusion of Na^+ and H_3O^+, but not the diffusion of other alcalis. Since lithium is present, a thin hydrated layer can form by the $Li^+ \rightleftharpoons H_3O^+$ or H^+ interdiffusion.

The preceding results show that in a confined environment with a high SA/V ratio, the durability of this simple borosilicate glass is enhanced by sodium ion saturation but there is little evidence for a silica saturation effect probably due to formation of colloïd precipitate. Limited renewal of water seems even to be detrimental to the chemical resistance of this glass <u>with this experimental procedure</u>.

A qualitative explanation of these observations can be outlined as follows. In the case of a confined environment, dissolution of glass leads to the precipitation of reaction products when the concentration of species in solution reaches saturation levels with respect to these secondary solids. The system is then composed of three reservoirs interacting as follows :

$$\text{Glass} \longrightarrow \text{leachant} \rightleftharpoons \text{precipitated material}$$

This irreversible dissolution leads to the complete transformation of glass into secondary material which is the stable solid phase of the system. Therefore, the rate of precipitation is that of extraction of dissolved species from the solution and thus governs the dissolution rate of the glass. When the entire piece of glass has been destroyed, the remaining system composed of leachant + precipitated material can then reach thermodynamic equilibrium.

Finally, the authors caution that the procedures inherent in all 5 cases involve hydration-desiccation cycles. Such cycles may also play a role in the dissolution process observed.

CONCLUSION

Leach tests on a simple borosilicate glass with various rates of water renewal show an alcali saturation effect but not a saturation effect for silica. This may be due to formation of precipitates or colloïds. This study showed that restricted water access apparently enhanced the degradation rate. Although such results are opposite to those obtained by Braithwaite (6) on a different glass composition, it is likely that the differences are due to the absence of multiple valence oxides in our simple glass. Understanding the way in which these oxides can affect gel formation and stability is of particular importance when trying to make meaningful predictions for the long term stability of HLW glass in a geological repository. The assesment of the long term durability in a confined environment should be made on a case by case basis until a clear understanding of the basic processes involved is achieved.

REFERENCES

1. J.C. Dran, Y. Langevin, M. Maurette, J.C. Petit, B. Vassent, in "Scientific Basis for Nuclear Waste Management", 4, S.V. Topp ed. (Plenum Press New York 1982), in press.
2. E.C. Ethridge, D.E. Clark, L.L. Hench, Physics and Chemistry of Glass, 20, (1979), 35.
3. E. Lue Yen-Bower, D.E. Clark, L.L. Hench, in "Ceramics in Nuclear Waste Management", T.D. Chikalla and J.E. Mendel ed. (Technical Information Center, U.S. Dept. of Energy 1979) pp. 41-46.
4. F. Lanza, in "Radioactive Waste Management and Disposal", R. Simon and S. Orlowski (Harwood Acad. Publ. 1980), p. 363.
5. A.E. Hugues, J.A.C. Marphes, A.M. Stoneham. AERE-R10189 (1981). J. Nucl. Materials (1982) in press.
6. J.W. Braithwaite in "Scientific Basis for Nuclear Waste Management", 2, C.T.M. Northrup ed. (Plenum Press New-York, 1979).
7. Z. Boksay, G. Bouquet, S. Dobos, Physics and Chemistry of glasses, Vol. 8 N° 4 (1967), pp. 140-143.
8. B.C. Bunker, G.W. Arnold, D.E. Day, E.K. Beauchamp, Personnal communication.
9. B.C. Bunker, G.W. Arnold, J.A. Wilder, Personnal communication.
10. A. Decarreau, S. Vernhet, RAST, Orsay-France, 1978, p 137.
11. G.J. McCarthy, W.B. White, R. Roy, B.E. Scheetz, S. Komarnemi, D.K. Smith, D.M. Roy, Nature, 273, (1978) 216.
12. S.N. Karkhanis, P.J.Melling, W.S. Fyfe, G.M. Bancroft, in "Scientific Basis for Nuclear Waste Management" 3, J.G. Moore ed. (Plenum Press New York 1981) pp. 115-122.

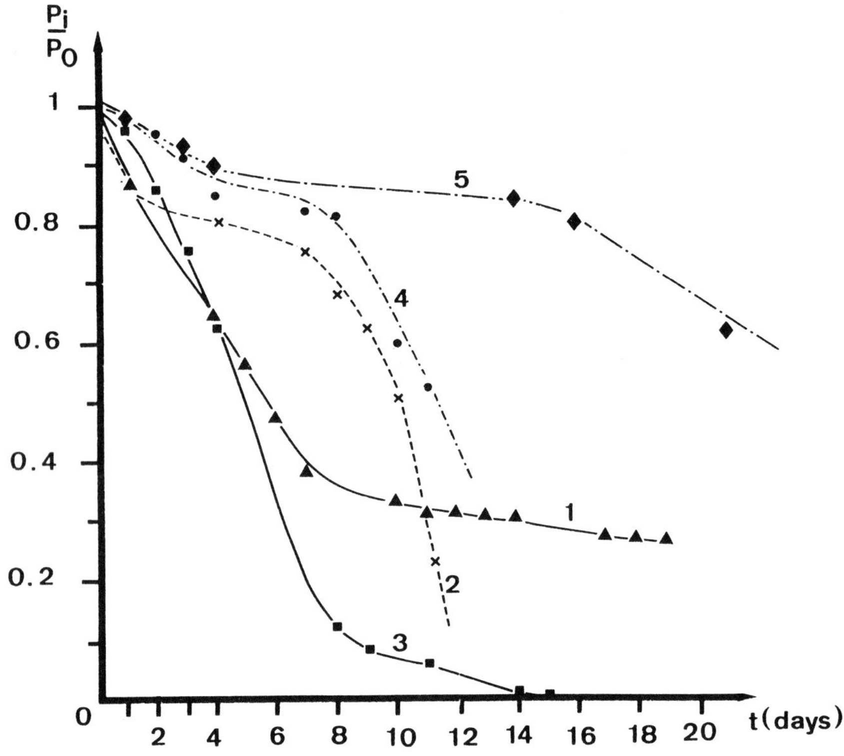

Figure 1 - Relative residual weight of the glass block versus time of experiment in various leaching conditions.

COMPARATIVE STUDY OF SEVEN GLASSES FOR SOLIDIFICATION OF NUCLEAR WASTES

J. L. NOGUES*, L. L. HENCH** AND J. ZARZYCKI*
*Laboratoire des Verres-Universite des Sciences et Techniques du Languedoc,
34060 Montpellier Cedex (France); **Ceramics Division, Department of Materials
Science and Engineering, University of Florida, Gainesville, Florida, USA

INTRODUCTION

The relative leaching behavior of seven alkali borosilicate glasses considered for immobilization of high level radioactive wastes was compared using a static 90°C leach test similar to MCC-1.[1] All compositions (Table I) were compatible with the low melting temperature, 1150°C, viscosity and other process variables required for the French AVM process. The quantity of simulated waste products was from 10.9–15.9 weight %.

EXPERIMENTAL PROCEDURE

Leaching times studied were 1, 3, 7, 14 and 28 days with ratios of glass surface area (SA) to solution volume (V) being $SA/V = 1.0 \ cm^{-1}$ and $0.1 \ cm^{-1}$. The pH of the D.I. water prior to leaching was 5.5. After leaching the solution pH was measured with a pH microelectrode and concentrations of Si^{4+}, B^{3+}, Al^{3+}, Mo^{6+} determined by ICP spectroscopy, Na^{+} by atomic absorption, and Fe^{3+} by atomic emission spectroscopy. Normalized, average leach rates were calculated as in the Marcoule Research Program. All sample surfaces were polished to a 600 grit surface with dry SiC paper and analyzed before and after leaching with infrared reflection spectroscopy (IRRS).[2,3] The IRRS spectra were normalized to a reflection intensity of 80 for the $1120 \ cm^{-1}$ Si-O-Si molecular stretching vibration of vitreous silica by using a shutter in the reference beam of a Perkin-Elmer 467 IR spectrometer.

RESULTS

Figure 1 compares the IRRS spectra of the seven glass compositions after 7 days leaching at 90°C, $SA/V = 0.1 \ cm^{-1}$. Glasses M1 and M2 show extensive reduction of intensity over the entire spectrum, with M1 being affected the most. This behavior is characterisitic of network dissolution following selective ion leaching.[3,4] This causes roughening of the surface, scattering of the incident IR beam, and a decrease in reflected intensity. Glass M3 is not altered by 7 days leaching, M4 and M6 show evidence of silica-rich film formation, and M5 and M7 show the beginnings of network damage of the surface.

After 14 days, the extent of surface attack progresses for all seven glass

TABLE 1

NUCLEAR WASTE COMPOSITIONS (WEIGHT %)

Glass / Oxide (%)	M 1	M 2	M 3	M 4	M 5	M 6	M 7
SiO_2	47.6	42.7	51.13	45.6	50.0	48.4	46.1
Al_2O_3	-	8.6	4.05	4.9	3	2.0	5.0
Na_2O	12.4	14.2	13.19	8.8	12.5	11.28	12.5
B_2O_3	18.6	17.8	13.56	22.0	14.41	18.47	14.2
Fe_2O_3	6.4	4.0	1.601	2.8	2.96	2.96	2.96
CaO	-	-	4.10	-	2	3.76	4.1
MgO	-	-	0.348	-	-	-	-
MoO_3	2.47	2.09	1.592	2.62	1.7	1.7	1.7
Li_2O	-	-	-	-	2.0	-	2.0
S.W.P.*	15	12.7	10.9	15.9	12.22	12.22	12.22

* Simulated Waste Products

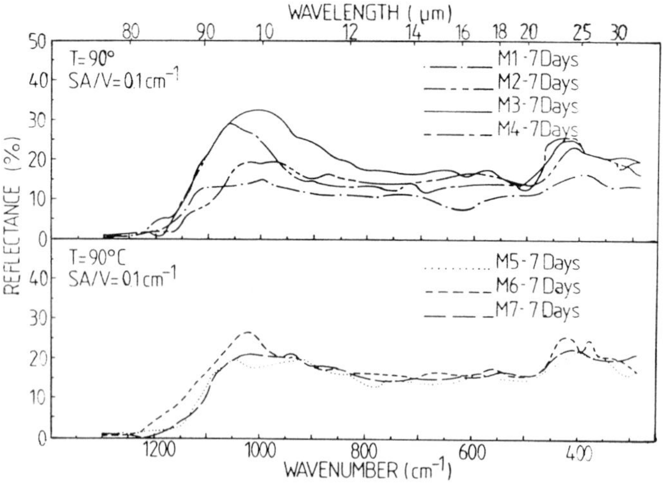

Fig. 1. IRRS after 7 days of corrosion (SA/V = 0.1 cm^{-1}).

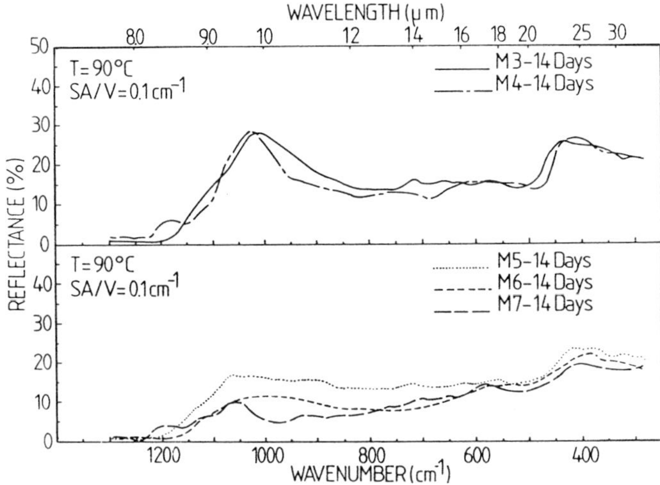

Fig. 2.   IRRS after 14 days of corrosion (SA/V = 0.1 cm^{-1}).

(Fig. 2). Glasses M1 and M2 are not included in Fig. 2 because the IRRS
reflection intensity is too low to be measured for those glasses due to rapid
surface attack. Glass M3 shows the least attack with the spectrum still nearly
the same as before leaching. M4 shows dealkalization since the intensity from
950 cm^{-1} to 800 cm^{-1} is reduced. Sharpening of the Si-O-Si stretching vibration at 1030 cm^{-1} indicates that a silica-rich layer has formed. Glass M5,
M6 and M7 show more surface attack. Twenty eight days is sufficient time for
all seven glasses to show surface deterioration by IRRS, Fig. 3. Glass M3 is
considerably more leach resistant than the other formulas, with its original
IRRS spectra only slightly affected by leaching. Results for leaching at SA/V
= 1.0 cm^{-1} are approximately the same as shown above except glass M7 shows
considerably more resistance to leaching.

The rate of glass surface attack is strongly dependent on the concentration
of network formers in the glass, Fig. 4. On the ordinate of Fig. 4 is plotted
the number of days required for a glass surface to be attacked sufficiently
for its IRRS spectrum at 950 cm^{-1} (Si-O-alkali region) to be decreased by 15%.
The abscissa is the sum of various combinations of glass network formers. The
graph shows that a critical concentration of SiO_2 + Al_2O_3 + Fe_2O_3 > approximately 55% (by weight) or SiO_2 + Al_2O_3 > 53% must be present in the glass for
the surface to be resistant to leaching.

214

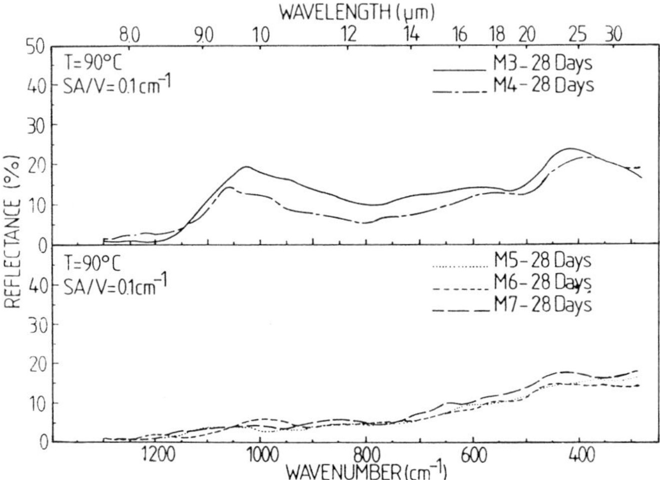

Fig. 3. IRRS after 28 days of corrosion (SA/V = 0.1 cm^{-1})

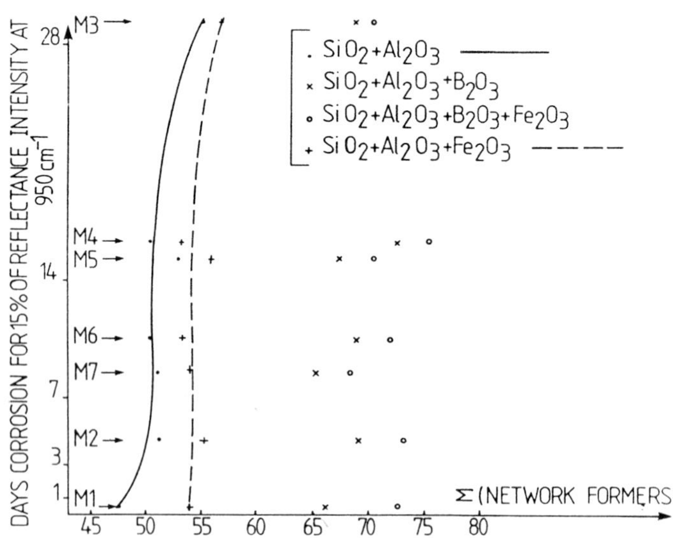

Fig. 4. Influence of network formers.

The time dependence of pH and leachant concentration of Si, B, and Na are shown in Figs. 5, 6, 7, and 8, respectively. All leachant solutions increased in pH with time, with M3 being the slowest and reaching the lowest pH value by 28 days. Leach rates for Si, B, and Na also showed considerable composition dependence with M3 being consistently lower. The cyclic behavior of the leach rates at short times is not fully understood at this time but is probably related to changes in dissolution and precipitation rates of cations that are very sensitive to pH fluctuations.

DISCUSSION

The relative resistance to 90°C static leaching at SA/V = 0.1 cm^{-1} for the seven compositions based upon three analytical criteria is:

(1) IRRS $\quad$ M3 > M4 > M5 $\cong$ M6 $\cong$ M7 > M2 > M1

(2) pH $\quad$ M3 > M4 > M6 $\cong$ M5 $\cong$ M7 > M1

(3) Solution Analysis $\quad$ M3 > M7 > M5 > M4 $\cong$ M6 > M2 > M1

Therefore, the final ranking of the seven glasses is:

(4) Overall Resistance $\quad$ M3 > M7 > M4 > M5 $\cong$ M6 > M2 > M1

Results obtained at SA/V = 1.0 cm^{-1} are generally the same but less conclusive due to the small solution volume which makes chemical analyses less accurate.

The above ranking showed that glass M3 performed well in all analyses. At the beginning of the leaching glass M7 had high average leach rates concurrent with a rapid decrease in IRRS intensity and increase in the solution pH. But after 14 days of leaching, the surface layer developed is thick enough and sufficiently impermeable to give very low average leach rates. The most important condition for the nuclear waste glass chemical durability is to have low leach rates ; glass M7 behaves in this manner. Glass M4 yields good results for IRRS analysis but the relatively high average leach rates show that the diffusion through the surface layer is rather easy. The other compositions (M5, M6, M2 and M1) give relatively poor IRRS and leaching rates corresponding to a lower chemical resistance. The worse glass composition of the series studied is glass M1. Average leach rates for the two best glasses, M3 and M7 after 28 days corrosion in 90° C water at SA/V = 0.1 cm^{-1} are :

$$L_{Si} \; (0,28) \cong 2.10^{-5} \; g.cm^{-2}.d^{-1}$$

$$L_{B} \; (0,28) \cong 2.10^{-5} \; g.cm^{-2}.d^{-1}$$

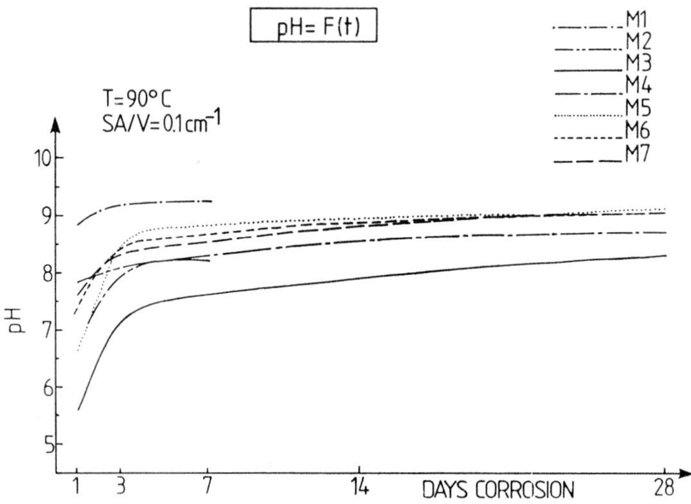

Fig. 5. pH = F(t).

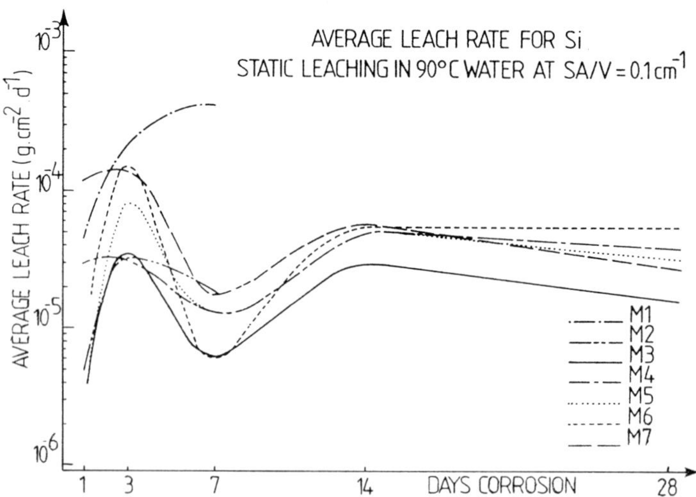

Fig. 6. Average leach rates for Si.

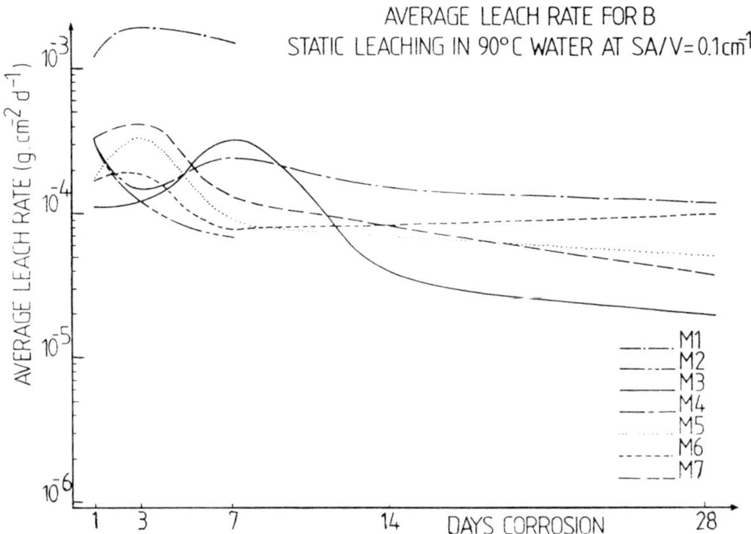

Fig. 7. Average leach rates for B.

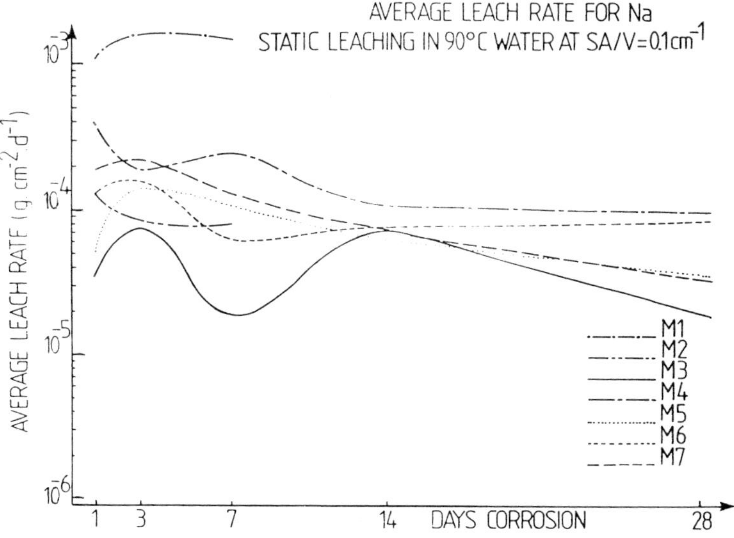

Fig. 8. Average leach rates for Na.

$$L_{Na}\ (0,28) \cong 2.10^{-5} g.cm^{-2}.d^{-1}$$

$$L_{Fe}\ (0,28) \cong 3.10^{-7} g.cm^{-2}.d^{-1}$$

CONCLUSIONS

With the range of glass compositions studied, it was not possible to determine the effect of each element on leaching behavior, however some conclusions regarding the general influence of the glass network formers can be made:

(1) The addition of Al_2O_3, between 2 and 5% (weight) results in a large increase in the chemical durability of the glass.

(2) The presence of Fe_2O_3, between 1.5 and 3% (weight) is necessary to develop with Al_2O_3 a second protective layer on top of the silica-rich film that results from rapid dealkalization.

(3) To obtain good durability the glass composition must contain more than 51% (weight) of (SiO_2 + Al_2O_3) and more than 54% (weight) of (SiO_2 + Al_2O_3 + Fe_2O_3).

(4) The best result was obtained with glass M3 which contains 55.18% (weight) of (SiO_2 + Al_2O_3) and 56.78% (weight) of (SiO_2 + Al_2O_3 + Fe_2O_3). High leach resistance and good surface stability was present even though this glass contained the largest quantity of Na_2O (13.19%) in the series.

(5) The difference between the results obtained at SA/V = 1.0 cm^{-1} and SA/V = 0.1 cm^{-1} shows the importance of understanding both the effects of glass composition and solution concentrations on the behavior of nuclear waste glasses.

ACKNOWLEDGMENTS

The authors gratefully acknowledge Mr. N. JACQUET-FRANCILLON, Commissariat a l'Energie Atomique, for providing the glasses and Mr. R. BONNIAUD, Mr. C. SOMBRET, Mr. E. VERNAZ and Mr. F. PACAUD, Commissariat a l'Energie Atomique, for their encouragements throughout this study and one of the authors (LLH) acknowledges financial support of the U.S. Department of Energy.

REFERENCES

1.   MCC-1 Static Leach Test (1980) Materials Characterization Center, Battelle Pacific Northwest Laboratory, Richland, Washington.
2.   Sanders, D. M., Person, W. B. and Hench, L. L. (1972) Appl. Spectroscopy, 26, 530-536.
3.   Clark, D. E., Pantano, C. G. and Hench, L. L. (1972) in Glass Corrosion, Books for Industry, New York, New York.
4.   Sanders, D. M. and Hench, L. L. (1973) J. Am. Ceram. Soc. 56, 373-377.

CHEMICAL STABILITY OF SIMULATED HLW FORMS IN CONTACT WITH CLAY MEDIA[+]

P. VAN ISEGHEM, W. TIMMERMANS, R. DE BATIST[x], Materials Science Department,
S.C.K./C.E.N., B-2400 MOL (Belgium); [x]Also Rijksuniversitair Centrum, Antwerpen, B-2020 ANTWERPEN (Belgium).

ABSTRACT

The corrosion stability of six simulated HLW forms proposed by several
European countries (five borosilicate glasses and one borosilicate
glass ceramic) in contact with different media relating to the clay
disposal has been investigated for periods up to 80 days and at a surface area
to solution volume ratio of 1 cm^{-1} under non de-aerated conditions. In the
reference medium distilled water the corrosion stabilities are largely determined by saturation effects for elements such as Si, Ca, Mg, Sr, Fe and U.
Si saturation is found to be enhanced by a large Al_2O_3 concentration in the
glass. In the clay-water mixture these saturation effects are much less predominant. Wet clay attacks the waste forms faster than the clay-water mixture,
although the corrosion rates tend to decrease with time in both clay media.
In general, no simple Arrhenius-type temperature dependence for the corrosion
behaviour is found for the interval between 40 and 200°C.

1. INTRODUCTION

In the Belgian nuclear program the use of borosilicate glasses for the conditioning of liquid high level waste produced by reprocessing plants is considered, in view of eventual disposal in deep geological layers. To evaluate
the suitability of clay as a disposal medium, possible interaction effects between high level waste glass and clay environments are being investigated so
that models for the long-term behaviour of glass in such environmental conditions can be established. S.C.K./C.E.N. contributes to this evaluation by
carrying out experimental studies of the stability of a number of borosilicate glass compositions in contact with clay environments.

Six simulated waste forms are investigated in this context as a part of the
current R & D programme of the Commission of the European Communities on Management and Storage of Radioactive Waste. Two of them, borosilicate glasses
SAN 60 2519 L_3C_2 and SM 58 LW 11 have been designed (in France and in F.R. of
Germany, resp.) for the incorporation of the HEWC and LEWC (high, low enriched
waste concentrates) stored at the Eurochemic plant at Mol. The borosilicate
glasses UK 209 (United Kingdom), SON 583020U_2 and SON 641920F_2 (France) and the
glass-ceramic C31 - 3 EC (F.R. Germany) were selected as reference waste forms.
They have been studied already during the previous joint CEC programme[1]. The
corrosion media investigated include distilled water as reference medium, a
clay-water mixture and wet clay. Tests were performed at different tempera-

---

[+]Supported in part by the Commission of the European Communities tures (40, 70, 90, 120, 150, 200°C) under non de-aerated conditions. It is clear that for the presently considered storage and disposal concept (i.e. a surface storage of several decades and a canister lifetime of several hundred years) the temperature at the interface glass-clay will be well below 100°C at first contact. Temperatures in excess of 90°C are therefore investigated mainly for their accelerating influence on corrosion. The influence of the pressure (the lithostatic pressure of the clay layer studied is about 40 bar) was not considered here, since it was found that pressure does not have a marked influence on the corrosion behaviour of borosilicate waste glasses[2]. In view of conditions likely to arise during geological disposal, a relatively large surface area to leachant volume ($SA.V^{-1}$) ratio was chosen (1 cm^{-1}).

In this contribution the discussion is restricted to the corrosion behaviour in pure water at 90°C and the effects of leachant composition and of temperature. The migration behaviour through the clay of the radioisotopes released from the glass is also being studied but will not be considered here.

2. EXPERIMENTAL

The compositions of the simulated waste forms are listed in table 1. Glasses SON58, SON64, UK209 and SAN60 and the glass-ceramic C31 (nucleated during 3 h at 620°C, and crystallized during 15 h at 825°C) were prepared by the laboratories which have developed them. Glass SM58LW11 was prepared in our laboratories, by premelting the frit during 3 h at 1400°C. After cooling, this frit was crushed and melted together with the waste oxide powders during 3 h at 1200°C. Both meltings were performed in a Pt-10% Rh crucible. After melting, glass SM58 was annealed during 3 h at 500°C.

The corrosion experiments are carried out in thermostats ($\leqslant$ 90°C) or autoclaves ($\geqslant$ 120°C). At temperatures $\leqslant$ 90°C thin glass plates (surface area about 5 cm^2, thickness about 1 mm, polished with 600 grit paper) were fixed in teflon cells, filled with about 5 cm^3 of the leaching medium to give a $SA.V^{-1}$ ratio of about 1 cm^{-1}. At temperatures $\geqslant$ 120°C, small polished cubic (6 mm x 6 mm x 6 mm) samples were used. The clay-water mixture was prepared by mixing 100 g of Boom clay (taken in the frozen state at 220 m depth under the Mol site, and stored in water environment at ground level) with 1 l distilled water at room temperature; pH values of the clay-water mixture are between 9.0 and 9.4. The anionic part of the water equilibrated with the clay consists mainly of PO_4^{3-} ($\approx$ 100 ppm), SO_4^{2-} ($\approx$ 30 ppm), NO_3^- ($\approx$ 10 ppm), Cl^- ($\approx$ 2 ppm) and C^- ($\approx$ 1 ppm). In this medium, the waste form samples were mounted on teflon supports, to prevent contact of the samples with clay sediments, which form at the bottom of the cells.

221

TABLE 1

COMPOSITIONS (IN WT %) OF THE HLW WASTE FORMS

	SON583020U$_2$	SON641920F$_3$	C31-3EC	UK 209	SAN602519L$_3$C$_2$	SM58LW11
Frit						
SiO$_2$	43.30	43.80	34.75	50.88	43.41	56.87
B$_2$O$_3$	18.80	17.15	4.14	11.12	17.00	12.28
Na$_2$O	9.30	11.40	1.12	8.30	10.67	4.63
Li$_2$O			0.98	3.99	5.00	3.74
Al$_2$O$_3$			10.28			1.15
MgO			1.44			2.05
CaO			3.84		3.50	3.83
BaO			14.48			
TiO$_2$			2.80			4.45
ZnO			4.88			
ZrO$_2$			0.80			
As$_2$O$_3$			0.48			
Waste Ox						
FP Ox	22.30	13.15	15.20	9.75	1.21	9.72
Fe$_2$O$_3$	0.60	5.86	1.47	2.73	0.33	1.17
Cr$_2$O$_3$	0.20	0.49	0.42	0.56		
NiO	0.10	0.20	0.23	0.36		0.11
MgO				6.34		
Al$_2$O$_3$	1.00	1.20		5.11	18.09	
Na$_2$O			2.20			
Gd$_2$O$_3$		5.86				
P$_2$O$_5$	0.60			0.23		
ZnO				0.40		
U$_3$O$_8$	3.80	0.89	0.46	0.06	0.02	

The autoclave tests were done by simply heating a stainless steel vessel of 1 l
containing the corrosion cells. Autoclave test duration was 1 or 7 days. For
the experiments at ⩽ 90°C, test duration was 3, 7, 14, 28, 42 days(in this case,
the samples were weighed at the different intervals, and then further corroded)
Most of the experiments were carried out in threefold. The results are interpreted in terms of weight loss (in case of the autoclave tests, after removal
of the weakly bound surface layer) and pH (measured at room temperature).

At 90°C, some corrosion tests were also conducted in distilled water with an
analysis sequence of 0.08, 0.33, 1, 3, 9, 27 and 80 days. At each time, the
sample was weighed (and not replaced in the corrosion cell) and analysed by
infrared reflection spectroscopy, while the pH and composition of the corresponding leachates were determined. The amounts of U and Si were analysed by
fluorimetry and spectrophotometry respectively. The other elements (B, Na, Li, Mg
Ca, Ba, Fe, Sr) were analysed by atomic absorption spectrometry.

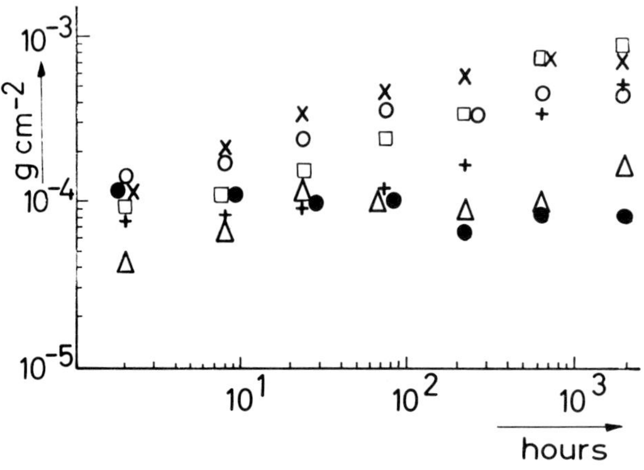

Fig. 1 : Weight losses (in g·cm^{-2}) as a function of time (in hours) in distilled water at 90°C and SA.V^{-1} = 1 cm^{-1} for the different waste forms (X SM58); O SON64; □ SON58; + UK209; ● C31; △ SAN60)

3. RESULTS AND DISCUSSION

3.1. Corrosion in distilled water at 90°C

For the simulated glasses SON, SM and UK, the cumulated weight losses, expressed per unit geometric surface area (fig. 1) increase with time for the 80 days covered untill now in this experiment. The weight of waste glass SAN decreases only during the first day and remains fairly constant afterwards. For all glasses, the corrosion rate based on weight losses seems to tend asymptotically to zero. For the glass ceramic C31, this initial weight loss is limited to the first 2 h, and the specific weight loss becomes zero between 2 h and 80 days.

IRRS spectra also clearly differentiate between waste forms SAN 60, C31 and the others. For waste forms SAN 60 and C31, the shape of the IRRS spectrum (see fig. 2a) remains relatively constant as a function of corrosion time. IRRS spectra for the other glasses (see fig. 2b) on the other hand reveal a pronounced growth and sharpening of the Si-O-Si stretching vibration ($\simeq$1050cm^{-1}) together with a decrease of the O-M^{+}(and O-M^{2+}) bands (between 960 and 700 cm^{-1}) with increasing corrosion time. From these IRRS spectra, one may conclude that the glass surfaces become enriched in Si^{3} (or depleted in alkali's).

An explanation for these observed differences in corrosion behaviour can be obtained from the chemical analysis of the leachate (fig. 3), which suggests that they are related primarely to saturation effects. In fig. 3, the specific weight losses, based on the specific elemental leaching are plotted, using the following formula : $\text{NSWL} = \dfrac{q}{Q/100} \dfrac{1}{S}$

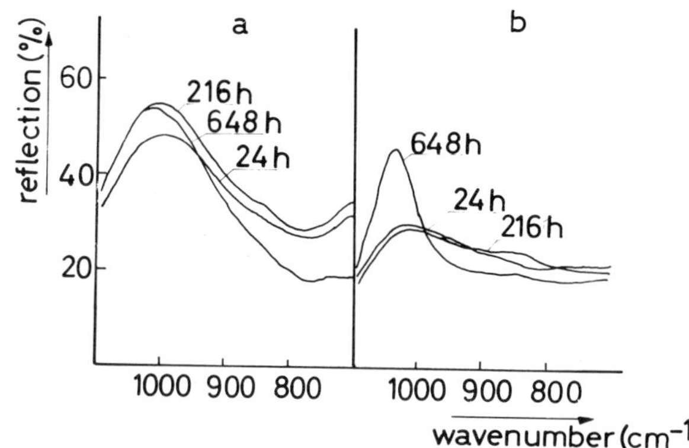

Fig. 2 : IRRS spectra for glasses SAN60 (a) and SON64 (b) as a function of the corrosion time.

where NSWL is the normalized specific weight loss, q the amount of the element in the leachate (in g), Q the weight percentage of the element in the glass and S the geometrical surface area of the glass (in cm^2). From fig. 3, two types of elements can be distinguished, namely soluble elements such as Si, B, Na and Li, and the other, less soluble elements such as Mg, Ca, Sr, Fe and U. Considering first the soluble elements it is found that for glasses SM (fig. 3a) and UK, for the alkali elements Na and Li as well as for B, initial leaching follows a $\sqrt{t}$ diffusion law during about 24 h. In later stages matrix dissolution (proportional with t) is observed. This changeover in corrosion mechanism occurs at a pH value of about 9.0 (fig. 4). These results agree with observations made elsewhere[4]. In glass SM, the amount of Si leached is proportional with time initially, and appears to saturate at about 130 ppm, whereas for glass UK saturation is not yet observed after 27 days. For glass UK, the leaching of Si can be described by a $t^{0.6}$ law.

For glasses SAN (fig. 3b), SON58 and SON64 (fig. 3c) description in terms of diffusion and dissolution does not seem to apply. The time dependence of the leaching of Na and B from glasses SON58 and 64 yields slopes between 0.5 and 1.0. In both glasses, the amount of Si leached is found to saturate at about 80 ppm. For glass SAN, saturation of Si (about 15 ppm), B (about 7 ppm) and Na (about 9 ppm) appears to limit corrosion. Comparison of these results suggests that indeed saturation effects (primarily of Si) play an essential role. Since the presence of Al in solution strongly reduces the Si solubility[5], it is conceivable that the concentration of Al_2O_3 in the glass also influences the corrosion stability of glasses in contact with small amounts of water.

As far as the less soluble elements are concerned, these are found in only minute quantities in the leachate. The observed saturation values are : about 1 ppm for Mg, about 4 ppm for Ca and about 0.1 ppm for Sr. These values agree fairly well with results obtained at Battelle[6]. Iron, whenever measurable,

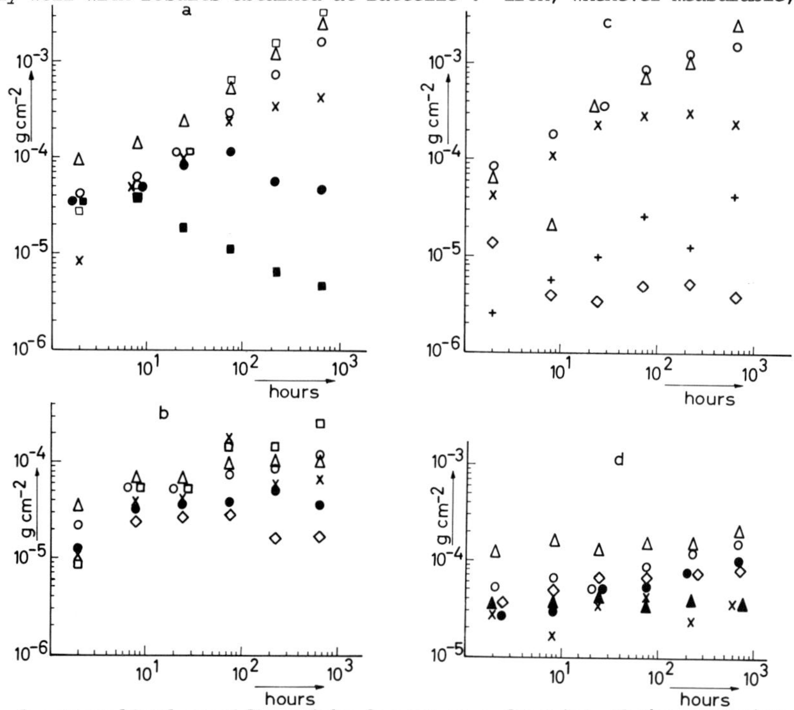

Fig. 3 : Normalized specific weight losses as a function of the corrosion time in distilled water at 90°C for four HLW waste forms; a : SM58; b : SAN60; c : SON64; d : C31. (x Si; o B; △ Na; □ Li; ■ Mg; ●Ca; ▲ Ba; ◇ Sr ; + U)

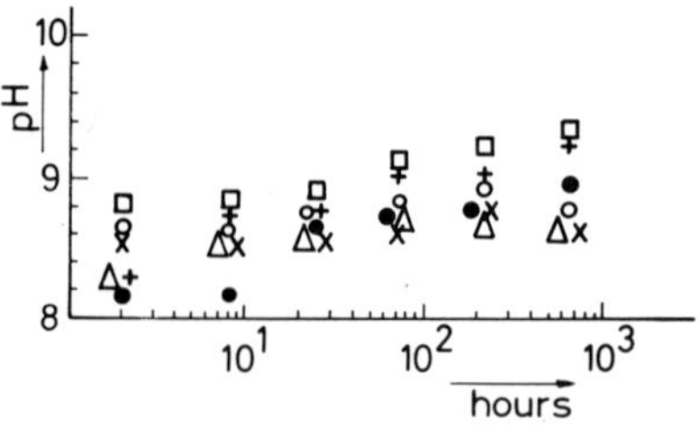

Fig. 4 : Evolution of pH with corrosion time (90°C, SA.V^{-1} = 1 cm^{-1}, distilled water) for the six HLW waste forms (△ SAN60; x SON58; o SON64; □ SM58; + UK209; ● C31)

saturates at a value below 0.1 ppm. Uranium, which is present to some extent
in glasses SON58 and 64, saturates at about 0.8 ppm for SON58 and reaches about
0.4 ppm for SON64 after 27 days. From the normalized weight losses shown in
fig. 3, it is clear that the glasses are leached incongruently. As it has been
observed[7] that elements such as Fe, U, Ca and Mg are trapped within the glass
surface layer, one may assume that for the corrosion times reported, stationary,
congruent leaching conditions have not yet been reached.

The corrosion behaviour of glass-ceramic C31 (fig. 3d) differs from that of
the borosilicate glasses. The amount of Na, B, Ca and Sr leached increases
very slowly with time, whereas the amount of Si and Ba leached does not increase further after 2 h. This suggests that the celsian crystalline phase
present in glass-ceramic C31 (Ba $Al_2Si_2O_8$) is more durable than the residual
glass. Analogous results were also found on a similar celsian glass-ceramic
CGC-HMI[8].

3.2. Effect of the corrosion medium

At the different temperatures investigated, the specific weight losses are
in general higher in clay media than in pure water (see table 2, listing results at 90°C). The corrosion rates, as calculated form the measured weight
losses, decrease with time in the clay media too. Moreover the large differences between the corrosion stabilities of the different waste forms which are observed in distilled water are strongly reduced in the clay media. These observations suggest that other factors than the intrinsic glass properties also play
a role in the corrosion in clay.

The prime reason for the higher weight losses measured in the clay-water
mixture seems to be the absence of saturation effects (at least for the exposure
times used thus far). Indeed corrosion experiments at 90°C in stationary distilled
water at a $SA.V^{-1}$ ratio of 0.1 cm^{-1} yield comparable specific weight losses as
obtained for the clay-water mixture at a $SA.V^{-1}$ ratio of 1 cm^{-1}. Therefore, the
cations and anions present in the clay-water solution do not seem to strongly
influence the corrosion stability. It is suggested that the apparent increase of
the solubility of glass components in the clay-water mixture is mainly caused by
the absorption of certain cations (e.g. Ca^{2+}, Mg^{2+}, Sr^{2+}) on clay particles
present in the solution.

The preliminary results available thus far indicate that wet clay corrodes
the waste forms faster than both distilled water and the clay-water mixture.
This increase in corrosivity as compared to the clay-water mixture might be due
to cation exchange reactions (e.g. $Cs_{waste\ form} \rightleftarrows Na_{clay}$[9]) between the waste

TABLE 2

SPECIFIC WEIGHT LOSS (SLW IN 10^{-5}gcm^{-2}) AND pH VALUES AFTER 3 and 28 DAYS AT 90°C IN DIFFERENT MEDIA

Medium	Time		SAN60	SM58	UK209	C31	SON58	SON64
Distilled water	3d	SWL	9.0	28.2	10.6	8.0	55.0	33.9
		pH	8.71	9.02	8.96	8.84	8.55	8.72
	28d	SWL	13.0	67.2	32.7	8.6	67.4	41.4
		pH	8.49	9.24	9.14	8.53	8.51	8.62
Clay – water mixture	3d	SWL	27.4	41.2	24.4	30.8	77.0	43.6
		pH	8.05	8.18	8.07	8.12	8.02	8.04
	28d	SWL	99.2	125.0	129.6	98.0	155.1	86.8
		pH	8.28	8.72	8.37	8.43	8.10	8.14
Wet clay	3d	SWL	23.6	65.9	53.7	68.9	173.1	77.9
	28d	SWL	259.0	335.0	185.1	213.0	524.4	265.1

form and the solid clay.

3.3. Effect of temperature (40, 70, 90, 120, 150, 200°C)

For corrosion times of one week in the three media considered, specific weight losses at 40°C are a factor 30 to 100 smaller than the corresponding specific weight losses at 200°C. The weight losses in distilled water of waste forms SAN and C31 (fig. 5) show an Arrhenius type behaviour between 40 and 200°C, yielding an activation energy E = 34.8 kj mol^{-1}. The other glasses behave differently, E values from 40 to 90°C being comprised between 53 and 35 kj mol^{-1}, while above 90°C values between 35 and 13 kj mol^{-1} are obtained. This complex temperature dependency is not surprising, since the corrosion mechanisms for glasses can change with temperature[3] and with time[10].

ACKNOWLEDGEMENTS

The authors wish to thank Dr. F. Lievens and his staff for the chemical analysis of the leachates, Mr. M. Rotti for the IRRS analyses, Mr. M. Borgers, Mr. G. Wilms and Mrs.R. Franckart-Vercauter for technical assistance during the corrosion tests, and Dr. P. Henrion for helpful discussions.

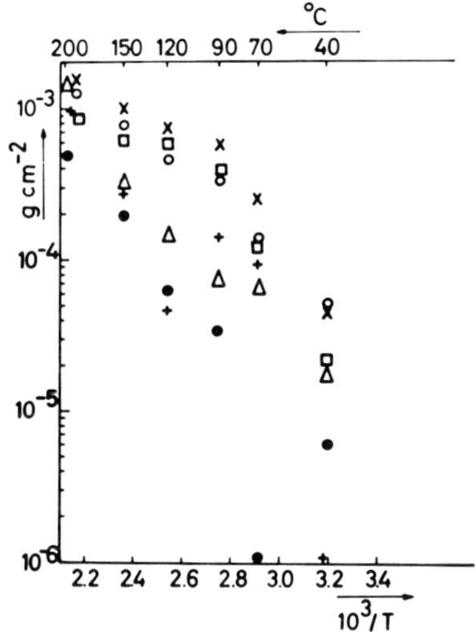

Fig. 5 : Arrhenius plot of the specific weight losses after 7 days corrosion in distilled water (△ SAN60; X SON58 o SON64; □ SM58; + UK209; ● C31)

REFERENCES

1. J.A.C. Marples et al. "Testing and evaluation of the properties of various potential materials for immobilising high activity waste", EUR 7138 EN (1981).
2. P. Van Iseghem et al. "Interaction of vitrified high level waste with clay environment", Proc. International Seminar on Chemistry and Process Engineering for High-Level Liquid Waste Solidification, Jülich (F.R. Germany), June 1981, in press.
3. D.E. Clark et al. "Corrosion of glass", Books for Industry (New York), 1979
4. L.L. Hench et al. "Physical Chemistry of glass surfaces", J. non Cryst. Sol. 28 (1978), 83 - 105
5. R.K. Iler "Effect of adsorbed alumina on the solubility of amorphous silica in water", J. of Coll. and Interf. Science, 43 (1973), no 2, 399 - 408.
6. H.T. Fullam "Solubility effects in waste-glass/demineralized-water systems", PNL 3614 (1981).
7. A. Barkatt et al. "Evaluation of chemical stability of vitrification media for radioactive waste products", Phys. and Chem. of Glasses, 22 (1981) no 4, 73 - 85.
8. R.O. Lokken "Leaching behaviour of glass ceramic nuclear waste forms", PNL 4022 (1981).
9. R.M. Garrels et al. "Solutions, minerals and equilibria", Harper and Row, (New York), 1965.
10.F. Lanza et al. "Methodology of leach testing of borosilicate glasses in water", Proc. First European Community Conference on Radioactive Waste Management and Disposal, Luxembourg (1980).

Mechanical Properties, Crystallization

CHARACTERIZATION OF MECHANICAL PROPERTIES OF NUCLEAR WASTE GLASSES

HERBERT RICHTER* AND PETER OFFERMANN**
*Fraunhofer-Institut f. Werkstoffmechanik, Rosastr. 9, D-7800 Freiburg;
**Hahn-Meitner Institut f. Kernforschung, P.O. Box 39 01 28, D-1000 Berlin

1. INTRODUCTION

Part of nuclear fuel cycle waste is highly dangerous, and must be safely isolated from people. Although the site of the final waste disposal must be the main safety barrier, the form of the waste and its properties are also important considerations.

Glass and glass ceramics are regarded as suitable nuclear waste solidification products; mechanical stability is one of the important properties of the material. Such mechanical stability is necessary on the one hand for the safe handling and transportation of nuclear waste glass, and on the other for its relevance to leaching processes. Leaching by water or by aqueous solutions must be regarded as the most dangerous process by which radioactive materials can be set free from the glass. Because the leaching rate is proportional to the total surface area of the materials exposed to the liquid, it is of utmost importance to keep this area as small as possible and to prevent any enlargement by cracking processes which can arise due to mechanically or thermally induced stresses.

2. TEST TECHNIQUES

It is known that flaws or cracks in glass can increase in size at stresses which do not result in instantaneous failure.[1] Further, it has been shown that the velocity v of such crack extension can be measured as a single-valued function of the fracture mechanics parameter K_I which characterizes the stress intensity at the crack tip.[2] This parameter, appropriately called the stress intensity factor, can be written in terms of applied load, flaw size and specimen geometry. It has been shown that at low crack velocities the function $v(K_I)$ has the empirical form[3]

$$v = A \cdot K_I^{\,n} \tag{1}$$

where A and n are empirical constants which depend mainly on material and environment. These constants characterize the susceptibility of the material to slow crack growth.

Given the crack velocity function $v(K_I)$ for a given material and environment and the stress to which the most severe flaw will be subjected, one can in principle, calculate the increase in material surface area with time.[3]

Another widely used fracture mechanics parameter is the "critical stress intensity factor" K_{IC}, called the "fracture toughness". It characterizes the K_I value at the onset of rapid crack growth and gives the upper loading limit at which spontaneous failure of the loaded specimen can be expected. For the present purpose the onset of rapid crack growth is assumed to lie at about $v = 1/mms$.

In addition to these fracture mechanics parameters it is important to have a measure of the strength which reflects the flaw size distribution at the stressed surface. The strength or the fracture stress of a material characterizes the average load capability of a given glass sample.

Young's modulus, E, and the coefficient of thermal expansion, α, are among other important parameters required, to estimate the build up of residual stresses due to temperature gradients which can develop during intermediate storage and final disposal of glass cylinders.

2.1. PRINCIPLES OF TEST TECHNIQUE SELECTION

In choosing methods for testing the crack propagation and strength behavior of nuclear waste glasses it was necessary to take into account the special test conditions in a hot cell. Though the investigations reported here were done using simulated nuclear waste glasses and could, therefore, be performed outside a hot cell, it was planned ultimately to use the same facilities for tests on real nuclear waste glasses within a hot cell. It was therefore proposed to choose economical and simple test methods because the preparation of material and samples and the handling of test apparatus within a hot cell is an expensive and complicated process.

2.2. $v\text{-}K_I$ CURVE

In order to measure the $v\text{-}K_I$ dependence in opaque brittle materials such as nuclear waste glasses under the conditions mentioned above, the double torsion method was selected[4,5]. In this specimen (Figure 1) the crack length, a, is proportional to the displacement, y, at the loading points at a given load, F; the crack velocity, $v = da/dt$, is then given by

$$v = (\dot{y}-y\dot{F}/F)/(F \cdot C) \qquad (2)$$

where $C = 6b_m^2 (1-v)/(bd^3E)$ is a specimen constant with E = Young's modulus and v = Poisson's ratio The value of K_I for the double torsion specimen is

$$K_I = F (CE/d)^{\frac{1}{2}} \qquad (3)$$

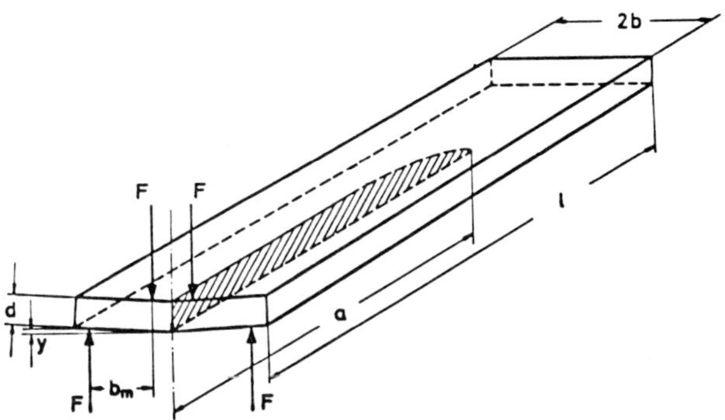

Fig. 1. Schematic representation of the double-torsion test configuration.

In contrast to most double torsion measurements reported in the literature which use constant displacement or constant load techniques, in the present experiments the load was increased continuously. By this method it is possible to measure a large range of crack velocities in one experiment. The load-point displacement, was measured using an inductive displacement transducer, and the corresponding load using a strain gage load cell. Both displacement and load were recorded continuously by a strip chart recorder and additionally stored in digitalized form on magnetic tape. Loading rate $\dot{F}$, and rate of displacement, $\dot{y}$, the quantities necessary to evaluate equation (2) were obtained by numerical methods.

2.3. FRACTURE STRESS

In order to measure fracture stresses the Hertzian indentation test was chosen because, in contrast to other strength tests, the test specimen is not broken catastrophically. It is then possible to obtain a set of strength data on the same specimen.[6] In our investigations about 50 indentation tests could be conducted on a plate of about 20 x 20 mm. In Hertzian indentation a sphere is pressed against the target specimen at increasing load F, until a more or less completely developed cone crack can be observed (Fig. 2). In this test the maximum radial tensile stress occurs at the contact circle and decreases away from the periphery according to

$$\sigma_{rr} = (1 - 2\upsilon) \cdot F/(2\pi r^2) \tag{4}$$

where r is the radial distance from the point of contact, and υ is Poisson's ratio

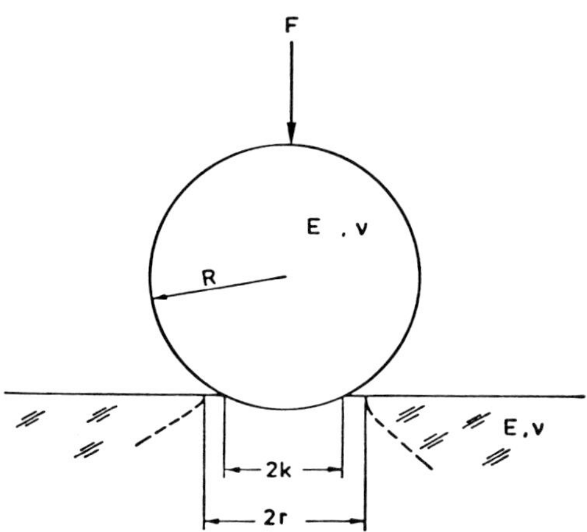

Fig. 2. Parameters of the Hertzian indentation configuration

The radius of the contact circle, k, is given by

$$k^3 = 3 \cdot F \cdot R \cdot (1 - \nu^2)/(2 \cdot E) \qquad (5)$$

where R is the radius of the spherical indenter and where it is assumed that the indenter and the target material have same values of Young's modulus, E, and Poisson's ratio, ν.

In order to take into account the strong inhomogeneity of the stress field great care was taken in the present experimental setup to be able to indicate the first evidence of fracture which is, generally, a shallow circular crack just outside the contact circle. In order to assure most favorable conditions for producing ring (as distinct from cone) cracking, a "stiff" loading system was devised. The indenter was accordingly pressed against the target specimen utilizing thermal expansion of an aluminum bar mounted into a rigid frame. The loading velocity was about $4 \cdot 10^{-6}$ mm/s. Most reliably the first evidence of fracture could be indicated by a quartz transducer which is sensitive to acoustic signals emitted during cracking. This transducer was placed between the loading sphere and a load cell which measures the load applied to the sphere. Immediately after the acoustic emission signal the specimen was unloaded. In this way in most cases ring cracks without cone cracking could be obtained.

To minimize roughness and friction effects[7] in the present study polished glass spheres were used as indenters. Most experiments were conducted using 3 mm diameter glass spheres; in a few experiments 4.5 mm diameter spheres were used. Most experiments were conducted in air as environment, a few tests were run under water. Immediately after each experiment the radius of the ring crack was measured at a magnification of 20. Inserting this radius, r, into equation (4) the fracture stress was calculated.

2.4. Young's modulus, coefficient of thermal expansion.

Young's modulus, E, was measured between room temperature and 500 °C by the resonance technique with the flexural mode of vibration, using bars of rectangular cross section of about 2 x 3.5 x 50 mm. The temperature range is expected to cover the temperatures that a glass cylinder might attain during intermediate storage and final disposal. The coefficient of thermal expansion was measured between 100 and 400 °C.

3.   RESULTS

3.1. v-K_I CURVES

The results of the v-K_I-measurements are plotted as log v versus log K_I. In Figure 3 the v-K_I curves for 4 different materials are shown. For comparison the curve for ordinary soda-lime silicate glass is included. The curves for the nuclear waste glasses cover the range in which v-K_I-data were found for all materials. It can be seen that the materials with special crystal structures (Diopside and Eukryptite) lie at the highest K_I-values, i.e. they show the greatest resistance to crack growth. Only the glass VG 98-3 seems to lie significantly below the v-K_I curve of soda-lime silicate glass. The occurrence of a plateau-like behavior in this glass can be ascribed to an influence of the environment, i.e. mainly of water vapor, on crack velocity. The parameters K_{IC}, n and log A, measured in the present investigations are compiled in Table 1. The units used to calculate log A are mm/s for the crack velocity, v, and N/mm$^{3/2}$ for the stress intensity factor, K_I. It can be seen that most materials are less susceptible to slow crack growth than soda-lime silica glass.

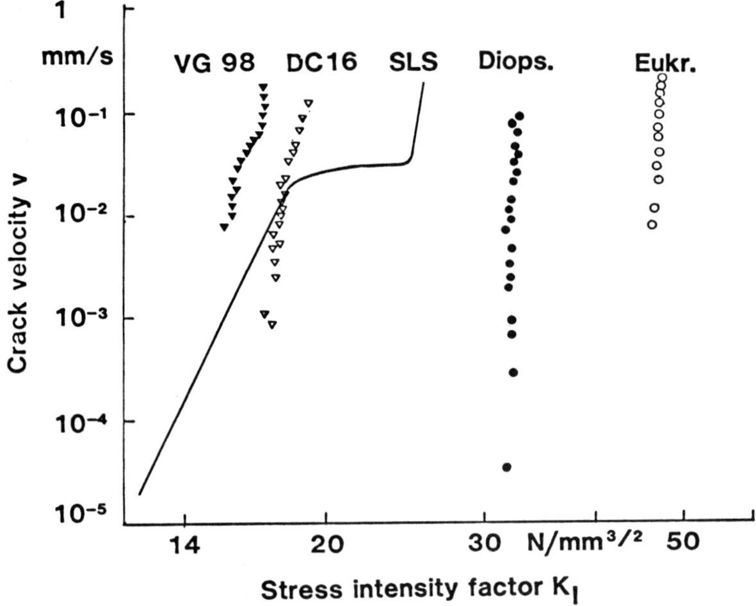

Fig. 3. v-K_I curves for soda-lime silicate glass and different waste glasses.

TABLE 1

FRACTURE TOUGHNESS AND CRACK GROWTH PARAMETERS OF NUCLEAR WASTE GLASSES

Material	K_{IC} [N mm$^{-3/2}$]	n	log A
Soda-lime-silicate glass	24	18.1	- 24.5
Soda-lime-silicate glass (remelted)	24	17	- 23.5
VG 98-3	17	36	- 45
UK 209	22	113	-152
DC 16	20	36	- 48
DC-0-3	33	96	-146
B-I-3	30	67	- 99
Diopside	33	107	-163
Eukryptite	50	151	-258
Perowskite	30	129	-192

3.2. FRACTURE STRESS

As mentioned above the fracture stress of a brittle material is to be regarded as a statistically distributed quantity. Results of the present measurements are, therefore, shown in a statistical plot using the Weibull distribution function.[8] It relates in its 2 parametric form the probability of failure, P, to the observed fracture stress, σ, by

$$\ln \ln (1/(1-P)) = m \ln (\sigma/\sigma_o) \tag{6}$$

where m and σ_o are characteristic constants.

In Figure 4 the strength distributions measured for the glass UK 189 with different surface finishings ares shown. A polishing process seems to decrease the surface strength compared to that of a surface generated in a fracture process. The surface which was in contact with a polished stainless steel surface during cooling down after melting shows the highest strength. This could be due to residual compressive stresses in the surface of the specimen. The fracture stress for a fire polished surface (not shown in Figure 4) lies at about the same values as those measured on the fracture surface. This example is more or less typical for the behavior of all materials investigated.

A compilation of these results listing median strength values, i.e. the strength at P = .5, is given in Table 2. In addition to the test conditions mentioned above also a measurement obtained under water is included in Table 2. Water has, as usual, a detrimental effect on the measured fracture stress.

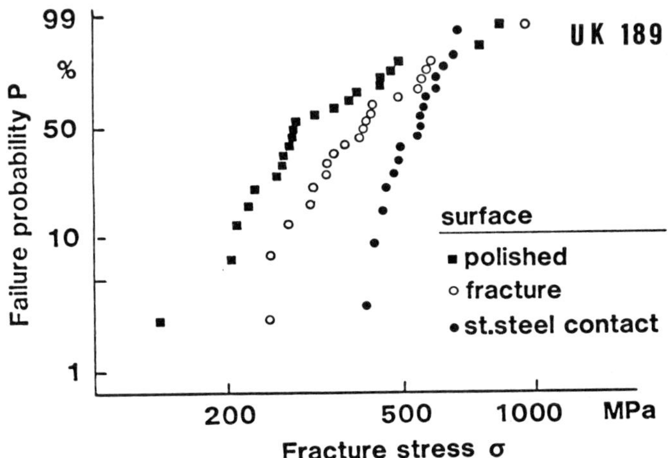

Fig. 4. Effect of specimen surface condition on the fracture stress distribution measured by Hertzian indentation.

TABLE 2

FRACTURE STRESS OF NUCLEAR WASTE GLASSES MEASURED BY HERTZIAN INDENTATION

Material	Fracture stress [MPa]				
	Surface condition				
	(0)	(1)	(2)	(3)	(4)
Soda-lime-silicate glass (SLS)	390				
VG 98/3		361	364	389	431
SON 58		428	443		531
UK 189		396	475		561
UK 209		397	406	559	598
DC 16		425			
DC-0-3		(550)			
DC-0-3 (under water)		(500)			
B-I-3		558			
Diopside		(700)			
Eukryptite		547			
Perowskite		(676)			

Surface condition: (0) as received float glass surface; (1) polished surface; (2) fracture surface; (3) flame polished surface; (4) surface obtained by cooling glass in contact with polished stainless steel.

In some glasses it could be observed that the ring crack disappeared during the microscopic observation; in some other cases no ring crack could be seen in the microscope though a marked acoustic emission signal had been observed. In these cases the radius of the contact circle was calculated according to equation (3) and this radius was used instead of the ring crack radius to calculate the fracture stress according to equation (2). This approach must obviously result in higher values of fracture stresses. In Table 2 parentheses are placed around values obtained from calculated contact circle radii.

The value of the strength of soda-lime silica glass measured by Hertzian indentation is about 390 MPa which is significantly higher than values obtained by commonly used strength test techniques such as 4 point bending, i.e. about 50 to 150 MPa. The high values obtained in the Hertzian test result from the fact that only a small area is stressed so that the probability to encounter a large flaw is accordingly small.

The values reported herein must, therefore, like all strength determinations be considered not to be intrinsic quantities, but must be compared only to values obtained under the same test and surface finishing conditions. In contrast to this restriction, the fracture mechanics values reported above can be considered to be independent on specimen size and test technique and can, therefore, be considered to be real material constants.

3.3. YOUNG'S MODULUS, COEFFICIENT OF THERMAL EXPANSION

Data for Young's modulus, measured between room temperature and 500 °C, and for the coefficient of thermal expansion, measured between 100 and 400 °C, are compiled in Table 3 and Table 4 respectively.

No significant irregularities were found for Young's moduli. The increase in slope of the coefficient of thermal expansion for the glass B-I-3 and DC-O-3 can be ascribed to the influence of crystal phases in these glasses.

Young's moduli and coefficients of thermal expansion lie higher than those known for soda lime silicate glass and borosilicate glass; this means that the nuclear waste glasses investigated, could be more susceptible to the development of thermal stress due to temperature gradients than the mentioned more normal glasses.

4. CONCLUSION

In order to provide data for a mechanical stability assessment of nuclear waste forms, fracture mechanics crack growth parameters, strength values, Young's modulus (between room temperature and 500°C) and the coefficient of thermal expansion (between 100 and 400°C) were measured.

As far as crack growth and strength is concerned the experimental results show that loading of borosilicate glass with waste oxides does not detrimentally influence the properties of the resulting glass; glass ceramics show an even improved behaviour.

Young's modulus and coefficient of thermal expansion also lie in the range of data known for ordinary glasses.

The results mean that the glasses investigated can be regarded to have a sufficient potential as a solidification product as far as their mechanical properties are concerned.

However, in order to make a complete assessment of the mechanical stability of glassy nuclear waste forms the stress distribution in the glass form under consideration and its flaw size distribution must be analyzed as well.

TABLE 3

EFFECT OF TEMPERATURE ON YOUNG'S MODULI OF NUCLEAR WASTE GLASSES

Material	Young's modulus [GPa]			
	23 °C	150 °C	300 °C	500 °C
DC-O-3	90.09	89.29	88.48	86.42
DC-16	83.14	77.75	76.91	71.63
UK 209	78.47	77.32	75.65	74.04
Diopside	95.99	93.89	93.06	
Eukryptite	90.49	89.36	88.32	85.99
Perowskite	83.91	80.36	77.06	76.61

TABLE 4

COEFFICIENT OF THERMAL EXPANSION AT DIFFERENT TEMPERATURES

Material	$\alpha \cdot 10^6$ K			
	20-100 °C	20-200 °C	20-300 °C	20-400 °C
SON-58	7.79	7.56	7.48	7.73
C31-3	8.21	7.95	7.91	8.23
UK 189	7.05	7.54	7.89	8.44
SM 58	8.05	8.17	8.41	8.88
UK 209	7.42	8.08	8.52	9.11
DC 03	7.69	8.47	9.96	>12.00
B1-3	8.11	9.02	10.77	10.56
DC 16	9.79	10.23	10.35	10.87
VG98/3	10.42	11.47	>12.00	>12.00

5. ACKNOWLEDGEMENT

Part of this work was supported by the Commission of the European Communities under Contract WAS-122-53 D.

6. REFERENCES

1. Baker, T.C. and Preston, F.W. (1946) J. of Appl. Physics, 17, 179-188.
2. Wiederhorn, S.M. (1967) J. of the Amer. Ceram. Soc., 50, 407-414.
3. Evans, A.G. and Wiederhorn, S.M. (1974) Int. J. of Fracture, 10, 379-392.
4. Outwater, J.O. and Gerry, D.J. (1966) Report on Contr. NONR 4219 Washington, D.C.
5. Evans, A.G. (1972) J. of Mater. Sci., 7, 1137-1146.
6. Lawn, B. and Wilshaw, R. (1975) J. of Mater. Sci., 10, 1049-1081.
7. Johnson, K.L. et al. (1973) Proc. of the Roy Soc. London, A334, 95-117.
8. Weibull, W.A. (1951) J. of Appl. Mechan., 18, 293-297.

SCIENTIFIC BASIS FOR RADIOACTIVE WASTE MANAGEMENT - V
Werner.Lutze, editor

FRACTURE APPRAISAL OF LARGE SCALE GLASS BLOCKS UNDER REALISTIC THERMAL CONDITIONS

FRANCIS LAUDE, ETIENNE VERNAZ, and MICHEL SAINT-GAUDENS
Départment de Genie Radioactif, C.E.A. Valrho, Site de Marcoule BP 171,
30200 Bagnols sur Ceze, France.

INTRODUCTION

The compositions of borosilicate glasses used to solidify fission product solutions are selected for their low leach rates. Fracturing of the glass blocks, caused primarily by thermal and residual stresses during cooling, increases the potential leaching surface area and the number of small particles, both of which are undesirable since long term container failure must be considered. It is therefore important to know the state of fracture of the glass occurring at different stages of its thermal history. A theoretical appraisal of the fracture state of the glass is presented, complimented by an experimental study using industrial scale glass blocks under realistic thermal conditions.

POSSIBILITIES AND LIMITS OF A THEORETICAL APPRAISAL OF FRACTURE

Stresses due to the cooling of a glass block[1]. When a glass block cools after casting, stresses appear because the surface cools faster than the core. The surface solidifies first around a more dilated core, which in turn solidifies later and compresses the surface. This stress can generally be avoided by annealing (i.e. the glass is brought to a sufficiently high temperature for a sufficiently long period of time to remove the stresses and then is slowly cooled). The residual stresses in the block are directly linked to the thermal gradients which exist at the moment of solidification. The lower the gradient $\frac{\partial \theta}{\partial x}$ the less the stress, in other words, the slower the cooling rate during solidification, the less the stress.

These stresses can also be created by blowing cold air over the hot glass, so that the surface is under compression (i.e. tempering). In addition to these residual stresses there are also transitory stresses during the cooling, the sign of which (tension or compression) depends on the increase or decrease of the local thermal gradient with time (sign of $\frac{\partial \left(\frac{\partial \theta}{\partial x} \right)}{\partial t}$ in x).

The cooling history of a fission product glass is intermediate between that of tempering and annealing. Independent of the technological problems posed, total annealing of the block is impossible, as the heat generated by the fission products imposes a thermal gradient between the core and the surface independent

of the cooling process. It is however important to reduce these stresses in
order to reduce or even eliminate fracturing of the block during cooling as
well as the residual stresses in the pieces which could cause ultimate fracturing by stress corrosion[2].

Thermal stresses which develop during the cooling of a glass block are extremely complex and the details difficult to analyze. The glass does not solidify at a given temperature, but there exists a temperature range above which
the stresses are almost instantly relaxed, and below which there is no relaxation. The extent of this temperature range depends on the cooling rate. For
example:

- 550°-500° for a tempered glass (taking into account the rapid cooling
       rate, relaxation becomes negligible below 500°).
- 550°-450° for an annealed glass (taking into account the cooling rate,
       relaxation becomes negligible below 450°).

Solidification begins when the surface reaches the upper limit of this temperature range and ceases when the core reaches the lower temperature limit.
For fission product glasses the cooling period is extremely slow, and the temperature range spans several hundred degrees. As it passes through these temperatures (taking several years), the thermal gradient of the block changes
creating new stresses called solidification stresses. These stresses are partially relaxed, where the temperature is sufficiently high. However, relaxation
for any point modifies the stress distribution in the whole block (including
points sufficiently cooled to be perfectly elastic) according to the relation

$$\bar{\sigma} = \frac{1}{V} \int \sigma(r, \, \theta, \, z) r \, dr \, d\theta \quad dz = 0 \tag{1}$$

Finally, when the core temperature is less than the minimum of the solidification field, the block behaves as an elastic material; at this point there is
a certain temperature gradient between the core and the surface. When this gradient disappears (after several hundred years), temperature equalization stresses (e.g. superficial compression, core tension) are generated. In the case of
an annealed glass, these are the only residual stresses.

Numerical calculus methods for stress calculation. It is possible to calculate the stresses completely at all points whatever their origin (e.g. solidification stresses, temperature equalization stresses, transitory stresses), as
long as the visco-elastic behaviour of the glass as a function of temperature
is known (law of relaxation). These methods have been successfully used by Lee
et al.[3] to calculate tempering stresses in flat glass.

A code perfected by CEA/DEMT for calculating thermal and mechanical stresses has been used with the "law of relaxation" of the type[4]:

$$\frac{d\sigma}{dt} = A\ (\theta)\ \cdot\ \sigma^2 \tag{2}$$

$$\text{with } A = A_o\ e^{(\alpha_1 \cdot\ \theta - \alpha_2)} \tag{3}$$

where A_o, α_1 and α_2 are constant, σ is the stress and θ is the temperature.

The calculation should be made at each stage for the whole volume, because the total deformation $\varepsilon(t)$ should satisfy equation (1).

The preliminary results are good. Figure 1 gives the evolution of the tangential component of stress on the surface of a cylinder with a heat generation of 26 W/l at the moment of vitrification. It is assumed that after 300 years heat generation is zero, as in the scenario of Fig. 2.

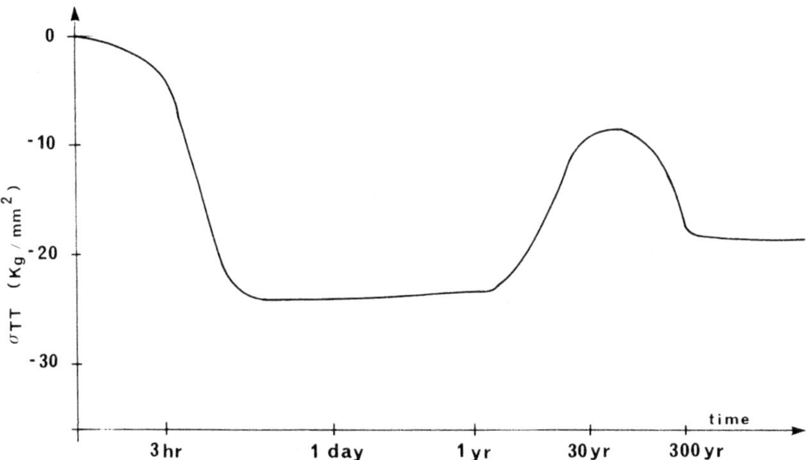

Fig. 1. Evolution of calculated stresses at the lateral surface of a glass block ($\emptyset$ 40 cm, H: 120 cm, heat generation 26 Wl^{-1})

In this case, the stresses σ_{TT} on the walls of the cylinder are always compressional. In fact $\dfrac{\partial \left| \dfrac{\partial \theta}{\partial x} \right|}{\partial t}$ is always negative, on the lateral surfaces, when the solidification field is reached. It is to be noted that due to the variation of the cooling scenario between two extremes (one open to the air, the other on a pedestal) tension components appear. The increase of σ_{TT} between t = 1 day and t = 3 years corresponds to solidification stresses. These stresses are partially

242

relaxed after 30 years. Between 30 and 300 years temperature equalization stresses increase and become permanent. Our program also makes it possible to consider composite structures in which the materials do not have the same Young's modulus, law of relaxation, coefficient of thermal dilation. So it is possible to calculate the effect of the metallic container on these stresses.

Resulting fracture rate. The determination of fracture rate, using the profile of the stresses in the block, is difficult, because fracturing depends on heat stresses and on their redistribution on the glass faults (e.g. fissures and inclusions). Fracturing occurs when, at a given point the stress intensity factor passes a critical value K_{IC}

$$K_I = \sigma \sqrt{a} \cdot f \text{ (geometry)} \tag{4}.$$

K_I depends not only on the stress, but also on a complex geometrical function of the block (f) and its flows (characteristic length, a)[5]. Experiments provide a better approach for linking the stress to the resulting fracture rate. Therefore the calculation of stress profiles hardly seems a realistic basis to estimate a fracture rate. On the other hand it is desirable to optimize the cooling histories in order to reduce the stresses. The method we propose is the following:

- selection of the critical cooling parameters (size and kind of container, heating or cooling during pouring)
- calculation of thermal profile in each case
- calculation of corresponding stresses (intensity and sign)
- optimization of cooling history to reduce stresses, taking technological constraints into account.

EXPERIMENTAL STUDIES

Method. The fracture state can be characterized by the granulometry of the pieces of glass or by the fracture ratio defined by

$$FR = \frac{\text{surface area of the cracked glass}}{\text{surface area of the uncracked glass}} \tag{5}.$$

Since it is difficult to determine the value of this characterization using theoretical calculations for stress, a series of experiments have been carried out. After pouring the inactive glass into graphite or stainless steel containers (diameter 300-400 mm), the critical stages of the thermal history are simulated, and then the glass is withdrawn from the mould. The pieces of glass are classified and the total surface area of the pieces is measured by comparison of the leach rates of the fractured glass to that of other cylindrical samples prepared under the same conditions.

Thermal history of glass blocks from a vitrification plant. First the glass is poured into a stainless steel container in the cell of the vitrification plant. The container can easily be cooled in a cylindrical water jacket. Temperatures reach equilibrium after one to two days. During this period the temperature decreases rapidly in the first few hours and the temperature gradients are important. After welding on the cover and decontaminating the walls (using pressurized water), the canisters are transferred to an interim surface storage site, consisting of concrete vaults ten meters deep, in which the containers are stored in a series of vertical pits.

After a brief forced-air cooling period, the natural convection created by stacks is sufficient. In this period the temperature of the glass decreases from a maximum initial temperature of $\leqslant 450^{\circ}$ C. The length of this interim storage period depends on whether or not natural convection can be used in the final repository. The following cases can be considered:

Short-term interim storage < 10 years, disposal in a compact way in a geologic formation with natural convection for 100-300 years. After closure the temperature rises slowly to a maximum of several tens of degrees C.

Interim storage of 30 years. Disposal in a less compact way in a geologic formation with immediate closure. In this case a maximum increase in temperature is reached in one to two years.

Simulation is of course limited to periods when the temperature variations are the most pronounced, particularly for the stages prior to ultimate disposal.

Cooling simulation before interim storage. This scenario represents the cooling of a glass block with a heat generation of 26 Wl^{-1} corresponding to PWR fission products solution vitrified after four years. During pouring and until temperature equilibrium is attained, the container is placed inside a cylindrical water jacket. By calculation the curves of decreasing temperatures can be obtained (Fig. 2).

244

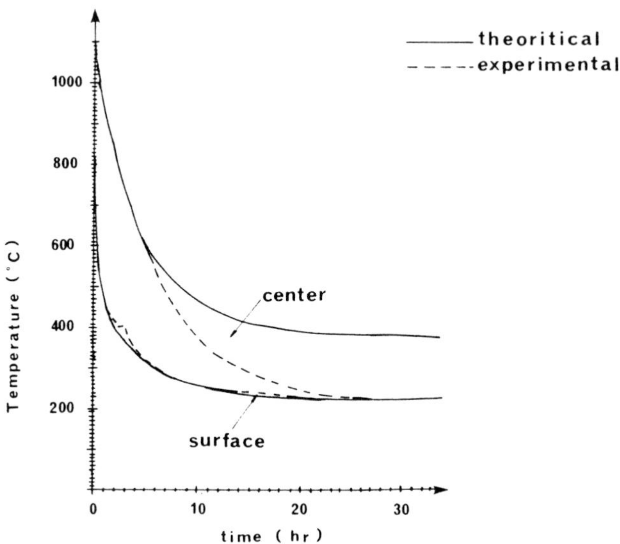

Fig. 2. Temperature evolution for cooling a glass block
($\emptyset$ 35 cm, heat generation: 26 Wl^{-1}) in the cell.

For this experiment, 130 kg of borosilicate glass was poured into a (35 cm
diameter) container and placed inside an electric furnace. It was then possible
to control the decrease of the surface temperature of the glass. The centerline
temperature of this inactive glass decreases slightly faster, but during the
first hours the test results are significant. Graphite and stainless steel containers were used for these experiments; with the graphite the glass is easily
withdrawn from the mould. The stainless steel container was especially designed
in four sections which were easily dismantled.

In both cases the glass was fractured into a relatively large number of
small pieces. To determine the surface area, the fractured glass was leached
in 94 liters of boiling water in a stainless steel vapour-heated tank. The
pieces of glass were permanently immersed, and a water renewal rate of 17 volume
percent per hour was produced by the reflux from a condenser.

Simultaneously with a smaller apparatus, three cylindrical (60 mm diameter)
glass samples were also leached. The ratio $\frac{\text{glass volume}}{\text{water volume}}$ and the renewal rate of
the water were the same. During the first fifteen days the curves of cumulative
sodium loss increase regularly and are similar.

TABLE 1

GRAVIMETRIC CLASSIFICATION OF GLASS FRAGMENTS

Class	< 1 g	< 10 g	< 100 g
Cumulative %	2,5	12,2	36,3
Class	< 500 g	< 1000 g	< 5000 g
Cumulative %	64,9	72,4	87,9

By comparing the cumulative sodium loss during this period the following fracture ratios were obtained:

graphite: FR = 10.3

stainless steel: FR = 9.0

Simulation of quenching due to water impact. If pressurized water is used to decontamine the walls of the container, a superficial fracturing of the block is observed. The higher the temperature of the wall, the more important the fracturing. At 50° to 100° C, the fracturing takes the form of a slight chipping along the periphery which is greatly accentuated over 200° C. To quantify this phenomenon, the glass containers were plunged into a small pool containing 1.2 m^3 of water at several different wall temperatures. Experiments were carried out to evaluate the maximal limit for the most drastic cases. A 320 kg glass block in a stainless steel container (diameter 0.4 m, height 1.1 m) was plunged into the water (core temperature = 800° C; the surface = 560° C). These temperatures represent the balance temperatures of a glass with a heat generation of 90 Wl^{-1} during the "before storage" stage. The temperature of the water rises by 25° C. The glass was found to be largely fractured with a large number of small particles. The fracture ratio measured by sodium leaching at room temperature was 16.6.

TABLE 2

GRAVIMETRIC CLASSIFICATION OF GLASS FRAGMENTS (FR = 16.6)

Class	< 1 g	< 10 g	< 100 g
Cumulative %	1,3	8,3	49,5
Class	< 500 g	< 1000 g	< 5000 g
Cumulative %	74,5	82,9	100

A more realistic experiment was carried out using a 130 kg glass block in a steel container (diameter 0.35 m) previously submitted to the "before storage" cooling scenario. The surface temperature was 220°C when plunged into the pool, and by using the same sodium leaching process the fracture ratio was found to be 12.

TABLE 3

GRAVIMETRIC CLASSIFICATION OF GLASS FRAGMENTS (FR = 12)

Class	< 1 g	< 10 g	< 100 g
Cumulative %	3,7	20,3	45,2
Class	< 500 g	< 1000 g	< 5000 g
Cumulative %	60,1	67	87,2

Simulation of placing a container into an interim storage pit. When the container is placed in the pit its walls are cooled by air having a temperature of 30°-80° C, whilst the wall temperature itself can reach approximately 200° C. In order to simulate this, a small insulated experimental pit was constructed in which it is possible to vary the air-flow rate. At this time no final results are available, but preliminary data indicate that the effects are slight.

Experimental reassembly of fractured glass. Glass fractured during the scenarios previously described was kept in its container at a temperature approximately equivalent to that of its softening point for a short time in order to avoid crystallization.

By maintaining the glass at 800° C for twenty four hours and cooling it at 5°C per hour below 650°C, the block was perfectly reconstituted, with no significant crystallization. When kept for six hours at 600° C, the reconstitution caused individual fragments to adhere to one another and avoid particle dispersion; however, the reconstituted block remained vulnerable to mechanic shock.

CONCLUSIONS

The theoretical studies show that it is possible to calculate the stresses created for different cooling histories, but it is difficult to evaluate quantitatively the state of fracture of a glass, because this depends on the presence and distribution of inclusions and faults.

The experimental studies concerning large glass blocks allow order of magnitude estimates of fracturing. It is difficult to avoid such fracturing, especially during the cooling period after casting in a hot cell.

Short periods of reheating may cause the physical reconstitution of the glass. The heat created by fission product decay may contribute to this process.

REFERENCES

1. Gardon, R. (1980) "Thermal tempering of glass" in Glass Science and Technology, Vol. 5: Elasticity and Strength in Glasses. Ed. by D.R. Uhlmann and N.J. Kreidl, p. 145.
2. Wiederhorn, S.M. and Bolz, L.W. (1970) J. Am. Ceramic Soc. 53, (10), pp. 543-548.
3. Lee, E.H., Rogers, T.G. and Woo, T.C. (1965) J. Am. Ceramic Soc., 48, pp. 480-487.
4. Adams, L.H. and Williamson, E.D. (1977) "Annealing and Strengthening in the Glass Industry". Ed. by D. Alexis, G. Pincus and Thomas R. Homes, p. 3.
5. Bui, H.D. (1978) "Mecanique de la Rupture Fragile", Ed. Masson.

AN ATTEMPT TO ASSESS THE LONG-TERM CRYSTALLIZATION RATE OF NUCLEAR WASTE GLASSES

NOËL JACQUET-FRANCILLON, FRANÇOIS PACAUD and PIERRE QUEILLE
Départment de Genie Radioactif, C.E.A. Valrho, Site de Marcoule BP 171,
30200 Bagnols sur Ceze, France

INTRODUCTION

The crystallization of fission product glasses is generally caused by maintaining these glasses at high temperatures (600° to 900° C) for several days or weeks in order to evaluate the effects of crystallization on leach resistance and mechanical properties.

The examination of glasses subjected to such thermal treatment involves the identification of the crystalline phases obtained[1] and the determination of T.T.T. diagrams[2]. Over the past few years it has appeared possible to measure the crystalline portion of these glasses by x-ray diffraction[3,4], by image analysis[4] or by measuring specific heat[6]. An exhaustive paper dealing with measurement problems has appeared recently[6].

At the C.E.A Marcoule, isothermal calorimetry has been used to measure the thermal release which accompanies crystallization in order to attempt to quantify the devitrification capacity of the different glasses tested and to establish the time-temperature relationships. Such data could allow for long-term, low temperature extrapolation of the behaviour of the glass during geological storage. The present work describes the results obtained using three non-radioactive glasses of increasing chemical complexity.

EXPERIMENTAL

The following three glasses were chosen for their varied but well known devitrification yields:
. lithium disilicate 2 $SiO_2 \cdot Li_2O$; 100 % volume in crystals.
. SAN 38 33 16 F' 1001 glass (for vitrification of MTR solution) 56 % vol. in
  crystals.
. SON 58 30 24 U2 glass (for vitrification of LWR solution) 5 to 6 % vol. in
  crystals.

Preparation. The three glasses were melted at 1300° C in a platinum crucible; the composition of the mixed constituents is given in Table 1.

TABLE 1

COMPOSITION (wt.%) OF THE GLASSES

Reference	SiO_2	Al_2O_3	B_2O_3	Na_2O	Li_2O	Ox.FP.	Diverse
$2\ SiO_2 \cdot Li_2O$	80.0				20.0		
SAN 383316 F' 1001	29.85	25.86	15.15	23.88		2.37	2.39
SON 583024 U2	42.17		23.16	9.17		22.00	3.50

The fused glasses are cast directly in the measuring crucibles (sintered alumina for the fission product glasses, graphite for the $2\ SiO_2 \cdot Li_2O$). Before measuring a sample is extracted of each type of glass and observed through an optical microscope by reflected and transmitted light.

Apparatus. A high temperature model SETARAM microcalorimeter was used. The apparatus essentially consists of two fluxmeters (measure and reference) placed in a thermostatically controlled bloc that can be heated and held constant at any temperature between ambient and 1000° C. The limit of detection for thermal effects is in the order of 0.85 joule h^{-1}. The samples are first positioned slightly above the fluxmeters at a temperature around the glass transition region, which makes it possible to trigger off the nucleation of the vitreous bulk (Table 2) and are then placed in their final position in the two fluxmeters.

TABLE 2

NUCLEATION TREATMENT OF THE GLASSES

Reference	First position	Glass transition temperature
$2\ SiO_2 \cdot Li_2O$	4h30' at 450° C	478° C
SAN 383316 F' 1001	30' at 440° C	460° C
SON 583024 U2	35' at 500° C	525° C

The energy released by devitrification after a period of time t, Qt, is calculated by integrating the curve of the thermal output q as a function of time

$$Q(t) = \int_{o}^{t} qdt = K \cdot S(t)$$

K = calorimeter constant,

S(t) = area below the curve at time t.

The exothermic reaction is considered to be over when $\frac{\Delta q}{\Delta t} < 0.85$ joule h^{-1}. If we consider that the total heat released, Qc, represents maximum devitrification, α_c, expressed in volume units, then after a time t

$$\frac{\alpha(t)}{Q(t)} = \frac{\alpha_c}{Q(t_c)}$$

$0 < \alpha(t)$ and $\alpha_c \lesssim 1$; α_c = maximum volume fraction which may potentially crystallize; $\alpha(t_c) = \alpha_c$ when $t = t_c$; t_c = reaction end time.

LITHIUM DISILICATE : $2 \, SiO_2 \cdot Li_2O$

This glass may be devitrified 100 percent: $\alpha_c = 1$. Hence, $\alpha(t)$, noted hereafter α, represents the total volume fraction crystallized after a time t at a given temperature.

$$\alpha = \frac{Q(t)}{Q(t_c)}$$

Six isothermic runs were carried out at different temperatures. The graphic representation of α as a function of time and temperature is given in figure 1.

TABLE 3

RESULTS FOR LITHIUM DISILICATE

Temperature (^{o}C)	Θ	500	532	542	562	576	580
Mass (g)	m	8.27	8.40	8.62	9.25	8.16	8.15
Time (s)x10^4	t_c	86.04	23.04	16.92	11.88	8.46	7.20
Total heat (joule)x10^3	Qt_c	2.541	2.597	2.629	2.956	2.330	2.527
Crystallized volume (cm^3)	V	3.38	3.43	3.52	3.78	3.33	3.33
Devitrification heat (joule·cm^{-3})	Qc_v	752	757	747	782	700	759

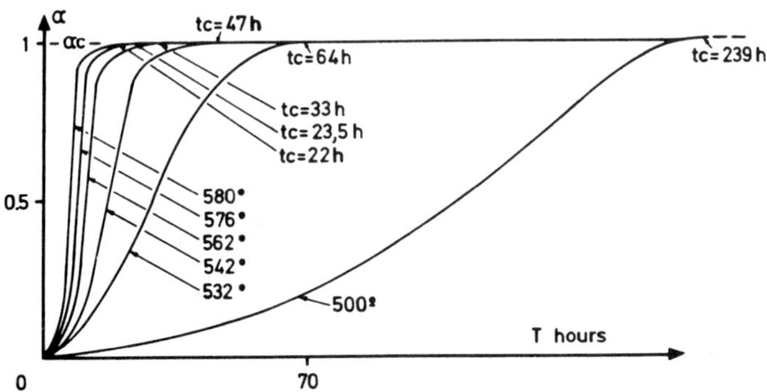

Fig. 1. α evolution for 2 $SiO_2 \cdot Li_2O$ as a function of time and temperature.

Discussion. The devitrification heat measured, 750 joule cm^{-3}, is higher than that found by measuring the dissolution heat[7], 611 joule cm^{-3}. On the other hand it differs little from that measured by temperature scanning or by placing the sample into position at constant temperature, 807 joule cm^{-3}. We have arbitrarily taken this value as corresponding to a 100 % crystallization of the material.

The devitrified volume fraction of a glass α, during isothermal treatment is usually given by the equation[8,9]:

$$\alpha = 1 - e^{-(\frac{t}{\tau})^k}$$

The calculation of k for two degrees of crystallization, 25 % and 75 %, and for each of the six test temperatures gives k_{ex} = 2.98 with σ_k = ± 0.21. It is thus experimentally verified that k = 3 and that bulk nucleation is mainly homogeneous with three-dimensional crystal growth[10]. The constant τ is independent of time and of α, and varies as a function of time according to a law of Arrhenius, as does the reaction end time, t_c, measured experimentally using the calorimeter (Table 4).

$$\log \bar{\tau} = \frac{K}{T} - K'$$

with K = 32,267 s K^{-1}, K' = 28.78 s. T is the temperature in Kelvin (K = E/R with E, the activation energy and R, the gas constant). If we pose $\frac{1}{\bar{\tau}} = e^{K'} \cdot e^{-\frac{K}{T}}$ we can calculate the activation energy, E, of the devitrification reaction. E was calculated to be 265 K Joule·$mole^{-1}$. This value is near to that proposed by other authors using DSC[11] or micro-DTA[12,13].

α may be expressed:

$$\frac{1}{3} \text{Log} \left[\text{Log} \left(\frac{1}{1-\alpha} \right) \right] = \text{Log } t - \frac{32,267}{T} + 28.78$$

$$(t \text{ in seconds, } T \text{ in K}).$$

Empirically, the reaction end time, t_c, measured isothermally by microcalorimetry, is linked to the temperature (Fig. 2):

$$\text{Log } t_c = \frac{19,473}{T} - 11.6826$$

TABLE 4

TIME-TEMPERATURE RELATION FOR 2 $SiO_2 \cdot Li_2O$

Temperature (^{O}C)	500	532	542	562	576	580
$\frac{10^3}{T(K)}$	1.2937	1.2422	1.227	1.1976	1.1779	1.1723
$\bar{\tau}(s) \times 10^4$	47.2	7.82	4.04	1.89	1.25	0.80
Log $\bar{\tau}$	13.065	11.267	10.607	9.847	9.433	8.982
Log t_c	13.665	12.348	12.039	11.685	11.346	11.184

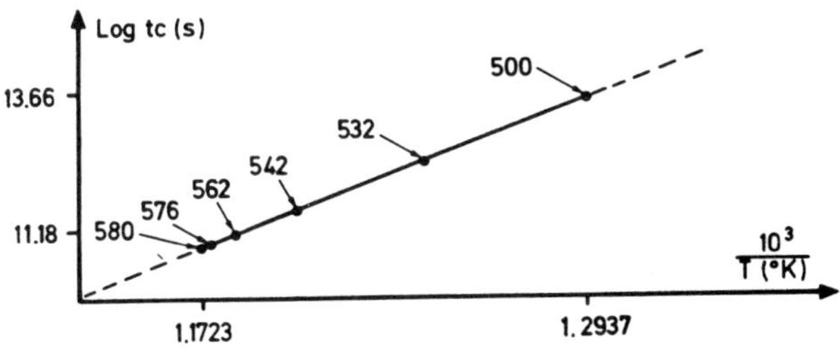

Fig. 2. Empirical time-temperature relation at maximum crystallization for 2 $SiO_2 \cdot Li_2O$.

The values which can be calculated for low temperatures, 300 K, lead to astronomically long periods of time, $4 \cdot 10^{15}$ years, comparable to those already put forward[14].

From a practical point of view, the devitrification heat measured by isothermal microcalorimetry is nearly the same, although slightly lower than that measured by temperature scanning.

Samples held isothermally at the same temperatures but without prior nucleation always lead to longer times for a given crystallization rate than measured by X-ray diffraction.

APPLICATION TO FISSION PRODUCT GLASSES

SAN 38 33 16 F' 1001 Glass. This test glass, suitable for the vitrification of alumina loaded solutions with a low fission product oxide content was chosen because it is the most easily crystallized of the glasses selected for thermal stability tests. A study of devitrification products reveals two crystalline phases: (1) Nepheline, $NaAlSiO_4$, is predominant and has a rapid growth rate. The volume fraction does not appear to go over 52 % even after several months of thermal treatments[15,16]; (2) probably a sodium aluminoborosilicate, a microcrystalline phase, developing after the appearance of the nepheline. The volume of this secondary phase was unmeasurably small even after heating the sample 92 hours at 706° C. Four isothermal runs were carried out at different temperatures (Table 5). The graphic representation of the ingrowth of crystalline material as a function of time and temperature is given in figure 3.

TABLE 5

RESULTS FOR SAN 38 33 16 F' 1001 GLASS

Temperatures (°C)	617	659	697	749
m(g)	21.46	15.30	18.95	12.82
t_c (s) x 10^4	13.68	7.92	5.04	3.24
Qt_c (joule) x 10^3	3.09	2.18	2.63	1.82

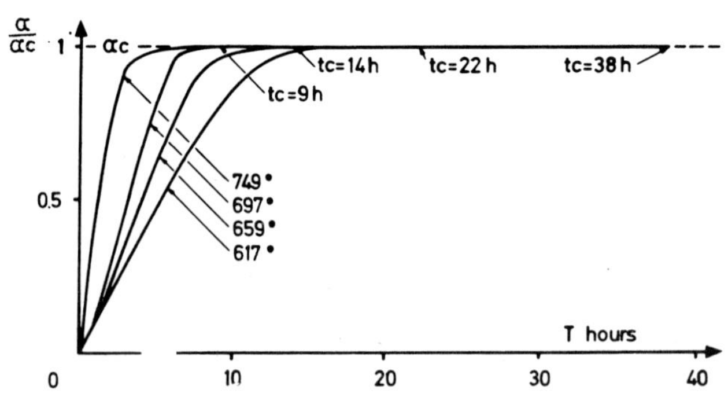

Fig. 3. α evolution for SAN 38 33 16 F' 1001 glass as a function of time and temperature.

Discussion. We may use the following relation[8,9]:

$$\frac{\alpha}{\alpha c} = \frac{Q(t)}{Q(t_{1}c)} = 1 - e^{-\left(\frac{t}{\tau 1}\right)^{k_{1}}}$$

Calculation of the coefficient k_1 using the method utilised for $2\,SiO_2 \cdot Li_2O$ gives $k_{1_{ex}} \simeq 1.46$ with $\sigma_{k_1} = \pm 0.20$ (Table 6). k differs little from 3/2, which is the value proposed[13] in the case of crystalline growth controlled by a diffusion mechanism with a fixed number of nuclei.

TABLE 6

TIME-TEMPERATURE RELATION FOR SAN 38 33 16 F' 1001 GLASS

Temperature ($^{\circ}$C)	617	659	697	749
$\dfrac{10^3}{T}$ (K)	1.1236	1.073	1.0309	0.9785
$\bar{\tau}1$ (s) x 10^4	2.21	1.69	1.36	0.475
Log $\bar{\tau}1$	10.003	9.735	9.518	8.466
Log t_{1c}	11.8263	11.2797	10.8277	10.3859

The time constant τ_1, as well as the end reaction time t_{1c}, obey an Arrhenius law.

$$Log\ \bar{\tau}_1 = \frac{K1}{T} - K'1$$

with $K1 = 10,201$ s K^{-1}, $K'1 = 1.296$ s.

The volume fraction $\dfrac{\alpha}{\alpha c}$ is given by the relation:

$$\frac{3}{2}\ Log\left[Log\left(\frac{\alpha c}{\alpha c - \alpha}\right)\right] = Log\ t - \frac{10,201}{T} + 1.296$$

The activation energy, E_1, of the reaction is 85 kilojoule mole^{-1}.
The experimental reaction end time, t_{1c}, is such that

$$Log\ t_{1c} = \frac{9,985}{T} + 0.5807$$

(T in K, t_{1c} in seconds).

As for $2\,SiO_2 \cdot Li_2O$, the volumetric devitrification heat is constant for the range of temperatures studied, and is close to that of $2\,SiO_2 \cdot Li_2O$, considering that the volume of devitrified glass is practically that of the nepheline for a short-term test (Table 7). Finally, as for the disilicate, the devitrification heat measured by DSC differs not more than 5 % of that found isothermally.

TABLE 7

VOLUMETRIC DEVITRIFICATION HEAT OF SAN 38 33 16 F' 1001 GLASS

Temperature ($^{\circ}$C)	617	659	697	749
Devitrified volume (cm^3)	4.29	3.06	3.79	2.56
Qc (joule cm^{-3})	720	712	694	711

SON 58 30 24 U2 Glass. This glass containing 22 % by weight of fission product oxides has been developed in the course of LWR solution vitrification
studies. It appears to be the most easily crystallized of the glasses in its
category, and has been the object of an in-depth study[4]. Four crystalline
phases have been identified after thermal treatment:

. $SrMoO_4$ - may contain lanthanides
. $(Ce,U)O_2$ - cerium is predominant
. $(Mn,Ni)_3O_4$ - spinel structure type
. a fourth phase of small volume which may be cristobalite.

It is to be noticed that due to the low content in Na_2O this glass has a tendency to show phase separation which is not evenly distributed in the matrix.
The sum of these devitrification products represents 10 % by weight or 5 to 6 %
of the volume.

Nine isothermal runs were carried out at temperatures from 586° tp 761°. The
graphical representation of the evolution of the percentage of the total crystallized phase is given in figure 4.

TABLE 8

RESULTS FOR SON 58 30 24 U2 GLASS

Temperatures ($^{\circ}$C)	586	586	600	620	655	656	671	689	761
m(g)	29.5	27.4	23.3	26.3	22.9	24.1	23.6	23.6	25.5
t_{2c} (s) x 10^4	25.5	25.2	23.0	19.4	18.0	18.0	16.2	14.4	7.75
Qc (joule) x 10^2	4.44	4.90	3.73	4.34	2.56	3.73	3.73	3.78	1.80

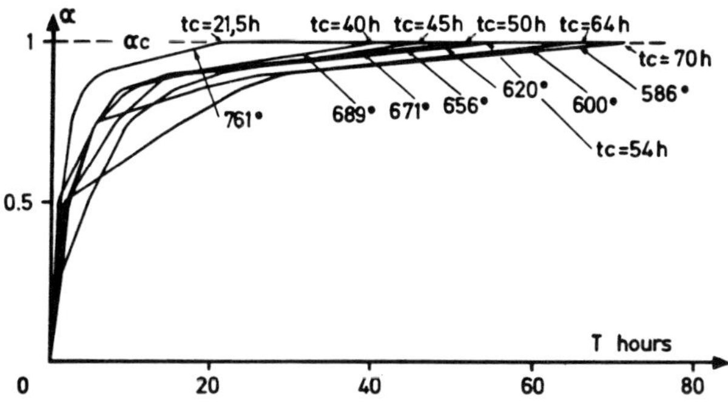

Fig. 4. α evolution for SON 58 30 24 U2 glass as a function of time and temperature.

Discussion. The use of previous equations[8,9] does not give a constant value for the coefficient k (0.42 < k < 1.06). Thus it is difficult to give a simple description of the evolution of the total crystallization rate of this glass. However, the reaction end time, t_{2c}, follows Arrhenius behaviour (Fig. 5).

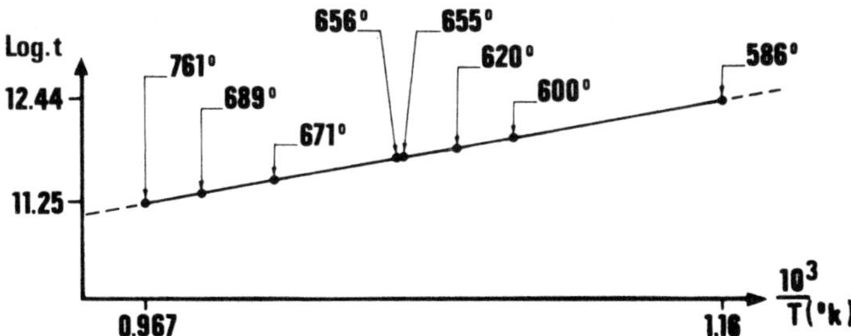

Fig. 5. Time-temperature relation for SON 58 30 24 U2 glass.

$$\text{Log } t_{2c} = \frac{5,463}{T} + 6.1264 \text{ (t in seconds, T in K)}.$$

The time which can be calculated from this relation is comparable to that found after thermal treatment at 760° C[4], and the volume fraction of the two samples measured by calorimetry and X-ray diffraction is also in the same range, 5 to 6 % volume. Table 9 gives the volumetric devitrification heat of SON 58 30 24 U2 glass calculated on this basis. Finally, exactly as for the disilicate and the SAN 38 33 16 F' 1001 glass, the devitrification heat measured by DSC differs not more than 10 % of that found isothermally.

TABLE 9

VOLUMETRIC DEVITRIFCTAION HEAT OF SON 58 30 24 U2 GLASS

Temperature (^{O}C)	585	586	600	620	655	656	671	689	761
Volume of de-vitrified glass (cm^3)	0.59	0.55	0.47	0.53	0.46	0.48	0.47	0.47	0.51
Volumetric devi-trification heat (joule cm^{-3})	752	891	794	819	556	776	793	804	353

CONCLUSIONS

There exists an empirical Arrhenius type relationship between time and temperature at maximum crystallization for temperatures above Tg; the apparent activation energy is lowered by an increase in the number of crystalline phases forming in the glass. A maximum crystallized volume fraction can be calculated by measuring the devitrification heat output either by isothermal calorimetry or by slow DSC. The volumetric devitrification heat is largely independent of the nature of the crystals formed. However, the microcalorimetric sensitivity limits this method to glasses which devitrify by more than 2 to 3 % of their volume in a short time.

The utilization of these empirically calculated relationships for the determination of a crystallization rate below the glass transformation temperature, T_g, should be done with great caution, because at this temperature other phenomena control changes of the glass structure such as devitrification.

REFERENCES

1. Bonniaud, R., Pacaud, F., Sombret, C. (1975), Verres et Réfractaires, Vol 29 n° 1, 25-33.
2. Dalton, J.T., Boult, K.A., Chamberlain, H.E., Marples, J.A.C. (1981), Proceedings of the International Seminar on Chemistry and Process Engineering for High-Level Liquid Waste Solidification - Kernforschungsanlage Jülich, Odoj, R., Merz, E. ed.
3. Turcotte, R.P., Wald, J.W. and May, R.P. (1981) Boston, MRS Symposia Proceedings, Vol. 6, Scientific Basis for Radioactive Waste Management, Topp, S. ed., Elsevier - North-Holland.
4. Morlevat, J.P., Uny, G., Jacquet-Francillon, N. (1981), Proceedings of the International Seminar on Chemistry and Process Engineering for High-Level Liquid Waste Solidification - Kernforschungsanlage Jülich, Odoj, R., Merz, E. ed.

5. Boulos, E.N., Depaula, R.P., EI-Bayoumi, O.H., Lagakos, N., Macedo, P.B., Moynihan, C.T., and Rekhson, S.M. (1980). Journal of the American Ceramic Society, Vol 63, 9-10, pp. 496-501.
6. Hinz, W. (1977) Journal of Non-Crystalline Solids, Vol 25 (1-3), pp. 217-260.
7. Temple, P., Zarzycki, J. (1975) Compte rendu Académie des Sciences Paris Tome 280 (23/6/75) Série C-1443 à 1446.
8. Johnson, W.A., Mehl, R.F. (1939) Trans. Am. Inst. Min. Metall. Engrs 135 p. 416.
9. Avrami, M., Journal of Chemical Physics - 7 (1939) p. 1103
                                              - 8 (1940) p. 212
                                              - 9 (1941) p. 177.
10. Hautojärvi, P., Vehanen, A., Komppa, V., Pajanne, E. (1978) Journal of Non-Crystalline Solids 29, pp. 365-381.
11. Matusita, K., Sakka, S. (1979) Physics and Chemistry of Glasses, Vol 20, No. 4, August 1979, pp. 81-84.
12. Matusita, K., Sakka, S., Matsui, Y. (1975) Journal of Materials Science 10, pp. 961-966.
13. Marotta, A., Buri, A. (1978) Thermochimica Acta 25, pp. 155-160.
14. Uny, E., Nauer, G. (1978) Journal of Non-Crystalline Solids 29, pp. 207-214.
15. Morlevat, J.P. (1980) CEA/CEN Grenoble - personal communication.
16. Morlevat, J.P. (1981) CEA/CEN Grenoble - personal communication.

SCIENTIFIC BASIS FOR RADIOACTIVE WASTE MANAGEMENT - V
Werner.Lutze, editor

QUANTITATIVE DETERMINATION OF CRYSTALLINE PHASES IN NUCLEAR WASTE GLASSES

R. H. FELD AND M. STAMMLER
Battelle-Institut e.V., Am Römerhof 35, D-6000 Frankfurt am Main

INTRODUCTION

In order to assess the longterm chemical and physical stability
of high level waste glasses it is important to know the amounts
of crystalline phases present.
In principal all information about the crystalline state can
be derived from its X-ray powder diffraction pattern,provid-
ed the spectral resolution is sufficiently high. Powder-
diffraction techniques are the only means available to
determine the concentration of crystalline phases in the
presence of equivalent amorphous components. In addition it
permits the determination of specific crystalline phases in
a multicomponent system.
To test the merits of the quantitative X-ray powder diffraction analysis for the characterization of nuclear waste glasses,
measurements were carried out to
- determine relatively small weight fractions of crystalline
  phases next to high concentrations of amorphous materials
- test whether components remain stable in their crystalline
  state, yield new crystalline compounds or dissolve in the
  glassmatrix
- determine the total crystallinity of nuclear waste products.
The "matrix flushing" method (1) was chosen since it was considered to be accurate, fast, and relatively easy to apply
(2).

MATRIX-FLUSHING

The original matrix flushing procedure required a binary
reference mixture of each phase i to be determined and
exactly the same amount of the flushing agent c which may
be any well crystallized compound not present in the sample.
After addition of a known weight fraction of flushing agent
to the sample, it should "flush out"mass absorption, diffract-
ing power and instrument geometry. The method was simplified,
according to (3) yielding the following equation:

$$K_i = \frac{x_c}{x_i} \cdot \frac{I_i}{I_c} \qquad (1)$$

K_i = relative intensity ratio

x_c = wt.fraction c

x_i = wt.fraction phase i

I_c = intensity of c

I_i = intensity of phase i

Thus n relative intensity ratios K_i could in fact be
determined using only one reference mixture with n + 1
phases, where one of them is the flushing agent.

After known weight fractions x_i and x_c have been mixed, only the strongest intensity I_i for each phase and the strongest intensity I_c of the flushing agent are being measured. To determine the weight fraction x_i' of the sample according to

$$x_i' = x_c' \cdot \frac{1}{K_i} \cdot \frac{I_i'}{I_c'} \qquad (2)$$

a known wt. percentage x_c' of flushing agent is added and the intensities I_i' and I_c' are measured.

Integral intensities are used rather than peak height intensities, as the latter are extremely sensitive to differences in line width arising from lattice distortions and crystallite size.

SAMPLES AND MEASUREMENTS

Although the simulated radwaste-glass mixtures were made for different purposes, they were found to be ideally suited as test substances.
The simulated waste contained 34.2 wt% ruthenium, a substance which does not change during processing; 17.3 wt% Zr which was expected to partly oxidise and dissolve in the glass matrix, and 32.2 wt% MoO_2 which could react with components of the glass. The remaining components were present at such low concentrations that they were below the detection threshold when diluted with 85-90 wt% of glass.

A lead borosilicate glass (BBS 10) and a lead silicate glass (BS Mn8) were used. Their compositions are shown in table 1. Pure BsMn8 did not show any crystalline phases, BBS 10 contained approximately 0,5 wt% of crystalline SiO_2.

Table 1: Composition of waste glasses

	BSMn 8	BBS 10
SiO_2	32 wt%	42 wt%
PbO	54 wt%	36 wt%
B_2O_3	–	12.6 wt%
CaO	–	1.8 wt%
MgO	–	0.9 wt%
Al_2O_3	–	6.7 wt%
MnO	14 wt%	

To prepare the samples 10 to 15 wt% of simulated radwaste was mixed with each glassfrit. Water or 0.7 m HNO_3 were used as suspension agents. The mixture was then formed into pellets of about 10 mm diameter.

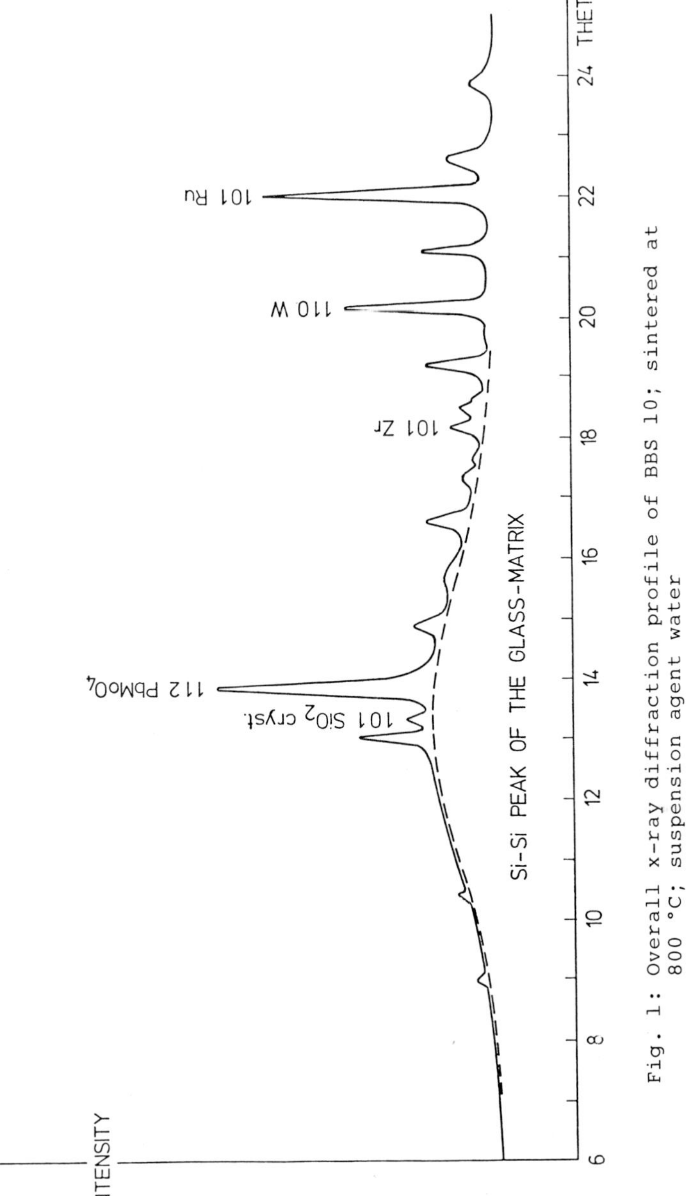

Fig. 1: Overall x-ray diffraction profile of BBS 10; sintered at
800 °C; suspension agent water

After drying at 500°C the samples were sintered at temperatures
between 800°C and 900°C for 10 min. One sample was tempered
at 600°C for 12h after sintering.
The samples were ground into a fine powder ($<$10 μ) using a
corundum mortar. Special care was taken to avoid prefered
orientation when filling the powders into the sample-holders.
The overall diffraction diagram (fig. 1 for BBS 10 with
15 wt% radwaste) shows that the strongest reflections are the
1 0 1 of ruthenium, the 1 0 1 of zirconium and 1 1 2 of
wulfenite ($PbMoO_4$). The latter orginates from the reaction
$PbO+MoO_3 \longrightarrow PbMoO_4$ under sintering conditions (4). In
BBS 10 the 1 0 1 reflection of quartz was present as well.

For the measurements a Siemens X-ray-powder diffractometer
with CuKα -radiation was used. The data were recorded by a
Siemens 300-310 computer. The strongest reflection of each
compound was measured by step scans (step width 0.02° (2 $\emptyset$)).
The recording time was four seconds per step for the measurement of the reference intensity ratios (table 2), as well as
for the measurements of the actual samples (Table 3).

DATA ANALYSIS

Data analysis for the samples and the reference mixtures was
carried out using programmes of the Siemens DIFFRAC 310 software (5). From the raw data approximate peak heights and full-
width at half maximum (FWHM) were determined. They served
as starting parameters for full matrix least square calculations using the Marquart (6) algorithm to match doublets of
$K_{\alpha1}/K_{\alpha2}$ lines with the analytical shape of Gaussian curves
to a series of measured points. The measured and the fitted
curves were recorded. After five refinement cycles excellent
agreement between the fitted and the measured profiles was
obtained. The free variable parameters were peak height, full
width at half maximum and the peak position. The integrated
intensities used for the quantitative determination of the
crystalline phases were computed from the calculated profiles.
The errors of the intensity ratios and the errors of the calculated weight fractions were determined by error propagation
from integral intensities errors, which were derived from counting statistics. The errors in the weighted samples were considered to be small enough to be neglected.

RELATIVE INTENSITY RATIOS

The 1 1 0-reflection of tungsten did not overlap with any of
the reflections in the sample - so tungsten was chosen as
flushing agent and between 1.5 and 2.6 et% of tungsten was
added to the samples during crushing.
Two reference mixtures were prepared, one containing 20 wt%
each of tungsten, ruthenium and wulfenite plus 40 wt% of quartz
the other containing 50 wt% of both tungsten and zirconium.
Measurements of the integral intensities from these mixtures
were used to obtain the relative intensity ratios (K_i) to an
absolute accuracy between 5 and 10% as given in table 2. They
served as the basis for the subsequent quantitative determinations.

Table 2: relative intensity ratios

reflection	relative intensity ratio (K_i)
W 110	1.0
Ru 101	0.67 ± 0.01
$PbMoO_4$ 112	0.75 ± 0.01
$SiO_{2\,Crist}$ 101	0.48 ± 0.01
Zr 101	0.44 ± 0.01

RESULTS AND DISCUSSION

The overall profile is shown for BBS 10, with 15 wt% rad waste
suspended in water and sintered at 800 °C (fig.1). The profiles
for the samples containing BSMn8 are similar, exept that no
crystalline SiO_2 could be detected.

The high amorphous background dominates in the entire spectrum
with a superimposed Si-Si peak having a maximum at a value of
approximately $26°$ 2θ.
In general the background intensities are relatively high
(Fig. 1; Fig. 2).

Neither MoO_3 nor ZrO_2-peaks are found in the spectra.

As expected for the strong peaks of ruthenium, tungsten and
wulfenite, the measured profiles agree excellently with the
fitted Gaussian curves (fig. 2a-c). But even the smaller
peaks of quartz and zirconium are well resolved and can be
fitted (fig. 2 d-e).
The results of the quantitative analysis for all seven
samples are compiled in table 3. For the three samples with
the frit BSMn8, all determined weight fractions agree well
with the expected values. The reaction of lead oxide with
molybdenum oxide to wulfenite appears to be quantitative
within the experimental error.
For BBS 10 different results were obtained. Only half of the
expected weight fraction of wulfenite has been found. As
BBS 10 contains some calcium oxide it is likely that
$Ca_xPb_{1-x}MoO_4$ is produced due to the complete miscibility
of the system $CaMoO_4-PbMoO_4$ (7).

Table 3a: Results of Quantitative X-ray Analysis for Glass BSMn8

sample	wt % rad waste	sintering temperature ^{o}C	suspension agent	wt % W added	reflection	integral intensity	wt % initial	crystal phase calculated from X-ray data	
1	10	800	H_2O	2.00	W-110	14231± 651			
					$PbMoO_4$-112	64654±2097	–	12.1	1.1
					MoO_3	–	3.2	4.8	0.4*
					Ru-101	15845± 715	3.4	3.3	0.4
					Zr-101	4598± 800	1.7	1.5	0.4
2	15	800	H_2O	1.99	W-110	14897±1081			
					$PbMoO_4$-112	57086±2162	–	10.2 ± 1.3	
					MoO_3	–	4.8	4.0 ± 0.5*	
					Ru-101	24357± 921	5.1	4.9 ± 0.6	
					Zr-101	5574± 879	2.5	1.7 ± 0.4	
3	15	800	0.7 m HNO_3	2.68	W-110	20100± 969			
					$PbMoO_4$-112	62986±2249	–	11.2 ± 1.1	
					MoO_3	–	4.8	4.4 ± 0.4*	
					Ru-101	22218±1089	5.1	4.4 ± 0.5	
					Zr-101	6425± 869	2.5	1.9 ± 0.4	

*wt-fraction of MoO_3 calculated from the value for $PbMoO_4$

Table 3b: Results of Quantitative X-ray Analysis for Glass BBS 10

sample	wt % rad waste	sintering temperature $^\circ$C	suspension agent	wt % W added	reflection	integral intensity	wt % initial	crystal phase calculated from X-ray data
1	15	900	H_2O	2.03	W-110	24370±1108		
					PbMoO$_4$-112	39082±2126	–	4.3 ± 0.5*
					MoO_3	–	4.8	1.7 ± 0.2
					SiO_2-101	2562±2109	–	0.4 ± 0.4
					Ru-101	38617±1148	5.1	4.7 ± 0.4
					Zr-101	6971± 791	2.5	1.3 ± 0.2
2	15	800	H_2O	2.14	W-110	25920±1104		
					PbMoO$_4$-112	46215±2110	–	5.1 ± 0.5
					MoO_3	–	4.8	2.0 ± 0.2*
					SiO_2-101	2247± 632	–	0.4 ± 0.2
					Ru-101	39133±1557	5.1	4.8 ± 0.5
					Zr-101	3890± 713	2.5	0.7 ± 0.2
3	15	900	0.7 m HNO_3	2.26	W-110	28959±1336		
					PbMoO$_4$-112	51862±2764	–	5.4 ± 0.7
					MoO_3	–	4.8	2.1 ± 0.3*
					SiO_2-101	3803± 598	–	0.6 ± 0.2
					Ru-101	36789±1295	5.1	4.3 ± 0.4
					Zr-101	6655± 725	2.5	1.2 ± 0.4
4	10	800/600 (12h)	H_2O	1.5	W-110	11586± 680		
					PbMoO$_4$-112	54986±1847	–	9.4 ± 0.9
					MoO_3	–	3.2	3.6 ± 0.3*
					SiO_2-101	7904± 902	–	2.1 ± 0.4
					Ru-101	20678±1193	3.4	3.9 ± 0.5
					Zr-101	3662±1026	1.7	1.1 ± 0.4

*wt-fraction of MoO_3 calculated from the value for PbMoO$_4$; the significantly lower value indicates that $Ca_xPb_{1-x}MoO_4$ (not determined) is formed.

Fig. 2: BBS 10, sintered at 800 °C, suspension agent water
—— fitted Gaussian curve
xxxx measured points

a) Wulfenit
(PbMoO$_4$)

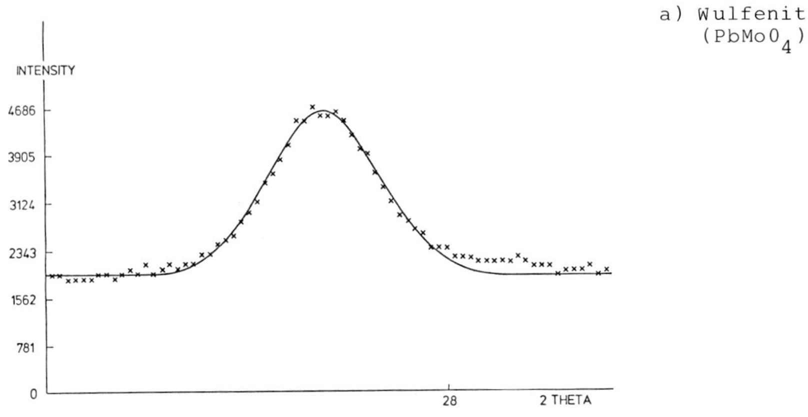

b) Tungsten

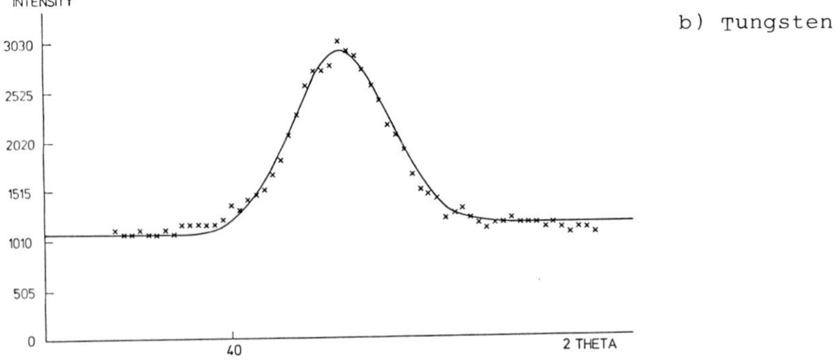

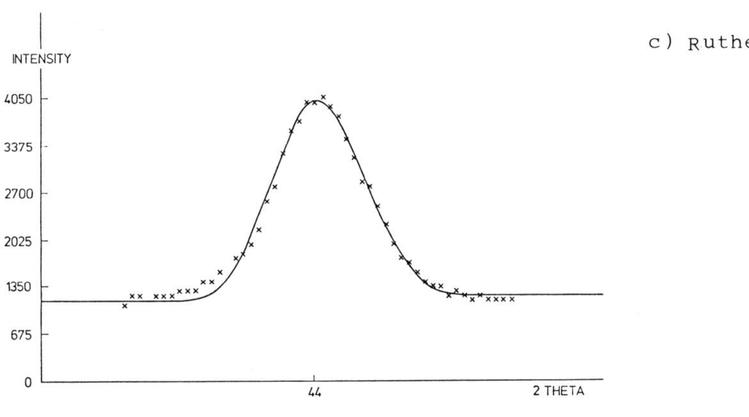

c) Ruthenium

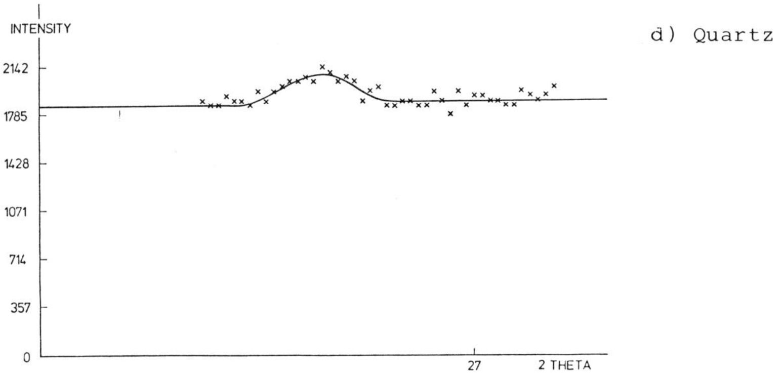

d) Quartz

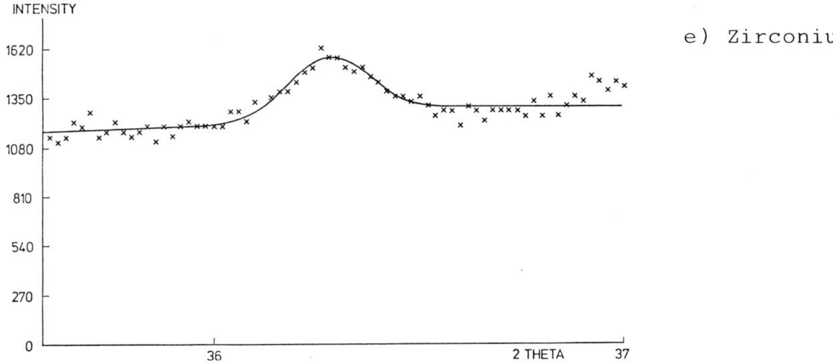

e) Zirconium

The ruthenium content remained unchanged in BBS 10 as well as
in BSMn8 when the experiments were carried out using water.
The lower values for ruthenium in those experiments where diluted
HNO_3 was used may be explained by the formation of volatile
ruthenium compounds. The experimentally determined lower values
for zirconium are believed to be due to the partial oxidation
to ZrO_2 during the sintering process and subsequent dissolution
in the glass phase. The small weight fraction of quartz found in
BBS 10 increased after the sample was tempered for 12 h at 600°C. In
general slight differences in the sinter temperature, the
content of simulated rad waste and the suspension agent did not
change the results significantly.
For BSMn8 the total crystallinity was calculated assuming that
the compounds which have not been measured remained unchanged.
For the samples with 15 wt% simulated rad waste the crystalline
content increased from 15 wt% to 19.3 ± 1.9 wt% using 0.7 m HNO_3
and to 19.9 ± 2.1 wt% using water.

CONCLUSION

It could be shown that it is possible to determine the weight
fractions of crystalline substances dispersed in a glass matrix.
The error is about 10% of the actual content for weight fractions
between 4 and 10 wt%. It is about 20 % for weight fractions between 1 and 4 wt% and up to 100 % for weight-fractions less
than 1 wt%. Stable phases like ruthenium, were in fact determined with the expected weight fractions. Chemical reactions
which may occur during sintering can be followed quantitatively.
This method appears to be well suited for determining the recrystallization behaviour of nuclear waste glasses. We learned
during the conference that a similar procedure is applied by
the CEA to characterize French HLW-glasses /8/.

ACKNOWLEDGEMENTS

We like to thank Dr. Helga Heide and Mr. Robens for the
assistance with the diffraction experiments and Mrs. Obers
for the sample preparation. The support by the Bundesministe-
rium für Forschung und Technologie is greatfully acknowledged.

REFERENCES

/1/   Chung, F.H., J. Appl.Cryst. 7, 519-531 (1974)

/2/   Alexander, L.E., Advances in the X-ray Analysis 20,
      1-13 (1976)

/3/   Chung, F.H., J. Appl.Cryst. 8, 17-19 (1975)

/4/   Schulien, S.; Eggersdorfer, R.; Stammler, M.;
      Wittekind, J.; Statusbericht 1981 des Projektträgers
      Universitätsforschung zum nuklearen Brennstoffkreislauf
      Kernforschungszentrum Karlsruhe Januar 1982

/5/ Siemens DIFFRAC 310 Manual

/6/ Marquardt, D.W. ; J.Soc.Industr. Appl. Math. $\underline{11}$,
     431-441 (1963)

/7/ Gmelin, Handbuch der Anorganischen Chemie 53, B2,
     S.265 (1976).

/8/ Morlevat, J.P., Uny, G. and Jacquet-Francillon,
     Seminar on Chemistry and Process Engineering for High
     Level Liquid Waste Solidification, Jülich, June 1.-5.,
     1981

Composition, Structure, Quality

SCIENTIFIC BASIS FOR RADIOACTIVE WASTE MANAGEMENT - V
Werner.Lutze, editor

EFFECT OF Fe_2O_3/ZnO ON TWO GLASS COMPOSITIONS FOR SOLIDIFICATION OF SWEDISH NUCLEAR WASTES

J. L. NOGUES* AND L. L. HENCH**
*Laboratoire des Verres, Universite des Sciences et Techniques du Languedoc,
34060 Montpellier Cedex (France); **Ceramics Division, Department of Materials
Science and Engineering, University of Florida, Gainesville, Florida, USA

INTRODUCTION

A recent study concluded that addition of Fe_2O_3 to a soda borosilicate
nuclear waste glass may significantly reduce damage by water attack due to
formation of a Fe-rich film on the glass surface.[1] However, differences in
SiO_2, B_2O_3, CaO, and concentration of fission products in previous glass
compositions make it impossible to ascribe the improved leach resistance solely
to Fe_2O_3 content. In the present work, leaching behavior of two glasses are
compared which differ only by the substitution of Fe_2O_3 for some of the ZnO
in the glass. Both glass compositions, Table 1, are compatible with the French
AVM process[2] and contain 9% (by weight) of simulated waste products characteristic of the Swedish nulcear waste program.

EXPERIMENTAL PROCEDURE

Leaching was performed at 90°C with deionized water in the MCC-1 Standard
Test for Leaching[3] at two ratios of glass surface area (SA) to solution volume
(V); SA/V = 1.0 cm^{-1} and 0.1 cm^{-1} for times of 1, 3, 7, 14 and 28 days.
Leachant solution pH was measured with a microelectrode and solution analysis
was performed by: ICP spectroscopy--Si^{4+}, B^{3+}, Mo^{6+}; atomic absorption--Fe^{3+};
and atomic emission--Na^+. Normalized average leach rates are reported. All sample surfaces were polished to a 600 grit surface with dry SiC paper and analyzed
before and after leaching with infrared reflection spectroscopy (IRRS).[4,5] The IRRS
spectra were normalized to a value of 80 for the 1120 cm^{-1} Si-O-Si molecular
stretching vibration of vitreous silica by using a shutter in the reference beam
of a Perkin-Elmer 467 IR spectrometer.

RESULTS

The substitution of Fe_2O_3 for ZnO in this pair of glasses results in ABS
41 (3.0% Fe_2O_3, 3.0% ZnO) having average leach rates nearly 3 times higher for
Si^{4+}, B^{3+}, Na^+ and Mo^{6+} than ABS 29 (0.6% Fe_2O_3, 6.0% ZnO), Figs. 1 and 2.

Although the normalized leach rates for Al^{3+} are the same for the two compositions, the normalized rates for Fe^{3+} release are three times higher for the

TABLE 1

NUCLEAR WASTE GLASS COMPOSITIONS (WEIGHT %)

OXIDE \ GLASS	ABS 29	ABS 41
SiO_2	52.0	52.0
B_2O_3	15.9	15.9
Al_2O_3	2.5	2.5
Na_2O	9.4	9.4
Fe_2O_3	0.6	3.0
ZnO	6.0	3.0
Li_2O	3.0	3.0
UO_2	1.66	1.66
S.W.P*	9	9

COMPOSITION OF SIMULATED NUCLEAR WASTE (WEIGHT %)

OXIDE	Cs_2O	SrO	BaO	Y_2O_3	ZrO_2	MoO_3	MnO_2	Ag_2O
WEIGHT %	9.78	2.89	5.11	1.67	14.22	18.11	8.56	0.12

OXIDE	SnO	Sb_2O_3	La_2O_3	Nd_2O_3	Pr_2O_3	Ce_2O_3	NiO	CdO
WEIGHT %	0.19	0.04	7.89	13.44	4.22	8.33	4.11	0.29

* Simulated waste products

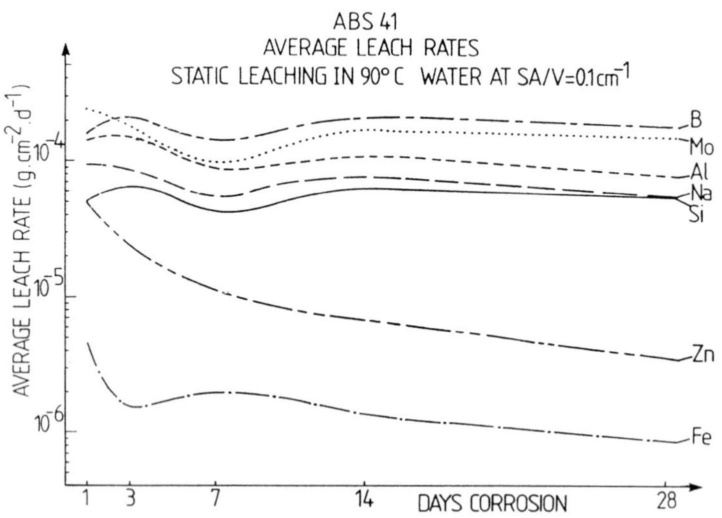

Fig. 1. Time dependence of leach rates for nuclear waste glass ABS 41.

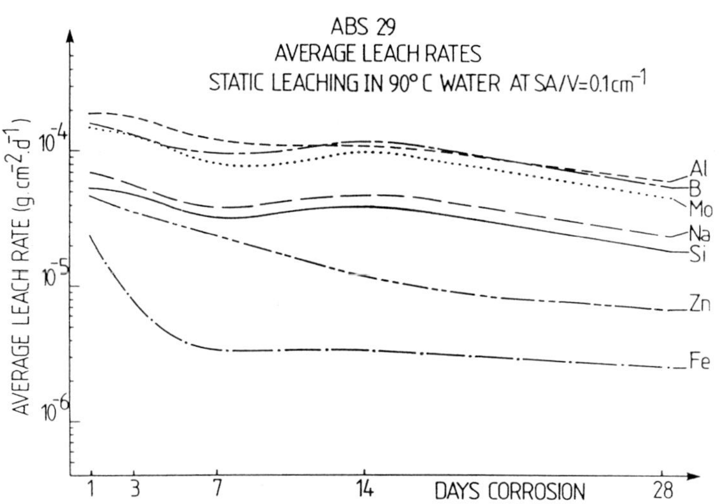

Fig. 2. Time dependence of leach rates for nuclear waste glass ABS 29

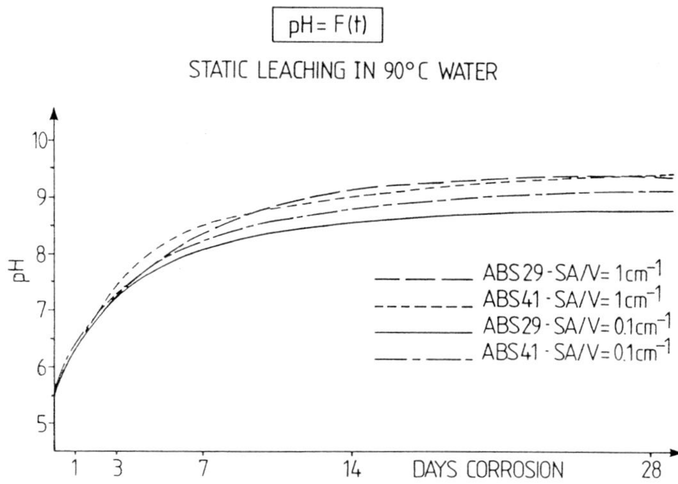

Fig. 3. Time dependence of leachant pH.

low Fe_2O_3 glass (ABS 29) than for high Fe_2O_3 glass (ABS 41). However, the values of leaching and elemental loss rates of Fe^{3+} are more than 30 times lower for both glasses than observed for the other elements. The difference in behavior of the two glasses cannot be attributed to variations in solution pH since little difference was observed between the glasses, Fig. 3.

After 1, 3 and 7 days leaching there is very little difference in the IRRS spectra for both glasses at either SA/V = 0.1 cm^{-1} or 1.0 cm^{-1}. However by 14 days (Fig. 4) the glass with the Fe_2O_3 substitution (ABS 41) is beginning to show evidence of more surface attack than the glass with high ZnO (ABS 29) at SA/V = 0.1 cm^{-1}. By 28 days the attack at SA/V = 0.1 cm^{-1} is much more severe for glass ABS 41 (Fig. 5). These results indicate that the surface is undergoing network breakdown and protective films have not been formed or are being destroyed. This difference in extent of surface damage is consistent with the leach behavior which shows glass ABS 29 having the lower leach rates for most species. However, at the higher value, SA/V = 1.0 cm^{-1}, the solution pH for both glasses is the same, and higher, and the IRRS spectra for both glasses shows little difference between them (compare Figs. 4a and 5a).

DISCUSSION

For this alkali zinc borosilicate glass composition, a mixture of ZnO and Fe_2O_3 in the glass is less effective in controlling release of ions in 90°C water than ZnO by itself. The previous study that reported enhanced leach resistance due to Fe_2O_3 compared a glass containing 6.4 weight percent Fe_2O_3. In the present study, ABS 29 contains 6 weight percent ZnO and exhibits good leach resistance. This suggests that a critical concentration of multivalence ions may be necessary for the surface to develop a protective second film on top of the SiO_2-rich layer that results from rapid dealkalization. Apparently if a mixture of multivalent ions are present, the concentration required to stabilize the protective secondary film is higher than if one species is concentrated in the surface film. This protective film is termed a Type III glass surface,[6] and was characteristic of the previously discussed nuclear waste glass containing 6.4 percent Fe_2O_3.[1]

At the higher SA/V = 1.0 cm^{-1} value, the mixed Fe + Zn ions are concentrated more quickly in solution, the secondary film forms more rapidly and the surface is protected more effectively than in the more dilute solution of SA/V = 0.1 cm^{-1}. These results show the importance of understanding both the effects of glass composition and solution concentrations on the behavior of nuclear waste glasses.

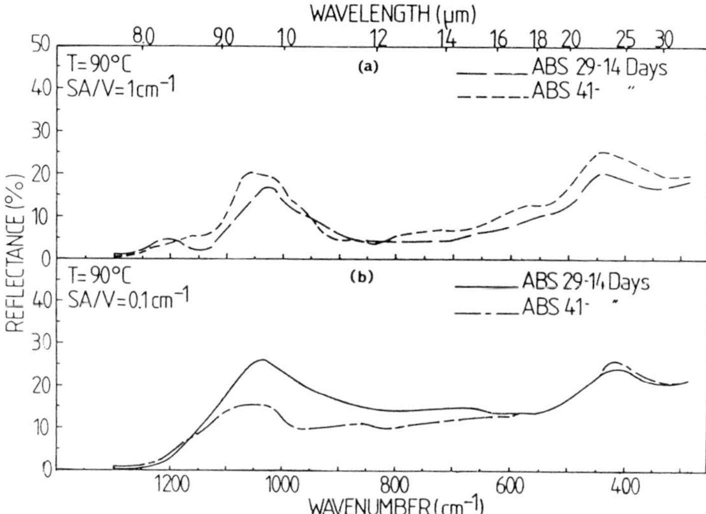

Fig. 4. IRRS spectra of ABS 29 and ABS 41 nuclear waste glasses after 14 days, 90°C static leaching.

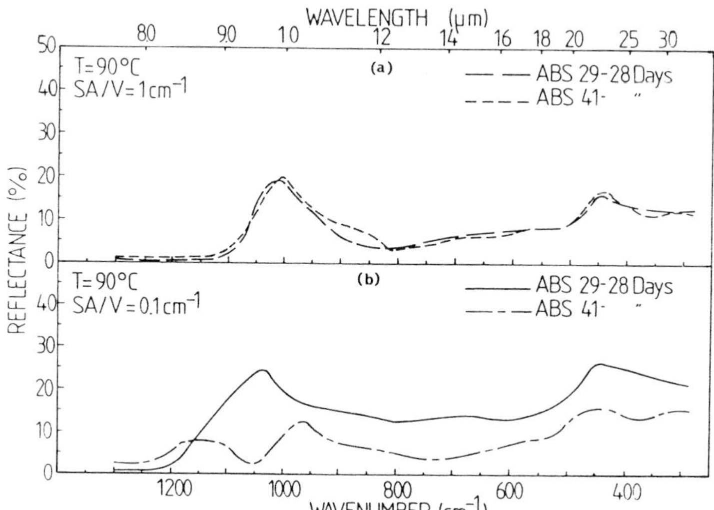

Fig. 5. IRRS spectra of ABS 29 and ABS 41 nuclear waste glasses after 28 days, 90°C static leaching.

CONCLUSIONS

Replacement of half of the ZnO in an alkali-zinc-borosilicate nuclear waste glass with Fe_2O_3 degrades leach resistance by approximately a factor of 3. Leach rates for the higher ZnO containing glass after 28 days are generally in the range of 5.10^{-5} $g.cm^{-2}.d^{-1}$ for B^{3+}, Al^{3+}, Mo^{6+}; 2.10^{-5} $g.cm^{-2}.d^{-1}$ for Na^+, Si^{4+}; and 2 to 7.10^{-6} $g.cm^{-2}.d^{-1}$ for Fe^{3+} and Zn^{2+}. The surface of both types of glass appears to be protected by dual protective layers, one rich in SiO_2 and a second very thin film rich in multivalent species. The second film that contains a mixture of Zn and Fe is less effective as a diffusion barrier and is less resistant to network breakdown than the film without the Fe^{3+}. This apparently is because a critical concentration of multivalence species is necessary to stabilize the second protective film. Differences in SA/V ratios affect the formation of the protective films and the rates of surface damage of the glass.

ACKNOWLEDGMENTS

The authors gratefully acknowledge Dr. T. Lakatos, Swedish Glass Research Institute for preparing the glasses and Dr. Lars Werme, SKBF/Project KBS, and Professor J. Zarzycki, Laboratoire des Verres, CNRS, Universite de Montpellier for their encouragement throughout this study. One of the authors (LLH) also acknowledges support of the U.S. Department of Energy.

REFERENCES

1. Hench, L. L., Urwongse, L. and Clark, D. E. (in press) "Surface Behavior of Two Nuclear Waste Glasses."
2. Bonniaud, R., Pacaud, F. and Sombret, C. (1975) Verres et Refractaires, Vol. 29-1.
3. MCC-1 Static Leach Test (1980) Materials Characterization Center Report, Pacific Northwest Laboratories, Richland, Washington,
4. Sanders, D. M., Person, W. D. and Hench, L. L. (1974) Appl. Spectroscopy, 28, 247-255.
5. Clark, D. E., Pantano, C. G. and Hench, L. L. (1979) in Glass Corrosion, Books for Industry, New York.
6. Hench, L. L. and Clark, D. E. (1978) J. Non-Crystalline Solids, 28, 83-105.

INVESTIGATION ON THE OXIDATION STATE AND THE BEHAVIOUR
OF MOLYBDENUM IN SILICATE GLASS

ANETTE HORNEBER[+]*, BOUBACAR CAMARA[+]** AND WERNER LUTZE[++]
[+]Institut für Werkstoffwissenschaften III, Universität Erlangen-Nürnberg,
Martensstr. 5, D 8520 Erlangen; [++]Hahn-Meitner-Institut für Kernforschung
Berlin GmbH., Glienicker Straße 100, D 1000 Berlin 39

INTRODUCTION

Knowledge about the oxidation state of molybdenum in glasses and glass ceramics which contain nuclear waste is important because of the danger of formation of water-soluble Cs-molybdate with Cs-137. In a previous paper[1], the behaviour of molybdenum in glasses with and without fission products was investigated with the aid of ESR and optical spectroscopy. The present paper continues this investigation and deals with the following aspects:

1) Oxidation state of Mo when organic reducing agents, i.e. formic acid,
   formaldehyde, tannic acid, are added to the batch.

2) Dependence of Mo "solubility" on the melting atmosphere.

3) Behaviour of the different oxidation states of Mo depending on the temperature and the basicity of the glass.

4) Redox interaction of Mo with Fe, Cr and Ti.

EXPERIMENTAL PROCEDURE

Glasses of the composition 74 % SiO_2, 20 % Na_2O, 6 % CaO , mass fractions,
and different amounts of MoO_3, Fe_2O_3, Cr_2O_3 and TiO_2 were investigated. Some
glasses had Cs_2O instead of Na_2O, substituted on a molar basis, so as to determine the influence of the basicity of the glass on the oxidation state of Mo.
Some phosphate glasses were investigated as well, in order to allow a comparison.

Rock crystal was used as raw material for SiO_2. The other substances were
p.a. quality. The glasses to be melted under reducing conditions were mixed
with the adequate amount of tannic acid. Melting took place in corundum crucibles in an electric furnace at 1673 K for 4 h. Glasses to be compared with
each other were melted simultaneously, so as to achieve identical melting conditions. For the ESR, the samples were ground, screened, and 100 mg of the

Present address: *) Verlag Schmid, D 7800 Freiburg i.Br., and **) BASF Farben
und Fasern AG., D 7122 Besigheim

fractions of 0.6 to 1 mm in diameter were used for each measurement. The spectra were recorded with the X-band Spectrometer 414 (Bruker Physik). The measurements were made at 298 K. For the optical spectra, polished plates were used.

RESULTS AND DISCUSSION

State of Mo in silicate glass

Molybdenum can occur in different oxidation states in glass, the most stable being Mo (VI). In[1] the assumption was made, that besides Mo (VI), Mo (III) (d^3, s = 3/2 or s' = 1/2) was present depending on the concentration of MoO_3 and on the melting atmosphere. The possibility of Mo (V) (d^1, s = 1/2) being present was not excluded. In the present work these investigations are examined again with an increased range of MoO_3 concentration.

Glasses melted under oxidizing conditions remain colourless. At an MoO_3 content of less than 0.4 %, no ESR spectrum was observed. Over 0.4 %, the spectrum figure 1, was obtained with g = 1.905 $\pm$ 0.001. The resonance intensity increases with increasing MoO_3 content. None of the glasses melted under oxidizing conditions showed optical band spectra. Most of the Mo is present as Mo (VI), only a low quantity as Mo (III) or Mo (V), which can be detected by ESR, but not by optical spectroscopy.

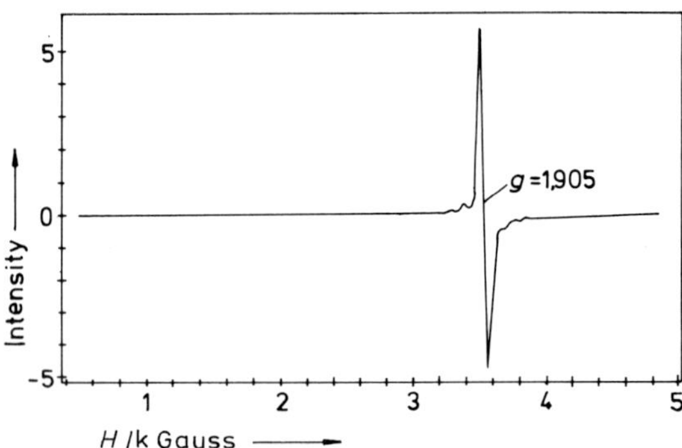

Fig. 1. Typical ESR-spectrum of Mo (III).

As opposed to the glasses produced under oxidizing conditions, the reduced glasses (2 g tannic acid per 40 g glass) show the typical spectrum of figure 1 at percentages as low as 0.2 % MoO_3. The intensity of the ESR spectrum increases rapidly with increasing Mo content, giving glasses of brown colour. At

1 % MoO_3 and above, the glasses show two absorption bands at 435 nm (22988 cm^{-1}) and 940 nm (10638 cm^{-1}) in the optical spectrum, as shown in figure 2.

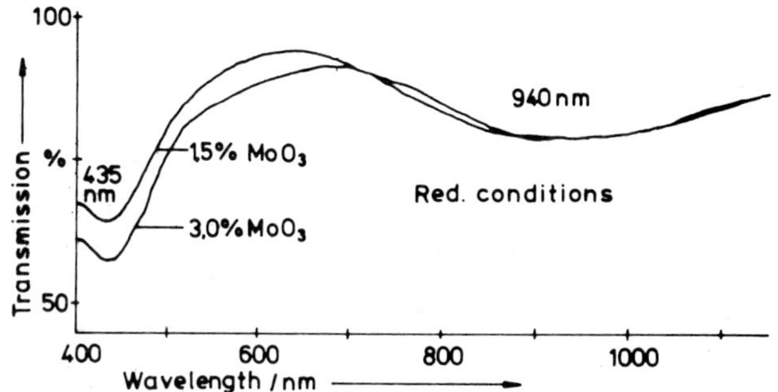

Fig. 2. Optical spectrum of MoO_3 containing glass melted under reducing conditions.

It is known that the relationship

$$g = 2.0023 - \frac{8\lambda}{\Delta} \qquad (1)$$

exists between the g-values as determined from the ESR spectrum and the ligand field absorption energy Δ, which can be used to determine the oxidation state of the respective ions. If, for example, g and Δ are determined from the experiment, the respective spin-orbit coupling constant λ, i.e. (λ:L·s), can be calculated and compared with the known value for the free ion. It should be noted that λ of the free ion is always greater than λ of the bonded ion. The following values were determined from the experiment: g = 1.905 and Δ = 22988 cm^{-1}. With equation (1), the value of λ = 279 cm^{-1} results.

For the two oxidation states Mo (III) and Mo (V) under discussion, the following values of the free ions are known[2]:

$$\text{Mo (III):} \quad = \quad 269 \ cm^{-1}$$
$$\text{Mo (V):} \quad = 1013 \ cm^{-1}$$

The value of 279 cm^{-1} determined experimentally is compatible with the spin orbit coupling constant of free Mo (III), but is very different from the Mo (V) value. A reduction of the spin orbit coupling constant from 1013 cm^{-1} (Mo(V)) to 279 cm^{-1} can be excluded. The fact of the experimental value being somewhat greater than the value of the free Mo (III) can be explained because equation (1) is approximate. The data from ESR and optical spectroscopy are indicative of Mo (III), as is the brown colour of the glasses.[3,4] Some information on the Mo (III) structure can be deduced from the absorption band at 435 nm (22988 cm^{-1}). Christ[5] found an absorption energy of 22650 cm^{-1} for Mo (III)

282

with 6 H_2O as ligands which shows a good coincidence with the observed value (22980 cm^{-1}). Therefore, it can be assumed that the band at 435 nm is produced by 6-fold coordinated Mo (III).

From the assumption that the absorption energy Δ of 22988 cm^{-1} corresponds to Mo (III) in octahedral coordination (O), the energy of Mo (III) in tetrahedral coordination (T) in these glasses may be calculated with the aid of the relationship: $\Delta_T = \frac{4}{9} \Delta_O$. For Δ_O = 22988 cm^{-1}, Δ_T is 10217 cm^{-1} (978 nm). The observed band at 940 nm (10638 cm^{-1}) probably corresponds to Mo (III). This would mean that Mo (III) appears in two different structures (coordinations) in the investigated glasses.

Basicity of the glass and oxidation state of molybdenum

Glasses in which Na had been substituted by Cs in the same mole proportion have a higher basicity. In these glasses, the fraction of Mo (III) - measured by the intensity of the resonance at g = 1.905 - is much smaller than in the Na glasses, for the glasses melted both under oxidizing and reducing conditions. In the case of glasses containing Cs it is difficult, even when increasing the amount of tannic acid, to reduce Mo (VI) to Mo (III) in larger quantities. An indication of the fact that the stability of the higher oxidation state of Mo increases with the basicity was also found with phosphate glasses: in the case of phosphate glasses free from alkali, Mo (III) was found, whereas Mo (V) was present in phosphate glasses containing alkali.

"Solubility" of molybdenum in silicate glasses

Molybdenum has low solubility in silicate and borosilicate glass.[6,7] With the aim of increasing the solubility in borosilicate glasses containing fission products, some authors[8] used metallic Si as a reducing agent. The tannic acid used in the present investigation avoids problems created by the use of Si. In oxidized glasses (without tannic acid), 2.5 % MoO_3 could be dissolved, giving transparent and homogeneous glasses. At 3 % MoO_3, white streaks appeared, and at 3.5 % MoO_3 white precipitations, which were identified by X-ray diffraction as Na_2MoO_4 and $Na_2MoO_4 \cdot 2 H_2O$. Probably the Cs-molybdate phase is also present in a similar form. On the contrary, glasses containing 3 % MoO_3 melted with tannic acid (2 g per 40 g glass) were of brown colour and without precipitations. This represents an increase of 20 % in solubility with respect to the glasses melted under oxidizing conditions.

Figure 3 shows the behaviour of Mo (III) as a function of the amount of tannic acid in the case of a glass containing 0.1 % MoO_3.

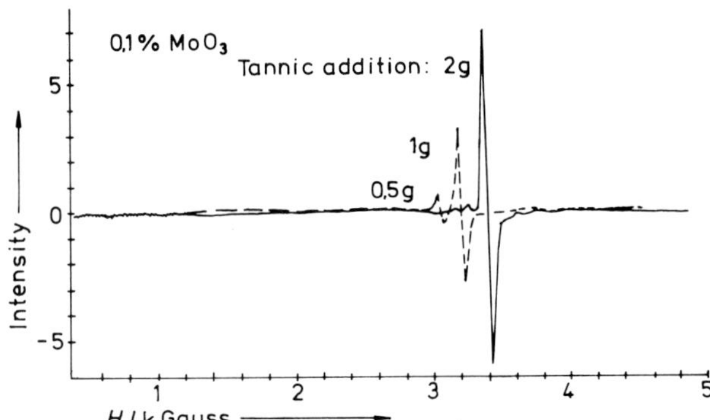

Fig. 3. ESR spectra of MoO_3 containing glasses with different amounts of tannic acid .

The reduction of Mo (VI) into Mo (III), as well as the solubility, increase with increasing tannic acid content. Formic acid and formaldehyde were used as reducing agents as well, the latter mainly because of the fact that they have already been used for the reduction of nitric acid in the HAW solution. But neither formic acid nor formaldehyde had a reducing effect during melting, thus having no influence on the oxidation state of molybdenum in the glass. Simultaneously, the influence of the melting temperature on the behaviour of Mo(III) was investigated. The results in the case of a glass containing 1 % of MoO_3 are shown in figure 4. As can be seen, Mo (III) decreases with increasing temperature, which demonstrates the low stability of this ion. At higher temperatures, more oxygen from the atmosphere dissolves in the glass melt, oxidizing Mo(III) to Mo (VI).

284

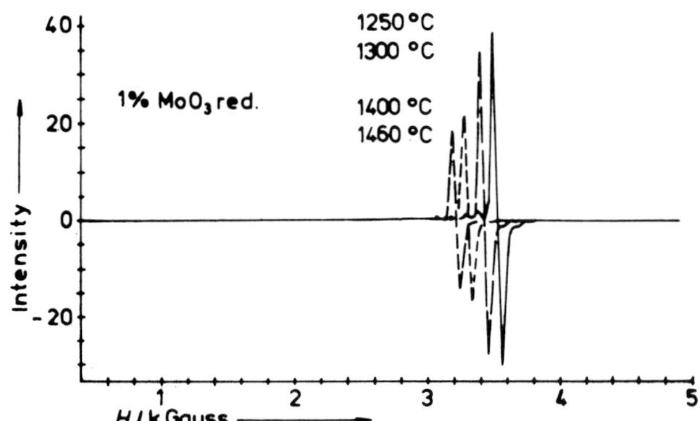

Fig. 4. ESR spectra of glasses melted at different temperatures under
reducing conditions.

Redox interactions between Mo and Fe, Cr and Ti

Mo - Fe. The knowledge about the redox interactions is of great importance
as Fe is present in HAW solutions. In[1,9], an interaction between Mo and Fe
was described. The present work continues the investigations in a wider range
of MoO_3 contents. For this purpose, glasses containing 0.05 % to 2.5 % MoO_3
and different amounts of Fe_2O_3 were investigated with the aid of ESR. The
following results were obtained: Under oxidizing conditions, glasses with
a small amount of MoO_3 (0.05 ... 0.4 % MoO_3) show no spectrum of Mo(III).
If Fe_2O_3 is added to these glasses the typical Mo (III) spectrum (figure 1) was
observed. This means that Mo (VI) is partially reduced into the Mo (III) state.
The intensity of this resonance increases at first with increasing Fe_2O_3 content,
then decreasing and finally disappearing. This means that Mo (III) is being oxidized. The same effect is also observed on glasses with higher MoO_3. This means
that in the redox reaction

$$Mo^{3+} + 3\ Fe^{3+} \rightleftharpoons Mo^{6+} + 3\ Fe^{2+} \tag{2}$$

the equilibrium can be displaced to the right or to the left according to the
actual value of the equilibrium constant $K = (a_{Mo^{6+}} \cdot a_{Fe^{2+}}^3)/(a_{Mo^{3+}} \cdot a_{Fe^{3+}}^3)$.
Apparently the interaction between the different ions cannot be neglected and
therefore the activities instead of concentrations have to be used. It should
be noted that the value of the Fe/Mo ratio at which the equilibrium is displaced
to the right by adding Fe_2O_3 decreases with increasing MoO_3 content, i.e. the
more Mo is present in the glass, the lower the Fe percentage necessary to oxidize Mo (III) into the Mo (VI) state. At 2.5 % MoO_3 it was difficult to reduce
Mo (VI) into Mo (III) by adding Fe_2O_3.

The diagram in figure 5 was obtained from the behaviour of the Mo(III) fraction upon addition of Fe_2O_3 (increasing or decreasing of the ESR intensity, resp.). This diagram shows the ranges of MoO_3 and Fe_2O_3 in which a reduction or oxidation can be expected in the case of the glasses which are the object of this investigation. The curve represents those Fe_2O_3 concentrations at which, for a given MoO_3 concentration, the equilibrium is displaced from left to right. With a low Fe_2O_3 content in the glass, Mo (VI) is reduced into Mo (III), a further increase of Fe_2O_3 produces the oxidation of Mo (III) into Mo (VI). A similar dependence of the equilibrium on the concentration was also found in the case of glasses containing Mn and Ce and Cr and Fe [10,11]. The influence of the Fe concentration on the Mo (III)/Mo (VI) ratio in glasses containing fission products should not be underestimated. In fact, the oxidation of Mo (III) by Fe (III) can reduce the solubility of Mo in the glass.

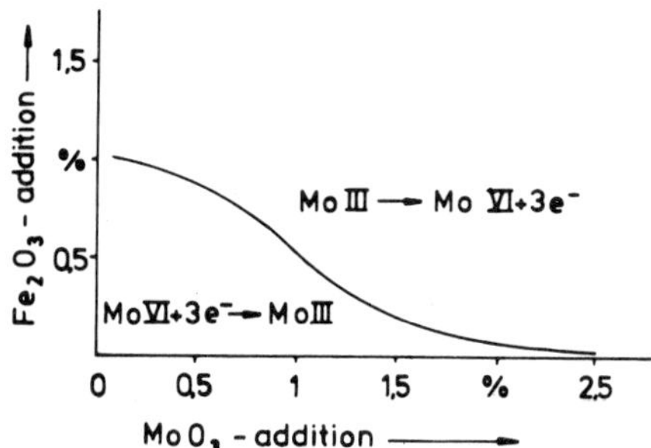

Fig. 5. Diagram showing the fractions of MoO_3 and Fe_2O_3 with which Mo is either oxidized or reduced.

Tibor Lakatos[12] investigated borosilicate glasses containing Cs and 1.6 % MoO_3 as well as 0.6 % Fe_2O_3. Si and SiC were used as reducing agents. The author did not find a difference between glasses melted under oxidizing and reducing conditions. We believe that this fact is due to the role of Fe in this relation, which probably compensated the reducing effect of Si and SiC on Mo(VI).

Redox reactions with Cr and Ti

In the glass melt, an intense and susceptible reaction takes place, which can be well observed by means of the spectroscopic behaviour of Mo (III) and Fe (III). These investigations were also conducted with Cr and Ti.

286

Mo and Cr. As an indicator for the behaviour of Mo, the ESR-spectrum can be
again used (figure 1),and the ESR-spectrum (figure 6) or the optical spectrum (figure 7) can be used in the case of chromium. As in the case of Mo, together with Cr (III), Cr (VI) is a common oxidation state. This oxidation state
has no ESR spectrum ($3d^{o}$). From the redox potentials of the two pairs Mo(III)
--> Mo (VI) and Cr (III) --> Cr (VI), the following redox reaction may be written:

$$Mo\ (III)\ +\ Cr\ (VI)\ \rightleftharpoons\ Mo\ (VI)\ +\ Cr\ (III) \tag{3}$$

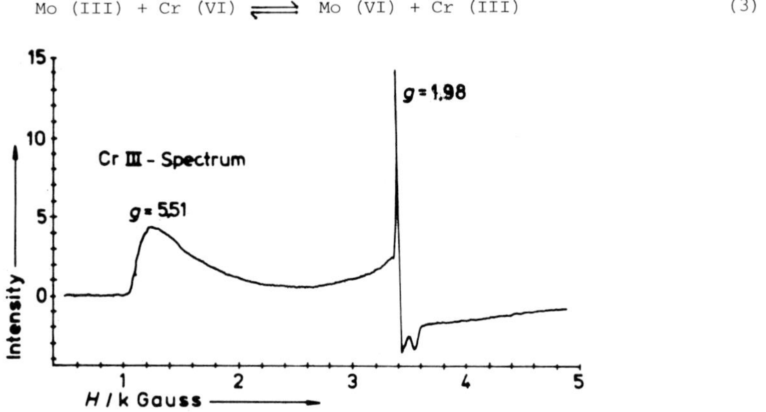

Fig. 6. Typical ESR spectrum of Cr (III).

For the study of this system, glasses containing Cr_2O_3 and MoO_3 were melted
under oxidizing and reducing conditions with the aid of tannic acid. Under oxidizing conditions, an addition of MoO_3 to glasses containing Cr produces only a
small influence on the ESR spectrum of Cr (slight decrease of the signal at
g = 1.98, little increase of the peak at 5.51). Even at 2.5 % MoO_3, the Mo(III)
peak was not observed. The optical spectrum[2] (figure 7) shows that in fact an
interaction according to reaction[3] has taken place after the addition of 2.5 %
MoO_3 to a glass containing 0.5 % Cr_2O_3. As the optical spectrum with the triplet
represents Cr (III)[13], the increase in intensity found on the three bands
(680 nm, 652 cm, 635 nm) means an increase of the Cr (III) concentration with
respect to the glass without Mo, and a displacement of the equilibrium according to equation (3) to the right, further decreasing the Mo (III) fraction,
which is low in oxidized glasses. This explains why the Mo (III) peak at
g = 1.905 could not be observed even at 2.5 % MoO_3.

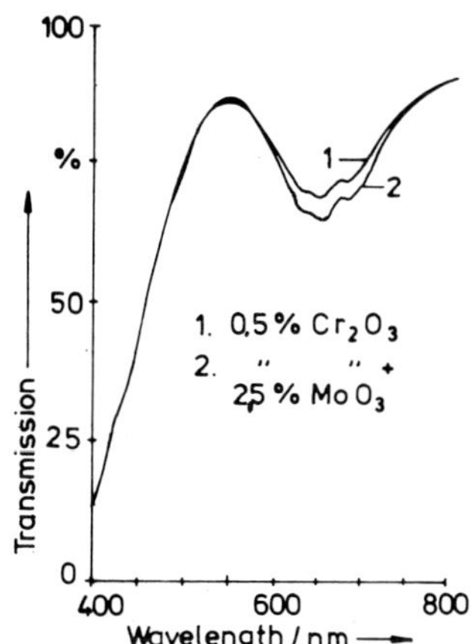

Fig. 7. Optical spectra of glasses containing Cr_2O_3 or Cr_2O_3 plus MoO_3.

In reduced glasses, the Cr (III) as well as the Mo (III) peaks could be observed (figure 8). This means that the oxidation states postulated in equation (3) in fact do co-exist.

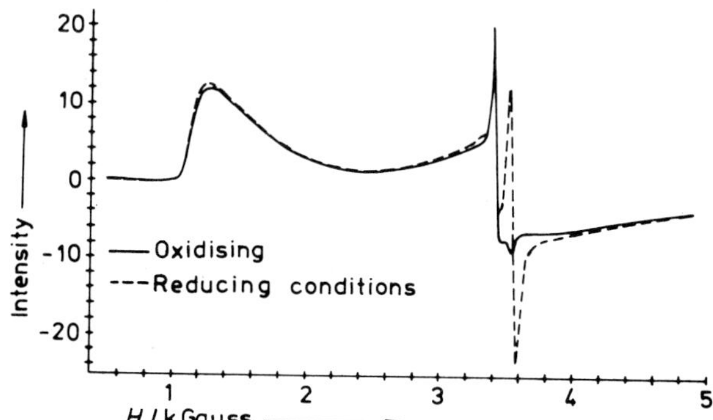

Fig. 8. ESR spectra of glasses containing Cr_2O_3 and MoO_3 melted under
different conditions.

<u>Mo and Ti.</u> Due to the high stability of Ti (IV), no redox interaction can be expected. Even ·at 4 % TiO_2 no significant interaction between Mo and Ti was observed.[14]

ACKNOWLEDGEMENT

The authors thank Prof. Oel for his interest in this work and for his valuable suggestions, and Dr. Bauer for the helpful discussions and comments.

REFERENCES

1. Camara, B., Lutze, W. and Lux, J. (1979) Scientific Basis for Nuclear Waste Management, Northrup, Jr., C.J.M. ed., Plenum Press, New York and London, Vol 2, pp. 93-102.
2. Fraga, S., Karwowski, J. and Saxena, K.M.S. (1976) Handbook of Atomic Data, Elsevier Scientific Publishing Company, Amsterdam-Oxford.
3. Landry, J. and Forunier, J.T. Arpa No. 306, Research Center, American Optical Company, Southbridge, MA.
4. Hartmann, H. and Schmidt, J.J. (1957) Z. Phys. Chem. N.F. 11, 234.
5. Christ, K. (1965) Dissertation: Spektralphotometrische und magnetische Untersuchungen an Molybdän (III)-Komplexverbindungen, Frankfurt/M.
6. Weyl, W.A. (1951) Coloured Glasses, Soci. of Glass Techn., Sheffield, 213.
7. Lutze, W., Borchardt, J. and Dé, A.K. (1979) Scientific Basis for Nuclear Waste Management, McCarthy, G.J. ed, Plenum Press, New York and London, pp. 69 - 81.
8. Singer, G. and Tomozawa, M. (1979) in Proc. Int. Symposium on Ceramics in Nuclear Waste Management, Chikalla, T.D. and Mendel, J.E. ed, Technical Information Center, pp. 193-197.
9. Lux, J. (1979/1980) Diplomarbeit, Universität Erlangen-Nürnberg, Institut für Werkstoffwissenschaften III.
10. Weyl, W.A. (1938) Sprechsaal 71, Jahrg. 9.
11. Stepanov, S.A., Zarubina, T.V., Iglatev, E.G. and Skorospelova, V.I. (1979) Sov. Journal of Glass Physics and Chemistry (Fizika i chimija stekla) 5,314.
12. Lakatos, T. (1979) Teknisk Rapport, Swedish Glass Research Inst. Växjö,79.
13. Camara, B. and Schaeffer, B. (1981) Ber. der Deutschen Keram. Gesellschaft 7, 519.
14. Horneber, A. (1980/1981) Diplomarbeit, Universität Erlangen-Nürnberg, Institut für Werkstoffwissenschaften III.

THE MATERIALS BALANCE - SCIENTIFIC FUNDAMENTALS FOR THE QUALITY ASSURANCE OF VITRIFIED WASTE

EDWIN SCHIEWER*, HARALD RABE*, SIEGFRIED WEISENBURGER**
*Hahn-Meitner-Institut für Kernforschung Berlin GmbH., Glienicker Straße 100,
D-1000 Berlin 39, **Kernforschungszentrum Karlsruhe GmbH., P.O.Box 3640,
D-7500 Karlsruhe

INTRODUCTION

In the vitrification of HLW the aim is to create a product, the stability (thermal, mechanical) of which is, to the greatest possible degree, independent of fluctuations in its composition. Process control during vitrification of HLLW is necessary in order to assure the quality of the final glass product. The chemical processes are controlled by quantitative analysis of all material streams entering and leaving the melting device. There are six material streams to consider: frit input, waste input, corrosion products (generated by the vitrification process), off-gas, recycled off-gas, glass output. Additionally, the capacity of the glass for buffering short-term fluctuations in the incoming streams must be known. Sufficient sampling during non-radioactive vitrification in addition to materials tests can provide the necessary data basis. The results are combined in a materials balance.

Under radioactive operation access to samples is rather limited and quality assurance must be derived from experience obtained during non-radioactive operation. A comprehensive understanding of the materials balance may provide a scientific basis for reliable quality assurance when sampling is limited.

EXPERIMENTAL

In the Kernforschungszentrum Karlsruhe a vitrification experiment lasting 230 hours was carried out using a non-radioactive HLLW-simulate in the ceramic melter K2 (contents = 700 kg of glass, surface of bath = 0.64 m^2, for further data see[1]). 3.8 t of glass were produced. A borosilicate glass rich in barium was used for these experiments. It has already been established that this glass was suitable for LEWC vitrification in a ceramic melter[2]. For technical procedure reasons the borosilicate frit was applied in the forms of powder and glass beads (in a ratio of 52.9 : 47.1). The compositions of the frit forms, determined experimentally, and the composition of the frit mixture, which was calculated from them, are listed in table 1.

The HLLW-simulate was adapted to the composition of the LEWC (low enriched

TABLE 1

COMPOSITION OF GLASS FRITS (wt.%)

Constituents	beads	powder	52.9 % powder + 47.1 % beads
SiO_2	45.99 ± 0.20	45.77 ± 0.60	45.78 ± 0.40
B_2O_3	9.15 ± 0.40	8.96 ± 0.40	9.05 ± 0.40
Al_2O_3	11.50 ± 0.40	12.31 ± 0.60	11.93 ± 0.50
Li_2O	1.96 ± 0.05	1.74 ± 0.15	1.84 ± 0.10
Na_2O	0.64 ± 0.20	6.86 ± 0.60	3.93 ± 0.40
K_2O	0.25 ± 0.05	–	0.12 ± 0.03
MgO	0.30 ± 0.05	0.20 ± 0.05	0.25 ± 0.05
CaO	9.47 ± 0.10	8.01 ± 0.30	8.70 ± 0.20
SrO	0.32 ± 0.05	0.35 ± 0.05	0.34 ± 0.05
BaO	16.50 ± 0.10	11.61 ± 0.60	13.91 ± 0.35
TiO_2	3.82 ± 0.30	4.07 ± 0.30	3.95 ± 0.30
SO_3	0.10 ± 0.02	0.12 ± 0.03	0.11 ± 0.03

TABLE 2

COMPOSITION OF LEWC-SIMULATE

Constituents	(wt.%)	Constituents	(wt.%)
Al_2O_3	7.97 ± 0.40	Nd_2O_3	2.82 ± 0.14
Li_2O	14.29 ± 0.71	TeO_2	0.35 ± 0.02
Na_2O	12.24 ± 0.61	ZrO_2	6.83 ± 0.34
SrO	0.40 ± 0.02	MoO_3	2.40 ± 0.12
BaO	0.84 ± 0.04	RuO_2	1.29 ± 0.06
SO_3	4.65 ± 0.23	Fe_2O_3	14.46 ± 0.72
Rb_2O	0.17 ± 0.01	NiO	3.65 ± 0.18
Cs_2O	1.09 ± 0.05	Cr_2O_3	2.38 ± 0.12
Y_2O_3	0.38 ± 0.02	P_2O_5	0.05 ± 0.00
La_2O_3	0.72 ± 0.04	NaF	14.48 ± 0.72
CeO_2	1.55 ± 0.08	ZnO	0.13 ± 0.01
Pr_2O_3	1.42 ± 0.07	MnO_2	5.44 ± 0.27

waste concentrate) stored at EUROCHEMIC in Mol, Belgium. It contained LiO_2 as an additive to obtain the viscosity of the glass melt required for this experiment. The nominal composition of the simulate is given in table 2. The actual composition was not determined. Due to impurities in the chemicals used, deviations from the required values are to be expected. These deviations were taken into account with $5_{rel.}$% in each case.

The dosage for the frit forms and the LEWC-simulate was to be determined so as to produce a glass with 15 $\pm$ 3 wt.% LEWC oxides. The maximum deviations of $\pm$ 3 % should only be the result of fluctuations in the dosage of the LEWC-simulate. During the first 85 hours of the experiment the average rate of production was 13 kg glass per hour. In the following 125 hours it was increased to an average of 18 kg and in the last 20 hours to almost 25 kg glass per hour.

The average length of time the glass was kept in the melter decreased for the above periods of time from 62 to 44 and to 32 hours respectively.

The loss of oxides, and fluorides, by evaporation in the melter was 33 $\pm$ 3 kg. 25 $\pm$ 3 kg of this were almost continiously recycled into the melter so that altogether only 8 $\pm$ 3 kg evaporation loss occurred, in other words 0.2 $\pm$ 0.1 % of the glass produced.

A thermocouple with a protective tube made of molybdenum was used to measure the temperature of the glass melt. After 149 working hours the thermocouple was found to be faulty. When the experiment had been completed it was discovered that the molybdenum tube had dissolved in the melt. In all an additional 1.15 kg Mo (= 1.73 kg MoO_3) entered the glass.

After the melter had been completely emptied the melt was found to have been covered by a layer of a substance referred to as "yellow phase".

RESULTS

Analysis of glass samples

Two to five glass samples were taken (during casting and from the cut canister) after 100 hours and after 226 hours of glass production. For each analysis 0.5 g of each sample were digested with 2 g Cs_2CO_3 in Pt crucibles. After acidification the solutions were quantitatively analysed by means of ICP atomic emission spectroscopy. At the time being, the elements Rb, Cs, Y, La, Pr, Nd, Ru, F, Te and P could not be determined. The calculated percentage of all these oxides in the glass (with 15 $\pm$ 3 % in weight LEWC oxides) is 2.63 $\pm$ 0.52 % in weight. These values were used for the materials balance.

The normalized analytical results are listed in table 3. Each is the mean

TABLE 3

MATERIALS BALANCE (NOMINAL AND EXPERIMENTAL COMPOSITIONS OF LEWC-GLASSES [wt.%])

Consti-tuents	Nominal values for glasses containing		Experimental yields (normalized) for glasses after		In view of corrosion products and yellow phase corrected nominal values for glasses containing	
	15 % LEWC	18 % LEWC	100 hours	226 hours	15 % LEWC	18 % LEWC
SiO_2	37.84 ± 0.34	36.24 ± 0.33	37.39 ± 0.70	35.33 ± 0.70	37.79 ± 0.34	36.21 ± 0.33
B_2O_3	7.47 ± 0.34	7.15 ± 0.33	7.42 ± 0.40	7.32 ± 0.40	7.46 ± 0.34	7.12 ± 0.33
Al_2O_3	11.23 ± 0.49	11.10 ± 0.48	11.17 ± 0.40	11.41 ± 0.40	11.27 ± 0.49	11.24 ± 0.49
Li_2O	4.02 ± 0.20	4.45 ± 0.22	4.29 ± 0.20	4.78 ± 0.20	4.01 ± 0.20	4.43 ± 0.22
Na_2O	7.24 ± 0.43	7.91 ± 0.44	7.50 ± 0.60	8.17 ± 0.60	7.23 ± 0.43	7.88 ± 0.45
K_2O	0.10 ± 0.03	0.09 ± 0.02	0.10 ± 0.03	0.13 ± 0.03	0.10 ± 0.03	0.09 ± 0.02
MgO	0.21 ± 0.04	0.20 ± 0.04	0.29 ± 0.05	0.33 ± 0.05	0.21 ± 0.04	0.20 ± 0.04
CaO	7.18 ± 0.17	6.87 ± 0.16	6.95 ± 0.50	6.28 ± 0.50	7.17 ± 0.17	6.84 ± 0.16
SrO	0.35 ± 0.05	0.35 ± 0.05	0.35 ± 0.05	0.36 ± 0.05	0.35 ± 0.05	0.35 ± 0.05
BaO	11.62 ± 0.31	11.17 ± 0.30	11.46 ± 0.30	11.22 ± 0.30	11.59 ± 0.31	11.13 ± 0.30
TiO_2	3.26 ± 0.26	3.12 ± 0.25	3.09 ± 0.25	2.89 ± 0.25	3.25 ± 0.26	3.11 ± 0.26
CeO_2	0.27 ± 0.01	0.33 ± 0.02	0.27 ± 0.02	0.32 ± 0.02	0.27 ± 0.01	0.33 ± 0.02
ZrO_2	1.20 ± 0.06	1.43 ± 0.07	1.30 ± 0.06	1.58 ± 0.06	1.26 ± 0.06	1.58 ± 0.08
MnO_2	0.95 ± 0.05	1.14 ± 0.06	0.95 ± 0.05	1.13 ± 0.05	0.95 ± 0.05	1.14 ± 0.06
Fe_2O_3	2.53 ± 0.12	3.04 ± 0.16	3.01 ± 0.15	3.68 ± 0.15	2.96 ± 0.15	3.69 ± 0.18
NiO	0.64 ± 0.03	0.77 ± 0.04	0.65 ± 0.02	0.84 ± 0.02	0.67 ± 0.03	0.85 ± 0.04
Cr_2O_3	0.42 ± 0.02	0.50 ± 0.03	0.41 ± 0.02	0.49 ± 0.03	0.38 ± 0.02	0.48 ± 0.02
SO_3	0.90 ± 0.06	1.07 ± 0.06	0.62 ± 0.08	0.69 ± 0.09	0.62 ± 0.05	0.79 ± 0.04
MoO_3	0.42 ± 0.02	0.50 ± 0.03	0.63 ± 0.04	0.48 ± 0.03	0.31 ± 0.02	0.39 ± 0.02
Sum of other const.	2.63 ± 0.13	3.15 ± 0.16	2.63 (theor.)	3.15 (theor.)	2.63 ± 0.13	3.15 ± 0.16
Sum	100.48	100.58	100.48	100.58	100.48	100.58

value of 4 analyses. No differences could be found in the composition within the contents of a single glass canister. The totals 100.48 (for 15 % LEWC) and 100.58 (for 18 % LEWC) take into account the fact that a very small proportion of the components referred to as oxides were in fact present in the form of fluorides. Table 3 also contains the nominal values for glasses with 15 and 18 wt.% LEWC oxides.

<u>Analysis of the "yellow phase"</u>

After the melter had been completely emptied the yellow phase was situated in the top part of the last canister. The amount was found to be 6.0 $\pm$ 0.2 kg, i.e. 0.16 % of the amount of glass produced.

For the purpose of analysis the yellow phase was separated into a part soluble in water, a part soluble in diluted (1n) hydrochloric acid, a part soluble in concentrated (5n) hydrochloric acid and an insoluble residue.

The soluble fractions were analysed by means of ICP atomic emission spectroscopy. The results were listed in table 4. From these and from the results of the X-ray diffraction analysis the composition of the yellow phase was determined: 51 wt.% alkali sulphates, 5 wt.% alkali chromates, 6 wt.% alkali molybdates, 19 wt.% $CaMoO_4$, 14 wt.% $Ba(Sr)CrO_4$ and 3 wt.% residue. The mean density of the yellow phase was 2.5 gcm^{-3}. Using this value and the surface of melt (0.64 m^2) a layer of yellow phase of some 3 mm thickness was calulated.

THE MATERIALS BALANCE

Of the six material streams: frit input, waste input, corrosion products (generated by the vitrification process), off-gas, recycled off-gas and glass-output, only the proportion of corrosion products is unknown. From the differences between the nominal and the actual values of the glass analyses (within the fluctuation range of the LEWC dosage) it is possible to draw conclusions about the causes and extent of the corrosion effects in the apparatus. Therefore, the mean LEWC oxide contents of the glasses must first be calculated.

As already mentioned the non-recycled loss of solids by evaporation was only 0.2 $\pm$ 0.1 %. These were mainly the LEWC components fluorine and RuO_2, F with about 0.14 %. Of the RuO_2 and F total amounts 4.0 % RuO_2 and 1.4 % F had been lost. There are no evaporation losses to be considered for the components determined in the glass samples (see table 3).

TABLE 4

COMPOSITION OF THE YELLOW PHASE (NORMALIZED VALUES [wt.%])

Constituents	1. part (sol.in H_2O)		2. part (sol.in 1nHCl)		3. part (sol.in 5nHCl)		4. part (insoluble)		Sum	
	wt.%	mol%	wt.%	mol%	wt.%	mol%	wt.%	mol%	wt.%	mol%
Basic oxides:										
Li_2O	4.30	11.35	0.17	0.45	0.00	–	–	–	4.47	11.80
Na_2O	19.07	24.27	0.05	0.06	0.01	–	–	–	19.13	24.34
K_2O	0.50	0.42	0.00	–	0.01	0.01	–	–	0.51	0.43
MgO	0.05	0.10	0.11	0.21	0.08	0.16	–	–	0.24	0.47
CaO	0.19	0.27	2.77	3.90	2.44	3.43	–	–	5.40	7.60
SrO	0.02	0.02	0.30	0.22	0.15	0.11	–	–	0.47	0.35
BaO	0.00	–	5.10	2.62	0.73	0.37	2.31	1.19	8.14	4.18
Basic oxides, total:	24.13	36.43	8.50	7.46	3.42	4.09	2.31	1.19	38.36	49.17
Acid oxides:										
SiO_2	0.15	0.20	0.38	0.50	0.06	0.07	–	–	0.59	0.77
B_2O_3	0.84	0.95	0.15	0.17	0.01	0.01	–	–	1.00	1.13
Al_2O_3	0.11	0.09	0.22	0.17	0.20	0.15	–	–	0.53	0.41
CrO_3	5.37	4.23	3.54	2.79	0.37	0.29	–	–	9.28	7.31
SO_3	29.71	29.28	0.06	0.06	0.16	0.16	1,21	1,19	31.14	30.69
MoO_3	4.17	2.29	7.14	3.91	6.80	3.72	–	–	18.11	9.92
P_2O_5	0.07	0.04	0.24	0.13	0.09	0.05	–	–	0.40	0.22
Acid oxides, total:	40.42	37.08	11.73	7.73	7.69	4.45	1.21	1.19	61.05	50.45
Other oxides, total:	0.04	0.03	0.47	0.30	0.08	0.05	–	–	0.59	0.38
Sum	64.59	73.54	20.70	15.49	11.19	8.59	3.52	2.38	100.00	100.00

<u>LEWC content of the glass produced</u>

The components best suited for use in the calculation of the mean LEWC oxide content in the glass are those with the best known input (either only frit or only LEWC) and those of which the contents in the glass is influenced to the least possible degree by the yellow phase or by corrosion processes. These include B_2O_3, TiO_2, CeO_2 and MnO_2. A comparison of the determined contents of these components with the nominal values for glasses with 15 to 18 % LEWC show that the LEWC content in the glass was 15 % after 100 hours and 18 % after the experiment had run for 226 hours.

<u>Effect of MoO_3 and the yellow phase on the materials balance</u>

As a result of the fact that a protecting tube was accidently dissolved altogether 1.73 kg MoO_3 entered the melt. Assuming that the rate at which the molybdenum was dissolved was constant there would be an additional MoO_3 input of 7.2 g MoO_3 per hour or an increase in the MoO_3 content in the glass of only 0.04 wt.%.

The contents of 0.62 % and 0.46 % MoO_3 measured after 100 and 226 hours respectively show that the MoO_3 from the protecting tube did <u>not</u> enter the melt at a constant rate.

If one takes into consideration the fact that the SO_3 contents of the samples under investigation are distinctly below the nominal values, the results could be explained by the presence of a phase containing MoO_3-SO_3. The "yellow phase" may have already formed after only a few hours of running due to the melt being saturated with SO_3 and the MoO_3 would have entered the melt via this yellow phase.

In the absence of other data it can be roughly estimated that the amount and composition of the yellow phase were constant between sampling after 100 hours and the end of the experiment. This would produce the following effects on the nominal values in the glass for the main components of the yellow phase:

a) Li_2O, CaO and BaO: no significant changes

b) Na_2O: decrease of 0.10 %

c) SO_3: decrease of 0.26 %

d) MoO_3: decrease of 0.15 %

e) Cr_2O_3: decrease of 0.06 %.

If these values are taken into account the calculated content for MoO_3 in the 100 hour sample does not coincide with the nominal value. The 100 hour sample should contain 0.31 % MoO_3. The deviation observed can be explained by the fact that the amount and composition of the yellow phase were not constant. These deviations are of less importance for the determination of the corrosion products of the input.

Corrosion products

The following corrosion products can be expected during the normal running of the melter:

a) from the melter material (ER216): Al_2O_3, ZrO_2, Cr_2O_3, (SiO_2),

b) from the electrodes and the outlet at the bottom (Inconel 690): Cr_2O_3, NiO, (Fe_2O_3),

c) from the dust removal wet scrubber and the pipes and containers: Fe_2O_3.

Melter material ER 216. The ZrO_2 contents of the glasses, which are increased compared with the nominal values, are most suitable for the determination of the corrosion rate of ER 216 (30 % Al_2O_3, 28 % ZrO_2, 28 % Cr_2O_3, 14 % SiO_2; density = 4.0 g cm^{-3}). Al_2O_3 and SiO_2 could not be determined precisely enough for this purpose.

For the total length of the experiment the ZrO_2 content of the glass is on the average 0.08 $\pm$ 0.03 % higher than expected. Therefore, the additional input is (3.0 $\pm$ 1.2) kg ZrO_2 or (10.7 $\pm$ 4.3) kg ER 216. The nominal values of the glasses for the remaining components of the melter material are increased by 0.09 % for Al_2O_3, 0.08 % for Cr_2O_3 and 0.04 % for SiO_2. Thus, the rate of corrosion for the 10 days experiment is (140 $\pm$ 60) μm·d^{-1}.

In corrosion experiments with similar glasses on a laboratory scale a corrosion rate of 120 μm per day was found. In these experiments Cr_2O_3 did not enter the glass melt, but adhered to the surface of ER 216. A similar effect was to be expected in the described experiment.

Electrode material Inconel 690. For the determination of the Inconel 690 input (60 % Ni, 30 % Cr, 10 % Fe; density = 8.1 g cm^{-3}) the NiO contents of the glasses, which are increased compared with the nominal values are used, as Cr_2O_3 and Fe_2O_3 might originate too from the corrosion of other components of the system.

The average increase in the NiO content of the glass is (0.04 $\pm$ 0.02) %. This results in a corrosion input of (1.2 $\pm$ 0.6) kg Ni or (2.0 $\pm$ 1.2) kg Inconel. The values for Cr_2O_3 and Fe_2O_3 are thus increased by 0.02 % or 0.01 respectively, i.e. the increases are barely detectable. The corrosion rate for Inconel (electrode surface = 60 dm^2) is therefore (40 $\pm$ 25) μm per day.

Container and pipe material (stainless steel). It may be assumed that Fe_2O_3 from the dust removal wet scrubber and the containers and pipes, in contact with the acid off-gas and its aqueous solutions, enters the melter as a corrosion product via recycling. As can be seen in table 3 after 100 running hours the glass samples contain 0.48 % (15 % LEWC) and after 226 hours 0.64 % (18 % LEWC) more Fe_2O_3 than originates from the LEWC. An analytical value (0.50 ± 0.10) % above the nominal values is taken as the mean value for the whole experiment. From this a corrosion input of (13 ± 2.6) kg Fe is obtained for the 10 day experiment. Assuming that this amount of Fe only comes from the wet scrubber (1.5 m^2) the corrosion rate will be (110 ± 22)μm per day.

SUMMARY AND CONCLUSIONS

For a 10 day vitrification experiment with simulated HLLW an almost complete materials balance could be achieved.

The dissolution of a molybdenum protecting tube could easily be recognised in the materials balance.

The corrosion product input can be calculated from the materials balance, and the corrosion rates for the corresponding materials can also be obtained. The corrosion rates for the 10 day experiment do not show any interference in the vitrification of the LEWC.

The maximum SO_3 content of the melt is approximately 0.70 %. Changes in the composition of the glass frit only slightly influence the solubility of SO_3 in the borosilicate glass. The saturation with SO_3 occurs as soon as the LEWC content in the glass reaches 14 %. In future experiments the HLW contents should be adapted to the greatest possible SO_3 content in the glass. Furthermore, the use of glass frits which are free of SO_3 should be observed.

If the solubility of the SO_3 in the glass has been exceeded, a second phase forms in the melter, which must increase in amount when the evaporation products are completely recycled, and as the duration of vitrification is extended. It cannot, therefore, be excluded that the degree of the corrosion of other individual materials will not be affected.

REFERENCES

1. Grünewald, W. and Weisenburger, S. (1981) International Seminar on Chemistry and Process Engineering for High-Level Liquid Waste Solidification, Jülich.
2. Heimerl, W.(1981) International Seminar on Chemistry and Process Engineering for High-Level Liquid Waste Solidification, Jülich.

CERAMICS

Published 1982 by Elsevier Science Publishing Co
SCIENTIFIC BASIS FOR RADIOACTIVE WASTE MANAGEMENT - V
Werner.Lutze, editor

INCORPORATION OF HIGH-LEVEL WASTES IN SYNROC: RESULTS FROM RECENT PROCESS ENGINEERING STUDIES AT LAWRENCE LIVERMORE NATIONAL LABORATORY*

J. H. CAMPBELL, C. L. HOENIG, F. J. ACKERMAN, P. E. PETERS, J. Z. GRENS
Lawrence Livermore National Laboratory, P.O. Box 808, Livermore, California, USA 94550

INTRODUCTION

In October 1981 SYNROC-D was selected as the reference alternate waste form to borosilicate glass for immobilization of defense wastes. A total of eight candidate waste forms competed in this selection process and the decision of which alternate waste form to choose was based primarily on performance properties.

The major advantage of SYNROC-D is that it gives a 10^3 improvement in actinide leach rates as measured by MCC-1 and 2 leach tests over borosilicate glass (SRL waste glass 131). Also the waste loading (as volume %) of SYNROC is three times that of the reference SRL glass giving SYNROC a marked economic advantage for interim storage, transportation, and repository storage[1,2]. These savings offset the additional costs associated with the processing facility[2], thus making the total system cost for SYNROC comparable to borosilicate glass.

In a recent series of papers, we have addressed the formulation[3], preparation[4], characterization[5,6], and performance testing[6,7] of SYNROC containing Savannah River Plant defense wastes. We have also recently published a brief description of the SYNROC process flow sheets and a schematic layout of an associate processing facility[8]. In this brief paper we summarize highlights from recent engineering research and development, in particular, results from fluidized bed calcination studies of SYNROC slurry.

SYNROC-C VS. SYNROC-D

The SYNROC composition originally proposed by Ringwood[9] was designed for storage of commercial reactor wastes. This material is called SYNROC-C (C for commercial). At LLNL we are working on a modification of this original form for the storage of defense wastes, i.e. SYNROC-D (D meaning defense).

There are two major crystal-chemical differences between SYNROC-C and SYNROC-D. First, an additional inert (i.e., contains no radionuclides) spinel phase is formed from the large quantities of aluminum and transition metals

---

* Work performed under the auspices of the U.S. Department of Energy by Lawrence Livermore National Laboratory under Contract W-7405-Eng-48.

300

metals (mainly Fe, Ni, and Mn) in the defense waste. Secondly, due to the presence of Na and Si in much of the defense waste, a silicate phase (nepheline) is used as a Cs host.

In regard to formulation, it is important to realize that only four additives (TiO_2, ZrO_2, CaO and SiO_2) are used in the preparation of SYNROC containing SRP defense waste (Fig. 1). Further, we have found that components in the waste sludge feed can vary by as much as ±50 wt% in composition without affecting either the quantity of SYNROC additives needed or the quality of the final product[6,10]. Our experience shows the SYNROC formulation to be tolerant of variations in feed composition and thus quite capable of handling reasonable feed stream upsets[10].

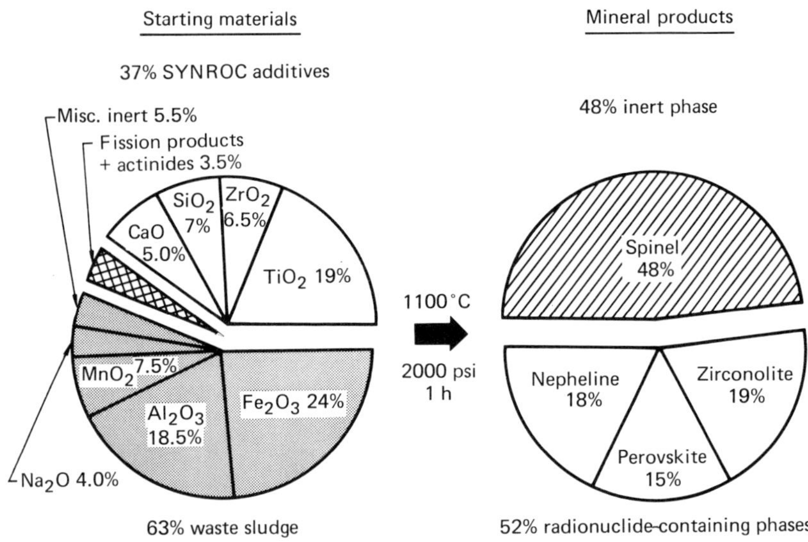

Fig. 1. Pie chart showing the wt% composition of the starting materials and synthetic mineral products during preparation of SYNROC-D containing SRP composite waste sludge.

STATUS OF SYNROC-D PROCESS DEVELOPMENT AT LLNL

The SYNROC-D process, at the present state (two years R&D) is less developed than the more mature (greater than 15 years) borosilicate glass process. This is expected. SYNROC is a new waste form and nearly all research to date has been directed toward lab synthesis and testing of the product. However, as a result of recent engineering R&D, significant simplification in SYNROC-D processing has occurred over the past 18 months.

In our initial lab process, (Fig. 2a) we blended and sometimes milled the additives with the waste slurry. After spray drying, this mixture was calcined and then reduced in two separate batch operations. Often the powder was reground prior to hot pressing to insure homogenity. The initial capacity was only about 50g/day.

In our first improvement of the process (Fig. 2b), we eliminated all grinding operations and now simply blend the waste sludge and SYNROC additives in a 55-gallon drum (for small batches) or a 350 gallon tank (for large batches). This slurry is fed to a large spray dryer having an output of approximately 5-10kg solids per hour. The spray dried powder is then simultaneously calcined and reduced in a large rotary calciner. This product can be densified by hot uniaxial pressing (HUP) or hot isostatic pressing (HIP).

More recently, we have demonstrated further improvement through use of a slurry-fed, fluidized-bed calciner as a replacement for the spray dryer and rotary calciner (Fig. 3c). This initial work was done in cooperation with the Idaho Chemical Processing Plant using a 10cm (4 in.) diam. fluidized bed. (Further details of this work are given below.) The fluidized bed will eventually increase our calcined powder production capacity to about 100 kg/day.

POWDER PROCESSING: FLUIDIZED BED CALCINATION

Calcined SYNROC-D powders ready for high-temperature consolidation need to meet the following specifications: (i) pycnometric density greater than 3.2 g/cm^3, (ii) reactant grain size less than or equal to 70 microns (with a mean approximately equal to 15 microns), (iii) tap density greater than 1.0 g/cm^3, and (iv) volatile content less than 1.0 wt%. Some control of the oxidation state of elements such as Fe, Mn, and U is also necessary. However, recent work at LLNL[10] and Rockwell Science Center[11] shows that the range of allowed oxidation state is quite large and thus control of this parameter is less critical.

Laboratory-scale experiments are currently underway at LLNL investigating the use of a slurry-fed, fluidized bed calciner. This calciner is similar in design to the units in operation at Idaho Chemical Processing Plant (ICPP) and offers several advantages for preparing SYNROC powders:

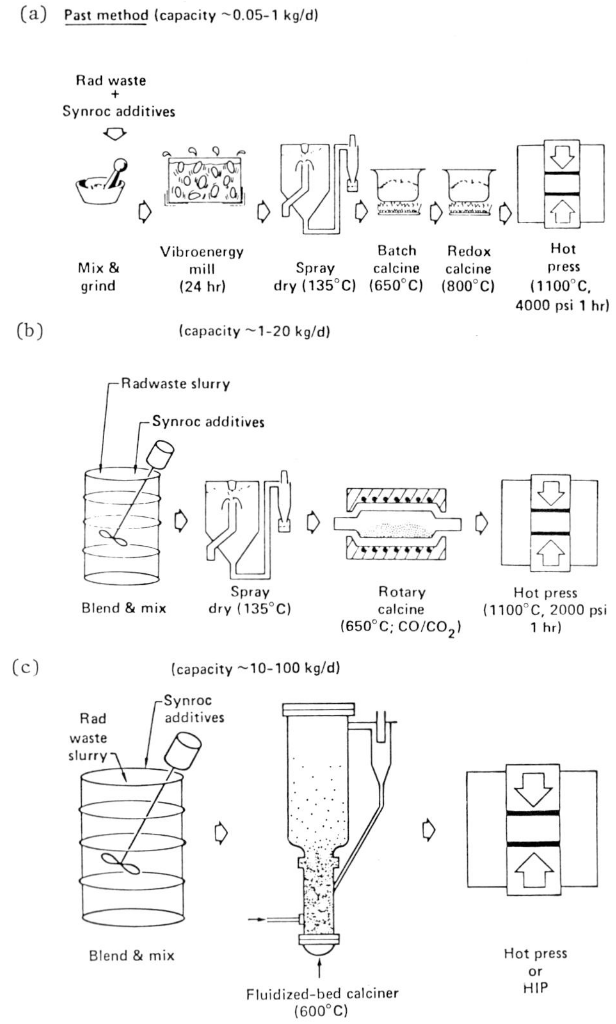

Fig. 2. Evolution of SYNROC processing from a laboratory (a) to various stages [(b) and (c)] of pilot-plant scale operation.

(1) High throughput with moderate-to-long residence times.

(2) High heat transfer rates and a broad range of operating temperatures.

(3) No internal moving parts.

(4) A fully-calcined product with high packing density (greater than 35% T.D.).

(5) Twenty years of operating experience on real radwaste feed (at ICPP).

A schematic diagram of our 10cm (4 in.) laboratory-scale fluidized bed calciner is given in Fig. 3. Fluidizing gas (air) is introduced through a bottom plenum and the waste slurry is sprayed into the bed through a nozzle driven by high pressure air. Another nozzle is used to introduce a kerosene/O_2 mixture that provides the heat for drying and decomposition reactions. We have also added a series of band heaters to the vessel walls that provide an alternate heat source for operations up to 500°C.

Typical calciner operating conditions for processing a SYNROC slurry feed are given in Table 1. These tests involved over 70 hours of operation on a 10cm (4 in.) diam. calciner at ICPP.

As mentioned above, one of the most desirable features of the fluidized bed calciner is the high packing density of the powder product. This high density is achieved by continuous deposition of a coating of product on a seed particle; the coating has a low porosity and consists of an intimate mix of microcrystalline grains of SYNROC additives and waste feed. Thus, although the product particle size is quite large (0.2-0.8mm), the reactant grains composing each particle are small enough (micron size) to give high reactivity during consolidation.

The fines produced during calcination (about 25% of the total product) are carried overhead and collected by a cyclone separator. These fines can be selectively introduced into the bed to provide the "seed" material for continued particle growth.

CONSOLIDATION VIA HOT ISOSTATIC PRESSING (HIP)

In HIP processing, the packing density of the powder is important because it impacts the design, size, and number of HIP cans needed to achieve a given throughput (Fig. 4). The expected plant throughput is about 1.45 metric tons SYNROC at 60 wt% waste loading. At a powder-packing density of 35%, this quantity of SYNROC can be loaded into a single can having an initial height of 1.47m (58 in.) and a diameter of 0.91m (36 in.) or two canisters 0.65m

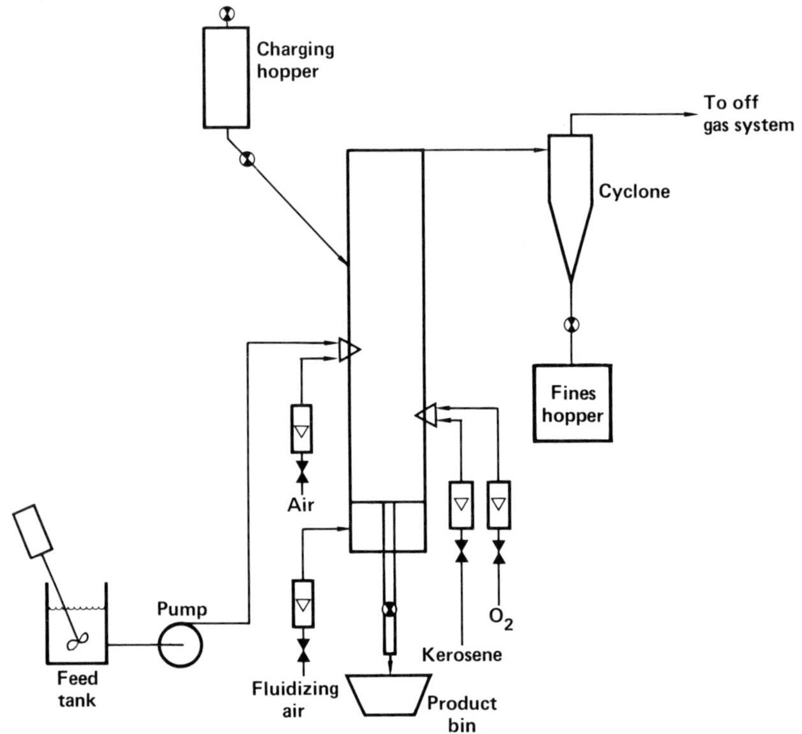

Fig. 3. Schematic of laboratory-scale fluidized bed calciner.

TABLE 1
SUMMARY OF TYPICAL OPERATING CONDITIONS FOR FLUIDIZED-BED CALCINATION OF SYNROC FEED (DATA FROM 10cm DIAM. UNIT AT IDAHO CHEMICAL PROCESSING PLANT.)

Bed cross-sectional area; (m^2)	7.85×10^{-3}
Slurry feed rate; (l/h)	2.0
Solids loading: dissolved; (kg/l)	0.27
suspended; (kg/l)	0.093
Calcining rate; $(kg\ oxide/m^2/h)$	43
Fuel feed rate (kerosene); (cm^3/h)	942
Oxidizer feed rate (oxygen); (m^3/h)	1.6
Fluidizing air (at 600°C, 1 atm); (m/s)	0.32–0.35
Bed pressure; (psig)	3.5
ΔP across the bed; (in. H_2O)	21–25
Bed temperature; (°C)	595–610
Product packing density[a]; (g/cm^3)	1.6–1.7
Particle (pycnometric) density (g/cm^3)	3.6–3.8

[a] loose packed, without vibration or compression

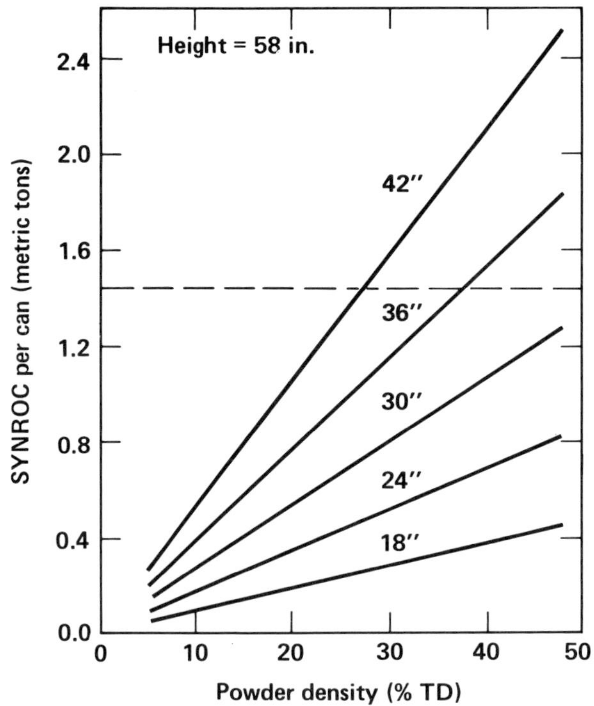

Fig. 4. Effect of powder packing density and canister dimension on throughput.

(26 in.) diam. by 1.47m (58 in.) high. HIP experience has shown that for powders with packing densities greater than approximately 50% theoretical (i.e.
less than 50% porosity), simple, cylindrical canister designs can be consolidated reliably. This is because the packed powder provides sufficient internal support to the metal canister to prevent buckling and distortion during
densification. In the event of packing densities less than 35% T.D. a special
bellows-type canister[12] is used which allows most of the compaction to take
place axially. A bellows canister is currently proposed for use in SYNROC-D
consolidation.

Recently Ringwood[13] has successfully densified SYNROC in bellows cans
measuring 25cm (10 in.) diam. x 60cm (24 in.) high; the final can sizes after
consolidation were about 20x20cm (8x8 in.). Work is currently underway at
LLNL on preparing two 30x30cm bellows canisters (dimensions are after consolidation), each containing 100kgm of SYNROC-D.

Estimates of Cycle Times for HIP Operation

SYNROC-D powder densifies rapidly at modest temperatures and pressures. For example, our experiments have shown that densification can be achieved in about 10 minutes at 4000 psi and 1100°C. Applying these results to large HIP canisters indicates that the time required for actual consolidation is small compared to the time required to preheat the canister to the necessary densification temperature.

We have calculated the cycle time for densifying a 0.65m (26 in.) diam. by 1.47m (58 in.) can containing SYNROC-D at 35% packing density. After consolidation this can is approximately 56cm (22 in.) diam.x79cm (31 in.) and contains 0.45 metric ton of SRP waste. The finished can is designed to fit the reference 61cm (24 in.) diam. waste canister. Our estimates show that it will take less than 24 hours to preheat the HIP canister to a center-line temperature of 800°C before charging to the HIP furnance. Large industrial HIP furnaces are usually loaded hot to maximize throughput (i.e., reduce cycle times). Once loaded into the HIP unit, it will take about 11 hours to process the canister and then cool to 800°C for recharging. This includes a three-hour hold period at maximum temperature. Therefore a single HIP unit can be projected to process two canisters a day. Our estimates agree well with actual industrial experience that shows large HIP systems can be operated on a 12-hour cycle under similar time-temperature-pressure profiles[8].

A schematic diagram of the envisioned SYNROC process (at this stage of development) is given in Fig. 5. It shows the use of a fluidized bed calciner to prepare SYNROC powder that is then fed to a storage hopper. Bellows-type canisters are filled, evacuated, sealed and preheated. The preheated canisters are loaded into a HIP unit where they are densified, then removed and cooled, and finally loaded into a waste storage container. After sealing, this container is decontaminated and transferred to the interim storage facility and then, ultimately, to an underground repository.

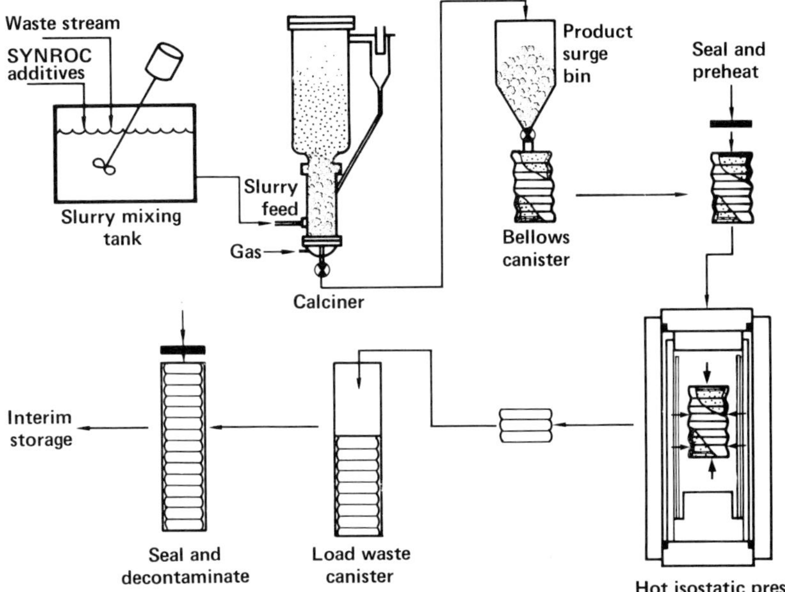

Fig. 5. Schematic of SYNROC-D process.

REFERENCES

1. Newby, D., (1981) Engineered Waste Package Conceptual Design Defense High-Level Waste (Form 1), Commercial High-Level Waste (Form 1) and Spent Fuel (Form 2) Disposal in Salt, Westinghouse Electric Corporation AESD-TME-3113.

2. McDonnell, W., (Dec. 1981) Savannah River Laboratory, Private Communication,

3. Ryerson, F. J., et al., (1981) Formulation Of SYNROC-D Additives for Savannah River Plant High Level Radioactive Waste, Lawrence Livermore National Laboratory, Livermore, CA, UCRL-53237.

4. Hoenig, C. L., et al., (1982) Preparation and Properties of SYNROC-D for Savannah River Plant High-Level Defense Waste, Lawrence Livermore National Laboratory, Livermore, CA, UCRL-53195.

5. Ryerson, F. J., (in progress) Characterization of SYNROC-D Samples Prepared for the Comparative Leach Test, Lawrence Livermore National Laboratory, Livermore, CA.

6. Campbell, J. H., et al., (1982) Properties of SYNROC-D Nuclear Waste Form: A State-of-the-Art Review, Lawrence Livermore National Laboratory, Livermore, CA, UCRL-53240.

7. Van Konynenburg, R. A., and Guinan, M. W., (1981) Radiation Effects in SYNROC-D, Lawrence Livermore National Laboratory, Livermore, CA, UCRL-86679.

8. Rozsa, R. B., and Hoenig, C. L., (1981) SYNROC Processing Options, Lawrence Livermore National Laboratory, UCRL-53187,.

9. A. E. Ringwood, et al., (1979) Geochem. J. 13, 14.

10. Campbell, J. H., et al., (1982) Flexibility in Formulating and Processing SYNROC-D: A Current Assessment, Lawrence Livermore National Laboratory, Livermore, CA, UCRL-87349.
11. Harker, A., (Feb. 1982) Rockwell Science Center, private communication
12. Larker, H. T., (1981) ASEA Inc., Robertsford, Sweden, private communication.
13. Ringwood, A. E., Australian National University, (March 1982) private communication.

Published 1982 by Elsevier Science Publishing Co
SCIENTIFIC BASIS FOR RADIOACTIVE WASTE MANAGEMENT- V
Werner.Lutze, editor

THE MICROSTRUCTURE OF SYNROC

D. R. COUSENS, S. MYHRA, J. PENROSE, R. L. SEGALL, R. St.C. SMART AND P. S. TURNER
School of Science, Griffith University, Nathan, Queensland 4111, Australia

INTRODUCTION

The Synroc concept was introduced by Ringwood in 1978[1] and, as is well known, the material now proposed for the disposal of high-level civilian nuclear waste consists primarily of hollandite ($BaAl_2Ti_6O_{16}$), perovskite ($CaTiO_3$) and zirconolite ($CaZrTi_2O_7$).

There is limited theoretical understanding of the dissolution kinetics of ceramics. The known excellent leach resistance of rutile does not of itself provide an explanation for the leach resistance of a mixture of titanates such as Synroc. Moreover, variable dissolution rates have been reported depending on the precise conditions of manufacture[2,3] and clearly the microstructure will be important in determining dissolution rates. For example, the presence of minor phases[4] will be significant if important elements such as Cs and Sr segregate into them and become more readily leached. More generally, studies of simple ionic[5] and semiconducting oxides[6,7] have shown that structural parameters are important in dissolution. Another aspect of any proposed high-level waste matrix which must be understood is the effect of irradiation by energetic particles. Again microstructural information is essential for a detailed analysis. Factors like phase size, grain size, phase distribution, the presence of minor phases and their location are all significant in interpreting radiation damage studies. Finally, and perhaps most significantly, the routine production of a reproducible ceramic waste form requires a knowledge of the processes of manufacture under real, non-equilibrium conditions. Microstructural studies are a major source of information on the details of the evolution of the final product from the complex mixture of starting materials.

The work is primarily on Synroc B (the material without radwaste) which should have the same structure as Synroc C (the material containing simulated waste) since the waste elements are supposed to partition themselves between the perovskite, hollandite and zirconolite[8]. However, work on Synroc C is also described.

EXPERIMENTAL

The research has been done on Synroc prepared at the Australian Atomic Energy
Commission Research Establishment via the oxide route. Fine powders of TiO_2,
Al_2O_3, ZrO_2, $BaCO_3$ and $CaCO_3$ are ball milled and the final product is made by
hot pressing[9]. The Synroc C studied had a waste loading of 10 wt percent.

Specimens have been examined by conventional transmission electron microscopy
at 100kV, by high resolution transmission electron microscopy using the JEOL
200CX operating at 200kV, by high voltage transmission electron microscopy at
1MeV using the EM7, by conventional scanning electron microscopy and by scanning
transmission electron microscopy with energy dispersive analysis. In addition
mechanically polished and chemically or thermally etched specimens have been
examined by optical microscopy.

The mechanically polished specimens for optical microscopy and conventional
scanning electron microscopy were prepared by the normal metallographic techniques to a fine diamond finish, and structure was then visible. No satisfactory
chemical etchant was developed; various reagents attacked the surface
differentially but not so that a one to one correspondence between the finely
divided phases and the optical microscope characteristics could be established.
Attack in distilled water at 200°C was as effective as any of the acid reagents.
Thermal etching was done in air at 1000°C and again revealed structure which
could not readily be related to the underlying microstructure. However, some
macroscopic inhomogeneity which has not as yet been satisfactorily related to
the fine scale microstructure was revealed both by chemical and thermal etching
in specimens examined by optical microscopy. All electron microscope specimens
were prepared by ion beam thinning using 5 kV argon ions. The electron microscope results obtained at 1 MeV were identical with those obtained at 100 kV,
suggesting that possible preferential thinning of particular phases was not
giving rise to unrepresentative results.

THE MICROSTRUCTURE OF SYNROC B

There are essentially three aspects of the microstructure of a multiphase,
multicomponent system - first the phases present and their volume fraction,
second the spatial distribution of the phases and third the grain size within
each phase.

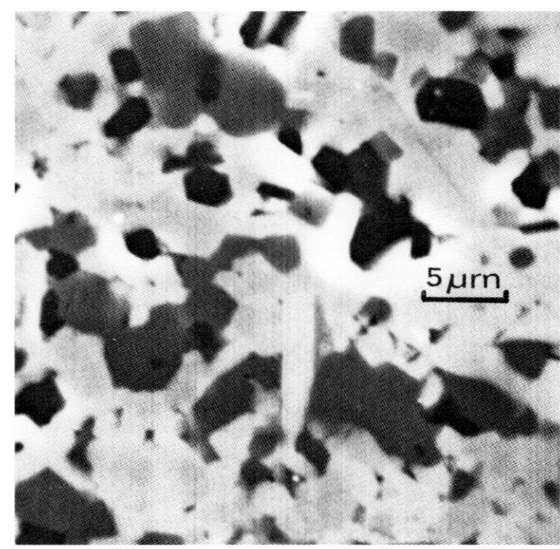

Fig. 1. Scanning electron micrograph in back-scattered electron mode of bulk specimen of Synroc B.

Figure 1 is a conventional scanning electron micrograph of a bulk specimen of Synroc B taken in the back-scatter mode. As will be seen below the resolution of this technique using bulk specimens is too poor to "see" individual phase regions. However, there are at least three distinguishable regions - light, mid and dark grey. These are not the major phases, indeed they are not single-phase regions. Nevertheless, one mid grey region is like another and they do represent a real aspect of the microstructure. We will return to this question after a more detailed examination of the phases present and of their distribution.

There are two fortunate occurrences which somewhat simplify the study of Synroc by transmission electron microscopy. First, as has been discovered by Headley[10], hollandite can often be recognised in the image without recourse to tedious diffraction analysis, by means of the complex structure (visible as narrow bands in bright field images), produced by the ion beam thinning process. Second, perovskite commonly has a characteristic appearance due to twinning and/or polytypism[11]. The latter is caused by the presence of aluminium, zirconium or barium in the calcium titanate.

On the other hand, the remaining major phase, the monoclinic zirconolite, is normally featureless in the image, but the large interplanar spacing of the {001} planes and the rather characteristic diffraction patterns containing this reciprocal lattice vector can be used as a "fingerprinting" device provided the grain size is sufficiently large for single crystal diffraction patterns to be obtainable.

In addition to these major phases, a calcium titanium aluminate (CTA) is found in titanate waste forms[12]. We have observed a phase whose lattice spacings and composition (from scanning transmission electron microscopy with energy dispersive analysis, STEM/EDS) are consistent with magnetoplumbite, $Ca(AlTi)_{12}O_{19}$. This phase is commonly in the form of single grains in the perovskite and hollandite rich areas. A cubic phase with a lattice parameter of approximately 1 nm has also been observed. It could be the pyrochlore structure reported by Tewhey et al.[13] from the X-ray work of G.S. Smith. These authors report a steadily decreasing proportion of this phase as the calcining temperature is increased. Metallic particles and a number of other unidentified phases which are distinct from those already mentioned have also been observed. Finally, we have found a modified zirconolite which is of some interest as it suggests an additional mode for the Synroc matrix phases of taking up radwaste elements. The phase has the composition of the ordinary zirconolite found in the Synroc as determined by the STEM/EDS method.

The new phase is an intergrowth of zirconolite as ordinarily observed and a very similar phase which matches on the {001} plane. Figure 2 shows the two regions in a high resolution transmission electron micrograph taken at 200 kV. On the left is zirconolite as ordinarily observed in Synroc B, on the right is the other phase which has an _apparent_ spacing exactly half as great. The _a_ and _c_ axes are perpendicular to the electron beam direction in Figures 2 and 3. It is likely that ordering has taken place in one phase, possibly favoured by some substitution of another element for particular titanium atoms, with a suppression of the forbidden 001 reflections in one phase leading to the observed contrast.

Fig. 2. High resolution transmission electron micrograph of the "modified
zirconolite" phase. On the left a spacing of 1.14nm is found, on the
right the *apparent* spacing is 0.57nm.

Fig. 3. "Modified zirconolite" showing the plates of the intergrown constituents
with widths sometimes of only one lattice parameter.

314

Figure 3 shows a detail of the intergrowth structure. Regions of only one or
two lattice parameters in thickness are observed as well as much wider regions.
Thus zirconolite, in addition to its ability to take up radwaste elements by
ordinary elemental substitution, has further flexibility in being able to change
to this modified structure. Some aluminium has been found, not surprisingly, in
the zirconolite in Synroc B and this may play a part in promoting the intergrowth structure. This new feature, of structural modification, needs further
study to ensure that there are no concommitant deleterious effects.

Synroc B is thus a complex multiphase system. For this reason and because of
batch to batch variations and sample to sample variations it is difficult to
make precise assertions about the details of the phase distribution. This difficulty is compounded by the inherent complexity of the structure and by detailed
electron microscopy required by the fine scale of the phases. Figure 4 is a low
magnification composite electron micrograph which indicates the nature of the
microstructure although it should be emphasized that other areas would have a
different microstructure. Figure 4 is thus representative but no one composite
picture of this extent could be accurately described as typical.

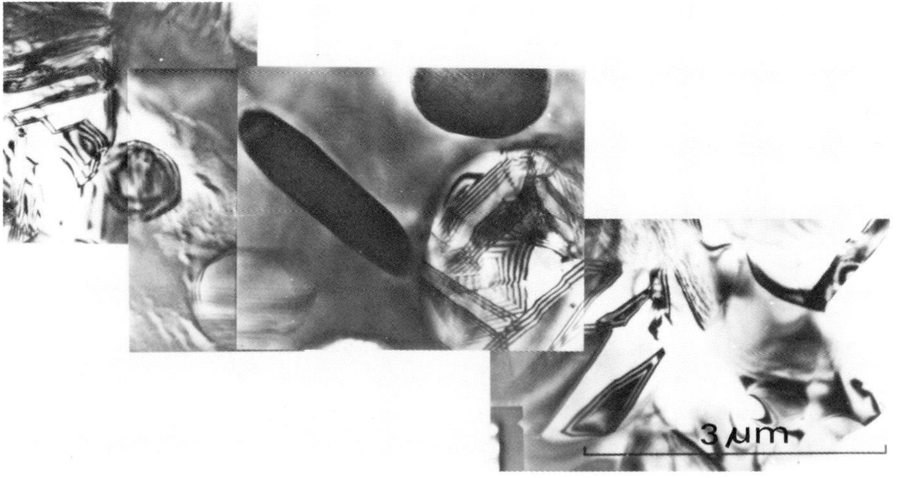

Fig. 4. Composite transmission electron micrograph of Synroc B taken at 1 MeV.
The typical twinned perovskite can be seen, hollandite, CTA and an unidentified
phase are also present in this micrograph.

The extent of the presence of the various minor phases is variable, depending significantly on the precise conditions of manufacture of the Synroc. We have not been able as yet to establish any systematic correlation between variables such as redox potential and hot pressing temperature and phase composition. We have not observed in any of the Synroc studied, any glassy phases. These would be readily detectable by transmission electron microscopy.

One further incidental observation is worth recording. Synroc is resistant to electron radiation damage. High electron fluxes (approximately 10^{23} electrons $m^{-2} s^{-1}$) are achieved in transmission electron microscopy. Significant damage was not observed in long exposures at voltages from 100 kV to 1 MeV. This comment does not apply to the hollandite where electron irradiation effects would be masked by the pre-existing ion beam damage.

A most striking feature of the grain size and the size of the single phase regions of the Synroc B are their very great variability. Grains of perovskite (twinned) larger than 20 μm in diameter have been observed. The average grain size of the three major phases is approximately 1 μm. The size of single phase regions is equally variable. Sometimes a region of hollandite which is a single grain of say 200 nm in diameter is found embedded in a large grain of CTA or zirconolite. In other areas a polycrystalline aggregate of one phase containing an estimated one hundred grains of average diameter nearly 1 μm is observed. All three major phases show this variability both in phase size and grain size. No particular association of one of the major phases with another has been detected nor have we established that any of the minor phases occurs exclusively within or next to one of the major phases. It does seem common for the modified zirconolite to adjoin, at least at one interface, ordinary zirconolite although this is certainly not always the case.

Returning to Figure 1 it is clear that the regions showing different contrast in conventional scanning electron microscopy are not the phases present but are on a considerably coarser scale. The back-scattered electron image contrast arises from differences in average atomic number. It seems that these regions, on the scale of tens of microns, are a consequence of events in the formation of the Synroc. If the five constituents in the pseudo-quinary Synroc system are nominally present in proportions a, b, c, d, e, there will inevitably be local variations a+δa b+δb and so on. It seems probable that this continuous variability leads to a small, discrete number of possible reaction sequences, depending on which elements are locally present in excess. Thus the scale of the variations seen in the conventional scanning micrographs depends on a combination of the diffusion length at elevated temperature and the effectiveness of mixing of the initial oxide powders.

THE MICROSTRUCTURE OF SYNROC C

The Synroc C we have studied is significantly different in microstructure from Synroc B. Both the size of the single phase regions and the grain size are smaller. Again both are variable but the average grain size is about 0.3 μm. The average phase size is similarly decreased compared with Synroc B although this comparison is complicated by the presence of additional phases in Synroc C. The modified zirconolite seen in Synroc B was also observed in Synroc C. There is a considerably higher density of small metallic particles in Synroc C.

The most striking difference is the presence in Synroc C of phases at grain boundary triple points and of precipitation within the grains. It is a little difficult to be precise about what constitutes precipitation in a multiphase system. In accord with the usual terminology in metallic alloys, we use the expression to mean a fine distribution of a minor phase within another phase. Figure 5 shows precipitation. In addition to the obvious precipitates, the perovskite (upper) grain also contains very fine precipitates. None of these fine precipitates has been identified as yet.

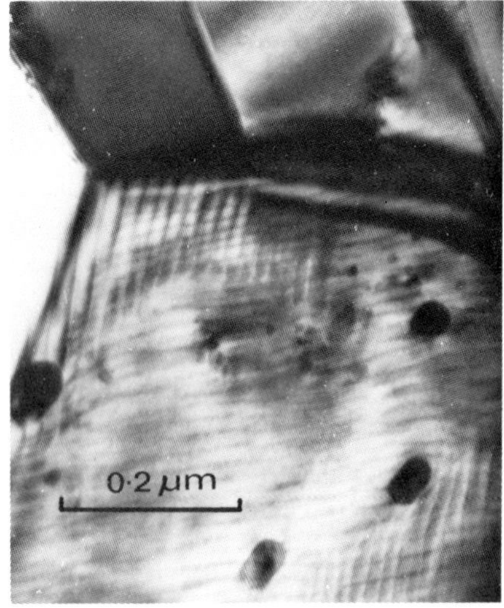

Fig. 5. Transmission electron micrograph of precipitation within the grains of Synroc C.

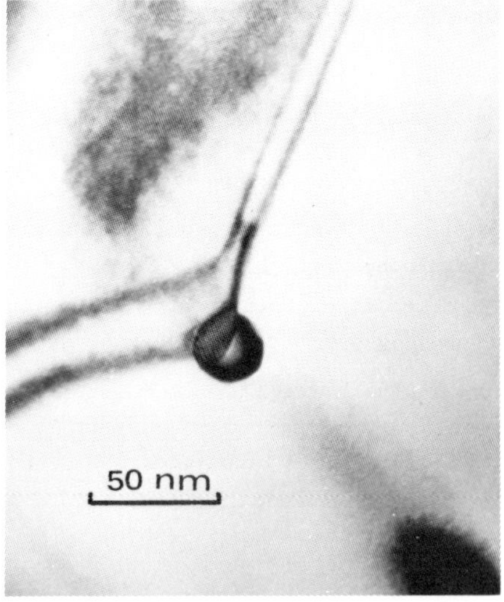

Fig. 6. Transmission electron micrograph of triple point precipitate in Synroc C.

Figure 6 shows an unidentified triple point precipitate. This is a common observation in Synroc C and has not been seen in Synroc B. At least three types of triple point phase has been observed, each having a distinctive morphology.

These features are significant in that it is clear that the added radwaste does not necessarily simply partition itself among the constituents of Synroc B. The nature and extent of such precipitation will evidently be a subtle function of the details of the fabrication process.

SUMMARY

An investigation of the microstructure of Synroc B prepared by the oxide route has shown that minor phases are present to a significant extent. Among these is a modified form of zirconolite which consists of an intergrowth of normal zirconolite with a derivative structure of the same phase. This provides an additional mechanism for the flexible uptake of foreign elements into zirconolite. Synroc C is a complex multiphase material and radwaste additions significantly affect the final form of the titanate ceramic. These modifications of Synroc B may be beneficial, the finer grain and phase size seem advantageous. On the other hand if there were segregation of a critical waste element into a soluble minor phase this could be a serious problem. In Synroc B and Synroc C the distribution of phases shows systematic minor variation on the scale of tens of microns, probably as the result of initial local compositional variations in the mixture of oxide starting materials.

ACKNOWLEDGEMENTS

We are very grateful to K.D. Reeve, E.J. Ramm, D.M. Levins and J.L. Woolfrey for the provision of Synroc specimens fabricated at the Australian Atomic Energy Commission Research Establishment, Lucas Heights. Our work has been made possible by financial support from the Australian Government under its National Energy Research Development and Demonstration Programme. We would like to thank Dr. J.L. Hutchison with whom one of us (RLS) is collaborating in the structural work on zirconolite described in the paper. One of us (RLS) would like to thank Professor Sir Peter Hirsch, FRS, for the provision of laboratory facilities in the Department of Metallurgy and Science of Materials, Oxford University, where a part of this work was done.

REFERENCES

1. A.E. Ringwood (1978) "Safe Disposal of High Level Nuclear Waste: A New
   Strategy" Australian National University Press.
2. A.E. Ringwood, V.M. Oversby, S.E. Kesson, W. Sinclair, N. Ware, W. Hibberson
   and A. Major (1981) Nuclear and Chemical Waste Management, $\underline{2}$, 287.
3. K.D. Reeve, D.M. Levins, E.J. Ramm, J.L. Woolfrey, W.J. Buykx, R.K. Ryan and
   J.F. Chapman (1981)Waste Management 1981 Symposium,Tucson,Arizona, R.Post ed.
4. A.E. Ringwood and S.E. Kesson (1979) "Immobilization of High Level Nuclear
   Reactor Wastes in Synroc", Int.Symp.Ceramics in Nuclear Waste Management,
   T.D. Chikalla and J.E. Mendel eds.
5. C.F. Jones, R.L. Segall, R.St.C. Smart and P.S. Turner (1981) Proc.Roy.Soc.
   A $\underline{374}$, 141.
6. N. Valverde and C. Wagner (1976) Ber. Bunsenges. Phys.Chem., $\underline{80}$, 330.
7. C.F. Jones, R.L. Segall, R.St.C. Smart and P.S. Turner (1978) J.C.S. Faraday
   I, $\underline{74}$, 1615.
8. A.E. Ringwood, S.E. Kesson, N.G. Ware, W. Hibberson and A. Major (1979)
   Geochem. J., $\underline{13}$, 141.
9. K.D. Reeve, D.M. Levins, E.J. Ramm, J.L. Woolfrey and W.J. Buykx (1982)
   "Scientific Basis for Nuclear Waste Management", $\underline{4}$, S. Topp, Ed., Elsevier-
   North-Holland, New York, in press.
10. T.J. Headley (1981) "39th Ann.Proc. Electron Microscopy Soc.Amer." Atlanta,
    Georgia, G.W. Bailey ed, p 114.
11. J.L. Hutchison and A.J. Jacobson (1975) Acta Cryst. B, $\underline{31}$, 1442.
12. A.B. Harker, C.M. Jantzen, P.E.O. Morgan and D.R. Clarke (1981) "Scientific
    Basis for Nuclear Waste Management" $\underline{3}$, J.G. Moore ed, Plenum Press,
    New York, p 123.
13. J.D. Tewhey, C.L. Hoenig, H.W. Newkirk, R.B. Rozsa, D.G. Coles and
    F.J. Ryerson (1981) "Alternative Nuclear Waste Forms and Interactions in
    Geologic Media", L.A. Boatner and G.C. Battle eds, Conf 8005107 p 4.

Published 1982 by Elsevier Science Publishing Co
SCIENTIFIC BASIS FOR RADIOACTIVE WASTE MANAGEMENT - V
Werner.Lutze, editor

SIMS DEPTH PROFILING STUDIES OF SPHENE-BASED CERAMICS AND GLASS CERAMICS LEACHED IN SYNTHETIC GROUNDWATER

P.J. HAYWARD, W.H. HOCKING, F.E. DOERN and E.V. CECCHETTO

Atomic Energy of Canada Limited, Whiteshell Nuclear Research Establishment, Pinawa, Manitoba, Canada, ROE 1LO

ABSTRACT

Glass ceramics and ceramics based on the mineral sphene ($CaTiSiO_5$) are being developed to host the wastes arising from possible future CANDU* fuel reprocessing. Results from leaching tests in deionized water and in synthetic groundwater indicate that these materials are highly durable. Secondary Ion Mass Spectrometry (SIMS) depth profiling of leached specimens suggests that leaching in the glass ceramics is predominantly confined to the glass phase. The high ionic strength and composition of the groundwater have a significant passivating effect on leaching and surface alteration phenomena, and encourage the precipitation of new phases on the ceramic surface. Leaching results, scanning electron microscope (SEM) observations and SIMS depth profile measurements are compared and discussed.

INTRODUCTION

Atomic Energy of Canada Limited is developing the technology to immobilize and isolate the high-level wastes that would arise from possible future reprocessing of irradiated uranium and thorium fuels from CANDU* nuclear reactors. The waste immobilization program includes the development of durable matrices to host these wastes, including boroaluminosilicate glasses, sphene-based glass ceramics and ceramics. All are being examined for thermal, mechanical, geochemical and radiation stability in the postulated hydrothermal environment of a flooded underground vault in the Canadian Shield [1].

Recent analyses [2] of deep groundwater samples from representative sites within the Shield indicate that saline groundwaters with high Na^+, Ca^{2+} and Cl^- and low K^+, Mg^{2+} and HCO_3^- concentrations are to be expected at depths greater than $\sim$ 500 m. Below $\sim$ 1000 m, the total dissolved solids in these waters may exceed 200 g.L^{-1}.

It has been suggested [3] that sphene, $CaTiSiO_5$, should be a thermodynamically stable phase under anticipated vault conditions. Thermodynamic calculations and leaching experiments using synthetic groundwaters and both natural and synthesized $CaTiSiO_5$ [4] appear to support this prediction. The purpose of the present work is to examine the nature of

---

* CANada Deuterium Uranium

the surface interactions that occur when sphene-based materials, containing simulated fission products and actinides, are leached in a synthetic saline groundwater of high ionic strength. Under these circumstances, it becomes impractical to rely on solution analyses alone for an understanding of the leaching processes. Analytical techniques such as Electron Spectroscopy for Chemical Analysis (ESCA) and Auger Electron Spectroscopy (AES), in combination with ion sputtering, or Secondary Ion Mass Spectrometry (SIMS) are, therefore, used to profile the outermost layers of the waste form chemically.

The relative advantages and drawbacks of these techniques for analyzing the corrosion behaviour of nuclear waste glasses have been summarized before [5-8]. The choice of SIMS depth profiling in the present experiments was mainly dictated by the wide range of ion concentrations that this technique can detect. This was essential for the present experiments because of the need to monitor major and trace (< 0.1 atom percent) elements simultaneously.

EXPERIMENTAL PROCEDURES

1. Sample Preparation

Target compositions for the ceramic, glass ceramic and synthetic groundwater are listed in Table 1. In all cases, the compositions were verified by chemical analyses.

The parent glass of the glass ceramic was melted at 1350°-1400°C. This was followed by casting, annealing at 750°C and cooling to ambient temperature. During cooling, the glass phase-separated to give a microstructure composed of 0.1-1.0 µm diameter titanosilicate droplets in an aluminosilicate matrix (see Figure 1a). On reheating to 1050°C, and holding at this temperature for one hour, the glass devitrified to give a glass ceramic, consisting of $CaTiSiO_5$ crystallites in an aluminosilicate glass matrix (Figure 1b).

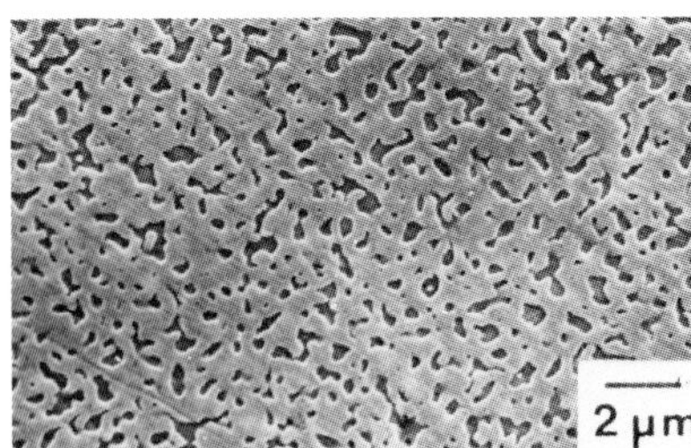

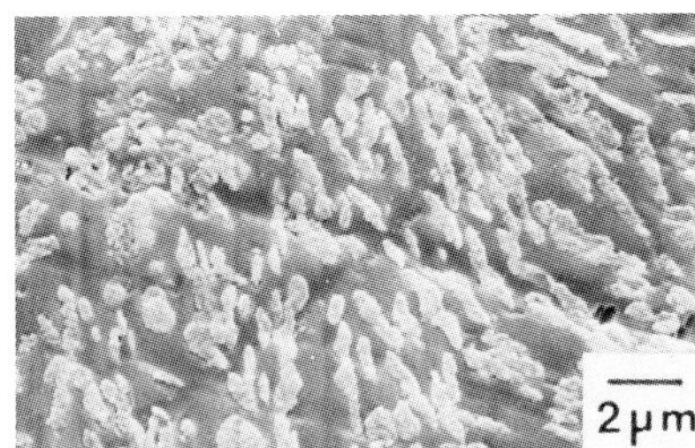

FIGURE 1: Scanning electron micrographs showing (a) phase-separated titanosilicate glass, and (b) sphene glass ceramic. Both samples etched for 20 seconds in 1% hydrofluoric acid.

The ceramic was prepared by melting a stoichiometric mixture of $CaCO_3$, TiO_2 and SiO_2, fritting in water, reheating to $1000^{\circ}C$ to crystallize the frit, and then grinding under butanol to give a fine (below 10 μm) powder. The dried powder was blended with a solution of the nitrates of Cs, Sr, La, Ce and U, plus additional SiO_2 and TiO_2, and redried. One percent stearic acid and three percent of wax binder were added in ether-carbon tetrachloride solution, the slurry was dried and granulated, and pellets 25 mm in diameter were pressed in a hardened steel die at 25 MPa. Sintering was accomplished by heating at 5°/minute to $1310^{\circ}C$ and holding at this temperature for 3 hours. The final ceramic consisted of interlocking $CaTiSiO_5$ crystallites, 5-20 μm in diameter, with an estimated closed porosity of $\sim$ 5 to 10%, and was impermeable to water.

2. Leaching Experiments

The synthetic groundwater experiments, reported previously [4], have been extended to 360 days, using the same experimental conditions as before. In all cases, crushed granite was added to the leachant to simulate a rock-groundwater environment. In some experiments, the pH was buffered in the range 8.0-9.0 by adding a powdered CaO-TiO_2-SiO_2 glass frit.

TABLE 1
Target compositions for the glass ceramic, ceramic and synthetic groundwater used in leaching/depth profiling experiments (s.s. = solid solution.)

	Oxides, weight %		mg L^{-1}
Components	Glass Ceramic	Ceramic	Groundwater
Ca	13.8	26.8	15000
Na	6.0	-	5050
Al	7.7	-	-
Ti	17.6	39.4	-
Si	50.9	29.6	15
Cs	0.5	0.8	-
Sr	0.5	0.5	20
La	1.0	1.0	-
Ce	1.0	1.0	-
U	1.0	1.0	-
Mg^{2+}	-	-	200
K^{+}	-	-	50
Cl^{-}	-	-	34260
SO_4^{2-}	-	-	790
HCO_3^{-}	-	-	10
NO_3^{-}	-	-	50

Phases Present		
$CaTiSiO_5$ s.s.	$CaTiSiO_5$ s.s.	pH at 100° = 6.37
Alumino-silicate glass		Total dissolved solids = 56 g.L^{-1}

3. Depth Profiling Experiments

Samples for depth profiling were prepared by fracturing coupons of each material. The fractured portions were leached in Teflon containers at 100°C under static conditions for 10 days, using either deionized water or synthetic groundwater plus 5-10 g of crushed granite. In each experiment, a sample surface area to leachant volume ratio of 0.1 cm^{-1} was employed. After 10 days, the samples were gently rinsed in deionized water and dried in air at 70°C.

The altered fracture surfaces of the specimens were analyzed using a Physical Electronics Industries Model 3500 SIMS II Spectrometer mounted on a 590A Scanning Auger Microprobe. Secondary ion mass spectra were excited by a 4.5 kV argon ion beam rastered over approximately 1 mm^2 of the specimen surface. Approximate calibration of the depth scale was derived from sputtering studies on standard films of tantalum oxide. Full details of the SIMS measurements will be published elsewhere.

The measured secondary ion depth profiles were converted into elemental composition profiles by employing the empirical, relative sensitivity factor method described by Newbury [9]. Suites of relative sensitivity factors, S, were calculated, using known bulk atomic concentrations, from the secondary ion yields measured for freshly fractured specimens of ceramic or glass ceramic. Under nominally identical conditions, the S values were reproducible only to within $\pm$ 50%: uranium showed the largest variation, no doubt reflecting its low ion current. The S values determined for the ceramic were in approximate agreement (within a factor of two) with those obtained for the glass ceramic. The relative sensitivity factors derived in this study were generally similar to those reported by McIntyre et al [7], except in the case of cesium, strontium and potassium, where they were 10^3, 30 and 20 times larger, respectively.

RESULTS AND DISCUSSION

Before considering the results in detail, it should be emphasized that the absolute values of the elemental concentrations recorded during depth profiling are subject to some uncertainty, since they assume that the relative sensitivity coefficients determined for each element in the unaltered material are also applicable to the altered surface layers. This is clearly an oversimplification, and could lead to serious errors where the ionization efficiency of a species is particularly sensitive to subtle matrix effects. Cesium seems to be an example of this effect. Previously quoted Cs leach rates for the glass ceramic in deionized water [4] are in the range 0.8-1.0 x 10^{-10} kg.m^{-2}.s^{-1}, i.e. slightly lower than those for Na. In the present SIMS results, however, there are consistent discrepancies between calculated cesium concentrations in the unaltered and the leached layers of both the ceramic and the glass ceramic, which could be taken as indicating very high cesium leaching losses. These results are probably misleading, and caused by extreme sensitivity of cesium ionization efficiency to matrix effects.

1. Ceramic Leached in Deionized Water.

The depth profile results for the ceramic leached in deionized water are shown in Figure 2. Apart from slight surface enrichment in Ce and La to a depth of ∿ 100 nm, little change in composition is seen as a function of depth. This is in general agreement with the low static leach rates published previously [4], and was confirmed by Scanning Electron Microscopy (SEM), where no visible evidence for leaching was observed.

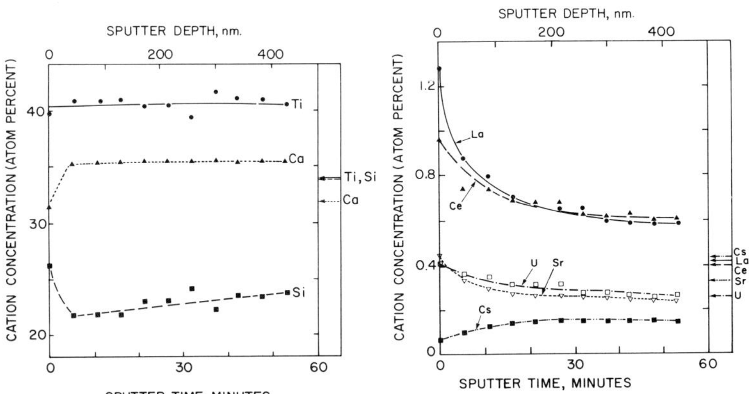

FIGURE 2:   Depth profiles of sphene ceramic leached in deionized water for 10 days at 100°C. Bulk elemental concentrations are shown in the right margins.

2. Ceramic Leached in Synthetic Groundwater.

Results of static leach tests for the ceramic in the synthetic groundwater are presented in Figure 3, expressed as specimen weight change versus time for two different pH ranges. The ceramic specimens exhibited initial weight gains, which were maintained in the experiments at pH 8.0-9.0, but not in the experiments at pH 6.0-7.0. This behaviour is in agreement with thermodynamic predictions for the H^+-H_2O-CaO-TiO_2-SiO_2 system, since the groundwater composition lies within the sphene stability field for the higher pH range, and in the rutile stability field for the lower pH range.

The SIMS depth profile of the ceramic specimen leached in synthetic groundwater (pH 6.0-7.0) showed clear evidence of the precipitation of a magnesium silicate deposit of ∿ 300 nm average thickness (see Figure 4). This feature was independently confirmed by ESCA, and correlates with the 0.02% weight gain found for this specimen after leaching. Below this deposit, there appears to be a layer depleted in Si and enriched in Ca. The Sr, La and Ce concentrations show a depth dependence similar to Ti, whereas the concentrations of U, Cs, Na and Al appear to be similar in the surface deposit and in the underlying layer.

324

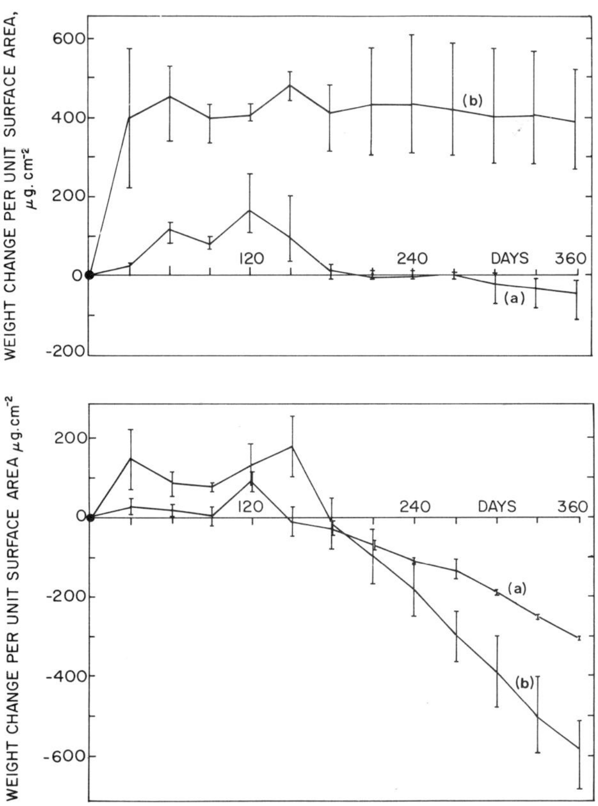

FIGURE 3:   Leaching results for (1) CaTiSiO₅ ceramic, and (2) sphene glass
ceramic. Static tests at 100°C, using (a) synthetic groundwater
plus crushed granite, pH 6.0-7.0, and (b) synthetic groundwater
plus powdered granite plus CaO-TiO₂-SiO₂ glass powder, pH
8.0-9.0.

SEM examination of the leached fracture surface of this specimen verified the SIMS and ESCA results. Figure 5(a) shows the unleached fracture
surface, and Figure 5(b) shows the same surface after 10 days of leaching
in the groundwater. A precipitated layer of a new phase is visible as a
lichen-like growth on the sphene crystals, together with the presence of
occasional NaCl crystals (white) in the photomicrograph. By subtracting
the energy dispersive x-ray spectrum obtained from an unleached sphene
grain from that of a leached sphene grain, it was confirmed that the
fibrous growth was a magnesium silicate phase.

A theoretical basis for this observation was provided by thermodynamic
calculations of the degree of saturation of the groundwater with respect to
common minerals, using the computer program SOLMNEQ [10]. The calculations

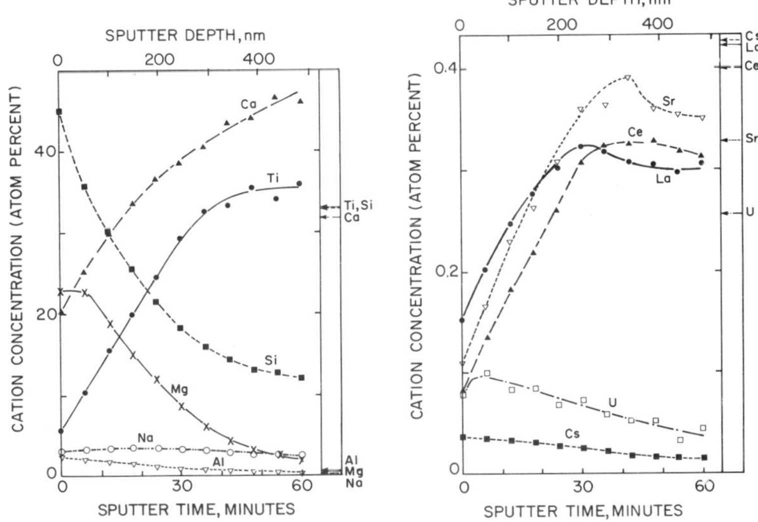

FIGURE 4: Depth profiles of sphene ceramic leached in synthetic ground water for 10 days at 100°C. Bulk elemental concentrations are shown in the right margins.

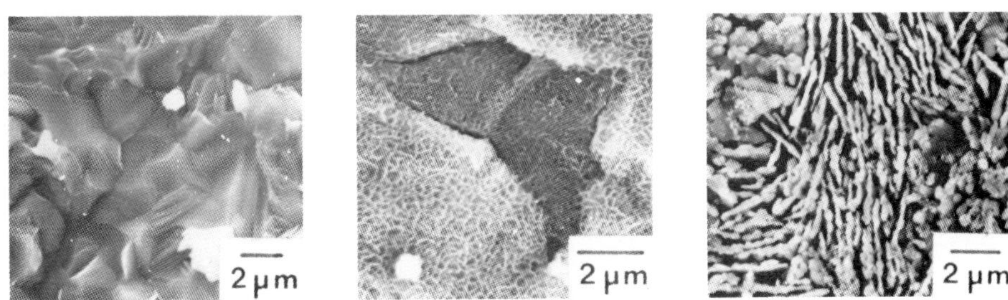

FIGURE 5: Scanning electron micrographs, showing
(a) sphene ceramic, unleached fracture surface
(b) sphene ceramic, fracture surface, leached for 10 days in synthetic groundwater at 100°C.
(c) sphene glass ceramic, fracture surface, leached for 10 days in deionized water at 100°C.

showed that the brine would be supersaturated with respect to certain magnesium-bearing silicate minerals, including talc and tremolite, and close to saturation with respect to chrysotile, $Mg_3[Si_2O_5](OH)_4$. The latter has a typical hydrothermal mode of formation, and a fibrous morphology [11], similar to that observed in Figure 5(b). The precipitated phase is, therefore, tentatively identified as chrysotile.

3. Glass Ceramic Leached in Deionized Water.

Previously quoted results of static leaching tests [4] showed leach rates for Na, Al and Si that were an order of magnitude or more higher than for Ca and Ti. The SIMS results, shown in Figure 6, indicate that a thick (>800 nm) surface layer, depleted in Na, Al and Si, and enriched in Ca, Ti, La and Ce, was produced after 10 days of leaching. The photomicrograph of the SIMS specimen after leaching, shown in Figure 5(c), clearly confirms that the aluminosilicate glass matrix has been preferentially dissolved.

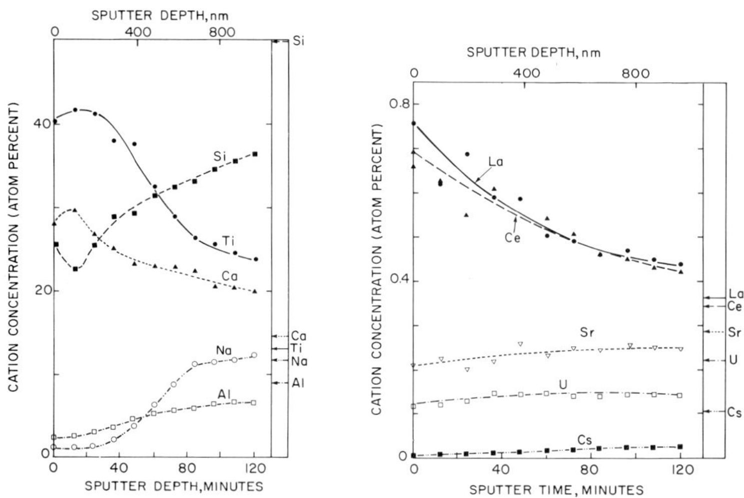

FIGURE 6: Depth profiles of sphene glass ceramic leached in deionized water for 10 days at 100°C. Bulk elemental concentrations are shown in the right margins.

4. Glass Ceramic Leached in Synthetic Groundwater.

Results from the static leaching experiments, shown in Figure 3(b), demonstrate initial weight gains for the specimens in the experiments with pH 8.0-9.0, which revert to weight losses after 150 days. The mean bulk leach rate between days 150 and 360 is 4.3×10^{-10} $kg.m^{-2}.s^{-1}$. In view of the lack of leaching of sphene under these conditions, this leach rate may be taken as approximating that of the aluminosilicate glass matrix. Similar behaviour was observed in the experiments with pH 6.0-7.0, with a mean bulk leach rate between days 150 and 360 of 1.4×10^{-10} $kg.m^{-2}.s^{-1}$. Using the same reasoning as above, therefore, it may be deduced that the rate of dissolution of the aluminosilicate glass phase increases in the higher pH range.

The SIMS results, shown in Figure 7, showed no major changes in composition with depth after 10 days of leaching at pH 6.0-7.0, except for slight surface enrichment in Si, K and Ce to a depth of $\sim$ 50 nm, and possible slight depletion in La, Ti, Ca and Na. The K enrichment may be due to ion exchange with Na in the aluminosilicate glass phase. No visible evidence for leaching was found using SEM.

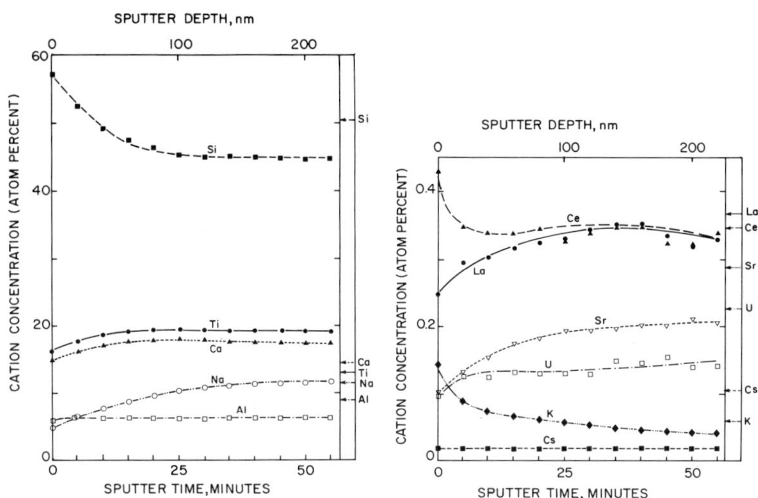

FIGURE 7: Depth profiles of sphene glass ceramic leached in synthetic groundwater for 10 days at 100°C. Bulk elemental concentrations are shown in the right margins.

SUMMARY

It is concluded from the SEM and SIMS results, and from the leachant analyses and weight change observations, that leaching and surface alteration phenomena in the glass ceramic are confined mainly to the aluminosilicate glass phase. The high ionic strength and composition of the groundwaters have a significant passivating effect on glass-ceramic leaching. Our results indicate that leaching data obtained with deionized water may be misleading in assessing radionuclide release rates in groundwaters.

The $CaTiSiO_5$ phase in both the glass ceramic and the ceramic undergoes little or no bulk dissolution in the synthetic groundwater, and fission product release is probably confined to ion-exchange reactions with the groundwater components at the material surface.

Further experience with depth profiling of these materials under different leaching conditions and with different SIMS operating parameters is required to improve confidence in the analyses for certain elements, particularly Cs and U.

328

ACKNOWLEDGEMENTS

The assistance of B. Phillips, J. Moulder, K. Smith and the staff of
Physical Electronics Industries, Eden Prairie, Minnesota, and of M.A.
Thompson, L. Brown, N. Pearson and R. Hamon of the Analytical Science
Branch for their help with SIMS, SEM-EDX and leaching solution analyses is
gratefully acknowledged.

REFERENCES

1. R.S. Dixon and E.L.J. Rosinger, editors, 1981 December, "Third Annual Re-
    port of the Canadian Nuclear Fuel Waste Management Program", Atomic Energy
    of Canada Limited Report, AECL-6821.

2. S.K. Frape and P. Fritz, 1981 January, "A Preliminary Report on the Occur-
    rence and Geochemistry of Saline Groundwaters on the Canadian Shield",
    Unpublished work, available from Atomic Energy of Canada Limited,
    Technical Record TR-136.

3. H.W. Nesbitt, G.M. Bancroft, S.N. Karkhanis and W.S. Fyfe, 1981, Scientific
    Basis for Nuclear Waste Management, Vol 3, ed. J.G. Moore, 131, Plenum Press.

4. P.J. Hayward and E.V. Cecchetto, 1981, in Proceedings of 4th Annual Confe-
    rence of the Materials Research Society on the Scientific Basis Underlying
    Nuclear Waste Management, Boston, Proceedings in press.

5. L.R. Pederson, M.T. Thomas and G.L. McVay, 1981, J. Vac. Sci. Technol., 18,
    732.

6. G.L. McVay and L.R. Pederson, 1981, Scientific Basis for Nuclear Waste Ma-
    nagement, Vol 3, ed. J.G. Moore, 323, Plenum Press.

7. N.S. McIntyre, G.G. Strathdee and B.F. Philipps, 1980, Surface Science, 100,
    71.

8. C.W. Magee, W.L. Harrington and R.E. Honig, 1978, Rev. Sci. Instrum., 49(4),
    477.

9. D.E. Newbury, 1979, Scanning, 3, 110.

10. Y.K. Kharaka and I. Barnes, (1973), "SOLMNEQ-Solution Mineral Equilibrium
    Calculations", U.S. Geol. Survey Comp. Centre. PB-215-899.

11. W.A. Deer, R.A. Howie and J. Zussman, 1966, "An Introduction to the Rock
    Forming Minerals", Longmans.

LEACHING OF NATURAL AND SYNTHETIC SPHENE AND PEROVSKITE

J.B. METSON, G.M. BANCROFT, S.M. KANETKAR, H.W. NESBITT, W.S. FYFE,
Centre for Chemical Physics and Department of Geology, University of Western
Ontario, London, Canada N6A 5B7 and P.J. HAYWARD, Atomic Energy of Canada Ltd.,
Pinawa, Manitoba ROE 1L0.

INTRODUCTION

Titanates and titanate-based ceramics are being considered as possible hosts
for fuel reprocessing wastes in several waste management programs, largely
because of their low leach rates. A variety of titanate based wasteforms is
being examined, including the hot-pressed product of the rutile microencapsula-
tion process[1] and the titanate-based SYNROC assemblage[2].

However, for relatively low waste loadings, and for waste vaults sited in
formations characterised by silica-rich groundwaters, a wasteform based on the
titanosilicate sphene ($CaTiSiO_5$) may offer a number of advantages. Reference
to the H^+-H_2O-CaO-TiO_2-SiO_2 activity diagram[3] shows sphene to be in, or close
to, thermodynamic equilibrium with a range of natural groundwaters. Analysis
of natural sphene samples show the structure has considerable ability to
tolerate a wide range of multivalent cations, which can substitute for Ca and
Ti in the lattice. Rare earth substitutions are particularly common and have
been reported at levels of 46 wt% Re_2O_3[4].

Atomic Energy of Canada Limited (AECL) is currently investigating the use of
sphene-based glass ceramics to host the wastes produced from possible reprocessing of irradiated CANDU* fuels[5]. We report here on the leaching behaviour
of natural and synthetic sphenes and a $CaTiSiO_5$ glass, in deionized water,
with varying pH and with Ca^{2+} and silica additions to the leachant. Leaching
studies on natural and synthetic perovskite in deionized water are also
reported.

EXPERIMENTAL

Crystalline sphene samples were obtained from two locations in Ontario,
North Crosby township and Lake Clear, Renfrew County. Sphene ceramics were
prepared by grinding, cold pressing and sintering a crystallised $CaTiSiO_5$

----

*CANada Deuterium Uranium.

330

frit. $CaTiSiO_5$ glasses were prepared by melting well mixed silica and titania
with calcium carbonate. After loss of CO_2, the melt was held at 1450°C for
1 hour, then cast in stainless steel molds. The cast cylinders were annealed
at 750°C before cutting and polishing into 3 mm thickness, 10 mm diameter
discs.

Perovskite leach results were obtained from a 120 mg single crystal from
Magnet Cove Arkansas and a 100 mg synthetic, flux-grown single crystal (98%
theoretical density).

In all cases, samples were leached in Teflon bottles at a temperature of
90°C. Limited sample quantities made the MCC-1 leach test method[6] impractical
for all but the sphene glass samples, so alternative methods were used. The
IAEA method[7] was used for all perovskite leach testing as small surface areas
are involved and solution replacement is the only practical way of assuring
sufficient leachant for analysis. The method also provides convenient
comparison with other workers who have studied perovskite leaching utilizing
this test.[8] For sphene leach tests, 2 ml samples were removed from the
leachant at intervals of 1,3,7,14,28 and, for some samples, 56 days and beyond.
An initial SA/V of 0.1 cm^{-1} was used and the total volume removed during
sampling was less than 10% of the initial leachant volume.

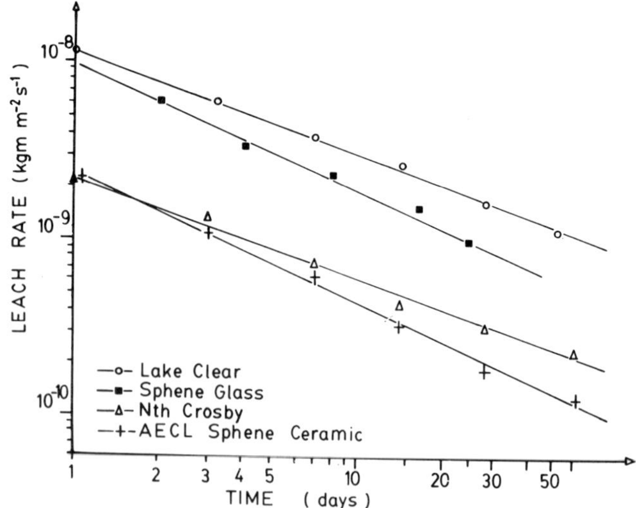

Fig. 1.   Bulk leach rates of two natural sphenes, a sphene
ceramic and a sphene glass, in deionized water, 90°C.

Bulk leach rates were calculated from Ca^{2+} analysis of the leachant using the formula:

$$L.R. = \frac{wt.\ Ca^{2+}\ leached}{leach\ time} \times \frac{1}{surface\ area} \times \frac{100}{\%\ Ca^{2+}\ in\ sample}$$

This assumes congruent dissolution although it will be demonstrated later in results from surface analysis of leached samples, that incongruent dissolution occurs. Titanium analyses gave values below Atomic Absorption resolution (0.1 ppm) and well below calcium levels in the deionized water leach tests. Thus the bulk leach rates shown (figure 1) represent maximum levels of attack.

Results and Discussion
Sphene leaching.

Results from the leaching of the three crystalline sphenes and the $CaTiSiO_5$ glass are shown in figure 1. The logarithm of leach rate follows an approximately linear decline with log time. For the North Crosby sample where leaching has been followed for 140 days, a plateau in calcium leach rate was observed at $\sim 3.5 \times 10^{-11}$ kg m^{-2} s^{-1}. This converts to a bulk leach rate $\sim 10^{-10}$ kg m^{-2} s^{-1} (kg m^{-2} s^{-1} = 1.16×10^{-4} g cm^{-2} d^{-1}).

A comparison of static (as described) vs. dynamic (by the IAEA method of solution replacement) leaching, using the AECL sphene ceramic, shows relatively small differences. Maximum separation occurs after 30 days when leach rates differ by a factor of 3. After 60 days the static leach rate appears to be leveling off, again at $\sim 10^{-10}$ kg m^{-2} s^{-1}, while the dynamic rate continues to decline. This suggests that the concentration of the leached species is influencing the leach rate after 60 days.

The effect of small Ca^{2+} concentrations on leaching was examined by measuring weight losses of sphene samples placed in solutions containing $10 \rightarrow 50$ ppm Ca^{2+}. Effects in the first week were minimal. However, beyond this, samples leached in deionized water lost weight rapidly, while in calcium-doped solutions, small weight gains were observed (figure 2). Thus even small concentrations of calcium, well short of those needed to attain calculated thermodynamic equilibrium levels for sphene , inhibit leaching. However, the weight gains observed bear no relationship to the level of calcium in the three experiments.

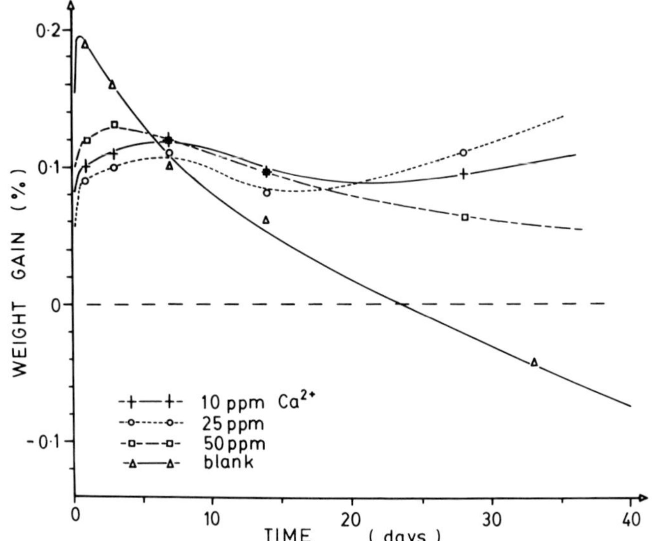

Fig. 2.   Weight gains of Lake Clear sphene discs in
solutions containing 10, 25 and 50 ppm Ca^{2+}, 90°C.

In a 500 ppm Ca^{2+} solution saturated with ground quartz, sphene caramics
showed small but consistent weight gains through 150 days leaching. This
leachant solution lies within the sphene stability region in the H^{+}-H_2O-CaO-
TiO_2-SiO_2 stability field diagram , and the results support the prediction
that there should be no net leaching of sphene in such a solution.

The pH dependence of the sphene leach rate was investigated with both
natural crystalline and $CaTiSiO_5$ glass samples. pH was regulated through HCl
and NaOH addition to minimise modification of the leachant. Both show
decreasing leach rates with increasing pH (figure 3). The observed pH
dependence is consistent with a mechanism of selective leaching of a surface
layer, where proton diffusion into the sphene lattice or exchange with
accessible cations is the rate determining step. Although TiO_2 solubility is
small, it is markedly higher in acidic solutions[9]. Thus, the observed pH
dependence could also be attributed to TiO_2 insolubility and reprecipitation on
the sphene surface. This reprecipitated layer would then inhibit further
leaching.

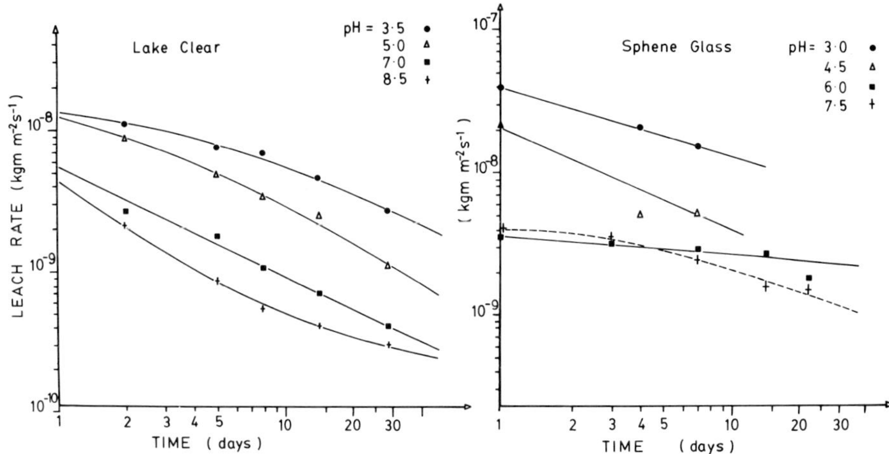

Fig. 3. The effect of pH on the bulk leach rate of Lake
Clear sphene and a sphene glass in deionized
water, 90°C.

The nature of the leached surface layer.

To examine leached surfaces, $CaTiSiO_5$ glasses were leached in deionized
water at 90°C and examined by X-Ray Photoelectron Spectroscopy (XPS) and
Secondary Ion Mass Spectrometry (SIMS). From the photoelectron spectra of the
2p levels of Ca, Ti and Si relative concentrations of these elements in the
top 5 → 10 nm of the glass were calculated.[10] Figure 4 shows the element
concentration ratios $\frac{n_{Ca}}{n_{Ti}}$ and $\frac{n_{Si}}{n_{Ti}}$ as a function of leaching time.

The larger than theoretical $\frac{n_{Si}}{n_{Ti}}$ is partly due to non-stoichiometry at the
surface, and this non-stoichiometry is also observed in the SIMS profiles
(figure 5). It is apparently due to some silica contamination of the surface
in the polishing of the discs. We observe a rapid decline in Si and Ca 2p
intensities within the first 12 hours of leaching, leaving a layer highly
enriched in TiO_2. After 8 days leaching, TiO_2 enrichment is still apparent.

SIMS data were reduced by normalising element counts to the bulk composition
which was assumed to have been reached in the sputtering of the blank sample.
Sputtering depths were assessed from interference microscope images of the
beam impact craters. SIMS profiles confirm the XPS observations of rapid
calcium and silica depletion. However the 8 day profile (figure 5) is markedly

different from the other spectra in a number of respects. This surface layer
is of considerable depth, extending at least 200 nm. A rapid sputtering rate
indicates the layer is of lesser density and the composition of the layer,
uniform throughout the sputtered profile, is very different from the bulk. We
therefore attribute these observations to the presence of a precipitated TiO_2
layer rather than the selectively leached layer observed on samples leached
for 1 to 24 hours.

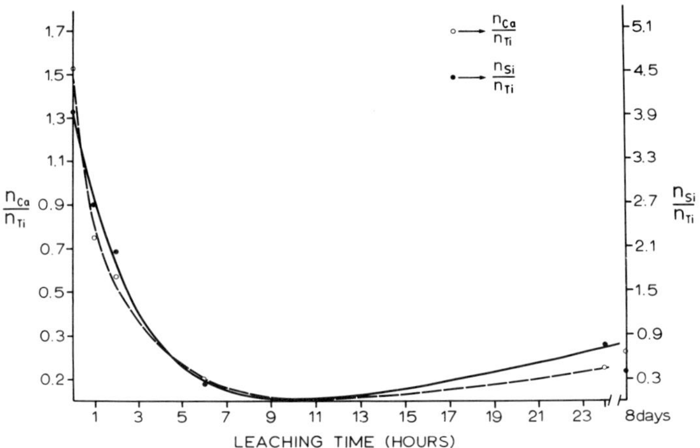

Fig. 4. Element ratios $\dfrac{n_{Ca}}{n_{Ti}}$ and $\dfrac{n_{Si}}{n_{Ti}}$ vs. leaching time from
x-ray photoelectron spectra of sphene glasses
leached in deionized water, 90°C.

A $CaTiSiO_5$ glass sample leached for 12 days in a 10 ppm Ca^{2+} solution,
gave similar XPS spectra to the 8-day sample. This sample was also examined
by X-ray diffraction, which identified anatase on the glass surface. This
supports selective leaching being followed by either the crystallisation of
the amorphous TiO_2 gel layer, or reprecipitation of TiO_2 from the leachant.
This process must occur after relatively short leach times to enable X-ray
characterisation of the phase after only 12 days leaching.

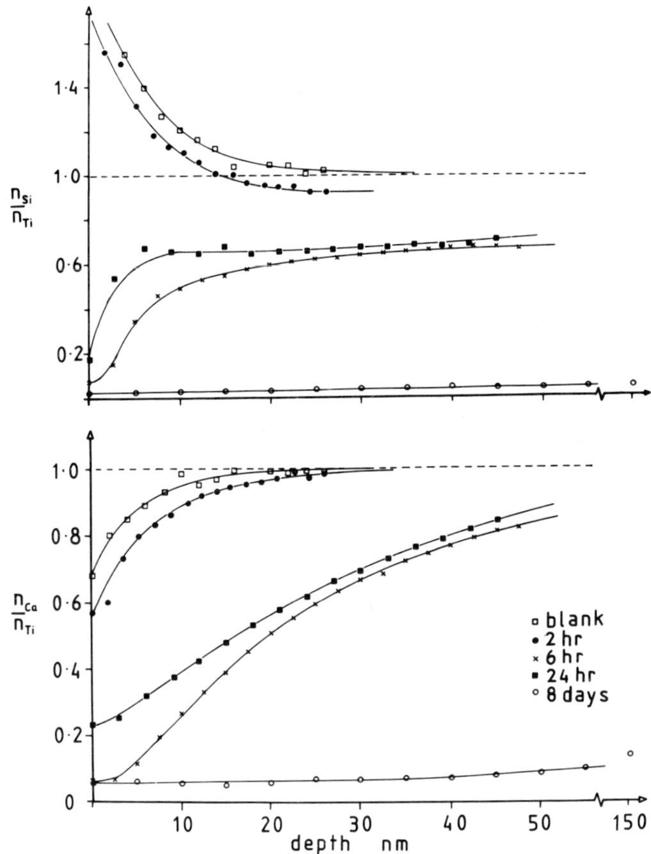

Fig. 5.   Element ratios $\dfrac{n_{Si}}{n_{Ti}}$ and $\dfrac{n_{Ca}}{n_{Ti}}$ vs. depth from SIMS profiles

of sphene glasses leached in deionized water, 90°C.

<u>Perovskite Leaching.</u>

Work on perovskite was limited to determining bulk leach rates (calculated
from calcium concentrations as before) for a natural and a synthetic single
crystal.   Leach rates using the IAEA method of solution replacement are shown
in figure 6, together with values for the AECL sphene ceramic leached under
similar conditions and reported values for perovskite grains from Magnet Cove[8].
The leach rates from Magnet Cove grains are an order of magnitude lower than
those from monolithic samples for reasons which are not immediately apparent.
Both results are based on geometric surface areas and an SA/V = 0.2 cm^{-1}.   The

suggestion of Ringwood et al.[8] that calcite impurities enhance the leach rate of the single crystal is unlikely as the single crystal used has been leached and repolished several times. Any selectively leached phase would be rapidly removed, and the consistent leach results suggest that any secondary phase is extremely minor. Examination by electron microprobe showed traces of apatite as the only identifiable inclusions in the crystal.

It has been proposed by Ringwood et al.[8] that TiO_2 rich surface layers occur on leached perovskites as we have observed on sphene glasses. From the similarity between the leach rates of perovskites and sphenes (figures 1 and 6), it appears that such a surface layer does not afford significantly more protection to perovskite, a titanate, than it does to sphene, a titanosilicate.

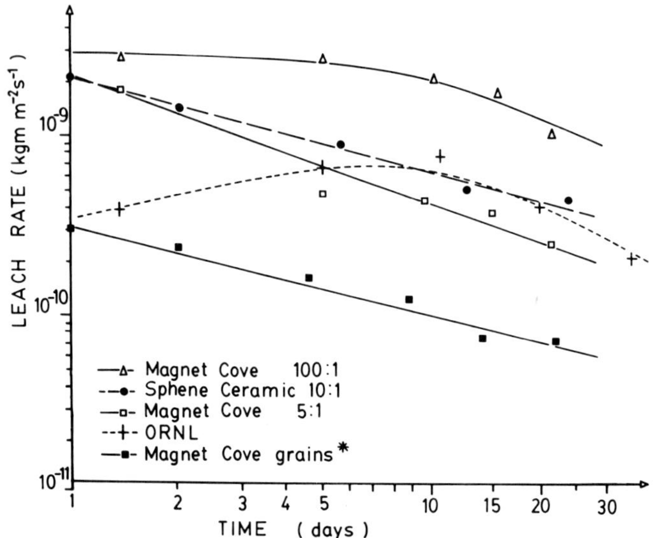

Fig. 6.   Bulk leach rates for perovskite samples in deionized water, 90°C. The ratios refer to the volume : surface area. * Ref. 8.

CONCLUSION

The leach rates of natural and synthetic sphenes in deionized water are initially between 10^{-8} and 10^{-9} kg m^{-2} s^{-1} and fall by an order of magnitude over 28 days. The three forms of sphene tested, i.e. natural crystalline and ceramic, and $CaTiSiO_5$ glass, show maximum leach rate differences of less than an order of magnitude over the 60-day time interval considered here. Perovskite leach rates from synthetic and naturally occurring single crystals,

show similar initial rates to the more resistant sphenes and a parallel
decline in leach rate with time.

Small but consistent weight gains are observed for ceramic samples leached
in quartz-saturated/500 ppm Ca^{2+} solutions, confirming the predictions of the
sphene/perovskite/rutile stability diagram produced by Nesbitt et al.[3]
Hayward and Cecchetto[5] have also observed weight gains in sphene leaching
experiments in a synthetic saline groundwater. This supports the validity
of using thermodynamic data to predict leach behaviour of crystalline
materials, at least in a relatively simple system such as H^+-H_2O-CaO-SiO_2-TiO_2.

Surface studies on a leached $CaTiSiO_5$ glass show rapid loss of Ca^{2+} and
Si^{4+} from the surface of the glass. After 8 days in deionized water at 90°C
the zone of calcium and silica depletion extends to a depth of at least 200 nm.
There is evidence that the initial selectively leached layer, as observed
after 1 to 24 hours leaching, has been replaced by a reprecipitated TiO_2 layer
of much greater depth after 8 days leaching.

The leaching of crystalline sphenes in deionized water could be expected
to follow the same path, as similar bulk leach rates results are observed.
For natural crystalline materials however, our experiments indicate that the
problem of obtaining a consistent surface on polished discs makes it
difficult to resolve whether the same behaviour occurs. Fractured surfaces,
used extensively in surface studies of glass leaching, minimise surface
modification[11] and offer better prospects for characterisation by surface
techniques.

ACKNOWLEDGEMENTS

We would like to acknowledge the assistance of Mr. I.A. Russell in the
experimental work in this paper. We also thank Dr. L.A. Boatner of Oak Ridge
National Laboratory for supplying the synthetic perovskite and AECL for the
financial support of this work.

REFERENCES

1. Forberg, S., Westermark, T., Larker, H. and Widell, B. (1979) Scientific
    Basis for Nuclear Waste Management, Vol. 1, ed. G.J. McCarthy, pp. 201-
    205.

2. Ringwood, A.E., Kesson, S.E., Ware, N.G., Hibberson, W.O. and Major, A.
    (1979) Geochemical Journal, 13, 141-165.

3. Nesbitt, H.W., Bancroft, G.M., Fyfe, W.S., Karkhanis, S.N. and
    Nishijima, A. (1981) Nature, 289, No. 5796, 358-362.

4. Exley, R.A. (1980) Earth and Planetary Science Letters, <u>48</u>, 97-110.

5. Hayward, P.J. and Cecchetto, E.V. (1982) Scientific Basis for Nuclear Waste Management, Vol. *6* (in press).

6. Strachan, D.M., Barnes, B.O. and Turcotte, R.P. (1981) Scientific Basis for Nuclear Waste Management, Vol. 3, ed. J.G. Moore, pp. 347-354.

7. Hespe, E.D. (1971) Atomic Energy Review, <u>9</u>, 1-12.

8. Ringwood, A.E., Oversby, V.M., Kesson, S.E., Sinclair, W., Ware, N., Hibberson, W. and Major, A. (1981) Nuclear and Chemical Waste Management, <u>2</u>, 287-305.

9. Imahashi, M. and Takamatsu, N. (1976) Bulletin Chemical Society of Japan, <u>49</u>, (6), 1549-1553.

10. Bancroft, G.M., Gupta, R.P., Hardin, A.H. and Ternan, M. (1979) Analytical Chemistry, 51, (13), 2102-2107.

11. McIntyre, N.S., Strathdee, G.G. and Phillips, B.F. (1980) Surface Science, <u>100</u>, 71-84.

RADIATION EFFECTS

Published 1982 by Elsevier Science Publishing Co
SCIENTIFIC BASIS FOR RADIOACTIVE WASTE MANAGEMENT - v
Werner.Lutze, editor

EFFECTS OF RADIATION DAMAGE AND RADIOLYSIS ON THE LEACHING OF VITRIFIED WASTE*

W.G.BURNS, A.E.HUGHES, J.A.C.MARPLES, R.S.NELSON and A.M.STONEHAM
Atomic Energy Research Establishment, Harwell, Didcot, Oxon OX11 0RA, U.K.

INTRODUCTION

Attack on the glass by leaching in water is the only probable process by which the long-lived radioactive isotopes incorporated in vitrified waste might be returned to the environment after future disposal. Radiation from the incorporated waste could increase the leach rate either by damaging the structure of the glass or by radiolysing the water near the glass surface to produce ionic species that attack the glass more readily.

A consideration of the various radiation damage processes leads to the conclusion that the most important contribution to damage to the glass comes from the recoil nuclei resulting from α-decay of actinides incorporated in the waste, together with the fission products[1-3]: these actinides are chiefly curium, americium and neptunium, together with any small amounts of plutonium and uranium that have not been removed for future use as nuclear fuel. The recoiling actinide nucleus from an α-decay has an energy of about 100keV and displaces over 1000 atoms in the glass; although the α-particle has a much greater energy ($\sim$5-6MeV) most of this is lost in ionising the atoms in its path and this ionisation is thought to have mostly only a transitory effect on the glass. The β-particles (electrons) emitted by the fission products likewise lose almost all their energy by ionisation.

The conventional way to test the ability of the glass to withstand the effects of α-decays is to dope samples of the glass with a short half-life α-emitter such as ^{238}Pu or one of the curium isotopes[1-8]. This subjects the glass to precisely the same process as will occur in the real waste but with a greatly increased dose rate. These experiments have not shown significant increases in leach rate. However, recent work in which the radiation damage has been produced with ion beams has suggested that after a critical dose of radiation a large increase in leach rate could occur, by up to a factor of fifty[9-11].

In addition, leach tests in a flux of γ rays have shown that radiolysis effects in the water can increase the leach rates of the glasses[12-14].

---

*Parts of this work were commissioned jointly by the UK Department of the Environment and the Commission of the European Communities as part of their Radioactive Waste Management Research Programmes. The results may be used in the formulation of UK Government policy but at this stage they do not necessarily represent such policy. The work is described in more detail elsewhere[3].

ION-BEAM IRRADIATIONS

Dran et al.[9-11] irradiated glasses and crystalline minerals with a beam of 200 keV Pb ions and showed that an enhanced leach rate occurred above a critical dose of about 5×10^{12} ions.cm^{-2}, which corresponded to the dose at which the zones damaged by each Pb ion overlapped; the zones had a diameter of about 100Å and a length of 500Å. Each incident Pb ion produced about 3000 displacements, giving 3×10^{21} displacements.cm^{-3} at the critical dose. Each α-decay has been calculated to produce 1380 displacements, so that the critical dose is $3 \times 10^{21}/1380 = 2.2 \times 10^{18}$ α-decays.cm^{-3}, equivalent to 8.5×10^{17} α-decays.g^{-1}. Dran et al arrive at a slightly larger equivalent number of α-decays by a somewhat different argument. The increased leach rates observed after this critical dose varied from zero to a factor of about 50 depending on composition with most of the increases being by factors of between 15 and 30. The leach rates were normally measured at 100°C in a 250g.ℓ^{-1} NaCl solution.

LEACH TESTING OF ^{238}Pu-DOPED GLASSES

Two sets of experiments have been carried out. The first was restricted to the glass composition UK189 (M5) doped with about 5 wt% ^{238}Pu (Table 1).

TABLE 1

GLASS COMPOSITIONS (in wt%)

Glass	FPO_x	SiO_2	B_2O_3	Na_2O	Li_2O	Al_2O_3	MgO	Fe_2O_3	U_3O_8
UK189 (M5)	9.6	41.5	21.9	7.7	3.7	5.0	6.2	2.7	0.1
UK209 (M22)	9.8	50.9	11.1	8.3	4.0	5.1	6.3	2.7	0.1
F.SON 58.3 0.20.U2	22.7	43.6	19.0	9.4				0.6	3.6
G.VG 98/3 (a)	15.5	41.8	10.5	22.2		2.3	0.4	0.7	1.2
G.CELSIAN B1/3 (b)	15.2	28.0	6.4	3.8	2.4	12.8	1.2	1.5	0.5

(a) Also CaO: 2.3, TiO_2: 3.5
(b) Also CaO: 4.0, TiO_2: 4.6, BaO: 14.8, ZnO: 3.6
All the glasses contain small amounts of other oxides

These samples have been stored for over 6 years and their leach rates have been measured at intervals by the Soxhlet technique[15], where the sample is exposed to frequently changed, freshly distilled water at 100°C. The results are given in Fig.1. It can be seen that only modest increases in leach rate have been observed even after 5.6×10^{18} α-decays per gram, equivalent to $\sim$ 1.4 million years for vitrified Magnox waste. The PuO$_2$ may not all be homogeneously dispersed in the glass, as will be the case for the actinide isotopes in real waste storage glass. One of these samples was also leached in a 250g.ℓ NaCl solution at 100°C as used by Dran et al. and again showed only a small increase in leach rate (Table 2). Note also that the leach rates in this strong solution were an order of magnitude smaller than those found for distilled water.

In the second series of experiments samples of four glasses and a glass ceramic were doped with 2.5 wt% $^{238}PuO_2$. The glass compositions are given in Table 1: they were suggested by UK, French and German participants in a comparative study arranged by the European Commission[4]. The glasses were again leach-tested at intervals by the Soxhlet technique at 100°C and the results are given in Table 3. Again only relatively small changes in leach rate were found.

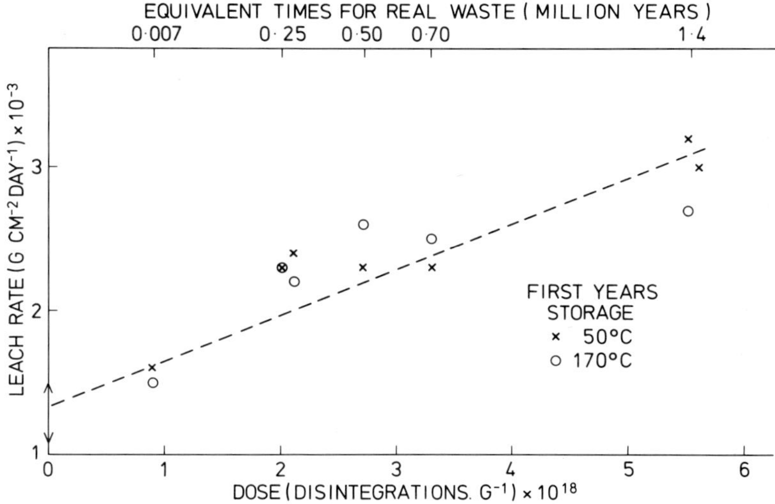

FIG.1   Increase in leach rate with dose for glass 189.   Note the suppression of the zero on the vertical axis.

TABLE 2

LEACH TESTS AT 100°C ON GLASS 189 IN NaCl SOLUTION

Sample	Leaching conditions	Leach rate (g.cm^{-2}day^{-1})
Irradiated	Soxhlet	2.7×10^{-3}
Control	Static distilled water	2.7×10^{-4}
Control	Static NaCl solution	3.1×10^{-5}
Irradiated	Static NaCl solution	3.7×10^{-5}

The NaCl concentration was $250g.\ell^{-1}$; although the conditions are described as 'static' there was considerable convective circulation.

The results on the irradiated sample were obtained after a dose of 5.5×10^{18} α-disintegrations per gram.

The equivalent storage times for the highest doses are given in the
last column of Table 3. These assume that the fuel is reprocessed 6 months
after it is taken out of the reactor. The wide differences between the equivalent storage times are due to the different reactor systems considered (UK:
Magnox, French and German glasses: LWR) and to the different assumed concentrations of fission products in the glasses (UK:7.9 wt%, French: 18 wt%,
German: 12.5 wt%).

TABLE 3

SOXHLET LEACH RATES $(mg.cm^{-2} \ day^{-1})$ AT VARIOUS DOSES

Glass	Initial leach rate	Leach rate after irradiation of 1.1×10^{18} α-disintegrations.g^{-1}	2.0×10^{18}	Equivalent time for real waste (years)
UK189 (M5)	1.33	1.91	2.95	500,000
UK209 (M22)	0.21	0.23	0.31	500,000
F.SON 58.30.20.U2	2.26	6.9	9.1	3,000
G.VG 98/3	2.21	2.91	3.03	10,000
G.CELSIAN B1/3	1.17	0.95	1.07	10,000

Thus, in all the simulations using α-emitters there is no sign of the large
effects observed by Dran et al. To attempt to reconcile this difference we
have looked at the effect of dose rate on the results and the possibility of
recovery occurring during irradiation.

RECOVERY DURING IRRADIATION

The damage caused by α-decays is almost entirely in the form of heavily
damaged zones around the track of the recoil nucleus. The build up of damage
thus consists essentially of the increase in the number of such zones within
the glass.

r = rate of damage (α-disintegrations $cm^{-3}sec^{-1}$)

v = volume of damaged zone (cm^{3})

F = fraction of volume of sample occupied by damaged zone.

In the absence of any recovery: $\dfrac{dF}{dt} = vr \ (1 - F)$ (1)

where 1-F is the probability that a new damaged zone occurs in a previously
undamaged volume.

Integrating, we have: $F = 1 - \exp \ (-vrt)$ (2)

We can now include a term which allows for the simultaneous recovery of the
damaged regions. This is assumed, for simplicity, to be a first order decay
of the damage so that equation (1) becomes: $\dfrac{dF}{dt} = vr \ (1 - F) - kF$ (3)

On integrating we have:
$$F = \left(\frac{vr}{vr+k}\right)(1 - \exp\{-[vr+k]t\}) \qquad (4)$$

Similar exponentially saturating behaviour is predicted as for equation (2) but the saturation value of F is reduced by a factor $vr/(vr+k)$

The forms of equations (2) and (4) fit curves of the density change of glasses with dose extremely well[3]. The value of v required is about 4×10^{-19} cm^3 which is consistent with the expected volume of a damaged zone. Using this value we can calculate values of vr appropriate to the case of real vitrified waste to these experiments with $^{238}PuO_2$-doped glasses and to the ion bombardment experiments. These are shown in Table 4 and emphasise the vast difference in damage rates between the real glass and the ion beam simulation experiments.

We have estimated k for glass 189 from the stored energy release curves and from the rate at which the density changes anneal out. Again assuming that the recovery is first-order, any irradiation induced property change ΔP (e.g. concentration of displaced atoms, change of sample density) that is assumed proportional to F should recover exponentially with time:

$$\Delta P = \Delta P_o \exp\,(-kt) \qquad (5)$$

Differentiating with respect to t we have

$$\frac{\partial(\Delta P)}{\partial t} = -\Delta P_o \cdot k\,\exp\,(-kt) \qquad (6)$$

$$\left(\frac{\partial(\Delta P)}{\partial t}\right)\cdot = -\Delta P \cdot k \qquad (7)$$

In the curves for release of stored energy, given in reference 3, the rate of heat release at any temperature is thus equal to k times the energy still to be released, given by the area under the curve above that temperature. For 189 glass this gave the values plotted in Figure 2.

TABLE 4

FREQUENCY OF DAMAGING EVENTS FOR VARIOUS SITUATIONS

Situation	$r\,(cm^{-3}s^{-1})$	$vr\,(s^{-1})$	$vr/(vr+k)*$
Radwaste, $10^2 - 10^6$y	10^6	10^{-12}	10^{-5}
Radwaste, first 10^2y	10^8	10^{-10}	10^{-3}
Simulation with α-emitter $\sim$ 1Ci.cm^{-3}	3.5×10^{10}	1.4×10^{-8}	0.12
200kev Pb ions ($\gtrsim 10^8$ ions $cm^{-2}s^{-1}$)	$\gtrsim 2 \times 10^{13}$	$\gtrsim 10^{-5}$	$\gtrsim 0.99$

*for $k = 10^{-7}s^{-1}$ see text

A small sample of glass 189 which had accumulated a dose of 1.25×10^{18} disintegrations per gram was annealed successively at progressively higher temperatures. Values of k were derived from the measured densities directly, using equation (5) and these are also plotted in Figure 2. It can be seen that the two different properties give very different values of k, suggesting that

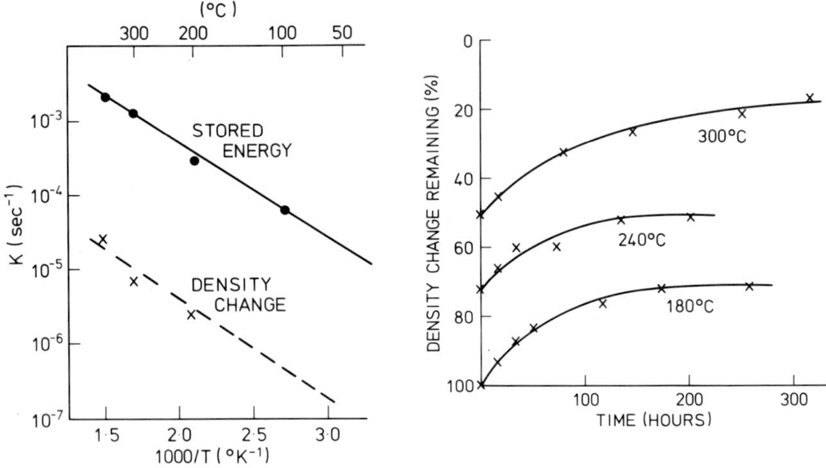

FIG.2 Estimates of the recovery rate FIG.3 Annealing of density changes
constant k for glass 189 in glass 209

the recovery is not a simple effect. Some evidence for this can be seen in the
'structure' in the stored energy release curves[4]. These effects would occur if
several different kinds of defect occur in the damaged glass which anneal out
at different temperatures and contribute differently to stored energy and
density changes. This was confirmed by measuring the density of an irradiated
sample of glass 209 after annealing at progressively higher temperatures (Fig.3)
At each temperature, part of the density change was quickly annealed out but
further annealing produced only a small change in the density. Much longer
heat treatment times will be needed to determine an effective value of k.
Nevertheless, it seems reasonable to conclude from Figure 2 that recovery is
roughly represented by a value of $k > 10^{-7}$ s^{-1} for temperatures of 25-200°C
which cover the simulation experiments and most storage and repository conditions. From Table 4 we can see that, with this value of k, recovery during
irradiation will be important in real vitrified waste, significant in simulations with α-emitters, but entirely negligible in ion beam simulations. It is
important to note that if vr/(vr+k) is much less than unity the damaged zones
never overlap and the critical dose envisaged by Dran et al will never be
reached.

This simple argument, although based on preliminary values of the parameters
shows that recovery must be taken into consideration in assessing the relevance
of simulation experiments to the real situation in vitrified waste. Results on

the same glass composition must also be compared since Dran et al have recently shown that the increase in the leach rate after ion-bombardment is sensitive to glass composition.

RADIOLYSIS OF THE LEACHANT

McVay et al.[12-14] have measured the leach rate of US glass 76-68 while the glass-leachant interface was being irradiated with ^{60}Co γ-rays with a dose rate of 2.4 Mrad.h^{-1}. They reported an increase of about X2 in the leach rate as measured by weight loss and some larger increases in the leach rate of individual elements. These increases were shown not to be due to damage to the glass by the γ-rays but were associated with the effects of radiation on the leachant. In some cases air was present and there the increases could be caused by the nitric acid produced by the irradiation of air in the presence of water. However, some increases in leach rate were found using an apparatus from which air was excluded[14] and these must have been caused by the radiolysis of the water.

We have calculated the concentrations of radicals produced by the irradiation of the water next to the glass surface, using the methods of radiation chemistry[16-19]. We calculated the yields of the species O_2^-, OH, H, e_{aq}^-, H^+, HO_2, H_2 and H_2O_2 produced by the ionisation of the leachant, using the known rates of secondary reactions[20] involving these species, un-ionised water molecules and possible solutes.

We have considered the following three cases: (i) 2.4 Mrad.h^{-1} γ-radiation as used by McVay et al; (ii) 4 Mrad.h^{-1} α-radiation, which is the dose to the leachant when the $^{238}PuO_2$-doped samples are leach tested. Here the water is assumed to be saturated with dissolved air; (iii) Radiation conditions that could occur in a repository assuming that the leachant contacts the glass directly at all times. This is a pessimistic assumption, neglecting the life of the canister, often estimated to be 1000 years. The dose rates to the leachant are given in Table 5.

TABLE 5

DOSE RATES TO THE LEACHANT IN A WASTE REPOSITORY

Time after vitrification (years)	Dose rates (Mrad.h^{-1})	
	Ionising radiation (β,γ)	α -radiation
4	0.53	2.7×10^{-2}
50	0.08	2.2×10^{-2}
500	~ 0	1.1×10^{-2}

These figures assume that the leachant is in contact with the radioactive glass.

The results for the first two cases are shown in Table 6. In the experiments where the glass was leached in a γ-flux, the radicals with the highest concentrations were O_2^- and OH, suggesting that one of these must have been responsible for the enhanced leach rate. There were also small changes ($\lesssim$ a factor of 2) in the H^+ and OH^- concentrations[3] but experiments where the leach rate was measured as a function of pH show that such small changes will have a negligible effect on the leach rate. In the second case where the $^{238}PuO_2$-doped samples were tested, no enhanced leach rate was found. Since for these samples the O_2^- concentration is higher than it is in the γ-irradiation experiment, it seems that the OH radical is the one that produced most of the increase in leach rate in the latter case.

TABLE 6

CONCENTRATIONS IN MOL.DM^{-3} OF REACTIVE RADICALS PRODUCED BY RADIOLYSIS

Radiation	H	OH	e_{aq}^-	O_2^-	H^+	OH^-
2.4 Mrad.h^{-1} γ	9.4×10^{-11}	4.2×10^{-9}	9.4×10^{-11}	3.8×10^{-8}	1.2×10^{-7}	8.3×10^{-8}
4 Mrad.h^{-1} α	1.1×10^{-13}	2.3×10^{-11}	4.1×10^{-14}	1.9×10^{-7}	1.3×10^{-7}	7.7×10^{-9}

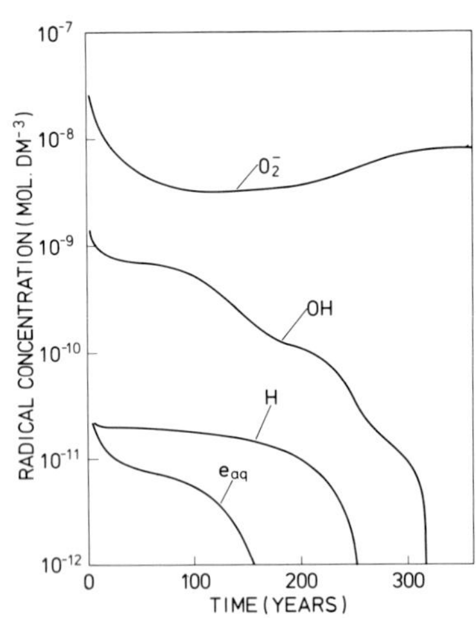

FIG.4 Concentrations of radiolytic species produced by radiolysis of water in contact with glass in a repository

The variation of the concentrations of these radicals under the conditions expected in a repository (but neglecting any shielding effect of the can) are plotted in Fig.4. Note that the concentrations quickly fall to values low compared with those found for the laboratory experiment under γ-irradiation. Also, we calculated that the ions NO_3^- and Cl^-, often found in groundwaters, can reduce the OH concentration by factors of up to several thousand. Thus we conclude that, in the absence of air, radiolysis of the leachant will not have a large effect on the leach rate in a real repository.

In water with the presence of air or nitrogen, as noted above, nitric acid is produced by irradiation. From experimental results[21,22] the nitric acid concentration N mol.dm^{-3} at time t hours is given by $N = 2C_o R\{1 - \exp(-1.45 \times 10^{-5}\, GDt)\}$, where D is the dose rate in Mrad h^{-1}, R is the ratio of the volume of air to the volume of liquid and the G value for the reaction in 1.9. C_o is the con-

centration of N_2 in the air in $mol.dm^{-3}$. This equation shows that a pH change
from 5.7 to 3.3, as observed by McVay and Pederson[14] in an experiment with air
present, would be consistent with a value of $R \simeq 1$.

However, a repository should be free of air pockets a few years after burial
(N.A.Chapman, private communication) and in any case the air would have to be
irradiated through the canister and a layer of water. Note also that when the
canister has failed - say after a few hundred years - the only important
radiation is from α-particles, which have a range of only about 40μm in water.
Under these conditions, nitric acid will only be produced if it can be formed
by α-radiation from air dissolved in the water. Rai et al[23] attempted to show
that such production was possible but their experiments were not completely unambiguous[24] because they did not exclude a gaseous phase entirely from the apparatus. On the contrary, Hall et al.[25] measured the leach rates in water containing dissolved air of glass samples doped with realistic amounts of Pu and
Am and obtained similar values to those of undoped samples, showing either that
no nitric acid was produced or only an insignificant quantity. Thus, under the
conditions in a repository it is unlikely that nitric acid formation will be a
problem.

ACKNOWLEDGEMENTS

The authors are grateful to Dr. G. L. McVay for discussing his results by
correspondence before publication, and to Dr. J. C. Dran for making available
preprints of recent papers.

REFERENCES

1. Hall, A.R., Dalton, J.D., Hudson, B. and Marples, J.A.C. (1976) in "Manage-
    ment of Radioactive Wastes from the Nuclear Fuel Cycle". IAEA-SM-207/24.
    Vienna II, 3-12.

2. Permar, P.H. and McDonnel, W.R. (1981) in "Proceedings of the 10th Symposium
    on Effects of Radiation in Materials" (American Society for Testing and Ma-
    terials, Philadelphia) to be published (available from National Technical
    Information Services CONF 800609-5, and as Dupont Report DP-MS-80-27).

3. Burns, W.G., Hughes, A.E., Marples, J.A.C., Nelson, R.S. and Stoneham, A.M.
    (1981) and (1982) AERE Harwell Report 10189 and J. Nucl. Mats. 107, to be
    published.

4. Marples, J.A.C. et al. European Research Reports (1981) 3 (3) 395.

5. Weber, J.E., Turcotte, R.P., Bunnell, L.R., Roberts, F.P. and Westsik, J.H.
    (1979) in "Ceramics in Nuclear Waste Management" Eds. T.D. Chikalla and
    J.E. Mendel (Technical Information Center, U.S. Department of Energy,
    CONF-790420) pp. 294-299.

6. Mendel, J.E., Ross, W.A., Roberts, F.P., Turcotte, R.P., Katayama, Y.B. and
    Westsik, J.H. (1976) in "Management of Radioactive Wastes from the Nuclear
    Fuel Cycle" (IAEA, Vienna) Vol 2, pp. 49-61.

7. Bibler, N.E. and Kelley, J.A. (1978) Dupont Report DP-1482 (Dupont Savannah River Laboratory, Aiken, South Carolina).

8. Scheffler, K. and Riege, U. (1977) KfK Report 2422 (Kernforschungszentrum, Karlsruhe).

9. Dran, J.C., Maurette, M. and Petit, J.C. (1980) Science $\underline{209}$, 1518-1520.

10. Dran, J.C., Maurette, M., Petit, J.C. and Vassent, B. (1981) in "Scientific Basis for Nuclear Waste Management" $\underline{3}$, J.G. Moore Ed. Plenum Press 449.

11. Dran, J.C., Langevin, Y., Maurette, M., Petit, J.C. and Vassent, B. (1982) in "Scientific Basis for Nuclear Waste Management" Vol.**6** (to be published).

12. McVay, G.L., Weber, W.H. and Pederson, L.R. (1980) ORNL Conference on Leach-ability of Radioactive Solids. Gatlinburg, Tennessee.

13. McVay, G.L. and Buckwalter, C.Q. (1980) Nucl. Technol. $\underline{51}$, 123-129.

14. McVay, G.L. and Pederson, L.R. (1981) J.Amer.Ceram.Soc. $\underline{64}$, 154-158.

15. Boult, K.A., Dalton, J.T., Hall, A.R., Hough, A. and Marples, J.A.C. (1978) AERE Harwell Report R-9188.

16. Burns, W.G. and Moore, P.B. (1976) Rad. Effects $\underline{30}$, 233-242.

17. Burns, W.G. and Moore, P.B. (1978) in "Proceedings of the Conference on Water Chemistry of Nuclear Reactors" (British Nuclear Energy Society, London) pp. 282-289.

18. Anbar, M. (1968) in "Fundamental Processes in Radiation Chemistry" Ed. P. Ausloos (Interscience, New York) pp. 651-685.

19. Appleby, A. and Schwarz, H.A. (1969) J.Phys.Chem. $\underline{73}$, 1937-1941.

20. Boyd, A.W., Carber, M.B. and Dixon, R.S. (1980) Radiat.Phys.Chem. $\underline{15}$, 177-185.

21. Wright, J., Linacre, J.F., Marsh, W.R. and Bates, T.H. (1956), Proceedings of the International Conference on the Peaceful Uses of Atomic Energy (United Nations, New York) Vol. 7, pp. 560-563.

22. Linacre, J.K. and Marsh, W.R. (1981) AERE Report R-10027 (Harwell).

23. Rai, D., Strickert, R.G. and Ryan, J.L. (1980) Inorganic and Nuclear Chemistry Letters $\underline{16}$, 551-555.

24. Burns, W.G., Hughes, A.E., Marples, J.A.C., Nelson, R.S. and Stoneham, A.M. (1982) Nature $\underline{295}$, 130-132.

25. Hall, A.R., Hough, A. and Marples, J.A.C., "Leaching of Vitrified High Level Radioactive Waste", Paper at this conference.

Published 1982 by Elsevier Science Publishing Co
SCIENTIFIC BASIS FOR RADIOACTIVE WASTE MANAGEMENT - V
Werner.Lutze, editor

STRUCTURAL EFFECTS OF RADIATION DAMAGE IN SILICA BASED GLASSES

A. MANARA, P.N. GIBSON
Joint Research Centre, 21020 Ispra (Varese), ITALY
and
M. ANTONINI
Istituto di Fisica and GNSM, Universita' di Modena, 41100
Modena, ITALY

ABSTRACT

The results of recent measurements of optical absorption,
etching rate and transmission electron microscopy in pure
silica and borosilicate glasses are reported and discussed.
At dose saturation conditions, the dependence of the optical
density associated with the production of single atomic
defects from the mass of the impinging particle shows a
marked saturation at masses $\geqslant$ 20 amu. The corresponding
etching rates increase by about 4 times with respect to unirradiated samples. In borosilicate glasses, the temperature
dependence of the threshold dose rate of electrons to initiate the nucleation of bubbles shows a marked increase from
about $300^{\circ}$to 600°K.

INTRODUCTION

In previous papers, we reported and discussed results concerning the behaviour of amorphous silica and of silica-based glasses incorporating various
oxides to simulate high-level nuclear waste /1 - 3/.

In those investigations, on one side the basic damage mechanisms were
examined in terms of production, saturation and thermal annealing of single
atomic defects; on the other side the more direct implications of α-recoil
damage on borosilicate and waste glasses were recorded in terms of density
variation, stored energy and microstructural modifications. From the first
class of experiments it was found that different kinds of particles introduce
different types of atomic defects at dose saturation levels which depend upon
the mass of the bombarding particle and upon the previous thermal and irradiation history of the sample. From the second class of experiments, heavily
electron damaged glasses revealed large microstructural modifications
consisting of bubble formation and phase segregation.

These results have raised interest to correlate the damage produced at the
basic level in pure silica with the larger modifications induced in more
complex glasses. In particular, a connection can be looked for between the
type and number of atomic defects present at dose saturation with the onset of
nucleation effects. For this purpose a careful analysis of the mass dependence
of saturation levels is necessary and a new particle of intermediate mass,i.e.

neon, has been therefore added to the previously adopted light and heavy ions
as a bombarding external source. The choice of neon has been made both to fulfill
this requirement and because neon seems to be among the first heavy ions which
at high energy can produce clearly visible nuclear tracks in silicate glasses
/4/. Further correlations might be found between the first stages of nucleation
of defects and the enhancement of chemical activity of the damaged regions /5/.
The connection might even be of more interest when considering that the deep
penetrations of high energy heavy ions adopted in our irradiations allow us to
examine the etching rate not only near the surface of the irradiated specimens,
as is the case with heavy low energy ions, but also within the bulk.

In the following, we first describe the most relevant experimental aspects
of the irradiations and of the post-irradiation tests. We then report on the
obtained results of optical absorption, etching rate and electron microscopy.
Finally we discuss the results within the frame of an integrated picture of the
observed irradiation effects.

IRRADIATIONS AND POST IRRADIATION TESTS

Optical absorption and etching rate measurements were performed in synthetic
amorphous silica, purchased from "Quartz and Silice", France, in the form of
5×5 mm^2 slabs, 0.2 mm thick. "In situ" electron microscope observations were
made in borosilicate glasses containing 55% wt. SiO_2, 17.5% B_2O_3, 6% Al_2O_3 and
21.5% Na_2O. This is one of the typical matrix compositions generally adopted
before waste storage. Samples were disks of 3 mm diameter, 3 um thin.

Heavy ion irradiations, including the new neon irradiations, were made at
the Variable Energy Cyclotron (AERE, Harwell, U.K.) using 46.5 MeV Ni^{+6} and
20.5 MeV Ne^{+3} ions, with the same facilities and procedures described elsewhere
/6/. Total doses ranged between 0.1 and 1 dpa at ion currents between 0.1 and
1 uA. Helium and electron bombardments were performed in the two Van De Graaf
accelerators of the J.R.C., Ispra both at 1.5 MeV. Much higher doses, up to 50
dpa, and dose rates, up to 3×10^{24} particles m^{-2} sec^{-1} were used during "in
situ" electron irradiations. As already pointed out /2/, these values are much
higher with respect to typical doses and dose rates to be expected during the
storage of actual waste. However, they have been adopted in order to determine
the initial stages of observable bubble growing.

Optical absorption spectra were recorded in the energy region from 3.5 to
6.5 eV, using the same procedure used in previous papers /3,6/. Some high
vacuum spectra were taken up to about 9 eV to find peak positions and intensities of the high energy bands. During in situ electron microscope irradiations,
temperature was varied from R.T. to over 1000°K, using a double tilt axis hole
entry hot stage with the specimen's chamber differentially pumped to below
1.5 10^{-5} Pa. Etchings were made using a 50% HF solution. Before etching, samples
were optically examined and some surface 'crazing' /7/ was found in neon irradiated samples at all doses. In order to measure the etching rate, a section of
the sample was masked off on both sides before immersion in the HF solution for
one minute. The amount of material etched away was measured using a 'Talystep
1' instrument which mechanically detected the step formed at the edge of the
masked off area.

RESULTS

Optical absorption. The optical absorption spectra give a direct indication
of the number of single atomic defects present in the glass after irradiation.
Typical spectra obtained at dose saturation conditions in amorphous silica
glasses irradiated with Ni, Ne and He ions are shown in figure 1. All data have
been normalized for convenience to the same depth of 5.5 um, that is to the range
of the ions of 1.5 MeV. From the observation of the figure it appears that the
general aspects of all the curves is the same. In particular, the two major
bands B_2 and E_1' already mentioned in previous papers /1,3,6/ are also present
at the same values of 5.05 and 5.8 eV, respectively, in neon irradiated samples.
The strength of the B_2 band is intermediate between those of alphas and Nickel,
confirming that this band, as well as the high energy band at 7.12 eV whose tail
underlies the spectrum, is present after all massive particle bombardments.

The production efficiency of all defects seems, however, to have some mass
dependence as is better shown in figure 2, where the peak heights obtained after
numerical deconvolution and fitting of the measured spectra, are plotted vs.
the mass of the impinging particles. All curves have some tendency for saturation when the mass is higher than that of neon, although the saturation levels
do not coincide, being lower for the E_1' centers and higher for the other two
centers. From this behaviour it can be deduced that only when the mass of the
incident particle is larger than about 20 amu, the response of the glass to
displacive bombardment at saturation is independent from the specific particle
which has created the damage. This circumstance should be taken into account
during light particle simulations of the recoil of α - emitting actinides. The
anomalous high value of the E_1' band in the case of proton irradiation is not
fully understood but it could be related to a chemical interaction between the
implanted hydrogen and the glass components.

"In situ" electron microscopy. The effects of 1 MeV electron irradiation at
high doses and relatively high currents previously reported /2,3/ and interpreted in terms of the formation of oxygen gas bubbles, have been systematically examined as a function of the irradiation temperature to detect the
temperature dependence of the onset of bubble nucleation. The results are shown
in figure 3 in terms of minimum electron fluxes necessary to initiate bubble
nucleation. It can be seen from the figure that this quantity drastically
decreases with temperature. Moreover in the last series of observations it has
been found that much smaller dose rates are indeed enough to provoke further
growing of bubbles.

Enhanced chemical etching. The etching behaviour before and after irradiation
is shown in figure 4. The etching rate is about 4 times higher in Ne and Ni
bombarded samples with respect to unirradiated materials. Its value increases
with dose up to saturation, attained at approximately 0.1 dpa. When the etched
depth roughly equals the range of the impinging particles, the slope of the
etching line decreases to about that of the unirradiated sample, clearly
indicating that the same rate is present all along ion paths. Proton irradiated
samples were also examined and a lower increase in etching rate was found under
saturation conditions.

352

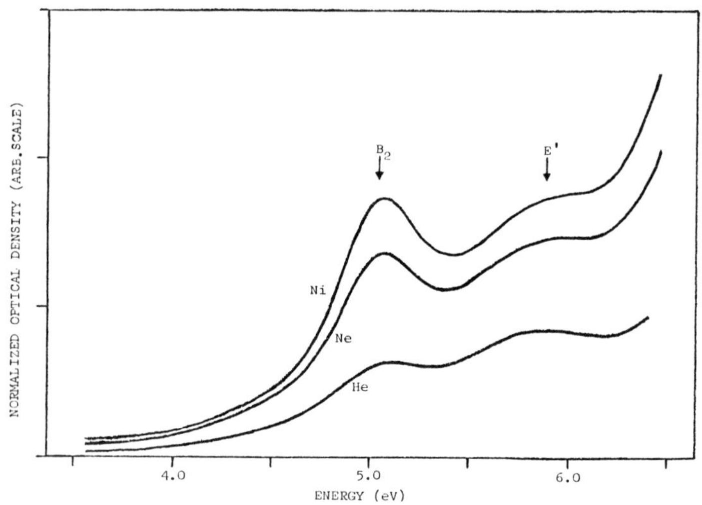

FIG. 1. ABSORPTION SPECTRA (SATURATED) OF Ni, Ne AND He ION IRRADIATED SILICA.

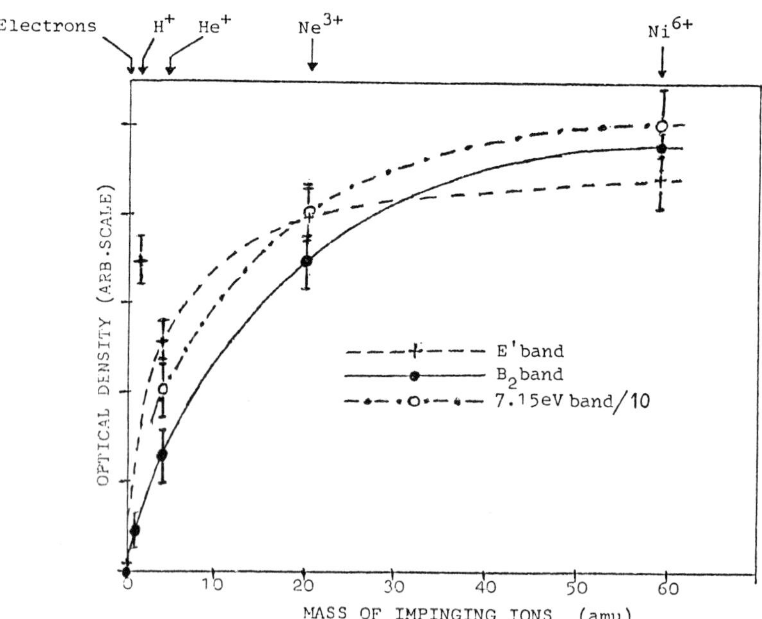

FIG.2 . SUMMARY OF THE HEIGHT AT SATURATION OF THE VARIOUS ABSORPTION BANDS IN
IRRADIATED SILICA AS A FUNCTION OF MASS OF IMPINGING ION.

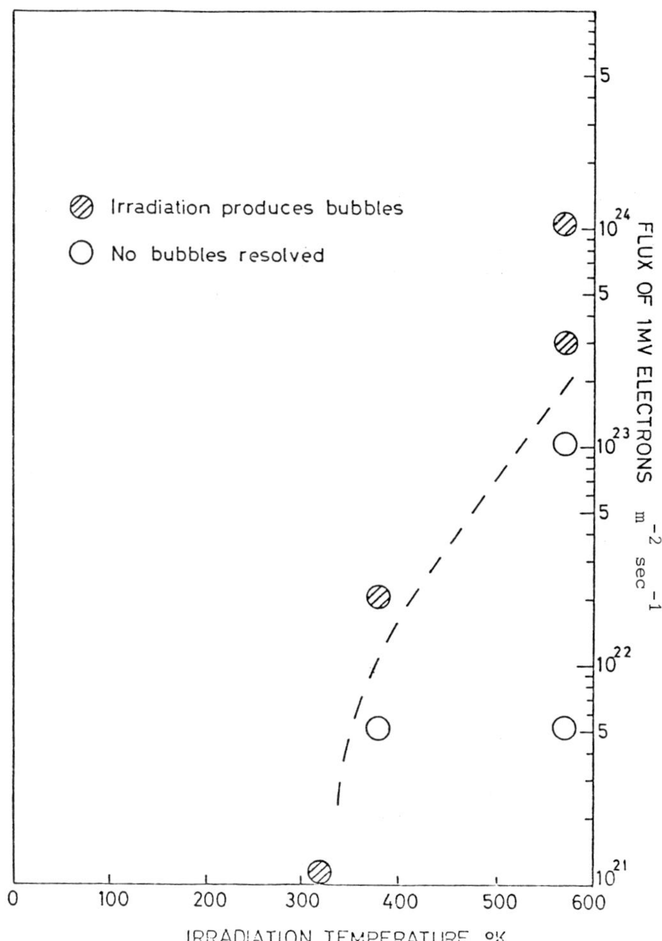

FIG.3. TEMPERATURE DEPENDENCE OF THE THRESHOLD FLUX FOR BUBBLE NUCLEATION IN BOROSILICATE GLASSES.

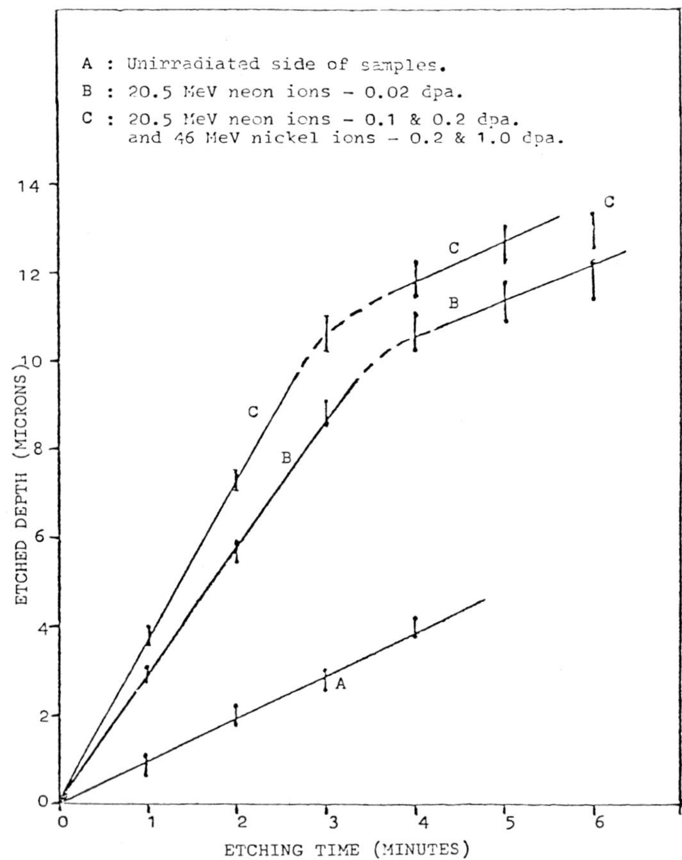

FIG. 4. ETCHING BEHAVIOUR OF HEAVY-ION IRRADIATED
SILICA SAMPLES COMPARED WITH AN UNIRRADIA-
TED SPECIMEN.

DISCUSSION

The presence of a saturation effect for the total maximum concentration of atomic point defects produced during irradiation at high masses can be interpreted as if a steady state condition is attained in terms of defect recombination during irradiation or assuming that defects produced at sufficiently high concentration precipitate into some form of aggregate. This precipitation would then occur as soon as the energy and the electrical charge deposition per unit length of penetration overcomes a threshold value. When comparing either of these alternative hypothesis, the second one seems more plausible. In fact the increase in etching rate agrees well with a model of accumulation and overlapping of etchable tracks in silicate glasses already proposed by Krätschmer to explain a similar effect observed at lower dose rates /8/. From the doses required to saturate our optical absorption spectra, an approximate track diameter has been calculated to be about 8 $\overset{\circ}{A}$ for Ne^{+3} and 25 $\overset{\circ}{A}$ for Ni^{+6} ions, agreeing reasonably with Krätschmer's calculations.

A recent theory of nuclear track formation relates the production of tracks with the precipitation of point defects to produce extended defects along the path of the particle /9/. These are kept separated by gap zones loaded with point defects. In this framework further more systematic measurements of enhanced etching and electron microscopy under different irradiation conditions could more definitively indicate the presence of extended defects even in pure irradiated vitreous silica.

ACKNOWLEDGMENTS

We wish to acknowledge Dr. S.N. Buckley, AERE Harwell, and Mr. S.A. Manthorpe who have provided the basic help during the electron microscope observations. We also wish to thank Mr. L. Mammarella for technical assistance during etching rate measurements.

REFERENCES

1. M. Antonini, P. Camagni, F. Lanza, A. Manara in: Scientific Basis for Nuclear Waste Management, J.M. Northrup Jr. ed. (Plenum Press, New York 1980) 2 pp. 127-133.

2. M. Antonini, S. Buckley, P. Camagni, P.N. Gibson, A. Manara in: Scientific Basis for Nuclear Waste Management, S.V. Topp ed. (Boston, Nov. 1981) to be published.

3. M. Antonini, P. Camagni, P.N. Gibson, A. Manara in: Radiation Effects in Insulators (Arco, Italy 1981) to be published in Rad. Effects.

4. R.L. Fleischer, P.B. Price and R.M. Walker, J. Appl. Phys. 36 3645 (1965).

5. J.D. Dran, Y. Langevin, M. Maurette and J.C. Petit in: Scientific Basis for Nuclear Waste Management, S.V. Topp ed. (Boston, Nov. 1981) to be published.

6. M. Antonini, P. Camagni, A. Manara, L. Moro, J. non Cryst. Solids, <u>44</u>, 321 (1981).

7. W. Primak, <u>The Compacted States of Vitreous Silica</u> (Gordon and Breach, New York 1975) p. 104.

8. W. Krätschmer in: <u>Nuclear Photography and Track Detectors</u>, Conf. Proc., (1975, Bucarest).

9. E. Dartyje, J.P. Durand, Y. Langevin and M. Maurette, Phys. Rev. <u>B 23</u>, 5213 (1981).

Published 1982 by Elsevier Science Publishing Co
SCIENTIFIC BASIS FOR RADIOACTIVE WASTE MANAGEMENT - V
Werner.Lutze, editor

NEAR-SURFACE LEACHING STUDIES OF Pb-IMPLANTED SAVANNAH RIVER WASTE GLASS

G. W. ARNOLD and C. J. M. NORTHRUP*
Sandia National Laboratories, Albuquerque, NM, 87185, USA
N. E. BIBLER
E. I. DuPont de Nemours and Company, Savannah River Laboratory,
Aiken, SC, 29801, USA

INTRODUCTION

Ion implantation into simulated nuclear waste glasses is a rapid means of producing near-surface energy deposition similar to that produced by α-recoil nuclei after long storage times (typically 10^3 - 10^6 years). For example, Dran, Maurette, and Petit[1] used 200 keV Pb-ion implantations in glass at a fluence of $5 \times 10^{12}/cm^2$ to produce surface damage. This fluence is equivalent to approximately 2×10^{18} alpha-decays/cm^3 which corresponds to approximately 10^6 years storage for glass containing Savannah River Plant (SRP) defense high-level waste (DHLW). These authors[1] found that this fluence value corresponded to a critical fluence (Φ_c) for enhanced etching (a factor of 20 increase as inferred by step-height changes) for several silicate glasses when etched in a NaCl solution at 100°C. This critical fluence value also corresponds very well with the fluence at which significant overlap of individual ion tracks occurs. In a subsequent paper,[2] Dran and his colleagues at Orsay have examined the etching problem in greater detail for a wide variety of glasses and etching conditions and have shown that the leaching of implanted glasses is a complex process and that the overall rate of dissolution depends upon the form of the hydrated layer and its swelling and subsequent removal. The rates at which these various steps occur determines the kinetics of the release of the glass constituents to the solution. Recently, Primak[3] has also observed etch rate increases due to ion implantation using a variety cf inert gas ions along with D^+ ions. The increases he observed, however, were approximately 7 times smaller than those observed by Dran, et al.[1] The reported increase in etch rate has drawn the attention of the Harwell group,[4] who have observed that significant increased leaching rate increases have not been noted in their glasses doped internally with alpha emitting actinides even when the α-decay dose exceeds that produced by a Pb-fluence of Φ_c. For example, a glass doped with Pu-238 has been under observation for an alpha decay dose of $> 1.3 \times 10^{19}$ decays/cm^3. Based on weight loss measurements,

---

*This work performed at Sandia National Laboratories supported by the U.S. Department of Energy under contract number DE-AC04-76DP00789.

the etch rate increased by a factor of <2. Similarly, preliminary measurements of the leach rate of ion implanted SRP waste glass based on solution analyses indicate no increase in leach rate up to doses equivalent to 3×10^{20} α-decays/cm^3.[5]

The present experiments with SRP simulated nuclear waste glass implanted with Pb ions (see the Experimental Section), used RBS and ERD to follow in detail the changes in composition which occur in the near-surface region upon leaching in deionized water at 90°C. Analyses of the leach solutions were made in an attempt to correlate the actual leach rates with the observed near-surface compositional changes. These experiments show that radiation damage can cause changes in the composition of the near-surface of the leached glass. We also find that a critical fluence is reached where abrupt changes of the surface elemental composition occur as a result of leaching. This fluence is near the value observed by both Dran, et al.[1] and Primak.[3] Solution analyses were not made for all the leaching experiments. However, those analyses which were made indicate that the amount of material actually leaving the glass is not significantly increased as a result of the radiation damage.

EXPERIMENTAL

Compositions of the frit and simulated DHLW are shown in Table I. The frit is currently the reference proposed by SRP to immobilize all the DHLW currently stored at SRP. Glasses were prepared by melting a dry mix (73 wt % frit and 27 wt % simulated waste) at 1150°C for at least three hours. To attain homogeneity, the glasses were then crushed and remelted at 1150°C. They were finally annealed at 500°C for 1 hour. Samples were then cut and polished for irradiation. Static leach tests were made using deionized water at 90°C. The surface area to leachate volume ratio (SA/V) was 0.05 cm^{-1}. Analyses for dissolved Si, Li, B and Na were made using inductively coupled plasma spectroscopy (ICP). Similar results were observed for both acidified and nonacidified leachates.

TABLE I

Composition of Reference Frit 131 and Simulated Defense
High-Level Waste (TDS-3A)[a]

Frit Component (wt %)	Simulated Waste Component (wt %)
SiO$_2$ (57.9), B$_2$O$_3$ (14.7), Na$_2$O (17.7) Li$_2$O (5.7), MgO (2.0), TiO$_2$ (1.0) La$_2$O$_3$ (0.5), ZrO$_2$ (0.5)	Fe$_2$O$_3$ (42.9), MnO$_2$ (12.3), zeolite (9.3), Al$_2$O$_3$ (8.6), NiO (5.3), SiO$_2$ (3.7), CaO (3.2), Na$_2$O (3.1), coal (2.1), Na$_2$SO$_4$ (0.5), Nd$_2$O$_3$ (9.3)

[a]Glass mix (dry basis) is 73 wt % frit and 27 wt % waste.

Implants were made in a few seconds with an Accelerators, Inc., implanter (15 - 250 keV). Single energy implants (207 keV) and multiple-energy (40-250 keV) Pb-implants were made; the single implants were made because of the previous experimental work by others[1,2] and experimental simplicity. The multiple-energy implants were made to approximate a uniformly damaged near-surface region. Fluences for the multiple-energy implants were chosen[6] so that the total energy deposition approximated that for 3×10^{18}, 3×10^{19}, and 3×10^{20} α-decays/cm^3. These are referred to as "low", "medium", and "high" dose implants throughout this paper. Some samples were also implanted with H to a fluence of 1×10^{16} H/cm^2 at either 100 keV or 45 keV. These H implants were used as a marker for monitoring the depth of compositional changes and material removal rates at the glass surface.

The near-surface elemental concentrations were determined by Rutherford backscattering spectrometry (RBS) and elastic recoil detection (ERD). For RBS, the probe particle was 2.8 MeV He$^+$ and the angle between surface normal and the incident beam was set at 55° for high resolution ($\sim$ 50 A). For elastic recoil detection (ERD),[7] 12 MeV C^{3+} was used to determine the position of the 100 kev hydrogen marker which initially was at a depth of $\sim$ 1 μm. Hydrogen and lithium profiles were obtained using 20 MeV Si^{4+} as the ERD probing beam.

RESULTS

Our results can best be presented and discussed in terms of the type of implantation (multiple or single energy) and the leaching times (5, 6 or 1.5 hours).

A. <u>Multiple Energy Implantation Followed by 5 and 6 Hr Leach Tests</u>

For these experiments, three samples were implanted (on both sides) each to different doses (low, medium, and high fluence) which are equivalent to 3×10^{18}, 3×10^{19} and 3×10^{20} α-decays/cm^3. The glasses along with an unimplanted sample were then leached by two 5-hour tests and then three 6-hour tests for a cumulative leach time of 28 hours. Leach rates for Si, B, Li, and Na were measured by solution analysis. The results are plotted in Figure 1 as a function of equivalent α-decays/cm^3. Based on solution analyses, no evidence for significant enhanced leaching due to implantation is observed in the data shown in Figure 1. If enhanced leaching occurred, it should be especially apparent in the first leach test where the leaching occurred primarily in the implantation-damaged surface. For a leach rate of 1 g/m^{2}d, approximately 800 A is leached in 5 hours, which is approximately the range of the highest energy Pb ions used in the implantation. In the first leach test, the maximum difference for any leach rate based on any of the four elements is within a factor of two of the respective leach rates for the unimplanted glass. No trend with radiation fluence is evident even though the largest two fluences exceed the equivalent α-decay dose produced by the Φ_c determined by Dran, et al.[1] In the second leach test, where some of the damaged layer is probably still being leached, the variation in the leach rates is still less than a factor of 2. Results of the fifth leach test are somewhat scattered also but indicate no effect of implantation. These results may display scatter because of the

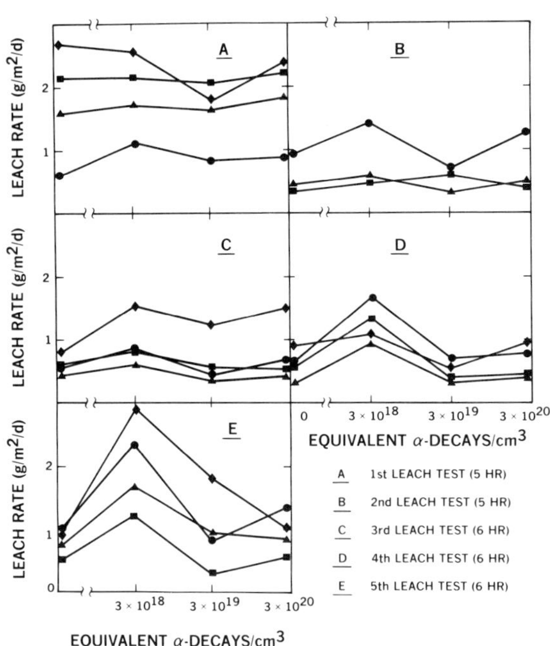

Figure 1. Leach rate in deionized water vs. equivalent α-decays/cm^3 for 5 leach tests of Pb-ion implanted SRP-DHLW glass. Leach rates are based on 4 elements dissolved in the leachate: ●, Si; ■, B; ▲, Li; ◆, Na (T = 90°C, SA/V = 0.05 cm^{-1}, ρ = 2.7 g/cm^3).

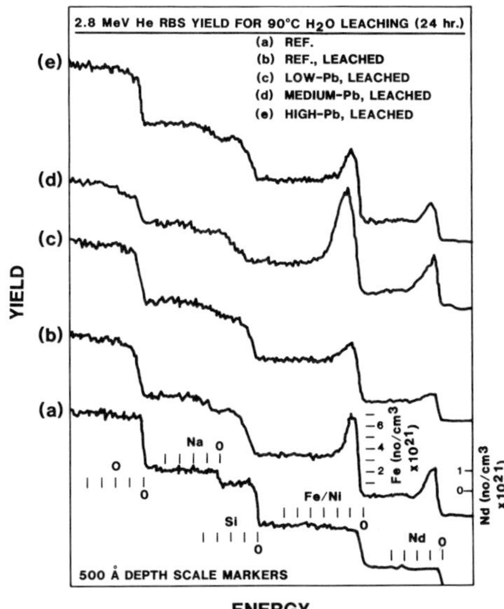

Fig. 2.
RBS yield vs. recoil energy.

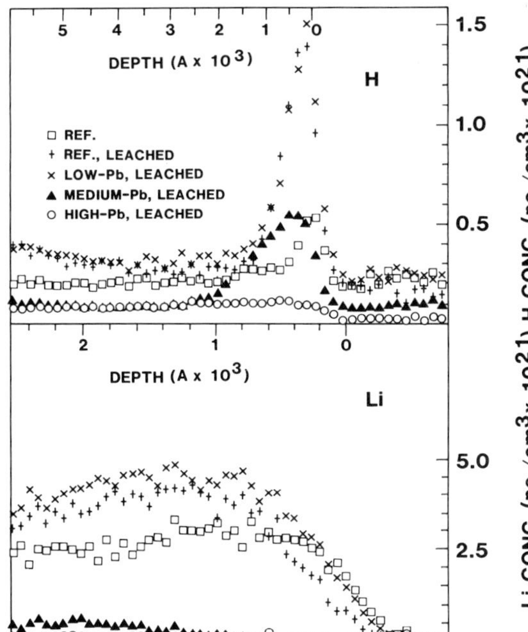

Fig. 3.
ERD yield (H, Li) vs.
depth.

instability and partial flaking off of the surface layer formed by the previous four leach tests.

After the fifth leach test, the RBS and ERD spectra of these four glasses were measured (Figures 2 and 3). The spectra indicate that implantation followed by leaching results in compositional changes in the near-surface of the leached glasses. Leaching at 90°C of the unimplanted glass produced surface peaks of Fe and/or Ni and of Nd (Figure 2). These distributions are 2.5 - 3 times the bulk values and peak $\sim$ 200 A from the surface and extend $\sim$ 1000 A into the sample. In addition, the Si and O edges show erosion to a depth of $\sim$ 4000 A due to a loss of $\sim$ 4 x 10^{16} Si/O atoms/cm^2. As seen from the data in Figure 2, the surface compositions of the implanted glasses after leaching differ for "low," "medium," and "high" Pb-fluence levels.

For the low-level implant, the major difference in RBS spectra, relative to the leached unimplanted sample, is that the Fe/Ni and Nd surface peaks are reduced by about 65%. For the medium Pb-fluence, leaching at 90°C reduces the near-surface Fe/Ni level from $\sim$ 6 x 10^{19}/cm^3 to 4 x 10^{19}/cm^3. The removal depth is >3000A. However, the surface Fe/Ni peak is enhanced about 25% and extends to greater depth ($\sim$ 1500 A). The Nd surface peak is also broadened and has a shoulder near the expected depth ($\sim$ 800 A) of the implanted Pb. Si and O both show greater losses than in the reference material with the additional loss coming from the ion-implanted layer. Rough calculations show $\sim$ 1400 Si/O atoms removed per incident Pb ion which is reasonable for the average number of displacements in the collison cascade. Release of these displaced atoms apparently occurs for critical energy deposition values as discussed below for single-energy implants. The alkali content for the medium- and high-Pb fluence levels has also been significantly altered. This is especially true for Li (Fig. 3) which is removed from depths >3000 A. Na is also preferentially removed from the implanted region but the effect is not as well defined (Fig. 2). No change in Li leach-rate is observed (Fig. 1) despite the apparent loss of Li (Fig. 3). Some loss of H could also have occurred due to a longer time interval between leaching and scattering measurements than was the case for the single-energy implants of (B) below.Leaching of the high-level Pb implant results in an RBS spectrum almost identical to that for the leached unimplanted sample. ERD measurements, however, show loss of Li as was observed for the medium-dose implant. On the basis of the ERD spectra for the medium and high fluence and the RBS spectra for the medium Pb-implant, Pb-ion fluences equivalent to

α-decay doses on the order of $10^{19}/cm^3$ produce signficant changes in the near-surface composition of leached SRP glass.

B. <u>Single Energy Implantation Followed by Four 1.5 Hr Leach Tests</u>

To obtain information at various stages of damage and/or leaching, experiments were performed using single energy implants (207 keV Pb) at fluences of 5 x 10^{11} - 1 x 10^{16} ions/cm^2. These glasses were also implanted with 100 keV H ions as a depth marker at ~ 10,000 A. Leach rates were determined after each of four 1.5 hr leach tests at 90°C. Again, the leach rates based on Si, B, Li and Na, did not vary by more than a factor of 2 to 3 over the entire fluence range. Values of the leach rates agreed with those determined in the earlier tests (nominally 1 g/m^{2}d, see Figure 1) even though the leach times were very short. A significant decrease in the Na leach rate was observed for all the glasses as the cumulative leach time increased. Figure 4 shows this decrease for the two unirradiated glasses and two irradiated glasses. The decrease in leach rate for Na as well as that for Fe, Ni and other elements in unirradiated glass has also been noted by Altenhein, et al.[8] Apparently, Na very near the initial surface of the glass is easily removed. From the data in Figure 4, it is clear that for these short leach times, the glasses implanted to Pb-ion fluences of 1 x 10^{14} and 1 x $10^{16}/cm^2$ are leaching similarly to the unirradiated glass. This result was also confirmed by the RBS and ERD spectra (not shown). Enhancement of Fe/Ni, Nd and Li at the surface were observed along with Na and Si depletion.

Measurement of the depth of the H marker by ERD using C^{+3} ions indicated that the marker had not moved even though Si and other elements were being leached from the surface. This could occur because the gel layer containing Fe, Ni, etc., largely remains and the net energy loss is primarily due to lower Z components which do not contribute significantly to the ERD measurement of the marker depth. The movement of other species into the near-surface region may also partially compensate by replacing the leached atoms in solution. A Si removal rate of 1 g/m^{2}d infers removal on the order of 200 A per 1.5 hr test for a total of 1200 A. Altenhein, et al.[8] have speculated that the enhancement of Fe, Ni and other elements may form a protective layer by forming insoluble oxides and hydrates.

C. <u>Additional 10 Hr. Leach Test of Single-Energy Implanted Glasses</u>

All the samples discussed above were leached an additional 10 hrs. Unfortunately, it was not possible to obtain solution analysis of thes leachates. Figures 5-7 compare the RBS spectra and the ERD profiles for the implanted samples after the 10 hr leach to those for unimplanted material.

364

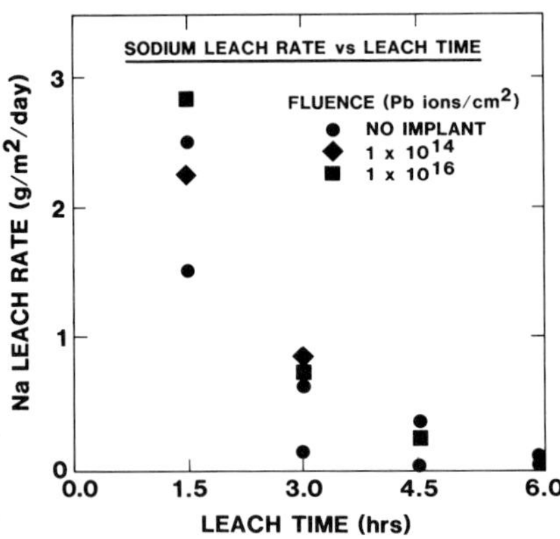

Fig. 4. Sodium leach rates vs leach time

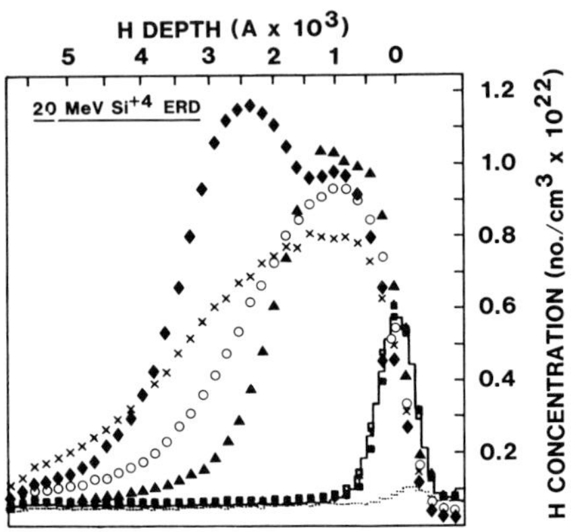

Fig. 5. ERD yield vs. depth for 207 keV Pb-implanted samples: ___ reference (no implant/leach), ... reference (leached), ■ 5×10^{11}/cm^2, □ 1×10^{12}/cm^2, + 5×10^{12}/cm^2, ▲ 1×10^{13}/cm^2, X 1×10^{14}/cm^2, ○ 1×10^{15}/cm^2, ◆ 1×10^{16}/cm^2.

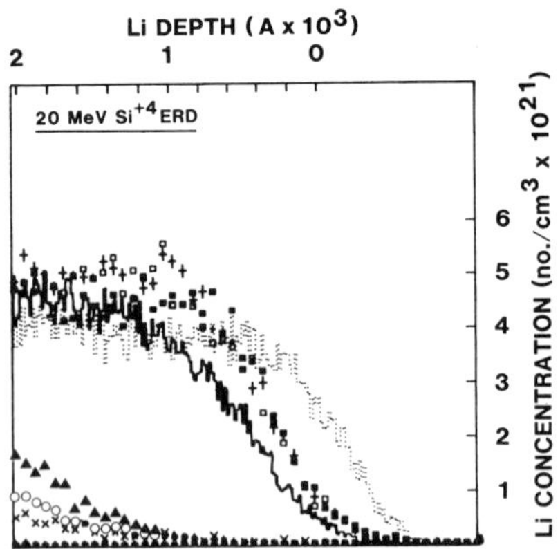

Fig. 6. ERD yield (Li) vs. depth for 207 keV Pb-implanted samples: ___ reference (no implant/leach), ... reference (leached), ■ 5×10^{11}/cm^2, □ 1×10^{12}/cm^2, + 5×10^{12}/cm^2, ▲ 1×10^{13}/cm^2, X 1×10^{14}/cm^2, ○ 1×10^{15}/cm^2, ◆ 1×10^{16}/cm^2.

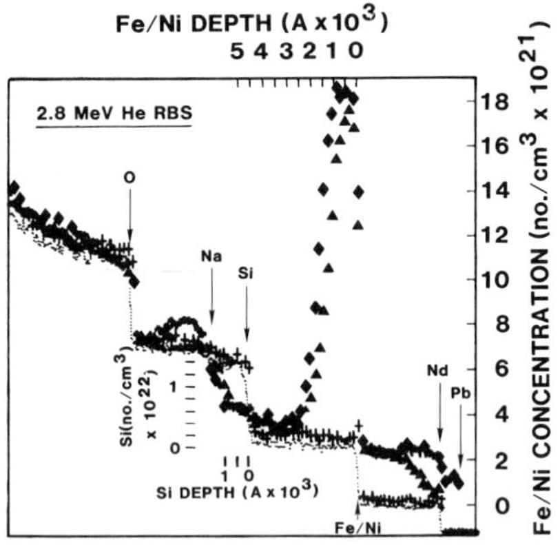

Fig. 7. RBS yield vs. depth for 207 keV Pb-implanted samples: ... reference (leached), + 5×10^{12}/cm^2, ▲ 1×10^{13}/cm^2, ◆ 1×10^{16}/cm^2.

Each series of spectra show large changes occurring for the same critical fluence in the range $\Phi_c = 5 \times 10^{12}$ to 1×10^{13} 207 keV Pb/cm^2. This fluence agrees with that observed by Dran, et al.[1] and Primak.[3] For this fluence interval, Figure 5 shows the movement of H into the glass to depths greater than 5000 A from the surface and concentrations an order of magnitude larger than bulk values. Measurements with 12 MeV C^{+3} ERD gave no evidence for the 100 keV H depth marker indicating that the physical surface of the glass had been removed to depths >1 μm by this 10 hr leaching step. The integrated H-distributions for samples where the fluence was Φ_c or greater, is about 3×10^{17} H/cm^2 compared to about 8×10^{16} H/cm^2 for those with fluences below Φ_c. Figure 6 shows a sharp loss of Li for fluence levels > Φ_c; for those with fluences below ϕ_c the Li levels are about 8×10^{16} Li/cm^2 while above Φ_c the level is between 1×10^{14} - 8×10^{15} Li/cm^2, integrated over the depths indicated on Figure 6. In contrast, it is also apparent that for fluences < Φ_c, near-surface Li concentrations are increased over leached un-implanted reference sample levels.

Figure 7 indicates that for fluences > Φ_c several important changes occur in the RBS spectra. The most apparent change is a large increase in the near-surface concentration of Fe/Ni to depths of ~ 5000 A. In addition, the Nd near-surface level increases by more than a factor of 2. It is also clear that the Si near the surface has been eroded severely. If one assumes that all of the Fe/Ni and Nd represent ions from the bulk glass pushed back by the moving surface, then material is removed to depths of ~ 1.1 μm which is con-sistent with the loss of the H ion depth marker. The levels of H gained and Li lost are consistent with a picture of H:Li ratio of 2-3 which is consistent with H$_2$O or (OH$_3$)$^+$ diffusion and exchange.

DISCUSSION

These results clearly demonstrate the existence of a radiation damage effect on near-surface compositional profiles for Pb-ion implantation of this boro-silicate nuclear waste glass after leaching. However, a corresponding increase in leach rates measured by solution analysis was not detected. The existence of a radiation damage effect itself is not surprising. It is well known that radiation damage alters many physical properties in glass and, in particular, etching rates in fused silica in dilute hydrochloric acid solution (see, e.g., Webb, et al.[9]). These authors[9] found an increase in initial etch-rate (0.3% HF solution) of 4-5 for H, C, and O ions implanted in SiO$_2$ glass. It should also be noted that a maximum occurs in initial etch rate for these ions at about the same energy deposition (2-4 $\times 10^{20}$ keV/cm^3) into atomic processes.[15]

The maximum is followed by a decrease in etch rate with increasing fluence. Similar behavior has been noted for changes in volume, density and refractive index,[10] lateral stress,[11] hardness,[12] ioninduced luminescence,[13] and thermo-luminescence[14] for SiO_2 glass. For SiO_2, the maxima in the induced changes occur at approximately the same energy deposition. These experiments have been discussed in a recent review paper on ion implantation effects in glasses.[15] The decrease in the rate of change of these properties, which is observed after the initial increase occurs at nearly the same energy deposition value and may be due to plastic flow. Such plastic flow results when the compaction of the glass in the implanted region is opposed by lateral tensile stresses in the bulk of the material. In the present experiments, the critical fluence, Φ_c, also occurs at $\sim 2 \times 10^{20}$ keV/cm^3 and may be related to this structural transformation. It should also be noted that the critical Pb-ion fluence corresponds to ion-track overlap for isolated tracks of ~ 45 A radius. This value is reasonably close to the 25-40 A radius for isolated Pb-ion tracks seen by Headley, et al.[16] in TEM measurements on Pb-implanted zircon, zircono-lite and Hollandite. The onset of significant compositional changes appears to occur much more abruptly than in the etching rate experiments of Webb, et al.[10] The sudden sharp compositional changes with Pb-fluence and leaching may reflect the initial high yield stress (and relatively low elasticity) of the glass which must be overcome to initiate plastic flow and consequent increases in leaching rates. In consonance with this view, Dran, et al.[2] have observed in their experiments with a number of glasses that the more radiation resistant glasses exhibit more abrupt increases in leaching rates.

The ion implantation process itself causes movement of weakly bound modifier ions in the glass structure towards the surface. This is especially the case for alkali ions in alkali silicate glasses and has been noted in earlier work.[17,18] The driving force for the alkali movement when glasses are bombarded with energetic ions is not fully understood, but may involve electrostatic forces and/or stress fields generated in the material near the end of the ion track.

We should comment on the possible practical implications of our observations. Burns, et al.[4] have discussed the implantation results of Dran, et. al.[1] as noted in the Introduction. The difference in dose rates seems to be a reasonable reconciliation of the wide variations in the leach rates found for the two sets of experiments. As we have already indicated, a possible interpretation of our results could be that the stresses generated in the damaged region may play a significant role in producing enhanced reaction rates because of a structural transformation of the glass. If this hypothesis should be borne out by further experiments, the implication would be a confirmation of the recognized need for reasonable homogenity in the nuclear waste glass. Otherwise, the non-uniform α-emitting segments in near-surface regions might make possible the generation of stresses which would allow more rapid dissolution in ground water than is presently measured or anticipated. There is at present no indication from ongoing research on real waste-containing glasses or those doped with α-emitters that this effect occurs.

CONCLUSIONS

These experiments on near-surface compositional profiles and solution analyses of Pb-ion implanted SRP nuclear waste glass allow the following conclusions to be made:

a. Pb-ion simulation of α-recoil damage in SRP glass, followed by leaching in deionized water at 90°C, clearly shows that radiation damage affects near-surface compositional profiles. In particular, large changes were observed in H, Li, Na, Si, Fe/Ni and Nd profiles.

b. For single energy (207 keV) Pb-implants, a critical fluence, Φ_C, of $5 \times 10^{12} - 1 \times 10^{13}/cm^2$ was determined for the onset of compositional changes in general agreement with results obtained by Dran, et al[1] and Primak.[3]

c. The critical fluence, Φ_C, implies an energy deposition (nuclear processes) of $\sim 2 \times 10^{20}$ keV/cm^3. This value corresponds approximately to that for ion-track overlap and to the value observed in many ion-implantation experiments[15] on glass at which observed physical properties begin to change in a manner suggestive of plastic flow or a structural transformation. Such changes could conceivably occur in real waste-containing glasses where segregation of waste has occurred. However, no reports of such an effect in such glasses have yet been noted.

d. A clear conclusion is that a need exists for further Pb-ion simulation experiments to verify and corroborate these preliminary experiments. In particular, solution analyses are needed to confirm the loss of material by leaching implied by the ERD and RBS measurements at the critical fluence. Although the overall leach rates may not be significantly affected by radiation damage, the compositional changes are sufficiently large as to suggest that phase separation and subsequent crystallization might occur.

e. Our work indicates that even though the surface compositions are affected, the overall leach _rates_ are not significantly altered by Pb-ion simulation of α-recoil damage. Solution analyses were not made after the 10 hr leach of the single-energy implants which showed loss of material at ϕ_C. It is therefore not known how the leach rate varied. If only slight changes occurred as for the multiple-energy implants, this might indicate that material flaked off or was redeposited from solution on the container walls. The results indicate, however, that the performance of SRP glass in the long-term immobilization of SRP defense high-level waste will not be adversely affected.

REFERENCES

1. Dran, J. C., Maurette, M. and Petit J. C. (1980), Science _209_, 1518.
2. Dran, J. C., Langevin, Y. Maurette, M., Petit, J. C. and Vassent, B., _Proceedings of the 4th International Symposium on the Scientific Basis for Nuclear Waste Management_, Materials Research Society Symposium, 11/15-20/81, Boston, MA, to be published.
3. Primak, W. (1982) Nuclear Sci. and Eng. _80_, 689.
4. Burns, W. G., Hughes, A. E., Marples, J.A.C., Nelson, R. S., and Stoneham, A. M. (1982) Nature _295_, 130.
5. Bibler, N. E., _Proceedings of the 4th International Symposium on the Scientific Basis for Nuclear Waste Management_, Materials Research Society Symposium, 11/15-20/81, Boston, MA, to be published.
6. Northrup, C.J.M., Arnold, G. W. and Headley, T. J., _Proceedings of the 4th International Symposium on the Scientific Basis for Nuclear Waste Management_, Materials Research Society Symposium, 11/15-20/81 , Boston, MA, to be published.
7. Doyle, B. L. and Peercy, P. S. (1979) _The Analysis of Hydrogen in Solids_, edited by R. L. Schwoebel and J. L. Warren, (DOE/ER-0026), NTIS, Springfield, VA, p. 92.
8. Altenheim, F. K., Lutze, W. and Malow, G. (1981), _Scientific Basis for Nuclear Waste Management_, Vol. 3, edited by J. G. Moore (Plenum Press), p. 363.
9. Webb, A. P., Houghton, A. J. and Townsend, P. D. (1976) Rad. Eff. _30_, 177.
10. Hines, R. L. and Arndt, R. (1960) Phys. Rev. _119_, 623.
11. EerNisse, E. P. (1974) J. Appl. Phys. _45_, 167.
12. Jensen, T., Lawn, B. R., Dalglish, R. L. and Kelly, J. C. (1976), Rad. Eff. _28_, 245.
13. Chandler, P. J., Jaque, F. and Townsend, P. D. (1979), Rad. Eff. _42_, 45.
14. Arnold, G. W. (1977) _Ion Implantation in Semiconductors_, edited by F. Chernow, J. A. Borders and D. K. Brice (Plenum Press), p. 275.
15. Arnold, G. W., Rad. Eff., to be published.
16. Headley, T. J., Arnold, G. W. and Northrup, C.J.M. (these proceedings).
17. Arnold, G. W. and Borders, J. A. (1977) J. Appl. Phys. _48_, 1488.
18. Borders, J. A. and Arnold, G. W. (1975) _Proceedings Second International Conference on Ion Beam Surface Layer Analysis_, Karlsruhe, Germany, 9/15-19/75.

Published 1982 by Elsevier Science Publishing Co
SCIENTIFIC BASIS FOR RADIOACTIVE WASTE MANAGEMENT - V
Werner.Lutze, editor

A STUDY OF RADIATION EFFECTS IN CURIUM-DOPED $Gd_2Ti_2O_7$ (PYROCHLORE) AND $CaZrTi_2O_7$ (ZIRCONOLITE)*

J.W. WALD[+] AND P. OFFERMANN[++] [+]Pacific Northwest Laboratory, P.O. Box 999, Richland, Washington, 99352, USA. [++]Hahn-Meitner-Institut für Kernforschung Berlin, Glienicker Strasse 100, Postfach 39 01 28, D1000 Berlin 39, Germany.

INTRODUCTION

Radiation effects studies in both glass and glass ceramic nuclear waste forms have identified a rare-earth titanate phase of the general formula $(RE)_2Ti_2O_7$, which is capable of acting as a host phase for actinides.[1,2] Ringwood and co-workers[3] have also proposed a structurally similar phase, zirconolite $(CaZrTi_2O_7)$, as one of the primary host phases in the SYNROC waste form. Data from these and other previous studies, as well as mineralogical information available on these titanate phases, have not provided an unambiguous interpretation of the effects of radiation damage relative to nuclear waste forms. This paper reports new laboratory data concerning radiation damage effects in both of these phases.

It is known that both pyrochlores and zirconolites can become x-ray amorphous as a result of self damage from actinide element decay. Ewing and Haaker[4] reviewed investigations of naturally occurring metamict minerals and concluded that titanate phases with similar crystalline structures as those studied in this work can be found metamict but may also be partially or completely crystalline. Limited leaching studies have indicated increases in release rates for metamict versus non-metamict phases.[5] Microcracking has also been observed in waste glasses[2] and the zirconolite phase[6] during damage ingrowth as a result of anisotropic swelling. Microcracking could lead to a decrease in mechanical strength and an increase in surface area available to potential leach solution.

The gadolinium titanate phase studied in the present work has the cubic (Fd3m) pyrochlore structure, as described more fully elsewhere,[7-10] with a reported unit cell constant of a=10.181 Å.[11] The structure consists of titanium-oxygen octahedra which share corners to form a continuous network, with gadolinium and extra oxygen anions occupying holes formed by the octahedral network. The TiO_6 octahedra form sheets parallel to {110} sets of planes. The pyrochlore structure has been related by some to a modified fluorite structure.[12,13] Zirconolite is very similar in structure to pyrochlore, containing (001) sheets of TiO_6 octahedra,[14] and has also been related to a defect fluorite structure.[15-17] The zirconolite structure is monoclinic (C2/c), however, with reported cell constants of a=12.4458 Å, b=7.2734 Å, c=11.3924 Å, β=100.533°.[14]

---

* This work was supported by the U.S. Department of Energy under contract DE-AC06-76RLO 1830.

The experimental results presented in this paper describe alpha decay self-damage studies in ^{244}Cm-doped $Gd_2Ti_2O_7$ and $CaZrTi_2O_7$ single-phase, polycrystalline materials. Property behavior was monitored at ambient temperature by x-ray diffraction and apparent density measurements during the initial damage ingrowth period (approximately 2-year time span). Microstructural examination was conducted by optical metallographic techniques after an x-ray amorphous state was reached. Results are compared with those of others on the ingrowth of damage effects in natural and synthetic zirconolites.

EXPERIMENTAL TECHNIQUE

Starting materials for the crystalline phases consisted of analytical reagent-grade gadolinium and zirconyl nitrates, calcium carbonate, and isopropyl titanate. Stoichiometric mixtures were wet milled in ethyl alcohol in an agate laboratory disc mill, air dried, and sieved to -100 mesh. These blended mixtures served as the precursor materials for the final actinide-doped crystalline phases.

The actinide species used in the doping work was ^{244}Cm with a half-life of 18.8 years. Curium was chosen because it was expected to easily substitute for gadolinium (3+) or as $2Cm^{3+}$ in place of (Ca^{2+} + Zr^{4+}) under the fabrication conditions used. Curium also provides damage levels in the range of 10^{18} to 10^{19} α-decays/cm^3 in reasonable time periods ($\sim$1.5 years) at low doping levels. Each crystalline phase was doped with 3 wt% ^{244}Cm in an inert atmosphere glovebox. Curium oxide powder was dissolved in hot nitric acid and then added to the precursor materials in the appropriate quantities. This slurry was dried on a hot plate and then processed to obtain the desired crystalline phases.

Preparation of the crystalline phase from the doped precursor typically involved a two-step firing operation, the first step being a bulk calcining operation at $\sim$725°C in air followed by cooling to room temperature and mechanical grinding and sieving to -200 mesh (74 μm). The second step involved cold pressing pellets of 1.3 cm diameter and firing in air at the desired conditions of time and temperature. The gadolinium titanate was pressed at 137.9 MPa (20,000 psi) and fired at 1400°C for 46 h, while the calcium-zirconium-titanate was pressed at 275.8 MPa (40,000 psi) and fired at 1325°C for 40 h.

X-ray diffraction (XRD) measurements were made on the samples starting at time "zero" (just after firing) and were taken at regular intervals until an x-ray amorphous state was reached. Two samples were required for this portion of the analysis so that surface contact could be made between the pellets during interim storage, thereby ensuring that the surface analyzed by XRD was typical of the bulk. A G.E. XRD-5/DIANO 8000 Diffractometer was used to slow-scan selected peaks of the crystalline phases and the externally fixed silicon standard. Solid pellets with a flat ground surface were temporarily fixed in an aluminum mount by the use of set screws and spacer pins to provide an exact alignment in the diffractometer for each analysis. The intent of the externally mounted silicon standard was to provide a reference material that would not become damaged by radiation during the course of the study. The NBS silicon (SRM-640) was mixed with a casting resin to form a thick paste which was filled and set in a cavity adjacent to the pellet sample.

Apparent density measurements were made by conventional immersion techniques, using water as the immersion fluid (similar to ASTM procedure C-693-74), each time XRD measurements were taken. An Al_2O_3 ceramic disc, prepared to approximate the geometric size of the test samples, was used as the measurement standard. Mass readings were made to ± 0.0001 g for the 2 g samples, giving a sensitivity of about $\pm 10^{-4}$ g/cm^3 and overall accuracy of about $\pm 10^{-3}$ g/cm^3. Pellet porosity was evaluated both by quantitative image analysis of microstructural cross sections and by water saturation during the density measurement. The greater fraction of measured porosity was found to be open porosity in both cases, with total porosity at 19% and 14% for $Gd_2Ti_2O_7$ and $CaZrTi_2O_7$, respectively.

RESULTS AND DISCUSSION

Figure 1 illustrates the volumetric swelling determined by x-ray diffraction and apparent density for both $Gd_2Ti_2O_7$ and $CaZrTi_2O_7$ as a function of cumulative dose. Bulk (macroscopic) swelling, determined by density measurement, was found to be similar to the calculated x-ray volumetric swelling at all doses for the $Gd_2Ti_2O_7$. The zirconolite phase ($CaZrTi_2O_7$), however, demonstrated a lower x-ray volumetric swelling than macroscopic swelling except at low accumulated doses. At a dose of 10^{19} α-decays/cm^3 in the zirconolite system, lattice swelling was about 2%, while macroscopic swelling was about 3%. This result is consistent with that reported by Clinard[5] in a similar evaluation. Both XRD and density data, in the present study, indicate that the cubic $Gd_2Ti_2O_7$ phase underwent the largest expansion in unit cell volume at $\sim 4.2\%$ compared with $\sim 2.1\%$ for $CaZrTi_2O_7$ at the last measurable dose levels.

The reason for the close agreement between macroscopic and lattice swelling in one structure and the large difference in the other is not entirely clear, but it may reflect the degree of disorder obtained in each structure as it goes toward the x-ray amorphous state. The degree of disorder may be small in the

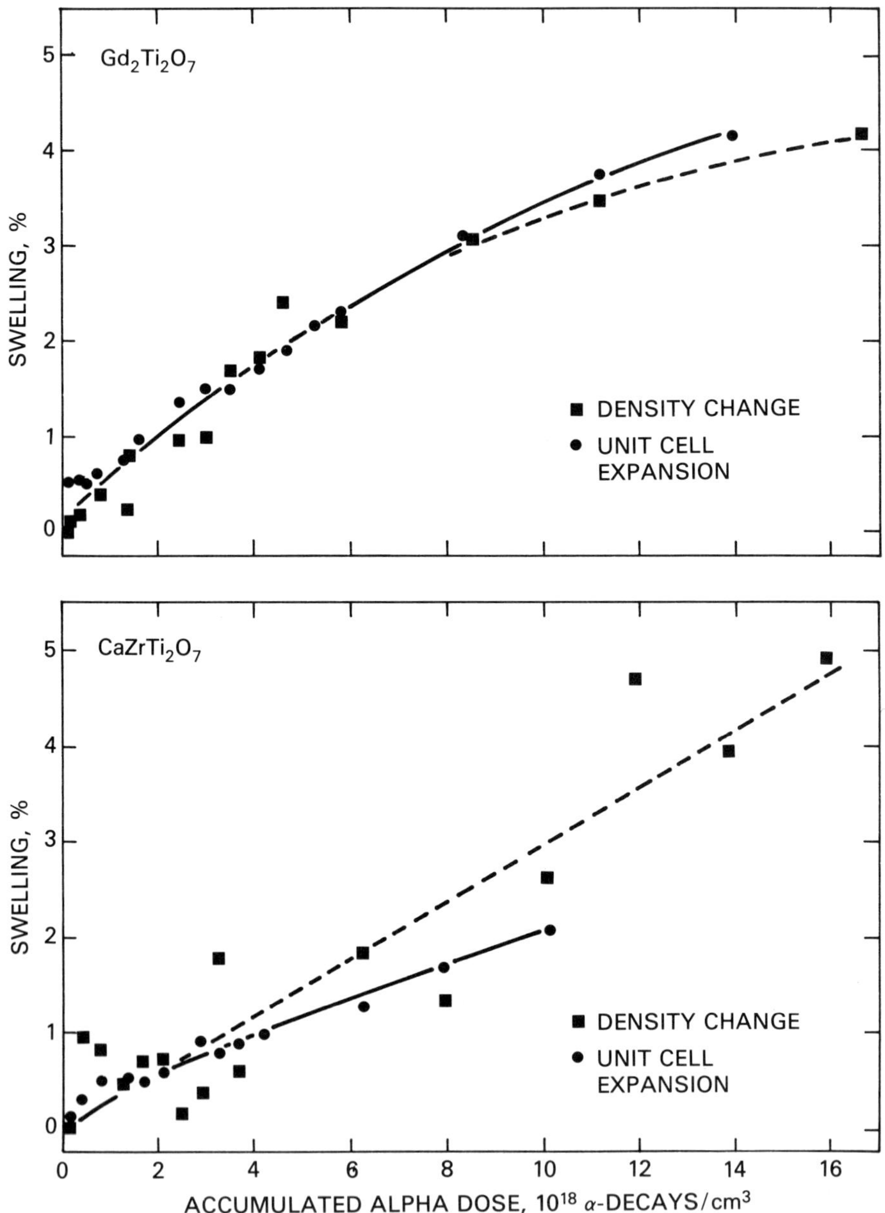

Fig. 1. Volumetric swelling for $Gd_2Ti_2O_7$ and $CaZrTi_2O_7$ measured by apparent density (macroscopic) and x-ray diffraction (microscopic).

$Gd_2Ti_2O_7$ phase, whereas radiation-induced structural disorder in the $CaZrTi_2O_7$ phase would likely be high.

Figure 2 illustrates the integrated intensity ratio changes of some of the main XRD peaks, referenced to the silicon standard. Both phases show general decreases in diffracted intensities at about the same rate. In the $Gd_2Ti_2O_7$ phase, no clear anisotropic decreases are evident, at least from comparing the (400) and the (440) XRD lines. The $CaZrTi_2O_7$ phase, on the other hand, exhibits rather marked anisotropic changes, evidenced partly in this figure.

Figure 3 more clearly shows the anisotropic behavior in the zirconolite phase as it undergoes the transformation to the x-ray amorphous state. Radiation-induced swelling occurs primarily in the c-axis direction with a minor change in the a and b axis direction. The monoclinic cell angle β (data not shown) did not vary measurably. The initial lattice constants were a=12.456 Å, b=7.271 Å, c=11.387 Å, and β=100.556°. At a dose of 10^{19} α-decays/cm^3, the "a" parameter expanded by about 0.3%; the "b" parameter did not change significantly; and the "c" parameter increased by about 1.6%. In spite of this relatively large aniso-tropic expansion ratio ($\Delta c/\Delta a \simeq 5$), no microcracking was observed. Figure 4 shows microstructural cross sections of both phases at a cumulative dose of $\sim 1.5 \times 10^{19}$ α-decays/cm^3. No microcracking was evident in either material, and both samples maintained their physical integrity during the course of the study.

Ringwood et al.[18] reported that in certain natural mineral zirconolites, with cumulative doses of $\sim 3.2 \times 10^{19}$ α-decays/cm^3, the zirconolite superstructure reflections completely disappeared, while the fluorite sublattice remained essen-tially intact. The implication in this finding was that high cumulative radia-tion doses caused only a degree of disorder in the structure, leaving a crystal-line substructure unaffected. In this study, though, the zirconolite phase appeared to retain its monoclinic symmetry as it became x-ray amorphous, at least as long as it was measurable. X-ray diffraction lines associated with both the zirconolite superstructure and the fluorite sublattice were observed to shift position and decrease in intensity at approximately equal rates. At the completion of the experiment, no x-ray diffraction lines were observed.

Changes in unit cell volumes for both phases were observed to follow an exponential relationship of the form $\Delta V/V_o = A[1-\exp(-BD)]$. The parameters A and B are constants; A is the saturation property change at infinite dose (D), and B is the rate of change. Table 1 lists the values for the constants A and B as determined by regression analyses of the XRD data from this work, along with the values for related crystalline phases studied in other work. The saturation dose values listed in the table were estimated from the previous equation by

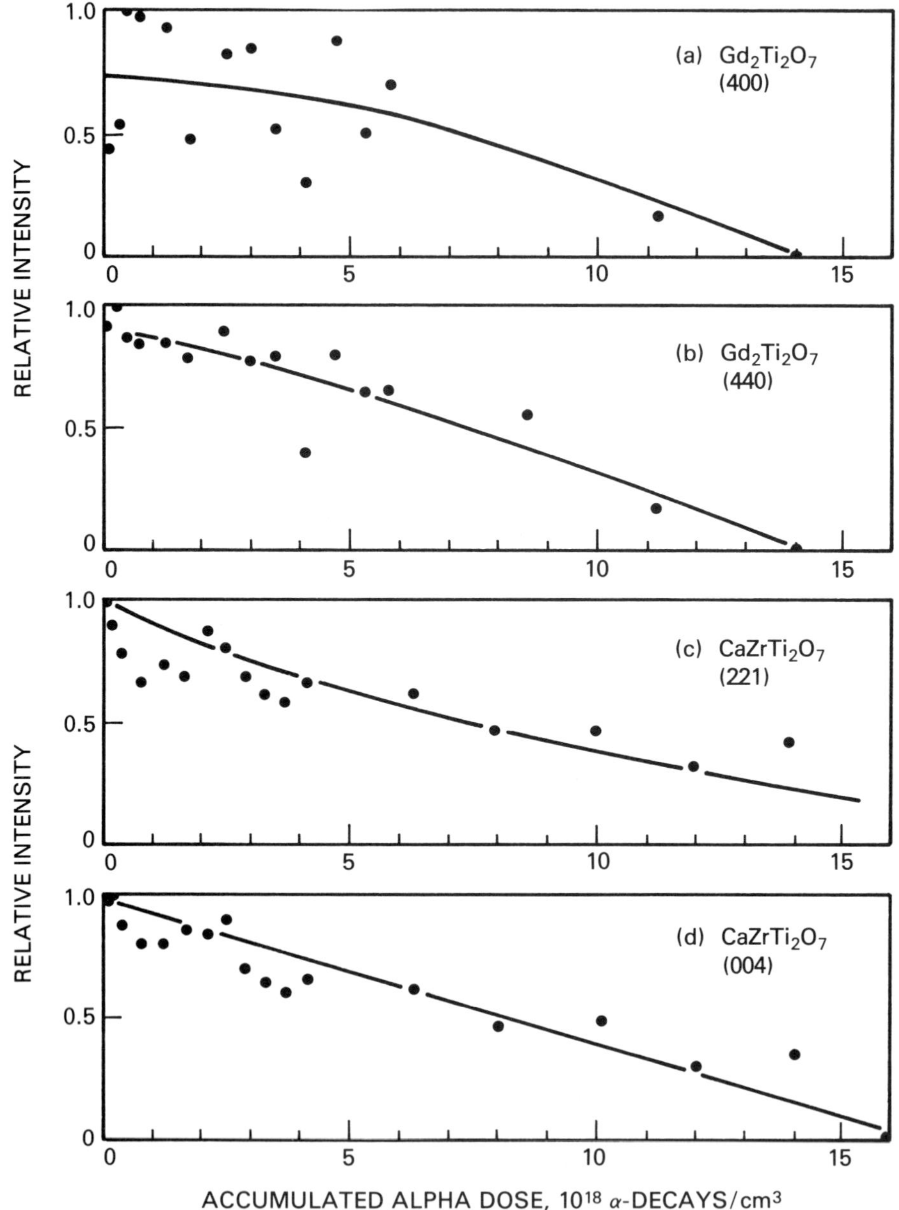

Fig. 2. Percent change in x-ray diffraction intensity of selected peaks for $Gd_2Ti_2O_7$ and $CaZrTi_2O_7$ referenced to the silicon standard

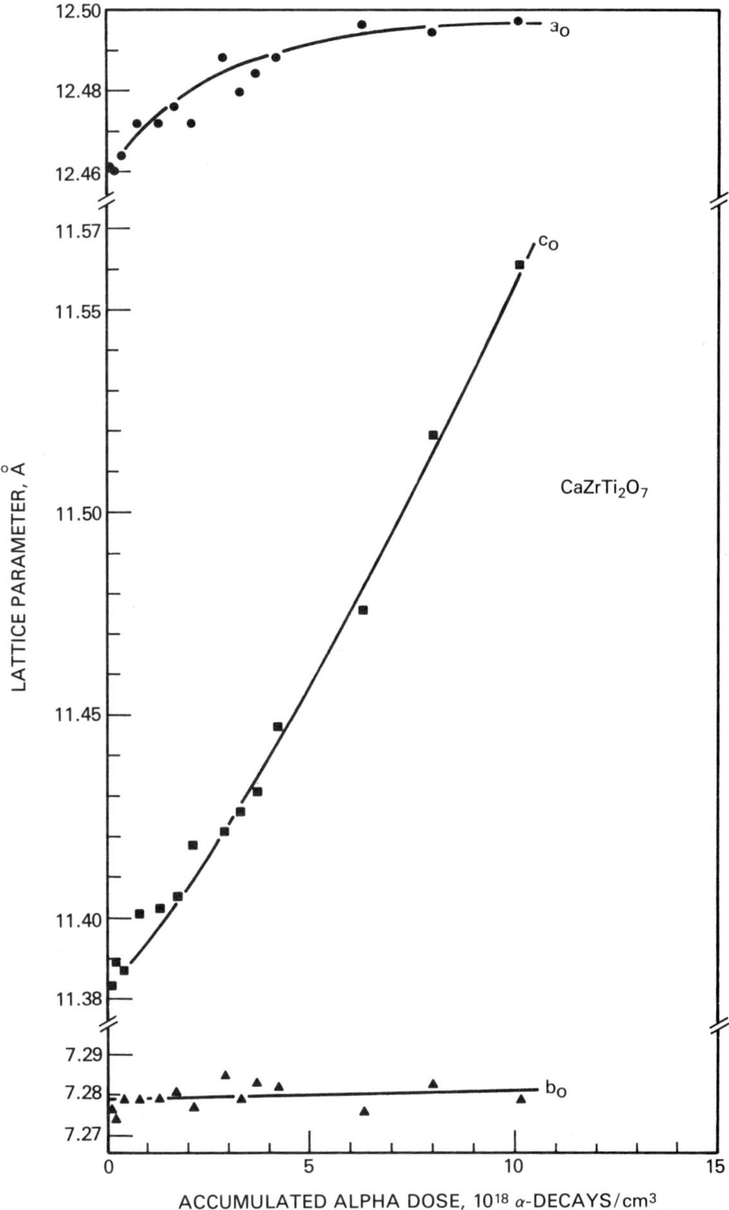

Fig. 3. Unit cell constant changes for monoclinic $CaZrTi_2O_7$ as a function of accumulated alpha dose.

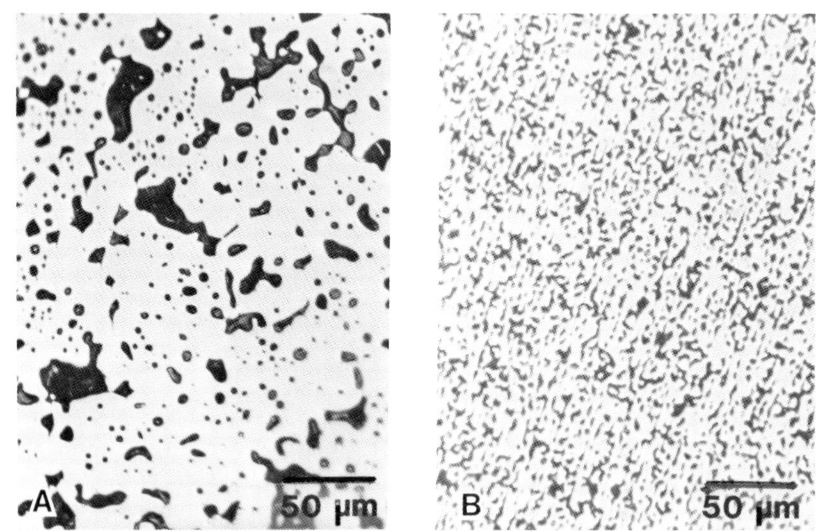

Fig. 4. Photomicrographs of (a) $Gd_2Ti_2O_7$ and (b) $CaZrTi_2O_7$ at a total accumulated alpha dose of $\sim 1.5 \times 10^{19}$ α-decays/cm^3.

TABLE 1

RADIATION DAMAGE INGROWTH PARAMETERS FROM X-RAY DIFFRACTION MEASUREMENTS

Phase	A[a]	B[a]	DOSE		Reference
			X-Ray Amorphous	Saturation (99%)	
$Gd_2Ti_2O_7$	0.054	1.01×10^{-19} cm^3	$\sim 1.7 \times 10^{19}$ $\frac{\alpha\text{-decays}}{\text{cm}^3}$	$\sim 4.6 \times 10^{19}$ $\frac{\alpha\text{-decays}}{\text{cm}^3}$	This work
$CaZrTi_2O_7$	0.076	2.81×10^{-20}	$\sim 1.6 \times 10^{19}$	$\sim 1.6 \times 10^{20}$	This work
$CmAlO_3$	0.096	7.88×10^{-20}	$\sim 1.4 \times 10^{19}$	$\sim 5.8 \times 10^{19}$	19
$ZrSiO_4$	0.0502	$\sim 3.2 \times 10^{-19}$ (b)	$\sim 2.4 \times 10^{19}$	$\sim 2.1 \times 10^{19}$	20
PuO_2	0.0096	4.27×10^{-19}	---	$\sim 1.1 \times 10^{19}$	21

(a) A & B are constants of the equation: $\Delta V/V_o = A[1-\exp(-BD)]$
(b) Calculated from values at saturation only

setting $\Delta V/V_o$ equal to 99% of its value at infinite dose. Many crystalline structures have demonstrated a similar exponential approach to radiation damage saturation even if they remained fully crystalline. For both the $Gd_2Ti_2O_7$ and $CaZrTi_2O_7$ phases studied here, expansion of the unit cell was not completely saturated before the onset of x-ray amorphization. This result is similar to that reported for the $CmAlO_3$ system (perovskite structure).[19] To give an additional example, saturation of lattice effects in the zircon structure occurs prior to actual x-ray amorphization, while in actinide dioxides such as PuO_2 with a fluorite structure saturation occurs with no tendency toward x-ray amorphization being reported. The compounds listed in Table 1 represent a range of generic structure types [pyrochlore, distorted pyrochlore (monoclinic), perovskite, zircon, and fluorite]. The data for these phases indicate that while the approach to saturation of damage ingrowth and the absolute magnitudes of the changes may vary measurably, the dose values required to reach these saturation values (assumed at 99%) are in the range of 10^{19} to 10^{20} α-decays/cm^3.

CONCLUSIONS

The present work shows that radiation damage ingrowth in $Gd_2Ti_2O_7$ and $CaZrTi_2O_7$ follows an expected exponential relationship of the form $\Delta V/V_o = A[1-\exp(-BD)]$ based on XRD lattice swelling. Macroscopic swelling was identical to lattice swelling for $Gd_2Ti_2O_7$ ($\sim$5.4%), while in $CaZrTi_2O_7$, macroscopic swelling was about 1.5 times greater than lattice swelling at the last measurable levels. Estimated values of unit cell volume and density swelling at saturation for $CaZrTi_2O_7$ are 7.6% and 10.3%, respectively. The $CaZrTi_2O_7$ phase exhibits anisotropic behavior in which the expansion of the monoclinic cell in the c-direction is over five times that of the a-direction. Despite this relatively large anisotropic expansion ratio, no microcracking was observed. Microcracking was not observed in $Gd_2Ti_2O_7$ either, and both materials maintained their physical integrity during the course of the study.

REFERENCES

1. Turcotte, R.P., et al. (1982) Am. Cer. Soc., PNL-SA-9869.
2. Weber, W.J., et al. (1979) Proc. Int. Symp. ACS, CONF-790420, 294-299.
3. Oversby, V.M. and Ringwood, A.E. (1981) Rad. Waste Mgmt. 1(3), 289.
4. Ewing, R.C. and Haaker, R.F. (1980) Nuc. Chem. Waste. Mgmt., 1, 51.
5. Ewing, R.C. (1981) in Proc. Conf. Alternate Nuclear Waste Forms and Interactions in Geologic Media, L.A. Boatner and G.C. Battle eds., CONF-8005107, 81.
6. Clinard, F. et al. (1981) Presented at Fourth MRS Symposium on Scientific Basis for Nuclear Waste Management, Boston, MA, 11/15-20/81 (published in Proceedings).
7. Aleshin, E. and Roy, R. (1962) J. Am. Cer. Soc., 45(1), 18.
8. Hogarth, D.D. (1977) Am. Min. 62, 403.
9. McCauley, R.A. (1980) J. Appl. Phys. 51(1), 290.
10. Nyman, H. et al. (1978) J. Sol. St. Chem., 26, 123.
11. Waring, J.L. and Schneider, J.L. (1965) J. Res. NBS, 69A,(3).
12. Pyantenko, Yu. A. (1960) Sov. Phys.: Cryst. 4(2), 184.

13. Barker, W.W., et al. (1970) in The Chemistry of Extended Defects in Non-Metallic Solids, L. Eyring and M. O'Keefe, ed. North-Holland Pub. Co. Amsterdam, 198.
14. Rossell, H.J. (1980) Nature, 283, 282.
15. Gatehouse, B.M., et al. (1981) Acta. Cryst. B37, 306.
16. Pyantenko, Yu. A. and Pudovkina, Z.V. (1964) Sov. Phys.: Cryst., 9(1), 76.
17. Pudovkina, Z.V. and Pyantenko, Yu. A. (1966) Akad. Neuk SSSR, Min. Muzei, Trudy, 17, 124.
18. Ringwood, A.E., Oversby, A.E. and Sinclair, W. (1980) Scientific Basis for Nuclear Waste Management, 2, Plenum Press, N.Y. 273-280.
19. Mosley, W.C. (1971) J. Am. Cer. Soc., 54(10), 475.
20. Holland, H.D. and Gottfried, D. (1955) Acta. Cryst. 8(6), 291.
21. Chikalla, T.D. and Turcotte, R.P. (1973) Rad. Effects, 19, 93.

DOSE-DEPENDENCE OF Pb-ION IMPLANTATION DAMAGE IN ZIRCONOLITE, HOLLANDITE, AND ZIRCON*

T. J. HEADLEY, G. W. ARNOLD, and C. J. M. NORTHRUP
Sandia National Laboratories, Albuquerque, New Mexico 87185, USA

INTRODUCTION

The long-term stability of nuclear waste forms is an important consideration in their selection for safe disposal of radioactive waste. Stability against long-term radiation damage is particularly difficult to assess by short-term laboratory experiments. Much of the displacement damage in high-level waste forms will be generated by heavy recoil nuclei emitted during the α-decay process of long-lived actinide elements. Hence, an accelerated aging test which reliably simulates the α-recoil damage accumulated during thousands of years of storage is desirable. One recent approach to this simulation is to implant the waste form with heavy Pb-ions.[1-6] If the validity of this approach is to be fully assessed, two important questions which have not yet been investigated must be answered. (1) Is the structural damage, including cumulative effects, similar for irradiation by Pb-ions and α-recoil nuclei in a given material? (2) Is the dose-dependence of the accumulated damage similar? The purpose of this investigation was to assess the extent of these similarities in selected materials. We utilized transmission electron microscopy (TEM) to characterize the radiation damage and measure its dose-dependence.

Synthetic crystalline zirconolite and (Ba,Ti)-hollandite, and natural crystalline zircon were implanted with Pb-ions over a wide range of equivalent α-doses and then characterized by TEM. Zirconolite and (Ba,Ti)-hollandite are component phases in titanate-based ceramic waste forms such as Sandia titanate[7,8] and SYNROC,[9] while zircon is a mineral whose radiation damage and metamictization from natural α-decay have been studied extensively. For comparison, natural samples of alpha-damaged zirconolite and zircon, which had received known α-doses, were also characterized by TEM. Natural samples of (Ba,Ti)-hollandite were not available for comparison.

---

*This work performed at Sandia National Laboratories supported by the U.S. Department of Energy under contract DE-AC04-76DP00789.

EXPERIMENTAL

Pb Implantations. The materials implanted were synthetic zirconolite ($CaZrTi_2O_7$), synthetic (Ba,Ti)-hollandite ($BaAl_2Ti_6O_{16}$) containing 4.7 wt. % Cs_2O, and highly crystalline zircon ($ZrSiO_4$) from Mud Tank, Australia. The materials were crushed and thin flakes were collected on holey carbon sub-strates on TEM support grids. These specimens were then implanted with Pb-ions in an accelerator using multiple energy (40-240 keV) implants to achieve uniform energy deposition with depth.[4] Maximum damage depth was ~ 600 Å so that only the thinner edges of flakes were damaged through their full thick-ness. The energy deposited in displacement processes for Pb-ion doses was converted to equivalent α-doses using the computer codes of Brice.[10] The samples were implanted to equivalent dose levels varying from 10^{17} to 5×10^{20} α/cm^3.

Naturally Damaged Zircons and Zirconolites. Four natural zircons were examined. Their respective total α-doses, calculated from U/Pb concentration dating, were 1.5×10^{18}, 8.9×10^{18}, 3.0×10^{19}, and 4.4×10^{19} α/cm^3. A more complete description of these zircons together with data on their leachability as a function of total α-dose is given in reference 20. Two zirconolites from Kaiserstuhl, Germany and Jacupiranga, Brazil were examined. Their total α-doses were 4.4×10^{18} and 4.9×10^{19} α/cm^3, respectively, calculated from their total U + Th concentration measured by electron microprobe analysis and the known age of their deposits.

Electron Microscopy. The samples were examined in a JEOL-200CX electron microscope using reduced beam intensities to avoid alteration of the damaged structures in the electron beam. The combined techniques of bright field imaging, high resolution dark field imaging, electron diffraction, and high resolution lattice imaging were used.

RESULTS

Pb Implantations. The evolution of structural damage with increasing dose was similar in all three implanted materials. For descriptive purposes, this evolution can be conveniently divided into a sequence of three stages with respect to effects observed in TEM images and/or electron diffraction patterns. This does not necessarily correspond to discrete changes in physical proper-ties, however. Figure 1a-c shows representative micrographs from each stage. In Stage I the damage appears as individual defect clusters ~25-40 Å in dia-meter and interpreted as cross-sections of individual Pb-ion tracks. Their num-ber density increases with increasing dose within this stage. No effects from implantation were observed in the electron diffraction patterns in Stage I. Stage II begins with the onset of significant overlap of the strain contrast images such that individual clusters can no longer be readily distinguished.

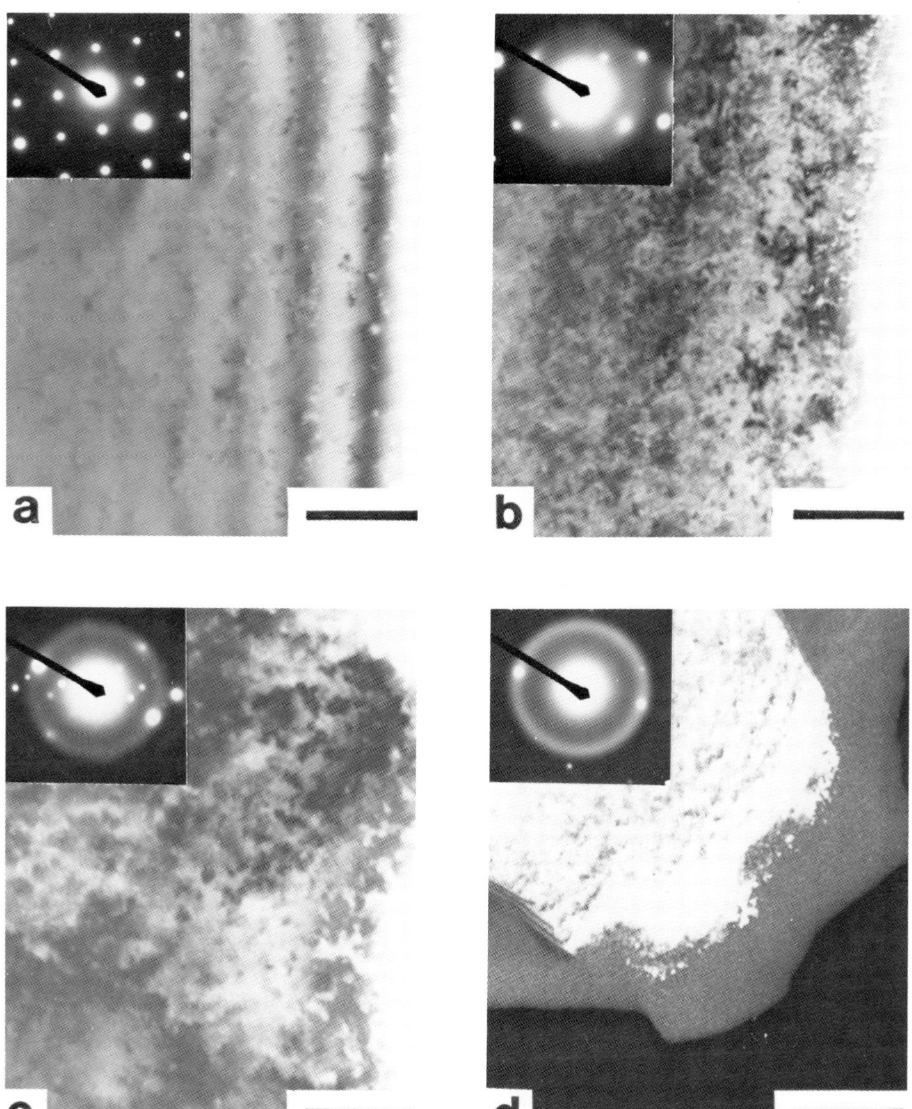

Figure 1. Pb-ion damage in (a) zircon, Stage I damage; 1×10^{18} α/cm^3. (b) zircon, Stage II damage; 5×10^{18} α/cm^3. (c) zirconolite, Stage III damage; 1×10^{19} α/cm^3. (d) zirconolite, metamict rim, dark field image; 1×10^{20} α/cm^3. Flakes oriented for strong diffraction. Bar = 50 Å in (a)-(c) and 0.2 μm in (d).

This is accompanied by the appearance in electron diffraction patterns of a broad halo of scattered intensity normally associated with the presence of amorphous material. The experimental fluence required for the onset of Stage II was consistent with expected values based on the overlap of tracks of $\sim$25-50 Å size. Stage III begins when substantial volumes of amorphous material can be detected by TEM imaging techniques,e.g.,by lattice imaging where the absence of lattice fringes delineated amorphous regions. This stage continues with the progressive destruction of crystalline remnants. At doses beyond Stage III, the materials are rendered TEM amorphous (metamict) as illustrated in Figure 1d. In this micrograph there is an amorphous rim surrounding the thin flake corresponding to the extent of the $\sim$ 600 Å implantation thickness. Beyond this rim, diffraction contours appear in the thicker, unimplanted part of the flake. The crystalline reflections in the diffraction pattern of Figure 1d arise from this thicker, unimplanted region and the dark field image was taken from one of these reflections. No crystalline remnants could be detected in the amorphous rims by high resolution techniques (as have been previously described for examining the amorphous nature of metamict minerals[11]).

Table I gives the dose-dependence of Stages I, II, and III for these three materials.

TABLE I. Dose-Dependence of Pb Implantation Damage

	Dose Range (α/cm^3)		
	Zirconolite	Hollandite	Zircon
Stage I	5×10^{17}–5×10^{18}	1×10^{17}–>1×10^{18}	5×10^{17}–>1×10^{18}
Stage II	>5×10^{18}–<1×10^{19}	5×10^{18}–<1×10^{19}	5×10^{18}–<1×10^{19}
Stage III	1×10^{19}–>2×10^{19}	1×10^{19}– 2×10^{19}	1×10^{19}–>2×10^{19}
TEM Amorphous	$\geq 5 \times 10^{19}$	>2×10^{19}	$\geq 5 \times 10^{19}$

In addition to lattice damage, the (Ba,Ti)-hollandite developed a modulated structure during Stage I. This observation is not important to the results of the present investigation, but we note that this effect is similar to that observed in bulk TEM specimens of (Ba,Ti)-hollandite thinned by ion bombardment with $\sim$ 7 keV Ar$^+$ ions.[12]

<u>Naturally Damaged Zircons and Zirconolites</u>. Alpha-radiation damage in the natural zircons and zirconolites was found to be similar in morphology and sequence to that from Pb implantation, i.e., isolated defects at lower doses progressing to overlapping images of defects and the appearance of diffraction halos and clearly amorphous material at higher doses. Representative

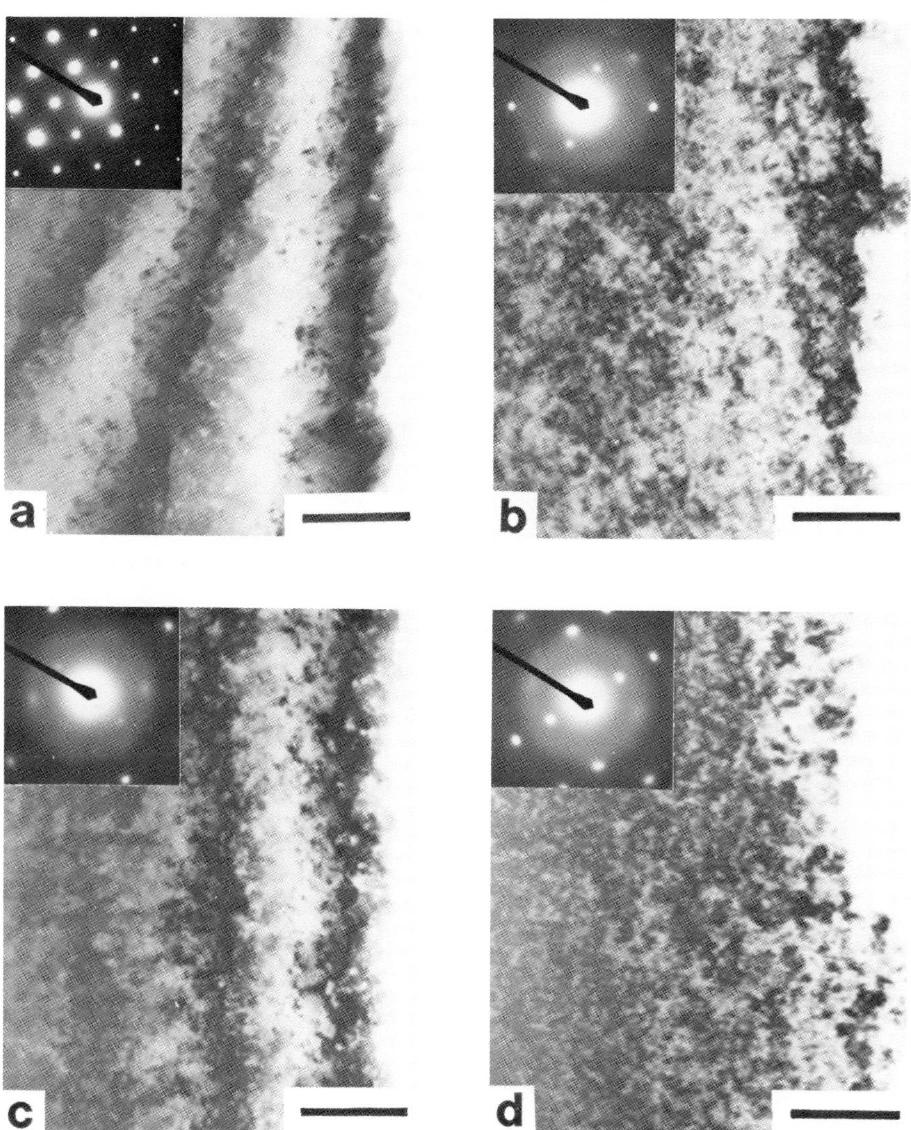

Figure 2. α-damage in natural zircons. (a) #RSD-9, Stage I damage; 1.5x10^{18} α/cm^3. (b) #W-77, Stage II damage; 8.9x10^{18} α/cm^3. (c) #DB-10, Stage II damage; 3.0x10^{19} α/cm^3. (d) #M-83-NM, Stage III damage; 4.4x10^{19} α/cm^3. Flakes oriented for strong diffraction. Bar = 500 Å.

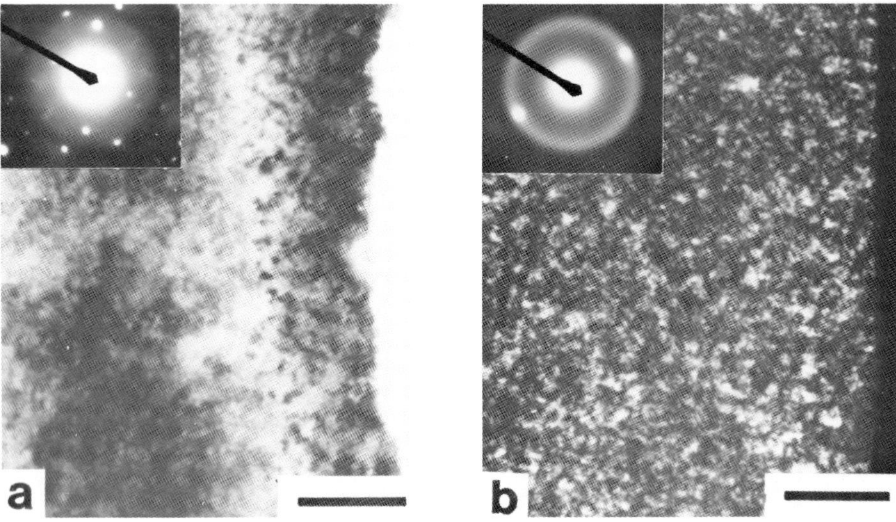

Figure 3. α-damage in natural zirconolites. (a) Kaiserstuhl sample, Stage II damage; 4.4×10^{18} α/cm^3. (b) Jacupiranga sample, advanced Stage III damage, dark field image, bright patches are crystalline remnants; 4.9×10^{19} α/cm^3. Flakes oriented for strong diffraction. Bar = 500 Å.

TABLE II. Extent of α-Damage in Natural Zircons and Zirconolites

Zircons	Observed Damage	Calculated*α-dose(α/cm^3)
RSD-9	near end of Stage I	1.5×10^{18}
W-77	near end of Stage II	8.9×10^{18}
DB-10	Stage II	3.0×10^{19}
M-83-NM	well into Stage III	4.4×10^{19}
Zirconolites		
Kaiserstuhl	beginning of Stage II	4.4×10^{18}
Jacupiranga	near end of Stage III	4.9×10^{19}

* α/gm were converted to α/cm^3 by multiplying by a constant density of 4.6 gm/cm^3.

micrographs are given in Figure 2 for the zircons and Figure 3 for the zirconolites, and these should be compared with Fig. 1. The structural damage in
these samples is categorized in Table II by the same 3-stage criteria for
damage used in Table I.

By comparing the calculated doses in Table II with the dose levels in
Table I, it is seen that there is remarkably good agreement between the dose-
dependence of α-recoil damage and Pb-ion damage with one exception. Zircon
#DB-10 with a known dose of 3.0×10^{19} α/cm^3 would fall at or beyond the end
of Stage III in Table I whereas TEM showed it to be damaged only into Stage II.
This discrepancy may be related to an inhomogeneous sampling for this zircon.

DISCUSSION

This investigation has shown that structural damage from Pb-ion implantation is similar in morphology and dose-dependence to that from α-damage for
the case of zircon and zirconolite. It is also instructive to compare the dose
required for metamictization by Pb implantation reported here with α-doses
received by metamict zircons and zirconolites reported in the literature. For
zirconolite, we find Pb implantation to a calculated equivalent dose of
$\sim 5 \times 10^{19}$ α/cm^3 to be sufficient for metamictization. Oversby and Ringwood[13]
report that Sri Lanka zirconolites having received total α-doses $> \sim 3.8 \times 10^{20}$
α/cm^3 are completely metamict. Clinard et al.[14] report that a Pu analog of
zirconolite, $CaPuTi_2O_7$, is x-ray metamict after a dose of 1.3×10^{19} α/cm^3.
These findings are in agreement within a factor of 8 with the calculated
dose required for metamictization by Pb-ions. Similarly, for zircon we find
Pb implantation to a calculated equivalent dose of $\sim 5 \times 10^{19}$ α/cm^3 to be
sufficient for complete metamictization. This in in good agreement with
reports[15,16] of natural zircons which are x-ray metamict after α-doses of
$\sim 4.8 \times 10^{19}$ α/cm^3 and $> \sim 4 \times 10^{19}$ α/cm^3, but zircons with doses as low as
1.1×10^{19} α/cm^3 have also been reported as metamict[17]. This is still within a factor of ~ 5 of the required equivalent Pb-dose.

The critical fluence required for metamictization by ion implantation in
these materials can be calculated and compared with the experimentally measured
dose. If metamictization is related to the beginning of overlap of individual
ion tracks, then a simple estimate can be made of the critical fluence following Morehead and Crowder[18]. Assuming a typical average displacement threshold
energy of 25 eV, and assuming that metamictization occurs when all atoms
within the core of the ion track are displaced, then

$$\phi_c = \overline{E}N_t(dE/dx)^{-1}cm^{-2}$$

where ϕ_c = critical fluence in ions/cm^2, $\overline{E}$ = average displacement energy = 25 eV, N_t = total number of target atoms $\simeq 9.2 \times 10^{22}$/cm^3 for these materials, and (dE/dx) = nuclear energy loss per unit path length. An estimate[18] for (dE/dx) is given by $7 \times 10^8 \rho z_1^{2/3} M_1/(M_1+M_2)$ eV/cm where ρ = density in g/cm^3, Z_1 = atomic number of implanted ions, and M_1 and M_2 are ion and target masses, respectively. M_2 in this case is the effective mass $\simeq 55$ for these materials. Then $(dE/dx) \simeq 5 \times 10^{10}$ eV/cm. As a check on this value, the tables of Gibbons et al.[19] can be used to give an estimated value of $\sim 4 \times 10^{10}$ eV/cm.

$$\text{Then } \phi_c \simeq 4.6 \times 10^{13} \text{ Pb ions/cm}^2.$$

The corresponding ion-track radius for complete overlap using this critical fluence is $R_o = 1/\sqrt{\phi_c}$

$$\text{and } R_o \simeq 1.5 \times 10^{-7} \text{ cm} = 15 \text{ Å (critical radius).}$$

The value of 30 Å for a critical diameter is in good agreement with the diameter($\sim$ 25-40 Å) of isolated defect clusters imaged in the TEM. The critical fluence of 4.6×10^{13} Pb ions/cm^2 is to be compared with the dose range for Stage II where significant overlap was observed in TEM images. The limits of Stage II were $>5 \times 10^{18}$ α/cm^3 to $<1 \times 10^{19}$ α/cm^3 which corresponds to $>2 \times 10^{13}$ Pb/cm^2 to $<3.5 \times 10^{13}$ Pb/cm^2, where the total Pb implanted at all implant energies is computed. The critical fluence of 4.6×10^{13} Pb/cm^2 is slightly greater than the upper limit for Stage II, but the agreement is quite good considering the assumptions made. On the other hand the agreement is better for Stage III where the limits correspond to

$$3.5 \times 10^{13} \text{Pb/cm}^2 < \phi_c = 4.6 \times 10^{13} \text{Pb/cm}^2 < 7 \times 10^{13} \text{Pb/cm}^2.$$

However the possible error involved in computing ϕ_c is large enough to allow either stage to be reasonable. The difference can also be related in part to the difficulty in establishing a dose for significant overlap from strain contrast images in the TEM.

A further consideration should be given to possible annealing effects during waste form storage. The comparison described here was for Pb implantation damage at room temperature and α-damage in natural samples which is presumed to have accumulated at ambient earth temperatures. We do not expect the accumulated α-damage in waste forms to anneal out during storage. The primary thermal cycle results from γ-radiation and should subside to < 100°C after several hundred years. However the α-damage becomes significant only after several thousand years storage. Hence the use of room temperature Pb implantation to simulate accumulated α-damage should be valid. Finally, we note that the similarity reported here between Pb-ion and α-recoil damage is confined to structural effects. Any chemical changes that may be associated with Pb-ion implantation[4] are not considered. Also for the fluences used in this study, the total amount of Pb deposited in these samples is insignificant.

CONCLUSIONS

The morphology of Pb-ion implantation damage in zircon, zirconolite, and (Ba,Ti)-hollandite, as observed in TEM images, can be described as occuring in three stages with increasing dose. Doses beyond the third stage render these materials TEM amorphous (metamict). The dose-dependence of the three stages in these materials has been determined. TEM of α-damaged zircons and zirconolites indicates that the damage sequence and dose-dependence is similar to those for Pb implantation. Thus it is concluded that Pb-ion implantation is a valid technique for simulating the structural effects of α-recoil damage in these materials. This provides reasonable confidence for using the technique for accelerated radiation-aging of waste forms for further post-irradiation experiments.

ACKNOWLEDGMENTS

R. G. Dosch of Sandia National Laboratories (SNL) supplied the synthetic zirconolite. The (Ba,Ti)-hollandite was provided by John Tewhey of Lawrence Livermore National Laboratories (LLNL). The undamaged natural zircon was supplied by R. C. Ewing of the University of New Mexico (UNM). The Kaiser-stuhl and Jacupiranga zirconolites were provided by V. Oversby of LLNL, and P. Hlava of SNL performed the electron microprobe analyses on these samples. The naturally damaged zircons and their total α-doses were from R. Zartman of the U. S. Geological Survey and were received through R. C. Ewing and R. F. Haaker of UNM. D. Wroble of SNL performed the Pb-ion implantations.

388

REFERENCES

1. J. C. Dran, Y. Langevin, M. Maurette, and J. C. Petit, _Scientific Basis for Nuclear Waste Management_, Vol. 2, C. J. M. Northrup, Ed., Plenum Press, New York (1980), p. 135.

2. J. C. Dran and J. C. Petit, _Science_, Vol. 209, (1980), p. 1518.

3. J. C. Dran, M. Maurette, J. C. Petit, and B. Vassent, _Scientific Basis for Nuclear Waste Management_, Vol. 3, J. G. Moore, Ed., Plenum Press, New York, (1981), p. 449.

4. C. J. M. Northrup, G. W. Arnold, and T. J. Headley, _Scientific Basis for Nuclear Waste Management_, Vol. 6 , S. Topp, Ed., Elsevier-North Holland, New York, (1982) in press.

5. J. C. Dran, Y. Langevin, M. Maurette, and J. C. Petit, _Scientific Basis for Nuclear Waste Management_, Vol. 6 , S. Topp, Ed., Elsevier-North Holland, New York, (1982) in press.

6. J. C. Dran, M. Maurette, J. C. Petit, and B. Vassent, _Scientific Basis for Nuclear Waste Management_, Vol. 6 , S. Topp, Ed., Elsevier-North Holland, New York, (1982) in press.

7. R. G. Dosch, in _Radioactive Waste in Geologic Storage_, S. Fried, Ed., Amer. Cer. Soc. Symp. Series No. 100, (1979), p. 129.

8. R. G. Dosch, A. W. Lynch, T. J. Headley, and P. F. Hlava, _Scientific Basis for Nuclear Waste Management_, Vol. 3, J. G. Moore, Ed., Plenum Press, New York, (1981), p. 123.

9. A. E. Ringwood, S. E. Kesson, N. G. Ware, W. Hibberson, and A. Major, _Nature_, Vol. 278, (1979), p. 219.

10. D. K. Brice, _Radiation Effects_, Vol. 6, (1970), p. 77.

11. T. J. Headley, R. C. Ewing, and R. F. Haaker, _Nature_, Vol. 293, (1981), p. 449.

12. T. J. Headley, _39th Annual Proc. Electron Microscope Soc. America_, G. W. Bailey, Ed., (1981), p. 114.

13. V. M. Oversby and A. E. Ringwood, _Radioactive Waste Management_, Vol. 1, (1981), p. 289.

14. F. W. Clinard, L. W. Hobbs, C. C. Land, D. E. Peterson, D. L. Rohr, and R. B. Roof, "Alpha Decay Self-Irradiation Damage in Pu^{238}-Substituted Zirconolite," LA-UR81-2401, Los Alamos National Laboratory, Los Alamos, NM, (1981).

15. H. D. Holland and D. Gottfried, _Acta Cryst._, Vol. 8, (1955), p. 291.

16. L. A. Bursill and A. C. McLaren, _phys. stat. sol._, Vol. 13, (1966), p. 331.

17. A. A. Krasnobayev, Y. M. Polexhayev, B. A. Yunikov, and B. K. Novoselov, _Geochem. Int._, Vol. 11, (1974), p. 195.

18. F. F. Morehead and B. L. Crowder, in _Ion Implantation,_ F. H. Eisen and B. L. Crowder, Eds., Gordon and Breach, (1971), p. 25.

19. J. F. Gibbons, W. S. Johnson, and S. W. Mylroie, _Projected Range Statistics,_ Halsted Press, (1974).

20. R. C. Ewing, R. F. Haaker, and W. Lutze, proceedings, this symposium.

LEACHABILITY OF ZIRCON AS A FUNCTION OF ALPHA DOSE

RODNEY C. EWING[+], RICHARD F. HAAKER[+] AND WERNER LUTZE[++]
[+]Department of Geology, University of New Mexico, Albuquerque, NM 87131 USA
[++]Hahn-Meitner Institute, Glienicker Strasse 100, 1000 Berlin 39, Federal
Republic of Germany

ABSTRACT

The variation in the leachability of naturally occurring zircons ($ZrSiO_4$) as a function of total alpha dose has been measured at $87^\circ C$. For calculated doses in the range of 10^{16} to 10^{18} alphas/gm there is an increase of one order of magnitude in the weight percent of zircon that is dissolved and an increase in the leach rate of the zircon 2.9×10^{-8} to 2.3×10^{-7} gm/cm^2 day. Totally metamict specimens (dose $\geq 10^{19}$ alphas/gm) have leach rates as high as 1.8×10^{-6} gm/cm^2 day.

INTRODUCTION

The effect of radiation damage on the leach rates of proposed high level radioactive waste forms is a factor that must be known before long term predictions of leaching behavior can be made. Since natural zircon is a compositionally simple and relatively common phase which occurs in crystalline and metamict forms, it is a convenient model for which the effect of radiation damage on leach rates can easily be measured. The purpose of this study is to report the variations in the leach rate of crystalline, partially metamict and metamict zircons for which the alpha-dose has been calculated based on U,Th and Pb isotopic age determinations, or estimated on the basis of x-ray diffraction analysis and density determinations.

There has been considerable discussion and controversy concerning radiation effects on the leachability of minerals and borosilicate glass and crystalline ceramic radioactive waste forms.[1,2,3,4,5] Since metamict minerals are often altered, microfractured, brittle and glassy, some mineralogists have assumed that such phases are less resistant than their crystalline counterparts.[6,7] Ringwood[8] citing the work of Pidgeon and others[9,10] has suggested that in phases susceptible to radiation damage, such as zircon, actinides would remain "tightly bound" even in the metamict state. Further work has shown that naturally occurring zirconolites have behaved as closed systems for U, Th and Pb and that there is only a slight increase in U leach rates with increased alpha dose.[11,12] Roy and Vance[13] have maintained that the solubility changes in a crystalline phase due to radiation effects are likely to be quite small.

In contrast, Fleischer[14], Kigoshi[15], and Eyal and Kaufman[16] have described the preferential loss of actinide, alpha-recoil nuclei in phases such as zircon, monazite, mica, feldspar, quartz and natural glass. Fleischer suggests that preferential loss of alpha-recoil nuclei via natural etching processes may be a factor in the discordance of radiometric ages.[14] Increased rates of dissolution are substantiated by the differential dissolution of altered and metamict zircon[17] and the increased solubility of zircons in a mixture of hydrofluoric and sulfuric acid as a function of degree of metamictization.[6] Also well documented is a dependence of fission track etching rates on the degree of radiation damage (fission track density) for zircons when $HF-H_2SO_4$ or the KOH-NaOH eutectic is used as the etchant.[18,19]

SPECIMEN DESCRIPTIONS

Samples 77-215, M-83-NM, DB-10 and RSD-9 were provided by R. Zartman of the U.S. Geological Survey. Descriptions of their geologic occurrence are referenced in Table 1. In general the samples consisted of individual crystals 50 to 75 microns in size (-200 +270 mesh). Individual crystals are colorless to light brown and may be complexly zoned and fractured. The crystals are usually prismatic, terminated by simple pyramids and basal pinacoids. Inclusions (e.g. quartz, feldspar and uraninite), identified by scanning electron microscopy, are not uncommon, and there may be variations of uranium concentrations among grains by a factor of two or three. Individual grains usually have uniform distributions of uranium but occasional high concentrations (probably UO_2) were noted. One should be careful to point out that the population of zircon crystals are heterogenous and do contain inclusions of other phases. Samples 3, 4, 17 and C1 were single crystals, purchased or provided by individuals. The crystals were ground until they were 50-106 microns (-140 +300 mesh) in size. Although subjected to careful electron microprobe analysis, no direct calculation of the total alpha-dose could be made. Estimates of the total alpha-dose were made on the basis of density determinations,[20] x-ray powder diffraction analysis and precession photographs. Sample 3 (from Thailand) is highly crystalline; back-reflections in the powder diffraction pattern show clear resolution of Cu $K_{alpha-1}$ and $K_{alpha-2}$ diffraction maxima. The density (4.61 gm/cm^3) suggests a dose of less than 2 x 10^{18} alphas/gm. Sample 4 (from Southwestern Minerals Co.) is partially crystalline; there are no back-reflections in the powder diffraction pattern. Precession photographs of sample 4 indicate that this zircon has normal unit cell dimensions but very broad and diffuse diffraction maxima. The

TABLE 1
AGES AND ALPHA DOSES FOR THE ZIRCON SPECIMENS

| | Concentration (ppm) | | | Age (millions years) | | | |
Sample No.	U	Th	Pb	$\frac{206_{Pb}}{238_U}$	$\frac{207_{Pb}}{235_U}$	$\frac{207_{Pb}}{206_{Pb}}$	$\frac{208_{Pb}}{232_{Th}}$
77–215	163.3	48.9	3.85	148	149	169	144
M–83–NM	556.6	188.3	396.9	3,009	3,210	3,340	3,359
DB–10	1,081	740	299	1,417	1,422	1,433	1,316
RSD–9	1,407.7	–	76.8	1,431	1,433	1,438	1,393

TABLE 1   Continued

Sample No.	Density (gm/cm^3)	Alpha Dose (alphas/gm)	Reference, Source or Locality
77–215	–	6.02×10^{16}	21, Zartman
M–83–NM	–	9.44×10^{18}	22, Zartman
DB–10	–	6.49×10^{18}	23, Zartman
RSD–9	–	3.30×10^{17}	24, Zartman
#3	4.61	$\leq 2 \times 10^{18}$ (est.)	Thailand
#4a	3.81	$\geq 1 \times 10^{19}$ (est.)	Southwestern Mineral Co.
#4b	3.81	$\geq 1 \times 10^{19}$ (est.)	Southwestern Mineral Co.
#17	4.64	$\leq 2 \times 10^{18}$ (est.)	Southwestern Mineral Co.
#C1	3.45	$\gg 1 \times 10^{19}$ (est.)	Raleigh Peak, Jefferson County Colorado
5b		0	flux grown zircon
7		0	sol-gel zircon

density (3.81 gm/cm^3) suggests a dose of greater than 10^{19} alphas/gm. Sample 17 (from Southwestern Minerals Co.) is highly crystalline; back-reflections in the x-ray powder diffraction pattern show clear resolution of Cu $K_{alpha-1}$ and $K_{alpha-2}$ diffraction maxima. The density (4.64 gm/cm^3) suggests a dose of less than 2×10^{18} alphas/gm. Specimen C1 (from near Raleigh Peak, Jefferson County, Colorado) is metamict and displays a single, broad diffraction halo. The very low density (3.45 gm/cm^3) suggests a dose of greater than 10^{19} alphas/gm.

In addition to the natural materials, two different snythetic zircon powders were leached. Zircon 5b was grown as single crystals from a $Na_2Mo_2O_7$ flux while zircon 7 was a poorly consolidated polycrystalline material prepared by a sol gel method[25]. Both synthetic zircons were highly crystalline and pure to the extent that no extra lines were observed in their x-ray powder diffraction patterns.

EXPERIMENTAL

The samples were placed in sealed fused quartz ampoules (except C1, which was pyrex) with a leachate solution of 5 weight percent $KHCO_3$. The samples weighed 3 to 20 mg (except C1, which was 978 mg) and were placed in 0.50 ml of solution (except C1, which contained 5 ml of solution). The samples were agitated by inversion twice a day and held at $87 \pm 2^{o}C$ for 9.9 days at a total pressure of less than two atmospheres The solutions were analyzed by an induction coupled plasma (ICP). Two ampoules were run for each sample (the average value is reported in Table 2); 0.3 ml of solution were removed from each ampoule and 2.7 ml of H_2O added to the leachate; the total 3.0 ml were analyzed by ICP, yielding eight consecutive measurements of 10 seconds each. The data and results listed in Table 2 are based on 168 individual measurements. The lowest value (0.027 ppm Zr in solution, Table 2) is 2.5 times above the detection limit for zirconium.

RESULTS AND DISCUSSION

It is interesting to note that for samples with approximately the same dose (Nos. 4 and C1, greater than 10^{19} alphas/gm), the experimental conditions (e.g. the amount of sample and leachate) do not have a significant effect on the result (percent $ZrSiO_4$ dissolved), Table 2.

TABLE 2
ZIRCON LEACHING DATA AND RESULTS

Sample No. and Mesh Size	Alpha Dose (alphas/gm)	%Saturation $(\frac{100 \times Dose}{1 \times 10^{19}})$	Surface[a] area (cm^2)
77-215 (-200+270)	6.02×10^{16}	0.8	1.91
M-83-NM (-200+270)	9.44×10^{18}	85.0	2.76
DB-10 (-200+270)	6.49×10^{18}	59.0	2.56
RSD-9 (-200+270)	3.30×10^{17}	3.0	5.68
#3 (-140+300)	$\leq 2 \times 10^{18}$ (est.)	≤ 20.0	1.76
#4a (-200+270)	$\geq 1 \times 10^{19}$ (est.)	≥ 100.0	0.61
#4b (-140+300)	$\geq 1 \times 10^{19}$ (est.)	≥ 100.0	2.31
#17 (-140+300)	$\leq 2.10^{18}$ (est.)	≤ 20.0	2.82
#C1 (-140+300)	$\geq 1 \times 19^{19}$ (est.)	≥ 100.0	124.0
5b (-140+300)	0	0.0	0.60
7 (-140+300)	0	0.0	1.45

Sample No.	Zr in solution[b] (ppm)	$ZrSiO_4$ dissolved (wt. %)	Bulk Leach Rate[c] (gm/cm^2 day)
77-215	0.027	2.42×10^{-3}	2.9×10^{-8}
M-83-NM	0.315	2.00×10^{-2}	2.3×10^{-7}
DB-10	0.116	7.88×10^{-3}	9.2×10^{-8}
RSD-9	0.0425	1.57×10^{-3}	1.9×10^{-8}
#3	0.193	1.63×10^{-2}	2.2×10^{-7}
#4a	0.54	1.50×10^{-1}	1.8×10^{-6}
#4b	2.70	1.12×10^{-1}	2.4×10^{-7}
#17	0.21	$.99 \times 10^{-3}$	1.5×10^{-7}
#C1	12.79	1.30×10^{-1}	2.1×10^{-6}
5b	0.095	1.94×10^{-2}	3.2×10^{-8}
7	0.243	1.03×10^{-2}	3.4×10^{-7}

a Surface areas were estimated using the assumption of spherical particles of 0.75 mm diameter for (-200+270) and 0.104 mm for (-140+300) mesh .

b Zr concentrations in the leachate after the 10 fold dilution was made.

c The bulk leach rates were calculated on the basis of Zr and the assumption of congruent dissolution.

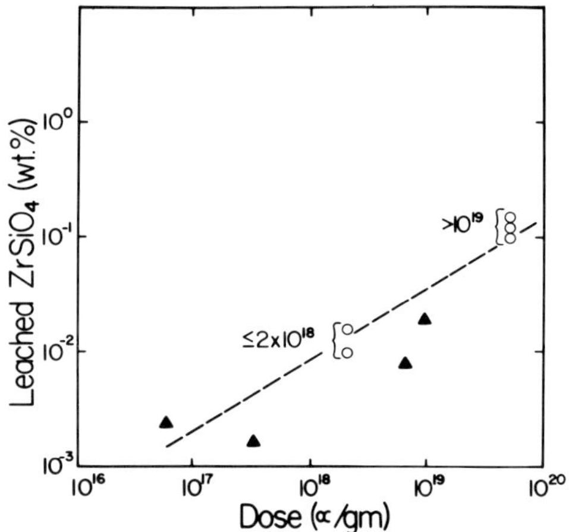

Fig. 1. Plot of wt. % ZrSiO$_4$ dissolved into solution (based on Zr in solution) vs. alpha-dose. For solid triangles the alpha-dose is calculated. For open circles the alpha-dose is estimated on the basis of density measurements and x-ray diffraction analysis.

For the first four samples (for which the alpha dose may be calculated from the isotopic data and ages), there is an increase in the weight percent zircon dissolved from 1.57×10^{-3} to 2.0×10^{-2} and an increase in the leach rate from 2.9×10^{-8} to 2.3×10^{-7} gm/cm^{2}day. For the remaining samples in which alpha dose is estimated to be greater than 10^{19} alphas/gm, the leach rate may be as high as 1.8×10^{-6} gm/cm^2 day. When all mineral samples are considered there is an increase in the weight percent of zircon dissolved of two orders of magnitude, Figure 1, and an increase in the leach rate of between one and two orders of magnitude, Figure 2. Synthetic zircon 5b (crushed single crystals) has a leach rate which is similar to those for zircons having an alpha dose of less than 10^{18} alphas/gm. Synthetic zircon 7 (a poorly consolidated powder having an average crystallite size of 5 microns)[25] has an order of magnitude higher leach rate than synthetic zircon 5b. Part of the difference in leach rates observed in this experiment may be due to differences in effective surface area.

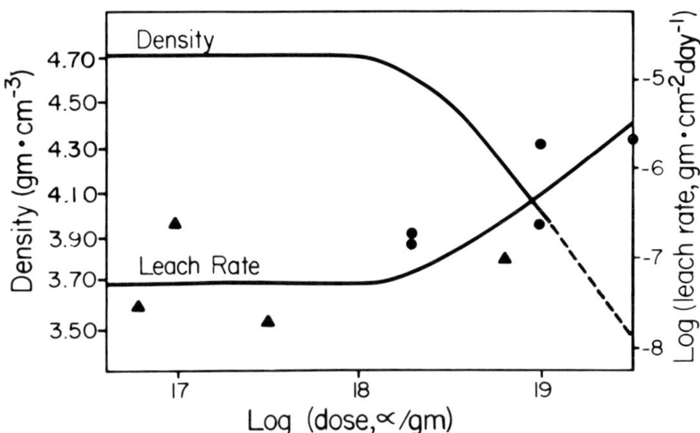

Fig. 2. The process of metamictization is not only accompanied by changes in physical properties, but also in leach rate. ▲ = Alpha dose calculated from radiometric age and U-Th content. ● = Alpha dose estimated from the relationship between density and alpha dose[20].

This work provides a quantitative basis for the statement that crystalline, radioactive waste forms may indeed experience an increase in their leach rates as a result of radiation effects caused by constituent radionuclides. The effect measured in these experiments may be in part due to increased surface area caused by microfractures or geochemical alteration of the more metamict zircons; however, the amount of zircon dissolved was not dramatically affected by differences in the surface areas of crystals from different size fractions (Table 2 no. 4). The increased surface area and microfracturing are not unique to zircon or metamict silicates, they might be expected to appear in radioactive waste forms which have accumulated a sufficiently high alpha-dose. The measured difference in leachability may in fact reflect differences in the atomic arrangment of the metamict state versus the prior, crystalline structure. Recent work[26,27,28] suggests that the structure of metamict silicates and complex Nb-Ta-Ti oxides involves not only the formation of a mosaic domain structure (50-100 A) with tilts of several degrees for partially metamict zircons, but the formation of substantial areas which are amorphous or consist of domains that are less than 10 A in size. Additionally, recent x-ray absorption fine structure spectroscopy (EXAFS) studies suggest distortion of the coordination polyhedra of major cations (e.g., Ti^{4+}) as compared to those found in the premetamict Nb-Ta-Ti oxide crystalline structures[29]. Both the TEM and EXAFS data suggest that there may be important structural changes that accompany the transition from the crystalline to metamict state, and these structural changes may account for the increased leachability. This report documents this effect in a single phase at a single temperature and pressure. The reported leach rates should not be interpreted as steady state values. Since the temperature dependence of the leach rates remains unknown it cannot be concluded that, under all conditions, metamict zircon will have a higher leach rate. The significance of the leach rate dependence on alpha-recoil damage (or metamictization) should be evaluated for individual crystalline phases which are components of proposed crystalline high level radioactive waste forms.

ACKNOWLEDGEMENTS

The authors are particularly indebted to R. E. Zartman for some of the samples used in this study and to Peter Schubert (HMI) for his efforts in preparing the samples. This work was in part supported by the U.S. Department of Energy and Battelle Pacific Northwest Laboratory through contract DE-AC-06-76RLO-1830.

Note added in proof: Please refer to the paper by T. J. Headley, G. W. Arnold, and C. J. M. Northrup (this volume) for a description (TEM) of samples RSD-9, DB-10 and M-83-NM.

REFERENCES

1. Dran, J. C., Maurette, M. and Petit, J. C. (1980) Science, 209, 1518.
2. Hirsch, E. H. (1980) Science, 209, 1520.
3. Dran, J. C., Maurette, M., Petit, J. C. and Vassent, B. (1981) in Scientific Basis for nuclear Waste Managment, vol. 3, Moore, J. ed., Plenum, New York, pp. 449–456.
4. Offermann, V. P. and Matzke, H. (1981) Atomwirtschaft, June, 350.
5. Weber, W. J., Wald, J. W. and Gray, W. J. (1981) in Scientific Basis for Nuclear Waste Management, vol. 3, Moore, J. ed., Plenum, New York, pp. 441–448.
6. Krasnobayev, A. A., Polezhayev, Y. M., Yunikov, B. A. and Novoselov, B. K. (1974) Geokhimiya, 2, 261.
7. Frondel, C. (1958) Systematic Mineralogy of Uranium and Thorium, Geological Survey Bulletin 1046, pp. 1–400.
8. Ringwood, A. E. (1981) Safe Disposal of High Level Nuclear Reactor Wastes: A New Strategy. Australian National University Press, Canberra, pp. 42–44.
9. Pidgeon, R. T., O'Neill J. and Silver, L. (1973) Fortschr. Min., 50, 118.
10. Pidgeon, R. T., O'Neill, J. and Silver, L. (1966) Science, 154, 1538.
11. Ringwood, A. E., Reeve, K. D. and Tewhey, J.D. (1981) in Scientific Basis for Nuclear Waste Management, vol. 3, Moore, J. ed., Plenum, New York, pp. 147–154.
12. Ringwood, A. E., Oversby, V. M., Kesson, S. E., Sinclair, W., Ware, N., Hibberson, W., and Major, A. (1981) Nuclear and Chemical Waste Management, 2, 287–305.
13. Roy, R. and Vance, E. R. (1981) Journal of Materials Science, 16, 1187.
14. Fleischer, R. L. (1980) Science, 207, 979.
15. Kigoshi, K. (1971) Science, 173, 47.
16. Eyal, Y. and Kaufman, A., Nuclear Technology, (in press); Eyal, this volume.
17. Krogh, T. E. and Davis, G. L. (1975) Annual Report of the Director of the Geophysical Laboratory, Carnegie Institution, 619.
18. Gleadow, A. J. W., Hurford, A. J. and Quaiffe, R. D. (1976) Earth and Planetary Science Letters, 33, 273.
19. Krishnaswame, S., Lal, D., Prabhu, N. and MacDougall, D. (1974) Earth and Planetary Science Letters, 22, 51.
20. Holland, H. and Gottfried, D. (1955) Acta Crystallographica, 8, 291.
21. Whetten, J. T., Zartman, R. E., Blakely, R. J. and Jones, D. L. (1980) Geological Society of America Bulletin, 91, 359.
22. Peterman, Z. E., Zartman, R. E. and Sims, P. K. (personal communication).
23. Zartman, R. E., Peterman, Z. E., Obradovisch, J. D., Gallego, M. D. and Bishop, D. T. (personal communication).
24. Zartman, R. E. (personal communication).
25. Haaker, R. F. and Ewing, R. C. (1981) Journal American Ceramic Society, 64.
26. Headley, T. J., Ewing, R. C. and Haaker, R. F. (1981) in Proceedings of the 39th Annual Meeting of the Electron Microscopy Society of America, 112.
27. Headley, T. J., Ewing, R. C. and Haaker, R. F. (1981) Nature, 293, 449.
28. Haaker, R. F., Ewing, R. C. and Headley, T. J. (in press) in Scientific Basis for Radioactive Waste Management, Vol.6 .
29. Greegor, R. B., Lytle, F. W., and Ewing, R. C. (1981) in Proceedings of the Eighth Annual Stanford Synchrotron Radiation Laboratory Users Group Meeting, SSRL Report No. 81-3, 15.

ISOTOPIC FRACTIONATION OF THORIUM AND URANIUM UPON LEACHING OF MONAZITE: ALPHA-RECOIL DAMAGE EFFECTS

YEHUDA EYAL
Department of Nuclear Engineering
Technion - Israel Institute of Technology, Haifa, Israel

INTRODUCTION

It has been proposed recently that high-level radioactive wastes from spent nuclear reactor fuels be converted into a crystalline phosphate form prior to their disposal in geological repositories.[1] This approach is based on the known durability in nature of the mineral monazite, $(Ce,La)PO_4$, which contains appreciable concentrations of thorium and uranium.[2] Of significance is the demonstrated resistance of monazite to excessive metamictization,[2] in spite of the fact that the mineral has been subjected since its formation to radiation from the radioactive disintegration of ^{238}U, ^{232}Th and their many intermediate decay products. Nevertheless, radiation effects on the leaching behavior of the material in possible groundwater environments must be studied in the development and testing of monazite-like phases for the long-term disposal of fission product and actinide wastes.

In a recent experiment,[3] three natural monazite specimens, in which radiation damage has accumulated over geological periods, were employed to investigate the relative dissolution rates of the actinide isotopes ^{238}U, ^{234}U, ^{232}Th, ^{230}Th and ^{228}Th. The minerals, containing 7.5-10.4 wt% ^{232}Th and 0.13-0.23 wt% ^{238}U, were subjected to a series of successive short-term leaching treatments in a bicarbonate-carbonate solution. Preferential dissolution by a factor of 1.1 to ∿10 of the radiogenic nuclides ^{234}U, ^{230}Th and ^{228}Th relative to their corresponding structurally incorporated isotopes ^{238}U and ^{232}Th has been observed. This isotopic fractionation was attributed to radiation damage caused by alpha-recoil atoms. Each alpha-recoil atom produces a trail of displaced atoms along a path a few hundred angstroms in length. This damage causes a local increase in the chemical reactivity of the mineral. The preferential transfer of the radiogenic isotopes to the solution may be explained by the preferential dissolution of the above damaged zones.

Here we report the continuation of the study for one of the three monazite samples (monazite M1 from New Mexico in the United States).[3] Whereas in Ref. 3 the dissolution included 0.2% of the total ^{232}Th radioactivity present in the mineral, here we present the leaching behavior up to dissolution of ∿1.8% of the total ^{232}Th radioactivity. This investigation provides evidence

that radiation effects largely control the dissolution of the mineral in
a well-developed stage of the leaching process. In particular, it may be inferred that etched alpha-recoil tracks provide a very effective means for the
solution to penetrate into deep layers of the crystal. This conclusion is
strongly supported by an additional experiment described here, that shows a
pronounced decrease in the leach rate of the mineral after laboratory heat treatment. The heat treatment has presumably caused a partial annealing of the
alpha-recoil damage. Evidence for lattice repair was obtained from X-ray diffraction measurements.

The thorium data further confirm the conclusion[3] that in the present monazite sample the alpha-recoil damage has persisted over a time range ($\sim 10^5$ yr)
that would be important for the bonding of the hazardous actinide nuclides and
their decay products in crystalline phases buried in deep geological formations.

MATERIALS AND METHODS

The present leaching study is, in part, a continuation of the experiment described in Ref. 3. The monazite specimen selected is sample M1 from New Mexico
in the United States. Prior to leaching the sample was ground to fine powder
with maximum particle size of ~ 110 μm. The leachant was an aqueous solution
containing 0.1 M $NaHCO_3$ and 0.1 M Na_2CO_3.

In Ref. 3, two independent series of experiments were performed. Each involved 1 g of the powder, and was conducted at room temperature. The leaching
procedure consisted of stirring the powder within 10 ml of the leachant for 3
min. The solution was then separated from the solid by centrifugation and filtration. The sample was in contact with the solution for a total of 14 to 22
min. The solid was immediately further leached by repeating the above leaching procedure four more times. In each of the five successive leaching treatments the leachant was a fresh 10-ml portion of the bicarbonate-carbonate
solution.

The further steps of the leaching process, performed in this work, involved
the immersion of the solid in the leachant for periods ranging from ~ 1 day to
36 days. During this part of the experiment the mineral and the liquid were
kept at rest, except for ~ 3 min mixing periods at the beginning and at the end
of each leaching treatment.

The radioactivities of ^{238}U, ^{234}U, ^{232}Th, ^{230}Th and ^{228}Th in each solution
were determined by alpha spectrometry preceded by chemical separation and
purification of uranium and thorium.[3]

EXPERIMENTAL RESULTS AND DISCUSSION

The transmutation schemes and the decay half-lives of the nuclides under investigation are as follows:

$$^{238}U \xrightarrow[4.5\times10^9\text{yr}]{\alpha} {}^{234}Th \xrightarrow[24\text{ days}]{\beta^-} {}^{234}Pa \xrightarrow[1.2\text{ min}]{\beta^-} {}^{234}U \xrightarrow[2.5\times10^5\text{yr}]{\alpha} {}^{230}Th \xrightarrow[8.0\times10^4\text{yr}]{\alpha}$$

and

$$^{232}Th \xrightarrow[1.4\times10^{10}\text{yr}]{\alpha} {}^{228}Ra \xrightarrow[5.8\text{ yr}]{\beta^-} {}^{228}Ac \xrightarrow[6.1\text{ h}]{\beta^-} {}^{228}Th \xrightarrow[1.9\text{ yr}]{\alpha} \quad .$$

The measured radioactivities and activity ratios in the mineral are presented in Table 1. The nuclides ^{230}Th and ^{234}U are in radioactive equilibrium with ^{238}U, and ^{228}Th is in radioactive equilibrium with ^{232}Th. Therefore, the mineral has been a closed system for at least the last $\sim10^6$ yr. The abundances of ^{232}Th and ^{238}U in the mineral are 10.37 wt% and 0.19 wt%, respectively.

The leaching data are presented in Table 2. The results of two independent experiments are reported separately. We focus attention primarily on the relative leach rates R and their variation as a function of cumulative percent of ^{232}Th activity leached. The latter quantity may describe perhaps most closely the dissolved portion of the overall monazite matrix, because the isotope ^{232}Th is one of the major nuclides in the composition of the sample. The value of R is given by the ratio of the fractions of activities leached. The latter quantities were determined with respect to the initial radioactivities in the mineral.

Perhaps the most important observation in the data of Table 2 is that the enhanced dissolution of both ^{228}Th and ^{230}Th relative to that of ^{232}Th is still maintained at an appreciable level at an advanced stage of the dissolution process. To understand this phenomenon, we recall that enhanced dissolution of the radiogenic nuclides at any given step of the dissolution process causes their rapid depletion from the layer of the mineral which is in contact with the solution. Therefore, continued enhanced dissolution of ^{228}Th and ^{230}Th relative to that of ^{232}Th indicates that the leachant penetrates into deeper layers of the crystal without dissolving much of the ^{232}Th radioactivity in the superficial layers of the solid in which ^{232}Th is now enriched. This may be explained by leachant penetration through etched alpha-recoil tracks.

Some comments are necessary about the ^{228}Th data presented in this work. The measured value of $R(^{228}Th/^{232}Th)$ may represent the correct leach rate of ^{228}Th relative to that of ^{232}Th only if the leach rate of ^{228}Th is similar to the leach rate of its 5.8 y parent, ^{228}Ra. Otherwise, the activity of ^{228}Th

402

TABLE 1

MEASURED RADIOACTIVITIES AND ACTIVITY RATIOS IN MONAZITE M1 (NEW MEXICO)[*]

Nuclide	Activity (dis/min.g^{-1})[a]	Nuclides	Activity Ratio
^{238}U	1432 (3)[b]	^{238}U/^{232}Th	0.057 (4)
^{234}U	1435 (3)	^{234}U/^{238}U	1.002 (2)
^{230}Th	1433 (5)	^{230}Th/^{232}Th	0.057 (5)
^{232}Th	25146 (3)	^{228}Th/^{232}Th	1.000 (2)
^{228}Th	25150 (3)	^{230}Th/^{228}Th	0.057 (5)

[*] The measurements involved totally dissolved samples (data from Ref. 3).
[a] Radioactivities (in units of disintegrations per minute per gram monazite).
[b] Uncertainties (in percent) are indicated in parentheses.

in the layers of the mineral which are in contact with the solution must be corrected for a net loss or a net gain by radioactive transmutations. The nuclide ^{228}Ra is a β^- emitter and its activity was not measured in the present experiment. However, we obtained the leach rate of ^{228}Ra relative to that of ^{232}Th in leaching treatment L_6 of experiment 1 (Table 2). This was done by observing the radioactivity of ^{228}Th in two separate portions of the leachate, one removed immediately after termination of the leaching process, the other removed ∿1 yr later. We obtain R(^{228}Ra/^{232}Th)=0.07 $\pm\ ^{0.12}_{0.07}$. Thus, the nuclide ^{228}Ra is largely retained in the solid phase. The observed radioactivity of ^{228}Th must be corrected, therefore, for the portion of activity produced during leaching by the decay of the excess amount of ^{228}Ra that is present in the surface layers of the mineral. Although the complete leaching behavior of ^{228}Ra is not yet known, we can readily obtain the lower limits of R(^{228}Th/^{232}Th) for all leaching treatments in Table 2. The results are presented in square parentheses adjacent to the observed values of R(^{228}Th/^{232}Th) in Table 2. The correction procedure involves the assumption that the ^{228}Ra radioactivity is fully retained in the mineral, while the radioactivity of ^{228}Th that is produced by the decay of ^{228}Ra during leaching is immediately dissolved in the leachant (via the dissolution of ^{228}Th itself, and, possibly, via a very rapid dissolution of the 6.1 h ^{228}Ac isotope). With this correction, the value of R(^{228}Th/^{230}Th) remains nearly constant at ∿1.5 over the range from ∿0.4% to ∿1.8% of total ^{232}Th activity leached.

As indicated in Ref. 3, the comparison of the dissolution rates of the long-lived ^{230}Th isotope with those of the short-lived ^{228}Th isotope (both nuclides are radiogenic and chemically similar) gives important information on the long-term stability of the alpha-recoil damage. We observe in Table 2 that the leach

TABLE 2

LEACHING DATA FOR MONAZITE M1 (NEW MEXICO) IN A BICARBONATE-CARBONATE SOLUTION[*]

Leaching Treatment	Leaching Time (days)	^{232}Th Activity (dis/min·g^{-1})[a]	^{228}Th/^{232}Th[c]	Relative Leach Rate R[b] ^{230}Th/^{232}Th[d]	^{238}U/^{232}Th[e]	^{234}U/^{238}U[f]
			Experiment 1			
L_{1-5}^{g}	0.06	50.9 (0.20)	2.77 [2.77]	3.46	4.99	1.09
L_6	21.8	146 (0.78)	1.48 [1.46]	1.62	4.48	1.04
Total	21.9	197 (0.78)				
			Experiment 2			
L_{1-5}^{h}	0.06	49.9 (0.20)	2.52 [2.52]	3.42	5.27	1.10
L_6	0.8	43.0 (0.37)	1.74 [1.74]	2.56	4.73	1.10
L_7	1.0	43.6 (0.54)	1.51 [1.50]	1.93	2.72	1.10
L_8	1.0	27.6 (0.65)	1.43 [1.42]	1.55	2.92	1.05[i]
L_9	1.0	24.0 (0.75)	1.55 [1.53]	1.38	2.30	
L_{10}	2.0	31.3 (0.87)	1.51 [1.49]	1.33	2.58	
L_{11}	2.0	35.6 (1.01)	1.35 [1.33]	1.08	1.73	
L_{12}	1.9	28.9 (1.13)	1.46 [1.43]	1.33	1.58	
L_{13}	3.1	26.2 (1.23)	1.57 [1.51]	1.19	2.27	
L_{14}	3.1	21.5 (1.32)	1.45 [1.38]	1.14	2.39	
L_{15}	5.9	27.3 (1.43)	1.67 [1.55]	1.27	2.37	
L_{16}	6.2	25.1 (1.53)	1.76 [1.61]	1.31	1.61	
L_{17}	21.8	31.1 (1.65)	1.97 [1.53]	1.57	2.67	
L_{18}	36.0	29.6 (1.77)	2.21 [1.41]	1.75	2.54	
Total	85.9	445 (1.77)				

[*]The results of two independent experiments are reported separately.
[a]Measured ^{232}Th radioactivity (in units of disintegrations per minute per gram monazite). The cumulative percent of ^{232}Th activity leached is indicated in parentheses. The uncertainties in experiment 1 are: 3%(L_{1-5}), 4%(L_6); in experiment 2 are: 4%(L_{1-5}), 5-8%(L_6-L_{18}).
[b]Relative leach rates, R, given by the ratios of fractions of activities leached.
[c]The lower limit of R(^{228}Th/^{232}Th), assuming complete retention of ^{228}Ra in the mineral, is indicated in square parentheses. The uncertainties in experiment 1 are 4%; in experiment 2 are 4-8%.
[d]The uncertainties in experiment 1 are 7%; in experiment 2 are: 8%(L_{1-5}), 13% (L_6 and L_7), ∿20% (L_8 - L_{18}).
[e]The uncertainties in experiment 1 are 7%; in experiment 2 are: 8% (L_{1-5}), 10-17% (L_6 - L_{18}).
[f]The uncertainties in experiment 1 are: 5%(L_{1-5}), 4% (L_6); in experiment 2 are: 7% (L_{1-5}), 8% (L_6 and L_7).
[g]Cumulative result for five successive 3-min leaching treatments (data from experiment 1 in Table 2 of Ref. 3).
[h]Cumulative result for five successive 3-min leaching treatments (data from experiment 2 in Table 2 of Ref. 3).
[i]Average value for leaching treatment L_8-L_{18}. The uncertainty is 5%.

rate of ^{230}Th is higher than the leach rate of ^{228}Th until $\sim$0.7% of the total ^{232}Th radioactivity is dissolved. This result indicates not only the stability of the alpha-recoil damage, but also the sensitivity of the leach rate to the size of the damaged crystal zone. The nuclide ^{230}Th is formed by two successive alpha transformations; the decay of ^{238}U to ^{234}Th and the decay of ^{234}U to ^{230}Th. The decrease in the leach rate of ^{230}Th relative to that of ^{228}Th in the next steps of the dissolution process may be explained by the depletion of ^{230}Th in the surface layers of the mineral during the early steps of the leaching process.

The uranium data in Table 2 show enhanced dissolution of ^{234}U relative to that of ^{238}U by a factor of 1.1 in the first step of the leaching process, and a tendency for decreasing amount of isotopic fractionation as the leaching process proceeds. The data also show a pronounced element fractionation; the dissolution rate of uranium is higher than the dissolution rate of **thorium** in all leaching treatments. This result, together with the ^{228}Ra data reported for leaching treatment L_6 of experiment 1 in Table 2, clearly indicates the importance of chemical effects in the leaching process.

We have studied the effect of heat treatment on the leaching behavior of monazite M1. In this experiment, 1 g of the powder was heated in an electric furnace under air atmosphere at 800°C for 6h. The sample was then slowly cooled to room temperature. We have noticed a weight loss of (0.55 ± 0.05)%, the nature of which was not identified. The sample was subjected to a series of successive short-term and long term leaching treatments in a bicarbonate-carbonate solution for a total of 300 days. The leaching procedures were the same as explained earlier. The results are presented in Table 3. By comparing these results with those listed in Table 2, we observe a pronounced reduction in the leach rate of the thorium isotopes after heat treatment. This result may be consistent with annealing of the alpha-recoil damage. It is known that fossil alpha-recoil tracks in various materials such as mica[4] can be annealed by heat treatments. However, the pronounced fractionation among the thorium isotopes, as well as the enhanced dissolution of ^{234}U relative to that of ^{238}U, clearly indicate that the annealing process is not complete (the ^{230}Th radioactivity and the radioactivity of uranium in some samples have not yet been measured with sufficient precision).

In an attempt to correlate the reduction in the leach rate of the thorium isotopes after heat treatment with effects imposed by lattice repair, we have performed powder X-ray diffraction measurements with heat-treated and unheated monazite samples. These measurements were performed under, as nearly as

TABLE 3

LEACHING DATA FOR HEAT-TREATED MONAZITE M1 (NEW MEXICO) IN A BICARBONATE-CARBONATE SOLUTION*

Leaching Treatment	Leaching Time (days)	^{232}Th Activity (dis/min.g^{-1})[a]	^{228}Th/^{232}Th[c]	Relative Leach Rate R[b] ^{230}Th/^{232}Th	^{238}U/^{232}Th[d]	^{234}U/^{238}U[e]
L^f_{1-6}	0.07	6.3 (0.025)	2.9 [2.9]	3.8[g]	45	1.10
L_7	24.0	5.6 (0.047)	2.7 [2.7]		36	1.03
L_8	24.1	3.5 (0.061)	3.8 [3.5]		16	
L_9	31.9	3.1 (0.074)	4.0 [3.6]		10.8	
L_{10}	38.0	3.0 (0.085)	4.5 [3.8]		7.2	
L_{11}	84.0	3.0 (0.097)	7.1 [5.3]			
L_{12}	98.0	2.9 (0.109)	12.4 [9.8]			
Total	300.1	27.4 (0.109)				

*The sample was heat-treated at 800°C for 6 h, then cooled to room temperature, prior to leaching.
[a]Measured ^{232}Th radioactivity (in units of disintegrations per minute per gram monazite). The cumulative percent of ^{232}Th activity leached is indicated in parentheses. The uncertainties are: 12% (L_{1-6}), ∿15% (L_7 - L_{12}).
[b]Relative leach rates, R, given by the ratios of fractions of activities leached.
[c]The lower limit of R(^{228}Th/^{232}Th), assuming the complete retention of ^{228}Ra in the mineral, is indicated in square parentheses. The uncertainties are: 23% (L_{1-6}) and ∿15% (L_7 to L_{12}).
[d]The uncertainties are: 14% (L_{1-6}), 17% (L_7-L_9), 25% (L_{10}).
[e]The uncertainties are 5%.
[f]Cumulative results for six successive 3-min leaching treatments.
[g]Average value for leaching treatments L_{1-6} to L_{10}. The uncertainty is ∿25%.

possible, similar experimental conditions. Two portions of the 2θ scans are presented in Figs. 1 and 2. We observe a sharpening of the diffraction maxima after heat treatment that can be explained by an increase in the effective crystallite size. This increase in the crystallite size was presumably caused by partial annealing of the alpha-recoil damage. It may be therefore concluded that alpha-recoil damage largely controls the dissolution rate of monazite in the present bicarbonate-carbonate solution.

The reason for the rise in the leach rate of ^{228}Th relative to that of ^{232}Th as a function of the leaching time (Table 3) is not clear. This rise cannot be fully explained by ^{228}Ra retention in the solid phase. A possible explanation may be offered if it is speculated that in the present heat treatment the process of partial lattice repair involved primarily the rearrangement of atoms located aside from the heavily damaged core of the alpha-recoil track. This kind of non-uniform rearrangement of the structure will provide much more

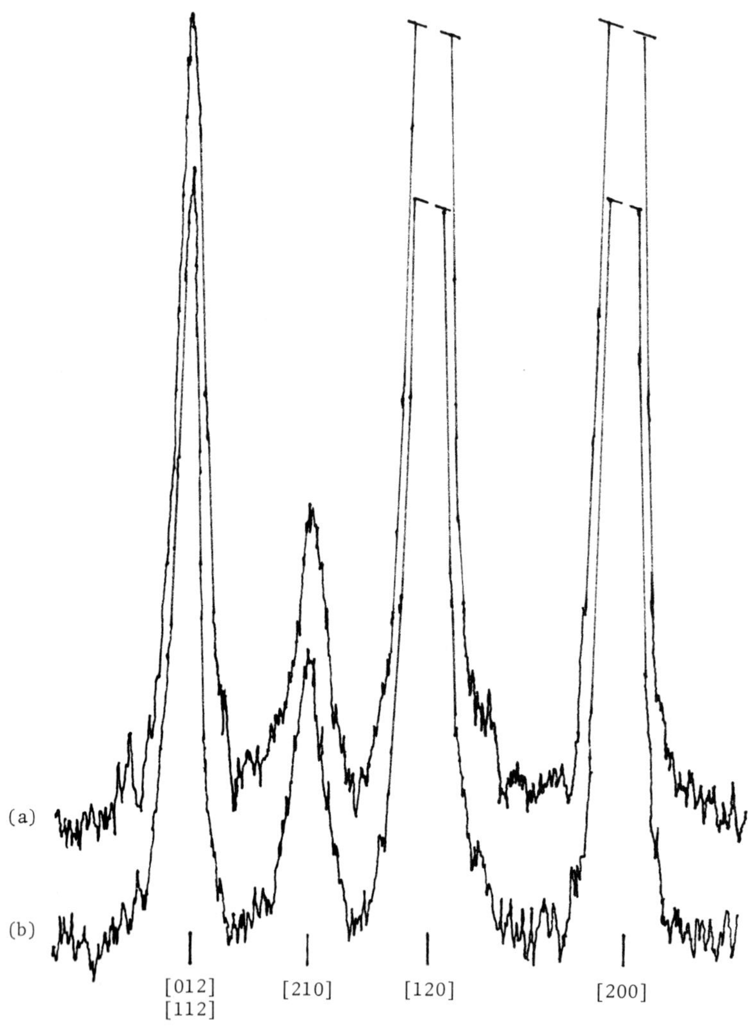

Fig. 1   The reflections of X-rays from the [200], [120], [210], [012] and [112] planes of monazite M1 powder (Cu K$_\alpha$ radiation, $2\theta \sim 26$-$32°$). (a) Unheated sample.   (b)   Sample heat-treated at 800°C for 6h.

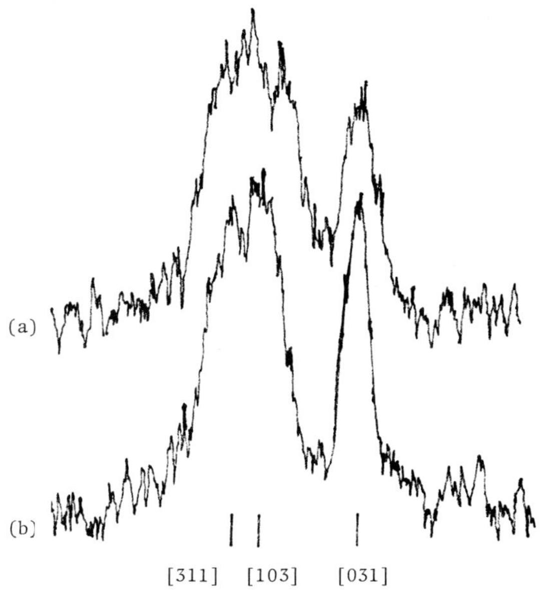

Fig. 2 The reflections of X-rays from the [031], [103] and [311] planes of monazite M1 powder (Cu K_α radiation, $2\theta \sim 40\text{-}43°$). (a) Unheated sample. (b) Sample heat-treated at 800°C for 6h.

protection, against attack by the solution, for primary matrix atoms (e.g., ^{232}Th atoms) than for alpha-recoil atoms and their decay products (e.g. ^{228}Th atoms) that are essentially always located at the end of the alpha-recoil track.

By comparing the uranium data in Tables 2 and 3, the leachability of uranium does not appear to be affected by the heat treatment in the early stage of the dissolution process. However, we have observed that $\sim70\%$ of the uranium activity measured in experiment L_{1-6} in Table 3 was released to the solution during the first 3-min leaching treatment. In the absence of heat treatment, $\sim35\%$ of the uranium activity reported in data L_{1-5} in Table 2 was released to the solution during the first 3-min leaching treatments. The former result may also be consistent with the occurrence of atomic movements during heat treatment; uranium is perhaps much less compatible than thorium with the monazite structure, as it may be inferred from the usually relatively very small abundance of uranium in monazite.

408

SUMMARY AND CONCLUSIONS

The present investigation includes a study of the relative dissolution rates of the actinide isotopes ^{238}U, ^{234}U, ^{232}Th, ^{230}Th and ^{228}Th upon leaching of monazite in a bicarbonate-carbonate solution. A preferential dissolution of the radiogenic isotopes ^{234}U, ^{230}Th and ^{228}Th relative to their corresponding structurally incorporated isotopes ^{238}U and ^{234}U has been observed in the early stages of the dissolution process as well as in an advanced stage of the leaching process. This isotopic fractionation may be attributed to radiation damage caused by alpha-recoil atoms.

The leachability of thorium in the present monazite sample is considerably reduced after heat treatment. The heat treatment has presumably caused a partial annealing of the alpha-recoil damage. Evidence for the occurrence of lattice repair is provided by powder X-ray diffraction measurements.

The present experiments may indicate that alpha-recoil damage may endanger the integrity of any crystalline phase in groundwater environments. In particular, the damage may reduce the retention of nuclides formed by alpha transmutations and their decay products in the solid phase relative to the retention of their corresponding isotopes bonded within the crystal structure. From thorium data reported in Ref. 3, and in the present investigation, it may be inferred that in monazite-like phases the overall alpha-recoil damage may increase nearly in proportion to the alpha-particle dose in the time range ($\sim 10^5$ yr) which is required for the effective isolation of actinide wastes.

REFERENCES

1. Abraham, M.M., Boatner, L.A., Quinby, T.C., Thomas, D.K. and Rappaz, M. (1980) Radioact. Waste Manage., 1, 181.
2. Ewing, R.C. and Haaker, R.F. (1980) Nucl. Chem. Waste Manage., 1, 51.
3. Eyal, Y. and Kaufman, A. (1982) Nuclear Technology (in press).
4. Huang, W.H., Maurette, M. and Walker, R.M. (1967) in Radioactive Dating and Methods of Low Level Counting, International Atomic Energy Agency, Vienna, pp. 415-429.

SCIENTIFIC BASIS FOR RADIOACTIVE WASTE MANAGEMENT - V
Werner.Lutze, editor

INVESTIGATION OF TITANIUM IN METAMICT Nb-Ta-Ti OXIDES USING THE EXTENDED X-RAY ABSORPTION FINE STRUCTURE TECHNIQUE

R. B. GREEGOR AND F. W. LYTLE
The Boeing Company, Seattle, Washington 98124
R. C. EWING AND R. F. HAAKER
Department of Geology, University of New Mexico, Albuquerque, NM 87131

INTRODUCTION

Recent proposals have suggested that radioactive wastes can be isolated as dilute solid solutions in a crystalline, titanate assemblage.[1] One titanate assemblage, SYNROC, consists of zirconolite ($CaZrTi_2O_7$), perovskite ($CaTiO_3$) and "hollandite" ($BaAl_2Ti_6O_{16}$) with additional accessory phases. There are two major problems in the evaluation of the long term stability of any crystalline wasteform such as SYNROC: 1) it is difficult to assess the long term stability of materials from short term laboratory experiments that are not necessarily valid simulations of complex geochemical processes, and 2) the corresponding titanate minerals are uncommon, making it difficult to study long term alteration and radiation effects on a significant number of specimens from different localities and geologic environments. There has even been considerable controversy concerning the stability of the reasonably common and simple phase, perovskite ($CaTiO_3$).[2]

An important approach in the evaluation of crystalline wasteforms is to study the long term stabilities of closely related mineral species which have become metamict (radiation damaged) and have been exposed to weathering processes for geologic periods of time. The documented metamictization and alteration effects can then be used as "benchmarks" for comparison with the results of short term laboratory leaching and irradiation experiments which have been designed to simulate long term effects. The Nb-Ta-Ti complex oxide minerals, containing appreciable rare earth elements (REE), U, and Th, are natural analogues for phases in proposed titanate radioactive wasteforms. Because of the geochemical similarities among Ti, Nb and Ta, a study of the metamict state and alteration effects in complex Nb-Ta-Ti oxides may yield data that is important in evaluating the long term stability of radioactive wasteforms.

The investigation reported here is concerned with the application of the extended x-ray absorption fine structure (EXAFS) and x-ray absorption near edge structure (XANES) spectroscopy to the study of the structure of the

metamict state. These techniques were used to obtain information on the
nearby atomic arrangement for the Ti site of the metamict Nb-Ta-Ti oxides of
euxenite and blomstrandine. X-ray absorption spectroscopy is unique in its
ability to provide such information about specific elements in these minerals
which are often non-crystalline.

DESCRIPTION OF SAMPLES AND EXPERIMENTAL DETAILS

The metamict Nb-Ta-Ti oxides of the type formula $A_x B_y O_z$ (A = REE, Fe^{+2},
Mn, Na, Ca, Th, U, Pb; B = Ti, Nb, Ta, Fe^{+3}) are generally assumed to include
the following minerals: euxenite, polycrase, priorite, blomstrandine,
aeschynite, pyrochlore-microlite, davidite, betafite, brannerite, columbite -
tantalite, samarskite, perovskite and zirconolite, as well as numerous variations. These complex oxides have caused problems of characterization since
their initial description in the early 1800's. Their complex and variable
compositions combined with their often metamict state and pervasive alteration have resulted in a contradictory nomenclature and inconsistent values
for many mineralogic parameters.[3] There are two distinct mechanisms that
alter complex Nb-Ta-Ti oxides, metamictization (alpha particle-recoil
nucleus radiation damage) and chemical alteration.

Minerals containing as little as 0.1 weight percent UO_2 or ThO_2 may become metamict. Accompanying the metamict transition are decreased density,
refractive index, and birefringence; and increased hydration, alteration
and microfracturing. As the crystal structure becomes increasingly disrupted,
x-ray diffraction maxima broaden, shift to lower values of 2θ and eventually
become indistinguishable from background. The x-ray diffraction pattern of
a metamict mineral is similar to that of a glass.[4,5] Recent transmission
electron microscope studies of metamict silicates and AB_2O_6 type Nb-Ta-Ti
oxides have demonstrated that the minerals are amorphous, at least to the
limits of resolution ($\sim$ 10 A).[6,7]

For EXAFS/XANES studies, samples of blomstrandine and euxenite were
selected from an array of specimens previously investigated by thermal
annealing and x-ray powder diffraction.[3] The composition of the samples
determined by electron microprobe analysis are given in table 1. The alpha
dose is estimated to be > 10^{18} α/gm.

These samples were taken to the Stanford Synchrotron Radiation Laboratory
(SSRL)[8] to perform EXAFS/XANES experiments. At SSRL the side station of
Beam Line VII (a Wiggler line) was used, with the synchrotron beam at a
current of approximately 60 mA and an energy of 3 GeV. A Si(220) double

TABLE 1

CHEMICAL ANALYSES OF METAMICT ORTHORHOMBIC AB_2O_6 - TYPE Nb-Ta-Ti OXIDES

	Blomstrandine R13	Blomstrandine R24	Euxenite R25
CaO	0.70	1.70	2.40
MnO	--	0.20	.46
Fe_2O_3	1.50	3.33	3.89
PbO	0.80	0.05	0.36
U_3O_8	3.80	4.40	12.12
ThO_2	5.40	5.80	2.85
Y_2O_3		12.70	11.80
Gd_2O_3		1.69	0.89
Dy_2O_3	30.9	2.15	1.82
Er_2O_3		0.02	2.11
Yb_2O_3		2.32	3.41
Ce_2O_3		0.25	0.23
Nd_2O_3	1.50	0.95	0.68
Sm_2O_3		1.34	0.56
Sum RE_2O_3	32.40	24.90	22.70
Nb_2O_5	19.00	17.72	23.19
Ta_2O_5	2.00	4.29	0.49
TiO_2	31.9	31.71	25.45
H_2O^-	0.40	--	--
H_2O^+	1.30	3.99	3.63
TOTAL	99.20	99.53	99.26

TABLE 2

EXAFS DETERMINED STRUCTURAL PARAMETERS

	Blomstrandine R13	Blomstrandine R24	Euxenite R25	Aeschynite*	Fersmite*
N	$6.4 \pm .7$	$5.3 \pm .4$	$5.3 \pm .4$	6.0	6.0
R (Å)	$2.01 \pm .05$	$2.03 \pm .05$	$2.03 \pm .05$	$2.03 \pm .09$	$2.02 \pm .21$
$\Delta\sigma^2$ (Å^2)	.00299	.00309	.00117	---	---

*Determined by x-ray diffraction[17,18]

412

crystal monochromator was used. The angle between the crystal faces was adjusted to detune the diffracted beam by 50% in order to reduce the harmonic content of the beam in the sample chamber. The beam in the sample chamber was approximately 1 mm x 20 mm, and the monochromator was moved in .25 eV steps in the vicinity of the Ti K-edge. Above the edge the step size was adjusted to provide a step of .050 $\pm$ .015 $\overset{\text{o}}{\text{A}}^{-1}$. The data were gathered at room temperature using a fluorescent x-ray detector similar to that described by Stern and Heald.[9] Nitrogen was used in the I_o and I chambers, with He in the sample holder area to enhance transmission of the x-rays and minimize background scattering in vicinity of the Ti K-edge. The Berkeley software[10] was used to gather the XANES/EXAFS data.

EXAFS RESULTS

The normalized EXAFS, Fourier transforms and inverse Fourier transforms for the samples studied are shown in figures 1a, b and c respectively.

The EXAFS oscillations in figure 1a can be described by single scattering theories[11,12,13,14] as

$$\chi(k) = \frac{1}{k} \sum_j \frac{N_j}{R_j^2} F_j(k) \exp(-2\sigma_j^2 k^2) \sin[2kR_j + \phi_j(k)]$$

In this representation, N_j is the number of atoms in the jth coordination shell, at distance R_j from the absorbing atom. $F_j(k)$ is the backscattering amplitude for an element in the jth shell, and σ_j^2 is the mean square relative displacement between the absorbing and scattering atoms along R_j. $\phi_j(k)$ is the total phase shift due to the interaction of the ejected photoelectron with the potential of the absorbing and scattering atom. Both $F_j(k)$ and $\phi_j(k)$ depend upon the kind of absorbing and backscattering atom. For oxygen neighbors the scattering amplitude function increases with decreasing k. Thus, information concerning the oxygen is contained at low k values between approximately 4 and 10 $\overset{\text{o}}{\text{A}}^{-1}$.

Evidence of oxygen coordination of the Ti is clearly evident in the Fourier transforms of EXAFS shown in figure 1b. The peaks in the radial distribution function correspond to the different coordination shells. The peaks indicated by the arrows in figure 1b are due to nearest-neighbor oxygen atoms. The position of this peak can be used to estimate the bond length by using a standard sample having a known R_j so that a ΔR_j correction can be made due to the effect of $\phi_j(k)$. The height and width of the peak are related to N_j and σ_j but precise determination of N_j or σ_j is difficult due

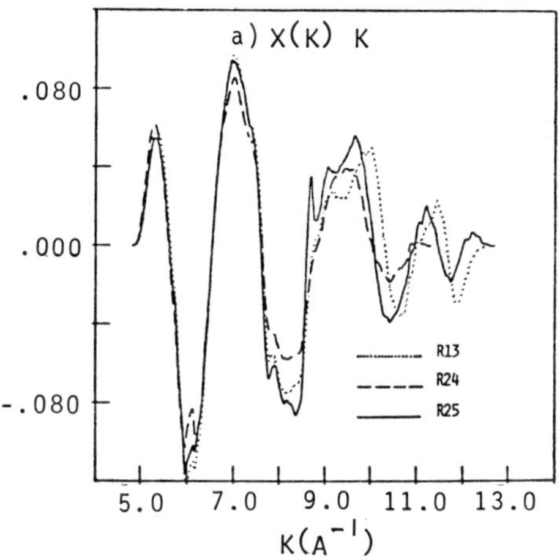
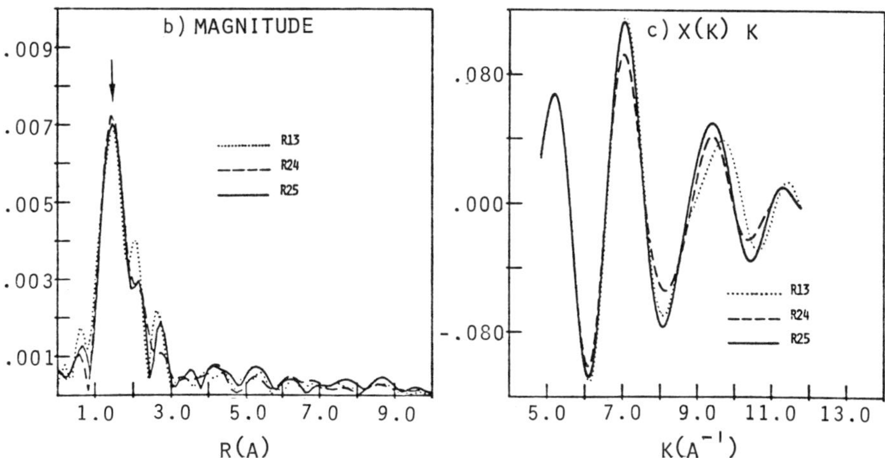

Fig. 1. The a) normalized EXAFS b) Fourier transform magnitude and c) inverse 1st shell transform of Blomstrandine (R13, R24) and Euxenite (R25).

414

to convolution effects of the Fourier transforms of $F_j(k)$ and the other $\chi(k)$ terms in R space.[15]

A more reliable method to determine N_j and σ_j is by isolating a given peak in the radial distribution function and performing an inverse Fourier transform. This procedure decouples the contribution of adjacent coordination shells so that the effect of only the coordination shell of interest is preserved. The resultant χ for one shell can be used in a least-squares fitting routine to determine N and σ. However, for systems having only one kind of atom in the coordination sphere of interest a ratioing technique can be implemented. The natural logarithm of the ratio of the inverse Fourier transforms of standard and unknown plotted as a function of k^2 has a slope which is proportional to $\Delta\sigma^2$ and an intercept proportional to N_s/N_u. Analytically this gives

$$\ln \frac{\chi_s(k)}{\chi_u(k)} = 2[\sigma_u{}^2 - \sigma_s{}^2]k^2 + \ln \frac{N_s}{N_u}\left(\frac{R_u}{R_s}\right)^2$$

which is of the form

$$y = mk^2 + b$$

where $m = 2\Delta\sigma^2$ and $b = \ln (N_s R_u{}^2/N_u R_s{}^2)$. Knowing N_s, R_u and R_s allows calculation of N_u for the unknown. The standard should be as chemically similar to the unknown as possible. Care must also be exercised in keeping the standards and unknowns thin or of the same effective thickness so that more accurate values of N_u can be obtained.

From the inverse transforms the number of oxygen neighbors about the Ti was evaluated using the "ln of the ratio" technique with anatase, rutile and GeO_2 as standards. As shown in table 2 this analysis yielded average oxygen coordination numbers of $5.3 \pm .4$ and $6.4 \pm .7$ for the two blomstrandine samples (R24, R13) and $5.3 \pm .4$ for the euxenite sample (R25). For the two blomstrandine samples we observed a decrease in the average titanium near-neighbor oxygen coordination and an increase in the structural disorder for the sample (R24) containing the greater percentage of U and Th. This is consistent with a model of radiation damage (creation of oxygen vacancies and structural disorder about the titanium site) due to alpha disintegration of the U and Th radionuclides. The EXAFS determined Ti-O average bond lengths for the peaks in the Fourier transforms are $2.02 \pm .05$ A for the blomstrandine and $2.03 \pm .05$ A for the euxenite. These values are within .01 A of the values of the crystalline counterparts of blomstrandine and euxenite.

XANES RESULTS

The use of XANES to probe chemical states of Ti and other elements has its physical basis in the absorption of x-rays which occurs upon photo-excitation and photo-emission of the metal atom core-level electrons. For x-ray photon energies near the core-level binding energy, transistions occur to bound states of the molecule containing the metal atom, giving rise to structure in the absorption spectrum in the vicinity of the absorption edge.

The near-edge spectrum of Ti and other transition metals is rich in information from which details of the coordination environment can be deduced. There is a distinction between the spectra for Ti in a highly symmetric octahedral environment, and the spectra for Ti in which inversion symmetry is broken, as in a tetrahedral environment. Figure 2 illustrates the XANES of the approximately octahedral environment for TiO_2 in the mineral forms rutile, anatase, and brookite. The near-edge spectra of these materials consist of three weak peaks lying below the absorption edge, plus weak features in the edge itself. The spectra (figure 3) for TiO_2 in SiO_2 glass, which has tetrahedral coordination,[16] exhibit a single strong pre-edge feature. This distinctive contrast between octahedral and tetrahedral phases of TiO_2 forms an important element in distinguishing Ti in octahedral form from Ti in tetrahedral, or any arrangement which breaks inversion symmetry.

As shown in figures 2 and 3 the pre-edge feature indicates that the Ti in the metamict samples does not have the same octahedral coordination as in anatase, rutile or brookite. The 3d feature is similar in appearance but not as intense as that usually observed for Ti in four-fold coordination. These results show that some fraction of the Ti is in an environment lacking inversion symmetry. From the shift in the Ti edges of the metamicts with respect to Ti metal it appears that the Ti is in a +4 oxidation state.

SUMMARY

A preliminary investigation of metamict orthorhombic AB_2O_6 complex Nb-Ta-Ti oxides has been conducted using the EXAFS/XANES technique. This investigation has demonstrated the feasibility of the technique to study complex geologic or simulated radioactive waste samples. The element specificity of the technique has allowed a detailed structural analysis of the Ti site in complex metamict Nb-Ta-Ti oxides. Although preliminary, these results suggest that significant changes in Ti coordination number

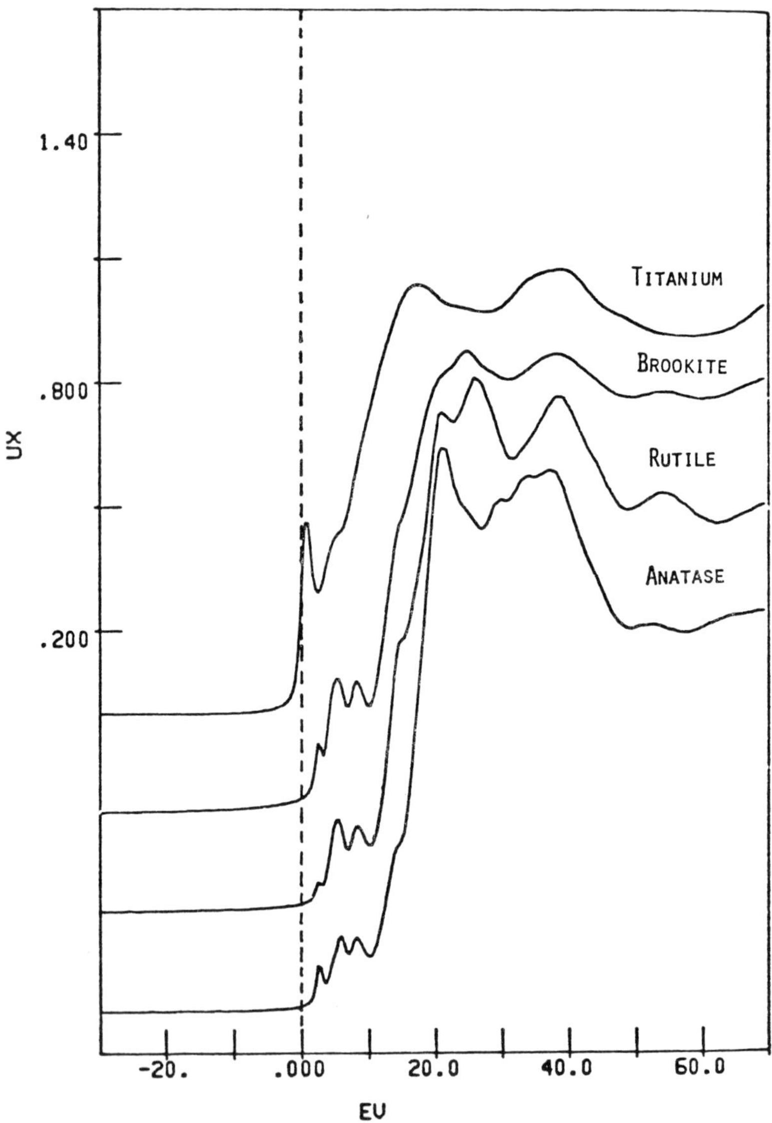

Fig. 2. Near-edge K-absorption spectra of Ti metal and three TiO_2 polymorphs in which Ti is approximately octahedrally coordinated. The zero of energy is the 1st inflection point of Ti metal.

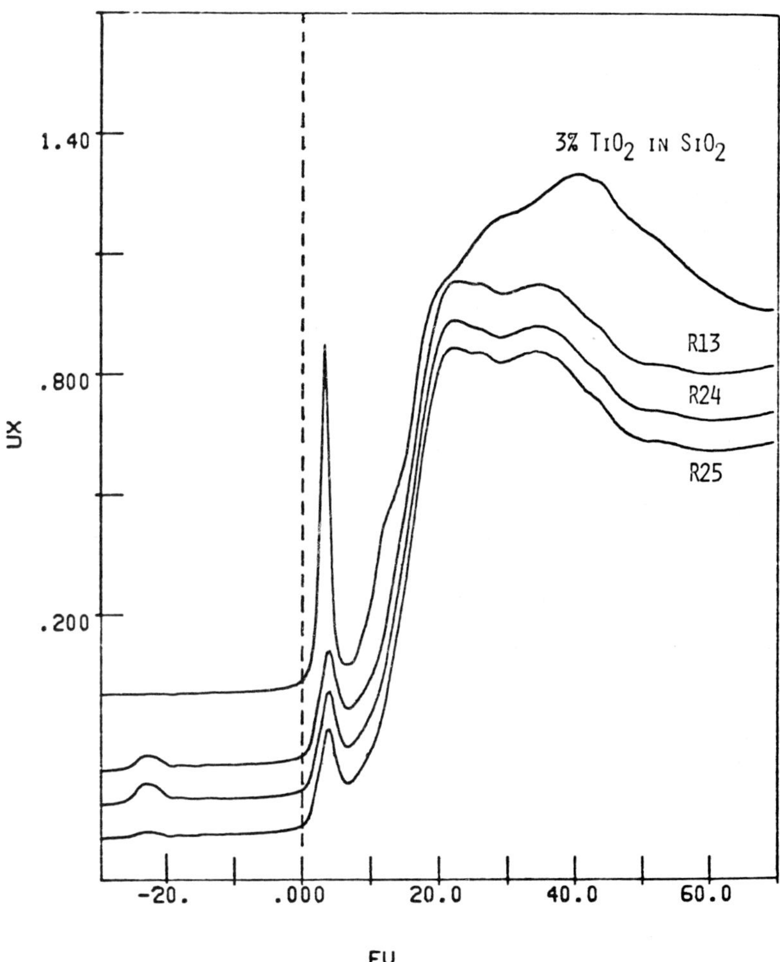

Fig. 3. Near-edge K-absorption spectra of 3% TiO$_2$ in SiO$_2$ in which Ti is tetrahedrally coordinated and the metamict minerals.

and geometry may take place during the metamictization process. This finding could have important implications for solid radioactive waste-forms.

ACKNOWLEDGMENTS

We are grateful to the excellent staff at SSRL for experimental support and to DOE and NSF for support of the facility. The research of two of us (R. B. Greegor and F. W. Lytle) was partially supported by NSF Grant DMR-8013706.

REFERENCES

1. Ringwood, A. E., Kesson, S. E., Ware, N. G., Hibberson, W., and Major, A (1979) Nature 278, 219.
2. Nesbitt, H., Bancroft, G., Fyfe, W., Karkhanis, S., Nishijima, A., and Shin, S. (1981) Nature 289, 358.
3. Ewing, R. C., Can. Min. (1976) 14, 111.
4. Haaker, R. F. and Ewing, R. C. (1979) in Ceramics in Nuclear Waste Management. Chickalla T. and Mendel J., ed 305.
5. Ewing, R. C. and Haaker, R. F. (1980) Chem. Waste Mgt. 1, 51.
6. Headley, T., Ewing, R. C. and Haaker, R. F. (1981) Nature 293, 449.
7. Ewing, R. C., Haaker, R. F., Headley, T. J. and Hlava, P. (1981) Proc. Fourth Int. Sym. on the Sci. Basis for Nucl. Waste Mgt. (in press).
8. Winick, H. and Bienenstock, A. (1978) Ann. Rev. Nucl. Part. Sci. 28, 33.
9. Stern, E. A. and Heald, S. (1979) Rev. Sci. Instr. 50, 1579.
10. Kirby, J. A. (1978), "Manual for Data Collection Program," Univ. Calif. Berkeley.
11. Sayers, D., Stern, E. and Lytle, F. (1971) Phys. Rev. Lett. 27, 1204.
12. Stern, E., Sayers, D. and Lytle, F. (1975) Phys. Rev. 11, 4836.
13. Lee, P., and Pendry, J. (1975) Phys. Rev. 11, 2795.
14. Lee. P., and Beni, G. (1977) Phys. Rev. 15, 2862.
15. Greegor, R. B. and Lytle, F. W. (1979) Phy. Rev. B. 20, 4902.
16. Sandstrom, D. R., Lytle, F. W., Wei, P. S. P., Greegor, R. B. Wong, J., and Schultz, P. (1980) J. Non-Crys. Sol. 41, 201.
17. Alexsandrov, V. B. (1961) Akad. Nauk SSSR Doklady, 149, 181.
18. Alexsandrov, V. B. (1960) Akad. Nauk SSSR Doklady, 132, 669.

CHARACTERIZATION OF REPOSITORIES

Salt

THE ISOLATION CAPABILITY OF A SALT DOME UTILIZED FOR HIGH-LEVEL WASTE DISPOSAL

J. HAMSTRA
Netherlands Energy Research Foundation, P.O. Box 1, 1755 ZG Petten N.H.,
The Netherlands.

INTRODUCTION

The overall objective of disposal of high-level and alpha bearing wastes in
deep geologic formations is to isolate these wastes from the biosphere for such
a period of time that a subsequent future release of radionuclides from the
buried wastes will not result in undue radiation exposures to man.

An effective long-term isolation of radioactive waste can, because of its
time scale, never be demonstrated. The performance of an underground disposal
concept can only be predicted, by evaluating and analysing the performance of
the overall system of man-made and natural barriers.

Model calculations can be used to perform these assessments. Up till now
most modelling was done for generic hostrock specific approaches. There are
basic differences in approach to achieve the required isolation of the wastes
in the distinct hostrocks under consideration. The diversities mainly de-
rive from differences in rock characteristics, in geohydrology and in response
to emplacement of the heat producing waste.

Crystalline rock has negligible plasticity. It therefore can develop fractures
that contain groundwater at the depth considered appropriate for disposal of
high-level waste. The isolation of the waste in a crystalline rock repository
therefore is assured by assigning an important barrier function to the waste
form, the waste container wall and a buffer material directly around the
emplaced waste containers. The disposal concept developed by the Nuclear Fuel
Safety Project (KBS) for determining the suitability of Swedish bedrock for
final storage of high-level waste is a good example of such a multi-barrier
disposal system that relies heavily on the man-made barriers.

Rock salt on the contrary can, because of its plastic properties, be
utilized as a potential host rock in such a way that its function as a natural
barrier becomes dominant in the overall disposal concept. It is this rock salt
and more in particular this salt dome specific potential that will be discussed
in this presentation.

In spite of its solubility in groundwater, rock salt is considered to be a
favourable host rock, because of
- its availability in large underground structures, such as salt domes,
- its viscoplastic properties that makes it essentially impermeable to liquids
  and gases,
- its compressive strength that in spite of its plasticity allows for excavations up to greater depths to remain accessible for disposal operations, and
- the ease with which it can be mined at relatively low costs.

It should be recognized that the size, shape and internal structure of a
salt dome will vary considerably from salt dome to salt dome. The complex
internal structure of a salt dome is a matter of concern, because the location
of the disposal area inside the salt dome needs to be adapted to an eventual
presence there of greater quantities less favourable non-halite evaporites,
such as anhydrite and carnallite.

All concepts for disposal of high-level waste in a salt dome aim at maintaining an isolation shield of hundreds of meters of undisturbed salt dome structure
over and around the disposal facility. Performance assessment studies prove that
this isolation shield is a major natural barrier against any future radionuclide
release.

The high isolation capability of a salt dome can only be made credible by a
careful and rational approach in developing that salt dome into a repository
area. That approach should first review the capability of the outercircumference
of a salt dome to serve as an isolation shield, then reconsider the disposal
mining impacts on that part of the salt dome and finally reconsider what
phenomena could on the long-term break the isolation shield around the sealed
and abandonned disposal facility.

THE PRESENT-DAY CONDITION OF THE ISOLATION SHIELD STRUCTURE

All salt dome disposal concepts aim at an appreciable mass of salt dome
structure overlying and surrounding the actual high-level waste disposal area.

Fig. 1 shows in cross-section the disposal concepts proposed for the salt
domes of Mors and Gorleben and a salt dome disposal concept model with an
isolation shield of 200 m thickness. This figure may illustrate how large the
isolation shield thickness may work out in reality, compared to the 200 m
thickness discussed in this review.

It is important to note that all natural processes or events that are to be
reconsidered on their future impact on the isolation capability of a salt dome

have been active somewhere some time period during the many millions to many
tens of millions years since the salt dome reached its full maturity. The
outer surface of the salt dome should therefore reveal eventual scars from
past impacts, if any.

Anhydrite sands or anhydrite caprock are for example often found at the
flanks and the top of salt domes as a residual evidence of dissolutioning
processes that evolved ever since the salt dome came into existance.

Neither the presence nor the absence of caprock at the flanks of a salt dome
is to be judged unfavourable for a salt dome of sufficient age. The presence of
residual anhydrite clearly contributes in retarding any future dissolutioning
of the underlying rock salt. The absence of an anhydrite layer can on the other
hand only be regarded as evidence that a variaty of geohydrological situations
experienced in the past all were unfavourable for effecting dissolutioning.
There is rational to assume that the geohydrologic situation around the salt
dome will in the near future remain within the range of geohydrological
conditions that were experienced up till the present and that thus the situation
will remain unfavourable for dissolutioning attacks.

Seismic reconnaissance will reveal the geometry of the salt dome with
sufficient accuracy to allow the reconnaissance development work to proceed
without penetrating the isolation shield area. The internal structure of the
isolation shield and its compositon according the different evaporites and
salt minerals can therefor not be established. From experience and knowledge
in salt mining in salt domes it can only be assumed that the stratigraphy
and the minerological composition of the isolation shield will be complicated
and strongly diversified.

The structural strength of the total isolation shield wall will not suffer
from that complexicity of its diversified stratigraphy of primary evaporites
and secondary salt minerals that resulted from metamorphism processes that
occured in the past.

Where these secundary salt minerals are found as fill-ups of old fractures
these scars from past tectonic movements should preferably be earmarked as
evidence of the selfhealing capability of rock salt more than that it should
be used as evidence that a total thickness of hundred of meters of salt dome
structure could loose its integrity.

Interbedded in the large masses of halite from different evaporite cycles
the relative thin potash beds and anhydrite banks and the local presence of
secondary salt minerals do not affect the overall structural capability of an

422

isolation shield of 200 m thickness nor its salt migration response to any
local strain differential.

Eventual anhydrite banks that were dragged along with the main upward salt
flow will have their main axis in the flow direction, that is mainly parallel
to the flanks. Only at a top of a salt dome that for long times was exposed to
dissolutioning processes, this vertical position of the anhydrite banks needs
special attention. Residual ridges of these vertical anhydrite banks may extend
through the flat top caprock up into an overlying aquifer. This is one of the
reconsiderations for the development of the network of roadways to be carried
out with great care. Major anhydrite banks and/or carnallite beds will only be
penetrated by roadways if judged necessary and justifiable. These crossing
will then anyhow be provided with a monolithic reinforced concrete liner and
with injection pipes to enable chemical or cement grouting in case this will be
required, primarily as a preventive measure. These liners are good examples of
engineered disposal mining features for maintaining the isolation shield in
good shape.

The existence in the isolation shield of local fissures, faults or cavities
of any importance is only imaginable when filled with a gas or fluid pressurized
up to the local prevailing lithostatic pressure.

There are quite some references to be found in literature of local fractures
observed in specific areas in rock salt. The point is to be stressed that none
of these fractures were of sufficient extension to create a passageway for
groundwater to enter deep into a salt dome. If these fractures ever served as
passageways for brine to migrate through the salt dome structure it only must
have been brine that stood under the prevailing lithostatic pressure and that
was being squeezed out of the salt dome.

In an undisturbed salt dome that reached its post-diapiric stage long ago
there is no compatibility for fractures all across the isolation shield area,
that are open towards an adjacent aquifer.

The hydrostatic pressure in a fracture that is open to an aquifer is insufficient to keep the fracture open against the lithostatic pressure that
prevails in the rock salt surrounding the fracture. The overpressure at the
outside of the fracture will establish a salt migration response, that easily
will overcome the very low formstability of the fracture and that will close
the fracture, the moment it would be formed.

It should thus be evident that a shield of 200 m of undefined salt dome
structure will anyhow be a solid natural barrier to start with. It then becomes

a prerequisite and a challenge for the disposal miner to limit his impacts on
the isolation shield to as few and as limited as possible.

DISPOSAL MINING IMPACTS ON THE ISOLATION SHIELD

The disposal mining impacts on the salt dome in general and the isolation
shield in particular can be limited to

- the two shaft penetrations through the top of the salt dome,
- the mine excavations required to get access to the interior of the salt dome
  and perform the actual waste disposal operations, and
- the decay heat production in the emplaced high-level waste packages and its
  accumulating heat load on the rock mass.

The two shaft penetrations are inevitable impacts on the integrity of the
isolation shield. Their construction, consisting of a welded steel cylinder
supported at the inside by a reinforced concrete tubbing and bedded at the outside in a layer of bitumen, is designed to withstand considerable strata
movements without losing its strength nor its watertightness. There is general
consensus between the experts in this specialized field of technology that this
shaft construction is qualified to survive any major damage during the operational period of the disposal mine.

Special backfilling and sealing provisions are foreseen for the shafts prior
to the abandonment of the repository. A careful working procedure can secure
that the previous shaft locations will become well healed scars in the isolation
shield.

The disposal mining excavations will, compared to the normal salt production
mining, be relative small in total excavated volume and low in extraction rate.

The disposal mining will only create a very limited rock salt creep response
in the direct vincinity of the mined openings and this will not affect the
structural stability of the isolation shield.

The high-level waste is to be emplaced in deep vertical disposal boreholes
from a depth of about 900 m down to about 1200 m. The spacing between these
deep disposal boreholes can be made so large that each 100 liter of high-level
waste will on the average be surrounded by some 10,000 m^3 rock salt. This large
spreading of the high-level waste canisters over the burial areas may be necessary to achieve rather stringent temperature limitations for the rock salt
such that no undue changes in certain hydrated salt minerals can occur.

The decay heat production in the emplaced high-level waste canisters that
causes the temperature rise of the rock salt in the disposal area will also

lead to a thermal expansion induced lithostatic pressure rise in that area.

For the non-retrievable deep borehole disposal concept the disposal bore-holes will be carefully plugged and sealed immediately after the emplacement of the canisters. The borehole convergence will then establish a confinement pressure on the emplaced canisters that is at least equal to the lithostatic pressure that prevails at their burial depth.

There thus will no longer be question of any pressure difference nor any vapour pressure difference towards an open borehole when evaluating the rock salt conditions directly around the emplaced high-level waste canisters. This is an important point because if some laboratory tests on rock salt samples and even in-situ tests in open boreholes fail to meet these equivalent condi-tions, their results thus become irrelevant for the in-situ disposal situation.

The pressure differences that will arise will cause salt creep where possible in the mine excavations and on the longer term also some thermal expansion movement at the top and on the flanks of the salt dome. It will be a design criterion to limit the total heat source such that the overall expansion movement at the top of the salt dome will be limited to an upward movement of less than 1,5 m in 150 years time. This movement is considered to be harmless to the structural stability of the isolation shield and also sufficiently slow for the overlying sediments to follow it.

It can also be a design criterion to plug and seal the high-level waste disposal boreholes in such a careful way that the seals will remain brine tight against the hydrostatic pressure that could prevail in a flooded disposal facility. Should an unforeseen flooding of the abandonned disposal facility on the longer term happen than there is still no rational to assume that the brine, if it should be squeezed out of the burial facility, could act as a transport medium for the radionuclides from the high-level waste to escape from their deep borehole confinement.

Summarizing disposal mining in a salt dome, that is carefully aimed at a true confinement of the high-level waste, need not to have any unfavourable effect on the original conditions in the isolation shield.

REVIEW OF PHENOMENA SUPPOSEDLY RELEVANT FOR BREACHING THE ISOLATION SHIELD

The phenomena often supposed to be relevant for developing a future breach in the ·isolation shield are an eventual future human intrusion into the salt dome, the sudden natural events such as earthquakes, and the slowly developing natural processes such as subrosion.

In short they can be reviewed as follows.

Future human intrusion

Future human intrusion into a salt dome that was used for nuclear waste
disposal purpose can not happen accidentally as long as a system of governmental
controls will continu to exist. In addition a thorough marking at the surface of
the disposal site is foreseen as a long-term warning. The disposal mining should,
because of its regional importance, be expected to be reminded by its markers and
their related story as long as humans will continu to live in that area.

Any future accidental intrusion can therefor only happen after such time
period that the radiation exposure level of a chance striking of some high-level
waste will have become too low to consider it a true risk [1].

A purposive future intrusion into a salt dome that was used for nuclear waste
disposal purpose is not considered in this context for the following reasons.

The resource value of the emplaced waste canisters is negligible in proportion to its mining costs. The salt itself will be of low interest because of the
continuing salt production in already existing mines nearby. If despite that
some adventurer knowingly wants to explore the old disposal facility this should
be considered an extremely costly affair that is undertaking at a voluntary risk,
against which no preventive measures can be effective.

Sudden natural events

The three sudden natural events that are often supposed to have an unfavourable effect on the isolation of the waste are a major earthquake, a meteorite
impact and a flooding of the area.

Earthquakes originate from a sudden release of a large amount of stored
deformation energy. The salt domes under consideration for nuclear waste
disposal are surrounded by sedimentary rocks and partly even overlain by unconsolidated sediments. Because of their very low to zero capacity for storage
of deformation energy these surroundings can never allow a large local seismic
centre to develop. If local underground displacements would ever occur they
therefor only will be very small compared to the total thickness of the
isolation shield [2].

For two reasons a major meteorite impact is not considered to be a realistic
phenomenon for breaching the isolation shield.

Firstly the direct consequences at the surface of the meteorite impact are

so disastrous that they fully will overcloud any future consequence of a radionuclide release from the repository, should that be initiated by that impact.

Secondly there is the very low to zero probability that a large meteorite
would hit the actual repository area.

A future flooding can only be supposed to be a long-term event for salt domes
situated in an embanked coastal region and related to a future rise of the sea
level. The sudden event then could arise from a burst through that embankment
at a time that the sea level has continued to rise above the land level. Thus
only an abandonned disposal site can be affected. Renewed sedimentation may
take place above the salt dome, the local geohydrology will change, but no
geohydrological situation will originate that is outside the variety of
situations experienced by the salt dome area in the past. Possible radiological
consequences of an unforeseen subsequent flooding of the abandonned repository
will be negligible compared to disastrous consequences of the preceding flooding
of the region itself.

Slowly developing natural processes

There are many slowly developing natural processes that could be considered.
Salt domes that have their top rock salt presently at about 200 m depth and
that have a post-diapiric low future rate of uplift can not be affected
seriously by any of them, except one.

The design criterion that the repository is to be arranged at larger depth
is a further guarantee that none of the surfactant phenomena can lead to a
breach in the isolation shield.

There then remains only the contact with groundwater leading to a so called
subrosion, that is considered capable even at greater depth to create a future
breach in the isolation shield.

Many salt domes show evidence of subrosion in the past, mainly at their flat
top surface in the form of a layer of residual anhydrite. Several salt domes
also have residual anhydrite sands or caprock at their flanks as a proof that
water circulation also existed along the flanks.

Literature data on past and present dissolutioning rates due to subrosion at
the top surface of German salt domes vary from an averaged 0.03 mm/y for many
salt domes in the past Malm-Wealden uplift period to an average 0.003 mm/y for
the Marne salt dome in the younger Lower Creteaceous calculated by Jaritz [3],
and from 0.04 to 0.004 mm/y for the salt dome of Salzgitter-Lebenstedt
evaluated by Preul [4].

If the geohydrological conditions around a salt dome are such that future
subrosion is a realistic assumption, then for certain these subrosion conditions have existed and thus been effective in the past for many million years
during those past geological time periods that had comparable to less
favourable hydrological circumstances. It is therefor also realistic to assume
that salt domes that have existed for tenth of millions of years and that
presently have their top at about 200 m depth can only continue to be exposed
to subrosion with a low dissolutioning rate in the order of hundreds to
thousands of a millimeter per year or meters to tenth of meters per million
years.

SUMMARIZING REMARKS

From the preceding review it can be concluded that a high isolation
capability of a salt dome that is to be used for nuclear waste disposal will
be dependant of the following prerequisites:
- the salt dome should have its top rock salt at about 200 m depth or more,
  should be sufficiently large in its dimensions and should have clearly reached
  its post-diapiric stage;
- the repository is to be developed carefully at greater depth and leaving a
  large isolation shield of at least 200 m thickness over and around the
  emplaced waste; and
- the abandonment and closure of the repository and the two shaft constructions
  is to be done carefully too.

The conclusion then is that the only phenomenon that could be considered
realistic in creating a future breach in the isolation shield will be a
continuous subrosion process. Its low dissolutioning rate in the order of one
meter every 100,000 years to one meter every million years is a good yardstick
for the high isolation capability of the salt dome.
The prediction then is that if the prerequisites can be fullfilled the
isolation capability will fall outside the scale of human chronology and
approach infinity.

REFERENCES

[1] Verkerk, B. (1981) Bijdrage tot de veiligheidsbeoordeling van het opbergen van radioactief afval in zout. Energiespectrum 5, nr. 5, p. 130.

[2] Leyedecker, G. (1980) Erdbeben in Nord-Deutschland. Z.dt.Geol.Ges., 131, p. 548, Hannover.

[3] Jaritz, W. (1980) Einige Aspekte der Entwicklungsgeschichte der Nordwestdeutschen Salzstöcke.
Z.dt.Geol.Ges., 131, p. 391 and 401, Hannover

[4] Preul, F. (1968) Die Subrosion am Salzstock von Salzgitter-Lebenstedt. Geol.Jb., 85, p. 809-816, Hannover.

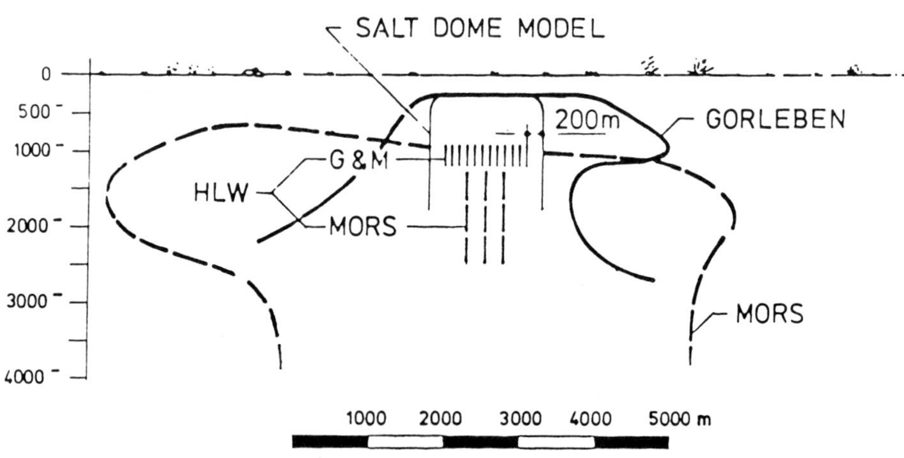

SALT DOME CROSS SECTIONS

FIG. 1

DISSOLUTION RATE OF SALT DOMES ON THE BASIS OF INTERPRETATION OF MEASURED SALINITY PROFILES

K.E. LINDSTRØM JENSEN
ELSAM, 7000 Fredericia, Denmark.

INTRODUCTION

When **radioactive** waste is disposed in a salt dome, it is important to evaluate the **hydrologic stability** of the dome. It depends on the dissolution rate of the dome, **which again** is determined by the transport and dispersion properties of the cap **rock and the** other formations surrounding the dome. The same properties are also **required** in safety assessment work for calculation of migration of radionuclides, **which** might be released from the dome to the surrounding strata.

In the **following analytical** solutions are described for calculation of salt concentrations **in ground** water corresponding to a certain dissolution rate of the dome. Using **these** solutions a method is described, which makes it possible to determine rate of dissolution, ground water velocity, and dispersivity in the strata just above cap rock on basis of measured salinity profiles above the dome.

The method is applied on the Danish salt dome Mors located in the northern part of Jutland. The cap rock is located at a depth of 600 m with limestone strata above. Interpretation of salinity profiles in the limestone just above cap rock gives a transverse dispersivity of 20 m in the limestone. The pore water velocity, here determined entirely on basis of measured salinity profiles, is found to be about 0.3 mm/y, which corresponds largely to that derived from measured permeabilities in the limestone. The corresponding average dissolution rate for the dome is determined to 0.0004 mm/y.

CALCULATION OF SALT CONCENTRATIONS IN GROUND WATER

This section treats the problem of how to relate a mass transport of salt from the dome to salt concentrations in the aquifer surrounding the dome.

Consider a model salt dome as shown in fig. 1. Salt is transported with a constant dissolution rate v_d (m/y) at the time $t = 0$ from the salt dome to the infinite porous medium above the dome through a rectangular dissolution area with edge lengths a(m) and b(m). The pore water velocity above the dome is

430

v(m/y) and elsewhere zero. The aquifer has porosity f and dispersion coefficients D_x, D_y, D_z (m^2/y).

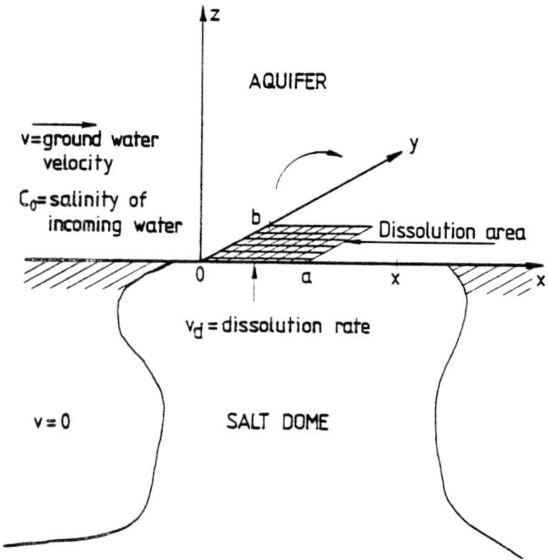

Fig. 1. Model salt dome.

The concentration of salt in the ground water $C(x,y,z,t)$ (kg/m^3) is determined by the solution to the partial differential equation

$$\frac{\partial C}{\partial t} = D_x \frac{\partial^2 C}{\partial x^2} + D_y \frac{\partial^2 C}{\partial y^2} + D_z \frac{\partial^2 C}{\partial z^2} - v \frac{\partial C}{\partial x} \tag{1}$$

It has been possible to apply an analytical approach for solution of this
equation resulting in a time dependent, analytical solution for $C(x,y,z,t)$ [1].
Above the dissolution area the concentration of salt in the ground water is
given by:

$$C(x,y,z,t) = \frac{2v_d \rho_s}{f\sqrt{\Pi}\,\sqrt{D_z}} \cdot$$

$$\left[\sqrt{t}\,\exp\left(-\frac{z^2}{4D_z t}\right) - z\sqrt{\frac{\Pi}{4D_z}}\left(1 - \text{erf}\sqrt{\frac{z^2}{4D_z t}}\right)\right] + C_0 \tag{2}$$

$$\text{for} \quad 0 \leq x \leq a$$
$$4\sqrt{D_y t} \leq y \leq b - 4\sqrt{D_y t}$$
$$0 \leq t \leq \frac{x}{v}$$

For t > $\frac{x}{v}$ the same equation can be used for the stationary solution by substitution of t by $\frac{x}{v}$. The density of salt ρ_s (kg/m³) is = 2165 kg/m³, and C_0 (kg/m³) is the salinity of incoming ground water from outside the dome.

Downstream the dissolution area the concentration is given by:

$$C(x,y,z,t) = 0 \qquad (3)$$

$$\text{for} \quad a \leq x$$
$$t \leq \frac{x-a}{v}$$

and

$$C(x,y,z,t) = \frac{2v_d \rho_s}{f\sqrt{\Pi}\ \sqrt{D_z}} \cdot$$

$$\left[\sqrt{t}\ \exp\left(-\frac{z^2}{4D_z t}\right) - \sqrt{\frac{x-a}{v}}\ \exp\left(-\frac{z^2 v}{4D_z(x-a)}\right) \right.$$

$$\left. - z\ \sqrt{\frac{\Pi}{4D_z}}\ \left(\text{erf}\ \sqrt{\frac{z^2 v}{4D_z(x-a)}}\ -\ \text{erf}\ \sqrt{\frac{z^2}{4D_z t}}\ \right) \right] + C_0 \qquad (4)$$

$$\text{for} \quad a \leq x$$
$$4\sqrt{D_y t} \leq y \leq b - 4\sqrt{D_y t}$$
$$\frac{x-a}{v} \leq t \leq \frac{x}{v}$$

For t > $\frac{x}{v}$ the stationary solution is obtained by substitution of t by $\frac{x}{v}$.

An example of calculated salt concentrations in the ground water corresponding to a salt dissolution rate of 1.2×10^{-4} m/y is shown in fig. 2.

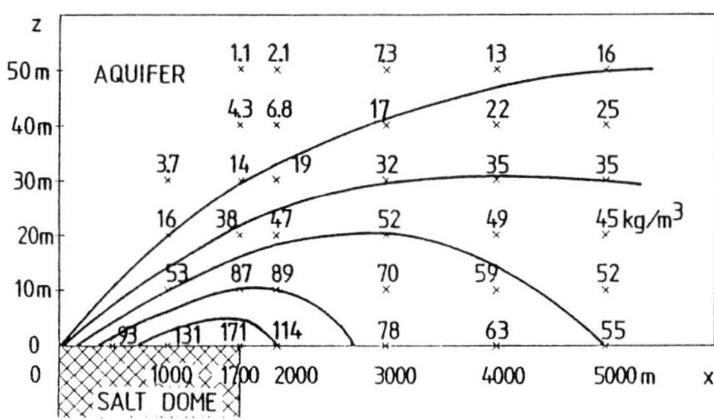

Fig. 2. Salt concentrations corresponding to a salt dissolution rate of 0.12 mm/y. Lines of constant concentration are indicated.

INTERPRETATION OF MEASURED SALINITY PROFILES

The Danish salt dome Mors located in the northern part of Jutland has been investigated as part of the Danish utilities' waste disposal project [2]. Fig. 3 shows a seismic interpretation of the Mors dome. The hydrological wells E1s and E2s are located close to the deep well Erslev 1, while E3s and E4s are situated close to Erslev 2.

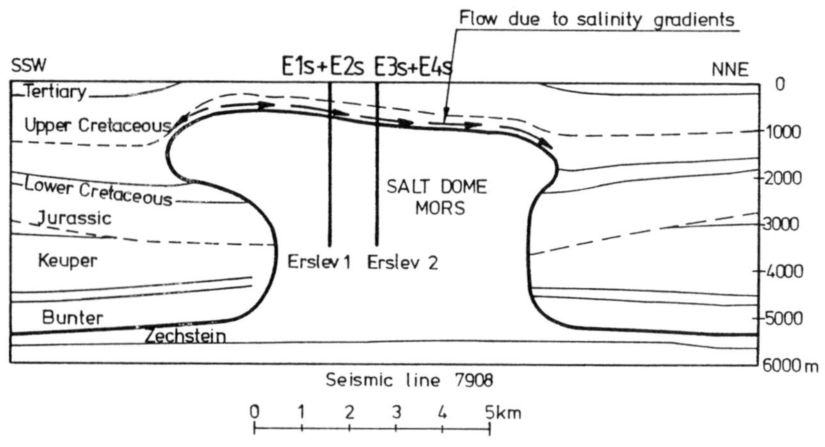

Fig. 3. Seismic interpretation of the Mors salt dome.

As shown in fig. 4 the measured salinity just above cap rock at Erslev 1 is about 150 kg/m³, which corresponds to half saturation concentration. Outside the dome, at a distance of about 10-30 km away, the salinity in the same depth is approx. 40 kg/m³. The considerably higher concentration above the dome indicates that dissolution has taken place in the past. The heavier saline water above the dome will move outwards and downwards following the contour of the dome as shown in fig. 3.

The salinity profiles above the dome depend on the dissolution rate and on the transport and dispersion of salt water in the limestone. This is used in the following to evaluate the dissolution rate, the ground water velocity and the dispersivity on basis of the measured salinity profile. Rearranging the stationary equation (2) gives

$$C(x,z) = (C(x,0)-C_0).$$

$$\left[\exp\left(-\frac{z^2}{4\alpha' x} \right) - z\sqrt{\frac{\Pi}{4\alpha' x}} \left(1 - \mathrm{erf}\sqrt{\frac{z^2}{4\alpha' x}} \right) \right] + C_0 \tag{5}$$

where α' is defined by

$$\alpha' = \frac{D_z}{v} \tag{6}$$

with D_z being defined by

$$D_z = \alpha v + D \tag{7}$$

where $\alpha(m)$ is the transverse dispersivity in the limestone, and $D(m^2/y)$ is the diffusion coefficient for salt in water = 0.025 m²/y. For the measured salinity profile at Erslev 1, x = 3.000 m, C(3000,0) = 150 kg/m³, and C_0 = 40 kg/m³. Equation (5) is now fitted to the measured salinity profile by variation of α'. As shown in fig. 4, α' = 30 m gives a reasonable fit to the measured profile.

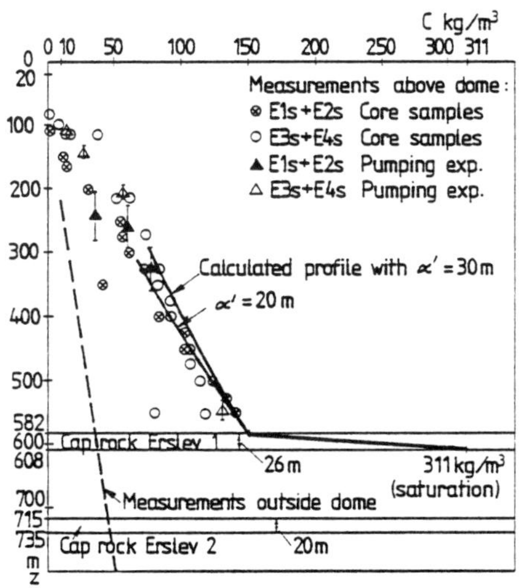

Fig. 4. Salinity profiles above and outside the Mors salt dome.

In the deep Erslev 1 and Erslev 2 drillings the cap rock has thicknesses of 26 m and 20 m respectively. In Erslev 2, a drill stem test was made directly in the cap rock in order to investigate the permeability of the cap rock. As it was almost impossible to extract water from the test, the cap rock has very little permeability. Therefore, the water in the cap rock must be almost without flow. The transport of salt from the dome up to the water-bearing strata above cap rock must then take place by diffusion in the cap rock at least at Erslev 1 and Erslev 2. Furthermore, logs in Erslev 2 indicate that the porosity of this deep lying cap rock is very low. An upper bound for the porosity is 0.1.

434

There is of course the possibility that the cap rock might be completely absent or might have a secondary porosity and permeability in areas, which have not been investigated. This situation is treated in the last section, where the cap rock is assumed to have heterogeneous properties.

Diffusion of salt in the cap rock gives

$$v_d = \frac{f_0 D}{\rho_s} \cdot \frac{\partial C}{\partial z}\Big|_{\text{cap rock}} \tag{3}$$

where f_0 is the porosity of cap rock. The diffusion of salt in the cap rock must be equal to the salt removed by the ground water above the cap rock. Elimination of v_d in the stationary equation (2) gives in combination with equation (8) the following equation for determination of the pore water velocity

$$v = \frac{2f_0 D}{f(C(x,0)-C_0)} \cdot \sqrt{\frac{x}{\Pi \alpha^1}} \cdot \frac{\partial C}{\partial z}\Big|_{\text{cap rock}} \tag{9}$$

By using v, the dispersivity may be determined from equation (6) and (7)

$$\alpha = \alpha^1 - \frac{D}{v} \tag{10}$$

The concentration gradient in cap rock at Erslev 1 is calculated to 6.2 kg/m^3/m from fig. 4. With $\alpha' = 30$ m and a porosity in the limestone above cap rock of $f = 0.23$ [2] equations (9), (8), and (10) give the following interdependent values for v, v_d, and α as functions of cap rock porosity f_0

Table 1.

f_0 Porosity	v (m/y) Pore water velocity	v_d (m/y) Rate of dissolution	α (m) Dispersivity
0.03	0.00207	2.1×10^{-6}	17.92
0.05	0.00346	3.6×10^{-6}	22.77
0.1	0.00691	7.2×10^{-6}	26.38

Measurements indicate [3] that the diffusion coefficient for salt in limestone (and perhaps in cap rock too) may be somewhat smaller than that used in the previous calculations. According to equation (9) a ten times smaller diffusion coefficient would reduce the pore water velocity by a factor of ten. This would make the pore water velocity, here determined entirely on basis of the measured salinity profile, to correspond to that based on the permeabilities measured in the limestone [2]. A ten times smaller diffusion coefficient would further reduce the dissolution rate by a factor of ten according to equation (8), whereas the dispersivity determined by equation (10) would remain unchanged. For a cap rock porosity of 0.05 the pore water velocity would then be about 0.3 mm/y, the dissolution rate 0.0004 mm/y, and the transverse dispersivity about 20 m. Variations in cap rock porosity will not significantly change these values.

TIME DEPENDENT SALINITY PROFILES

The greatest dissolution took place during the Lower Cretaceous about 100 mill. years ago, where the salt was in direct contact with the sea. Most of the cap rock was formed at that time. In the period following the Lower Cretaceous more and more chalk was deposited, which together with the subsequent compaction of the original porous sediments has decreased the dissolution rate of the dome.

The time dependent variation of the salinity profile for a decrease in the dissolution rate is illustrated in the following. In fig. 5 the stationary salinity profile is shown as determined by the stationary equation (2) and corresponding to a dissolution rate of 8.5×10^{-6} m/y. At t = 0 the dissolution rate is decreased from 8.5×10^{-6} m/y to 4×10^{-6} m/y. The associated time dependent variation of the salinity profile appears from fig. 5. The profiles are determined by superposition of the stationary profile by profiles determined by the transient equation (2) and corresponding to a dissolution rate of $(4 \times 10^{-6} - 8.5 \times 10^{-6})$ m/y.

Basic data:
v=0.003m/y   α=20m
f=0.23 C_0=0
D=0.025m²/y
a=x=3000m

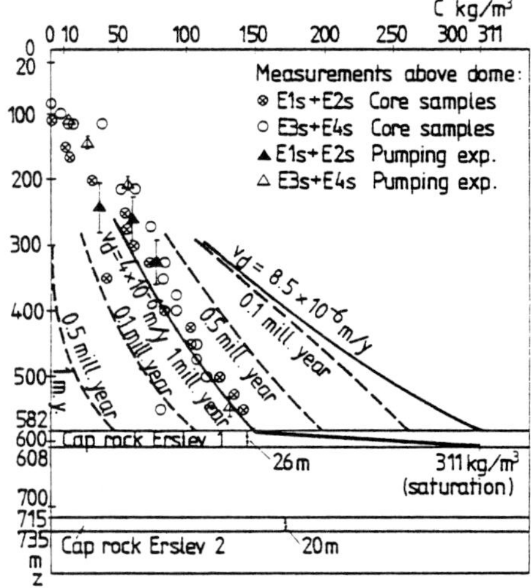

Fig. 5. Time dependent variation of salinity profile.

The time dependent variation of the salinity profile, when the dissolution rate is further decreased from 4×10^{-6} m/y to zero, is also shown in fig. 5. As it appears, the time involved with the changes in the salinity profile is 1 mill. year. A ten times smaller diffusion coefficient will further increase this time by a factor of ten. The slow ground water velocity in the limestone just above the dome implies therefore that the present measured salinity profiles might be the result of dissolution and migration of salt during the last 10-20 mill. years.

HETEROGENEOUS PROPERTIES OF CAP ROCK

The measured salinity profile at a specific location is not only dependent on the dissolution rate at that point. It depends also on the preceding dissolution rate at points located upstream to that point. This circumstance is used to evaluate the dissolution rate in areas without measurements located upstream to points with measured salinity profiles.

In the previous interpretation of the measured salinity profiles it is assumed that the cap rock has homogeneous properties, and that dissolution takes place from all points located upstream to the point of measurement. However, the cap rock might have heterogeneous properties in areas without measurements. In those areas the cap rock might have varying thickness, porosity, and permeability, or it might be completely absent.

An extreme case is considered in the following. It is shown, how the measured salinity profile might be interpreted as the result of salt dissolution along a confined area located upstream and apart the measured profile. With reference to fig. 1 the dissolution area is assumed to extend from 0 to a, with the measured profile located at $x \overset{>}{-} a$. Rearranging the stationary equation (4) gives

$$
C(x,z) = \frac{(C(x,0)-C_0)}{(\sqrt{x} - \sqrt{x-a})} \cdot
$$

$$
\left[\sqrt{x} \exp\left(-\frac{z^2}{4\alpha'x} \right) - \sqrt{x-a} \exp\left(-\frac{z^2}{4\alpha'(x-a)} \right) \right.
$$

$$
\left. - z \sqrt{\frac{\Pi}{4\alpha'}} \left(\operatorname{erf} \sqrt{\frac{z^2}{4\alpha'(x-a)}} - \operatorname{erf} \sqrt{\frac{z^2}{4\alpha'x}} \right) \right] + C_0 \tag{11}
$$

For chosen values of x and a is α' determined by fitting of equation (11) to the measured salinity profile. For an assumed pore water velocity of v = 0.003 m/y is α then determined by equation (10).

Rearranging the stationary equation (4) gives also

$$C(a,z) = \frac{(C(x,0)-C_0)}{(\sqrt{x} - \sqrt{x-a})} \cdot$$

$$\left[\sqrt{a}\, \exp\left(-\frac{z^2}{4\alpha'a} \right) - z\sqrt{\frac{\Pi}{4\alpha'}}\left(1-\mathrm{erf}\,\sqrt{\frac{z^2}{4\alpha'a}} \right)\right] + C_0 \tag{12}$$

Equation (12) determines the salinity profile $C(a,z)$ in the water leaving the dissolution area shown in fig. 1, which after subsequent transport and dispersion downstream ends up with the salinity profile at the location x. The associated dissolution rate can now be determined by solving for v_d in the stationary equation (2) or (4).

For chosen values of x and a, equations (11), (10), (12), and (2) or (4) give an interdependent set of values for α, $C(a,0)$, and v_d

Table 2

x(m)	a(m)	α(m)	$C(a,0)(kg/m^3)$	v_d (m/y)
3000	2500	2.7	210	3.2×10^{-6}
2000	1500	4.7	230	5.0×10^{-6}
1000	800	22	218	9.8×10^{-6}
200	200	442	150	4.7×10^{-5}
200	97	92	311	7.8×10^{-5}

Each of the situations in table 2 ends up with a salinity profile at the location x, which fits the measured profile. It appears from table 2 that dispersivity and dissolution rate increases, as the dissolution area moves closer to the measured salinity profile. The greatest dissolution rate occurs, when the water leaving the dissolution area is saturated just above the dome, as for the case where there is no cap rock, or when there is a water movement in the cap rock itself. For an upper bound value of α equal to about 100 m the last case in table 2 gives an upper bound for the corresponding dissolution rate of 7.8×10^{-5} m/y, which is about twenty times greater than the previous dissolution rate assuming diffusion in the cap rock. Even for the case with a heterogeneous cap rock it takes about 100,000 years to dissolve 10 m of salt.

REFERENCES

1. Lindstrøm Jensen, K.E., (1979) Transport of Radionuclides in Groundwater, ELSAM. E-K No. K79/190A, pp. 1-21.
2. ELKRAFT and ELSAM, (1981) Disposal of High-Level Waste from Nuclear Power Plants in Denmark, Salt Dome Investigations, vol. 1-5.
3. Carlsen, L., Bo, P. and Batsberg, W. (1981) Diffusion Coefficients in Chalk Samples from Erslev, Mors, Denmark, Risø National Laboratory, Denmark, pp. 1-7.

Published 1982 by Elsevier Science Publishing Co
SCIENTIFIC BASIS FOR RADIOACTIVE WASTE MANAGEMENT - V
Werner.Lutze, editor

DISSOLUTION OF EVAPORITES AND ITS POSSIBLE IMPACT ON THE INTEGRITY OF THE WASTE ISOLATION PILOT PLANT (WIPP) NEW MEXICO, USA

SIAVOSH M. ZAND

Environmental Evaluation Group, New Mexico Health and Environment Department
(At present, U.S. Geological Survey, 345 Middlefield Road, Menlo Park, CA
94025 USA).

INTRODUCTION

The purposes of this report are: (1) to discuss the manifestation of dissolution processes (such as land subsidence and method of formation of breccia
pipes and brine reservoirs) in the bedded salt of southeastern New Mexico in
the vicinity of the WIPP site, (2) to postulate conditions under which
breaches of the repository might occur, (3) to discuss radiological consequences of such breaches, and (4) to identify areas that need further
investigation.

The extent and activity of the dissolution processes both now and in the
future are of importance in evaluating the geologic adequacy of the WIPP
site. The dissolution of evaporites produces dissolution residues and may
lead to collapse features such as breccia pipes. Dissolutions may also be
responsible for brine reservoirs. The dissolution of evaporites begins by unsaturated water dissolving the soluble materials and continues as long as
unsaturated water and evaporites are in contact. The residual solid matrix is
weaker structurally and more porous than the original evaporite bed. This
porous residual matrix gradually crumbles and consolidates under the overburden weight forming a new layer. All these processes may occur simultaneously
at a given time at different locations within the Basin. The resulting residual formations may provide a significant pathway of the radionuclides to the
biosphere if a nuclear repository that is located within the bedded salt is
subjected to dissolution. Mathematical models of radionuclide transport are
utilized to evaluate the radiological consequences of the nuclear waste repository breach. Hydrogeologic and geochemical properties of the residual
formations are of crucial importance in these models. The cumulative effect
on overburden materials of the compaction of the dissolved evaporites is
represented by the land subsidence on the ground surface.

The dissolution of evaporites may be divided into two broad classes: (a)
shallow dissolution, and (b) deep dissolution. Most scientists agree on the

mechanisms that control shallow dissolution and on its possible impacts on the integrity of the repository. However, deep dissolution and its possible impact on the integrity of the repository is a matter of controversy.

SHALLOW DISSOLUTION

Shallow dissolution is the dissolving of surficially exposed evaporites by climatic precipitation and surface water. The evaporite formations subjected to shallow dissolution in the Basin are the Rustler Formation and upper part of the Salado Formation. The most pronounced remnant of this process is Nash Draw two miles to the west of the outer boundary of the WIPP site. The leading edge of the dissolution is referred to as the "dissolution front."

Bachman[1] suggests that intermittent dissolution of the Salado Formation near its western boundary began at least as early as 140 million years ago and is presumed to be continuing today. He further states that, if dissolution continues to be no more severe than during the past 500,000 yr, there is no reason to believe that the repository horizon will be threatened by dissolution for another 500,000 yr. Bachman also concludes: "Dissolution is presently an active process over the Capitan aquifer system along San Simon Swale about 32 km (20 mi) east of the WIPP site. However, this aquifer system does not underlie the site and there is no presently known aquifer system or process of dissolution in this area which is undermining the site."

The dominant variables influencing shallow dissolution are spacial and temporal variations of climatological variables such as precipitation and temperature; tilting of the basin; and chemical characteristics and solubility of the exposed evaporites in the path of the shallow dissolution. Bachman and Johnson[2] estimated that the average rate of shallow dissolution front movement towards the repository is vertically about 15cm (0.5 ft) of salt per 1000 yr and horizontally about 13 m/1000y (43 ft/1000 yr).

The closest known shallow dissolution front is 460 m (1500 ft) vertically above and 3.2 km (2 mi) horizontally to the west of the repository[3]. Assuming time invariancy and linearity of the processes, it will take 3.25 million years for the front to breach the repository, Figure 1. This period of time is several orders of magnitude longer than the 24,000 yr half-life of Plutonium-239 and the associated detrimental effects will be negligible.

DEEP DISSOLUTION

Deep dissolution is the dissolving of evaporites by unsaturated water deep within a geologic formation. It may be divided into two groups: (a) local or point source dissolution; and (b) regional or blanket dissolution. Deep dissolution may be closely related to the formation of breccia pipes and brine reservoirs. The brine density flow may be the main mechanism for the connection of an aquifer to the salt beds of the Salado, thus producing deep dissolution.[4] The unsaturated water of pressurized aquifers below the evaporites can rise through fractures and/or interfacial boundaries dissolving the salt and form denser brine that sinks back into the aquifer. The local or point source deep dissolution at its extreme can be represented by surfaced breccia pipes or breccia chimneys. The regional or blanket dissolution includes the dissolution of evaporites in large areas of the Basin forming more permeable strata and possibly brine reservoirs.

The extent to which deep dissolution is active at present and could be active in the future is a matter of controversy among geologists. Anderson[4] suggests that deep dissolution may be an active process throughout the Basin wherever evaporites exist and where there is a source of unsaturated and pressurized water as well as a pathway and driving force to move the water into the evaporites. Anderson[4,5,6] believes that the brine density flow mechanism through faults and fractures connecting the Delaware Mountain Group (DMG) the Capitan Reef and Castile formation is the main mechanism for deep dissolution. He considers the thinning of halite beds in some localities and the Total Dissolved Solids (TDS) concentration gradient in the artesian groundwater of the DMG aquifers as partial evidence to support his hypothesis. In summary, Anderson[5] believes: "The horizon selected for the repository, ..., is the one with the most active history of dissolution. If dissolution proceeds as it has in the past, the WIPP site will be breached at the horizon of the repository before the overlying salt is removed by surface ground-water flow. Hence, estimates of site stability based upon the rate of movement of a surface dissolution front (U.S. Department of Energy, 1980, p. 7-98) may not be pertinent."

Bachman[1,6] however, believes that the mechanism of brine density flow, given the recent past geohydrologic and climatic conditions in the region, is not adequate to explain all the relevant features in the Delaware Basin. In addition, Bachman does not believe this mechanism alone can account for the removal of large quantities of brine in the subsurface within the Basin due to the tightness of underlying aquifers. Bachman and other scientists[1,7] disagree with Anderson's hypothesis based on their belief that: a) the halite

442

beds in the Castile Formation were originally deposited in individual
subbasins within the larger Delaware Basin protected by relatively impermeable
enclosing beds of anhydrites; and b) that if there has been any dissolution it
occurred through surface erosion when the beds were nearer the surface after
Permian and before Cretaceous time.

Local Dissolution — One manifestation of local dissolution may be the
generation of a breccia pipe. There are two distinct theories on the formation of breccia pipes. Stanton[8] relates their formation primarily to deep
dissolution processes and believes they are formed gradually by subsidence of
overlying formations in cavities generated by the deep dissolution. According
to this interpretation, a breccia pipe caused by collapse should contain breccia which has been displaced downward from higher strata and the diameter of
the pipe should decrease upward. The second theory by Kopf[9,10] suggests tectonic action as the primary triggering mechanism for the development and
reactivation of some breccia pipes. According to this hypothesis, termed
"hydrotectonics" some breccia pipes are pressure relief tubes formed by power-
ful upward hydraulic injection of muddy tectonic breccia which then undergoes
reversible surges generated by fluctuations in volume of chambers along an
underlying regional undulatory thrust fault or decollement in which the tubes
are rooted. If the reversible surges continue, the top of the pipe may become
higher and, if it perforated the upper plate, will form a broad low mound of
mud on the land surface called a "mud volcano". Later subsidence may leave a
cone- or basin-shaped topographic depression in the center of the mound.

It is interesting to note that breccia pipes formed by Kopf's hydrotectonic hypothesis may be recognized by one or more of the following characteristics: (1) part of a breccia pipe may contain a mixture of fragments,
some of which have been displaced upward and others downward from their
original stratigraphic position; (2) the deformation of any particular stratum
increases toward the pipe; (3) the diameter of parts of a breccia pipe may
vary and appears to reflect variations in solubility and tectonic competence
of strata locally forming the walls of the pipe; and (4) preceding periods of
local uplift and subsidence of the Earth's surface activated by renewed
reversible surges within a directly underlying hydrotectonic pressure-relief
tube may be recognized and dated by comparing the normal thickness of sedimentary deposits in the region with the thickness of the same strata exposed in
the walls of the pipe — strata that were later penetrated by the top of the
upward-growing pressure-relief tube (Kopf, written commun., 1981).

<u>Regional Dissolution</u> — Regional or blanket dissolution includes the dissolution of the evaporites in large areas of the Basin and may result in formation of brine reservoirs. It is important to note that in both theories the development of brine reservoirs may occur during the initial phase in the formation of breccia pipes. In Stanton's theory[8], a brine reservoir may be the nucleus of the initial cavity that causes gradual subsidence of the overburden materials. In Kopf's hydrotectonic theory[9][10] the intersection of near-vertical fractures may predetermine the site for relief of anomalously high positive and negative fluid pressures generated along a part of a presumed deeply underlying active thrust fault. Once formed, the pipe may act as a vertical aquifer which penetrates underlying and overlying relatively flat lying aquacludes, and allows brines to move vertically from one aquifer to another. If the pipe penetrates all overlying strata, the brines could reach the earth's surface. Continued thrust faulting may reactivate part or all of the overlying pressure relief tube at any time in the distant future. The hypothetical consequences of events leading from the dissolution of the evaporites to the formation of a brine reservoir is depicted in Figure 2. The assumption in Figure 2 is that the unsaturated water from pressurized aquifers within the Basin flows through a fracture (B) and/or along an interfacial surface toward layers of more soluble evaporites. Upon dissolving these exposed layers the more dense water moves downward and the brine density flow occurs. It is also assumed that there will be a critical cross section (A) due to heterogenity of the evaporite layers or pecularities of flow boundary layer phenomena. This assumption is not necessary for the formation of a brine reservoir. However, it will help to demonstrate the dynamic nature of various processes within the evaporites breached with unsaturated water. Any possible obstruction of the critical cross-sections A and B will curtail brine density flow and will generate a brine reservoir. If the obstructions do not occur the process will be generating regional dissolution (extensive areas that behave like aquifers). The obstruction could be caused by the precipitation of returning denser brine and/or by collapse of the upper strata inside the critical cross-sections. The brine within the partly or completely isolated volume become more saturated with available constituents of the exposed evaporites. The enlargement of the brine reservoir may be renewed if the obstruction is removed by dissolution of the collapsed materials in the critical cross-section (A) by the brine or by some other mechanism.

<u>Radiological Consequences of the Breach of the WIPP as a Result of Deep</u>

<u>Dissolution</u>—Agreement does not exist among geologists on deep dissolution and its impact on the integrity of the WIPP repository. In the absence of such agreement, one should assume that the worst case event will occur and calculate the radiological consequences. One way to evaluate the possible radiological consequences of the breach of the repository as a result of deep dissolution is to assume a breach does occur and then to quantify the effects by using mathematical models. In the following paragraphs, such scenarios are considered.

The leading edge of the regional deep dissolution is called the dissolution front. The dissolution front may be in the form of fingers or wedges penetrating the stratiform evaporite. At the time of the closure of the repository, one can assume that the closest of these dissolution wedges is at a distance of X meters from the repository perimeter, moving towards it at a long-range average rate of r meters per year. This indicates that it takes t_1 = X/r years for the dissolution wedge to reach the repository after its closure Figure 3. After time t_1, the brine penetrates the repository and starts to dissolve, mobilizing and transporting the radionuclides towards the biosphere. The period of time associated with dissolution and/or mobilization of the radionuclides in the wastes is t_2. The time for the mobile radionuclides to go through the regional dissolution remnants is assumed to be t_3, while t_4 is the period of time the radionuclides move from the outflow of the dissolution remnants to the biosphere. Therefore, $T = t_1 + t_2 + t_3 + t_4$ is the total period of time, after the closure of the repository, at which the radionuclides will be present in the biosphere.

In the extreme condition, if path AB in Figure 3 is already well developed, then immediately after the closure of the repository the dissolution and mobilization of the radionuclide starts. After the dissolution of radionuclides, according to the brine density flow mechanism, the radionuclide-laden brine should be moving downward toward the DMG aquifers. However, there are sets of circumstances under which the radionuclide-laden brine may be moving upward toward the ground surface.

While moving upward, if the brine became intercepted by the Magenta and Culebra aquifers of the Rustler Formation, the port of its entry into the biosphere would be the outcrops of these aquifers in Nash Draw[12] and/or the Pecos River at Malaga Bend. This type of liquid breach scenario is already addressed in part by the Department of Energy[13][14] and the Environmental Evaluation Group.[3][15] Efforts should be made to refine the model to include: a) the effects of the fractures in the dolomitic Magenta and Culebra; b) the possibility of having much more permeable abandoned mine tunnels in potash as part of the radionuclide flow path to the biosphere; c) the significance of the outcrops of Magenta and Culebra aquifers in the Nash Draw 11 km (7 mi) closer to the repository than the outlet to the Pecos River at Malaga Bend; and d) the effects of long-term climatic and hydrologic variations. If the radionuclide-laden brine is pressurized similar to some brine reservoirs, or a catastrophic collapse occurs similar to a breccia pipe, then it is conceivable that the brine might move all the way upward to the ground surface. Various other credible scenarios may be visualized in this class of the breach of the repository.

Another possible scenario that may be developed when path AB in Figure 3 is already well developed at the time of the closure of the repository is that the deep dissolution remnant has formed a new flow path, unknown and undetected by our explorations (Figure 4). It is expected that this flow path would flow northeasterly as DMG aquifers do and would have a length of at least 15 km. This length is equivalent to 2/3 the flow path length of the Culebra and Magenta aquifers from the WIPP to Pecos River before entering the biosphere. The hydrogeological and geochemical properties of this new flow path might be similar to the Magenta and Culebra aquifers of the Rustler formation since both are the remnant of past dissolution processes. The two basic differences are the radionuclide flow path from the WIPP repository through the geosphere to the biosphere and the fracture permeability. The fracture permeability in the postulated aquifer would be most likely smaller than its corresponding value in the Magenta and Culebra because of the higher overburden weight. Therefore, the radiological health consequences of this breach of the repository might be similar to that estimated for transport through the Magenta and Culebra aquifers.[3,14,15]

Another extreme case consideration is to assume that the hydrogeological characteristics of the postulated flow path are similar to the Rustler-Salado interface aquifer. This possibility appears much less likely due to the fact that the repository is located completely within the Salado formation. This case is not addressed here but should be quantified when more data become available. Another possible pathway within the geosphere for the release of radionuclides to the biosphere as a result of deep dissolution-caused breaching of the repository is the movement of radionuclide-laden brine through the Bell Canyon aquifer. The minimum horizontal distance of the flow path from

446

the repository within the Bell Canyon aquifer is 15 km. Using Darcy's law, porosity of 0.16, hydraulic conductivity of 0.041 cm/day (0.016 ft/day), and hydraulic gradients of 2.2×10^{-3}, one obtains a travel time of 600,000 years. Even by not taking credit for any other time delays, this period is twenty-five times the half-life of Plutonium-239.

CONCLUSION

The radiological consequences of a breach of the WIPP TRU waste repository, as a result of dissolution processes were shown to be in two broad classes. Most of the possible scenarios that could happen as a result of the deep dissolution processes were shown to be addressed by the scenarios already presented elsewhere or are being investigated at the present. The remaining possible scenarios, except one, showed that the possible release of radionuclides through the geosphere to the biosphere would take periods of time orders-of-magnitude longer than the half-life of Plutonium-239. The one not specifically addressed, a postulated hydrologic pathway with the characteristics of the Salado/Castile interface aquifer, appears unlikely. However, it should be quantified when more data become available. To obtain a better understanding of the radiological health effects of the deep dissolution-caused breach of the repository, emphasis must be placed on better evaluation and refinement of the existing mathematical models and associated parameters.

In the majority of the existing scenarios the most significant pathway of the radionuclide through the geosphere to the biosphere is the dolomitic Magenta and Culebra aquifers. Because these formations are fractured, mathematical modeling of radionuclide transport is subject to a large degree of uncertainty. Other uncertainties that relate to this pathway which need further study are: a) the possibility of having more permeable abandoned mined potash tunnels as a part of the radionuclide flow path and b) the outcrops of the Magenta and Culebra formations in Nash Draw, 11 km (7 mi) before the outlet to the Pecos River at Malaga Bend. The problems associated with the quantifications of these uncertainties as well as the effects of long-term climatic and hydrologic changes and hydrogeological data requirements of the models should be further addressed.

Acknowledgement — Appreciation is expressed to R. W. Kopf (retired, U.S. Geological Survey) who kindly reviewed and commented on his hydrotectonic theory in this report.

REFERENCES

1.   Bachman, G. O., 1980, Regional Geology and Cenozoic History of Pecos Region Southeastern New Mexico: U. S. Geological Survey Open-File Report 80-1099.

2.   Bachman, G. O., and Johnson, R. B., 1973, Stability of Salt in the Permian Salt Basin of Kansas, Oklahoma, Texas, and New Mexico, with a Section on Dissolved Salts in Surface Water, by F. A. Swenson: U.S. Geological Survey Open-File Report USGS-4332-4.

3.   Neill, Robert H., et al., eds. (1979) Radiological Health Review of the Draft Environmental Impact Statement (DOE/EIS-0026-D) Waste Isolation Pilot Plant, U. S. Department of Energy, (EEG-3).

4.   Anderson, R. Y., Kirkland, D. W., 1980, Dissolution of Salt Deposits by Brine Density Flow, Geology, Vol 8, pp 66-69.

5.   Anderson, R. Y. (1981) Deep Seated Salt Dissolution in the Delaware Basin Texas and New Mexico, N.M. Geological Sociey Sp. Publ., No 10, pp 133-145.

6.   Environmental Evaluation Group (1980) Geotechnical Considerations for Radiological Hazard Assessment of WIPP: A Report of a Meeting Held on January 17-18, 1980 (EEG-6).

7.   Chaturvedi, Lokesh (1980) WIPP Site and Vicinity Geological Field Trip: A Report of A Field Trip to the Proposed Waste Isolation Pilot Plant Project, in Southeastern New Mexico, June 16 to 18, 1980, (EEG-7).

8.   Stanton, R. J. (1966) The Solution Brecciation Process, the Geological Society America, Bull., Vol. 77, pp 843-848.

9.   Kopf, R. W. (1979) Tectonic Model for Development of Breccia Pipes, The Geological Society of America, Program Abstracts, Vol. 11, no. 3(2), p 138.

10. Kopf, R. W., Unpublished, "Principles of Hydrotectonics and its Relevance Plate Tectonics."

11. Kopf, R. W. (1981) Hydrotectonics: Principles and Relevance: EOS, Vol. 62, No. 45, p. 1047.

12. Bachman, G. O. (1981) Geology of Nash Draw, Eddy County, New Mexico, U. S. Geological Survey, Open-File Report 81-31.

13. Department of Energy (1980) Final Environmental Impact Statement, Waste Isolation Pilot Plant.

14. D'Appolonia (1981) Modeling Verification Studies, Long-term Waste Isolation Assessment, Project No. NM87-648-701.

15. Wofsy, Carla (1980) The Significance of Certain Rustler Aquifer Parameters, for Predicting Long-Term Radiation Doses From WIPP, (EEG-8).

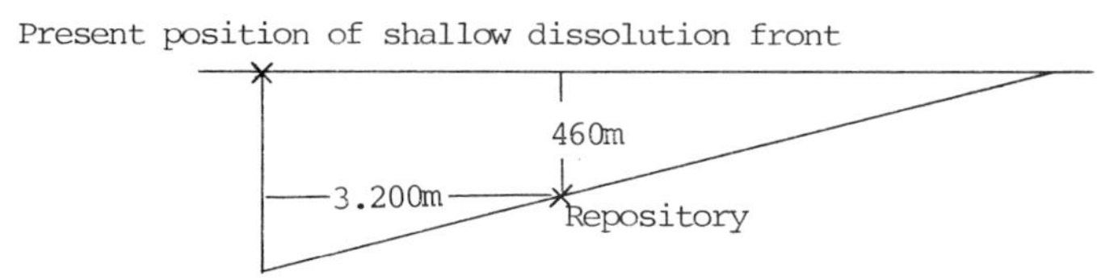

FIGURE 1 - Advancement of shallow dissolution front towards the repository.

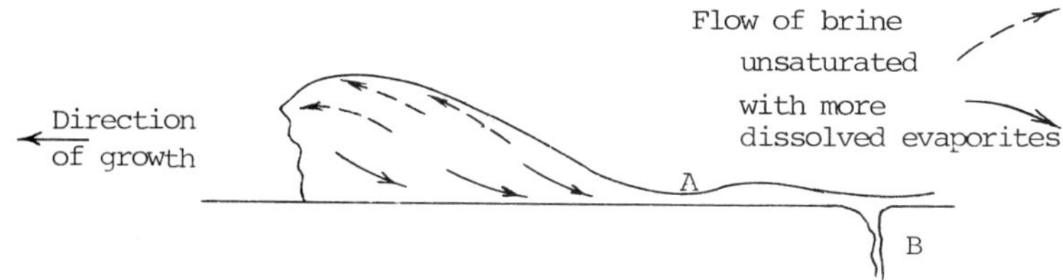

FIGURE 2 - Possible formation of a brine reservoir - a hypothetical sketch.

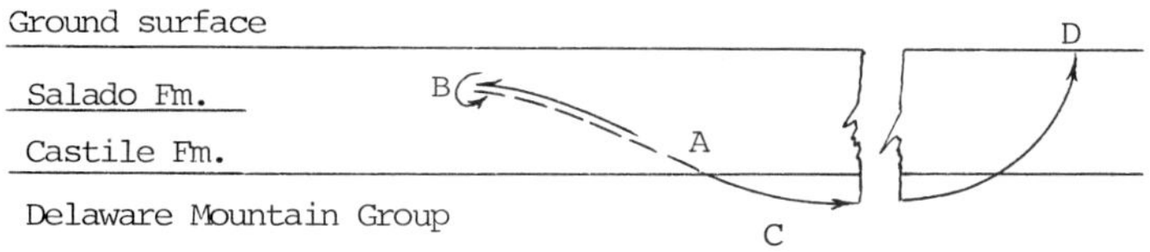

(A) dissolution wedge at the time of closure of the repository (B) radionuclide-laden brine entering the DMG (C) and the biosphere (D)

FIGURE 3 - A hypothetical generalized radionuclide flow path to the biosphere.

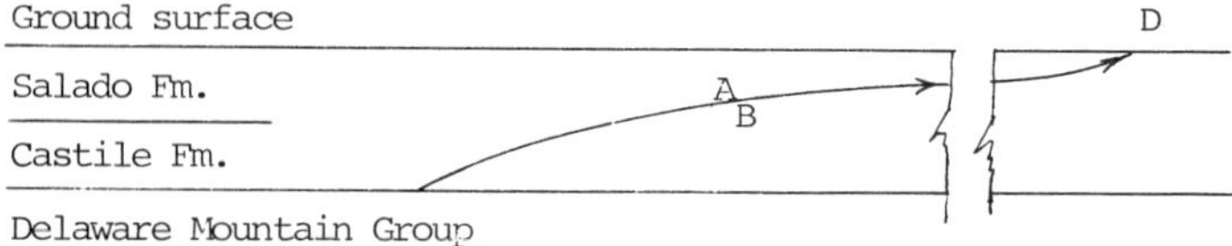

A, B and D are as in Figure 3

FIGURE 4 - A hypothetical direct radionuclide flow path to the biosphere.

DEFORMATION–DISSOLUTION POTENTIAL OF BEDDED SALT, WASTE ISOLATION PILOT PLANT SITE, DELAWARE BASIN, NEW MEXICO

ROGER Y. ANDERSON

Department of Geology, University of New Mexico, Albuquerque, New Mexico, USA

INTRODUCTION

Deep-seated collapse and brecciation is a feature common to bedded salt in a number of evaporite basins. The origin of these features is poorly understood. The relatively young geologic age of uplift and exposure of evaporites in the Delaware Basin, and a still-active hydrologic system, provides a unique opportunity to examine the pathways of water movement related to collapse features and salt dissolution. In addition, salt deformation structures are closely associated with brine, dissolution, and collapse. This report briefly discusses evidence for deep-seated dissolution, considers the relationship between deformation, dissolution, and brine, and examines the potential for dissolution at the WIPP (Waste Isolation Pilot Plant) site with respect to stages in the deformation-dissolution process.

The Delaware Basin, which contains the Upper Permian Castile and Salado evaporites, was uplifted, faulted on the western margin (Fig. 1A), and tilted 19m/km to the east-northeast during the latter part of the Cenozoic, probably 4-6 my ago. This event exposed the tilted limestone (reef) and sandstone aquifers beneath the overlying anhydrite and salt beds (Table 1). Meteoric waters entered the exposed aquifers in the uplifted western and southern margins of the basin. The aquifers became charged and pressurized, with the pressure surface high within the evaporites; almost at the surface. Salt has been removed from beneath and within the evaporite body, apparently by means not related to ordinary near-surface groundwater flow. This report considers how this may have happened.

EVIDENCE FOR DEEP-SEATED DISSOLUTION

The role of the reef aquifer in removing salt from the overlying Salado Formation can be observed in a chain of deep, isolated depressions that have developed above the inner margin of the deeply buried reef along the east side of the basin[1] (Fig. 1A). These depressions contain Pleistocene alluvial fill, with the northernmost depression along the east side having developed in the last 60,000 years. Collapse associated with these depressions has occurred at

TABLE 1

UPPER PERMIAN AND YOUNGER STRATIGRAPHIC UNITS IN THE DELAWARE BASIN

Age	Unit	Thickness (m)	Lithology	% Dissolved
Pleistocene	Gatunia Fm.	0 - 200	Sandstone	
Pliocene	Ogallala Fm.	0 - 50	Sandstone	
Triassic	Chinle Fm.	0 - 240	Mudstone	
	*Santa Rosa Fm.	40 - 90	Sandstone	
Permian	Dewey Lake Fm.	60 - 180	Siltstone	
	*Rustler Fm.	60 - 180	Anhy., Dol. Halite	
	Salado Fm.	400 - 600	Halite, anhydrite	
	Upper			55
	Middle			72
	Lower			
	Castile Fm. (restricted to basin, inside the reef)			
	Anhydrite IV	0 - 120		
	Halite III	0 - 90		65
	Anhydrite III	100		
	Halite II	50		49
	Anhydrite II	30		
	Halite I	90		42
	Anhydrite I	60		
	*Capitan Fm.	490	Limestone	
	*Delaware Mt. Gr.	900	Sandstone	

*Aquifer

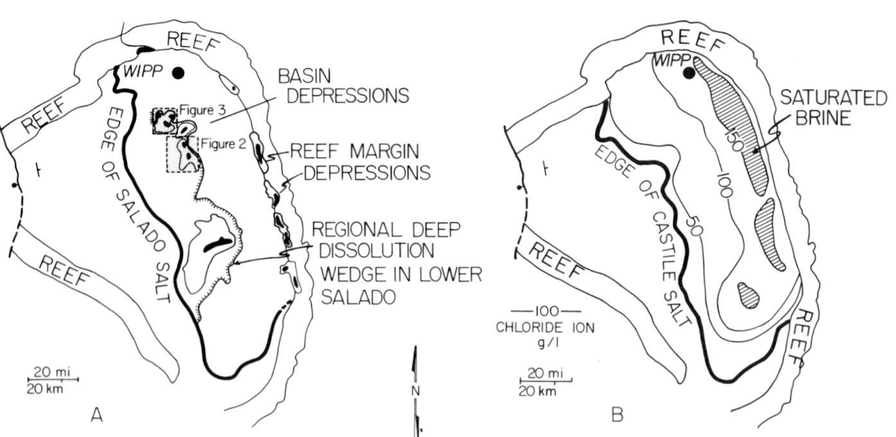

Fig. 1(A). Delaware Basin, showing deep collapse depressions along reef-margin and within the basin.

Fig. 1(B). Delaware Basin, showing salinity gradient in basin aquifer beneath Castile salt. Modified from Hiss[6].

two separate locations in historic time. Salado salt is missing from the center of these depressions, indicating that they root above the reef. Isolated collapse breccia chimneys, originally described by Vine[2] were cored during WIPP exploration and they also root in, or possibly just above, the reef.

Dissolution requires a dynamic hydrologic system. Gravity constraints will not permit the lifting of heavy brine generated near the base of these isolated depressions, some 800 m below the surface, to an elevation where the brine would be free to discharge in near-surface ground water flow. A mechanism of brine density flow has been suggested[3] which proposes that the heavy brine has been drained downward through fractures in the anhydrite which normally separates the salt from the aquifer. In the early stages, water moving upward through fractures and to the salt under artesian pressure, may have been a pathway of water flow which aided dissolution and collapse. In the later stages, the funnel-like action of the collapse itself appears to have gathered additional water to be drained downward into the reef aquifer[1]. Brine flows have been directly observed draining brine from beneath collapsed caprock in Gulf Coast salt domes[4], suggesting that sufficient permeability for brine flow exists in other geologic settings.

The pattern of deep-seated dissolution and collapse within the basin is more complex. A similar chain of isolated deep depressions, containing Pleistocene fill, occurs about half-way across the basin and somewhat parallel to the eastern chain of depressions (Fig. 1A). These depressions are most interesting and unusual because acoustic log correlations of anhydrite marker beds within the salt show that the missing salt responsible for the collapse has been removed from the Lower Salado, in the middle of the evaporite body (Fig. 2). Anhydrite marker beds of the Middle and Upper Salado as well as the overlying Rustler Formation (Fig. 3) are lowered into these depressions. Marker beds adjacent to the depressions define normal undisturbed salt beds, indicating that salt has not moved laterally.

Regionally, the isolated centers of collapse are integrated into a front or "wedge" of expanded dissolution and in places this wedge has extended down-dip, beneath overlying salt, for distances up to 30 km (Fig. 1A). Preference for dissolution in the Lower Salado is reflected in the total amount of salt that has been removed from the basin (Table 1).

The same gravity constraints that apply to water in the isolated reef-margin depressions also apply to the depressions in the basin. Heavy brine generated near the base of these depressions cannot be lifted to near-surface drainage and it must have followed other pathways. There appear to be two

452

options. Either the brine moved farther down dip through the most permeable beds in the Lower Salado or, more likely, Anhydrite III at the base of the Salado, or it drained downward through fractures into the underlying aquifer. There is probably insufficient permeability in the Lower Salado to allow for the down-dip drainage of brine, although drainage through Anhydrite III, fractured by underlying salt deformation, cannot be excluded as a possibility. Modeling studies related to WIPP, which use existing permeability data, suggest that too little brine can be moved in the basin aquifer. However, the isolated character of the deep depressions suggest that brine movement was downward.

The pressure surface in the Delaware Mountain Group aquifer is higher than for the Capitan (reef) aquifer indicating that water movement should be from the basin aquifer into the reef[5]. The salinity gradient in the basin aquifer closely parallels the distribution of overlying salt, with saturated brine occurring along the low, northeastern margin of the aquifer where it drains into the reef[6]. The concentration becomes progressively lower toward the western edge of salt and the 50 g/l chlorinity contour near the western edge of salt (Fig. 1B) is closely parallel to the zero salt line, suggesting an equilibrium condition whereby migration of brine in the aquifer is balanced in a dynamic system by contributions from the overlying salt.

The episode of water incursion into the evaporites and the brine reservoirs might also be explained if a sufficiently large thermal event had affected the area. The most likely time for such an event, consistent with regional tectonic history, was some 30-35 my ago, when the basin was involved with regional fracturing and a basaltic dike was employed. The salt structures and brine, however, post date such an event and regional heat flow during brine emplacement was probably low, as it is at present.

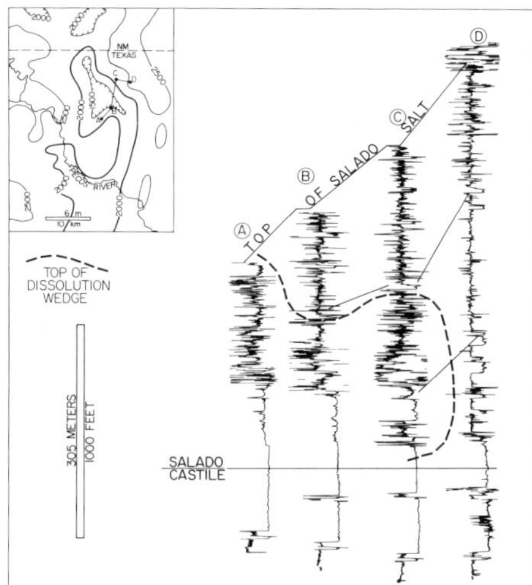

Fig. 2. Acoustic log correlations. Collapse of Rustler is proportional to salt removed from Lower Salado. See Fig. 1(A) for location. Map modified from Hiss[6].

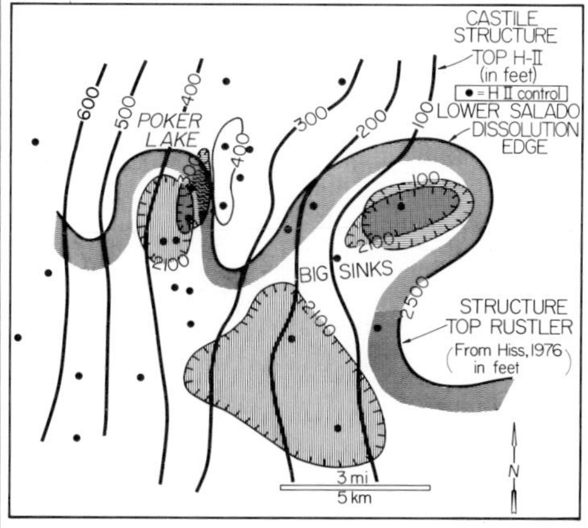

Fig. 3. Structure map of Castile and Rustler formations. Note that lowering of Rustler coincides with depressions deeper in the Castile. See Figs. 1(A) and 4 for locations.

SALT DEFORMATION AND BRINE ASSOCIATIONS

Anomalous, localized, thick and thin areas of Castile halite are scattered throughout the basin, with apparently a greater concentration near the basin margin (Fig. 4). Generally, it is the lower halite (Halite I) that has been deformed, but salt structures are also common in Halite II[7]. Structural deformation is complex in the so-called "disturbed zone" of anomalous seismic reflection data, which extends into the northern part of the WIPP site. Here, many seismic lines suggest a discontinuous "blocky" structure, rather than discrete salt anticlines. Closer spacing of boreholes in the basin would no-doubt reveal many additional structures. The pattern that emerges, however, is one of localized, isolated, and slightly elongate structures, sometimes flanked on one side or the other by similar-size areas of anomalously thin Castile salt. The pattern resembles that of structures in other basins where gravity-induced salt flow has occurred.

Substantial reservoirs of brine are sometimes found near, or associated with, salt deformation structures (Fig. 4). Brine flows of 20,000

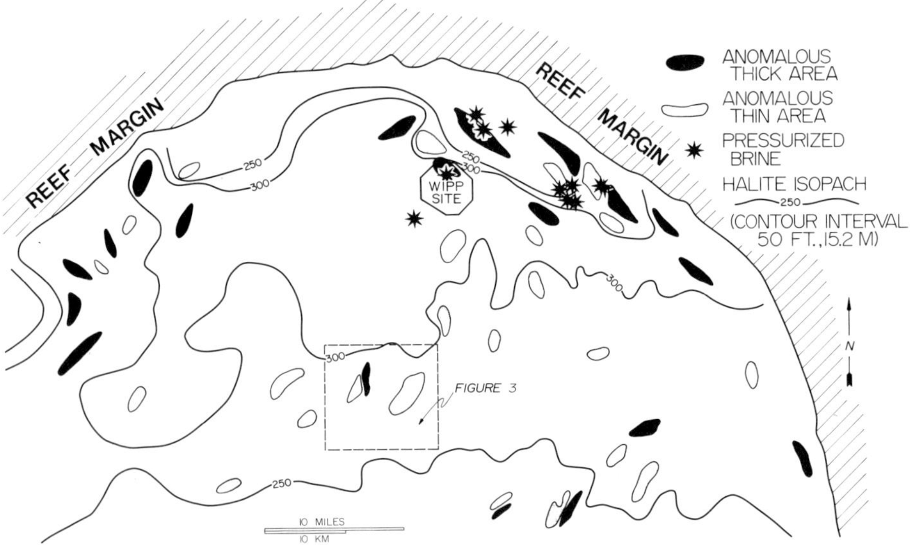

Fig. 4. Northern Delaware Basin, showing location of Halite I salt deformation structures and brine reservoirs. Structures plotted exceed 30 m departure from regional thickness of halite.

barrels per day and reservoir pressures of about 2000 psi have been reported[8]. The brine generally occurs in fractured, brecciated, and partially replaced anhydrite beds of the middle and upper Castile (Anhydrites II and III). The brine reservoirs appear to be marginal to areas of thickened salt, and some of the reservoirs, including the largest flows, are associated with thin salt and structural depressions where Castile salt has been almost completely removed.

High water pressures associated with evaporites are transient phenomena[9]. This implies that dynamic deformation, as well as water recharge, are required to account for present reservoir pressures. The young age of salt deformation is also fixed by Cenozoic regional microfolds that have themselves been involved in later salt deformation[7]. Inasmuch as brine has moved into some of the structures, water emplacement must be contemporaneous with, or post-date, salt deformation. An attempt to date the brine in one reservoir has yielded a minimum age of 570,000 years and a probable age of 800,000 years[8]. No Permian formation water has been identified[10].

ROLE OF WATER IN SALT DEFORMATION

The brine in the salt deformation structures, and the similar age of emplacement of both, raises the possibility of a genetic association and suggests a means whereby large quantities of geologically young water might gain access to the middle of the evaporites. Cores collected from ERDA 6 and WIPP 11 in the "disturbed zone" contain large clear crystals and masses of halite replacing original polycrystalline halite and anhydrite and filling voids (Fig. 5). At ERDA 6 the clear salt is apparently in contact with brine. The occurrence of this type of salt in the middle of the deformed salt in the WIPP 11 structure, suggests that there may have been a general episode of water penetration into the salt, coincident with salt flowage. At ERDA 6, wide fractures in anhydrite beds are now filled with clear halite (Fig. 6) indicating that there have been previous episodes of brine movement in places where there is no current brine storage.

The occurrence of previous water movement within the anhydrite and halite may help explain a discrepancy between laboratory estimates of yield strength for polycrystalline halite and the observed behavior of salt. Ductile flow and thermally induced gravity movement requires a burial depth of some 5000-8000 m[11]. The lower Castile salt was under about 1200 m of overburden at the time of deformation, never buried much deeper (Table 1), and never subjected to a significant regional thermal event. Talbot, after studying the movement of salt glaciers following the dampening by rains[12], suggests that "softening" of salt by water lowers the effective viscosity of polycrystalline halite by 3 - 9 orders of magnitude[13].

When softened by water, salt may have no yield point, and thermal convection may be initiated at shallow burial, especially in situations such as found in the Castile Formation, where thick anhydrite beds form an insulating blanket over the salt. The softening effect is most effective for fresher waters and not observed in saturated solutions[14]. This means that the most plausible source for water, if the "Joffee Effect" is involved, is from the basin aquifer beneath deformed salt, with the water introduced to the salt sometime after initial charging of the aquifer. This water, driven by artesian pressure, would reach the salt through fractures in the lower anhydrite. Presumably, the availability of fracture networks determine the localization and shape of the deformation (Fig. 4).

Once water has contacted salt and deformation is initiated, a feedback response results in additional fracturing of anhydrite beds and thick anhydrite laminae in the salt and further water penetration. Tilting and "foundering" of the proportionately heavier anhydrite beds may have provided

456

pathways for the upward movement of water to the fractured anhydrite at the top of the Castile (WIPP 12 brine reservoir). This hypothesis, and the observed water pressures, mean that brine reservoirs have varying degrees of interconnection to each other and to the source aquifer.

Fig. 5. Clear halite replacing re-crystallized halite and anhydrite in WIPP 11 core. Depth is 945 m.

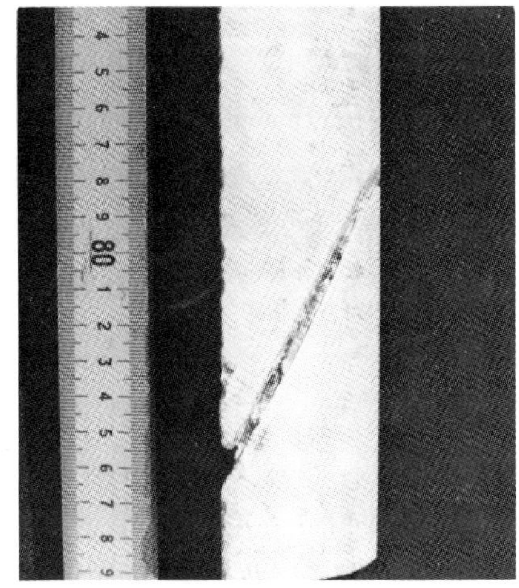

Fig. 6. Open fracture in anhydrite filled with halite. ERDA 6, Anhydrite II, Depth is 830 m.

CONTROLS ON EXPANDING DISSOLUTION

The deep collapse depressions near Poker Lake and Big Sinks (Figs. 1A, 3) provide a clue as to how structural deformation in the Castile may have controlled the later development of expanding dissolution outward from local centers in the Lower Salado. The close spacing of boreholes in the Poker Lake oil field defines Castile structure and reveals that the larger collapse depression in the Rustler Formation lies almost directly above thin Castile salt and localized salt deformation (Fig. 3). Another collapse depression in the Rustler, to the east of Poker Lake near Big Sinks, overlies another area of anomalously thin Castile salt. This alignment suggests that the structures below may have provided the conduits required to drain brine from the overlying expanded area of dissolution in the Lower Salado.

It is not necessary to invoke the softening of salt by aquifer waters to develop a link between Castile structure and the missing salt above. This mechanism, however, does provide pre-existing fractures and upward water movement that may later be the same pathways for descending brine drainage. If, however, brine drainage was down dip through Anhydrite III, improved fracture permeability, in the anhydrite above local areas of salt deformation, might provide the initial pathways for water movement, for the introduction of water to the reservoirs, and for later expanded dissolution.

The proposed models suggest that deep-seated dissolution has progressed eastward across the basin in a series of stages:

1. Charging of the artesian basin aquifer with meteoric waters and water penetration of fractures; salt flowage, possibly induced by water-softening; development of fracture permeability in anhydrites and conduction of water into brine reservoirs. Possibly some brine drainage and minor deep collapse.

2. Dissolution of lower salado through fractured zones above Castile structures; drainage of brine downward into basin aquifer, or possibly down dip; collapse of Salado, Rustler, and overlying beds and accumulation of alluvial fill.

3. Expanded dissolution in Lower Salado is incorporated into regional front or "wedge"; ultimate integration of Lower Salado dissolution with surface drainage.

DISSOLUTION POTENTIAL AT THE WIPP SITE

The missing salt in the Lower Salado (Fig. 2) and the fact that dissolution has progressed eastward selectively means that the site will most likely be breached at the horizon of the repository before overlying salt is removed by near-surface dissolution processes. The rate at which this type of dissolution is progressing in the basin cannot be predicted because it does not appear to move as a continuous front, but rather, moves ahead to predisposed areas which are later incorporated into the front.

A "disturbed zone" of structural complexity extends into the site area from the north. The WIPP 12 borehole, 1.6 km north of the center of the site, encountered a thickened section of Halite I, a thinned Anhydrite III, and a reservoir of brine that flowed about 5000 m^3 at the surface. Correlation of Castile beds in boreholes in the "disturbed zone" indicate that there are probably offsets in the basin aquifer. One seismic survey[15] suggests faulting at the top of the aquifer.

If the stages of dissolution identified here are correct, then the site must be considered an area predisposed to the deep-seated type of dissolution. How far the process has progressed beyond the present regional front is not

458

known. There are several structural features, identifyable in geophysical logs, that suggest that Lower Salado salt is locally missing to the east of the regional front. One of these is a depression in the Middle Salado marker beds 3.2 km north of the center of the site[15]. Another larger depression, 8 km southeast of the center of the site, is missing the Lower Salado salt in the acoustic log. In addition, two large late Pleistocene sinks occur within the basin and east of the regional front. If any of these features is the result of the removal of Lower Salado salt by dissolution, then the process has reached stage 2 in the vicinity of the WIPP site.

In summary, salt is missing from the horizon of the repository, at localities not too distant from the site, and by a largely unknown dissolution process. Repository evaluation is difficult without knowing how these conditions developed. If the explanations offered here are even partially correct, the problems may be generic to bedded salt, depending upon the particular geologic setting of the basin.

REFERENCES
1.   Anderson, R. Y. (1981) New Mexico Geol. Soc., Special Publ. 10; pp. 133-145.
2.   Vine, J. D. (1970) American Assoc. Petroleum Geologists, Bull., 44, pp. 1903-1911.
3.   Anderson, R. Y. and Kirkland, D. W. (1980) Geology 8, pp. 66-69.
4.   Rezak, R. and Bright, T. S. (1981) Geo-Marine Letters, 1, pp. 97-103.
5.   Hiss, W. L. (1975) Univ. Colorado, Ph.D. Dissertation, pp. 1-396.
6.   Hiss, W. L. (1976) New Mexico Bur. Mines & Mineral Resources, Map 7.
7.   Anderson, R. Y. and Powers. D. W. (1978) New Mexico Bur. Mines and Mineral Resources, Circular 159, pp 79-83.
8.   Register, J. K. (1981) U. S. Dept. of Energy. TME 3080, pp. 1-14.
9.   Hubbert, W. K. and Rubey, W. W. (1959) Geol. Soc. America, Bull. 70, pp. 115-166.
10. Lambert, S. J. (1978) New Mexico Bur. Mines, Circular 159, pp. 33-38.
11. Gussow, W. C. (1968) American Assoc. Petroleum Geologists, Mem. 8, pp. 16-52.
12. Talbot, C. J. (1980) Geologica Ultraiectina, Rijksuniversiteit, Utrecht, Special Publ. 1, Geology and Nuclear Waste Disposal, pp. 53-67.
13. Talbot, C. J. (1982) Personal communication.
14. Ode, H. (1968) Geol. Soc. America, Special Paper 88, pp. 543-595.
15. Powers, D. W., Lambert, S. J., Shaffer, S. E., Hill, L. R. and Weart, W. D. (1978) Sandia Laboratories, Albuquerque, New Mexico, SAND 78-1696, 1.

Published 1982 by Elsevier Science Publishing Co.
SCIENTIFIC BASIS FOR RADIOACTIVE WASTE MANAGEMENT - V
Werner.Lutze, editor

MINERALOGICAL AND GEOCHEMICAL FACTORS INFLUENCING THE FINAL DISPOSAL OF HLW IN THE STASSFURT HALITE

HERMANN GIES
Gesellschaft für Strahlen- und Umweltforschung mbH München, Institut für Tief-lagerung, D-3300 Braunschweig, FRG

STRATIGRAPHY AND MINERALOGY

As the thickest and primarily monomineralic saline horizon in the Federal Republic of Germany, investigation of the Zechstein-2-Series, Staßfurt-Halite (Na2) is being emphasized for the final disposal of HLW. This horizon is of particular importance, especially since it is the base of the whole saline formation in northern Germany (Fig. 1).

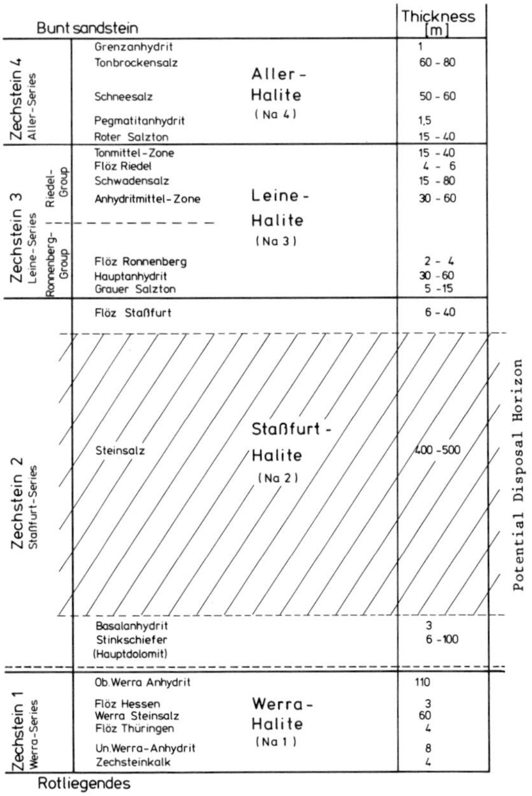

Buntsandstein

				Thickness [m]
Zechstein 4 (Aller-Series)		Grenzanhydrit		1
		Tonbrockensalz	Aller-Halite (Na 4)	60 - 80
		Schneesalz		50 - 60
		Pegmatitanhydrit		1,5
		Roter Salzton		15 - 40
Zechstein 3 (Leine-Series)	Riedel-Group	Tonmittel-Zone		15 - 40
		Flöz Riedel		4 - 6
		Schwadensalz		15 - 80
		Anhydritmittel-Zone	Leine-Halite (Na 3)	30 - 60
	Ronnenberg-Group	Flöz Ronnenberg		2 - 4
		Hauptanhydrit		30 - 60
		Grauer Salzton		5 - 15
		Flöz Staßfurt		6 - 40
Zechstein 2 (Staßfurt-Series)		Steinsalz	Staßfurt-Halite (Na 2)	400 - 500
		Basalanhydrit		3
		Stinkschiefer (Hauptdolomit)		6 - 100
Zechstein 1 (Werra-Series)		Ob.Werra Anhydrit		110
		Flöz Hessen	Werra-Halite (Na 1)	3
		Werra Steinsalz		60
		Flöz Thüringen		4
		Un.Werra-Anhydrit		8
		Zechsteinkalk		4

Rotliegendes

Potential Disposal Horizon

Fig. 1. Stratigraphy of the "Germanischer Zechstein"

460

The thicker chloride parts of the oldest precipitation sequence, the Werra
Series (Z1), were deposited only on the border seas of the Zechstein ocean.
The overlying Na2, which is overlain again by the carnallitic potash seam
'Staßfurt', consists of a deep sequence of 8 - 20 cm thick halite banks which
are separated by sulfate layers of only a few millimeters thickness. This
sulfate is chiefly anhydrite ($CaSO_4$) in the thickest parts of Na2, the so-
called 'Hauptsalz' ('Main Salt'), whilst in the stratigraphically higher
regions, polyhalite ($K_2MgCa_2(SO_4)_4 \cdot 2H_2O$) and finally kieserite ($MgSO_4 \cdot H_2O$)
are increasingly admixed. This corresponds to a general Mg-increase when
approaching the overlying potash seam.

The general stratigraphic structure and the resulting stratigraphic division
are based in part on mineralogy and petrography   and   on textural differences
which are subject to certain local and regional changes. The characteristic
features and the stratigraphic development as mentioned above are, however,
generally present. For an exact description of the facies of the most variable
uppermost Na2-horizon, extensive sampling was carried out at the Asse, the
results of which may principally be regarded as an example for larger parts
of the depositional area. Thus,   samples from drill cores, channel samples
from stopes and average running samples of 1.5 m are investigated mineralogically
and geochemically. The mineralogical analysis of a larger number of individual
samples had already shown[1] that only four minerals (halite, anhydrite, polyhalite and, in very small amounts (< 1,0%), kieserite can occur). Minor and
irregular quantities of carnallite are first detected at the base of the
potash seam.

Microscopic investigations[2] carried out on a larger number of samples
showed that the sulfate layers generally consist of up to > 90 vol. % of
anhydrite. Polyhalite is found mainly in the upper horizons in variable
quantities. It only approaches 50% of the sulfate portion in the exceptional
case. The polyhalite is normally concentrated on the upper and lower boundary
of the anhydrite layers. Additional mineral components in the sulfate layers
may include halite, but argillaceous admixtures are most common.

Particularly in the boundary region of the anhydrite layers a more inti-
mately intergrown texture of coarse-grained, mostly tabular anhydrite crystals
(up to 500 μ in grain size) was observed, which could lead to an increased
strength of these boundary layers. Kieserite is found only in the highest
stratigraphic horizons of Na2 by means of chemical analysis, possibly being
only a component of included brine.

GEOCHEMICAL CHARACTERISTICS AND WATER CONTENTS

Geochemical analysis can also be used for stratigraphic purposes. Fig. 2 shows the distribution of elements [+] in the cross section of a 350 m deep core drilled through the upper region of Na2, started approximately 15 m below the boundary to the potash seam 'Staßfurt'.

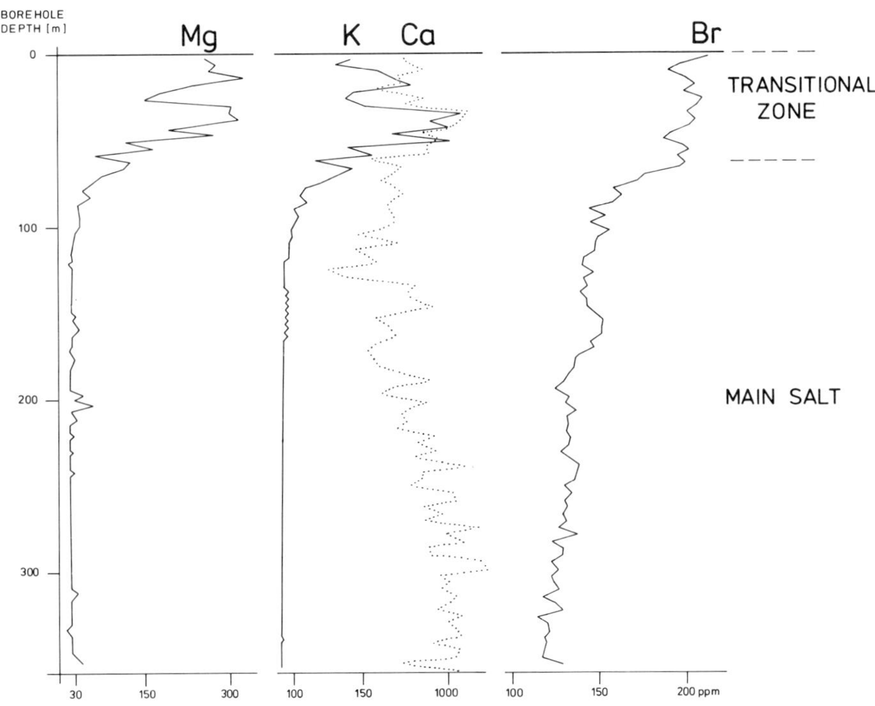

Fig. 2.   Variation of some elements in a drill profile through the Upper
Staßfurt-Halite

[+] The geochemical analyses have been performed by H. Bachmann in the Institute for Sedimentary Petrography, University Heidelberg.

It is evident that the upper region of Na2 is distinguished by significantly higher K-, Mg-, and Br-contents and that their reduction to lower values may be used as a stratigraphic boundary between the Main Salt and the transitional region to the overlying carnallite seam.

The same is true of the water contents[3] which mainly represent hydration water from polyhalite (cf. Fig. 3, foot-note). The difference between both stratigraphic horizons is expressed here in a most striking manner, in that the uppermost Na2 has an average H_2O-content of 0.25 wt. % , while the Main Salt has 0.04 wt. % only.

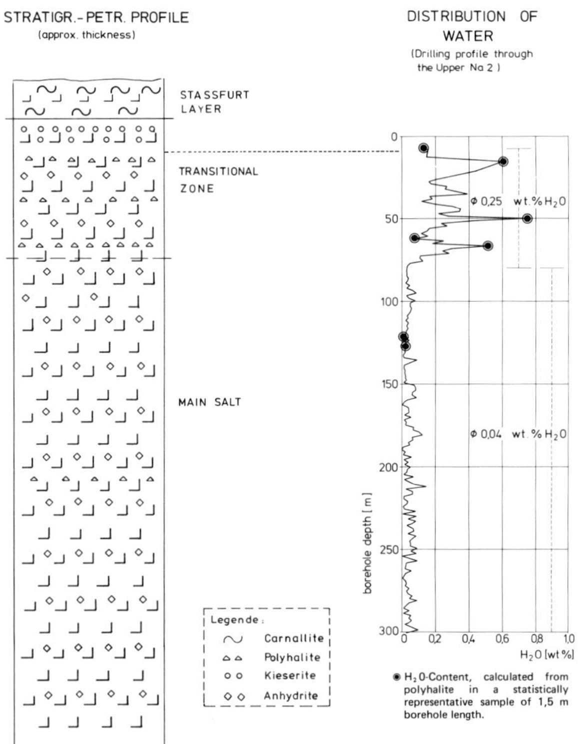

Fig. 3. Correlation of stratigraphy and water-contents of the Staßfurt-Halite

According to the results of the investigation of the conditions for the thermal liberation of the hydration water [3], the same dependence on the confining pressure may be assumed for polyhalite (and kieserite) as has been proved for carnallite [4]. This means that under conditions of continued rock pressure no liberation of water of hydration occurs. According to Winske [5], however, a stress relaxation is to be expected up to 1 m radius around the disposal boreholes. Thus, essentially the H_2O-quantities within that range must be considered. Assuming their release, which further depends on the permeability of the salt, the maximum amount of water expected for given water contents would be the following:

at an average H_2O-content of   a maximum of water released per meter borehole

0.25 wt. %	of 23.0 1 H_2O (Transition Zone)
0.05 wt. %	4.6 1 H_2O (Main Salt)

CONSEQUENCES

Because of the water content of the Transition Zone which may potentially enter a borehole, it is recommended that this zone be excluded from HLW-disposal activities.

It is also necessary to maintain a distance from the stratigraphically overlying carnallite seam, which guarantees that it is not heated to more than 135° C [4] in order to avoid a thermal liberation of H_2O. With these requirements met it should be possible to maintain a maximum temperature of 200° C in the repository. This requires, according to Fig. 4, a safety distance from the carnallite of $\geq$ 40 m [6]. This would also insure that the stratigraphic transition zone, assuming a maximum width of 40 - 60 m, would be excluded from the HLW-disposal region.

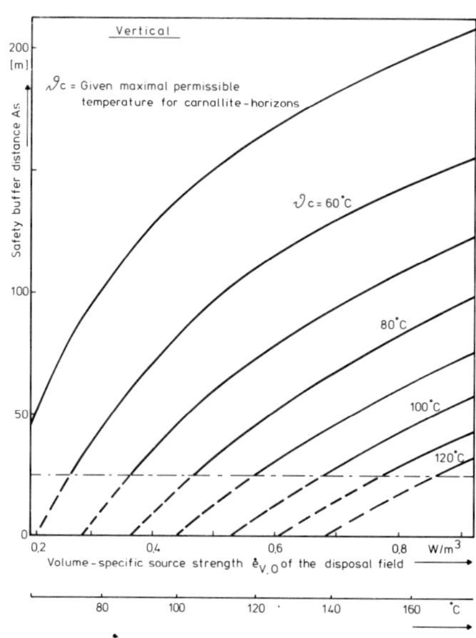

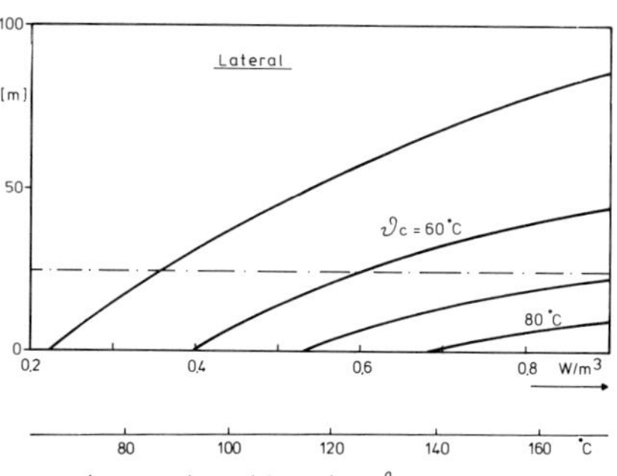

Fig. 4. Vertical and lateral safety buffer as a function of the maximal permissible temperature of the carnallite as well as the volume-specific disposal field source at the time of emplacement.
source diameter D=0.3m
source age t_{RE}=10a; source length L=50m

The much thicker Main Salt, situated below the Transition Zone, has, according to the data obtained so far, an average composition of

$$93 - 96 \quad \text{vol.\%} \quad \text{halite;}$$
$$4 - 7 \quad \text{vol.\%} \quad \text{anhydrite and}$$
$$\text{max.} \quad 1 \quad \text{vol.\%} \quad \text{polyhalite}$$

(and practically no kieserite, sylvite or other
minerals such as clays).

The halite brine that would result from such a composition, if it is a brine at all, would contain at $200°C$ up to 55 wt.% of soluble components. Due to their high solubility, the maximum of 1% Mg-sulfates (mainly polyhalite) would be dissolved, but, despite the high NaCl-concentration, the $CaSO_4$-solubility should not exceed 0.654 wt.% at any time.[7] The pH-values of such brines are generally < 6.0 and may be reduced once more to ≤ 2 at rising temperatures and due to radiolysis.

A separate paper in this volume[8] deals with the gases included in salt and their thermal release.

REFERENCES

1. Gies, H. and Rothfuchs, T. (1981) International Seminar on Chemistry and Process Engineering for High-Level Liquid Waste Solidification, Jülich, pp. 844-872, Odoj, R., Merz. E. ed.
2. Diem, W. (1981) Masters Thesis Techn. Univ. Clausthal, pp. 1-157.
3. Jockwer, N. (1981) Diss. Techn. Univ. Clausthal.
4. Kern, H. and Franke, J. (1980) Glückauf Forschungshefte $\underline{6}$, 252-255, Essen.
5. Winske, P. (1981) in 'Admissible Thermal Loading of Rocks'. Study for the Commission of the European Comm., pp. 1-401, in press.
6. Albers, G. et al. (1981) 1. Semi-Annual Report 1981 of GSF for the CEC, available at GSF/Inst. f. Tieflagerung, Braunschweig.
7. Marshall, W. et al. (1964) J. Chem. and Eng. Data $\underline{9}$, 187-191.
8. Jockwer, N. (1982) MRS Symposia Proceedings, Vol. 11, W. Lutze, ed.

GAS PRODUCTION AND LIBERATION FROM ROCK SALT SAMPLES AND POTENTIAL CONSEQUENCES ON THE DISPOSAL OF HIGH-LEVEL RADIOACTIVE WASTE IN SALT DOMES

NORBERT JOCKWER
Gesellschaft für Strahlen- und Umweltforschung mbH, Institut für Tieflagerung,
Theodor-Heuß-Str. 4, D-3300 Braunschweig, Federal Republic of Germany

1. INTRODUCTION

Rock salt from the north German salt domes which are to be used for disposing high level radioactive waste consist, along with the major mineral halite, the minor minerals anhydrite, polyhalite and kieserite and of some trace minerals such as carbonates, clay, bitumencomponents. Small amounts of free water - and gas components either are adsorbed on the crystal boundaries or form inclusions with a size of less than one micrometer up to some millimeters.

In the surroundings of an emplacement borehole with high level radioactive waste the rock salt will be heated to about 200 °C and the water of the hydrated minerals as well as the water and gas adsorbed on the crystal boundaries and present as inclusions will be liberated into the intergranular spaces. At the elevated temperature carbondioxide,different hydrocarbons, hydrochloric acid and sulphurdioxide may be generated by thermal decomposition of the carbonates, clay, bitumen and other trace minerals.

As a result of γ-radiation gases such as hydrogen, oxygen, hydrochloric acid and carbondioxide will be generated by radiolysis.

2. RESULTS

2.1 GAS GENERATION AND LIBERATION AT ELEVATED TEMPERATURE

Some hundreds of rock salt samples from the Asse mine and about 50 from three other north German mines out of different depths and different stratigraphic layers have been investigated to determine the mineralogical composition and the water and gas contents.

Table 1 shows the minerals which have been found in 136 salt samples from statigraphic layers which might be suitable for disposing of high level radio-active waste. The amounts of these minerals, in addition to the halite, range from 0.1 to 5 wt %. In some small layers the concentration of the minor minerals reach 20 or even 30 wt %. The content of clay, carbonates and bitumen is much less than 0.1 wt %.

TABLE 1

Minerals being found within 136 salt samples from the Stassfurt and Leine horizon

Mineral	being found in ... samples
halite	136
anhydrite	73
polyhalite	106
kieserite	19
sylvin	3

Primarily, as a result of the content of the hydrated minerals rock salt contains small amounts of water. The investigation of some hundreds of salt samples has shown that 55 % of these samples have a water content less than 0.1 wt % and 75 % less than 0.2 wt %. (Jockwer [1;2;3]). The amount of the water adsorbed at the crystal boundaries and in fluid inclusions is much less than the water derived from the hydrated minerals. At elevated temperature in the range of 200 °C this water will be liberated and then migrate on the crystal boundaries and through microfissures as a result of a gradient of total pressure, partial pressure, concentration and temperature.

To investigate the liberation of the water, the loss in weight at different constant temperatures between 90 and 630 °C has been measured in a thermoanalyser whose oven has been rinsed with absolute dry nitrogen. Figure 1 shows the loss in weight over a 50 hour heating period versus temperature for this period for three salt samples of different mineralogical composition. It indicates that using an absolute dry gas stream no minimum temperature for the liberation of the water exists. The different steps are caused by different release velocities. The loss in weight above 300 °C is caused by thermal cracking of some of the minor and trace minerals, the liberation of gas components, and submicroscopic fluid inclusions. Above 600 °C rock salt begins to sublime.

The thermogravimetric investigation of rock salt compared to water determination by either the Karl-Fischer method or the moisture analyser of Du Pont has shown that besides water, gas components also are liberated below 200 °C. Table 2 shows the results of 20 rock salt samples from the Asse salt mine taken from 7 different locations. Samples from the same locations have an alphabetic index. The different components have been determined using a gas chromatograph while heating the samples up to 500 °C and rinsing the oven with very pure nitrogen.

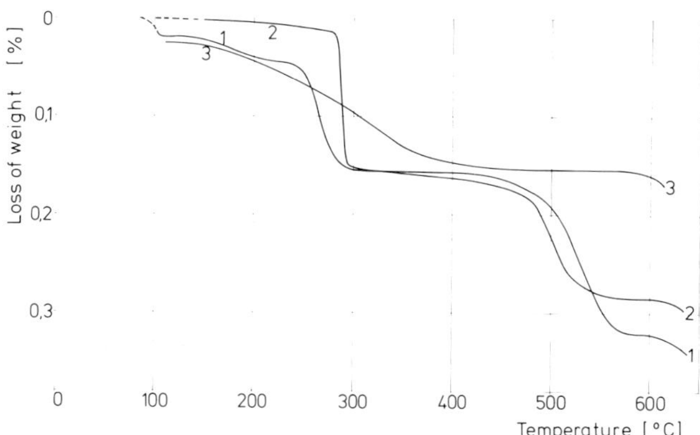

Fig. 1. Loss in weight from three salt samples as a function of temperature. Curve 1: Polyhalitic salt with 0.12 % hydration water from polyhalite and 0.035% adsorbed water. Curve 2: Kieseritic salt with 0.147 % hydration water from kieserite and no adsorbed water. Curve 3: Synthetic salt with all water adsorbed at the crystal boundaries.

None of the samples have shown a detectable liberation of gas at room temperature when rinsing them with very pure nitrogen. Only H_2S could be smelled during grinding. The thermally liberated SO_2 was not detectable quantitatively but qualitatively.

Figure 2 shows the intensity of the liberated components H_2S, HCl, CO_2, CH_4, H_2O and SO_2 versus the time and temperature as measured with a mass spectrometer. The ground rock salt sample of 4.33 grams was heated up at 1 °C/min to 450 °C, then the temperature was held constant for 100 hours. This figure shows that gas liberation, especially of the components HCl, CO_2 and SO_2, begins above room temperature. At about 250 °C the water of the hydrated minor minerals polyhalite and kieserite is released rapidly. The crystal lattice of the rock salt is destroyed and therefore a greater liberation peak of all components occurs. Holding the temperature constant at 450 °C with release dropping results in further liberation of the components H_2S, HCl, CO_2 and SO_2 to zero after about 100 hours. During the whole measurement the oven in which the salt was heated and baked was rinsed with 1 l/h pure helium. If the oven is not rinsed, secondary chemical reactions between the different liberated gas components take place, in particular the production of H_2S becomes lower while that of SO_2 increases.

TABLE 2

Thermally liberated gases from rock salt samples during heating to 500°C

Sample No.	HCl [ppm]	H_2S [ppm]	CO_2 [ppm]	gaseous hydrocarbons [ppm]
1a	130	0.27	>300	60
1b	35	0.27	54	1.5
1c	35	0.05	28	7
1d	27	0.04	93	5
2a	51	0.12	24	2
2b	45	0.1	20	2
2c	20	0.75	119	5
2d	22	0.07	14	4
3a	14	0.7	>300	30
3b	20	4.3	183	5
3c	17	0.45	73	5
3d	22	0.3	9	5
4a	15	0.5	–	4
4b	15	0.35	10	4
5a	2	1.2	8	5
5b	30	0.6	5	6
6a	0.2	0.15	10	5
6b	4	0.2	5	4
7a	4	2.6	270	10
7b	3.5	1	5	6

1ppm = 1 mg gas per 1 kg salt

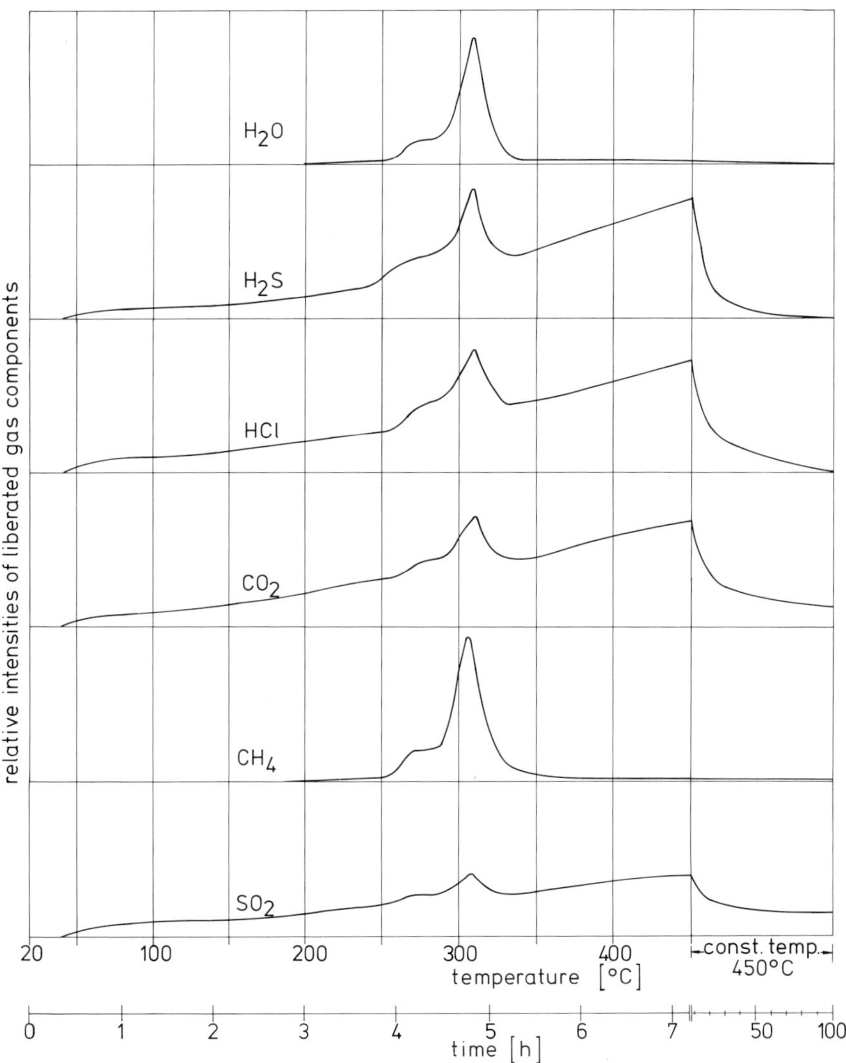

Fig. 2. Liberation of the gas components H_2S, HCl, CO_2, CH_4, H_2O and SO_2 out of a ground salt sample versus temperature and time.
Mineralogical composition: 74.3 wt % halite, 20.5 % polyhalite, 5.2 wt % anhydrite.
Total gas content: 4.3 ppm H_2S, 20 ppm HCl, 183 ppm CO_2, 35 ppm hydrocarbons, 1.26 wt % H_2O, amount of SO_2 not measureable.

472

2.2 GAS GENERATION BY RADIATION

High level radioactive waste which is to be disposed in rock salt has a medium energy of 0.7 MeV. With the density 2.2 g/cm³ of the rock salt the absorption coefficient is:

$$\mu \; = \; 0.187 \; \text{cm}^{-1}$$

according to the absorption law

$$I \; = \; I_0 \cdot e^{-\mu x}$$

about 95 % of the radiation will be adsorbed within a layer of 15 cm and 99.5 % within a layer of 30 cm around the emplacement borehole. This indicates that essentially all effects due to radiation and radiolysis will take place within less than half a meter around the disposed high level waste.

In order to determine the generation of gas components by radiolysis as a result of γ-radiation, rock salt samples of different mineralogical composition have been radiated in gastight glass phials. These phials had a diameter of 2 cm, length of 10 cm. Half of the volume was filled with 10 grams of ground salt. To get rid of the air in the residual volume and the gas components adsorbed on the crystal surfaces, the phials were evacuated and flooded with pure helium twenty times. Before sealing, they were filled with 1 bar helium. The radiation of these phials took place in the core of a reactor during 60 hours. The adsorbed γ-dose was about 10^7 rad and the temperature of the samples during radiation was about 70 °C. As a result of the radiation the ground salt became slightly blue to dark blue. Some crystals changed their color to yellow or brown.

After taking the phials out of the reactor core they were sent for gas determination. The duration between radiation and gas determination was about one week. Due to the gas generation by radiolysis a slight overpressure existed in the phials.

Table 3 shows the mineralogical composition and the water content of 7 salt samples and table 4 the gas components which were found in the residual volume of the phials originally filled with pure helium. HCl, CO_2, CO, Cl_2, H_2S, SO_2, and hydrocarbons other than CH_4, were not found in this residual volume. But as the amount of the radiated salt (about 10 g) was relatively small, further investigations concerning the generation of gas components were necessary.

In order to determine the gas components adsorbed at the crystal boundaries, the open phials were rinsed with very pure nitrogen first at room temperature

and then when they were heated with the salt samples to 250 °C. Table 5 shows
the liberated gas components and their amounts for the two conditions.

TABLE 3

Mineralogical composition and water content of the radiated salt samples

Sample	Mineralogical composition	water content
A	99 wt % kieserite	13 wt %
B	99 wt % carnallite	38.8 wt %
C	halite anhydrite polyhalite kieserite	0.5 wt %
D	halite anhydrite polyhalite kieserite	0.03 wt %
E	halite anhydrite polyhalite kieserite	0.8 wt %
F	92.5 wt % halite; 7.5 % anhydrite	0.09 wt %
G	95.6 wt % halite; 6.4 wt % anhydrite	0.05 wt %

TABLE 4

Gas components determined in the residual volume of the radiated phials

Sample	hydrogen (H_2) [ppm (v/v)]	oxygen (O_2) [ppm (v/v)]	methane (CH_4) [ppm (v/v)]
A	> 500		
B	2000		
C	2200		
D	1500	< 50	30 - 150
E	2800		
F	1400		
G	2200		

1 ppm (v/v) = 1 µl per 1 l

Oxychlor acid was not found in any sample with the detection limit at about
100 ppm.

474

TABLE 5

Liberation of HCl and CO_2 from the radiated salt samples during rinsing with very pure nitrogen

with * at room temperature; without * at about 250 °C

Sample	H Cl [ppm]	CO_2 [ppm]
A*	–	–
A	< 0.2	–
B*	< 0.2	60
B	100	–
C*	–	–
C	1.5	–
D*	3	100
D	0.6	20
E*	< 0.2	–
E	< 0.2	–
F*	1.5	40
F	0.6	–
G*	0.6	–
G	0.2	–

1 ppm = 1 mg per 1 kg

These results of the gas generation and liberation of radiated salt samples are preliminary since the amount of the salt in the phials was comparatively small. Further investigations with phials in which about half a kilogram salt can be radiated are planned. In addition the analysis of the different gas components will be improved.

Nevertheless these first results show that hydrogen, hydrochloric acid and carbondioxide are generated by radiation while the amount of oxygen is comparatively low. It might be possible that the oxygen from the radiolytic decomposition of the water of hydration has reacted with the hydrocarbons which leads to increased amounts of carbondioxide.

3. CONSEQUENCES

These preliminary results of gas generation at elevated temperature up to 200 °C and radiation up to 10^7 rad indicate that in disposing of high level radioactive waste in rock salt, the components H_2S, CO_2, O_2, H_2 and CH_4 will be liberated into the emplacement borehole. This presumes that the in-situ rock salt conditions will indeed be comparable to the conditions during the laboratory tests.

The components H_2S, HCl and O_2 can corrode the waste container and might interact with the solidified waste products. Therefore it is necessary to investigate what will happen to the solidified waste products when exposed to these components.

If the borehole is rinsed and is absolutely tight, the components H_2O, CO_2, H_2 and CH_4 will generate pressure in the borehole, in the intergranular spaces and between the crystal boundaries. This elevated pressure may increase to the fracture pressure of the host rock which then may lead to an uncontrolled release of the gas components.

To avoid destruction of the containment by corrosive gases and their interaction with the solidified waste products and to avoid fracturing of the host rock due to elevated pressure, rinsing the borehole until the majority of the gas components are liberated might be one technical solution. Afterwards the borehole can be sealed tight. It is not yet possible to calculate the time required to rinse the borehole.

ACKNOWLEDGEMENT

This investigation is being made within a research contract with the European Atomic Energy Community and Gesellschaft für Strahlen- und Umweltforschung mbH München.

REFERENCES

1. Jockwer, N. (1980): Laboratory investigation on the water content within the rock salt and its behavior in a temperature field of disposed high level waste. - Scientific basis for nuclear waste management. Boston 1980.

2. Jockwer, N. (1981): Untersuchungen zu Art und Menge des im Steinsalz des Zechsteins enthaltenen Wassers sowie dessen Freisetzung und Migration im Temperaturfeld endgelagerter radioaktiver Abfälle. - Dissertation, Technische Universität Clausthal.

3. Jockwer, N. (1981): Transport phenomena of water and gas components within rock salt in the temperature field of disposed high level waste. - Workshop on near-field phenomena in geologic repositories for radioactive waste. - Seattle 1981.

THERMOMECHANICAL IN-SITU EXPERIMENTS AND FINITE ELEMENT COMPUTATIONS

ALEXANDRA PUDEWILLS, ROLAND MÜLLER, EKKEHARD KORTHAUS, RAINER KÖSTER
Institut für Nukleare Entsorgungstechnik, Kernforschungszentrum Karlsruhe,
Federal Republic of Germany

INTRODUCTION

In the ultimate storage of high level wastes in rock salt thermomechanical phenomena caused by the temperature rises play an important role for
elaborating the storage concept. Important factors are both large space
effects concerning the repository and the geological formation and small
space phenomena influencing to a considerable extent the interaction
between waste blocks and the host rock. While the large space aspects are
largely inaccessible to an experimental investigation, the small space
effects, especially the convergence behavior of individual heated bore-
holes, are eligible for a systematic experimental and computational study
allowing verification of available computational methods and thermomechanical material laws for rock salt.

As has been shown by previous theoretical studies, the closure of heated
boreholes depends on the creep behavior of salt and on the stress boundary
conditions prevailing at the place of the experiment. Since we know from
experience that both parameters cannot be determined accurately enough,
it is reasonable to perform experiments repeatedly at different test places in order to obtain a realistic evaluation of the possible accuracy of
computational predictions of the overall behavior of the heated salt,
especially of the borehole closure. For this purpose three in-situ experiments were performed in ASSE, using a convergence measuring probe specifically developed for such investigations, and evaluated with the help of
thermomechanical model computations.

DESCRIPTION OF THE EXPERIMENTS

The experiments were carried out in ASSE at 775 m depth in older halite.
The novel mobile convergence measuring probe ("Standard probe") was successively installed in three boreholes of 185 mm diameter, drilled down to
about 5 m depth from a tunnel, and the closure of the borehole taking
place during a heating phase of about 100 days was observed.

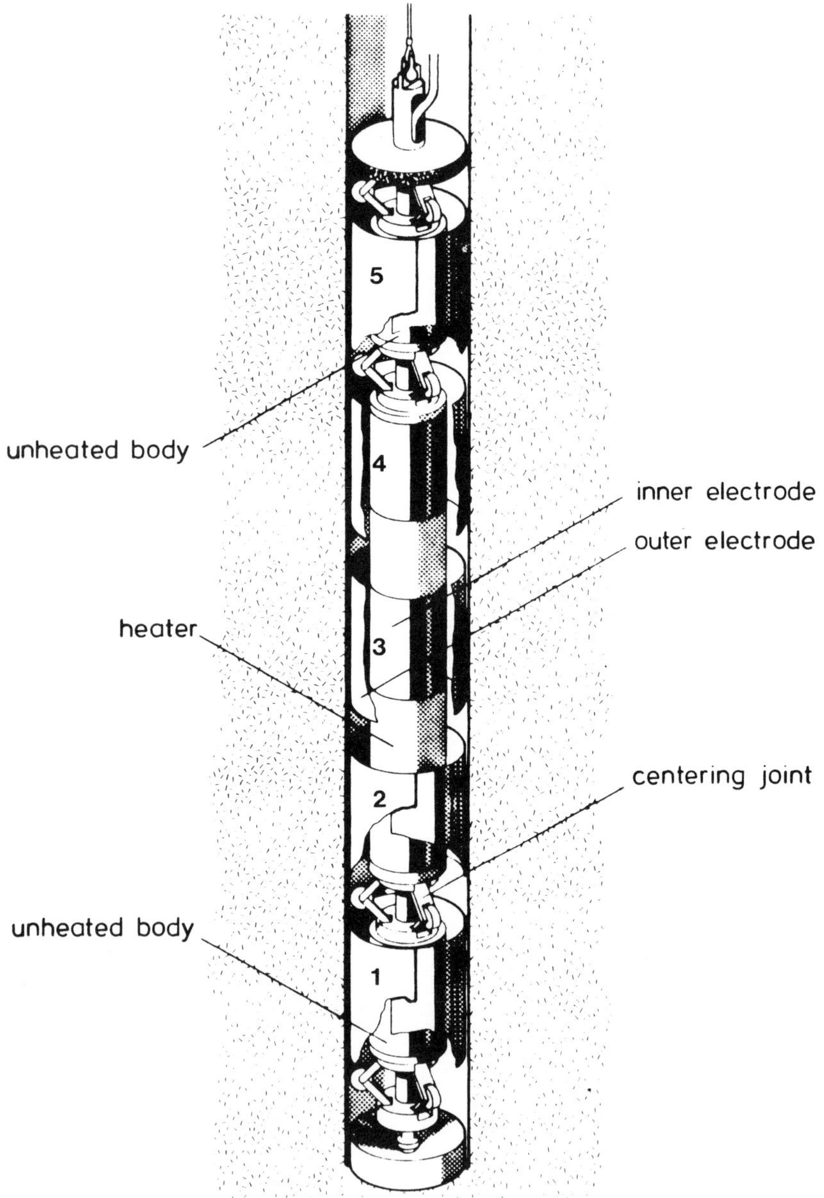

Fig.1 Schematic drawing of the convergence measuring probe

The main components of the probe are an electric heater and an in-line
convergence measuring device based on the measurement of the electric
capacitance between annular electrodes placed on the probe body and at
the borehole wall (Fig. 1). A number of thermocouples are installed on the
probe body and on the external annular electrodes for temperature monitoring. The probe electronic system consists of a controlled power supply
to the heater and an automatic data acquisition system for five convergence probes, 12 temperature measurement points, as well as for heater
power and voltage.

The convergence measuring technique applied here was chosen because
other usual techniques of length or distance measurement are not applicable at all, or applicable with difficulties, under the conditions prevailing here (high temperatures and temperature gradients, corrosive
atmosphere).

Three tests were made with heater outputs of 1.2, 1.3 and 1.4 kW,
leading to maximum salt temperatures of 140-170°C at the borehole wall
(tests 1-3). The test conditions were not optimum because experiments 1
and 3 were separated by only a 5 m distance and were relatively close to
a travel road passing laterally, which might have caused some disturbance
of the initial stress conditions in the salt not taken into account by
the computation models used. Moreover, since the ventilation system of
the test gallery was variable, the boundary conditions for the temperature
calculations were not well defined.

MODEL CONCEPTS AND METHODS OF COMPUTATION

To evaluate the experiments, two-dimensional axi-symmetric models were
used in which also the test section was approximated by a cylindrical
space. A more accurate definition of the actual geometry would have
called for very expensive three-dimensional computations.

The ASYTE/KA code /1/ specifically developed for the investigation of
repository structures was used to calculate the development of temperature and the thermomechanical analyses were made with the commercial
Finite Element program ADINA /2/. Calculation of the temperatures has
already been demonstrated for similar heater experiments performed in
ASSE /1/.

The geometric model used for the thermomechanical calculations and
the division into finite elements are shown in Fig. 2. The very fine

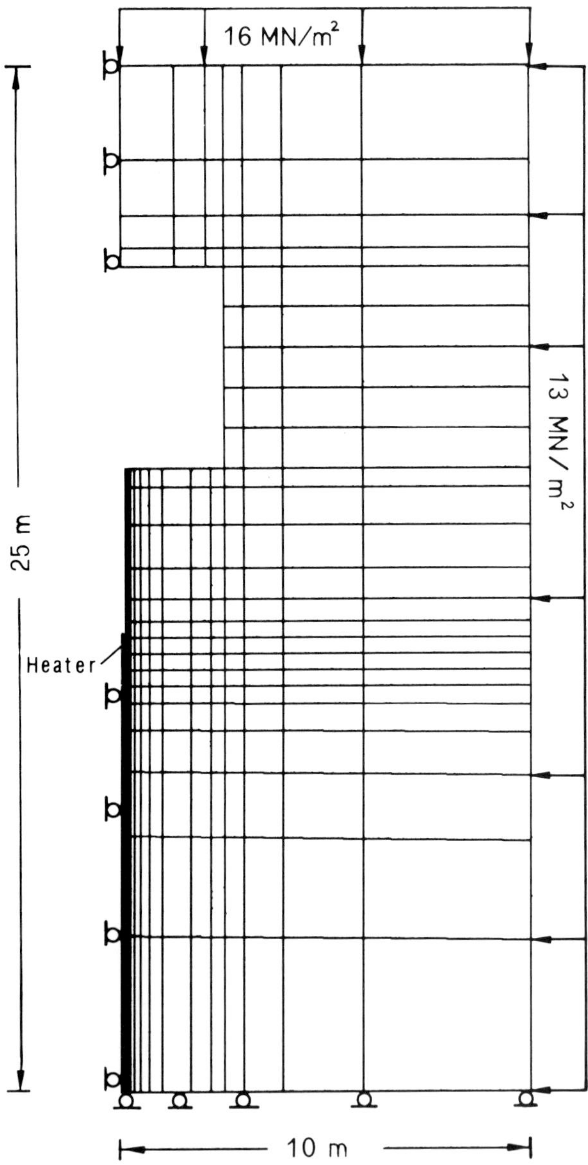

Fig.2 Finite element mesh and boundary
conditions for heater experiments

radial dividion at the borehole was necessary in order to deal with the
great temperature gradients in the immediate vicinity of the heater
(see Fig. 3).

The complex thermomechanical material behavior of the rock salt was
described as a thermoelastic-plastic material model with secondary creep.
The elastic-plastic short-term behavior was modeled by a simple bilinear
relation which, however, approximates well the strain behavior of rock
salt observed in the laboratory, including the dependence of temperature.

The time dependent creep strains were calculated with a creep law derived from the laboratory test on ASSE rock salt /3/:

$$\dot{\varepsilon} = A \cdot \exp(-Q/RT)\, \sigma^5$$

with activation energy Q = 54,21 kJ/mol

    A = 0.18 MPa^{-5}/d and R = 0.008314 kJ/mol/K

    σ = effective stress

The elastic properties of the rock salt assumed, to be homogeneous and
isotropic, were chosen as follows:

    Young's modulus E(T) = (20 k/ (T-273 K)$^{0.1}$. 7000 MPa

    Poisson ratio ν = 0.3

The rock pressure acting on the external faces of the model is required
as a boundary condition for the computations. While on the upper bounding
surface a vertical stress component of 16 MPa was defined, corresponding
to the lithostatic pressure estimated to prevail at 775 m depth, the horizontal component acting on the external faces was assumed to be 16 MPa
and 13 MPa (corresponding to a lateral pressure coefficient λ = 0.8) in
order to delimit the actual conditions.

The computations were made in two steps. In the first step the stress
condition prevailing in the salt at the beginning of the heating phase
is determined, which is characterized by the relocation of stress originating in the cavities. This is followed by the calculation of the thermomechanical events occurring in the neighborhood of the probe under the
influence of the space and time dependent heating of the salt.

RESULTS AND DISCUSSION

By the example of test 2, with the average heat output (1.3 kW) Fig. 3,
shows the calculated temperature distribution in the mid plane of the
probe at different points of time. For the same test, the influence of the

lateral pressure boundary condition was studied first. The comparison
shown in Fig. 4 between experiment and calculation at two measurement
positions revealed a very good agreement for a horizontal stress of 13 MPa,
while at σ_R = 16 MPa the convergences calculated were clearly too high.
Therefore, to evaluate the other two tests, only the value of 13 MPa was
used, since it can be considered to be more realistic.

Fig. 5 to 7 show the results of computations so obtained, compared
with the experimental data for three tests. In the tests 1 and 2 the results are lacking for the central measurement position (electrode). The
reason because is that salt efflorescences pushed inwards the annular
electrode and thus falsified considerably the measured values. The reason
for the practically disappearing convergence at the lower electrode in
test 1 has not yet been clarified.

The agreement between computation and experiment is very remarkable
altogether. But the assumed horizontal stress of 13 MPa can only be confirmed from this agreement if the creep law, which exerts a major influence on the borehole closure, can be validated in quantitative terms. To
illustrate further the computed results, the closures along the borehole
wall at different points of time have been represented in Fig. 8.

CONCLUSIONS

The experiments described here on heat-induced borehole closure and
their evaluation by model computations have shown that with the methods
of computation and material laws used, a relatively good prediction can be
made of the thermomechanical effects in the ultimate disposal of high
level wastes in rock salt, at least with respect to near field and relatively short time spans. Regarding a verification of the creep law and
the model, it would be necessary to make accompanying measurements of the
initial stresses at the place of the experiment, as well as laboratory
measurements on the creep behavior of the respective salt.

Since the measurement of absolute stresses in salt has posed problems,
an attempt should also be made to determine the initial stress condition
at the place of the experiment by large-space computations.

Acknowledgement. This work was carried out in cooperation with the European Atomic Energy Community within the scope of the indirect project
"Management and Storage of Radioactive Waste".

REFERENCES

1. Korthaus, E. et al. (Febr. 1979), Untersuchungen zur Temperaturentwicklung bei der Endlagerung hochradioaktiver Abfälle, Teil I und II,
   Zeitschrift Atomwirtschaft Atomtechnik, Nr. 2
2. Bathe, K.J. (Sept. 1978), A Finite Element Program for Automatic
   Dynamic Incremental Nonlinear Analysis. Massachusetts, Inst. of
   Technology.
3. Wallner, M. et al. (1979): Ermittlung zeit- und temperaturabhängiger
   mechanischer Kennwerte von Salzgesteinen.- Proc. 4. Int. Congr. Rock
   Mech., Montreux 1979, Vol. 1, 313-318

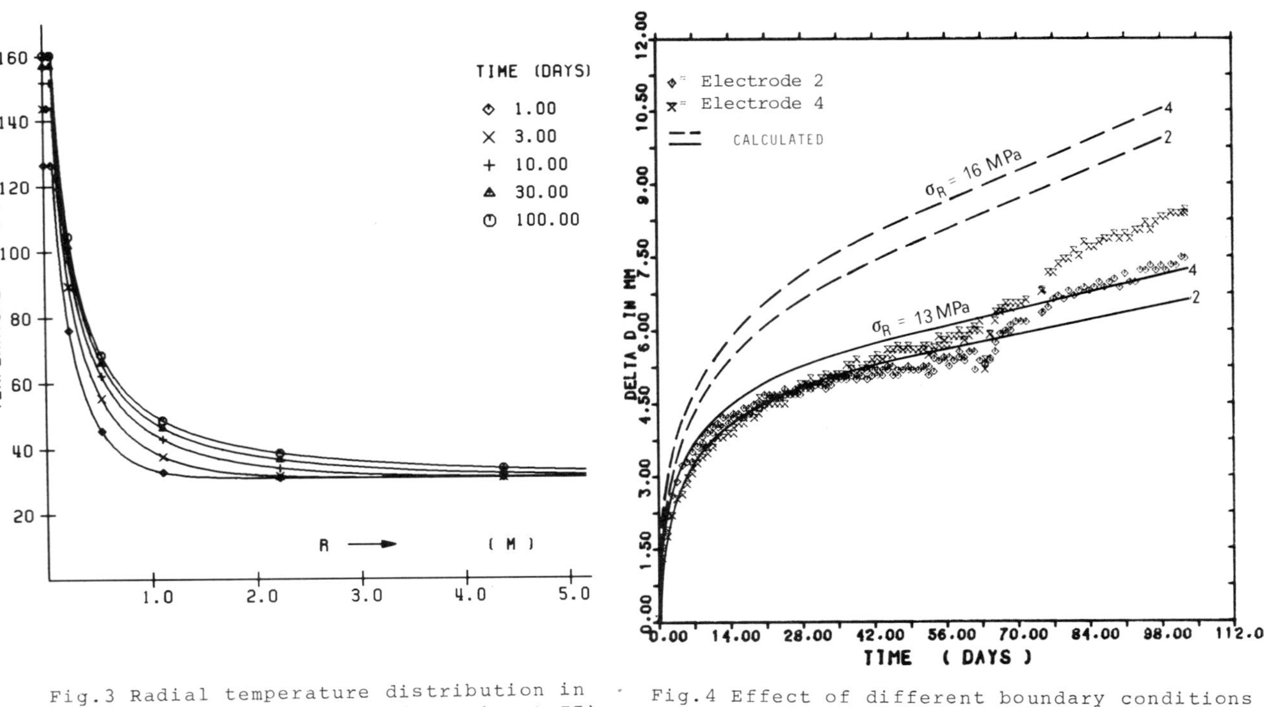

Fig.3 Radial temperature distribution in
the heater midplane (Experiment II)

Fig.4 Effect of different boundary conditions
on borehole closure

Experiment II   (Diameter change Delta D)

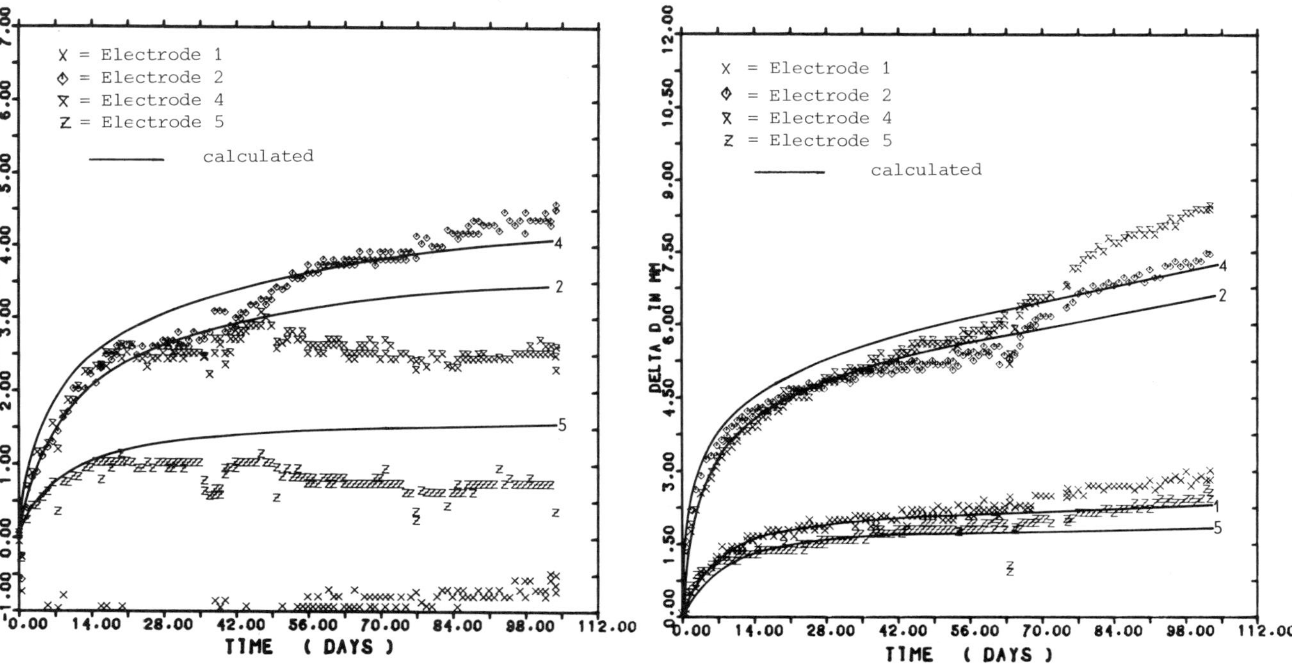

Fig.5 Comparison of measured and computed borehole closure for experiment I

Fig.6 Comparison of measured and computed borehole closure for experiment II

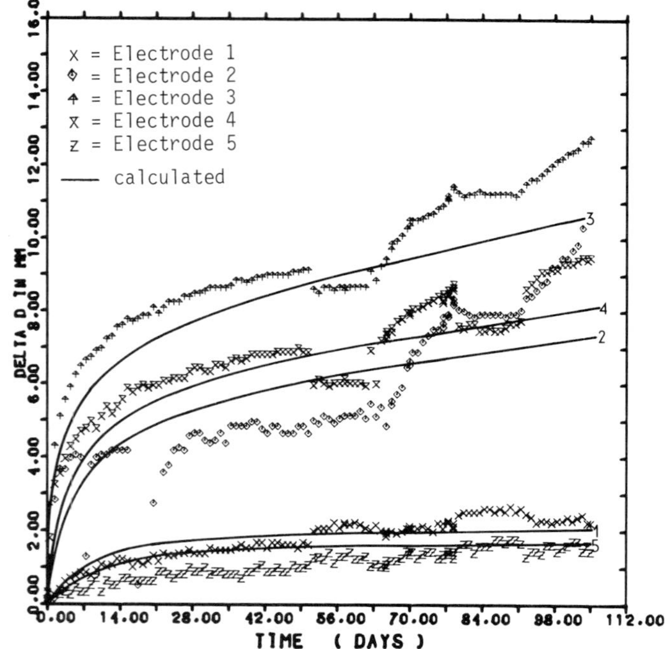

Fig.7 Comparison of measured and computed
borehole closure for experiment III.

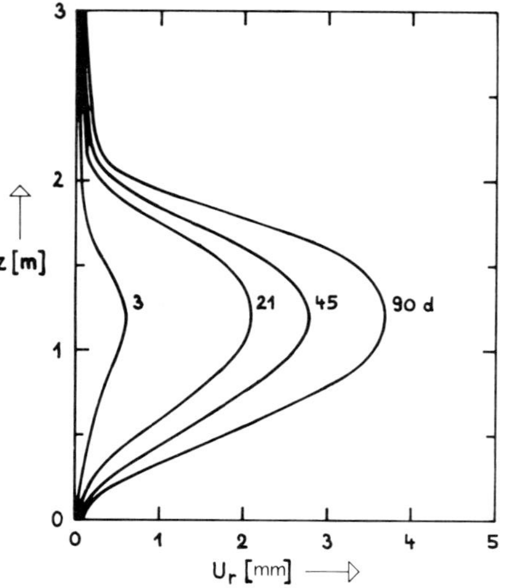

Fig.8 Radial displacement (Ur) of
borehole wall as function of vertical position (Z)
(Experiment II)

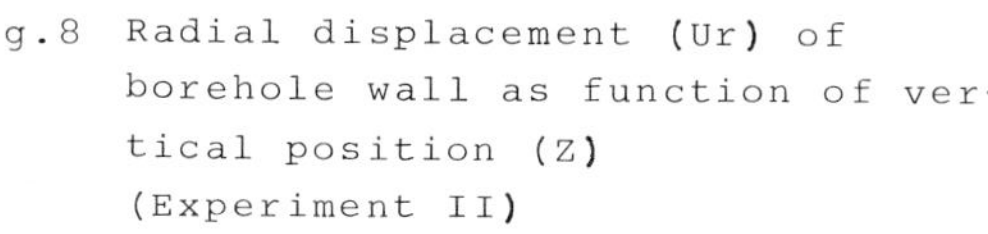

INVESTIGATIONS OF THE SOLUTION BEHAVIOUR OF NaCl IN THE QUINARY SYSTEM NaCl-KCl-MgCl$_2$-MgSO$_4$-H$_2$O AT DIFFERENT TEMPERATURES

R.CONRADT, H.ENGELKE and A.KAISER

Fraunhofer-Institut für Silicatforschung, Würzburg, Germany

INTRODUCTION

The German concept of high level waste final storage by the aid of waste glasses is based on rock salt formations as deposits. Investigations consider the accidental case of water entrance into the deposit where the waste forms then are exposed to highly concentrated salt brines at elevated temperatures and pressures.

In order to achieve standard corrosion conditions for the investigations of waste forms, a set of salt brines of definite composition has been generally accepted as possible corroding agents. The agreement is based on a model for the process of a hypothetical water entrance saying that in the initial period the entering water approaches a solution equilibrium at lower temperatures with the solid phases containing sodium, potassium and magnesium sulfates and chlorides. In the rock salt environment of the deposit higher temperatures will occur due to the radioactivity of the waste forms. That is why an additional uptake of NaCl by the "standard brines" has to be taken into account. However, data concerning this fact are not available. Therefore the intent of the investigations was to determine the additional amount of NaCl which can be dissolved at high temperatures in the so-called Q-solution mentioned below.

EXPERIMENTAL SET UP

The investigations of the quinary system up to 200°C require a transparent apparatus able to stand pressures of $\approx$16 bar. The apparatus used is a simplified version of an experimental equipment described for other investigations [1]. Through an inlet N$_2$ is led into the Duran-50 glass container to build up a pressure high enough to prevent its liquid content from boiling. A teflon tube is inserted into the mixture to provide sampling through a teflon valve.

The container is heated in a bath of silicon oil fitted with a thermocouple for temperature control and a calibrated thermometer. Both the oil bath and the mixtures in the reaction volume can be stirred by a magnetic stirrer. Heating devices around the inlet and outlet regions avoid condensation.

EXPERIMENTAL RESULTS

The starting point of the investigations was the invariant point Q of the quinary system $NaCl-KCl-MgCl_2-MgSO_4-H_2O$ at $55^{\circ}C$ according to D'Ans [2]. The composition of this solution (Q-solution) is given in table 1.

TABLE 1

COMPOSITION OF THE Q-SOLUTION AT $55^{\circ}C$ IN g/100 g

NaCl	1,45
KCl	4,75
$MgCl_2$	26,80
$MgSO_4$	1,40
H_2O	65,60
	100,00

The other mixtures investigated contain the same amounts of KCl, $MgCl_2$, $MgSO_4$ and H_2O but increasing amounts of NaCl. (Thus, the total weight becomes > 100 g for these mixtures.) Every mixture was directly weighed into the glass container and heated within ≈0,5 h close below the temperature at which complete dissolution was expected. Then this temperature was approached by very slow heating (≈0,1 K/min). The dissolution temperature was determined by direct observation supported by the use of a light source, performing three heating and cooling cycles with a reproducibility of some 0,1 K. The temperature thus determined was defined as the solution temperature of the investigated mixture. The (slightly lower) temperature of recrystallization during cooling was not respected because of the nucleation lag phenomena involved.

Table 2 presents the solution temperatures of mixtures containing different amounts of NaCl.

TABLE 2

SOLUTION TEMPERATURES OF Q-SOLUTION WITH VARYING AMOUNTS OF ADDITIONAL NaCl

additional NaCl (g)	total content of NaCl (g)	solution temperature (°C)
0	1,45	58 60
1,05	2,5	82 87
1,55	3,0	100
2,05	3,5	114
2,55	4,0	129
3,05	4,5	145
3,55	5,0	157 155
5,05	6,5	187

Solution temperatures obtained from different runs carried out under identical conditions show remarkably greater differences than during a single run. The deviations are estimated to be $\pm$ 3 K. This may be due to condensation phenomena inside the container and the fittings, to a not absolutely constant hydrate water content of the salts used for the preparation of the mixtures and finally to some loss of water from the system through small leakages due to the teflon fittings at higher temperatures and pressures.

After the determination of the solution temperatures the mixtures were cooled down to 80°C and the precipitates were allowed to settle down. Then most of the liquid phase was removed, the remaining substance was filtered and washed with small amounts of ethanol. The precipitates of the 3,5 g, 4 g, 4,5 g, 5 g, and 6,5 g NaCl containing mixtures were analyzed by X-ray diffraction. In all cases halite (NaCl) and sylvite (KCl) could be detected. Hints to the presence of a magnesium sulfate hydrate were found for the 5 g and 6,5 g mixtures, but not for the others.

SUMMARY

The experiments show that the Q-solution investigated is able to take up notable amounts of NaCl at temperatures up to 200°C. The data presented can be used to enhance the simulation of the accidental case of water entrance to a rock salt deposit. It must however be noted that the experimental procedure presumes a rather quick equilibration between the phases involved. All cases in which longer periods of time are needed to establish a thermodynamic equilibrium are not covered. The data are valid for the experimental conditions specified. There are some experimental observations requiring further investigations to make sure whether thermodynamic equilibrium is reached or not under these conditions.

REFERENCES

1. W.Strohmeier, A.Kaiser (1976) Journal of Organometallic Chemistry, 114, 273-279.
2. D'Ans, J., "Die Lösungsgleichgewichte der Systeme der Salze ozeanischer Salzablagerungen", Berlin 1933.

Published 1982 by Elsevier Science Publishing Co
SCIENTIFIC BASIS FOR RADIOACTIVE WASTE MANAGEMENT - V
Werner.Lutze, editor

INVESTIGATION OF DIFFUSION OF UO_2Cl_2 IN SATURATED NaCl SOLUTIONS AT VARIOUS TEMPERATURES

HARRY MURSO[*], BODO PLEWINSKY[**], DIETER LEOPOLD AND GÜNTER MARX
Free University of Berlin
Institute of Inorganic and Analytical Chemistry
Fabeckstraße 34-36, 1000 Berlin 33 (Dahlem), Germany

INTRODUCTION

Even in the early 1960's it had already been decided to place high-level radioactive waste in the salt formations of the Federal Republic of Germany. However a major concern is the formation of highly corrosive brines in the event of water filling the repositories.[1] In order to estimate the risks of waste storage in salt repositories it is necessary to investigate the physico-chemical properties of the waste materials and of those radionuclides extracted into brine. The extraction of radionuclides from borosilicate glasses or spent fuel elements has been discussed[2], taking into consideration the two transport mechanisms: diffusion and convection.

Diffusion measurements of actinides in saturated NaCl solutions have been made in order to determine which mechanism is rate determining. The diffusion coefficients of UO_2Cl_2 in saturated NaCl solutions were measured by use of an ultracentrifuge at 25, 40 and 50 $^\circ$C. The results are compared with those determined for the binary system UO_2Cl_2 - H_2O. Measurements of viscosity and molar mass were made in order to elucidate the difference in diffusion velocities in the systems studied.

EXPERIMENTAL DETAILS

Diffusion coefficient and molar mass determinations were carried out with a Beckman Instruments, Model E analytical ultracentrifuge, fitted with a standard schlieren optical system and an absorption optical system connected to a photo-electric scanner. The apparatus, methods of operation and data handling systems are described fully in references [3, 4, 5].

---

* Part of theses

** to whom correspondence should be addressed,

new address: Bundesanstalt für Materialprüfung (BAM),

Unter den Eichen 87, 1000 Berlin 45, Germany

492

Viscosity measurements were carried out with the aid of an automatic capillary viscosimeter (Ubbelohde type).

DIFFUSION IN TERNARY SYSTEMS

A ternary system - e. g. UO_2Cl_2 dissolved in saturated NaCl solutions obeys Fick's law as follows:[6]

$$H_2O \ (1) \ - \ UO_2Cl_2 \ (2) \ - \ NaCl \ (3),$$

$$-J_2 \ = \ D_{22}\mathrm{grad}c_2 \ + \ D_{23}\mathrm{grad}c_3,$$

$$-J_3 \ = \ D_{32}\mathrm{grad}c_2 \ + \ D_{33}\mathrm{grad}c_3.$$

J_i is the current density of component i and equals the amount of substance i being transported perpendicularly across a plain of reference of unit area in unit time. In such a system, four diffusion coefficients are required in order to quantitatively describe the diffusion behaviour. There are two main-term diffusion coefficients, D_{22} and D_{33}, and two cross-term diffusion coefficients, D_{23} and D_{32}. In this case, the main-term diffusion coefficient D_{22} describes the molecular transport of UO_2Cl_2 caused by its own concentration gradient. In addition, the NaCl concentration gradient can also cause the diffusion of UO_2Cl_2. The extent of this coupling effect is indicated by D_{23}. Of the three optical systems available with the Beckman ultracentrifuge (schlieren, Rayleigh interference and absorption) only the Rayleigh interference system permits the determination of all four diffusion coefficients simultaneously. For such an optical system, Creeth and Gosting[7] have developed expressions for the treatment of mixed solutes, although these were restricted to noninteracting solute flows. However, Albright and Sherrill[8] have recently presented theories applicable to more general cases of ternary diffusion.

Nevertheless, the use of absorption optics for investigating ternary systems has the one great advantage in that this optical system permits a _direct_ determination of the main-term diffusion coefficient, and for this reason _absorption_ optics were employed in this study.

DIFFUSION IN THE BINARY SYSTEM UO_2Cl_2 - H_2O

From a practical point of view, the determination of the diffusion coefficient of UO_2Cl_2 in pure water is of little value in estimating the risks of storing spent fuel in salt formations. On the other hand, such measurements are invaluable for the interpretation of data in ternary systems. For this reason diffusion coefficients of UO_2Cl_2 were measured for aqueous solutions at 25 and 40 $^{\circ}$C. At the former temperature absorption optics were employed for

UO_2Cl_2 concentration of 2.3×10^{-2} mol l^{-1} or less. Higher concentrations were investigated with the schlieren optical system. For measurements at 40 ^{O}C, only the schlieren optics were employed. The results are illustrated in Fig. 1. The function $D = D(c)$ at 25 ^{O}C is similar to the curves for the $Th(NO_3)_4$ and $UO_2(NO_3)_2$ systems.[9] The change in diffusion coefficient of UO_2Cl_2 with concentration is also shown for measurements at 40 ^{O}C. Upon comparing the curve measured at 25 ^{O}C with that measured at 40 ^{O}C, it can be seen that at the latter temperature the diffusion coefficients are much higher. For most binary systems there exists an exponential relationship between the diffusions coefficient and temperature;[10] the results of Fig. 1 support such a relationship.

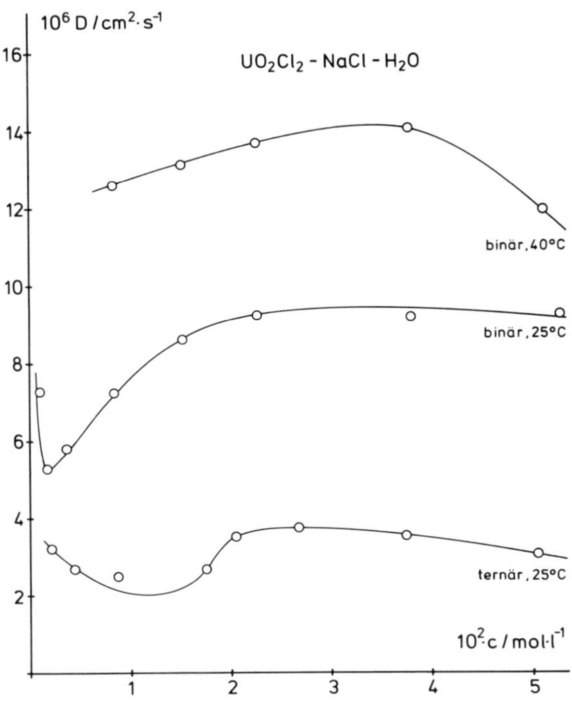

Fig. 1. Diffusion coefficient of UO_2Cl_2 as a function of the UO_2Cl_2 concentration.

DIFFUSION OF UO_2Cl_2 IN SATURATED NaCl SOLUTIONS

The main-term diffusion coefficient of UO_2Cl_2 was measured in a saturated NaCl solution (c(NaCl) = 5.2 mol l^{-1}) using absorption optics. The results obtained at 25, 40 and 50 $^{\circ}$C can be seen from Fig. 2. It is significant that for the ternary system the values of the diffusion coefficients are much smaller than for the binary system. It is thus evident that the diffusion processes of UO_2Cl_2 in saturated NaCl solutions are much slower than those in water. In order to explain such observations, viscosity measurements were carried out for both binary and ternary systems at 25, 40 and 50 $^{\circ}$C.

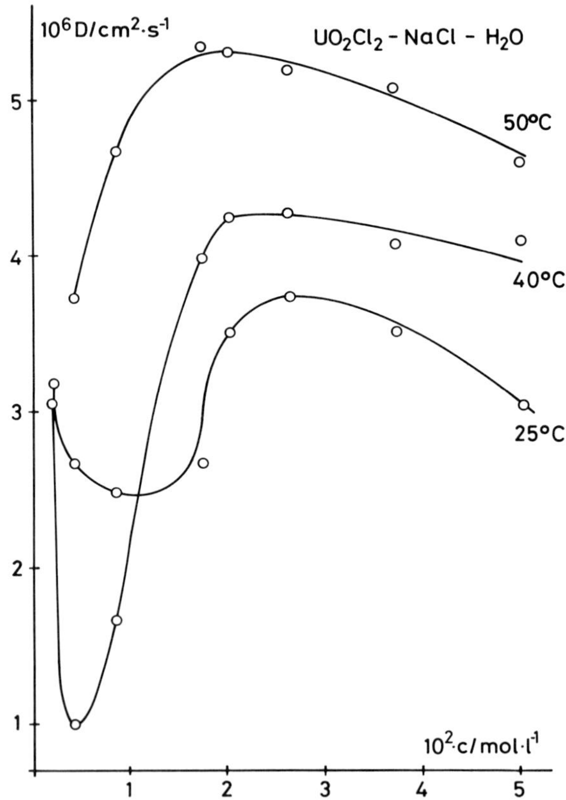

Fig. 2. Main-term diffusion coefficient of UO_2Cl_2 in saturated NaCl solution as a function of the UO_2Cl_2 concentration.

Results indicated that in the absence of NaCl the viscosities were approximately half those than when NaCl was present. Molar mass determinations (in saturated NaCl; $c(NaCl) = 5.2$ mol l^{-1}) were carried out using the equilibrium sedimentation method at 25 $^{\circ}C$ only and with the absorption optical system. The variation in the weight average molar mass, M_w, as a function of the initial concentration is shown in Fig. 3. The dimer content (as $(UO_2)_2(OH)_2^{2+}$)[11] was found to increase from 48% of the mass at a uranyl chloride concentration of 1.5×10^{-3} g cm^{-3}, to 70% of the mass at the plateau region. Interferometric measurements carried out by Rush et al.[12] on the UO_2Cl_2 - NaCl - H_2O system by use of an ultracentrifuge have not indicated the presence of any polymeric uranyl cations over the NaCl concentration range 0.01 $c(NaCl)/\text{mol } l^{-1}$ 0.1. Our results prove that the dimerization of uranyl chloride is caused by the high NaCl content.

Therefore in saturated NaCl solutions small diffusion velocities for UO_2Cl_2 are due to high viscosities and the increased radius of the uranyl chloride particles.

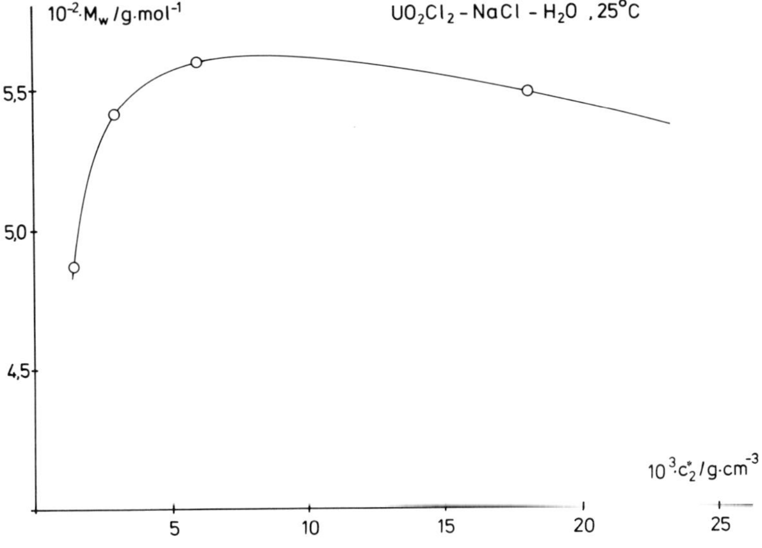

Fig. 3. The weight average molar mass M_w as a function of the initial UO_2Cl_2 concentration.

496

Fig. 2 shows that the temperature dependence of the UO_2Cl_2 diffusion coefficient is small in saturated NaCl solutions. At these temperatures the minima of the curves D = D(c) can be seen at low UO_2Cl_2 concentrations. The most pronounced minimum occurs at 40 ^{O}C and it can be seen that at this temperature, the values of the diffusion coefficient are smaller than at 25 ^{O}C. The changes in the diffusion coefficient are primarily a result of the temperature-dependancy of either the viscosity of the solution or the solute molecular dimensions. Separate determinations have indicated no anomalous behaviour in the viscosity versus concentration curves for UO_2Cl_2 over the 25 - 50 ^{O}C temperature range. The curious behaviour illustrated in Fig. 2 is therefore assumed to be due to a solute size effect brought about by molecular aggregation.

ACKNOWLEDGEMENTS

We wish to express our gratitude to Konrad Mündlein for his help and assistance during the course of this work.

We gratefully thank the Fonds der Chemischen Industrie for financial assistance.

REFERENCES

1. Closs, K. D., ed. (1980), "Vergleich der verschiedenen Entsorgungs-
   alternativen und Beurteilung ihrer Realisierbarkeit",
   Kernforschungszentrum Karlsruhe (KfK 3000).
2. Malow, G. (1981) in Proc. Int. Sem. on Chem. and Process-Eng. of High-
   level liquid waste solidifications, Odoj, R. and Merz, E. ed, Jülich,
   Germany.
3. Chervenka, C. H.(1973) A Manual of Methods for the Analytical Ultracentrifuge, Beckman Instruments, Palo Alto.
4. Fujita, H. (1975) Foundations of Ultracentrifugal Analysis, New York.
5. Mündlein, K. (1982) Studienarbeit TFH Berlin.
6. Haase, R. (1963) Thermodynamik der irreversiblen Prozesse, Darmstadt.
7. Creeth, J. M. and Gosting, L. J. (1958) J. Phys. Chem. 62, 58.
8. Albright, J. G. and Sherrill, B. (1979) J. Solution Chem. 8, 201.
9. Plewinsky, B., Rösel, B., Binder, Ch., Feldner, K.-H., Murso, H. and
   Marx, G. (1982) PBE-Berichte in press.
10.Longsworth, G. (1954) J. Phys. Chem. 58, 770.
11.Baes, C. F. and Mesmer, R. E. (1976) The Hydrolysis of Cations, New York.
12.Rush, R. M., Johnson, J. S. and Kraus, K. A. (1962) Inorg. Chem. 1, 378.

THE IMPORTANCE OF NEAR-FIELD PHENOMENA FOR NUCLIDE RELEASE FROM A FLOODED SALT
DOME REPOSITORY

RICHARD STORCK
Technische Universität Berlin, FRG

INTRODUCTION

Safety studies for final disposal of nuclear waste at the Gorleben site in
Germany are performed by the project "Safety Studies Entsorgung (PSE)" supported by the Federal Ministry for Research and Technology. The aim of the project
is to do consequence assessments for selected site-specific scenarios. The restriction to consequence assessments, instead of performing an overall risk assessment, is based on the expectation that the safety of a given repository site
can be demonstrated by showing that the consequences of each scenario, which
might reasonably occur, are acceptable.

The consequences of a given release scenario can be influenced by a number of
phenomena in the near field (inside the salt dome) and the far field (outside
the salt dome). Especially in the near field, the number of phenomena and their
interactions can be rather high, so that a limited investigation must restrict
itself to the most relevant phenomena. The identification of relevant phenomena
is a stepwise procedure which has to be carried out for each type of scenario.
This paper deals with the first step in identification of relevant near field
phenomena for the scenario "penetration of brine into a backfilled HLW-repository, directly after its closure". By use of parameter variation, the relevant
phenomena for the release of Tc-99 from the salt dome are identified.

The restriction to the calculation of nuclide quantities released from the
salt dome and to the consideration of just one nuclide can be justified by two
arguments. First, the consequences of water intrusion scenarios are dominated
by long lived nuclides, of which Tc-99 is one of the most important[1]. Second,
the radiation dose to the most exposed individual is approximately directly
proportional to the released amount of nuclides[1], so that the phenomena relevant to the release of Tc-99 can be addressed as relevant near field phenomena
for the consequences of this scenario.

DESCRIPTION OF THE REPOSITORY FOR HLW AND THE BARRIER SYSTEM

The repository, planned at the Gorleben site, will be a one-level system
situated 800 m below surface in homogeneous rock salt. The high level waste
will be disposed of in bore-holes, each containing 250 canisters and the waste

498

of 216 t_{HM} reprocessed fuel. The bore-holes will not be backfilled but sealed
with a salt concrete mixture. The disposal drifts, each allowing access to 6
bore-holes will be backfilled with crushed salt and sealed with a salt concrete
mixture at both ends.

In the case of groundwater penetration into the whole repository directly
after its closure, the sealings cannot retard the water from the waste canister,
because they are not designed for a hydrostatic pressure difference of about
10 MPa. The present canisters are only designed for transportation loads and
not for loads due to hydrostatic or lithostatic pressure. However, well designed canisters are under discussion promising lifetimes of several hundred
years. To evaluate the importance of high canister lifetimes for the described
disposal system, the canister lifetime will be handled as a variable.

The resulting barrier system for a repository with a capacity of 1400 t_{HM}/y
filled up over a period of 50 years is shown in figure 1. The multipliers
indicate the number of barriers in one barrier above it (e.g. 6 bore-holes per
disposal drift). There are two types of bore-holes which result from the convergence process before water penetration. The first one still has the annular
gap between the canister and the salt, and it is called the latest bore-hole.The
second one has time to close the annulus by creep of the salt, and it is called
the older bore-hole. The relationship between the number of drifts with later
and older bore-holes is determined by the closure time of the annulus in the
bore-hole compared to the operation time. Closure times between 7 and 16 years
have been calculated[2,3] and a value of 12.5 years has been chosen arbitrarily.

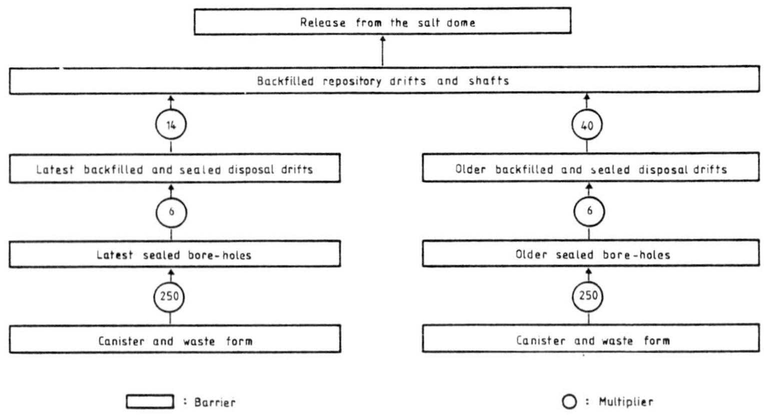

Fig. 1: Barrier system for the HLW-repository

BARRIER-MODELLING

The described barrier system for HLW-disposal in bore-holes comprises four different types of barriers, which have to be modelled for calculating the nuclide release to the overlying strata. The different types of barriers are: a) the waste package consisting of the canister and the waste form, b) the sealed bore-holes, c) the backfilled and sealed disposal drifts and d) the backfilled repository drifts. For each of them, a barrier model has been established taking into account various physical and chemical effects which are outlined in the next chapters.

Waste Package

The waste package consists of the canister and the waste form. For the canister it is assumed, that it only delays the contact between brine and waste form during its lifetime. Once contact is established, any barrier effect from the canister is neglected.

Leaching of the waste form is modelled by a constant leach rate ($R_o=10^{-3}$/y) for a reference temperature (T_o=473 K), corrected by a temperature function which follows normal Arrhenius behaviour (Q = 46 kJ/mol[4]; I_i: inventory of nuclide i).

$$R_{w,i}(t) = R_o (T_o) \cdot I_i (t) \cdot \exp \left(\frac{Q}{R} \left(\frac{1}{T_o} - \frac{1}{T(t)}\right)\right) \tag{1}$$

A reference leach rate of 10^{-3}/y will result in a total mobilisation after 10^3y if the temperature is kept at the reference temperature. Utilizing calculated temperatures[5] at the interface of the canister and rock salt, the mobilized amount of Tc-99 after 10^3 y will be reduced to 18 % of the initial inventory. The temperature function reaches its minimum of 0.27 % after 10,000 years. Total mobilization of the waste will then take place after 100,000 years.

The void volume in the sealed bore-hole and the nuclide transport to the disposal drift will be low enough, so that the leaching will be controlled by the solubility of the nuclides. Solubility numbers for Tc-99 in brine of 10^{-7} mol/l have been mentioned[9]. Assuming a safety factor of 1000, a conservative estimate of 10^{-4} mol/l has been used.

Bore-Hole

The bore-hole consists of two parts: The waste area with a height of 300 m and the sealing area with a height of 10 m. The two areas are modelled quite differently.

The waste area contains a hollow space initially as an annular gap between the column of the waste packages and the host rock and within the waste packages themselves. The hollow space within a waste package is present as a cavity at the top of waste and as cracks in the vitrified waste, both resulting from cooling. After penetration of brine into the hollow space of the waste area and the start of the leaching process, a uniform distribution of nuclides in the brine is assumed due to natural convection caused by high temperature gradients in that area.

The sealing area is assumed to be permeable because the sealing material is not designed for hydrostatic pressure differences of 10 MPa, which are possible after flooding of the repository. The sealing material is assumed to be a porous medium which allows nuclide transport from the waste area to the drift by three mechanisms: 1) diffusion, 2) forced convection due to creep of the salt, and 3) natural convection. For all of these, the delayed release to the drift, resulting from the transport through the sealing material, is neglected.

<u>Diffusion</u>: Nuclide release by molecular diffusion through the void volume of the sealing material can be calculated using an effective diffusion coefficient which is equal to the diffusion coefficient of ions in water ($D = 10^{-9}$ m^2/s) times the porosity of the porous medium ($\eta_B = 0.2$ at $t = 0$). The release rate by diffusion is given by

$$R_{B,i}^{D} = \frac{D \cdot \eta_B(t) \cdot A_B}{L_B} \, (c_{B,i}(t) - c_{D,i}(t)) \qquad (2)$$

where A_B is the cross section of the sealing (0.126 m^2) and L_B the height of the sealing (10 m). The concentration difference is that between the waste area in the bore-hole (B) and the disposal drift (D). The porosity of the sealing will be changed by the creep of the salt as explained later on.

<u>Forced Convection</u>: Creep of rock salt in the waste area causes nuclide release by forced convection. Creep of the salt can be characterized by a convergence rate (k) which is the relative change in volume per unit time. A convergence rate for the annulus of $6 \cdot 10^{-4}$/y has been calculated for flooded repository conditions[2]. The convergence rate is rather constant in time and results in closure of the 5 cm annulus after 1000 years, if the flooding occurs directly after disposal of the waste.

After closure of the annulus, the creep of the salt causes a decrease of the hollow space in the waste packages. A convergence rate for this phase has been assumed to be $3 \cdot 10^{-5}$/y for an initial porosity of 0.1 (fraction of hollow space in the waste package to the volume of the waste package). In the course of time, the convergence rate will be changed by the change in porosity of the waste form as explained later on.

The nuclide release rate by forced convection ($R^F_{B,i}$) can be calculated from the volume flow rate ($\dot{V}^F_B$) due to the convergence in the waste area (V_W: volume of the waste packages = 21.2 m³ at t = 0, V_A: volume of the annulus = 16.5 m³ at t = 0).

$$\dot{V}^F_B (t) = k_B (t) \cdot (V_W(t) + V_A(t)) \tag{3}$$

$$R^F_{B,i} = \dot{V}^F_B (t) \cdot c_{B,i} (t) \tag{4}$$

Natural Convection: Natural convection can be caused by the heat production, the gas production by radiolysis and by salt dissolution processes and will result in brine exchange between the bore-hole and the disposal drift.

Darcy velocities for the core region of porous waste forms have been calculated as a function of permeability ($K_B = 10^{-12}$ m² at t = 0) and heat production ($\dot{q}$ = 821 W/m at t = 0)[6]. Assuming this flow pattern to the sealing material and neglecting the effects from radiolysis and salt dissolution, the volume flow rate ($\dot{V}^N_B$) and the nuclide release rate by brine exchange from natural convection ($R^N_{B,i}$) can be calculated

$$\dot{V}^N_B (t) = 1.44 \cdot 10^{10} \, W^{-1}a^{-1} \cdot R_W \cdot K_B (t) \cdot \dot{q} (t) \tag{5}$$

$$R^N_{B,i} = \begin{cases} 0.0 & : \dot{V}^N_B \leq \dfrac{\dot{V}^F_B}{2} \\[2em] (c_{B,i}(t) - c_{D,i}(t)) \cdot (\dot{V}^N_B - \dfrac{\dot{V}^F_B}{2}) & : \dot{V}^N_B > \dfrac{\dot{V}^F_B}{2} \end{cases} \tag{6}$$

The permeability and the heat production will be changed by the convergence and the radioactive decay respectively. The radius of the waste form (R_W=0.15m) is kept constant.

The nuclide release from the bore-hole to the drift by the three effects mentioned above, will be influenced in a complex manner by the convergence. This context is demonstrated in the schematic representation of the barrier model for the sealed bore-hole in Figure 2.

502

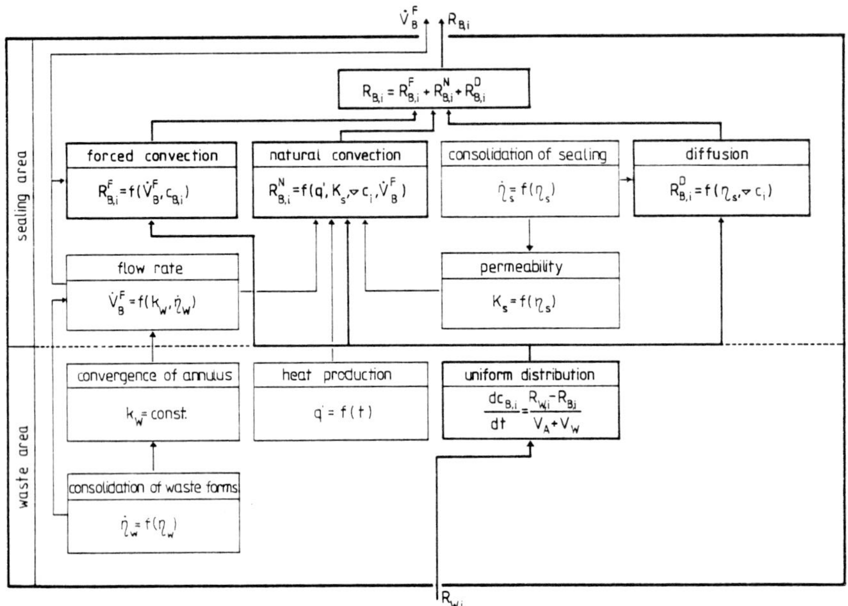

Fig. 2: Barrier-model for the sealed bore-hole

<u>Effect of Convergence:</u> The convergence in the bore-hole area will effect the porosity and the permeability of the bore-hole sealing as well as the porosity (remaining hollow space) in the waste forms. Simultaneously the convergence rate will be influenced by the porosity of the porous media, which is discussed first.

Based on analytical investigations, considering the consolidation of a single spherical void in a rock-salt-sphere under constant pressure conditions[7], the convergence rate (k) can be described by a reference value (k_o) for a reference porosity (n_o) corrected by a porosity function

$$k\ (n) = k_o\ (n_o)\ \cdot \frac{n}{n_o}\ \left(\frac{1 - n_o^{0,2}}{1 - n^{0,2}}\right)^5 \tag{7}$$

As an example: Beginning with a reference porosity of 0.4, a porosity decrease to 0.2 lowers the convergence rate by a factor of 24.

The convergence rate as a function of porosity effects the porosity itself. The porosity change causes a change in permeability (K), which can be described by a reference permeability (K_o) at a reference porosity (n_o) corrected by a porosity function

$$K(\eta) = K_o(\eta_o) \cdot (\frac{\eta}{\eta_o})^3 \cdot (\frac{1-\eta_o}{1-\eta})^2 \tag{8}$$

As an example: Beginning with a reference porosity of 0.4, a porosity decrease to 0.2 lowers the permeability by a factor of 14.

Disposal Drift

Similar to the bore-hole, the disposal drift consists also of two parts: the backfill area with a cross section of 6 m x 6.6 m (width x height) and a length of 280 m, and the sealing areas with the same cross section and a length of 15 m at both ends of the disposal drift. These two areas are also modelled quite differently, but similar as in the case of two areas of the bore-hole.

The backfill area is a porous medium with a relatively high porosity (initial value of 0.4), where thermally induced convection results in a uniform distribution of nuclides in the brine after its release from the bore-holes.

The sealing area has to be assumed permeable for the same reason as the bore-hole sealings, so that nuclide transport to the open drifts is possible by the same three mechanisms: 1) diffusion, 2) forced convection due to volume flow from the bore-holes and due to backfill consolidation, and 3) natural convection due to temperature gradients.

Diffusion: Nuclide release by molecular diffusion through the void volume of both sealings, can also be described by equation (2) where A is the cross section of the two sealings (80 m²) and L the length of one sealing (15 m). The concentration difference in this case is that between the repository drifts and the disposal drift. The porosity of the sealings will also be changed by the creeping of the rock. Convergence rates for this case as well as for the bore-hole sealings are assumed to be equal to those in the backfill area for a reference porosity of 0.4. Due to lower porosities in the sealing area, the initial convergence rate is a factor of 25 lower.

Forced Convection: Nuclide release from the backfill (V_D: volume of backfilled drift = 11,100 m³ at t = 0) is caused by the volume flow from all the bore-holes in that drift ($\Sigma\, V_B^F$) and by the consolidation of the backfill material

$$R_{D,i}^F = \dot{V}_D^F (t) \cdot c_{D,i} (t) \tag{9}$$

$$\dot{V}_D^F(t) = K_D(t) \cdot V_D + \Sigma \, \dot{V}_B^F \tag{10}$$

The convergence rate for the backfilled area k_D is assumed to be 10^{-4}/y for a reference porosity of 0.4 which is equal to calculated values for open drifts under flooded repository conditions but at normal rock temperature[1]. The volume of the backfilled drift itself as well as the porosity of the backfill will be changed by the consolidation process as defined by the definition of the convergence rate and equation 7.

Natural Convection through the drift sealings and therefore brine exchange between the repository drifts and the disposal drift can be caused by a horizontal temperature gradient along the length of the sealing. Darcy velocities have been assessed analytically for narrow horizontal cavities as a function of permeability ($K_D = 10^{-12} \text{m}^2$ at $t = 0$), temperature gradient (∇T_D) and height of the cavity ($H_D = 6.6$ m)[1]. They can be transformed to a volume flow rate by use of the cross section of the drift ($A_D = 40 \text{ m}^2$)

$$\dot{V}_D^N = 7.4 \cdot 10^{10} \text{ m}^{-1} \text{ K}^{-1} \text{ a}^{-1} \cdot \nabla T_D(t) \cdot A_D \cdot H_D \cdot K_D(t) \tag{11}$$

and nuclide transport is then given by a relationship similar to equation (6). The permeability will be changed as a function of porosity. Temperature gradients have been taken from reference 5 and will be changed as a function of time.

Repository Drifts

The backfilled repository drifts and the backfilled or sealed shafts are included in the fourth barrier-model. All former drifts and the shafts (10^6 m^3) are assumed to be backfilled up to a porosity of 0.4. Contaminated brine in the void volume is pressed out by the convergence process ($k_O = 10^{-4}$/y, see equation 7). Contamination of brine is modelled as uniform and hence, delay effects by distribution in the void volume or transport through the shaft are neglected.

RESULTS

The described barrier-models are integrated into a barrier-system as shown in figure 1. This barrier-system has been numerically evaluated for a period of 10^5 years. Some results are shown in figures 3 and 4 as volume flow rates or nuclide release rates at several points in the barrier system. The relationship in volume flow rates from latest bore-holes by forced and natural convection indicates, that there is no brine exchange in the time period between 10^2 and 10^3 years. The nuclide release from such bore-holes in that time period is dominated by the convergence process, whereas the leaching process is just replacing the nuclides, released by diffusion.

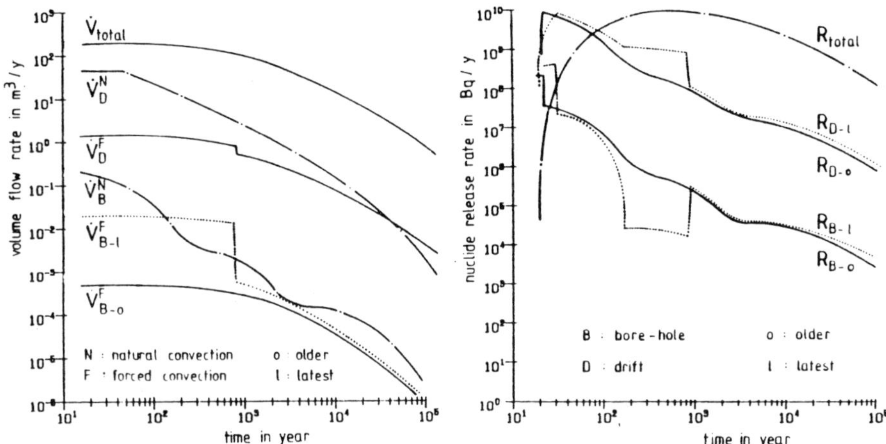

Fig. 3: Volume flow rates   Fig. 4: Nuclide release rates

The leaching of nuclides in both types of bore-holes is controlled almost from the beginning (2y, 10y) by the solubility. The release of nuclides from the bore-holes is dominated by natural convection over $3 \cdot 10^4$ years (with an interruption by forced convection for latest bore-holes) and by diffusion after $3 \cdot 10^4$ years. The release of nuclides from the drifts is dominated by natural convection over 5000 years and by forced convection afterwards.

The effectiveness of the barrier system can be characterized by the released amount of nuclides accumulated over 10^5 years divided by the amount of nuclides, which has been disposed of. This value is about 0.2 % for the whole system. A corresponding value for the release from bore-holes is 1.3 % for a latest and 0.8 % for the older bore-holes. Hence, there is no significant safety related effect from the annulus.

The effect of container lifetime on the accumulated released amount is shown in figure 5. It can be seen, that even container lifetimes of 10^3 years reduces the release just by a factor of 10. This statement is valid for solubility controlled leaching as well as for unrestrained leaching, where the absolute value for the released amount is a factor of 20 higher.

The relative importance of other effects can be described by the relative change of the above mentioned released amount divided by the relative variation of characteristic input variable. An importance value of 0.2 for example means, that after a change of 100 % at the input variable , the accumulated released amount will be changed by 20 %. The importance of several effects and input parameters have been quantified as follows

container lifetime	0.200
leach rate	0.030
temperature dependent leaching	0.002
solubility limit	0.900
leach rate (no solubility limit)	1.000
temp. dep. leaching (no solubility limit)	1.800
permeability of bore-hole sealings	0.700
permeability of drift sealings	0.100
diffusion through bore-hole sealings	0.025
diffusion through drift sealings	0.000
convergence rate	0.100
porosity dependent convergence	0.400

For estimating the absolute importance, the possible variation of the parameter has to be considered. The uncertainty in the solubility for instance is by 2 or 3 orders of magnitude, which changes the released amount in the same size. On the other hand, the temperature dependent leaching (no solubility limit) is of less absolute importance because even in case of neglecting the temperature effect, the released amount changes by a factor of about 2.

Therefore the relevant effects, which have to be investigated in more detail are: solubility controlled leaching, natural convection through bore-hole sealings and the convergence process as a function of porosity. Also bore-hole sealings with low permeability seem to be a more worthwile research target than container longevity.

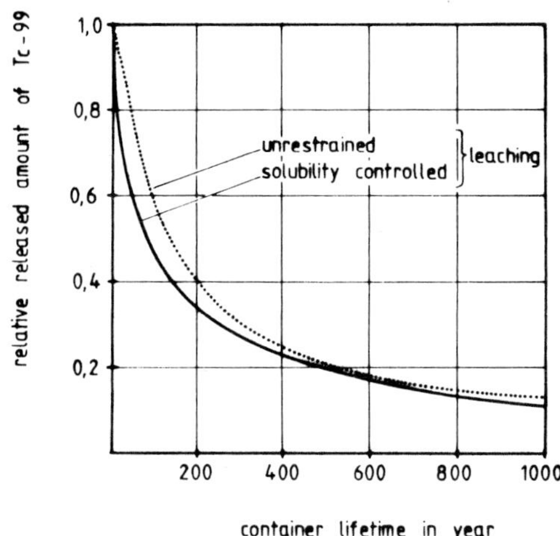

Fig. 5: Released Tc-99 vs. container lifetime

REFERENCES

1. Projekt Sicherheitsstudien Entsorgung, (1981) Zusammenfassender Zwischenbe-
   richt, Hahn-Meitner-Institut Berlin.
2. Pudewills, A. (1982) private communications, Gesellschaft für Kernforschung
   Karlsruhe.
3. Physikalisch-Technische Bundesanstalt, Braunschweig, (1980) private
   communications.
4. Malow, G. et al.(1978) Testing and Evaluation of the Properties of Various
   Potential Materials for Immobilizing High Activity Waste; Report EUR 6213 EN.
5. Delisle, G. (1979) Berechnung der Aufheizung einer Salzformation durch ein
   HAW-Endlager, Bundesanstalt für Geowissenschaften und Rohstoffe, Hannover,
   Archiv-Nr. 81837.
6. Storck, R., Brüggemann, R., Hossain, S. and Podtschaske, T. (1981) The
   Modelling of Barrier Effects from Bore-Holes in Safety Studies for Salt Dome
   Repositories, NEA/OECD, Workshop on Near Field Phenomena, Seattle,
   Aug. 31-Sept. 3.
7. Butcher, B.M. (1980) Creep Consolidation of Nuclear Depository Backfill
   Material, SAND - 79 - 2212.
8. Fair, G.M. and Hatch, L.P. (1933) Fundamental Factors Governing the Stream-
   line Flow of Water through Sand, Am. Water Works, pp. 1551-1556.
9. Elsam, Elkraft (1981) Disposal of High-Level Waste from Nuclear Power Plants
   in Denmark (Elsam: Denmark, 7000 Fredericia; Elkraft: Denmark,
   2750 Ballerup).

Igneous and Metamorphic Rocks

Published 1982 by Elsevier Science Publishing Co
SCIENTIFIC BASIS FOR RADIOACTIVE WASTE MANAGEMENT - V
Werner.Lutze, editor

DIFFUSION IN CRYSTALLINE ROCKS

K. Skagius and I. Neretnieks
Department of Chemical Engineering
Royal Institute of Technology
S-100 44 STOCKHOLM
Sweden

SUMMARY

Laboratory experiments to determine the sorption and the rate of diffusion of cesium and strontium in pieces of granite have been performed. The effective diffusivity, $D_p \cdot \varepsilon_p$ was found to be $1 - 2 \cdot 10^{-12}$ m^2/s for both cesium and strontium.

The diffusion of non-sorbing species in granites and other rock materials have been studied in laboratory scale. The non-sorbing species were iodide, tritiated water, Cr-EDTA and Uranine. In granites the effective diffusivities were determined to be $0.7 - 1.3 \cdot 10^{-13}$ m^2/s for iodide and $1.3 - 1.8 \cdot 10^{-13}$ m^2/s for tritiated water.

Electrical resistivity measurements in salt water saturated rock cores have been performed. The resistivity is measured in the saturated core and in the salt solution with which the core has been saturated. The ratio between these two resistivities has a direct relation to the ratio of the effective diffusivity for a component in the rock material and the diffusivity in free water for the same component.

The results from the electrical resistivity measurements and the experiments with diffusion of non-sorbing species are in fair agreement. The effective diffusivity for cesium and strontium (sorbing species) are, however, more than ten times higher than expected from the results of diffusion of non-sorbing species and the electrical resistivity measurements. This is interpreted as an effect of surface diffusion.

INTRODUCTION

Questions related to the final disposal of the nuclear wastes are studied intensively in many countries. The Swedish concept for a final repository for radioactive waste is to emplace the waste in crystalline rock. Canisters containing the waste are placed in holes in stable rock at about 500 m depth. The nuclides may, however, eventually leak out from the canisters and be transported with the moving ground water in fissures in the bedrock. To estimate the velocity of the moving nuclides in the fissures it is important to know the interaction mechanisms between the nuclides and the rock. The nuclides may sorb on the fissure surfaces and they may also diffuse through the stagnant fluid in the micropores of the rock and sorb on the micropore surfaces as well. This may have a strong impact on the retardation of the radionuclides. (1)

Three types of experiments in the laboratory scale to determine the sorption and the diffusion of nuclides in crystalline rocks have been performed.

o Sorption and diffusion experiments with cesium and strontium in two
   different granites.

o Diffusion experiments with non-sorbing species (iodide, tritiated water,
   Uranine and Cr-EDTA) in different rock materials.

o Electrical resistivity measurements in salt water saturated rock cores.

SORPTION OF STRONTIUM AND CESIUM ON GRANITE

Laboratory investigation

The granites in these experiments are from Finnsjön (quartz-granodiorite) outside Forsmark on the east coast of Sweden and from the Stripa mine (quartz-monzonite) in central Sweden.

The Finnsjö granite was in the form of cylindrical plates of 41 mm diameter and 5 mm thickness. They were taken from a drill core from about 100 m depth.

The Stripa granite was in the form of square plates 35 x 35 x 5 mm. These pieces were sawed out from a larger drill core from about 340 m depth in the Stripa mine.

Before starting the adsorption experiments, the granite pieces were saturated with "synthetic" ground water (the composition proposed by B. Allard (2)) by the following method: The pieces were heated at 90° C in vacua for three days. The dried pieces were then placed above a pan of synthetic ground water at ambient temperature in a vacuum chamber. A pressure close to the boiling point of the water (about 25 mm Hg) was maintained for several hours, and then the samples were dropped into the water.

After being in contact with the synthetic ground water for about a week ten pieces were put in a rack of teflon with a distance of 4 mm between the pieces. Because of this distance both sides of the plates would be in contact with the solution in the adsorption experiment. The rack of teflon with the granite pieces was placed in a bottle, and 1,100 ml of the solution containing strontium or cesium or both was added. Altogether 5 experiments were performed, strontium adsorption on Finnsjö granite and Stripa granite, cesium adsorption on Finnsjö granite and on Stripa granite and one experiment with simultaneous adsorption of cesium and strontium on Finnsjö granite. The initial con-

centration of strontium was about 10 ppm and of cesium about 15 ppm. The solid
to liquid ratio in the experiments with Finnsjö granite was 178.2 g/1,100 ml
and in the experiments with Stripa granite 165.4 g/1,100 ml.

All the bottles were placed in a shaking bath and the water in the bath
was temperated at 25° C. At intervals approximately following a geometric
progression, small fractions (1 ml) of the solutions were taken out, and the
concentration of cesium and strontium was measured by atomic absorption
spectrometry.

Determination of the sorption capacity and the rate of sorption

In the model below the granite pieces are treated as porous bodies. The
nuclides diffuse through the micropores and are then sorbed on the inner
surfaces of the solid. The diffusion equation which describes this sorption
process for one-dimensional diffusion is

$$\varepsilon_p \cdot \frac{\partial c_p}{\partial t} + \rho_s \cdot \frac{\partial q}{\partial t} = D_p \cdot \varepsilon_p \cdot \frac{\partial^2 c_p}{\partial x^2} \tag{1}$$

where ε_p is the porosity of the rock pieces, $c_p(x,t)$ and $q(x,t)$ are concentrations in the micropore fluid and in the solid phase respectively,
D_p is the pore diffusivity and x is the length coordinate. Local chemical
equilibrium is assumed at every point inside the pieces. The relation
between the concentration c_p and q within the pieces can be described by
the following equilibrium relation, the Freundlich isotherm.

$$q = k_f \cdot c_p^{\beta} \tag{2}$$

In these sorption experiments on pieces of granite no separate
determination of the equilibrium relation i.e. the Freundlich isotherm
(Eq. (2)) has been made. In an earlier investigation with adsorption and
diffusion of cesium and strontium in crushed Finnsjö and Stripa granite
(3), it was found that the isotherm was near linear ($\beta \approx 1$) for strontium
adsorption on both Finnsjö and Stripa granite. For cesium adsorption,
however, the isotherm was found to be non-linear, with the Freundlich
exponent $\beta = 0.54$ for Finnsjö granite and $\beta = 0.66$ for Stripa granite.

In the experiments with larger pieces of granite we assume that the
Freundlich exponent β has the same value as in the experiments with crushed
granite. From that assumption and the equilibrium concentrations the
constant k_f in the Freundlich isotherm can be calculated.

Table I shows the values of the parameters in the Freundlich isotherm
for the adsorption experiments on pieces of granite. In the experiments with

simultaneous adsorption the exponent β is assumed to be the same as in the experiments with a single species. As a comparison the corresponding values for adsorption on the largest particles (4-5 mm) (3) are included in Table I. The constant k_f is in all cases larger for the crushed particles than for the pieces. The difference is largest for cesium. This means that the adsorption capacity is higher for the particles, and it might be an effect of micro-fissures and cracks induced during the crushing of the granite.

Table I: Parameter values in the Freundlich isotherm
and estimated effective diffusivities.

		Adsorption on granite pieces			Adsorption on granite particles (3) d_p =4-5 mm		
		β	k_f	$D_p \cdot \varepsilon_p \, (m^2/s)$	β	k_f	$D_p \cdot \varepsilon_p \, (m^2/s)$
Stripa	Strontium	1	$8.9 \cdot 10^{-4}$	$1.3 \cdot 10^{-12}$	1	$1.0 \cdot 10^{-3}$	$1.0 \cdot 10^{-12}$
	Cesium	0.66	$1.1 \cdot 10^{-3}$	$1.0 \cdot 10^{-12}$	0.66	$3.9 \cdot 10^{-3}$	$1.0 \cdot 10^{-12}$
Finnsjö	Strontium	1	$2.7 \cdot 10^{-3}$	$2.2 \cdot 10^{-12}$	1	$6.9 \cdot 10^{-3}$	$24 \cdot 10^{-12}$
	Cesium	0.54	$9.8 \cdot 10^{-3}$	$1.6 \cdot 10^{-12}$	0.54	$3.0 \cdot 10^{-2}$	$16 \cdot 10^{-12}$
Finnsjö simultaneous sorption	Strontium	1	$3.7 \cdot 10^{-3}$	$3.1 \cdot 10^{-12}$			
	Cesium	0.54	$6.1 \cdot 10^{-3}$	$1.0 \cdot 10^{-12}$			

To estimate the diffusivities one needs to solve the diffusion equations. Crank (4) gives the analytical solution to the equations for a linear isotherm ($\beta = 1$). For the non-linear isotherm the equations have been solved numerically by a computer program (TRUMP) developed at the Lawrence Livermore Laboratories (5). The effective diffusivities are presented in Table I.

Fig. 1 shows the fit of the diffusion model to the experimental uptake curve for cesium and strontium on Finnsjö granite. Also the fit of the model with the Freundlich isotherm found for the crushed particles (d_p = 4-5 mm) (3) are presented. If equilibrium is not obtained after 10,000 hours, as we assume it does, then the diffusivity $D_p \cdot \varepsilon_p$ lies somewhere between the values for equilibrium after about 10,000 hours and equilibrium according to the crushed particles. The diffusion model gives, however, a better fit to the experimental values for the former case.

DIFFUSION OF NON-SORBING IONS IN ROCK MATERIAL

Experimental

In these experiments the diffusion of tritiated water, iodide, Cr-EDTA and Uranine through different rock materials was determined. The rock materials were granites from Finnsjön and Stripa, gneiss from Karlshamn at the east coast of Sweden and gabbro from Vipängen just outside Uppsala.

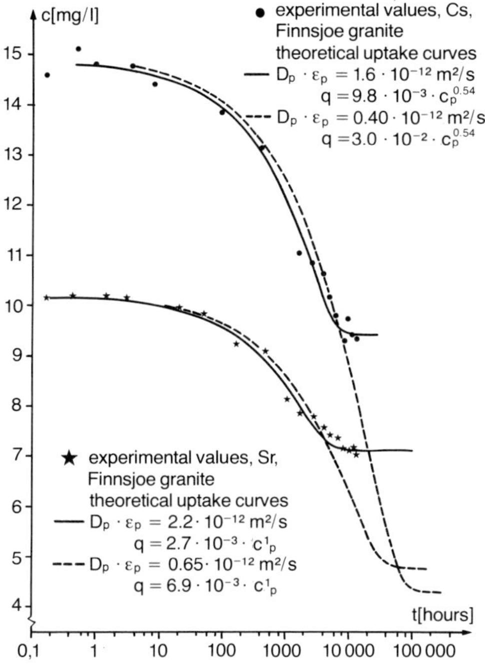

Fig. 1: The fit of the diffusion model to the experimental uptake
curve for strontium and cesium on Finnsjö granite.

The experiments were performed as follows: The rock piece was fixed
between two chambers. One of the chambers was filled with a solution
containing the diffusing component and the other chamber was filled with
distilled water. In one of the experiments with tritiated water synthetic
groundwater was used instead of distilled water. The concentrations of
tritiated water, iodide, Cr—EDTA and Uranine were ~50µCi, 1 mol/1,
9.9 g/1 and 10 g/1 respectively. The concentration increase in the chamber,
that from the beginning was free from the diffusing component, was
determined by taking out samples after different times and measuring the
concentration in the samples. Each time a sample had been taken out the
same volume distilled water (or synthetic groundwater) was added to the
chamber to keep the volume constant.

Determination of the diffusivities

The rate of transfer of diffusing substance through a porous plate of crossectional area A, can be described by Fick´s law of diffusion,

$$N = -D_e \cdot A \cdot \frac{dc}{dx} \tag{3}$$

where N is the flowrate of diffusing substance, c the concentration of diffusing substance, x the distance in the diffusing direction and D_e is the effective diffusion coefficient. When steady state is reached the flowrate of diffusing substance is given by the equation

$$N = -D_e \cdot A \cdot \frac{dc}{dx} = D_e \cdot A(C_1 - C_2)/\ell \tag{4}$$

where ℓ is the thickness of the plate and C_1 and C_2 are the concentrations on each side of the plate.

At steady state the concentration increase at the low concentration side of the plate is constant with time, and the flowrate can be calculated. In all the experiments performed the concentration on the low concentration side was negligible compared with the concentration on the high concentration side. By using Eq. (4) the effective diffusivity D_e, which in this case is an effective pore diffusivity $D_p \cdot \varepsilon_p$, of the components in the different rock materials was determined. The result is presented in Table II. Where two values are given, there are two results.

Table II: Effective diffusivities for diffusion in rock materials.

	Effective diffusivity, $D_p \cdot \varepsilon_p$ (m^2/s)			
	Finnsjö granite	Stripa granite	Gneiss	Gabbro
Tritium diffusion	$1.8 \cdot 10^{-13*}$ $1.3 \cdot 10^{-13}$			
Iodide diffusion	$8.4 \cdot 10^{-14}$ $7.4 \cdot 10^{-14}$	$1.3 \cdot 10^{-13}$ $1.2 \cdot 10^{-13}$	$1.4 \cdot 10^{-13}$ $5.0 \cdot 10^{-14}$	$<4.0 \cdot 10^{-16}$ $<4.0 \cdot 10^{-16}$
Cr-EDTA diffusion	$6.9 \cdot 10^{-15}$			
Uranine diffusion	$2.3 \cdot 10^{-15}$			

* Synthetic ground water is used in the experiment

ELECTRICAL RESISTIVITY MEASUREMENTS IN SATURATED ROCKS, IN ORDER TO DETERMINE THE DIFFUSIVITY

<u>Theory</u>

Direct measurements of diffusion in low porosity materials are very time consuming. It is therefore of interest to find a method which can reduce the experimental time. There are some indications that electrical conductivity and molecular diffusion may depend in the same way on the formation factor $\varepsilon_p \cdot \delta_D / \tau^2$ (6). This means that

$$\frac{D_p \cdot \varepsilon_p}{D_v} = \frac{\varepsilon_p \cdot \delta_D}{\tau^2} = \frac{R_o}{R_s} \tag{5}$$

where R_s is the resistivity of the salt water saturated rock sample and R_o is that of the salt water.

The concentration of the salt water solution must not be too low because then the pore surface conductivity might influence the results (7).

Experimental

Rock samples were saturated with salt solution (NaCl) with the same method as in the experiments with cesium and strontium. After saturation the samples were dried with a piece of paper on the outer surfaces except on the surfaces that would be in contact with the electrodes. The electrodes were polished copper plates. Between the electrodes and the rock sample a filter paper, saturated with the same salt solution as the rock sample, was placed. The resistance in the rock sample was measured by a conductivity meter (type Wheatstone bridge) at a frequence of 50 Hz. The resistance in the salt solution was measured with the same electrodes and conductivity meter. The electrodes were placed on each side of a transparent PVC cylinder. The cylinder was filled with the salt solution, and the resistance measured.

Two initial investigations were performed. The purpose with these was to decide suitable concentration of the salt solution and the length of the rock samples (distance between the electrodes). These initial investigations were made on Finnsjö granite rock samples.

Experimental results

The first initial investigation showed that the formation factor $\varepsilon_p \cdot \sigma_D / \tau^2$ $(= R_o / R_s)$ is dependent on the concentration at concentrations below 1 mol /1. That might be an effect of pore surface conductivity. Concentrations of 1 mol /ℓ and higher gives a factor that is almost independent of the concentration. For the subsequent measurements the concentration 1 mol /1 was chosen.

The other initial investigation showed that the shortest cores (ℓ = 10 mm) gave a somewhat higher value to the formation factor than the longer ones (ℓ = 30–100 mm). The reason for that might be that the shortest cores have more pores going through the whole piece than the longer ones. An other possible explanation is that some of the conductivity occured on the cylindrical outer surfaces of the shortest cores.

The results from electrical resistivity measurements in Finnsjö and Stripa granite, gneiss and gabbro could be found in Table III. The formation factor has about the same value for the Finnsjö and the Stripa granite. Gneiss gives a higher value, and gabbro a somewhat lower value than the granites. The difference between the lowest and the highest value for the same rock material is about 1.5-2 times.

DISCUSSION AND CONCLUSIONS

In all the experiments that have been performed the formation factor has been calculated either from determined values of the pore diffusivity or from measured electrical resistivities. If this factor for a certain porous material is known, as well as the diffusivity of a component in free water, then the effective diffusivity for the component in the porous material can be predicted.

In Table III the values of the formation factor from all the experiments are summarized. For Finnsjö granite the diffusion experiment with tritiated water, the diffusion experiment with iodide and the electrical resistivity measurements give results that are in fair agreement. The diffusion experiments with Cr-EDTA and Uranine give values that are lower.

In the sorption experiments with cesium and strontium the factor is about an order of magnitude higher, or more for strontium, than in the iodide diffusion experiment. This indicates that there is another mechanism of migration in addition to pore diffusion. Similar observations have been made in studies of cesium and strontium migration in clays (8).

The iodide diffusion experiment and the resistivity measurements in Stripa granite give values of the formation factor that are in good agreement. The factor determined in the experiments with cesium and strontium is markedly higher also for Stripa granite.

In the experiments with gneiss there are differences. Gneiss is a rather inhomogeneous material and there are probably variances in the material properties between different pieces of gneiss. This could be an explanation to the different results between the iodide diffusion experiment and the resistivity measurements and also to the differences in the results from the same type of experiment.

Table III: Values of the formation factor for different rocks and
determined with different methods.

Type of experiment		Finnsjö granite	Stripa granite	Gneiss	Gabbro
Diffusion exp.		$5.3-6.2\cdot10^{-5}$			
Tritiated water					
conc. range 10^{-9} M		$7.4-8.6\cdot10^{-5*}$			
Sorption exp.	Sr	$170-220\cdot10^{-5*}$	$98-130\cdot10^{-5*}$		
conc. range 10^{-4} M	Cs	$66-78\cdot10^{-5*}$	$41-49\cdot10^{-5*}$		
Simultaneous	Sr	$230-300\cdot10^{-5*}$			
sorption	Cs	$43-49\cdot10^{-5*}$			
Diffusion exp. Cr-EDTA		$2.0\cdot10^{-5}$			
conc. range 10^{-2} M					
Diffusion exp. Uranine		$0.5\cdot10^{-5}$			
conc. range 10^{-2} M					
Diffusion exp. Iodide		$4.9\cdot10^{-5}$	$7.6\cdot10^{-5}$	$7.6\cdot10^{-5}$	$<0.03\cdot10^{-5}$
conc. range 1 M		$5.1\cdot10^{-5}$	$8.7\cdot10^{-5}$	$3.5\cdot10^{-5}$	$<0.03\cdot10^{-5}$
Resistivity measurements		$8.3-13\cdot10^{-5}$	$6.7-10\cdot10^{-5}$	$14-25\cdot10^{-5}$	$2.2-4.3\cdot10^{-5}$
conc. range 1 M					

* Synthetic ground water is used in the experiment.

In the experiments with gabbro the results do not agree at all. It is
probably the iodide diffusion experiment that gives a much too low value.
The resistivity measurements indicate that there are pores in the material
where diffusion could take part. Maybe the iodide reacts chemically in some
way with the rock material.

Experiments where the diffusion in low porosity materials is determined
are often very time consuming. The results from the experiments here
indicate that electrical resistivity measurements, which is a faster
method, can give approximate values of the diffusivity. It must be
noted, however, that this is only valid when the transport mechanism
consists of only pore diffusion. When surface conductivity is large the
method may overestimate the effective diffusivity.

NOTATION

A	diffusion area	m^2
c	concentration in bulk fluid	mol/ℓ, cpm/ℓ, mg/ℓ
c_p	concentration in pore fluid	mg/ℓ
D	diffusion coefficient	m^2/s

D_e	effective diffusion coefficient	m^2/s
D_p	pore diffusivity	m^2/s
D_v	diffusivity in water	m^2/s
k_f	parameter in Freundlich isotherm	
ℓ	thickness or length of a piece	mm
N	rate of transfer	mol/s, cpm/s, mg/s
q	concentration in the solid material	mg/kg
R_o	electrical resistivity in salt water	Ωm
R_s	electrical resistivity in salt water saturated rock sample	Ωm
t	time	seconds, hours
x	length coordinate	
β	exponent in Freundlich isotherm	
δ_D	constrictivity for diffusion	
ε_p	porosity of the materials	
ρ_s	density of the solid material	kg/m^3
τ	tortuosity	
$\varepsilon_p \cdot \delta_D / \tau^2$	formation factor	

REFERENCES

1. Neretnieks I., Diffusion in the Rock Matrix: An Important Factor in Radionuclide Redardation? J. Geophys. Res., 85, 1980, p. 4379.
2. Allard B., Beall G.W., Sorption of Americium on Geologic Media, Journ. Environm. sci. and Health, 6, 1979, p. 507-518.
3. Skagius C., Svedberg G., Neretnieks I., A Study of Strontium and Cesium Sorption on Granite, Report PRAV 4.26, April 1981, (Accepted for publication in Nuclear Technology).
4. Crank J., The Mathematics of Diffusion, 2nd ed., Oxford University Press 1975.
5. Edwards A.L., TRUMP: A Computer Program for Transient and Steady State Temperature Distribution in Multidimensional Systems, National Technical Information Service, National Bureau of standards, Springfield Va., USA, 1972.
6. Klinkenberg L.J., Analogy between Diffusion and Electrical Conductivity in Porous Rocks, Geol. Soc. Am. Bull., 62, 1951, p. 559.
7. Brace W.F., The Effect of Pressure on the Electrical Resistivity of Water-Saturated Crystalline Rocks, J. Geophys. Res., 70, 1965, p. 5669.
8. Eriksen T., Jacobsson A., Ion Diffusion through Highly Compacted Bentonite, KBS Technical Report, 81-06.

Published 1982 by Elsevier Science Publishing Co
SCIENTIFIC BASIS FOR RADIOACTIVE WASTE MANAGEMENT - V
Werner.Lutze, editor

DIFFUSION IN THE MATRIX OF GRANITIC ROCK. FIELD TEST IN THE STRIPA MINE

Lars Birgersson, Ivars Neretnieks
Department of Chemical Engineering
Royal Institute of Technology
S-100 44 STOCKHOLM Sweden

SUMMARY

A migration experiment with the objective to investigate the existence of a connected pore system in "undisturbed" rock has been performed in the Stripa mine at the 360 m-level.

Since all three tracers (Cr-EDTA, I^- and Uranine) that were used in the experiment migrated at least 11 cm into the rock under a three month period, this experiment indicates the existence of a connected pore system in "undisturbed" rock.

BACKGROUND

The radionuclides migrating with flowing water may be considerably retarded if they can diffuse into the rock and sorb on the surfaces of the microfissures in the rock matrix (1).

At present there are a series of laboratory experiments being performed with the purpose to determine diffusion coefficients for various tracers in granite, but these experiments are not carried out in "undisturbed" rock. It cannot be ruled out that the reduction of the rock stresses which occur when samples are taken out have induced the microfissures. It is thus necessary to make experiments in rock in the natural stress environment. It is also necessary to know if diffusion can occur through the fissure filling material, since the radionuclides must pass through it before they can penetrate into the rock matrix.

This experiment was performed in the Stripa mine at the 360 m-level. That will give nearly the same conditions as for the planned nuclear waste storage.

INFLUENCE OF THE STRESSFIELD

Near drillholes and drifts ,the rockstresses will be changed compared to "undisturbed" rock. A general rule in these cases is that the rock stresses are changed about 2 hole diameters out from and below the hole. That is, outside these 2 hole diameters essentially "undisturbed" rock exists (2).

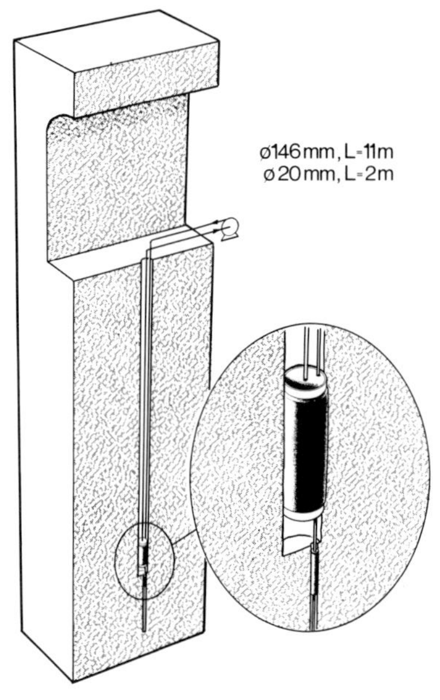

Figure 1. Drilling dimensions
and packer positions

Since the diameter of the drift where the experiment has taken place was approximately 5.5 m, a 11 m deep 146 mm hole was drilled. At this distance (11 m) from the drift the changes in the rock stresses due to the drift can be neglected , i.e. essentially "undisturbed" rock is reached. However, the existence of the 146 mm hole will cause a further change in the rockstresses approximately 0.3 m (2 hole diameters) outward and below.

Thus, in the bottom of the 146 mm hole a 20 mm hole (approximately 2 m long) was drilled. This 20 mm hole will cause a change in the rock stresses approximately 4 cm outward, but outside this disturbed zone and 0.3 m below the larger hole essentially "undisturbed" rock is reached.

If tracers can migrate from the little hole past the disturbed zone and into "undisturbed" rock, this experiment will indicate the existence of a connected pore system in "undisturbed" rock.

EXPERIMENTAL ALTERNATIVES : DIFFUSION - FLOW AND DIFFUSION

The most satisfactory way of doing this kind of experiment is to do a diffusion experiment with no over pressure in the injection hole. It is, unfortunatly, not possible to separate the radial concentration profile due to diffusion from that due to flow. Diffusion and flow may be of the same order of magnitude. The problem can be illustrated by following figure of the little hole in the bottom of the big hole.

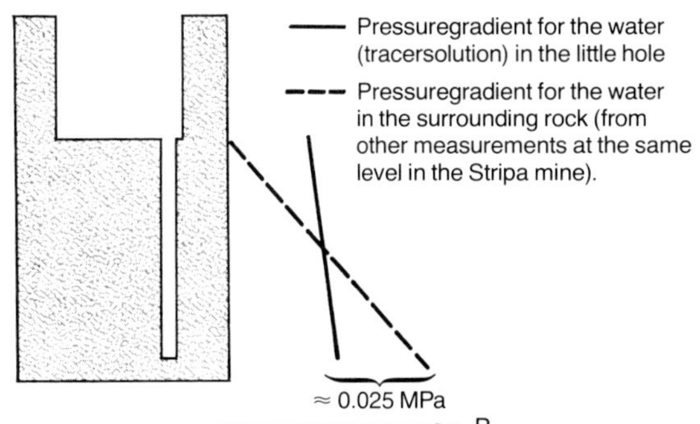

According to other measurements at the same level in the Stripa mine (3) ,the pressure difference is approximately 0.025 MPa per meter of injection hole.

Figure 2. Pressure gradients for injection
hole and surrounding rock

The significance of this pressure differance is determined by the hydraulic conductivity of the rock. If the hydraulic conductivity is $\geq 10^{-13}$ m/s, the radial flow distance because of the pressure difference is of the same order of magnitude as the expected diffusion distance. This means that the concentration profile that will arise in a diffusion experiment because of the concentration gradient will be more or less displaced by the radial flow because of the different pressure gradients.

According to literature values (4, 5, 6), the matrix (unfractured rock) has a hydraulic conductivity of $10^{-13} - 10^{-12}$ m/s. Because of these literature values it was decided to carry out the experiment with an over pressure of≈1 MPa and a contact time of≈3 month.

The chosen over pressure was high enough to eliminate the problem with the different pressure gradients, but small enough to avoid influencing the rock (2).

CONCENTRATION PROFILES: FLOW AND DIFFUSION

For diffusion-convection calculations of concentration profiles of non-sorbing tracers, four parameters influence the shape of the curves. The parameters and their expected values are:

D_p - diffusivity, $\approx 10^{-10}$ m^2/s (7).

ε_p - porosity, 0.345 % (8).

K_p - hydraulic conductivity for the rock matrix, 10^{-13} - 10^{-12} m/s (4, 5, 6).

t_c - contact time.

The equations that predict the transport distance for radial diffusion and flow are:

Diffusion equation: $\dfrac{\partial c}{\partial t} + V_r \dfrac{\partial c}{\partial r} = D_p \dfrac{1}{r} \dfrac{\partial}{\partial r} \left(r \dfrac{\partial c}{\partial r}\right)$ Eqv. (1)

Radial flow equation: $V_r = \dfrac{const.}{r}$ Eqv. (2)

(The Notation is given at the end of this paper.)

Initial and boundary conditions used imply: No tracer in rock at start and constant concentration in little hole at all times thereafter.

EXPERIMENTAL DESIGN

After drilling the holes, one small packer was placed in the little hole and one big packer was placed in the bottom of the big hole (see figure 1). This system, with inflow through a nylon tube in the bottom of the little hole and outflow through a tube just above the little hole ensures a good circulation during the injection of tracers.

After the installation of the packers, the water pressure and the water flow into the little hole were monitored. The water pressure was found to be 0.73 MPa and the waterflow into the little hole approximately 7 ml/h, which indicated that some waterbearing fissure(s) was/were intersecting the little hole.

After circulating tracers for a while, to get constant concentration in the system, the injection was initiated. A pressure of 1.63 Mpa (i.e. 0.90 Mpa over pressure) was used during the whole injection time.

OVERCORING AND SAMPLING

After about 3 month the injection was ended. The packers were taken up and the little hole was overcored. The core from the overcoring had a diameter of 132 mm and was about 2.5 m long, with the injection hole (20 mm) at the side of the core. The core were cut into 5 cm long cylinders. From these a number of sampling cores (ϕ 10 mm) were drilled at different distances from the injection hole. These sampling cores were leached with distilled water. The tracer concentration in the distilled water was determined.

TRACERS AND ANALYTICAL METHODES

Since the objective with this experiment is to investigate the existence of a connected pore system in "undisturbed" rock, it was decided to inject a mixture of non-sorbing tracers.

The following tracers were used:

Tracer	Analytical method	Injection concentration
Cr-EDTA	Atomic absorption	$\approx$ 5 000 ppm
[Na-Fluorescein]		
Uranine	Spectrophotometer	$\approx$ 20 000 ppm
I^-	Ion selective electrode	$\approx$ 100 000 ppm

The concentration of tracers in the injection mixture was high enough to follow the concentration profile down to $c/c_0 = 0.01$ for all tracers. The accuracy was about +/- 0.1 for all tracers at the lowest concentration.

CORE DESCRIPTION

The core from the overcoring of the injection hole was intersected by several fissures. Since there was an inflow of water into the little hole , at least one of them must have been waterbearing. Therefore a number of samples were taken close to the fissures, in order to see if there was any indication of tracers having migrated first into the fissure by water flow, and then through the fissure filling material and into the rock matrix by diffusion.

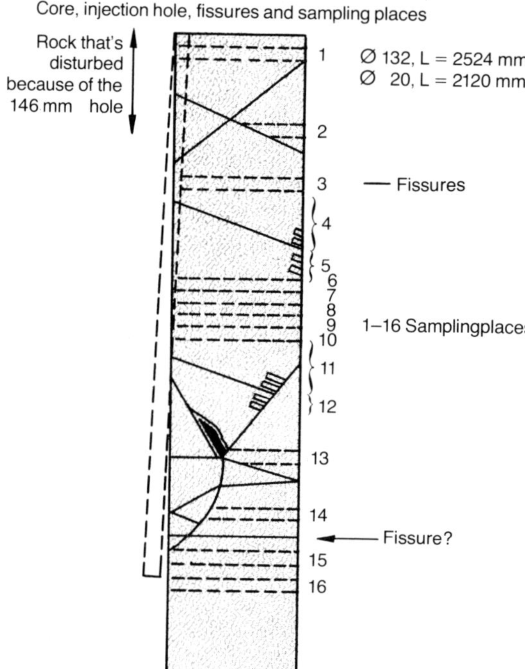

Figure 3. Core description

Comments on the sampling places:

3,6-10,15 and 16 - Investigation of the concentration profile in the rock matrix.

4,5,11 and 12   - Samples taken far from the injection hole and close to a fissure. Could indicate diffusion through fissure filling material.

13 and 14   - Samples taken behind a fissure. If tracer is found here it must have passed through the fissure filling material.

CONCENTRATION PROFILES IN CORE

The concentration profiles in the pieces where the migration had taken place in the rock matrix (i.e. "far" from any fissures), showed the same results. All tracers have passed the disturbed zone (approximately 4 cm) and migrated into "undisturbed" rock.

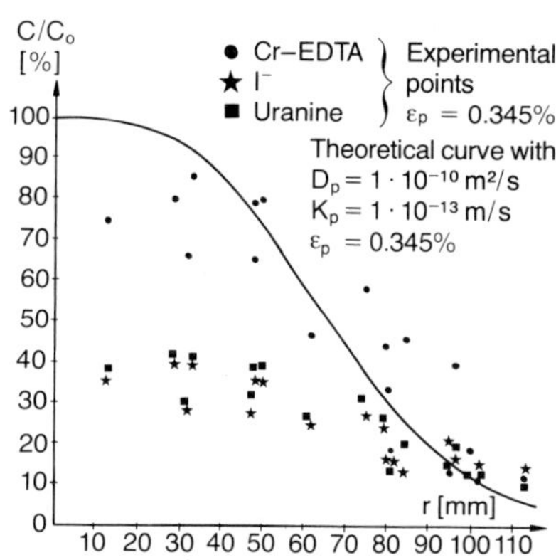

As an example, the tracer concentrations from piece No.10 and a theoretical curve (9) are shown below. (Diagram 1) All experimental points are based on a uniform porosity of 0.345% in the rock matrix.

All three tracers passed the disturbed zone and migrated into "undisturbed" rock.

The Cr-EDTA concentration profiles by simple convection-diffusion migration without chemical interaction.

Diagram 1. Tracer concentration vs distance from injection hole for piece No. 10 and a theoretically calculated curve.

INFLUENCE OF FISSURES

Since there was at least one waterbearing fissure intersecting the little hole, the investigation of points 4,5,11 and 12 could have indicated migration through the fissure filling material and into the rock matrix if the concentration had been higher close to the fissure and lower a little way out. Unfortunately there is not enough data to show this.

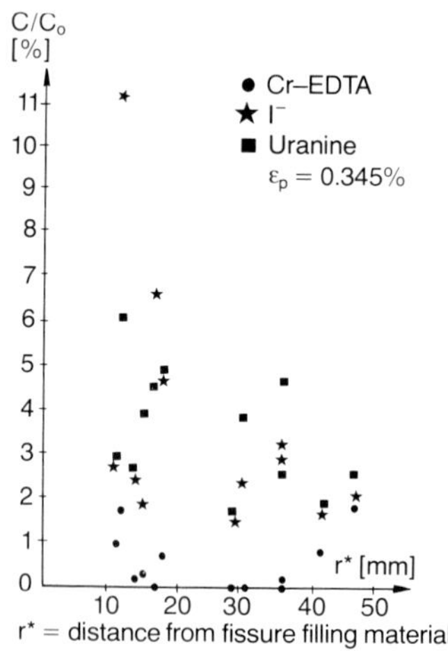

From the two pieces taken behind a fissure, the samples from piece No. 13 showed no or very low concentration. But samples from piece No. 14, showed that tracers had migrated into that piece. (Diag.2)

Since no (or very little) tracer was found in piece No. 15, probably because the bottom of the little hole was filled with granite particles from the drilling, the tracers must have migrated through the fissurefilling material (see fig 3).

Diagram 2. Tracer concentration vs distance from fissure for piece No. 14.

At least two of the tracers have passed through the fissure filling material. The fissure filling material seems to decrease the migration rate somewhat.

DISCUSSION

From the tracer concentrations shown in Diagram 1, it can be seen that the profile for Cr-EDTA can be explained with a theoretical curve, while c/c_o doesn't reach higher than 40-50 % for I$^-$ and Uranine. The magnitude of c/c_o for a sample is determined by the porosity and the calculations of c/c_o for all three tracers are based on a porosity of 0.345 %.

The porosity in our granite has been measured in three different ways.

- Mercury penetrometry using approximately 1000 bar. This measurement gave a porosity of 0.345 % (8).

- Comparing the weight for dry and wet granite. These measurements gave approximately the same porosity as the method above.

- Impregnating a sample with 1 M NaI and leaching in distilled water. This method gave a porosity of approximately 0.15-0.20 %, i.e. approximately half as much as the methods above.

If the concentration profiles for I^- and Uranine are calculated with the porosity that was obtained from the impregnating-leaching experiment with I^- (0.345 %/2) instead of 0.345 %, the curves for I and Uranine will reach the same level as the curve for Cr-EDTA.

At present, we have no explanation for the fact that concentration profiles must be calculated with different porosities for different tracers, in order to fit a theoretical curve.

CONCLUSION

The conclusions from this experiment are:

- Tracers have migrated through the disturbed zone and some distance into "undisturbed" rock.

- Tracers have passed through fissure filling material.

The results indicate that it is possible for tracers (and therefore radionuclides) to migrate from a fissure, through fissure filling material, and into the undisturbed rock matrix.

NOTATION

c	concentration in liquid	mol/m^3
c_o	injection concentration	mol/m^3

528

D_p	diffusivity in water in pores	m^2/s
K_p	hydraulic conductivity	m/s
L	length	m
p	pressure	Pa
r	radial distance	m
t	time	s
t_c	contact time	s
v_r	radial velocity	m/s
ε_p	porosity	m^3/m^3
ϕ	diameter	m

REFERENCES

1. Neretnieks, I., Diffusion in the Rock Matrix: An important Factor in Radionuclide Retardation? J. Geophys. Res., 1980, 85, p 4379.
2. Stephansson, O., Personal communication, Department of Mining Engineering, Colorado School of Mines, Golden, 1981.
3. Wilson, C., Macropermeability Experiment in Stripa (draft report), Earth Science Div., Lawrence Livermore Laboratory, 1981.
4. Brace, W.F., Walsh, J.B., Frangos, W.T., Permeability of Granite under high Pressure, J. Geophys. Res., 1968, 73, p. 2225.
5. Freeze, R.A., Cherry, J.A., Groundwater, Prentice-Hall, 1979.
6. Heard, H.C., Trimmer, D., Duba, A., Bonner, B., Permeability of Generic Repository Rocks at Simulated In Situ Condition, Lawrence Livermore Laboratory, UCRL-82609, April 1979.
7. Skagius, C., Neretnieks, I., Diffusion in Crystalline Rocks, presented at International Symposium on the Scientific Basis for Nuclear Waste Management, Berlin, 1982.
8. Müller-Vonmoos, M., Personal communication, Institut für Grundbau und Bodenmechanik, Eidgenössische Technische Hochschule Zürich, 1981.
9. Edwards, A.L., TRUMP: A Computer Program for Transient and Steady-State Temperature Distributions in Multidimensional Systems, Lawrence Livermore Laboratory, 1969.

Published 1982 by Elsevier Science Publishing Co
SCIENTIFIC BASIS FOR RADIOACTIVE WASTE MANAGEMENT - V
Werner.Lutze, editor

MIGRATION IN A SINGLE FRACTURE

Harald Abelin, Jard Gidlund, Ivars Neretnieks
Department of Chemical Engineering
Royal Institute of Technology
S-100 44 STOCKHOLM, Sweden

SUMMARY

It has been decided to investigate flow and sorbtion in a readily identifiable fracture which can be excavated for a detailed examination of flow path and sorbtion sites. The investigation is performed in the Stripa mine, 360 m below ground, where there is a natural water flow towards the drift. The bedrock is granite.

A method of tracer injection into a fracture, either as a step or a pulse, and of collection of water samples under anoxic atmosphere has been suggested and tested in a preparatory investigation. The introduction of tracers can be done either under natural pressure or by injection with over pressure.

An injection of Rhodamine-WT, Na-iodide and Na-Fluorescein with over pressure has been performed. It has been found that Rhodamine-WT is influenced in some way along the the flow path.

BACKGROUND

In the KBS (Swedish nuclear fuel supply co/division KBS) report (1), it is proposed that the final repository for radioactive waste should be at 500 m depth in crystalline rock. The safety analysis for this repository is based on the assumption that if and when any radionuclides are leached from the waste , the majority of the important radionuclides will interact chemically or physically with the bedrock and will be considerably retarded. This retardation and interaction depends upon the velocity of water, the sorbtion rates and equilibria of the reactions as well as the surface area of the rock in contact with the flowing water.

Most studies are based upon the assumption that the flow can be described as porous media flow. This might be true for very large distances where the flow would encounter a multitude of channels and some averaging may be conceivable on the scale considered. However, no large scale tracer tests have been performed in fissured crystalline rock with known flow paths. Transport over short distances, i.e in the near field of a canister, most probably occurs in individual fissures. On an intermediate scale where more than a few fissures conduct the flow, well type tracer tests alone cannot give the detailed information needed to understand dispersion and sorbtion phenomena in fissured rock. It has therefore been decided to investigate flow and sorption in readily identifiable fissures which can be excavated for a detailed examination of flow paths and sorption sites.

Several investigations on migration in single fractures are under way both in the laboratory and in the field. The laboratory runs are done with migration distances of up to 0.3 m (see Neretnieks et al (2)) and the field experiments with a migration distance of up to 30 m (see Gustavsson and Klockars (3)). In the present investigation a migration distance of about 5 m is used. The fracture will be excavated afterwards.

PURPOSE

The study to be performed has the following main objectives:

o To observe the movement of nonsorbing and sorbing tracers under controlled and well defined conditions in a real environment.

o To interpret the movement of tracers in such a way that the results become useful for the prediction of radionuclide migration.

o To obtain a basis for comparing laboratory data on sorption with observations in a real environment.

o To develop good techniques for small volume sampling of water and fracture surfaces with sorbed tracers.

o To gather experience with stable tracers before using radioactive tracers.

EXPERIMENTAL DESIGN

This investigation started in 1980 and will end in late 1983. Injection of sorbing tracers will start in mid 1982.

The Stripa mine is well suited for performing tracer tests in a single fracture as well as in a network of interconnected fractures. Old waterbearing fractures have been found in and near the drifts now in use. As the drifts are well below the water table, the fractures have been conducting water for a very long time and thus are as well "equilibrated" as we can reasonably achieve in a sorption experiment.

The technique to be used in this investigation is to locate a suitable fracture, drill an injection hole which intersects the fracture at a distance of about 5 m from the face of the drift, and several sampling holes in the fracture plane (see fig 1). Tracers are introduced into the fracture from the injection hole either by injection with an overpressure or by circulation in the compartment through which natural flow occurs. The natural water flow is towards the drift. Groundwater with tracers is collected in sampling holes. By having a series of sampling holes the transverse dispersion of the tracers can be observed in addition to the axial dispersion.

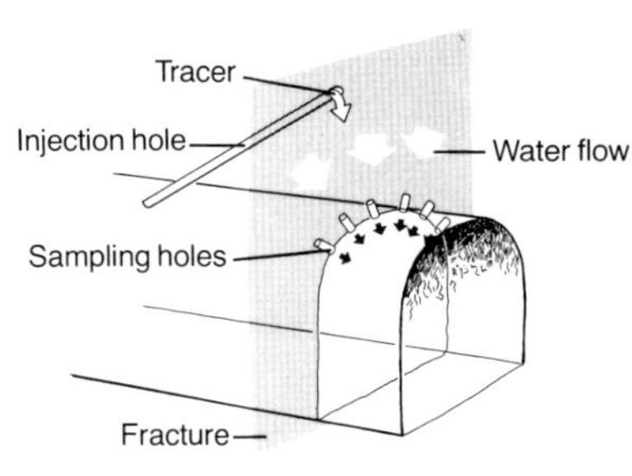

Figure 1. Drift with intersecting fracture

Injection and sampling techniques

The injection hole will be sealed off just below and above the fracture with two straddle mechanical packers, see figure 2. In the preparatory investigation an inflatable packer was used. In order to reduce the injection compartment volume a PVC filling with a diameter near that of the borehole was used. The introduction of tracers in the fracture could be done in two different ways:

1. By circulating tracers in the injection compartment under natural pressure.

2. By injecting tracers with a certain over pressure. The injected volume and the injection flow are measured.

Each sampling hole has a mechanical packer (see figure 2), with a funnel shaped top. It is possible to purge the sampling holes with nitrogen, in order to maintain the redox integrity of the samples. The water from the sampling holes is collected with a fractional collector which can be kept under an anoxic atmosphere.

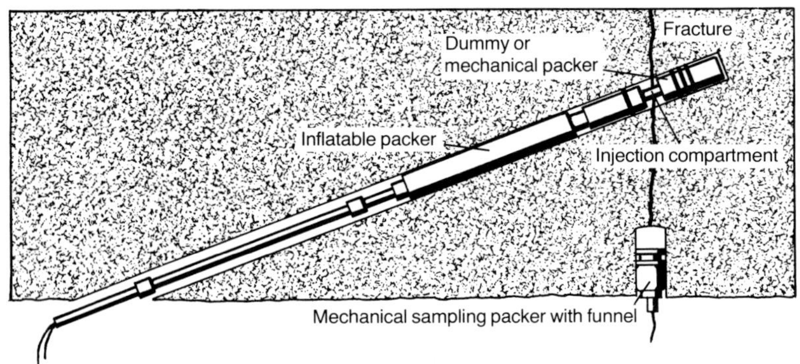

Figure 2 Injection and sampling packers

A more detailed describtion of the equipment is given in Abelin and Neretnieks (4).

Tracers

Stable tracers will be used throughout in the field experiments. Except for very few nuclides, notably Np, Pu and Tc all the objectives stated can be achieced with less effort. "Hot" laboratory experiments are run in parallel in a supporting investigation, Erikson (5).

Four classes of tracers will be used:

o Particles   0.4 micrometer plastic pellets are used to simulate the movement of particulate matter.

o High molecular weight   Blue Dextran M=2 000 000 or Albumin is used
  tracers   to simulate the movement of high molecular weight organic nonsorbing matter.

o Nonsorbing tracers   Various dyes, Bromide, Iodide.

o Sorbing tracers   Cs (I),Sr Ba (II),Eu Nd (III),Th (IV), *(V),
                        U (IV,VI).

*No suitable non-active tracer found.

The nonsorbing tracers are tested for sorbtion on crushed granite and materials from the equipment as well as stability in time. The solubility of the sorbing tracers in Stripa groundwater will be tested before use.

The analysis of the particles, the high molecular weight tracers, and the nonsorbing tracers will be done by spectrophotometer, ionselective electrode, and atomic absorbtion analysis. The sorbing tracers in water and granite will be analysed by neutron activation analysis.

PREPARATORY INVESTIGATION

To test the methods for injection and sample collection a preparatory investigation was run. A suitable fracture for this was found at the 360 m level in the Stripa mine in the so called "ov1" drift.

Figure 3 shows the site for the preparatory investigation with its three main equipment components: 1. Injection equipment

2. Packer and sampling devices

3. Fractional collector with anoxic box

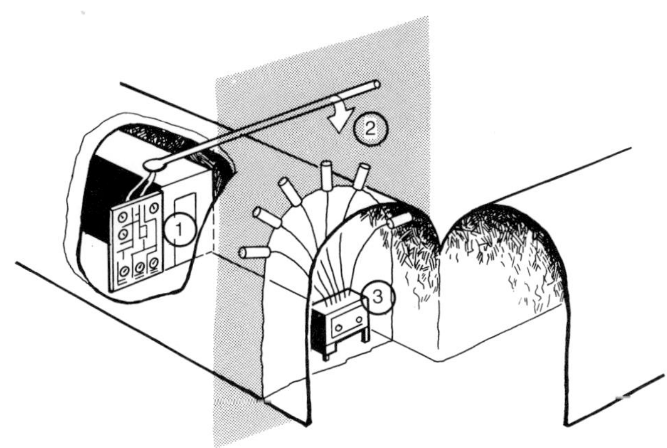

Figure 3: Test site at the 360 m level

534

A more detailed description of the equipment used in the preparatory investigation is given in (4).

Data on water flow, pressure and test for connection

After the injection hole was drilled, the water flow from the injection hole was monitored. A small mechanical packer was inserted near the end of of the injection hole. The water flow was found to be approximatly 100 ml/h. The natural pressure in the injection hole was found to be 0.28 MPa.

The sampling holes are at a distance of about 0.7 m from each other and situated in the same fracture. From what can be seen from table 1 there seems to be a considerable channeling in the fracture. After the water flow was monitored, Na-fluorescein was injected into the fracture to test for connection between the injection hole and the sampling holes. It took less than 20 h for the first tracer to reach sampling holes No. 4 and No. 5. Injection was continued and tracer eventually arrived in hole No. 3 as well. No tracer was found in the remaining three holes. After this test, flushing of the fracture with groundwater was attempted, however, the tracer could be detected in the flowing water for a very long time.

Table 1. Water flow from sampling holes

Hole no	water flow ml/h
1	0(approx)
2	0.2
3	5
4	83
5	28
6	0

Test run with Na-Fluorescein and Rhodamine-WT simultaneously

Na-fluorescein and Rhodamine-WT were injected simultaneously for 90 h. The pressure used was 0.5 MPa. The results from this test run can be seen in diagrams 1,2. Diagram 1 shows the breakthrough curves for hole No. 4 and No. 5. It can be seen that Rhodamine-WT appears later and more diluted in both holes. Diagram 2 shows the difference in breakthrough time and dilution between hole No. 4 and No. 5.

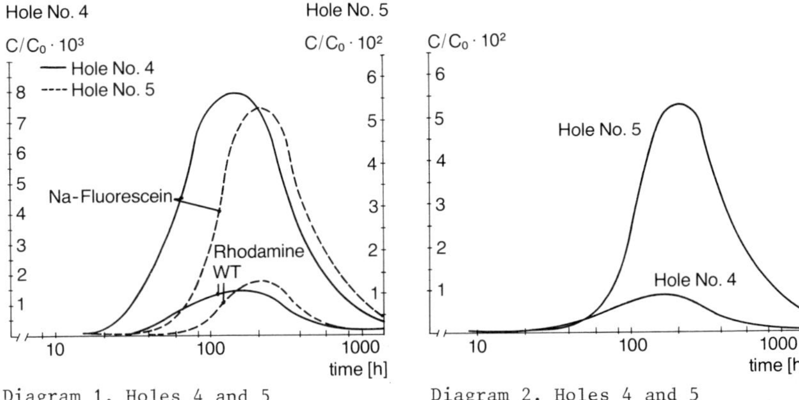

Diagram 1. Holes 4 and 5   Diagram 2. Holes 4 and 5

Tentative interpretation of the preliminary investigation

The very different flow rates to the various collecting holes indicate
a considerable channeling. The mean travel time for Na-Fluorescein is
about 100 and 200 h for holes No.4 and 5 respectively. From this, the
known flow rates in the collection holes (83 and 28 ml/h), the collection
length along the perimeter (about 0.7 m), and hydraulic head and geometry,
the hydraulic conductivity and equivalent fracture width for flow can be
determined.

$$K_{pf} = 0.5 \ \ln\left(\frac{r_2}{r_1}\right) \ (r_2^2 - r_1^2)/t_w \ (h_2 - h_1) \qquad (1)$$

$$v_{r1} = 0.5 \ \frac{1}{t_w} \left(\frac{r_2^2}{r_1} - r_1\right) \qquad (2)$$

$$d_f = Q_r / l_{r1} \ v_{r1} \qquad (3)$$

K_{pf} is the hydraulic conductivity of the fracture m/s

t_w is the mean water residence time (assumed to be s
equal to Na-Fluorescein residence time)

h_2, h_1 pressure head in point 2 and 1 respectively m
(28 m and 0 m)

v_{r1} water velocity at point 1 (fracture outlet in m/s
drift)

r_1, r_2 radial distance from center of drift m
(6 m and 2.25 m)

d_f fracture width for flow m

Q_r water flow rate m^3/s

l_{r1} collecting lenght (0.7m) m

The results are shown in table 2 below

Table 2. Calculated fracture widths

Hole No	K_{pf} [m/s]	d_f [m]	d_1 [m]
4	$1.4 \cdot 10^{-6}$	$1.8 \cdot 10^{-3}$	$0.0013 \cdot 10^{-3}$
5	$0.7 \cdot 10^{-6}$	$1.2 \cdot 10^{-3}$	$0.001 \cdot 10^{-3}$

An equivalent fracture width for laminar flow in a parallel walled fracture with the same hydraulic conductivity can also be determined:

$$d_1 = k_{pf} 12 \, v_k/g \qquad (4)$$

This is also shown in table 2.

ν_k is the kinematic viscosity of water $1 \ 10^{-6}$ Ns/m^2
g is the gravitational constant 9.81 m/s^2

It can be concluded that this type of laminar flow resistance is not the main cause of the total resistance.

Rhodamine-WT is influenced in some way along the flow path as the peak height is lower than that of Na-Fluorescein. Rhodamine-WT thus cannot be used to characterize the water residence time without a knowledge of the interaction mechanisms.

Experience gained during the preparatory investigation

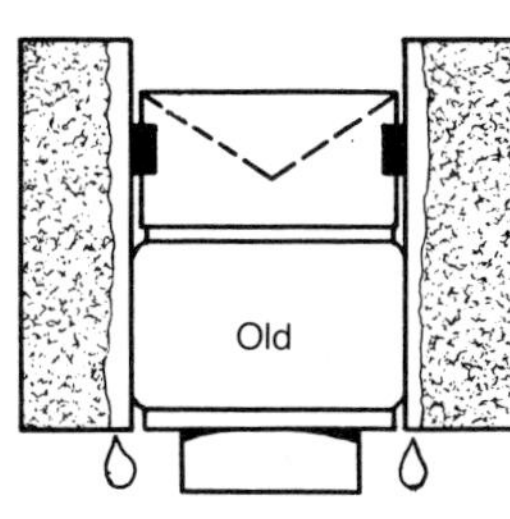

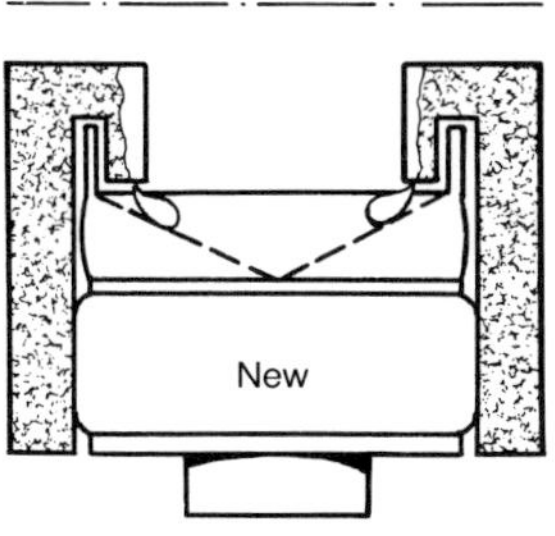

Figure 3 Sampling packer

The described method of injection and water sampling has been found to function well, and will be used in the main investigation with the following modifications. The number of injection holes has been increased from one to five, in order to reduce the risk of intersecting a less conductive part of the fracture.

During the preparatory investigation leakage occurred around the sampling packers, most likely due to capillarity of the fracture. The sampling packers have been redesigned (see fig 3) in order to minimize this leakage. The injection hole packers have been changed from a simple inflation, to a combination mechanical/inflation system in order to reduce the stagnant water volume in the injection compartment. The injection compartment has been further modified to allow the removal of small volume water samples.

538

To determine the recovery of injected tracer, all water from the
sampling holes will be collected. A few sampling holes will be drilled in
adjacent fractures, to check other possible flow paths in addition to the
main fracture.

REFERENCES
1.   "Handling of Final Storage of Unreprocessed Spent Nuclear Fuel",
      Vol II Technical KBS Report (1978).
2.   Neretnieks, I., Eriksen, T., Tahtinen, P., "Tracer Movment in a
      Single Fracture in Granitic Rock some Experimental Results and
      their Interpretation", Accepted for publication in Water Resources
      Research 1982
3.   Gustavsson, G., Klockars, K-E., "Studies on Groundwater Transport
      in Fractured Crystalline Rock under Controlled Conditions using non
      Radioactive Tracers", Technical Report KBS 81-07.
4.   Abelin, H., Neretnieks, I., "Migration in a Single Fracture
      Preliminary Experiments in Stripa", Internal Report KBS 81-03.
5.   Eriksen, T., Dep. Nucl. Chem, Royal Inst. Techn. Stockholm,
      Personal communic. 1982.

Published 1982 by Elsevier Science Publishing Co
SCIENTIFIC BASIS FOR RADIOACTIVE WASTE MANAGEMENT - V
Werner.Lutze, editor

MODEL FOR NEAR FIELD MIGRATION

Andersson, G., Rasmuson, A. and Neretnieks, I.
Department of Chemical Engineering
Royal Institute of Technology
S-100 44 STOCKHOLM, Sweden

SUMMARY

A model is proposed which describes the transport to and from a waste canister in a repository. The model includes flow and diffusion in the fractures in the bedrock as well as diffusion in the backfill. Calculations have been made on the inward transport of corrosive agents to a buried canister from the surrounding bedrock. The oxidants are transported from the flowing groundwater through the backfill to the surface of the canister, were they are assumed to react instantaneously with the canister material. The rock is modeled as a discrete fracture system. The fractures are assumed to be evenly spaced and the groundwater movement is described by potential flow. The model is three dimensional.

INTRODUCTION

In the Swedish concept for disposal of radioactive waste, the spent fuel is buried in mined caverns. The waste is sealed in canisters and buried in stable crystalline bedrock. The repository consists of a system of tunnels, excavated 500 m below ground level. Holes are drilled in the floor of the tunnels in which the canisters are emplaced. The space between canister and bedrock is filled with bentonite clay. The purpose of the clay barrier is to increase the transport resistance of migrating species, mechanically protect the canister and act as a sorbent for some radionuclides.

In this paper we only deal with the inward transport of corrosive agents (oxidants), such as oxygen for canisters made of copper & lead and sulfides for copper.

An important factor is the groundwater flow in the bedrock. There are three quantities determining the water flow in fractures, the hydraulic conductivity of the bedrock, the hydraulic gradient and the porosity of the bedrock . The velocity in a fracture is given by:

540

$$V_\infty = \frac{K_p \cdot i}{\varepsilon} \tag{1}$$

Measurements in the Swedish bedrock at large depths indicate a K_p-value of 10^{-9} m/s or less (1). In the following we will use $K_p=10^{-9}$ m/s and i=0.01 m/m. (The notation is listed at the end of the paper.)

In the backfill the flow will be very slow, due to the very low hydraulic conductivity ($K_p < 10^{-13}$ m/s (1)). The transport by flow as compared to transport by diffusion may be estimated in the following way. By using equation (1) the time to penetrate the backfill with water may be estimated. With the distance 0.40 m and porosity 0.25 it will take about 3 million years.

By using the equation for instationary diffusion (2), the time for a migrating species to penetrate the barrier can be calculated:

$$\frac{D \cdot t}{z^2} = 1 \tag{2}$$

D is the diffusivity in the pores, t is the time and z is the thickness of the backfill. A typical D-value for small molecules is $0.8 \cdot 10^{-10}$ m^2/s (8). This gives a diffusional time of approximately 6 years for 5% penetration.

CONCEPTUAL MODEL

The canister with the buried waste will corrode only if oxidants can be transported to its surface. Since the oxidizing reaction is assumed to be very fast the mass-transfer rate will govern the corrosion. The oxidants are transported by groundwater to the outer surface of the backfill. From there they diffuse through the backfill to the canister, were they react instantaneously (3). It is possible to model the mass transport as if there were two serial coupled transport resistances between the canister and the flowing groundwater. The first is due to the diffusion resistance in the groundwater. The second is the diffusion resistance in the back-fill. A number of approximations are made in the mathematical model. The most important are:

a) The fracture planes in the rock are assumed to be perpendicular to the axis of the canister. The fractures are assumed to be evenly spaced.

b) The canister is assumed to be of infinite length (i.e. end effects are neglected).

c) The groundwater flow in the fracture is described by potential flow.

d) Stationary conditions prevail.

e) Influence of temperature gradients are neglected.

f) Transport of water and oxidants in the rock matrix are neglected.

In figure 1 it is shown how a fracture conducts water past the compacted backfill.

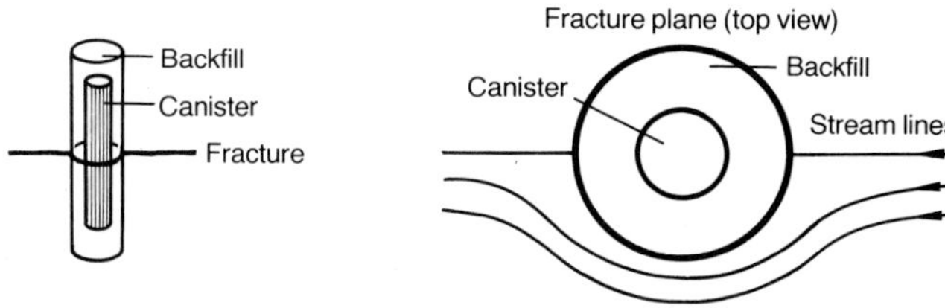

Figure 1. Waterflow around the backfill

We have decided to deal with the transport of oxidants under conditions of steady-state. This is easily justified by considering the total amount of oxidants that can be accumulated in the backfill: compared to the amount of canister material available for reaction, it is vanishingly small. The instationary phase is therefore neglected.

MATHEMATICAL MODEL

Assumptions a) and b) makes it possible to perform the calculations for only one fracture. This can be done by reasons of symmetry. Planes of symmetry will arise between each fracture, which makes every one of them behave identically. (See figure 2).

542

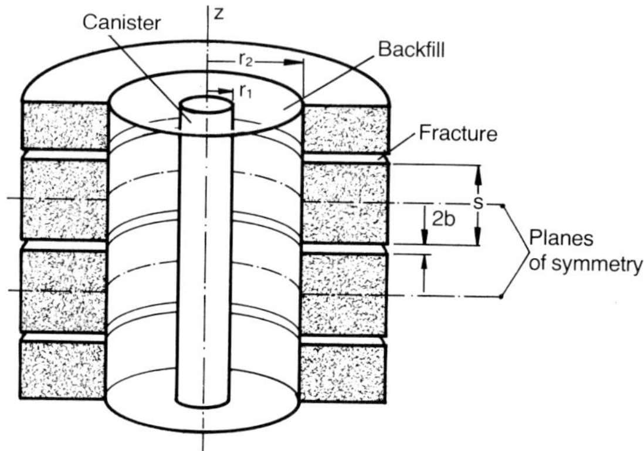

Figure 2. Definition sketch of modelled system: single canister, backfill and idealized model of fissured rock

TRANSPORT IN THE FRACTURE

The water flow and masstransfer have been modeled by using the Navier-
Stokes and conservation of mass (continuity) equations. The equation of
continuity includes diffusion of the solved components. Some simplifying
assumptions have been made to get a less complex system. The rock is
assumed to consist of impermeable blocks, separated by the fractures. The
flow in the fracture is described by potential flow.

The masstransport occurs only in the fractures. The transport
mechanisms are convection and diffusion. Changes in concentration of the
migrating components are assumed not to influence the material properties
of water. The equation for water flow and transport of oxidants are
accordingly decoupled. First the flow field in the fracture is calculated
without considering the concentration of oxidants. Then it is possible to
calculate the transport of oxidants, with known velocities and flow
directions.

Under conditions of steady-state the mass transport in the flowing
groundwater of a plane fracture is given by:

$$D_v \left[\frac{1}{r} \frac{\partial}{\partial r} \left(r \frac{\partial c}{\partial r} \right) + \frac{1}{r^2} \frac{\partial^2 c}{\partial \theta^2} \right] - v_r \frac{\partial c}{\partial r} - v_\theta \frac{\partial c}{\partial \theta} = 0 \qquad (3)$$

were $v(r)$ and $v(\theta)$ are known. This is the usual equation of continuity in
cylindrical coordinates for a dissolved component. The first two terms
stand for diffusion in the r- and θ-directions. The other two give the

convective transport in these directions. In this case r denotes the radial distance from the center of the canister and θ is the angle, which is taken as zero at the stagnation point.
(The notation is given at the end of the paper.)
Boundary conditions

$$c = c_2(\theta) \qquad r = r_2 \qquad 0 < \theta < \pi \tag{4}$$

$$c = c_o \qquad r \rightarrow \infty \qquad 0 < \theta < \pi \tag{5}$$

$$\frac{\partial c}{\partial \theta} = 0 \qquad r > r_2 \qquad \theta = 0 \tag{6}$$

$$\frac{\partial c}{\partial \theta} = 0 \qquad r > r_2 \qquad \theta = \pi \tag{7}$$

Condition (4) implies that the concentration on the outer surface depends upon one spatial coordinate. This gives the coupling to the conditions in the backfill. It simply states that the concentration in the groundwater at the surface of the backfill is equal to the concentration in the porewater at the surface.

The coupling between conditions in the fracture and conditions in the backfill is treated below. Then $c_2(\theta)=0$. In the numeric calculations the backfill and the fracture are treated as a connected system with varying material properties. The amount of oxidants which is transported into the backfill per unit time is given by:

$$N = -2 \, D_v \, r_2 \, 2b \int_0^\pi \frac{\partial c}{\partial r}\bigg|_{r=r_2} d\theta \tag{8}$$

The evaluation of the integral in equation (8) is made with the aid of equation (3). In this treatment the velocity componentes $v(r)$ and $v(\theta)$ are assumed to be independent of the z coordinate.

Futhermore, the friction against the outer surface of the backfill is neglected. This can be done because the velocity profile in the r-direction in the thin fracture is much steeper than the concentration profile.

FLOW FIELD IN THE FRACTURE

In order to solve equation (3) expressions for the velocity components of the flow field, $v(r)$ and $v(\theta)$, must be obtained. This has been done by using potential flow theory.

544

$$\nabla^2 \phi = 0 \tag{9}$$

This is the Laplace-equation. It is pertinent at this point to summarize the various assumptions made in arriving at equation (9). These are:

* Constant fracture width.
* Constant viscosity and density.
* Laminar flow (low velocities)
* Stationary conditions.

Equation (9) is also valid for stationary flow in a homogeneous, isotropic porous medium. The equation is therefore applicable also to fractures with such filling materials. The interesting case for us is potential flow around a cylinder. In this case equation (9) is solved using complex functions. The velocity components are given in rectangular coordinates by Bird et. al. (4), which have been transformed to cylindrical:

$$v_r = v_\infty \left[- \left(1 - \left(\frac{r_2}{r}\right)^2 \cos(2\theta)\right) \cos(\theta) + \left(\frac{r_2}{r}\right)^2 \sin(2\theta) \sin(\theta) \right] \tag{10}$$

$$v_\theta = v_\infty \left[\left(1 - \left(\frac{r_2}{r}\right)^2 \cos(2\theta)\right) \sin(\theta) + \left(\frac{r_2}{r}\right)^2 \sin(2\theta) \cos(\theta) \right] \tag{11}$$

DIFFUSION IN BACKFILL

The transport through the backfill is also described by the equation of continuity. Due to the very low permeability of the backfill, the convective part may be neglected. The fracture is modeled in two dimensions only , but in the backfill the geometry is three-dimensional.

$$D_L \left[\frac{1}{r} \frac{\partial}{\partial r} \left(r \frac{\partial c}{\partial r}\right) + \frac{\partial^2 c}{\partial z^2} + \frac{1}{r^2} \frac{\partial^2 c}{\partial \theta^2} \right] = 0 \tag{12}$$

c now denotes concentration in backfill. The terms represent, from left to right, diffusion in the r-, z- and θ-directions.

The boundary conditions used describe the case were the concentration on the canister surface is zero. Equation (8), but with D_L and c, applies for transition from flowing water to the backfill. Futhermore, the symmetry of the system is utilized. Due to space limitations the boundary conditions are not shown.

NUMERICAL TREATMENT

The model equations have been solved with a computer program (TRUMP), developed at Lawrence Livermore Laboratories, USA (5). The program uses the Integrated Finite Difference method. In this method the three dimensional space is subdivided into a number of small volume elements, which can be chosen quite arbitrarily in size and shape. A nodal point is defined in each element.

CALCULATED EXAMPLE

In order to simplify the comparison between the two barriers and to visualize the results, the concept of masstransfer resistance is introduced. This resistance is an analogy with electrical resistance and can be treated with the same formulas. The resistance in the backfill is defined by:

$$N = \frac{1}{R_L} \Delta c_L \tag{13}$$

The resistance between the flowing groundwater and the outer surface of the backfill is defined by:

$$N = \frac{1}{R_V} \Delta c_V \tag{14}$$

Δc_V denotes the difference in concentration between undisturbed groundwater far away and the concentration at the interface between water and backfill. The latter is averaged around the hole. It is now possible to introduce a total resistance (R) defined by:

$$N = \frac{1}{R} \Delta c \tag{15}$$

where $\Delta c = \Delta c_L + \Delta c_V$ and $R = R_V + R_L$

Values on R_V, R_L and R from a computed example for one fracture is given in Table 1. The results obtained using the above described numerical scheme are compared with the results from a simplified model (6). The latter is also described in (9).

Fracture width (mm)	$R_L \times 10^{-3}$ (years/m^3)		$R_V \times 10^{-3}$ (years/m^3)		$R \times 10^{-3}$ (years/m^3)	
	Simplified model	This model	Simplified model	This model	Simplified model	This model
0.1	0.21	–	4.6	4.2	4.8	–
1.0	0.15	0.17	1.5	1.3	1.6	1.5

Table 1. Resistances for <u>one</u> fracture.

By using the total resistances in Table 1 the total amount of oxidants transported to a buried canister can be calculated. In the Swedish concept the canister has a length of 5 m. This means that the canister intersects 5 fractures (fracture spacing 1 m).

Fracture width (mm)	Q_{eq} (1/year)	N_{S^2} (mg/year)	N_{Cu} (mg/year)
0.1	1.0	7.3	39
1.0	3.1	22	116

Table 2. Equivalent water flowrate Q_{eq} (9), transport rate of S^2 and corroded amount of copper.

The following values have been used in the computations.

$$D_L = 7.8 \cdot 10^{-11} \ m^2/s, \ D_V = 3.9 \cdot 10^{-9} \ m^2/s$$

$$S = 1 \ m, \ V_\infty = 3.15 \ 10^{-4} \ m/year,$$

$$r_1 = 0,375 \ m, \ r_2 = 0.75 \ m, \ \Delta c = 7 \ g/m^3$$

DISCUSSION AND CONCLUSIONS

A numerical model for transport in the near field area has been developed. The model takes into account flow and diffusion in the fracture and diffusion through the backfill. The masstransfer resistance between the groundwater and the outer surface of the backfill dominates the total resistance. The resistance in the backfill is of minor importance. The calculated corrosion rates in table 2 are very small. With the higest value, 116 mg/year, it would take more than 4 million years to corrode 10 mm of the 200 mm thick copper canister.

In the computed example there is a very good agreement between this model and the simplified model. The migration in the near-field can therefore, under conditions of steady-state, be deduced with the simplified model.

The numerical model has the capacity to include the effects of oxidants emmanating from radiolysis and release and migration of radio nuclides. Work on these mechanisms is in progress.

NOTATION

b	half-width of fracture	(m)
c	concentration	
Δc, Δc_L, Δc_V	concentration drop, total, in backfill, in groundwater resp.	(kg/m^3)
D_L, D_V	diffusivity in backfill and in groundwater resp.	(m^2/s)
i	hydraulic gradient	(m/m)
K_p	hydraulic conductivity	(m/s)
L	canister length	(m)
N	mass flow	(kg/s)
Q_{eq}	equivalent water flow rate	m^3/s (1/y)
r	radial coordinate	(m)
R, R_L, R_V	masstransfer resistance, total, in backfill, in groundwater resp.	(s/m^3)
t	time	(s)
v_∞	velocity in fracture at infinite distance from canister	(m/s)
v_r	velocity component in the r-direction	(m/s)
v_θ	velocity component in the θ-direction	(m/s)
z	longitudinal direction of the canister	(m)

Greek letters

ε	porosity	(m^3/m^3)
ϕ	potential	(m)
θ	angular coordinate	(rad)

REFERENCES

1. "Handling and Final Storage of Unreprocessed Spent Nuclear Fuel", Vol. II Technical KBS Report (1978).
2. Crank, J., "The Mathematics of Diffusion", 2:nd ed., Oxford Press, (1975).
3. Neretnieks, I., "Transport of oxidants and radionuclides through a clay barrier", KBS Technical Report No 79, (1978).
4. Bird, et al., "Transport Phenomena", Wiley & Sons, Inc, (1960).
5. Edwards, A. L., "TRUMP: A Computer Program for Transient and Steady State Temperature Distribution in Multidimensional Systems", National Technical Information Service, National Bureau of Standards, Springfield, Va., USA, (1972).
6. Neretnieks, I., "Transport Mechanisms and Rates of Transport of Radionuclides in the Geosphere as related to the Swedish KBS Concept", Underground Disposal of Radioactive Wastes Vol. II, International Atomic Energy Agency, Vienna (1980).
7. Rasmuson, A., Narasimhan, T. N. and Neretnieks I., "Chemical transport in a fissured rock: Validation of a numerical model", Accepted for publication in Water Resources Research (1982).
8. Skagius, C., Neretnieks, I., "Diffusivity Measurements of Methane and Hydrogen in Wet Clay", KBS Technical Report No. 86, (1978) (In Swedish).
9. Neretnieks, I., "Leach Rates of High Level Waste and Spent Fuel-Limiting Rates as Determined by Backfill and Bedrock Conditions". This publication.

Published 1982 by Elsevier Science Publishing Co
SCIENTIFIC BASIS FOR RADIOACTIVE WASTE MANAGEMENT - V
Werner.Lutze, editor

MODEL FOR FAR FIELD MIGRATION

Anders Rasmuson and Ivars Neretnieks
Department of Chemical Engineering
Royal Institute of Technology
S-100 44 Stockholm, Sweden

ABSTRACT

A transport model of radionuclide migration in fissured rock is
presented. It includes advection and hydrodynamic dispersion in the
fissures and diffusion and sorption in the rock blocks. 1 and 2D
analytical and 3D numerical solutions of the model have been developed. A
large number of calculations for the approximate range of variation of the
input parameters have been performed. A large impact of hydrodynamic
dispersion and micropore diffusion on the amount of radioactive material
reaching the biosphere is found.

INTRODUCTION

Radionuclides which escape from a final repository for high level waste
in deep geologic media, will have to migrate through the bedrock to reach
the biosphere. In the Swedish concept, the final repository for high level
waste will be in crystalline rock at 500 m depth (1). To this end
moderately fissured crystalline rocks such as the Swedish granite and
gneiss were investigated. The bedrock has a low hydraulic conductivity.
Water flow occurs in fractures in the rock. Radionuclides may interact in
several ways with the bedrock. Radionuclides may migrate into the
micropores of the rock through the mechanism of molecular diffusion (2).
Many nuclides will sorb (e.g. adsorption, ion exchange) on the pore
surfaces.

In the present paper a transport model of the process is presented. The
model considers advection and hydrodynamic dispersion in the fissures and
diffusion and sorption in the rock blocks. The model has been solved
analytically for single decaying species and simple boundary conditions
(3-5). A 3D numerical model has been developed to account for complex
geometry as well as chain decay (6, 7).

MATHEMATICAL MODEL

In the theoretical analysis the rock is regarded as a double-porosity medium consisting of porous blocks separated by fissures. Since the permeability of solid rock is very low water flow is assumed to take place in fissures only. However, transport of dissolved constituents to the interior of the rock blocks takes place by molecular diffusion.

Two simplified geometrical models of the fissured rock were used:

i) set of parallel fractures

ii) cubic system of orthogonal fractures

In the latter case, to model internal diffusion, the cubic blocks used in the hydraulic model are approximated by spheres having the same surface-to-volume ratio as a cubic block (4). Then $r_o = 0.5$ S.

In mathematical terms, the process is governed by two coupled transport equations, one for the fissures and one for the blocks.

The migration of the ith member of a radionuclide chain is described (the one-dimensional case is shown only) by:

$$\frac{\partial c_f^i}{\partial t} + U_f \frac{\partial c_f^i}{\partial z} - D_L \frac{\partial^2 c_f^i}{\partial z^2} = -\alpha N^i - \lambda_d^i c_f^i + \lambda_d^{i-1} c_f^{i-1} \tag{1}$$

$$K^i \frac{\partial c_p^i}{\partial t} = D_p \varepsilon_p \left(\frac{\partial^2 c_p^i}{\partial r^2} + \frac{\beta}{r} \frac{\partial c_p^i}{\partial r} \right) - K^i \lambda_d^i c_p^i + K^{i-1} \lambda_d^{i-1} c_p^{i-1} \tag{2}$$

$$N^i = -D_p \varepsilon_p \left(\frac{\partial c_p^i}{\partial r} \right)_{r=interface} \tag{3}$$

(The notation is listed at the end of this paper.)

The first equation is the mass balance of the ith member in the water in the fissures. The terms in this equation represent accumulation in water in the fissures, convective transport, transport by axial dispersion, accumulation in the blocks and disappearance of i by decay and appearance of i from decay of the preceding chain member.

In the second equation, the terms give sorption on the interior surfaces + accumulation in the pore fluid, diffusion in the pore fluid, and radioactive decay , respectively.

Equation (3) is the coupling condition between (1) and (2). α and β are geometrical parameters taking the values

$$\left.\begin{array}{l} \alpha = 1/b \\[2mm] \beta = 0 \end{array}\right\} \quad \text{set of parallel fractures}$$

$$\left.\begin{array}{l} \alpha = -\,3(1-\varepsilon_f)/\varepsilon_f r_o \\[2mm] \beta = 2 \end{array}\right\} \quad \text{spherical rock blocks}$$

The entity α is the interfacial rock surface per unit fracture volume. By convention, the coordinate within the rock is taken positive inwards for a set of parallel fractures and outwards (from the centre) in the case of spherical rock blocks. This causes the sign difference in α.

Assuming known values of hydraulic conductivity, hydraulic gradient and fissure spacing; fissure width, fissure porosity and average fracture velocity are calculated according to models proposed by Snow (8):

$$(2b)^3 = \phi\; \frac{12\mu}{\rho g} \cdot SK_p \tag{4}$$

$$\varepsilon_f = \omega(2b/S) \tag{5}$$

$$U_f \varepsilon_f = K_p i \tag{6}$$

where:

$$\left.\begin{array}{l} \phi = 1.0 \\[2mm] \omega = 1.0 \end{array}\right\} \quad \text{set of parallel fractures}$$

$$\left.\begin{array}{l} \phi = 0.5 \\[2mm] \omega = 3.0 \end{array}\right\} \quad \text{cubic system of orthogonal fractures}$$

SOLUTION OF EQUATIONS

Analytical solution

The analytical solution was derived for a single decaying species. The description of internal diffusion assumed that the blocks may be regarded as spheres. The boundary conditions where those of a semi-infinite system initially free of nuclide, and in which the inlet ($z = 0$) nuclide concentration suddenly is increased to c_o at time t_o, and then decreased to 0 again at $t_o + \Delta t$.

The analytical solution is presented in detail in Rasmuson and Neretnieks (3, 4). The solution is:

552

$$c_f/c_o = e^{-\lambda_d t}\left[u(z,t - t_o)\,H(t - t_o) - \right. \tag{7}$$
$$\left. -u(z,t - (t_o + \Delta t))\,H(t - (t_o + \Delta t))\right]$$

$$u(z,t) = \tfrac{1}{2} + \frac{2}{\pi}\int_o^\infty \exp\left[f(\delta,R,Pe,\lambda)\right] \tag{8}$$
$$\sin\left[g(\delta,R,Pe,y,\lambda)\right]\frac{d\lambda}{\lambda}$$

f and g are complicated functions of the variable of integration λ and of the following dimensionless parameters:

$$\delta = (3D_p\varepsilon_p/r_o^2)\,(z/mU_f) \qquad \text{bed length parameter}$$

$$R = K/m \qquad \text{distribution ratio}$$

$$Pe = zU_f/D_L \qquad \text{Peclet number}$$

$$y = 2D_p\varepsilon_p t/Kr_o^2 \qquad \text{contact time parameter}$$

The bed length parameter may be thought of as a ratio of one time, specifying the moving fluid, to another time, specifying the diffusion in particles. The distribution ratio gives the relative magnitudes of the capacity of the particles and the capacity of the fluid in the fissures. The Peclet number is a measure of the relative amount of hydrodynamic dispersion. A method for evaluating the infinite integral in equation (8) is given by Rasmuson and Neretnieks (4).

Analytical solutions for more complicated situations including transient adsorption (9), chemical reaction at the intrapore surfaces (10) and lateral dispersion from a circular disc source (11) have also been obtained. They are not discussed further.

Numerical solution

To be able to treat more complicated problems, the numerical model TRUMP (12) was tested. This code is based on an Integrated Finite Difference Method (13) and solves the advective diffusion equation in three dimensions. In the present study the program was originally used to calculate transport of a single radionuclide in fractured rock including advection and longitudinal dispersion in the fractures and diffusion and sorption in the matrix. The model was validated against the analytical solution (6).

Subsequently, the program has been extended so that calculations on radionuclide chain migration can be done (7). The extended version was given the acronym TRUCHN. In each time step, calculations for the members in the chain are done in succession, beginning with the first member in the chain. The decay-term for a parent nuclide is a source term for a daughter and is included as such in the calculation for the daughter. Since the mother nuclide is independent of the daughter nuclide, we may perform n single nuclide migration calculations for a n-member chain. However, the maximum allowed time step is governed by the nuclide where the concentration changes are largest. In practice, the modification added a number of DO-loops from 1 to ICHAIN, where ICHAIN is the number of members in the chain. The algorithm was also changed to directly include the decay terms.

INPUT PARAMETERS

Halflives and mass sorption coefficients were taken from the Nuclear Fuel Safety Project study (1). Mass sorption coefficients are converted to volume sorption coefficients K by multiplying by the rock density $\rho_p = 2700$ kg/m^3.

D_p is the diffusivity in the water in the pores and is related to the diffusivity in pure water by $D_p = D_v \delta_D/\tau^2$. Here δ_D/τ^2 is a geometric factor. The diffusivity of strong electrolytes in water at ambient temperature is $D_v \approx 1 - 3 \cdot 10^{-9}$ m^2/s. The ratio δ_D/τ^2 is expected to be somewhere in the interval 0.15 to 0.6 (2) and the porosity ε_p has been found to be 0.003 - 0.005 for the granites in this study. Expected values for the effective pore diffusivity $D_p \varepsilon_p$ would then be from $4 \cdot 10^{-13}$ m^2/s to $10 \cdot 10^{-12}$ m^2/s. This was experimentally confirmed by Skagius et. al (14). In the present study we used $D_p \varepsilon_p = 10^{-12}$ m^2/s.

554

The hydraulic conductivity used was in the range $10^{-7} - 10^{-9}$ m/s. The low permeability value was observed in several boreholes in granite and gneiss in south Sweden (15). The fissure spacing was taken as 1 - 50 m. This is substantiated by the borehole investigations. In the calculations we used a hydraulic gradient of 0.001 - 0.01 m/m.

The Peclet numbers used in the examples were based on a recent compilation of about 50 field measurements by Lallemand - Barrès and Peaudecerf (16). The data of Lallemand - Barrès and Peaudecerf have Peclet numbers ranging from 0.5 to 50, with most of the data around 5.

Times for canister penetration of 40, 5000 and 100,000 years have been used. The dissolution times were 30,000 and 500,000 years.

SOME RESULTS

Using the analytical solution a large number of calculations were made for the 23 most important radionuclides in a repository for spent fuel (5). The concentration in the flowing water is observed at a distance z downstream. This concentration is obtained from equation (7). The fraction of the original inventory of a nuclide arriving each year at distance z is then $c_f/\Delta t c_o$. For every case the breakthrough curves for the 23 nuclides at a given distance were calculated versus time. Figure 1 gives typical examples for z = 1000 m. The resulting curves for all those nuclides which have c_f/c_o larger than 10^{-9} at any time are shown. It may be seen that the impact of longitudinal dispersion is severe.

To illustrate the impact of matrix diffusion on the arrival times of the members of a radionuclide chain, a number of numerical calculations, using the code TRUCHN, were done for the two chains U-238 $\rightarrow$ Th-230 $\rightarrow$ Ra-226 and Pu-239 $\rightarrow$ U-235 $\rightarrow$ Pa-231 (7). Typical results for the U-238 chain are depicted in Figures 2-3. Other variables being equal, the distance from the source is a hundred times greater in Figure 3. In the figures two additional sets of curves are given. These curves were computed with the computer program GETOUT (17). One set of curves gives the result for surface sorption (penetration depth 10^{-4} m) while the other set gives the result for complete penetration of the rock matrix. As may be seen, the differences in first arrival times are very large. The arrival times, in the surface and volume sorption cases, differ with as much as four orders of magnitude. The corresponding times for instationary diffusion are located between these extreme values. In Figure 3 the curves for instationary diffusion and volume sorption nearly coincide. This is due to the long contact times.

DISCUSSION AND CONCLUSIONS

The analytical and numerical codes developed provide viable tools for use in nuclear waste evaluation studies. Using the codes a large number of calculations have been done. The influences of matrix diffusion and hydro-dynamic dispersion are given special attention. The case of instationary diffusion into the matrix is compared with the cases of instantaneous surface and volume sorption. The radionuclide arrival times to a certain distance will in general vary over many orders of magnitude between these cases. Longitudinal dispersion has a large impact on the early arrival of many radionuclides - distances travelled before decay may differ by orders of magnitude. For further results the reader is referred to the recent reports by Rasmuson and Neretnieks (5) and by Rasmuson et al. (7).

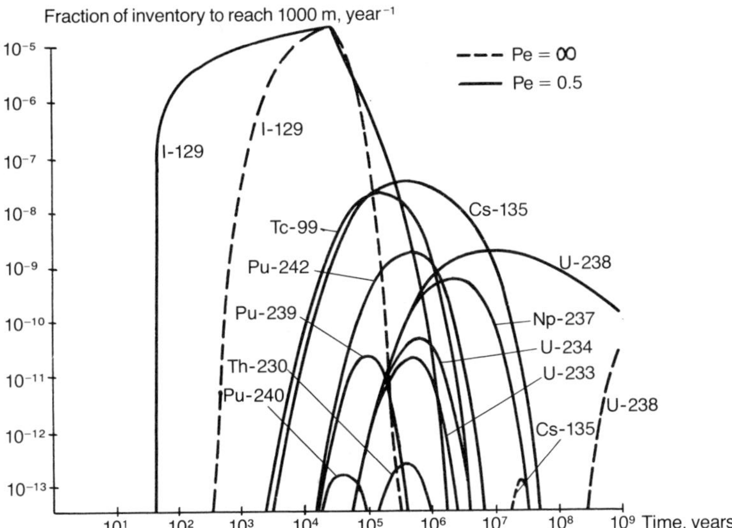

Figure 1: Analytical solution. t_o=40 years, Δt=3·10^4 years, K_p=10^{-9} m/s, i=0.01 m/m, S=50 m, $D_p \varepsilon_p$=10^{-12} m^2/s.

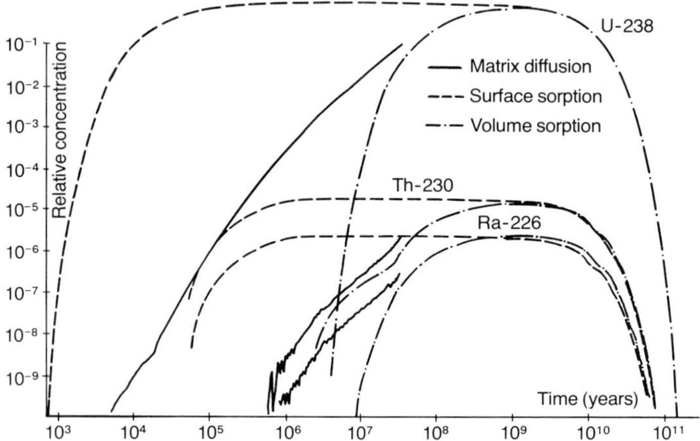

Figure 2: Numerical solution. S=1 m, short transport time, U238 - Ra226.

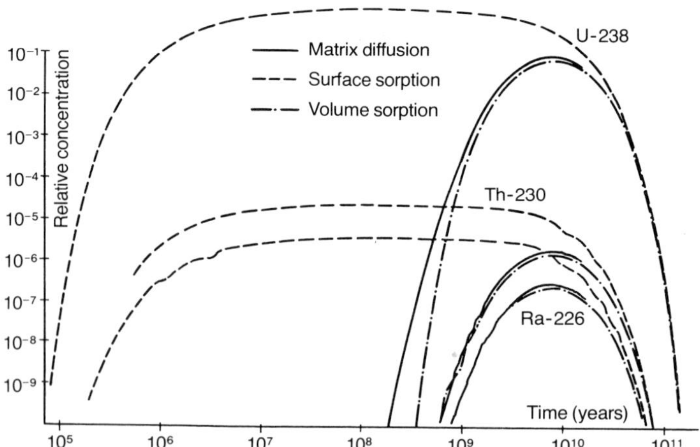

Figure 3: Numerical solution. S=1 m, long transport time, U238 - Ra226.

NOTATION

b	half width of fissure	m
c_f	concentration in liquid in fissures	mol/m^3
c_p	concentration in liquid in microfissures	mol/m^3
c_o	inlet concentration in the liquid	mol/m^3
D_L	longitudinal dispersion coefficient	m^2/s
D_p	diffusivity in water in pores	m^2/s
D_v	diffusivity in water	m^2/s
g	gravitational constant	m/s^2
H	Heaviside's step function	
i	hydraulic gradient	m/m
K	volume equilibrium constant	m^3/m^3
K_p	hydraulic conductivity	m/s
m	$= \varepsilon_f/(1-\varepsilon_f)$	
N	molar flux at the fracture-solid interface	$mol/m^2,s$
Pe	$=zU_f/D_L$, Peclet number	
r	radial distance from center of spherical particle or distance into rock from fissure surface	m
r_o	effective spherical radius	m
S	fissure spacing	m
t	time	s
t_o	time of canister penetration	s
Δt	time for dissolution of waste	s
U_f	average velocity of water in fissures	m/s
z	distance in flow direction	m

Greek letters

δ_D	constrictivity for diffusion	
ε_f	porosity of rock fissures	
ε_p	porosity of rock matrix	
λ_d	decay constant of radionuclide	s^{-1}
μ	viscosity of water	Ns/m^2
ρ	density of water	kg/m^3
τ	tortuosity	

Superscript

i	i^{th} chain member

REFERENCES

1. KBS - Nuclear Fuel Safety Project: Handling and final storage of unreprocessed spent nuclear fuel, vol. II. Technical report Stockholm, 1978.

2. Neretnieks, I.: Diffusion in the rock matrix: An important factor in radionuclide retardation? J. Geophys. Res. $\underline{85}$, 4379 (1980).

3. Rasmuson, A and I. Neretnieks: Exact solution of a model for diffusion in particles and longitudinal dispersion in packed beds. AIChE J. $\underline{26}$, 686 (1980).

4. Rasmuson, A. and I. Neretnieks: Migration of radionuclides in fissured rock: The influence of micropore diffusion and longitudinal dispersion. J. Geophys. Res. $\underline{86}$, 3749 (1981).

5. Rasmuson, A. and I. Neretnieks: Migration of radionuclides in fissured rock - Results obtained from a model based on the concepts of hydrodynamic dispersion and matrix diffusion. Manscript for pub.(Oct. 1981).

6. Rasmuson, A., T.N. Narasimhan and I. Neretnieks: Chemical transport in a fissured rock: Verification of a numerical model. Accepted for publication in Water Resources Res. (June 1981).

7. Rasmuson, A., A. Bengtsson, B. Grundfelt and I. Neretnieks: Radionuclide chain migration in fissured rock - The influence of matrix diffusion. Manuscript for publication (October 1981).

8. Snow, D.T.: Rock fracture spacings, openings and porosities J. Soil Mech. Found. Div. Amer. Soc. Civil Eng. $\underline{94}$ (SM1) 73 (1968).

9. Rasmuson A: Exact solution of a model for diffusion and transient adsorption in particles and longitudinal dispersion in packed beds; AIChE J. $\underline{27}$, 1032 (1981).

10. Rasmuson A: Transport processes and conversion in a fixed-bed catalytic reactor; Chem. Engng. Sci. $\underline{37}$, 411 (1982).

11. Rasmuson A: Diffusion and sorption in particles and two-dimensional dispersion in a porous medium; Water Resources Res. $\underline{17}$, 321 (1981).

12. Edwards, A.L., TRUMP: A computer program for transient and steady state temperature distributions in multidimensional systems. National Technical Information Service, National Bureau of Standards, Springfield Va. 1972.

13. Narasimhan, T.N and P.A. Witherspoon: An integrated finite difference method for analyzing fluid flow in porous media. Water Resources Res. $\underline{12}$, 57 (1976).

14. Skagius, C., G. Svedberg and I. Neretnieks: A study of strontium and cesium sorption on granite. Accepted for publication in Nuclear Technology (1981).

15. KBS - Nuclear Fuel Safety Project: Handling of spent nuclear fuel and final storage of vitrified high level reprocessing waste: Supplementary geologic studies, Report, Stockholm 1979. (Swedish)

16. Lallemand-Barrès, A., and P. Peaudecerf: Recherche des relations entre la valeur de la dispersivité macroscopique d'un milieu aquifère, ses autres caractéristiques et les conditions de mesure, Bull. Bur. Rech. Geol. Min. Fr., $\underline{4}$, 277, 1978.

17. Grundfelt, B.: Nuclide migration from a rock repository for spent fuel, Tech. Rep. 77, Nucl. Fuel Safety Proj., Stockholm, 1978. (Swedish)

LEACH RATES OF HIGH LEVEL WASTE AND SPENT FUEL. - LIMITING RATES AS DETERMINED BY BACKFILL AND BEDROCK CONDITIONS.

Ivars Neretnieks
Department of Chemical Engineering
Royal Institute of Technology
S-100 44 Stockholm, Sweden

SUMMARY

The release rate of radionuclides from a degraded canister containing radioactive waste is controlled by diffusion through the backfill and by diffusion into the water flowing past the canister. In low porosity low permeability bedrock like the Swedish granites the amount of water which can be contaminated may be very small. Two sample cases are calculated. One uses the conditions for a repository for vitrified high level waste, the other applies to a repository for spent fuel. The small amount of water which can carry the waste in combination with the low solubilities for the major longlived actinides limit the release rate from the repositories to very small values.

INTRODUCTION

The present study is based on the conditions which will prevail in a repository for radioactive waste in crystalline rock at large depth as in the Swedish Nuclear Fuel Safety Project (KBS) concept.

The canisters with either spent fuel (SF) or high level waste (HLW) are emplaced in holes in the floor of tunnels at 500m depth. The bedrock is crystalline e.g. granite or gneiss and has a low hydraulic conductivity (KBS 1977 Geology) $K_p < 10^{-9}$ m/s. The backfill consists of either compacted bentonite in the holes or a mixture of bentonite and quartz sand. The tunnels are backfilled with the latter mixture.

The bentonite, when wet, has a strong swelling pressure and very low hydraulic conductivity $K_p < 10^{-12}$ m/s (Pusch 1978). In the KBS concept the maximum temperature rise will be to less than 80° C on the canister surface. The canisters are designed not to corrode for very long times. If and when a breach occurs, the near field temperature is very near the ambient temperature in the bedrock $< 25^{o}$ C.

560

The present study is concerned with the transport of species to and from the degraded canister when all the spent fuel or high level waste is exposed to the groundwater in the bedrock. The study concentrates on the rate of release of radionuclides to the flowing water. One of the aims of the study has been to devise a method whereby input data to a farfield migration model may be obtained.

CONCEPTUAL MODEL AND MAIN MECHANISMS

The basic approach to the problem has been to determine how much of the water which flows past the repository will exchange species with the canisters. The exchange is limited by the diffusional resistance in the nearly stagnant water in the backfill as well as the diffusional resistance in the slowly moving water in the bedrock. As the canisters are spaced fairly far apart it is assumed that they do not influence each other.

The basic scenario is the following. At some time the canister material has degraded. The surface of the waste is wetted and the waste reacts "instantaneously" with the water, dissolving all constituents of the waste up to the solubility limit of each constituent. Some constituents like iodide will dissolve entirely in the water, others like silica of the glass, uranium of the spent fuel or nuclides like thorium will attain their equilibrium concentration in the water in contact with the waste.

There will be an instationary phase when the species diffuse out into the backfill and into the moving water. This is relatively short in relation to the total leach times of interest for the majority of the important nuclides (Neretnieks 79) and is not further discussed here. During the ensuing stationary period the species are transported away from the fuel at a rate permitted by the rate of diffusion through the backfill and by the rate at which the flowing water can carry them away. The concentrations of the species in water in contact with the surface of the waste is at the solubility limit of the species.

It is conceivable that all nuclides will not be released to the water at a rate large enough to keep the concentration at the solubility limit. This may occur if the release is dominated by the dissolution rate of the matrix of the waste (congruent dissolution). This might be the case for glasses high in silica or for spent fuel where most of the waste (95 %) consists of UO_2. The nuclides may only be released at the rate at which the silica or uranium is dissolved and transported away.

Both   the congruent dissolution case and the case limited by individual solubilities will be treated.

The problem is structured in the following parts

o Flowrate of water in the near field of the canister
o Diffusion in the water flowing outside the backfill
o Diffusion in the backfill
o Coupling of diffusional resistances

MATHEMATICAL MODEL

Water flow around the canister

The water flow in the vicinity of the canister will be somewhat larger than in the undisturbed rock far from the caverns where the rock is not influenced by mining activities. Assuming that the bedrock can be treated as a porous medium and that the rock is much more permeable near the canister than far away, it can be shown that the flux through an infinitely long cylindrical hole is twice that in the undisturbed rock (Tietjens 1960). The same applies to flow in a fissure intersecting a hole perpendicular to its axis.

The presence of more fissured rock near the backfilled holes acts as an empty hole in the above description. The presence of an impermeable area – the backfill – inside the permeable area does not influence the flow pattern in the rock. There are no detailed data on the size and orientation of the fissures in the rock. In the following the flowrate in the rock surrounding the backfill is taken to be twice that in the far field rock.

The flowrate of water in the bedrock in the far field can be determined if the hydraulic gradient and conductivity is known, the flowrate is $u_o=$ K_p i and the velocity of the water is $u_p=2u_o/\varepsilon_R$ in the near field.
(The notation is given at the end of the paper.)

Mass transfer to the water flowing past the backfill

Two cases are treated. In the first case we assume that the bedrock around the backfill can be treated as a porous medium surrounding an infinitely long impermeable medium (the backfill). The assumption of an impermeable backfill is a good approximation for the bentonite clay which has a very low hydraulic conductivity compared to the bedrock. The transport by diffusion of solved species is much faster than by flow (Neretnieks 1977). Transport by flow in the backfill is therefore neglected.

562

In the second case we assume that the bedrock consists of individual fissures intersecting an infinitely long clay cylinder at regular intervals. All water flow takes place in the fissures.

The mass transfer to the flowing water is treated as follows: The water which approaches the backfill has a known concentration c_∞ of the species of interest. As the water approaches and flows past the backfill it picks up (or is depleted of) the species which diffuse in the water from (or to) the outer surface of the backfill. In the following only outward transport will be described. Inward transport may be described in the same way. During its passage past the backfill the flowing water will pick up more and more of the diffusing species. When the water leaves the vicinity of the backfill, it has taken up an amount of species which is mainly dependent on the residence time of the water at the surface of the backfill and the diffusivity of the species. A rigorous treatment of this case is given by Andersson et al (1982). A simplified approach was taken by Neretnieks (1978, 1980).

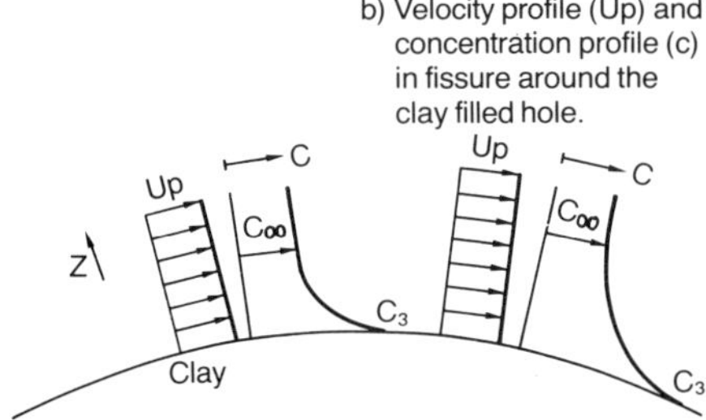

Figure 1. Velocity and concentration profile at the backfill interface

The velocity profile is assumed to be flat at the rock backfill interface as the fissured bedrock dominates the flow resistance. The residence time of the water at the outside of the backfill is approximated by $t_{res} = \pi r_3 / u_p$.

Assuming that the penetration depth of the diffusing species is short compared to flow length the mass N transferred during this time can be obtained by solving the diffusion equation for the appropriate initial and boundary condition. The solution is given in detail by Bird et al (1960 p. 537).

The results can be written

$$N = 2\pi L r_3 \, \varepsilon_R \, (c_\infty - c_3) \sqrt{\frac{4D_w}{\pi t_{res}}} \tag{1}$$

Mass transfer in the backfill

The distinction between the case where the bedrock is a "porous" medium and the case where it is a sparsely fissured rock becomes important when the migration in the backfill is to be treated. In the first case the diffusing species can enter the water in the bedrock everywhere along the backfill/bedrock interface, whereas in the second case the species must find the widely spaced individual fissures.

Bedrock is a porous medium

The diffusional transport through a cylindrical barrier can be written (Bird et al 1960, p. 287).

$$N = D_1 \; 2\pi L \; (c_1 - c_2)/\ln \; (r_2/r_1) \qquad (2)$$

If backfill material can be injected into the porous bedrock to some distance r_3 (or into the individual fissures) there will be an additional transport resistance. This can be written analoguous to equation 2.

$$N = D_2 \; \varepsilon_R \; 2\pi L \; (c_2 - c_3)/\ln \; (r_3/r_2) \qquad (3)$$

D_2 is the effective diffusivity in the backfill injected into the pores of the bedrock.

As it is implicitly assumed (not quite true) that c_1, c_2 and c_3 are constant around the cylindrical circumferences, c_2 and c_3 can be eliminated from equations 1, 2 and 3 to give

$$N = 2\pi L \; \cfrac{(c_1 - c_\infty)}{\underbrace{\cfrac{\ln \; (r_2/r_1)}{D_1} \cdot \frac{S}{\delta}}_{R_1} + \underbrace{\cfrac{\ln \; (r_3/r_2)}{D_2 \; \varepsilon_R}}_{R_2} + \underbrace{\cfrac{1}{r_3 \; \varepsilon_R \sqrt{\dfrac{4D_w}{\pi t_{res}}}}}_{R_3}} \qquad (4)$$

The introduction of S/δ is explained below.

The three terms in the denominator are proportional to the resistance to mass transfer in each section R_1, R_2 and R_3 respectively. They can be used to compare the importance of the various diffusional barriers.

Bedrock is a fissured medium

A case where the bedrock is fissured with equally spaced horizontal fissures is also modelled.

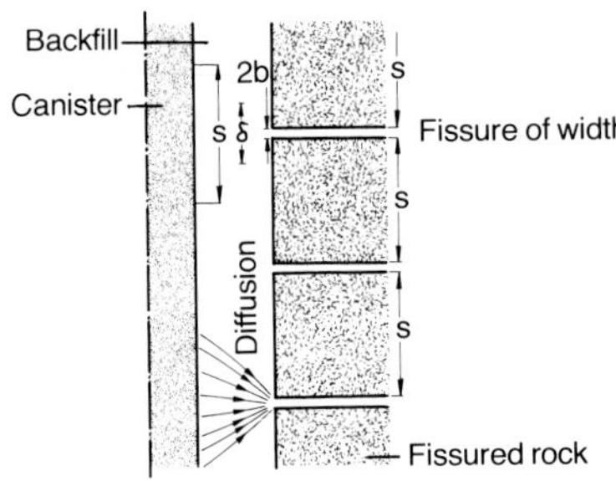

Figure 2 shows a crossection of the hole with a canister.

The diffusing species from the degraded canister will have an effectively decreasing crossectional area to migrate in until it reaches the fissures.

Assuming that all transport takes place between a circle segment 2b wide and an outer segment S wide at a distance S - 2b, an approximate average width for mass transport is obtained from the expression for diffusion between two concentric cylinders.

$$\delta = \frac{S - 2b}{\ln \frac{S}{2b}} \quad \overset{2b \ll S}{=} \quad \frac{S}{\ln \frac{S}{2b}} \tag{5}$$

Thus in the case of fissured rock only a part of the backfill length = δ/S is effectively utilized for mass transport.

The resistance in the backfill R_1 then increases by a factor S/δ as compared to the case where the rock is a "porous" material. For a porous medium $\delta = S$.

Andersson et al (1982) have shown that the approximations leading to equation 5 do not introduce grave errors.

The transport rate N of a species may also be expressed as an equivalent water flow rate Q_{eq}. which would increase in concentration from c_1 to c_∞.

$$N = Q_{eq} (c_1 - c_\infty) \tag{6}$$

APPLICATION TO LEACHING OF RADIOACTIVE WASTES

Specification of a repository case

Table 1 gives basic data for the cases studied. They have been taken from the Swedish KBS Studies (1977, 1978). Based on these data the equivalent water flow rate Q_{eq} and the resistances to mass transfer are calculated by eq. 4. The results are given in table 2.

$u_o=0.1$ $1/m^2$, yr, $S=1$ m, $2b=0.1$ mm, $D_w=2\cdot10^{-9}m^2.s^{-1}$, $D_1=D_w/50$, $D_2=D_w/10$, $r_1=0.3$ m (HLW) and 0.375 m (SF), $r_2=r_3=0.5$ m (HLW) and 0.75 m (SF), $L=1.5$m (HLW) and 4 m (SF).

Table 1. Basic data for the computed cases. Data are based on the Swedish KBS studies (1977, 1978).

In the central case for HLW the equivalent water flowrate is 0.14 l/yr per canister. This is the flowrate which will transport a leached species by increasing its concentration from c_∞ to c_1. This can be compared to the average flowrate of water through the repository. Each canister occupies 150 m^2 with a tunnel spacing of 25m and canister spacing of 6 m.   On the average then the water flowrate is 15 l/yr. Only Q_{eq} = 0.14 l/yr of this water will pick up the leached species with a concentration increase from c_∞ to c_1. For spent fuel canisters it will be 0.45 l/yr .

It is interesting to note that $R_3=16.7$ R_1 which means that only 6 % of the mass transfer resistance is due to diffusional resistance in the back-fill and 94 % is due to diffusional resistance in the small crossection area of the fissures in the bedrock.

<u>Leach rates</u>

The release rate is obtained from the equivalent water flow rate times the concentration increase. The release rates from a repository for HLW with 3000 canisters (equivalent to 3000 tonnes fuel) is given in table 3 and the release rate from a repository for 7000 canisters of spent fuel (10000 tonnes) is given in table 4.

DISCUSSION AND CONCLUSIONS

The release mechanisms described in this treatment are not dependent on the kinetically controlling factors nor on if the waste is degraded and increased in surface.

A low porosity bedrock may considerably limit the flowrate of water to which a leaking canister can deliver radionuclides. The major limiting factor for mass transfer is the small crosssections of the fissures in which the species migrate. Only a limited amount of water will be contaminated before it has passed the vicinity of the canister. The very low permeability backfill adds only little to the mass transfer resistance as the crossection for diffusion is large. Even a small injection distance of backfill into the fissures may add considerably to mass transfer resistance as can be seen from line 4 table 2.

Nuclides with a high solubility will be hindered very little while nuclides with a low solubility may have very low leach rates. The actinides Pu and Th have so low solubilities that the maximum leach rate from a whole repository can be expected to be less than 1 Ci. The release values would apply for oxidizing as well as reducing conditions for these nuclides. The release rate for uranium is determined by the concentration of carabonate in the water. The release rate for Np will be very high in an oxidizing environment if Np solubility were determining. Oxidizing conditions will prevail due to radiolysis (Christensen 1982). It seems probable, however, that Np and Pu, which are incorporated in the UO_2 matrix will not be released faster than the leach rate of UO_2. Furthermore the Np if released in oxidizing conditions would precipitate when the water turns reducing a little distance from the repository as it encounters ferrous iron in the bedrock.

	u_o	$2b$	S	R1/R1	R2/R1	R3/R1	Q_{eq}	
	$1/m^2 \cdot yr$	mm	m	–	–	–	l/yr	
	0.1	0.1	1	1	0	16.7	0.14	Central case (HLW)
	0.1	1	1	1	0	7.1	0.45	Variation 2b
HLW	1	0.1	1	1	0	5.3	0.43	Variation u_o
	0.1	0.1	1	1	77	15.3	0.027	Backfill injected 10 cm
SF	0.1	0.1	1	1	0	10.1	0.45	Central case (SF)

Table 2. Determined equivalent water flowrate Q_{eq} for leaching of HLW and SF and relative resistances to mass transfer of the various barriers. R_1/R_1, R_2/R_1 and R_3/R_1 are resistances in backfill around the canister, in backfill in the fissure, and in the flowing water in fissure respectively normalized to the resistance in the backfill around the canister.

	Solubility g/1 10^3 (ppm) (Allard 1981)	Inventory HLW (KBS 1977) g/canister (10^5 yr old)	Leach rate fraction/yr	Activity release μCi/3000 canisters·yr
SiO_2 b)	120-20	$75 \cdot 10^{\,3}$	$1.9 \cdot 10^{-7}$	-
U-238	485	897	$1.4 \cdot 10^{-4}$	129
Pu-242	2.4	256	$1.3 \cdot 10^{-6}$	3 940
Np-237 a)	0.007	837	$1.2 \cdot 10^{-9}$	2.1
Th-230	0.024	0.25	$1.3 \cdot 10^{-5}$	196
Pu239	2.4	10.7	$1.3 \cdot 10^{-6}$	61 900

Table 3. Leach rates of some important actinides from high level waste

a) Reducing conditions   b) Freeze (1979)

	Solubility g/1 10^3 (ppm) (Allard 1981)	Inventory SF (KBS 1978) g/canister (10^5 yr old)	Leach rate fraction/yr	Activity release μCi/7000 canister·yr
U-238	485	$1.33 \cdot 10^6$	$3.1 \cdot 10^{-7}$	948
Pu-242	2.4	466	$2.3 \cdot 10^{-6}$	29 500
Np-237a)	0.007	2 170	$1.45 \cdot 10^{-9}$	16
Th-230	0.024	64	$1.68 \cdot 10^{-7}$	1 470
Pu-239	2.4	325	$2.3 \cdot 10^{-6}$	464 000

Table 4. Leach rate of some important actinides from spent fuel

a) Reducing conditons

NOTATION

b	half width of fissure	m
c	concentration	g/m^3
D_w, D_1, D_2	diffusivity in water, backfill and backfill injected in bedrock porosity	m^2/s
i	hydraulic head	m/m
K_p	hydraulic conductivity	m/s
L	canister length	m
N	rate of mass transport	g/s
Q_{eq}	equivalent water flow rate	m^3/s (l/yr)
r	radius	m
R	resistance to mass transfer	s/m^3
S	fissure spacing	m
t	time	s
t_{res}	time for water flow past backfill	s
u_o	water flux in bedrock	$m^3/m,s$
u_p	water velocity	m/s
δ	average width	m^3
ε_R	porosity of bedrock = 2b/S	-

Subscripts for r and c:
1 at canister interface
2 at emplacement hole cylindrical surface
3 at interface between flowing water in fissure and backfill

REFERENCES

1. Allard B. Solubilities of Actinides in Neutral or Basic Solutions. Proc. Actinides 81. Asilomar Sept. 10-15 1981. In press.
2. Andersson G., Rasmuson A., Neretnieks I. Model for Near Field Migration. This publication.
3. Bird R.B., Stewart W.F., Lightfoot E.N. Transport Phenomena. John Wiley N.Y. 1960.
4. Christensen H, Studsvik Energiteknik AB, Studsvik Sweden. Personal Communication 1982.
5. Freeze A. R., Cherry J. A., Groundwater, Prentice Hall Inc. 1979.
6. KBS-1977. Handling of Spent Nuclear Fuel and Final Storage of Vitrified High-level Reprocessing Waste. Vol. IV Safety Analysis. KBS-Report, Stockholm 1977.
7. KBS-1978. Handling and Final Storage of Unreprocessed Spent Nuclear Fuel, Vol. II, Technical. KBS-Report, Stockholm 1978.
8. Nerenieks I. Retardation of Escaping Nuclides from a Final Repository. KBS-Technical Report 30, 1977.
9. Neretnieks I. Transport of Oxidants and Radionuclides through a Clay Barrier. KBS-Technical Report 79, 1978.
10. Neretnieks I. Transport Mechanisms and Rates of Transport of Radionuclides in the Geosphere as Related to the Swedish KBS concept. IAEA publ. SM-243/108 IAEA Vienna, 1980.
11. Pusch R. Highly Compacted Bentonite as a Buffer Substance. KBS-Technical Report 74, 1978.
12. Tietjens O. Strömungslehre. Springer 1960.

Published 1982 by Elsevier Science Publishing Co
SCIENTIFIC BASIS FOR RADIOACTIVE WASTE MANAGEMENT- V
Werner.Lutze, editor

AQUEOUS PHASE DIFFUSION IN CRYSTALLINE ROCK

M.H. Bradbury, D. Lever and D. Kinsey
Harwell Laboratories, Oxfordshire, OX11 ORA, England

1.   INTRODUCTION

One of the options being considered for the disposal of radioactive waste is deep burial in crystalline rocks such as granite. It is generally recognised that in such rocks groundwater flows mainly through the fracture networks so that these will be the "highways" for the return of radionuclides to the biosphere. The main factors retarding the radionuclide transport have been considered to be the slow water movement in the fissures over the long distances involved together with sorption both in man-made barriers surrounding the waste, and onto rock surfaces and degradation products in the fissures.

Recent theoretical studies[1,2] have indicated the importance also of diffusion into pores and micro-fissures in the rock matrix as an additional retardation and dispersion mechanism, particularly for species only weakly sorbed. This would delay the migration of radionuclides moving with the ground-water along major fissures or fractures. Since these fissures are likely to be widely spaced (a 10 m spacing may be typical) and the radionuclide pulse very long, diffusion will occur over extremely long periods of time and could retard the radionuclides by several orders of magnitude.

Because of the potential importance of this mechanism in radionuclide migration, a programme was started to measure diffusion coefficients of soluble species in water through rock. Diffusion coefficients have been measured in a number of granites, as a typical rock, using the very weakly sorbed iodide ion as tracer. In addition, measurements on weakly sorbed nuclides have yielded values of the sorption coefficient where other techniques are inadequate.

2.   EXPERIMENTAL PROCEDURE

Diffusion was measured through cylindrical samples of rock up to 75 mm in diameter and 20 mm thick at ambient temperature - small temperature variations had no discernible effect. Samples were sealed into perspex holders which were sandwiched vertically between the two perspex halves of the cell, the sealing being provided by two fully trapped 'O' rings. One half provided the reservoir for the diffusing species, and the other a stirred measurement cell.

The reservoir was filled with 0.1 M KI and the measurement cell to an equal depth with an isotonic solution of KNO_3. Prior to assembly the rock was saturated with the KNO_3 solution and the seal between rock and holder checked

570

by applying a static head of one metre for a few days.

Iodide concentrations in the measurement cell were followed as a function of time using standardised ion specific electrodes. An example of the variation in concentration with time is shown in figure 1. The linear portion can be fitted by the asymptotic solution of the diffusion equation (see section 3):

$$\frac{C_2}{C_1} \cdot \frac{V\ell}{A} = D_i t - \frac{1}{6} \alpha \ell^2 \quad \ldots \ (1)$$

where D_i is the intrinsic diffusion coefficient, C_1 is the inlet (reservoir) concentration, C_2 is the outlet (measured) concentration (with $C_2 \ll C_1$), A is the cross-sectional area of the rock sample, ℓ is the sample thickness, V is the solution volume in the measurement cell and t is the time. α, the rock capacity factor, is given by

$$\alpha = \varepsilon + \rho K$$

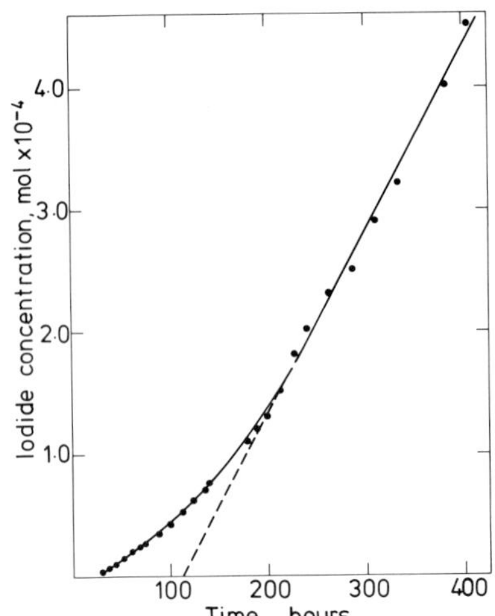

Fig. 1. Iodide ion diffusion in Ossian granite.

with ε the open porosity, ρ the density and K the sorption coefficient which can be concentration dependent. D_i and α can be obtained from the slope and intercept of the linear (steady state diffusion) portion of the plot.

In addition some elution experiments have been performed to obtain a second measurement of the rock capacity factor, α. For this a rock sample was taken immediately after a diffusion experiment had finished and placed in a known volume of KNO_3, then the iodide concentration was measured until no further change was recorded (after about 3 weeks). The value of α was obtained directly from the amount of iodide eluted.

3.   DIFFUSION IN POROUS MEDIA

In an isothermal system at uniform pressure, diffusion takes place in the direction of decreasing concentration, and the rate of transport per unit area of cross-section, F, is given by Fick's first law

$$F = -D\frac{\partial C}{\partial x} \qquad \ldots (2)$$

where D is the diffusion coefficient. Fick's second law

$$\frac{\partial}{\partial x} \left(D\frac{\partial C}{\partial x}\right) = \frac{\partial C}{\partial t} \qquad \ldots (3)$$

gives the rate of change of concentration at a point in a one-dimensional system.

These two equations are widely used but care must be taken in modifying them for porous media. Equation (2) is retained in superficially the same form, with an "intrinsic diffusion coefficient" D_i, which is a property of the three-component system: porous solid, solute and solvent. For continuity at interfaces, C is still defined as being per unit volume of solvent. The second law is modified by the introduction of the factor α, defined above, which relates the concentration per unit volume of the medium to that in the water. In most common situations the intrinsic diffusion coefficient is independent of distance, x, and then (3) is replaced by

$$D_i \frac{\partial^2 C}{\partial x^2} = \alpha \frac{\partial C}{\partial t} \qquad \ldots (4)$$

In this form it is tempting to use the "apparent diffusion coefficient", $D_a = D_i/\alpha$. In simple situations when only the concentration is to be calculated, this is satisfactory. However, the intrinsic diffusivity is essential for flux calculations, so it is recommended that D_i and α are kept separate as in (4). Certainly authors should make clear when they use diffusion coefficients whether they mean the intrinsic or apparent coefficient, and to date they have not always done this.

Solutions of (4) in various geometries have been widely published[3]. For the case here, a porous slab initially at zero concentration, with constant inlet concentration C_1 at $x = 0$, and outlet concentration C_2 ($C_2 \ll C_1$) at $x = \ell$, the solution is

$$\frac{C(x,t)}{C_1} = 1 - \frac{x}{\ell} - \frac{2}{\pi} \sum_{n=1}^{\infty} \frac{1}{n} \sin\left(\frac{n\pi x}{\ell}\right) \exp\left(-\frac{D_i n^2 \pi^2 t}{\ell^2 \alpha}\right) \qquad \ldots (5)$$

The total quantity diffused through the slab, $Q(t) = C_2(t)V$, is given by

$$\frac{Q}{A\ell C_1} = \frac{D_i t}{\ell^2} - \frac{\alpha}{6} - \frac{2\alpha}{\pi} \sum_{n=1}^{\infty} \frac{(-)^n}{n^2} \exp\left(-\frac{D_i n^2 \pi^2 t}{\ell^2 \alpha}\right) \qquad \ldots (6)$$

As t increases, the exponential terms in (6) rapidly decrease leaving the asymptotic solution (1), which is a straight line of slope D_i/ℓ^2, with an intercept on the time axis at $t = \alpha\ell^2/6D_i$.

4.   RESULTS

A summary of the diffusion results for granites from 4 different regions of the United Kingdom is given in Table 1. These were available commercially and have no direct significance for waste disposal. The table also gives values of α obtained from elution experiments. For the Ossian granite all of the values of D_i and most of the values of α lie within a factor of 2.5 of each other.

Table 1. Summary of Iodide Ion Diffusion Results in 4 United Kingdom Granites

	D_i $m^2 s^{-1}$	α	α (elution)	ψ
Ossian Granite	1.4 E-12	1.2 E-2		7 E-4
	1.4 E-12	1.6 E-2		7 E-4
	1.4 E-12	2.2 E-2	1.5 E-2	7 E-4
	1.6 E-12	1.2 E-2	1.0 E-2	8 E-4
	9.4 E-13	7.4 E-3		5 E-4
	7.5 E-13	3.1 E-3	2.1 E-3	4 E-4
	6.8 E-13	6.5 E-3		3 E-4
	6.4 E-13	5.7 E-3		3 E-4
Scottish Low-land Granite	4.0 E-13	2.2 E-3		2 E-4
	3.7 E-13	2.7 E-3		2 E-4
Skene Complex Granite	2.8 E-13	1.2 E-3		1 E-4
Cornish Carnmenellis Granite	2.5 E-14	7.7 E-4		1.2 E-5
	3.2 E-14	8.8 E-4		1.6 E-5

The present technique only yields a value of α, and the contributions from the porosity ε and sorption cannot be separated unless the porosity is measured by an independent technique. For a few samples this was done, but the results for the sorption of iodide ions were inconclusive, as the value of ε sometimes exceeded that of α. At this stage therefore, it can only be stated that ρK for iodide ions is very small and probably less than 10^{-2}.

For the other two Scottish granites (lowland and Skene complex) the measurements obtained so far indicate they have similar, but slightly lower values of D_i and α. However, a much more marked difference exists between the Scottish and Cornish (Carnmenellis) granites. For the Cornish granite D_i and α are factors of 10 and 3 lower respectively. This difference is not altogether surprising since the conditions of formation of the two types of granite were different[4]. The presence of degradation products in the pores could account for lower values of D_i and α.

In figure 2 three curves are shown. The full curve was constructed using (6) and the values of D_i and α obtained from the slope and intercept of the experimental curve. The broken curves show the effect of varying D_i by 20% above and below the experimentally determined value. Figure 3 gives the corresponding curves with α varied by ± 20%. The differences arising from sample to sample variation are clearly greater than the precision of the experimental technique (Table 1).

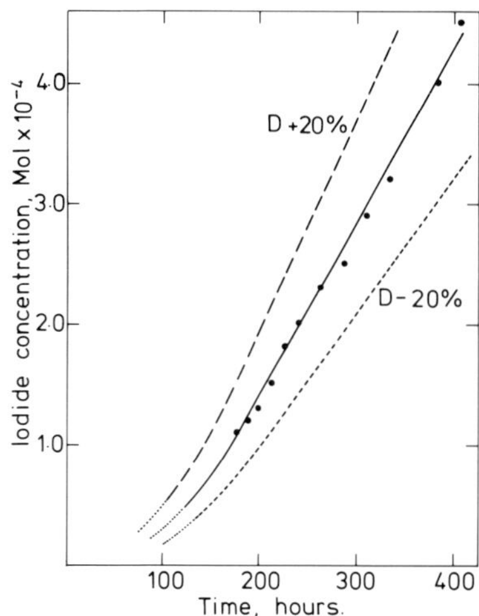

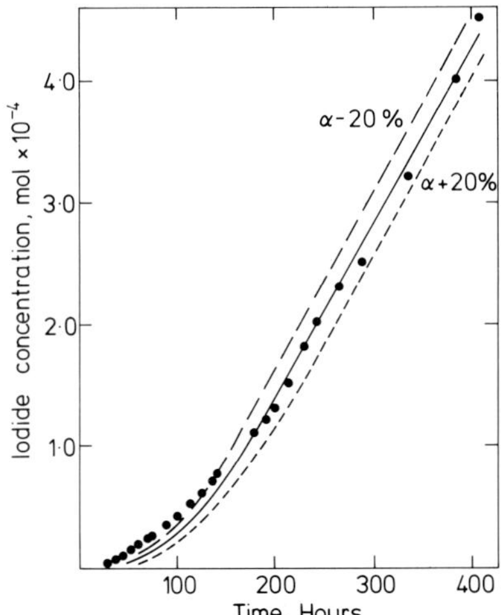

Fig. 2. Comparison of experimental results with theoretical curves of constant α with D_i varied by ± 20%.

Fig. 3. Comparison of experimental results with theoretical curves of constant D_i with α varied by ± 20%.

This type of measurement provides a good method for obtaining K values for weakly sorbed species ($K < 1$ ℓ kg^{-1}), which would otherwise be difficult to determine. As an example, figure 4 gives some preliminary results for diffusion in Ossian granite of technetium as the pertechnetate ion. This gives

$D_i = 7 \times 10^{-13}$ $m^2 s^{-1}$ and $\alpha = 0.156$. Taking a porosity value $\varepsilon = 7 \times 10^{-3}$ yields a K for pertechnetate of 0.06 ℓ kg^{-1}.

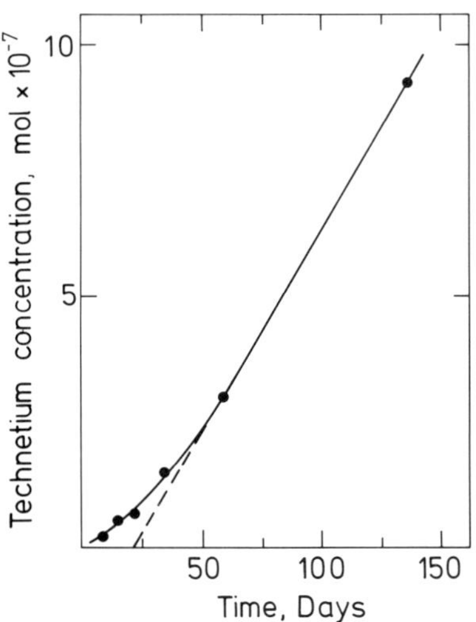

Fig. 4. Technetium, as the pertechnetate ion, diffusion in Ossian granite.

5. DISCUSSION

Intrinsic diffusion coefficients for other species

Experimental plots like those in figure 1 allow the intrinsic diffusion coefficient D_i and rock capacity factor α to be calculated for given species. It is important to note that D_i is derived independently of α as it is found from the steady state diffusion through the sample. The diffusion coefficients for ions in free water D_o may be measured, or derived from conductivity data[5]. D_o is related to D_i by the relation

$$D_i = D_o \, \psi \qquad \qquad \dots (7)$$

where ψ, the diffusibility, is a property of the porous medium, depending on the pore structure rather than the nature of the diffusing species. Diffusibilities are given in Table 1 based on D_o (Iodide) $= 2 \times 10^{-9}$ $m^2 s^{-1}$.

Knowledge of the diffusibility for a particular granite type allows D_i to be calculated for other species from their values of D_o, provided there are no "size factors" influencing diffusion, as could well be the case for colloids. With this condition in mind, it may nevertheless be instructive to give a first estimate of the intrinsic diffusion coefficient for colloids in granite. For colloids of radius 50 nm $D_O \sim 5 \times 10^{-12}$ m^2s^{-1}[6], so if $\psi \sim 5 \times 10^{-4}$, a value of D_i (colloid) $\sim 2.5 \times 10^{-15}$ m^2s^{-1} is obtained.

<u>Measurement of sorption coefficients for weakly sorbed species</u>

It is very difficult to measure by standard techniques the sorption coefficients for weakly sorbed species ($K < 1$ ℓ kg^{-1}), as small changes in concentration must be measured and the presence of even low concentrations of a different strongly sorbed form of the same nuclide would interfere. For example, Tc(IV) is strongly sorbed by many rocks, whereas Tc(VII) is very weakly sorbed. Oxidation or reduction from one valency state to the other can occur readily, so that Tc(IV) commonly interferes with sorption measurements on Tc(VII).

Measurements of α by the diffusion method is however unambiguous. For example, for Tc(VII) the presence of Tc(IV) as a small fraction of the Tc(VII) concentration affects only the accuracy with which the reservoir concentration is known, as only the more mobile Tc(VII) will reach the measurement cell within the time scale of the test.

It should be noted that although an independent measurement of the porosity is needed to estimate values of K, it is the directly measured value of α which is the parameter required in nuclide migration calculations.

<u>Granite capacity factor, α</u>

α is a measure of the capacity of the rock for the diffusing species. It is the sum of a physical contribution, the porosity ε, and a chemical contribution ρK. In the diffusion experiment the value of α derived from the intercept is a steady state value; however, in the transient region in figure 5 the theoretical curve, calculated from the experimental D_i and α values, lies significantly below the experimental points. The effect of varying α at constant D_i is seen in figure 3, where it can be seen that a decrease in α results in an increase in concentration. If the discrepancy shown in figure 5 is associated with α, then two explanations may be considered.

<u>K is a function of concentration</u>. In the transient region the concentration in the pore water across the granite is everywhere less than the concentration established during steady state diffusion, except at the reservoir surface, where it is fixed. Measurements of sorption as a function of concentration show that K decreases with increasing concentration[7]. Therefore, in the transient region a higher value of K, and thus of α, is expected. However, an increase in α results in a decrease in C (figure 3) and so this

576

explanation does not agree with the observations in figure 5.

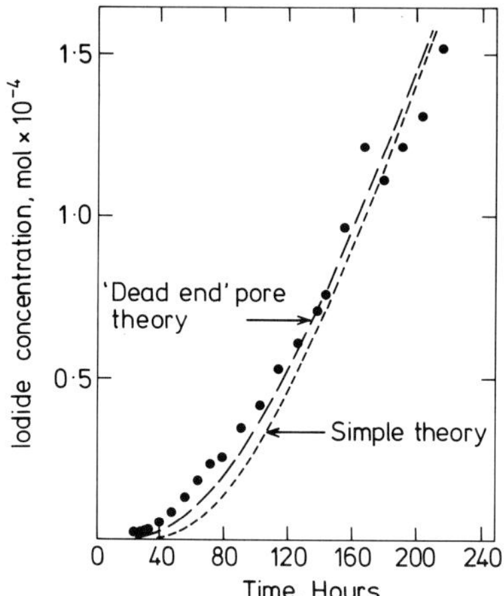

Fig. 5. A comparison of experimental results and theoretical curves in the transient region.

Interconnecting and dead-end pores. In real materials the porosity ε is made up of two contributions: one from readily accessible "transport" pores and the other from less accessible "interconnecting" and "dead-end" pores. Equilibrium between the concentration in the two sets of pores will only be reached after some time. In the short timescale before equilibration, in the transient region, there is a lower effective porosity but the porosity then apparently increases to its full value in steady state diffusion. The effect of a lower effective porosity, i.e. lower α, leads to an earlier breakthrough of the diffusion front and a concentration above that predicted using the steady state α value. This explanation is consistent with the observations.

These ideas can be incorporated into a simple mathematical model. The rock pores are divided into the two types: the transport pores, with capacity α^+ and concentration C^+, and the less accessible and dead-end pores with capacity α^* and average concentration C^*. The equation for diffusion through the transport pores is

$$\alpha^+ \frac{\partial C^+}{\partial t} = D_i \frac{\partial^2 C^+}{\partial x^2} - \alpha^* \frac{\partial C^*}{\partial t} \qquad \ldots \ (8)$$

The last term represents the transfer from transport to dead-end pores. This is given by

$$\frac{\partial C^*}{\partial t} = k(C^+ - C^*) \qquad \ldots (9)$$

The transfer coefficient k is approximately given by constant$.\frac{D_p}{\zeta^2}$, where $D_p = \frac{D_i}{E}$ and is a typical dead-end pore diffusion length. k^{-1} is a time scale, and is the relaxation time for equilibrium between the two sets of pores. Models like this have been used to describe transport by flow through porous media[8], whereas here it describes transport by molecular diffusion. For large times or large k, $C^+ \cong C^*$ and (8) reduces to

$$(\alpha^+ + \alpha^*) \frac{\partial C^+}{\partial t} = D_i \frac{\partial C^+}{\partial x^2} \qquad \ldots (10)$$

which is the same as (4) with $\alpha = \alpha^* + \alpha^+$.

Equations (8) and (9) can be solved in a similar manner to the original equation (4). It is found that steady state diffusion again yields D_i and α $(= \alpha^* + \alpha^+)$. Two adjustable parameters, α^*/α^+ and k^{-1}, can be varied to improve the fit in the transient region. A comparison of the results from experiment, the simple theory (6), and from the theory described here is given in figure 5. The values used are $\alpha^*/\alpha^+ = 1.0$ and $k^{-1} = 15$ hours, corresponding to a dead-end pore diffusion length $\zeta = 0.5 - 0.6$ cm. Though the fit to the experimental points in the transient region is still not particularly good, it does represent an improvement on the simple theory and can certainly be further ameliorated. A more detailed presentation of the model and additional results will be given in a later paper[9].

ACKNOWLEDGEMENTS

This work has been commissioned by the Department of the Environment, as part of its radioactive waste management research programme. The results will be used in the formulation of Government policy, but at this stage they do not necessarily represent Government policy. This work was supported by the European Atomic Energy Community in the framework of its R&D programme on Management and Storage of Radioactive Waste.

The authors would like to thank Dr S.J. Hemingway for permission to use some results of his computational work in figure 5.

578

REFERENCES

1.   Glueckauf, E. (1980) Harwell Report AERE-R 9823.

2.   Neretnieks, I. (1980) J. Geophys. Res., 85, 4379.

3.   Carslaw, H.S. and Jaeger, J.C. (1959) Conduction of Heat in Solids, Oxford.

4.   Bromley, A. Private communication.

5.   Harned, H.S. and Owen, B.B. (1958) The Physical Chemistry of Electrolytic Solutions, 3rd ed., Rheinhold.

6.   Thomason, H.P., Avery, R.G. and Ramsay, J.D.F. (1982) Harwell Report AERE-R 10479.

7.   McKinley, I.G. and West, J.M. (1981) IGS-NERC Report ENPU 81.

8.   Coats, K.H. and Smith, B.D. (1964), Soc. Pet. Eng. J. Trans. AIME, 231, 73.

9.   Bradbury, M.H., Hemingway, S.J. and Lever, D.A. To be published.

PERMEABILITY MONITORING TECHNIQUE OF THE NEAR-FIELD AND FAR-FIELD INTERFACE OF UNDERGROUND OPENINGS

ALEXANDER T. JAKUBICK[†] AND VICTOR DE KOROMPAY[††]
[†]Ontario Hydro, 800 Kipling Ave., Toronto, M8Z 5S4; [††]119 Angell Ave.,
Beaconsfield, Québec, H9W 4V6, Canada

INTRODUCTION

In July/August 1981 and February/March 1982 we sampled water for isotope analysis from water-bearing fractures of two mines in Elliot Lake. Permission for sampling was given to Ontario Hydro by Denison Mines Ltd and Rio Algom Ltd, Quirke Mine in conjunction with the CANMET Mine Safety Tests.[8] The purpose of the investigation was to find out whether there is a relationship between the water present in some fractures in a deep mine and the shallow groundwater in the far-field of the mine. The investigation was done in co-operation with University of Waterloo.

The results of the analyses for ^{18}O and ^2H contents are expressed as the $\delta‰$ notation and refer to the standard SMOW* (Figure 1). The ^{18}O and deuterium contents are in the same range as estimated for the Sudbury area by P. Fritz.[1] Compared to the global meteoric water line, the samples taken in the Denison Mine and Rio Algom Mine have an isotope composition close to precipitation and shallow groundwater. They have, however, lower ^{18}O and ^2H content than the surface water samples of the area. Also the mine water is slightly to the left of the meteoric waterline. Thus it could be speculated that the water is a brackish, high ^2H content, deep groundwater which was highly diluted by the local meteoric water.[2] The high salinity of mine water (see Table 1) supports this theory. As the isotope composition of the water in the near excavation zone changed with time it was assumed that this is the place where the mixing occurs.

Therefore, the conclusion can be made that shallow groundwater is entering the deeper workings of the mine. A hydrogeologically possible pathway is through the Quirke Lake overthrust which intersects Denison Mine between shaft No. 1 and No. 2. Below the fault the water is probably conducted by local fracture systems where mixing with deep groundwater occurs.

---

* SMOW = Standard Mean Ocean Water

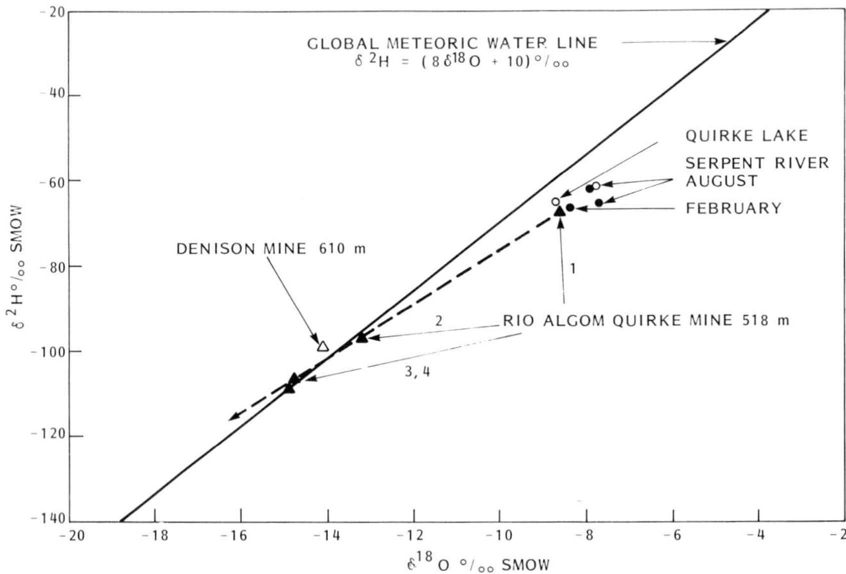

FIG. 1 – COMPARISON OF $\delta^{18}O$ AND $\delta^2 H$ VALUES OF WATER SAMPLES FROM SURFACE WATERS LOCATED ABOVE TWO MINES IN ELLIOT LAKE AND FROM WATER-BEARING FRACTURES IN THESE MINES. 1,2,3,4 INDICATE THE SAMPLING TIME AFTER THE FRACTURE WAS HIT (12, 24, 36 AND 108 HOURS, RESP.)

TABLE 1

COMPARISON OF CHEMICAL COMPOSITION OF SURFACE WATER (SERPENT RIVER) AND MINE WATER (RIO ALGOM QUIRKE MINE) IN ELLIOT LAKE

Constituent	Surface Water ppm	Mine Water ppm
pH	9.3	6.7
alk-OH as $CaCO_3$	15	<5
alk-HCO_3 as $CaCO_3$	34	22
Cl^-	42	1500
NO_3^-	160	17
SO_4^{2-}	510	<0.2
Ca^{2+}	208	554
Mg^{2+}	6.2	84
Na^+	23	515
K^+	51	13

Radiogeochemical experiments in laboratory showed that the sorption of ^{239}Pu and ^{241}Am on high quartz content rock surfaces is low (Figure 2)[3]. From the combination of these findings we concluded that the near-vault fractures may become an important factor in migration between the near-field and the far-field of geologic repository as the mixing processes there will strongly effect the radionuclide source term. It is for this reason that we need a method to assess the permeability of the near-excavation zone.

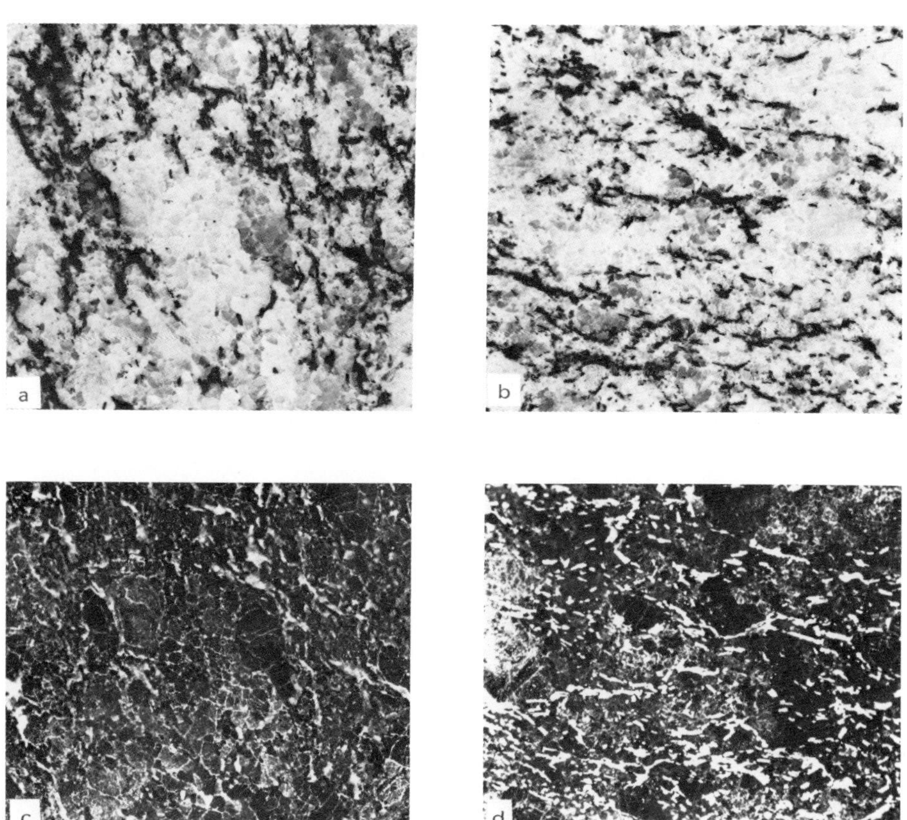

Fig. 2. Selective sorption of ^{239}Pu and ^{241}Am, resp. on high quartz content surfaces (granite from Tengboche, Nepal; Composition: quartz, feldspar, biotite). (a) and (b) original surfaces, (c) ^{239}Pu autoradiograph, (d) ^{241}Am autoradiograph. After 96 hours of contamination and subsequent decontamination only the biotite retained some of the transuranium nuclides.

HIGH RESOLUTION FRACTURE MONITORING

Testing Technique. The measurement of permeability of the tested media to a vacuum-induced air flow is the basis of the method.[4] The advantage of use of vacuum are the safety of operation (unlike injection, evacuation causes no pressure buildup in the roof of undergound openings) and the convenience of vacuum measurement compared to measurement of pressure. Figure 3 shows the testing assembly for 38.1 mm (1-1/2 inch) and 76.2 mm (3 inch) holes. Diamond drilled or percussion holes are used for the test. The section to be tested is isolated by a mechanical straddle packer and evacuated through a flow line. The discharged air is lead through a flow meter assembly. The rock face pressure and temperature are measured at the same time. The sealing performance of the packers can be checked by remeasuring the test section after alternating packer release and inflation.

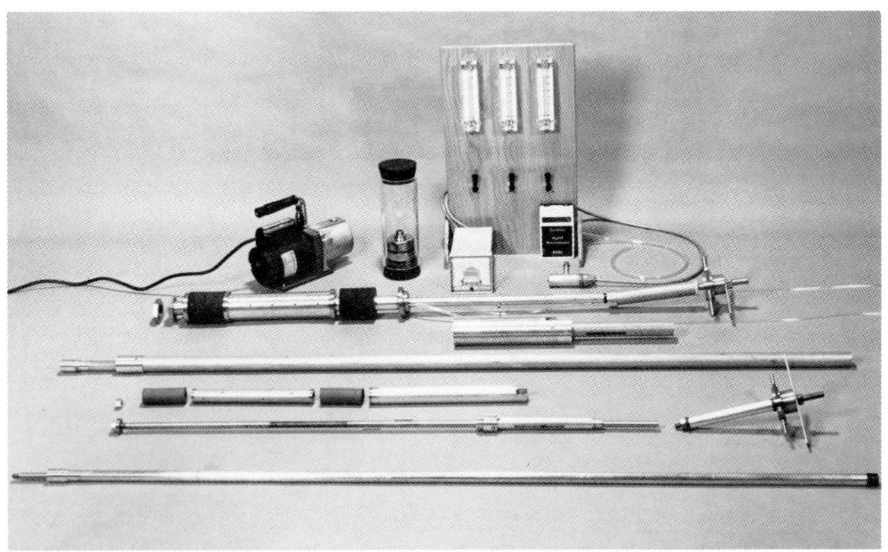

Fig. 3. The vacuum rock testing device consisting of 38.1 mm (1-1/2 inch) and 76.2 mm (3 inch) mechanical straddle packer assembly, setting tubes with inserted setting rods, setting tools, vacuum pump, pressure and temperature gauges and flowmeters.

The evacuated holes in the rock provide only information about the two ratios p/t and p/Q where p, Q, t are the pressure, air flow and time. There are two ways in which the ratios p/t and p/Q may be determined:

(1) monitoring the pressure recovery after evacuation $Q = \Delta p V/t$

(2) pressure/flow equilibration $Q = Q_p(p_a - p_p)$,

where Q is the flow due to the rock leakage, Δp the pressure change, V the volume of the test section, Q_p the effective pumping rate, p_a the achievable vacuum, p_p the end pressure of the pump. Low absolute pressure measurements that are characteristic for dense matrix or tight fractures produce as their final output only the values p_a and p/t, and no measurable flow. The permeability can be obtained independent of the flow rate if the pressure recovery is recorded. The value so obtained is not affected by boundary or flow conditions of the fracture. In case of flowing fractures, where flow is measured directly, the permeability can be derived from the flow equations when assumptions are made about the extent of the air drainage area. However, this pressure/flow test analysis may be incorrect because of erroneous boundary conditions.

When simultaneous flow of air and water occurs $k = k_{abs}K_{rw}(S_a)$ in which $k_{rw}(S_o)$ is the average relative permeability of water. This situation requires the additional measurement of the prevailing air-water ratio during the test and determination of the ratio k_{ra}/k_{rw} as a function of pressure.

<u>Test hole pressure response</u>. We use a partially linearized form of the flow equation for evaluation of the pressure recovery in the vacuum hole:

$$\frac{1}{r}\frac{\partial}{\partial r}\left(r\frac{\partial m(p)}{\partial r}\right) = \frac{n\mu c}{k}\frac{\partial m(p)}{\partial t} \tag{1}$$

where r is the radial distance, m(p) pseudo pressure, n the porosity, μ the viscosity and c the isothermal compressibility. The linearization of Equation 1 in respect to viscosity and compressibility of air by use of the Kirchhoff integral transformation was first suggested by Al-Hussainy et al[5]. This procedure is normally referred to as the pseudo-pressure solution technique. The reason for adapting this approach is because of its theoretical advantages. In using it one does not have to be concerned about the pressure range in which it is applicable, as is the case when using the pressure squared solution technique.[6,7,8] The special case of the solution applicable for pressure recovery analysis in explicit form is:

$$\frac{kh}{Q_1}\frac{1}{1422T}\Delta m(p_s) = \frac{1}{2}\ln\frac{t_1 + \Delta t}{\Delta t} + m_D(t_D) - \beta_D(t_D) \tag{2}$$

584

where h is the test section thickness, Q_1, the evacuation flow rate, T the absolute temperature, p_s static pressure, t_1 evacuation time, Δt pressure recovery time; $m_D(t_D)$ and $\beta_D(t_D)$ are constants which can be evaluated when the skin effect and the non-Darcy flow are of interest. We do not include them in our present analysis as they are not required for the evaluation of the pressure recovery for k.

The Horner plot[9] of $m(p_s)$ versus $\ln((t_1 + \Delta t)/\Delta t)$ for the recorded data is linear for intermediate values of Δt and the slope of this line is $m = 1422\ T\ Q_1/kh$ from which k can be calculated. This simplified approach of D.R. Horner was originally designed for liquids, however it can be applied to gas systems just as well.[10]

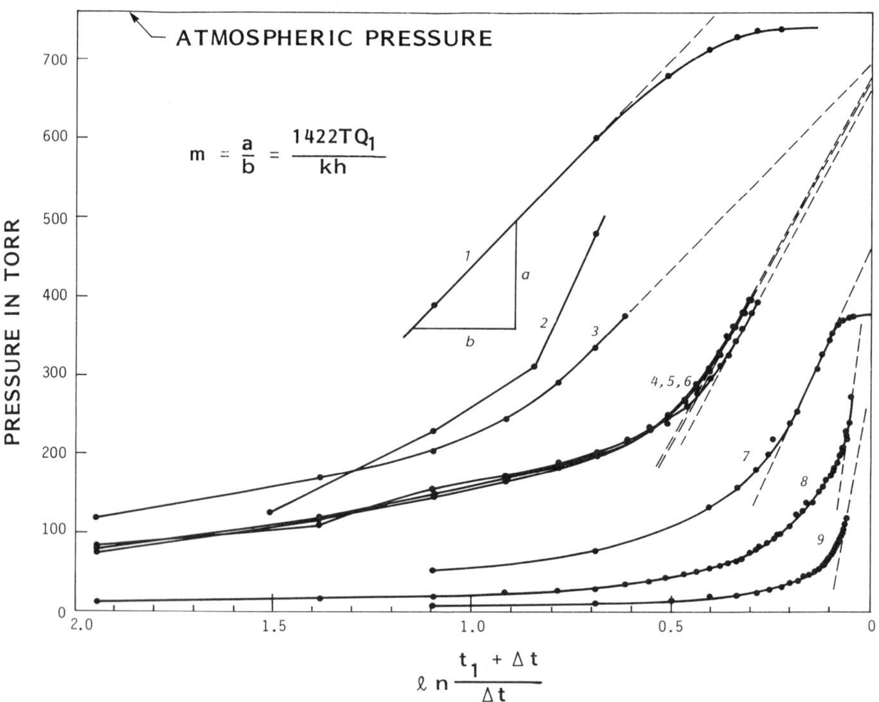

Fig. 4. Pressure response in a test hole after evacuation (Rio Algom mine 1700 feet). The curves are in order of decreasing permeabilities. t_1 is the evacuation time (t_1 = 60 s), Δt the pressure recovery time, T the absolute temperature, Q_1 the evacuation flow rate, k the permeability, h the test section thickness.

<u>Application</u>. In conjunction with the CANMET mine safety tests[11] in situ measurements were performed at 610 m and 518 m depths (Denison Mine and Rio Algom Mines, respectively). The tested formation was the ore-bearing quartzite of Elliot Lake. EX holes were drilled in the mine roof to a depth of 8.35 m. About 15 drillholes were tested. The testing was done in 25 cm sections. Up to the depth of 50 to 75 cm, a blast damaged zone of increased fracturation was encountered. Behind that zone the fracture spacing increased and only a few test sections dominated the permeability of the rock.

The Horner plot of pressure response curves from a borehole at the 518-m depth (Rio Algom, 7 level, test slope borehole No. 4) is in Figure 4. The intermediate section of the pressure recovery curves is, as predicted by equation (2), linear. At large values of Δt the pressure recovery follows the curved solid lines. The final vacuum is the average pressure within the bounded volume being drained. In the particular case of a continuous fracture, the vacuum will be completely lost and the normal atmospheric pressure regained (curve No 1). The curves which are asymptotic to a pressure value below the atmospheric level indicate finite fractures. If no fracture is present in the test section the initial vacuum achieved is higher (curve No. 9) than in the case of a discontinuous fracture. Curves No. 4, 5 and 6 demonstrate the repeatability of the test. Curve No. 9 is an illustrative example for an unfractured section of the rock.

Laboratory measurements of matrix permeability on cores gave 2 to 10 microdarcy for the tested rock. About the same values were obtained in situ by the vacuum hole method when tight zones were tested. The fracture permeabilities in the near-excavation zone were up to 10^5 times higher.

CONCLUSIONS

At this stage only rough preliminary estimates were tried. Currently we are working on a more accurate and dependable method of interpretation. Sets of curves are being estimated to show the effect of fracture discontinuities on the vacuum testing of rock; the transition from Darcian to non-Darcian flow is being tested. Master curves will serve, in the future, the comparison of field measurements with characteristic curves either for various values of k or for various $\bar{p}$. Even with the presented preliminary results the vacuum testing is a valid technique for assessing permeability of the near excavation zone.

Of complementary importance to the physical testing of the near-excavation zone is the evaluation of the origin of groundwater in water-bearing fractures intersecting the excavation which can only be achieved by isotope hydrogeochemical means.

Other areas of use of the vacuum hole technique are: Fracture detection for effective rock bolting; Evaluation of rock blast damage; Measurement of grouting effectiveness.

ACKNOWLEDGEMENTS

We would like to thank D. Hedley of CANMET Elliot Lake Laboratory for giving us the opportunity to participate in the mine safety tests, J.W. Roxburgh, N. Whitton and M. Bates of Rio Algom Ltd. for supporting the work, and A. Rickaby of Denison Mines Ltd. for providing access to the underground workings. We would also like to thank P. Fritz of University of Waterloo for the isotope analyses of water samples. Technical assistance throughout the measurements was given by R. Klein, Ontario Hydro. At Ontario Hydro the work was supported by the Research and Design and Development Divisions, as a part of the technical assistance program to AECL.

REFERENCES

1. Fritz, P., Reardon, E.J. (1979) Isotope and Chemical Characteristics of Mine Water in Sudbury Area, AECL TR-35.
2. Frape, S.K., Fritz, P. (1982) The Chemistry and Isotopic Composition of Saline Groundwater from the Sudbury Basin, Ontario. Can. Journal of Earth Sciences, in press.
3. Jakubick, A.T., Kahl, I. (1981) Plutonium and Americium Sorption in Sandy Soils and Granite. IAEA Coordinated Research Programme Meeting, Keble College, Oxford.
4. DeKorompay, V. (1980) A Vacuum Hole Monitoring System to Detect Fractures in Mine Roof. Div. Rep. Can. Met., Canada, MRP/MRL 80-87 (TR).
5. Al-Hussainy, R., Ramey, H.J., Jr. and Crawford, P.B. (1966) The Flow of Real Gases Through Porous Media. J. Pet. Tech., Trans. AIME, May, 624-636.
6. Dake, L.P. (1978) Fundamentals of Reservoir Engineering. Elsevier, Amsterdam.
7. Dumoré, J.M. (1974) Drainage Capillary Pressure Functions And Their Computation From One Another. Soc. Pet. Eng. J., October, 440.
8. Coats, K.H., Dempsey, J.R., Henderson, J.H. (1971). The Use of Vertical Equilibrium in Two Dimensional Simulation of Three Dimensional Reservoir Performance. Soc. Pet. Eng. J., March, 63-71.
9. Horner, D.R. (1951) Pressure Build Up in Wells. Proc. Third World Petroleum Congress, Brill, E.J. Ed., Leiden, 11, 503.
10. Tracy, G.W. (1956) Why Gas Wells Have Low Permeability, Oil & Gas Journal, August 6.
11. Hedley, D. (1982) Cooperative Mine Safety Research Program. Mining Research Laboratory, Canada Centre for Mineral and Energy Technology (CANMET), Elliot Lake.

Published 1982 by Elsevier Science Publishing Co
SCIENTIFIC BASIS FOR RADIOACTIVE WASTE MANAGEMENT – V
Werner.Lutze, editor

HYDROTHERMAL CONDITIONS AROUND A RADIOACTIVE WASTE REPOSITORY

ROGER THUNVIK[+] AND CAROL BRAESTER[++]
[+] Royal Institute of Technology, S-100 44 Stockholm, Sweden
[++] Israel Institute of Technology, Haifa 32000, Israel

INTRODUCTION

The possibility of permanent burial of radioactive waste from nuclear power plants, is studied in Sweden at the KBS (Nuclear Fuel Safety) – project. Definite repository sites have not yet been selected, but the general principles of construction regarding the layout have been devised (KBS[1]).

The feasibility of a prospective site for radioactive waste disposal is highly dependent on the geohydrological conditions. Heat emitted by the decaying waste will increase the temperature of the rock, changing groundwater density gradients and creating convective currents.

Under certain conditions water particles passing through the repository may reach the ground surface. It is therefore of significant interest in the safety analysis to predict pathlines and travel times of water particles, should any of the waste canisters be breached and the groundwater be contaminated.

The solutions presented illustrate the effect of heat released from a hypothetical repository on the groundwater movements around the repository.

THE FLOW MODEL

The prospective sites for radioactive waste repositories in Sweden are fractured hard rock formations. The fractured rock is conceptualized as a configuration of interconnected fractures surrounding the solid blocks.

It is assumed in the present investigation that the fractured rock formation in consideration can be treated by the continuum approach.

The water volume in the fractures is small in comparison with the solid rock volume. As a consequence, one may assume that thermal equilibrium between fluid and rock takes place instantaneously.

The governing equations for simultaneous flow of fluid and heat, derived from the basic conservation laws and Darcy´s law, are the equation for the conservation of fluid mass:

$$\phi\rho^f(c^f+c^r)p_{,t} - \phi\rho^f\beta T_{,t} - (\rho^f\frac{k}{\mu}ij\,(p_{,j} - \rho^f g_j))_{,i} = 0 \tag{1}$$

588

and the equation for the conservation of thermal energy:

$$((\rho C)^* T)_{,t} - (\lambda^* T_{,i})_{,i} + (\rho^f c^f (\tfrac{k}{\mu}ij \, (p_{,j} - \rho^f g_j))T)_{,i} = 0 \tag{2}$$

where $*$ denotes the equivalent properties of the composite, solid and fluid.

Water density and viscosity are considered functions of pressure and temperature through the equations of state:

$$\rho^f = \rho^f(p,T), \quad \mu^f = \mu^f(p,T) \tag{3}$$

Eqs. (1) to (3) form a system of coupled non-linear partial differential
equations.

METHOD OF SOLUTION

The equations are solved with the appropriate boundary conditions by the
Galerkin finite element method. The flow domain is discretized in a graded size
mesh of eight node quadrilateral elements.

According to the Galerkin method, one introduces the trial functions

$$p \cong \Psi_J p_J , \quad T \cong \Psi_J T_J \tag{4}$$

where $\Psi_J = \Psi_J(x_i)$ are basis functions, chosen to satisfy the essential boundary
conditions. The same basis functions are used to represent the variations in the
material properties over the elements.

Making use of the orthogonality conditions in Galerkin's method, one obtains
for the fluid mass conservation equation

$$<\phi\rho^f(c^f+c^r)p_{,t},\Psi_I> - <\phi\rho^f\beta T_{,t},\Psi_I> - <\rho^f\tfrac{k}{\mu}ij \, (p_{,j}-\rho^f g_j))_{,i},\Psi_I> = 0 \tag{5}$$

and for the thermal energy conservation equation

$$<((\rho C)^* T)_{,t},\Psi_I> - <(\lambda^* T_{,i})_{,i},\Psi_I> + <(\rho^f c^f (\tfrac{k}{\mu}ij \, (p_{,j}-\rho^f g_j))T)_{,i},\Psi_I> = 0 \tag{6}$$

Applying Green's theorem to all second order terms and using the finite difference approach to the time derivatives, one obtains a non-linear system of
algebraic equations, which is solved by an iterative procedure. For more details
regarding the method of solution as well as the flow model, the reader is directed to the report by Thunvik and Braester[2].

THE FLOW DOMAIN, BOUNDARY AND INITIAL CONDITIONS

The study area is a vertical cross-section of 3 km. lateral extent and 1.5 km. thickness. It is perpendicular to the longitudinal axis of the repository tunnels. The repository of 1 km. lateral extent is located at 500 m. depth below the ground surface.

The following three cases are investigated:

1) a repository located below a horizontal ground surface,

2) a repository located below the crest of a hill and

3) a repository located below a hillside

Because of the abundant precipitation in Sweden, the water table is usually located close to the ground surface level. A fairly conservative assumption is that the water table and the ground surface coincide. Seasonal variations in temperature are neglected. Thus, in the present investigation the upper boundary is considered to be an isobaric and an isothermal boundary.

Below a certain depth, the location of an impervious bottom has, in general, little influence on the groundwater flow. In this investigation an impervious bottom is located at 1500 m. below the ground surface. Constant temperature is prescribed at the bottom boundary according to the geothermal gradient.

The vertical sides of the flow domain are considered impervious to fluid and adiabatic.

For the water, the initial condition is the flow imposed by the slope of the upper boundary. For the heat, the initial condition is the natural geothermal gradient of 30°C/km.

It is assumed that the repository is entirely loaded at the same time.

INPUT PARAMETER VALUES

The parameter values used in the calculations are presented in table 1.

Table 1. Parameter values used in the calculations

Porosity	0.003	Fluid thermal	
Permeability	$10^{-14}\,\mathrm{m}^2$	conductivity	0.6 W/(mK)
Rock compressibility	$10^{-11}\,\mathrm{Pa}^{-1}$	Fluid specific heat	4180 J/(kgK)
Fluid density	998 kg/m^3	Rock density	2700 kg/m^3
Dynamic viscosity	0.001 Pas	Rock thermal	
Fluid compressibility	$10^{-10}\,\mathrm{Pa}^{-1}$	conductivity	3.5 W/(mK)
Fluid thermal volume		Rock specific heat	800 J/(kgK)
expansion	$0.0018\,\mathrm{K}^{-1}$	Thermal load	5.25 W/m^2

The values of fluid density and viscosity are given as reference values corresponding to a temperature of 20°C. The thermal load is assumed to be generated by 40 year old high level waste and the given value corresponds to a uniformly distributed load over the area of the repository.

The decay of the thermal load is considered to be:

$$Q(t)/Q(0) = 0.882 \exp(-7.327\text{x}10^{-10}t) + 0.118 \exp(-4.396\text{x}10^{-11}t) \tag{7}$$

Part of the previously given values are the result of field investigations at some test sites (Jeffry et al.[3], Carlsson et al.[4], Ekman et al.[5], Hult et al.[6]).

RESULTS

The results of the calculations, pathlines, flow times of water particles, groundwater fluxes and isotherms are displayed graphically in figs. 1 to 3.

Groundwater flow patterns without heat emission have also been calculated (Thunvik and Braester[2]). A comparison with the results presented here shows that the heat released from the hypothetical radioactive waste repository modifies the initial fluxes and subsequently also the flow times of the water particles.

The temperature reaches its maximum after about 50 years but the heat released from the repository continue to exert influence on the flow field for many thousands of years.

In case 1, the groundwater is initially at hydrostatic equilibrium. The pressure gradient and the geothermal gradient are parallel. This means, that in this case the groundwater movements are induced only by the heat released from the repository. As a result of the heat emission, one obtains a flow pattern characterized by a region of upward movements at the centre of the repository, rotational movements at the edge and downward movements at some distance from the repository (fig. 1).

Water particles starting from the centre of the repository reach the ground surface after approximately 650 years.

The calculations have been carried out for a period of about 1900 years. The convection currents being created at the edge of the repository will actually vanish within the period subject to the pathline trace.

In case 2, the repository is located symmetrically below the crest of the hill. Initially, the groundwater flow is governed by the slope of the water table. Two different values of the slope of the hillsides, 1/1000 and 1/100, are considered.

In the case of a slope of 1/1000, the effect of the heat is to create convection currents around the edges of the repository. As a result, the flow

times become longer for water particles being traced from the edges of the repository, but considerably shorter for the particles being traced from the
centre of the repository. The shortest exit time is that of the water particles
being traced from the centre of the repository and it is about 3800 years
(fig. 2). Without heat release, the shortest exit time is that of the particles
being traced from the edge of the repository and it is only 2550 years.

In the case of a slope of 1/100, the heat has a minor influence on the flow
times, and the dominating factor is the groundwater flow induced by the slope
of the water table. Generally, the flow times from the repository were prolonged by about 30 per cent with the effect of the heat from the repository.
The shortest exit times are those of the particles passing through the edges of
the repository: 290 years with heat emission, and 250 years without.

Because the repository is situated below an inflow area, this case represents, in general, a desirable location of a radioactive waste repository.

In case 3, with a hillside of 1/1000, the groundwater flow pattern is significantly modified by the heat, particularly in the region close to the repository where the groundwater is pushed up. This reduces the exit time from about
900 years without heat emission to about 625 years. Without heat emission, the
shortest exit time is that of the particles starting from the right edge of the
repository, while with it the shortest exit time is that of the particles
starting from a point at a distance of 300 m. to the left of the centre of the
repository (fig. 3).

These examples show that heat released from a radioactive waste repository
may have a significant impact on the flow regime around the repository.
Although the hypothetical repository in the presented examples causes a very
moderate increase in the temperature of the surrounding rock, it significantly
affects the flow patterns and consequently the flow times for water particles
from the repository to the ground surface. It may also be concluded from the
present investigation, that heat released from a radioactive waste repository
may increase as well as decrease the travel times, depending on the conditions
in and around the repository.

ACKNOWLEDGEMENT

The investigation reported here was carried out in the course of the
research sponsored by the Swedish KBS - Nuclear Fuel Safety - project.

592

NOMENCLATURE

Symbol   Description

c   compressibility
C   specific heat capacity
g   acceleration of gravity
k   permeability
p   pressure
Q   heat flow rate
t   time
T   temperature
x_i Cartesian coordinate

β coefficient of thermal volume expansion of the fluid
λ thermal conductivity
μ dynamic viscosity
ϕ porosity
Ψ basis function

$$\langle f(x), g(x) \rangle = \int_V f(x)g(x)\, dv, \text{ inner product}$$

superscripts

f   fluid
r   rock
*   equivalent medium

subscripts

i,j   indices used for Cartesian tensor notation,
         repeated indices indicate summation over
         these indices (i=j=1,2,3)

$p_{,t}$ partial time derivative of p

$p_{,j}$ gradient of p

I   node index

REFERENCES

1. KBS (1978), Handling and Final Storage of Unreprocessed Spent
    Nuclear Fuel, Volume II.

2. Thunvik, R., and Braester, C., (1980), Hydrothermal Conditions around
    a Radioactive Waste Repository, SKBF-KBS-TR-80-19.

3. Jeffry, J.A., Chan, T., N.G.W., and Witherspoon, P.A., (1979),
    Determination of In-situ Thermal Properties of Stripa Granite,
    LBL-8423, University of California.

4. Carlsson, L., et al., (1980), Kompletterande permeabilitetsmätningar
    i Finnsjöområdet, SKBF-TR-80-10 (in Swedish).

5. Ekman, L., and Gentzschen, B., (1980), Komplettering och sammanfatting
    av geohydrologiska undersökningar inom Sternöområdet, Karlshamn,
    SKBF-KBS-TR-80-01 (in Swedish).

6. Hult, A., Gidlund, G., and Thoregren, U., (1978), Permeabilitetsmätningar,
    KBS-TR-61 (in Swedish).

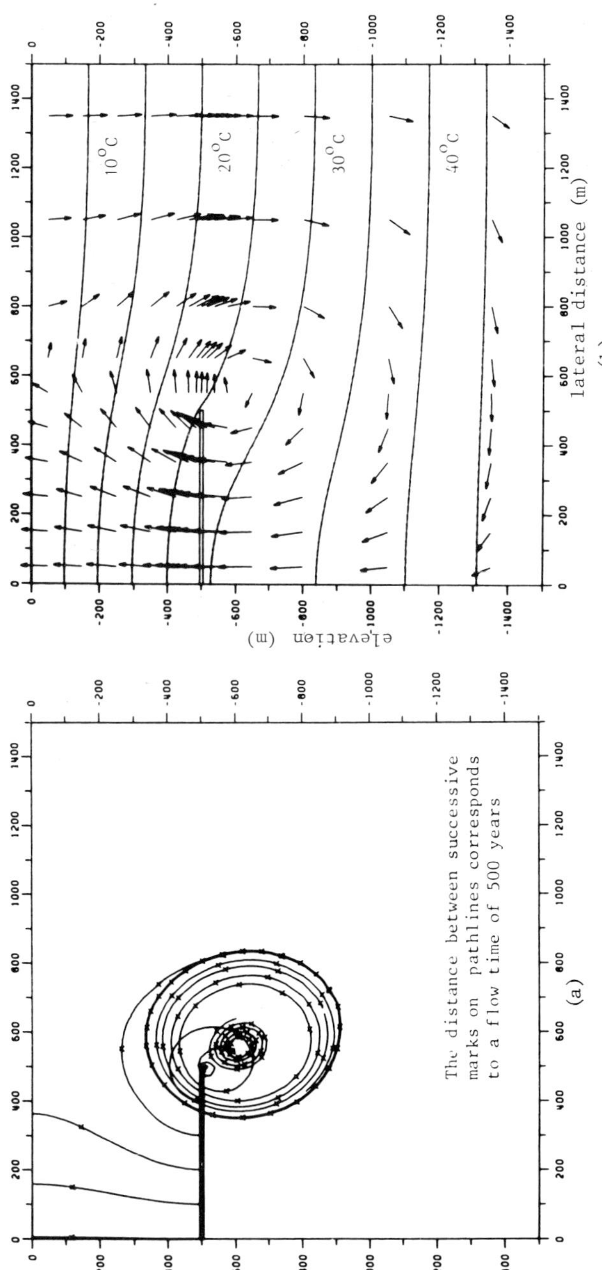

Figure 1. Graphical display of the results of the calculations for a repository located below a horizontal ground surface (a) pathlines and flow times (b) groundwater fluxes and isotherms after 1900 years heat release.

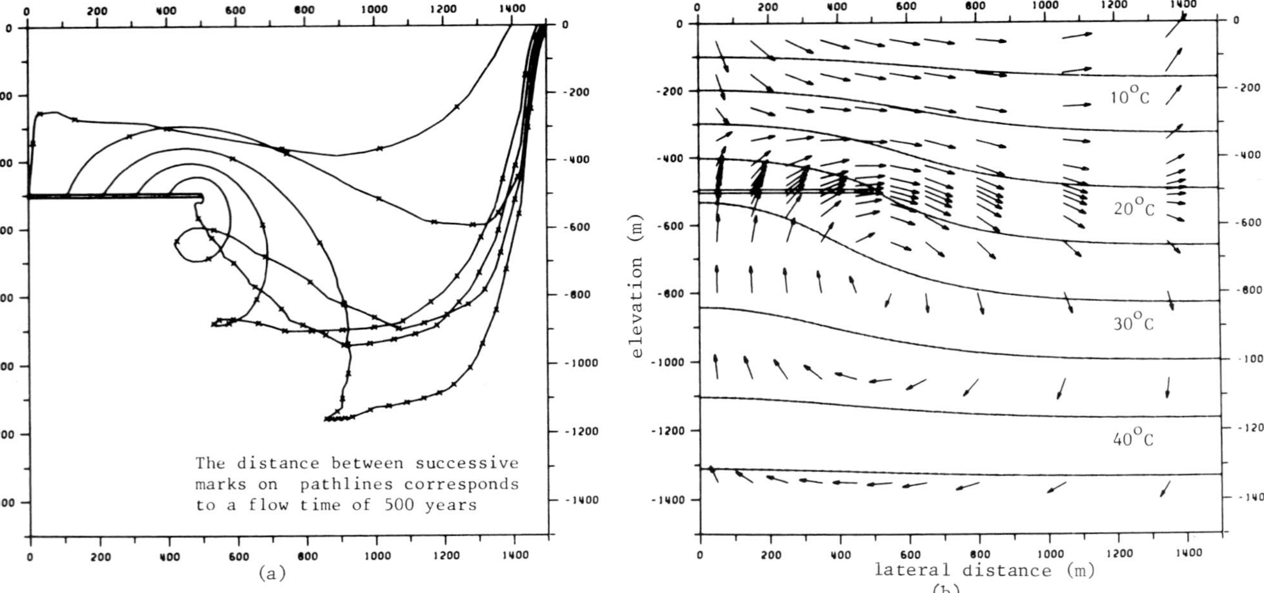

Figure 2. Graphical display of the results of the calculations for a repository located below the crest of a hill, hillside slope 1/1000, (a) pathlines and flow times (b) groundwater fluxes and isotherms after 1900 years heat release.

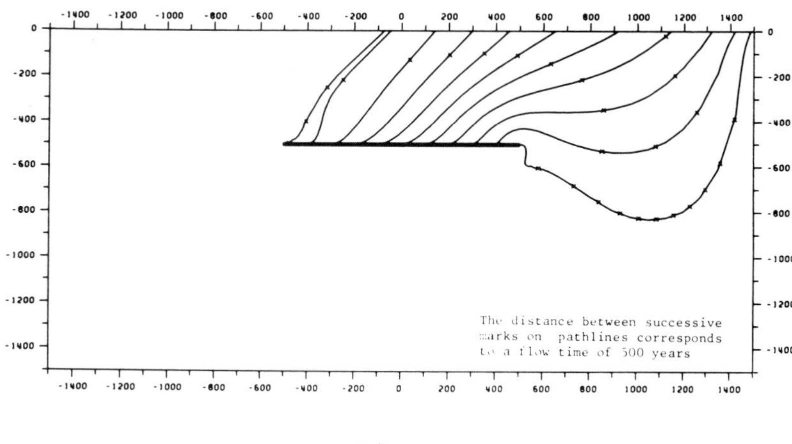

(a)

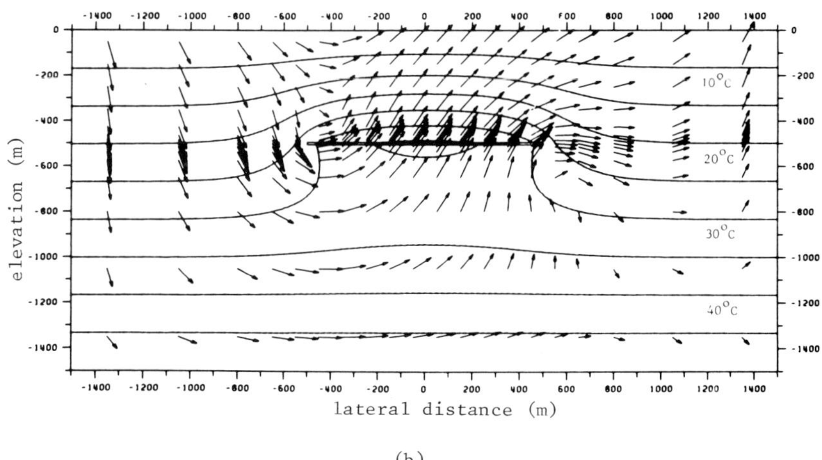

(b)

Figure 3. Graphical display of the results of the calculations for a repository
located below hillside with a slope of 1/1000 (a) pathlines and flow
times (b) groundwater fluxes and isotherms after 680 years heat
release.

Published 1982 by Elsevier Science Publishing Co
SCIENTIFIC BASIS FOR RADIOACTIVE WASTE MANAGEMENT- V
Werner.Lutze, editor

MODELING APPROACH TO DETERMINE SHORT- AND LONG-TERM THERMAL
AND THERMOMECHANICAL EFFECTS OF WASTE EMPLACEMENT IN A
REPOSITORY IN BASALT

TERRY F. LEHNHOFF*, K. THIRUMALAI[+], AND ALAN D. KRUG[+]
*University of Missouri - Rolla, Rolla, Missouri, 65401 USA;
[+]Basalt Waste Isolation Project, Rockwell Hanford Operations,
Richland, Washington, 99352 USA

INTRODUCTION

The Columbia River basalts, which underlie a large portion of the Pacific
Northwest of the United States of America, are being investigated as one of the
candidate media for a nuclear waste repository. The Basalt Waste Isolation
Project (BWIP)[1] of Rockwell Hanford Operations (Rockwell) is conducting these
investigations for the U.S. Department of Energy (DOE). Since the inception of
the program in 1976, a number of studies have led to the selection of a reference repository location and the start of construction of an exploratory
shaft.[1-3]

Development of reliable numerical-modeling procedures to predict the response of a geologic repository for nuclear waste emplacement is the key to
assessing its performance. The ability to predict the response of a repository to a desired level of accuracy depends largely on the answers to the
following key questions:

- Are adequate numerical techniques available to model the thermomechanical
  and hydrochemical behavior of basalt with a sufficient degree of accuracy?

- Are adequate methods and techniques available to measure appropriate rock
  and rock mass properties with an accuracy which is compatible with the
  numerical methods?

The performance of a repository in the United States will be assessed in
terms of the ability of the geologic repository to meet the safety criteria
established by the U.S. Nuclear Regulatory Commission (NRC), U.S. Environmental
Protection Agency, and DOE. The numerical-modeling techniques used to assess
the performance will provide guidance for the design of a repository. The
models developed to assess the repository performance should, therefore, be
capable of producing quantified predictions of short- and long-term repository
response. Integrated methods will be necessary to examine the response of

598

canister-, room-, repository-, and regional-scale environments. The range of
these scales and examples of the types of parameters required for the model
input are illustrated in Figure 1.

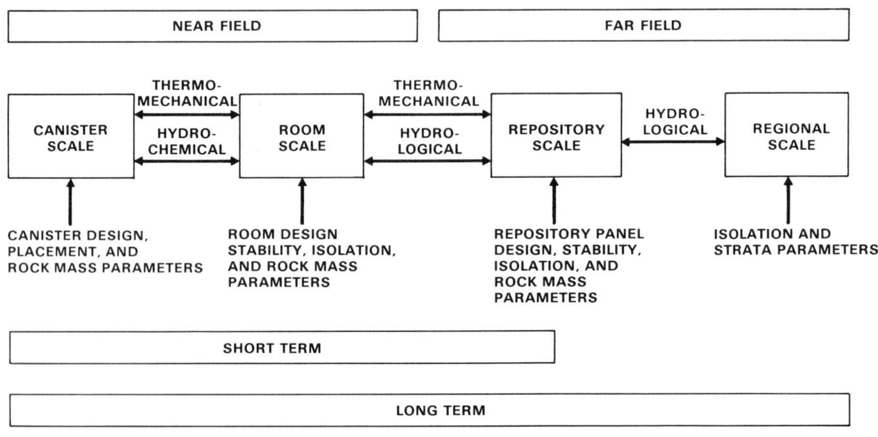

Fig. 1. Numerical-modeling scales and parameters.

Near-field studies relate to the time in the life of a repository when
canister- as well as room-scale analyses are of primary concern. The purpose
of these studies is to assess the performance of the design with regard to its
ability to safely receive and contain nuclear waste until the rooms are backfilled and sealed. A recent NRC rule requires, in effect, that repository
rooms must remain stable and accessible to facilitate retrieval of the canisters
from their placement holes during the 50 to 110 years of repository operation.

Far-field studies deal primarily with the period when the repository is
sealed and decommissioned until that time when the waste is no longer dangerous. Repository- and regional-scale analyses are of interest during this time
frame. The objective of these studies is to assess the ability of the repository to isolate nuclear waste in such a manner that radionuclides are effectively blocked from reaching the biosphere or until they are reduced to safe
levels. While hydrological concerns extend for many thousands of years, the
potential contribution to increased permeability due to thermal expansion and
thermal fracturing will peak in a much shorter time. Thus, the solid mechanics
studies of thermal expansion and thermal fracturing will be of interest in the
short term.

BASALT WASTE ISOLATION PROJECT APPROACH TO DEVELOP MODELS

The principal focus of BWIP-modeling investigations is in the basalts at the Hanford Site in the Pasco Basin near Richland, Washington. Basalt, in general, is characteristically a jointed and fractured rock. Preliminary field measurements to date, however, indicate that major portions of the deep basalt flows are highly impermeable to groundwater flow because of mineral infilling and large lithostatic pressure.[2] For near-field considerations, the intraflow structures in jointed basalt have a governing influence on the rock mass-property parameters and their response to the repository environment. The typical intraflow structures of a basalt flow are shown in Figure 2. The flow consists of: a ropy to brecciated vesicular flow top; an upper colonnade with large (0.7 to 2.2 m diameter) irregular columns; an entablature of small (0.2 to 0.9 m diameter) hackly to regular columns; and a lower colonnade of well-formed to irregular, large (0.5 to 1.5 m diameter) columns.

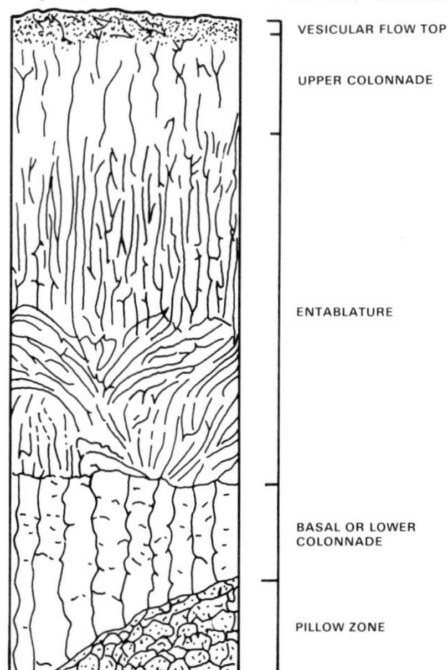

Fig. 2. Typical intraflow structures present in a Grande Ronde Basalt flow.

During the initial phases of the program when the rock properties as well as waste package configuration, room geometry, repository location, geometry, etc., could not be well defined, simplified mathematical models and less-complex

scoping-analysis procedures were used. These approximations provided a good
basis for engineering judgment and facilitated practical design alternatives.

Much of the early work was done with closed-form solutions or boundary-
element methods. These techniques are seen as the only reasonable and practical
approach to scoping studies. The large number of parameter variations necessary
for conceptual design of a repository preclude the initial application of elaborate and detailed finite-element methods. The thermomechanical analysis completed at the BWIP has progressed through much of the scoping phase and is now
entering the detailed analysis and design phase in some areas. Methods for detailed analysis are being demonstrated and many uncertainties are being clarified. Final repository-design studies will warrant the special effort necessary
to produce extensive finite-element analyses. The resulting finite-element models then permit analysis of design details and can expose various problem areas
for inspection and final evaluation. Nonlinear effects of all types can be
evaluated to determine if concerns exist as real problems. The detailed finite-
element modeling will contribute to the basis for making rational correct decisions that will result in a safe repository in basalt for storing nuclear waste.

NUMERICAL COMPUTER PROGRAMS AND MODELS

A list of some available computer programs and their capabilities is shown in
Table 1. Several of these programs are being applied on various aspects of the
BWIP analyses. Since there are several available programs, it has been, for the
most part, possible to select programs based on the preference of the individuals performing the analysis. These programs can then be verified by applying
them to problems with known solutions which have the essential elements of the
real problems to be solved. This should be obvious, since many of the larger
programs are capable of treating a broad spectrum of problems. They could,
conceivably, be functioning correctly in one area and not in another. Solutions
of the same problems using different programs can provide additional confirmation. Verification, while desirable and necessary, can never be a substitute
for good engineering judgment on the part of those doing the analysis.

In any large geotechnical project there will be problems that cannot be
handled by available computer programs. It will, occasionally, be necessary to
develop special-purpose programs to study correctly the impact of some unusual
but anticipated phenomena, such has been the case during the BWIP analysis. On
the other hand, in many areas, these tools of analysis are capable of allowing
detailed refinements that are not appropriate because the material properties

are not known to the necessary extent. Of questionable value is an attempt to
explore the realm of nonlinear analysis, anisotropy, or other complex processes
without providing statistically accurate values of material properties to support the analysis refinements. Thus, the importance of statistically dependable
material-property experiments cannot be overemphasized.

TABLE 1

EXAMPLES OF CODES AND CAPABILITIES FOR REPOSITORY ANALYSIS.

Geometry

Code Providing Basis for Model	Plane	Axisymmetric	3-Dimensional
ADINA & ADINAT	A	A	A
ANSYS	A	A	A
DBLOCK	A		
FINEL	A	A	A
HONDO II	A	A	
MARC	A	A	A
SANCHO	A	A	
SPECTROM 11 + 41	A	A	D
STEALTH	A	A	D

Phenomena Couplings

Code Providing Basis for Model	Thermal	Mechanical	Fluid Flow	Inertial	Thermal-Mechanical	Thermal-Fluid Flow	Mechanical-Fluid Flow	Thermal-Mechanical-Fluid Flow
ADINA & ADINAT	A	A		A	A			
ANSYS	A	A		A	A			
DBLOCK		A		A				
FINEL	A	A		A	A			
HONDO II		A		A				
MARC	A	A		A	A			
SANCHO		A		A				
SPECTROM 11 + 41	A	A		A				
STEALTH	A	A	D	A	A	D	D	

Kinematics

Code Providing Basis for Model	Small Strain/Small Displacement	Small Strain/Large Displacement	Large Strain/Large Displacement
ADINA & ADINAT	A	A	A
ANSYS	A	A	
DBLOCK	A	A	A
FINEL	A	A	A
HONDO II	A	A	A
MARC	A	A	
SANCHO	A	A	A
SPECTROM 11 + 41	A		
STEALTH	A	A	A

Constitutive Models

Code Providing Basis for Model	Linear Thermal	Nonlinear Thermal	Thermal Anisotropy	Linear Elastic	Nonlinear Elastic	Plastic	Viscous	Mechanical Anisotropy	Ubiquitous	Slip Surfaces	Cracking	Block Movement	Work Softening
ADINA & ADINAT	A	A	A	A	A	A	A	A	D		A		A
ANSYS	A	A	A	A	A	A	A	A		A			
DBLOCK				A						A		A	
FINEL	A	A	A	A	A	A	A	A	A	A			A
HONDO II				A		A				A			
MARC	A	A	A	A	A	A	A	A		A			
SANCHO				A		A	A		A	A	A		
SPECTROM 11 + 41	A	A	A	A	A	A		A	D	D			
STEALTH	A	A	A	A	A	A	A	A	D	D			A

NOTE: A - Available; D - Under Development.

While the process for near- and far-field analysis seems to fit a neat, continuous pattern where one set of analyses feed another, it is in fact a complex
matrix of these processes that will ultimately lead to the final repository
design. The following examples of studies completed recently or in progress
at Rockwell are evidence of this matrix concept.

ROCK MASS PROPERTIES AND MODELS

The accuracy and adequacy of a model to predict the response of basalt for
repository conditions depends, to a large extent, on the accuracy of the rock

properties that can be measured in situ. Laboratory values of mechanical property data can provide statistically reliable numbers but, in some cases, do not adequately represent field conditions.

Extensive studies to determine material properties for use in the various analysis techniques have been undertaken by the BWIP. Laboratory values of thermal and mechanical properties have been determined for the intact rock in the region of the proposed repository. Tests were designed and are in progress to obtain field measurement of mechanical properties of jointed basalt. The block test is a typical example of such a field test to determine deformation modulus and thermal-conductivity data from a representative volume of basalt ($2 \times 2 \times 2 \ m^3$). The block contains typical discontinuities of the intraflow structure of the Pomona basalt. The field experiment is being carried out in two steps. Step one consisted of the tests conducted after the first horizontal slot was completed and instrumented flat jacks were placed in the slot. Step two consisted of isolating a complete $2 \times 2 \times 2 \ m^3$ block containing enough joints to more adequately represent the mass properties of the basalt.

Completion of the first step of the jointed block test has allowed the determination of the rock mass modulus in the vertical direction. The U.S. Bureau of Mines' gages set in boreholes above the slot have contributed, with the flat jack instruments, to the determination of a mass modulus in the direction along the columnar structure of the Pomona basalt. Analytical solutions for the displacements, coupled with measured deformations, indicate a modulus of elasticity of 41 GPa, which is less than half the 85 GPa determined for intact laboratory-scale samples. The general agreement of this result with the mass modulus obtained from the estimated rock mass-classification procedure has provided support for using mass-modulus values based on the classification procedure until the appropriate tests can be performed. An example of the predicted and measured temperature distribution in the block is shown in Figure 3. The 1.57×10^{-3} W/m°C value of rock mass thermal conductivity from the block and the laboratory-determined value of thermal conductivity (1.56) are quite similar. The thermal conductivity and deformation modulus have provided a basis for obtaining good predictions of canister-scale response.

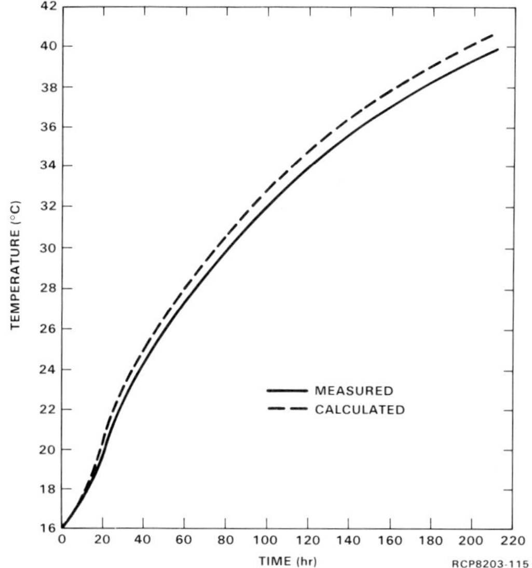

Fig. 3. Measured versus calculated temperatures for a typical thermocouple-block test.

CANISTER-SCALE MODELING

The values of mass modulus and thermal conductivity have been used for canister-scale modeling verification at the Near-Surface Test Facility (NSTF) with a full-scale electrical heater experiment. Electric heaters were placed in boreholes in a basalt flow (Pomona) at the NSTF. The heaters simulated a 1-kW canister loading, representing 5-year-old spent nuclear fuel. The power was increased in two steps up to 5 kW during a 2-year test period to examine the response of jointed basalt to an extreme range of loading conditions. Instrumentation was installed in the heater test area to measure temperature, displacement, and stress distribution as a function of time.

Structural features of the basalt imply that nonlinear and anisotropic analysis will be required to adequately predict thermal and mechanical response using numerical techniques. A fracture-joint spacing of 8 to 10 cm in the zone of the full-scale heater test suggests a potential for closing of joints during the early stages of loading, followed by more typical continuum-type deformation. A nonlinear or perhaps at least a bilinear approximation of the

stress-strain curve would be anticipated as a result of the two different
stages of deformation. Some experimental evidence supports the bilinear
stress-strain curve concept. The columnar structure of the colonnade section
of the basalt would result in differences in material property values in at
least two directions; i.e., along the axis of the columns the properties would
be expected to be different from the direction perpendicular to the axis of
the columns. The possibility of a need for several elastic constants exists
to describe the 3-dimensional anisotropic behavior of the basalt. Physically,
the nonlinear and anisotropic nature of the basalt rock is evident. The
extent to which these factors should be developed in a thermomechanical
analysis was the subject of a theoretical predictive study.

The logical first step, a linear axisymmetric numerical thermo-elastic analysis using average material properties, was determined to provide adequate
accuracy for most near-field predictions of temperatures, displacements, and
stresses. Good agreement (Figure 4) was obtained between the linear finite-
difference solution for temperatures and measured temperatures using average
material properties for Pomona basalt. Good agreement is also shown (Figure 4)
between vertical displacements from a linear finite-element solution and measured vertical displacements. The agreements between linear finite-element
solutions and measured horizontal displacements are shown in Figure 5. The
impact of joints on the displacements is being analyzed. The comparison is not
as good for stresses because of instrumentation performance and the present
limitation on interpretation of stress measurements made at specific points.

IN SITU STRESS PREDICTION

In another part of the matrix, the in situ stress conditions are being
studied. Drilled cores from deep boreholes in the Pasco Basin have exhibited
the core disking phenomena. Core disking generally indicates a possibility of
high horizontal in situ stresses. This finding at Hanford has led to detailed
finite-element modeling of the core-disking process as well as measurement of
the in situ stresses. Axisymmetric finite-element models were prepared to
analyze the state of stresses in the core while it was being removed from the
ground (Figure 6). In situ stresses, bit forces, and forces due to the pressure
of the drilling mud were applied to the model. The in situ stress ratio was
varied over the range of possible values and a nonlinear fracture analysis was

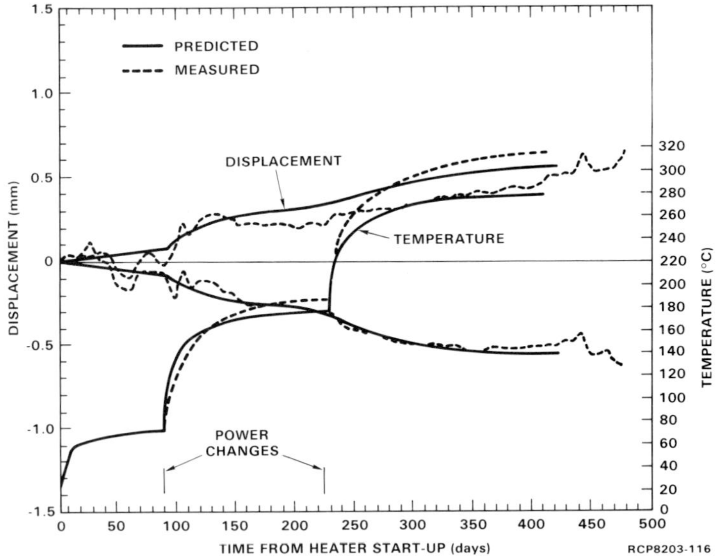

Fig. 4.   Temperature and vertical displacement relative to the midplane.

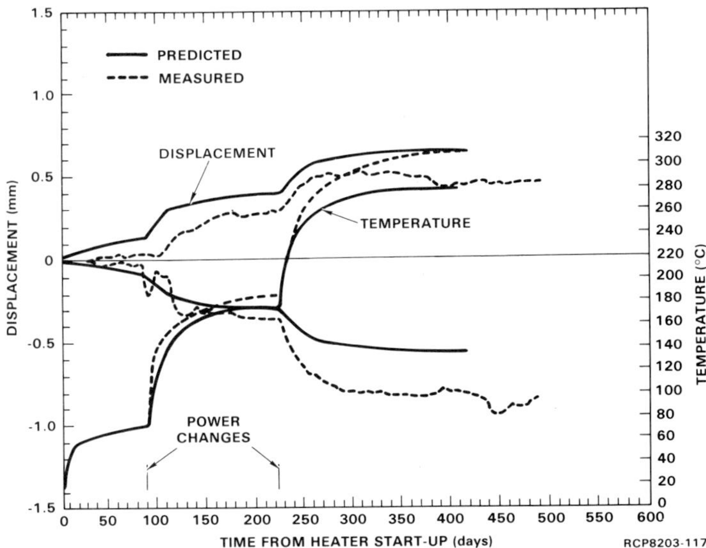

Fig. 5.   Temperature and horizontal displacement relative to the midplane.

performed. Using average material properties for the basalt at the repository horizon, a stress ratio of 1.6 to 2.0 was predicted. Hydraulic fracturing tests were performed at the repository horizon to measure the in situ stress values. The measurements confirmed a value of 2 for the horizontal to vertical stress ratio. As further confirmation of the axisymmetric finite-element model, a fracture analysis of the core during the drilling operation was used to successfully predict the shape and orientation of the disks from actual core holes.

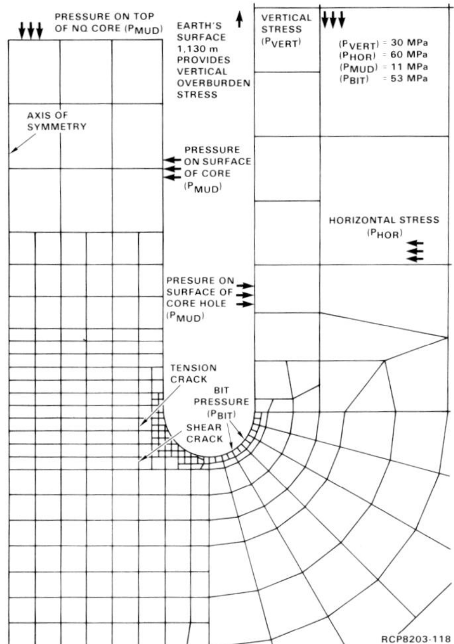

Fig. 6. Axisymmetric finite-element model of core during drilling.

ROOM-SCALE MODELING

As a result of the 2:1 stress ratio prediction, some modification of the room-canister geometry used in the conceptual design was necessary. A series of scoping studies which utilized closed-form and boundary-element solutions led to a new room-canister-scale geometry concept for more detailed finite-element analysis. The proposed room geometry, with horizontally opposed canisters of commercial high-level waste or spent fuel is shown in Figure 7. Also shown in this figure is a schematic of the elements used to model the proposed room-scale

configuration. Because of the increased complexity of the geometry, a
3-dimensional model was selected for the study. In order to present a clear
visual image, all of the elements in the model are not shown.

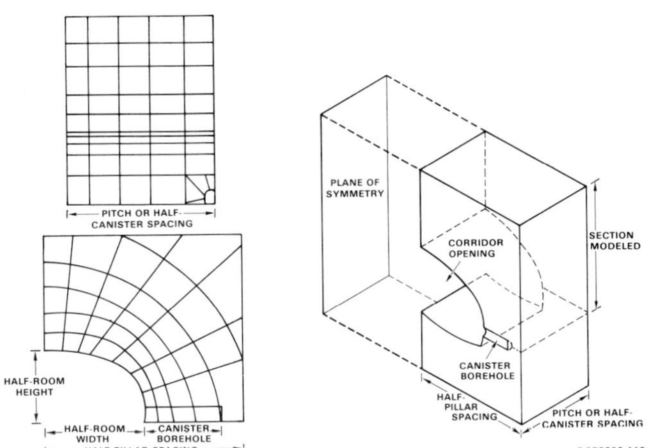

Fig. 7. Finite-element model of present room and canister borehole.

REPOSITORY- AND REGIONAL-SCALE ANALYSIS

In the far-field repository and regional aspects of the design, average
material properties for large regions are adequate. However, since a greater
area is used in the analytical model, it will be necessary in the final design
to have reasonable estimates of the properties of materials located quite far
from the repository. Closed-form, boundary-element, and finite-element methods
have provided excellent analysis of the larger-scale problems, as shown in
Figure 8. The final repository design will utilize finite-element models and
thermomechanical analysis coupled with analysis of the movement of groundwater.
The thermal gradient from the repository will provide a driving force comparable
to the natural hydrologic driving forces which move the groundwater. In the
early years of operation, the groundwater movement will have an effect on the
near-field thermal conditions around the repository. However, in the long term,
the slow migration of groundwater will have minimal potential to shift the overall temperature distribution.[2] The minimal shift in temperature distribution
will permit symmetrical finite-element models.

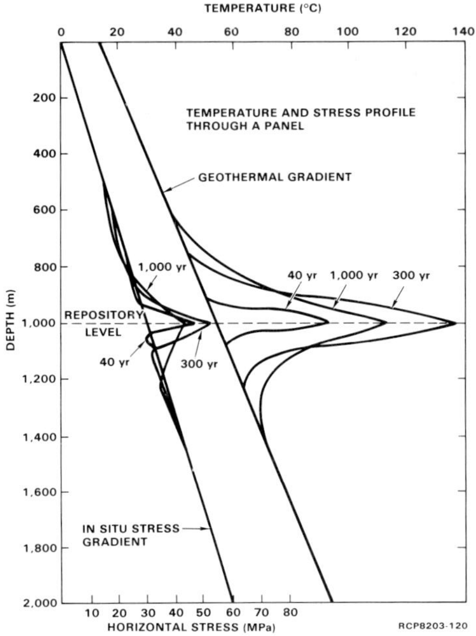

Fig. 8.   Temperature and stress in a panel of a repository in basalt.

The unique geology of the Hanford Site can utilize the thermomechanical contributions to potential decreases in permeability in the near field of the repository. The predictive studies have shown temperatures and stresses to be too small to cause additional fracturing of the rock in the vicinity of the repository (Figure 8). Further, there are numerous natural barriers to brittle fracture propagation. In the far field, the many layers of basalt between the repository and the surface, including flow-top regions and interbeds, will act as potential barriers to crack propagation. In the near field, the regularly spaced and infilled joints provide a good potential to inhibit fracture propagation. As an example, an NQ core with a diagonal pre-existing joint is shown in Figure 9. The horizontal cracks were induced as the core was being drilled. They obviously could not pass the barrier provided by the joint. The high density of joints could present a permeability problem. However, at Hanford much of the thermal expansion in the near field is likely to be absorbed by the closing of joints and any subsequent fracturing is expected to be minimal, thus providing a potential to decrease permeability.

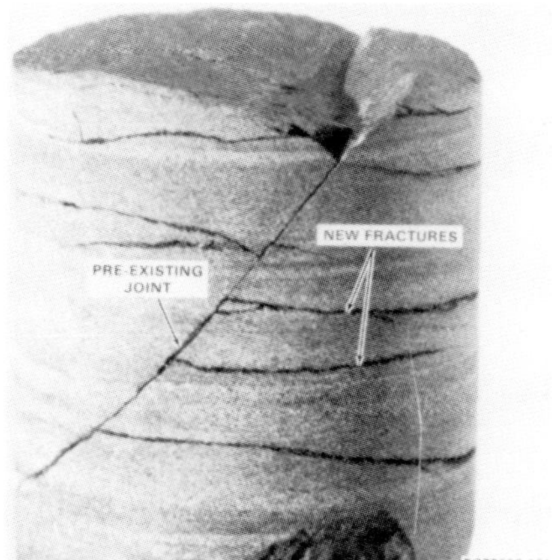

Fig. 9. Pre-existing joints as barriers to fracture propagation.

EMPHASIS FOR MODELING THERMOMECHANICAL EFFECTS

Investigating the results from BWIP-modeling analyses answers the two questions presented earlier. The available computer programs are adequate for most repository-related modeling and analysis. Effective application of those programs depends on the accuracy of available material properties. Extensive application of present as well as improved methods is required to obtain the material properties necessary to increase the accuracy of numerical-modeling studies. This is especially true in near-field analyses. Modeling refinements will be appropriate in each successive stage of updating the geometry of the repository room- and canister-placement configuration, as well as each time that significant changes in the material properties have been obtained.

REFERENCES

1. RHO-BWI-80-100, November 1980, Basalt Waste Isolation Project Annual Report - Fiscal Year 1980, Rockwell Hanford Operations, Richland, Washington.

2. RHO-BWI-C-58, October 1979, G. Hocking and C. M. St. John, Numerical Modeling of Rock Stresses within a Basaltic Nuclear Waste Repository - Annual Report, Fiscal Year 1979, Rockwell Hanford Operations, Richland, Washington.

3. RHO-BWI-C-86, September 1980, G. Hocking, J. R. Williams, P. Boonlualhor, I. Mathews, and G. Mustoe, Numerical Prediction of Basalt Response for Near-Surface Test Facility Heater Tests #1 and #2, Rockwell Hanford Operations, Richland, Washington.

Performance Assessment

A COMBINED ANALYTICAL MODEL FOR PERFORMANCE ASSESSMENT OF THE WASTE PACKAGE/ GEOLOGIC MEDIUM SYSTEMS

ASIT K. RAY[†], SHYAM NAIR[†], AND H. ERIC NUTTALL[††]
[†]Department of Chemical Engineering, University of Kentucky, Lexington, Kentucky, 40506-0046, USA; [††]Department of Chemical Engineering, University of New Mexico, Albuquerque, New Mexico, 87131, USA

INTRODUCTION

Geologic waste isolation systems currently under consideration for long term containment of high-level radioactive waste is based on a set of sequential barriers to release of radionuclides. Recently, Klingsberg and Duguid[1], and Pigford[2] have reviewed in detail the multiple-barrier disposal concept. These barriers are the waste form, canister, buffer, overpack, backfill and geologic media. Each of these barriers acts to retard ground water penetration to the repository as well as migration of radionuclides to the biosphere.

The possibility of water contamination is one of the major concerns from these wastes. Once a repository is filled with radioactive waste and sealed, the contamination can take place by the following sequence of events:

Groundwater intrusion into the repository
Canister failure and initiation of leaching
Transport of radionuclides from the repository to groundwater
Transport of radionuclides to biosphere

For an overall performance assessment of the geologic isolation systems, one needs to take the above sequence of events and interdependent functions of the barriers into consideration. Several numerical codes are currently available for performance assessment, but analytical solutions are needed to check accuracy of these codes. Analytical solutions for radionuclide transport in porous media have been developed by Lester et al.[3] and Burkholder et al.[4]. In this study we have developed a three dimensional model which incorporates the sequence of events preceding migration of nuclides to biosphere.

The model is based on the following assumptions:

(i) Canister failure occurs before groundwater invades the repository.

(ii) Leaching of radionuclides begins when the repository is saturated by groundwater.

(iii) Leaching is solubility controlled. Thus, at the canister/backfill interface the concentration of radionuclides in water is equal to equilibrium concentration.

612

    (iv)   All the engineered barriers are treated as a single homogeneous medium, having same physical properties, and the migration of radionuclides through the barriers is diffusion controlled.

    (v)   Radionuclides migrating out of the waste package are transported through groundwater by diffusion as well as by convection, and the geomedium surrounding the waste package is assumed to be infinite in expanse in all directions.

THEORETICAL ANALYSIS

(a) <u>Ground Water Intrusion Time</u>. A simple but crude estimate of the time, t_i, needed for groundwater to saturate the repository may be obtained from Darcy's Law

$$v_o = - \frac{K}{\mu} \nabla p, \tag{1}$$

where K is the permeability of engineered barriers, μ is the viscosity, ∇p is the pressure gradient, and v_o is the superficial velocity of water.

The intrusion time, t_i, can be calculated from a material balance across the barriers,

$$t_i = \frac{(b^2 - a^2)\varepsilon}{2bv_o}, \tag{2}$$

where b is the radius of the barriers, a is the radius of the canister, and ε is porosity of the barriers.

At a confining pressure of 18 MPa (lithostatic pressure at 800 m targeted depth)[5], and for the following physical characteristics of the barrier,

$$\varepsilon = 0.1, \ a = 0.15 \ m, \ K = 0.1 - 1 \ \mu \ darcy \ and$$
$$thickness \ of \ backfill \ material, \ b - a = 0.25 - 1 \ m.$$

we calculate

$$t_i = 10 - 100 \ years.$$

In our calculation we will assume that the leaching of radionuclides begins after 100 years from the time of implantation.

(b) <u>Leaching Rate and Migration of Nuclides Through Engineered Barriers.</u> Permeabilities of backfill materials are low enough to prevent significant transport of radionuclides by convection. Thus, the transport by molecular diffusion is going to be the dominant mechanism. In this situation the concentration distribution of radionuclides in the engineered barriers may be described by the following partial differential equation

$$D_b \left(\frac{\partial^2 C_b}{\partial r^2} + \frac{1}{r} \frac{\partial C_b}{\partial r} \right) - R_b \lambda C_b = R_b \frac{\partial C_b}{\partial t}, \tag{3}$$

where C_b is the concentration of radionuclides, D_b and R_b are the effective diffusivity and the retardation factor, respectively, and λ is the decay constant.

The above partial differential equation is subject to the following initial and boundary condition

$$t = 0, \quad C_b = 0 \quad \text{(no radionuclides present initially)} \tag{4}$$

$$C_b = C_{eq} \text{ at } r = a \; . \tag{5}$$

and

$$D_b \frac{\partial C_b}{\partial r} + K_L C_b = 0 \quad \text{at} \quad r = b, \tag{6a}$$

or

$$C_b = 0 \quad \text{for} \quad \frac{K_L b}{D_b} \rightarrow \infty \tag{6b}$$

Boundary condition (5) states that at the interface of the canister and the saturated backfill material the concentration of radionuclides in water is equal to the equilibrium composition, C_{eq}. For boundary condition (6a) it is assumed that at the backfill/geomedia interface the dissolved nuclides are transferred to flowing groundwater by a mixing mechanism which depends on the mass transfer coefficient, K_L, and the driving force, $C_b(b,t) - 0$. The transport of nuclides in geomedia by convection as well as by diffusion is expected to be significantly higher than the transport by diffusion in the backfill. Thus, Eq. (6a) may be simplified to Eq. (6b).

The partial differential equation can be solved by applying integral transform technique[6] to give

$$C_b(r,t) = \frac{C_{eq} \ln(r/b)}{\ln(a/b)} + \frac{2C_{eq}}{\pi} \sum_{n=1}^{\infty}$$
$$\frac{\{R_b \lambda + D_b \beta_n^2 \exp\,[-(D_b \beta_n^2 + R_b \lambda) t / R_b]\}}{\beta_n^2 (D_b \beta_n^2 + R_b \lambda) N_n J_o(\beta_n b) Y_o(\beta_n a)} \left[\frac{J_o(\beta_n r)}{J_o(\beta_n b)} - \frac{Y_o(\beta_n r)}{Y_o(\beta_n b)} \right], \tag{7}$$

where β_n's are the positive roots of the transcendental equation

$$\frac{J_o(\beta a)}{J_o(\beta b)} - \frac{Y_o(\beta a)}{Y_o(\beta b)} = 0, \tag{8}$$

and

$$N_n = \frac{2}{\pi^2} \frac{1}{\beta_n^2 J_o^2(\beta_n b) Y_o^2(\beta_n b)} \left[1 - \frac{J_o^2(\beta_n b)}{J_o^2(\beta_n a)} \right]. \tag{9}$$

614

The leaching rate of nuclides from the canister can be obtained from Eq. (7) as

$$L(t) = -D_b \left. \frac{\partial C_b}{\partial r} \right|_{r=a} = - \frac{D_b C_{eq}}{a \ln(a/b)} + \frac{4D_b C_{eq}}{\pi^2 a}$$

$$\sum_{n=1}^{\infty} \frac{\{R_b \lambda + D_b \beta_n^2 \exp[-(D_b \beta_n^2 + R_b \lambda)t/R_b]\}}{\beta_n^2 (D_b \beta_n^2 + R_b \lambda) N_n J_o^2 (\beta_n b) Y_o^2 (\beta_n a)} \tag{10}$$

Total leach time, T, can be calculated from the following differential equation

$$\frac{dN}{dt} = - \lambda N - L(t) A_c \tag{11}$$

with initial condition

$$N = N_o \exp(- \lambda t_i) \tag{12}$$

where N is the inventory of nuclides at time t, N_o is the inventory at the time of implantation, and A_c is the surface area of the canister.

From the solution of Eq. (11) we obtain the following relation for the leach time

$$N_o e^{-(T+t_i)} + \frac{D_b A_c C_{eq}(1-e^{-\lambda T})}{a \lambda \ln(a/b)}$$

$$- \frac{4D_b A_c R_b C_{eq}}{\pi^2 a} \sum_{n=1}^{\infty} \frac{\{1 - \exp[-D_b \beta_n^2 + R_b \lambda)T/R_b]\}}{\beta_n^2 (D_b \beta_n^2 + R_b \lambda) N_n J_o^2 (\beta_n b) Y_o^2 (\beta_n a)} = 0 \tag{13}$$

The leakage rate of nuclides from the waste package to flowing water is given by

$$W(t) = -D_b \left. \frac{\partial C_b}{\partial r} \right|_{r=b} \cdot 2\pi bnL$$

$$= - \frac{2\pi D_b nLC_{eq}}{\ln(a/b)} + \frac{8D_b nLC_{eq}}{\pi}$$

$$\sum_{n=1}^{\infty} \frac{\{R_b \lambda + D_b \beta_n^2 \exp[-(D_b \beta_n^2 + R_b \lambda)t/R_b]\}}{\beta_n^2 (D_b \beta_n^2 + R_b \lambda) N_n J_o^2 (\beta_n b) Y_o(\beta_n a) Y_o(\beta_n b)} \tag{14}$$

where L is the length of the canister and n is the number of canisters.

(c) <u>Transport of Radionuclides by Flowing Water</u>. In this section we formulate and solve the basic transport problem involving the radionuclides migrating out of the repository. We assume that the groundwater flows at a constant velocity U_x, and the flow is parallel to the x-axis. The concentration distribution of radionuclides are described by the following partial differential equation

$$R \frac{\partial C}{\partial t} = D\nabla^2 C - U_x \frac{\partial C}{\partial x} - R\lambda C + S(x,y,z,t) \tag{15}$$

where

$$\nabla^2 = \frac{\partial^2}{\partial x^2} + \frac{\partial^2}{\partial y^2} + \frac{\partial^2}{\partial z^2},$$

R is the retardation factor of the geomedia surrounding the repository.

The source term, $S(x,y,z,t)$, in Eq. (15) describes the release of radionuclides migrating out from the waste package. Since the lateral dimensions of the waste package is small compared with the infinite expanse of geomedia, we may treat the waste package as a line. Thus, the source term may be written as follows

$$S(x,y,z,t) = m(t)\delta(x-0)\delta(y-0)[U_{\ell_1}(z) - U_{\ell_2}(z)] \tag{16}$$

where

$$m(t) = -D_b \left.\frac{\partial C_b}{\partial r}\right|_{r=b} 2\pi b, \tag{17}$$

$$\ell_1 - \ell_2 = nL \tag{18}$$

and $\delta(x-x_o)$ represents the Dirac delta function. The function $m(t)$ may either be obtained using Eq. (14) or from any arbitrary, time-varying, mass-release function describing the waste package performance.

Applying the triple Fourier integral transform we obtain the following solution for the concentration distribution

$$C(x,y,z,t) = \frac{1}{8\pi D} \int_o^t \frac{m(\tau)}{(t-\tau)} \exp[-\lambda(t-\tau)] \exp\left\{-\frac{[x-U_x(t-\tau)/R]^2+y^2}{4D(t-\tau)/R}\right\}$$

$$\left\{\text{erf}\left[\frac{\left|z-\ell_1\right|}{2\sqrt{D(t-\tau)/R}}\right] - \text{erf}\left[\frac{\left|z-\ell_2\right|}{2\sqrt{D(t-\tau)/R}}\right]\right\} d\tau \tag{19}$$

The discharge rate at a downstream distance x can be calculated from the following relation

$$Q(x,t) = \int_{-\infty}^{\infty}\int_{-\infty}^{\infty} \left[-D\frac{\partial C}{\partial x} + U_x C\right] dydz = \left(\frac{\pi}{4R^3D}\right)^{1/2} \int_o^{\infty} \frac{m(\tau)}{(t-\tau)^{3/2}}$$

$$[x + U_x(t-\tau)/R] \cdot \exp[-\lambda(t-\tau)] \exp\left\{-\frac{[x-U_x(t-\tau)/R]^2}{4D(t-\tau)/R}\right\} \left\{nL \ \text{erf}\left(\frac{nL}{4\sqrt{D(t-\tau)/R}}\right)\right.$$

$$\left. + 4\sqrt{\frac{D(t-\tau)}{\pi r}} \left[\exp\left[-\frac{Rn^2L^2}{16D(t-\tau)}\right] - 1\right]\right\} d\tau \tag{20}$$

MODEL APPLICATIONS

To examine the behavior of the solutions developed above and to elucidate the effects of various parameters on radionuclide transport we have performed a number of numerical calculations using a test case. The physical parameters used in the test case are:

$$a = 0.15 \text{ m} \qquad b = 0.40 \text{ m} \qquad L = 3 \text{ m}$$
$$n = 1 \qquad R_b = 10^3 \qquad R = 10^2$$
$$D_b = 3.0 \times 10^{-2} \text{ m}^2/\text{yr} \qquad D = 10^{-1} \text{ m}^2/\text{yr} \qquad T = 10^7 \text{ yr}$$
$$\text{and } U_x = 0.10 \text{ m/yr}$$

Physical properties and initial inventories of nuclides used in our calculations are listed in Table 1.

TABLE 1

PHYSICAL PARAMETERS RELATED TO RADIONUCLIDES

Isotope	Half-life (yr)	Equilibrium Concentration (Ci/m^3)	Initial Inventory (Ci)
^{79}Se	7.0×10^4	10^{-9}	3.0×10^1
^{129}I	1.6×10^7	10^{-3}	2.0×10^5
^{235}U	7.1×10^8	10^{-6}	3.0×10^0
^{237}Np	2.2×10^6	10^{-6}	2.7×10^3
^{243}Am	7950	10^{-8}	1.0×10^3

The computed results of physical significance are leaching rates, leakage rates from the repository, maximum concentration level, and discharge rate at a downstream location. These results are plotted in Figures 1 through 4.

Figure 1 shows leaching rate as a function of time. For solubility controlled leaching the rate decreases in the beginning and then attains a steady value. For the test case studied here the steady state is obtained at about 1000 years after saturation of the repository. Because of relatively high solubility, isotope ^{129}I leaches at a higher rate than the other isotopes. Figure 2 shows the leakage from the repository to the surrounding geo-media. The leakage rate increases with time before reaching a steady value at about 1000 years.

The other calculated results of importance are the maximum concentration level in groundwater and total discharge rate at a downstream location. Figures 3 and 4 provide such information. The maximum concentration level

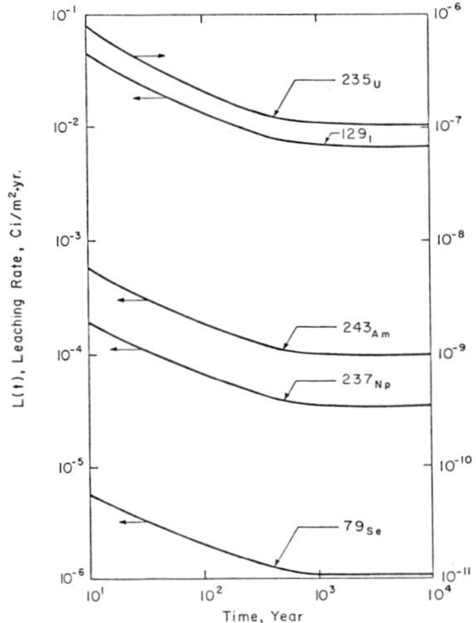

Fig. 1. Leaching rate as a function of time. L(t) Eq. 10

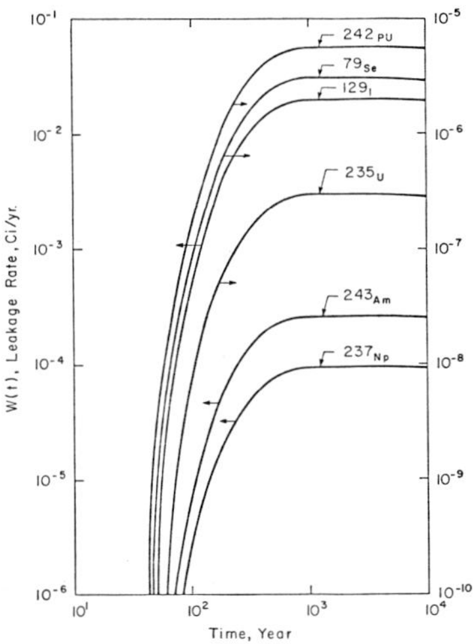

Fig. 2. Leakage rate from waste package as a function of time. W(t) Eq. (14)

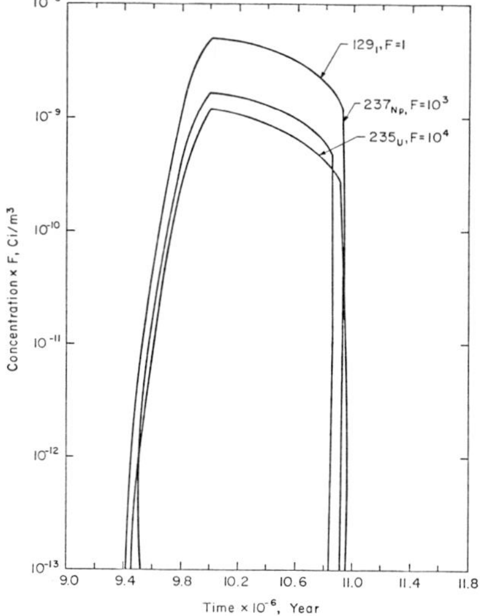

Fig. 3. Maximum concentration as a function of time at x = 10 km. C(10,0,0,t) Eq. (19)

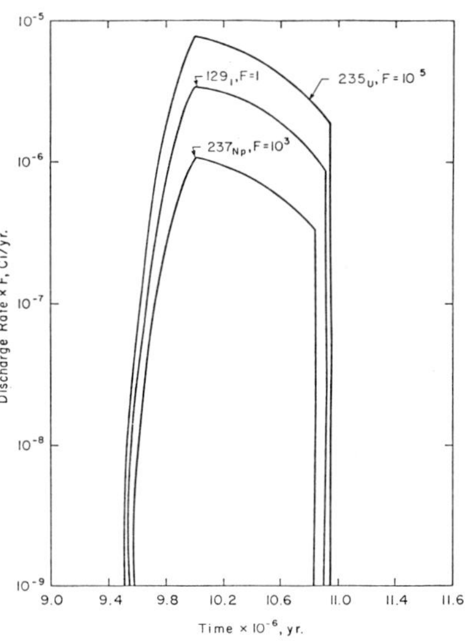

Fig. 4. Total discharge rate as a function of time at x = 10 km. Q(10,t) Eq. (20)

618

at a downstream location will exist on a line passing through the center of the canister and parallel to the x-axis. In Figure 3 we have plotted concentration as a function of time on such a line at a location, where x = 10 km. The results show that ^{79}Se and ^{243}Am completely decays out before reaching that location. Of the isotopes studied ^{129}I shows maximum concentration level ($\sim 5 \times 10^{-3}$ μ Ci/m^3), which occurs at about 1×10^7 years. Discharge rate calculations at x = 10 km are shown in Figure 4. The profiles are similar to the concentration profiles.

CONCLUSIONS

For sparsely soluble isotopes leaching rates are probably controlled by solubility limits[7]. For a situation like this backfill provide a significant resistance to escape of nuclides from the repository. Isotope ^{129}I shows significant concentration level at a 10 km downstream distance. For this isotope the backfill thickness of 0.25 m used in the test case is not sufficient to reduce the concentration to an insignificant level. Materials for backfill and thickness of the backfill probably have to be chosen from the consideration of nuclides like ^{129}I. The three dimensional transport model used here provides a relatively more accurate estimates of concentration levels and discharge rates than the one dimensional model.

REFERENCES

1. Klingsberg, C. and Duguid, J. (1982) American Scientist, 70, 182-190.
2. Pigford, T. H. (1982) Chemical Engineering Progress, 78, 18-26.
3. Lester, D. H., Jansen, G., and Burkholder, H. C. (1975) AIChE Symposium Series, 71, 202.
4. Burkholder, H. C. and DeFigh-Price, C. (1978) AIChE Symposium Series, 74, 91-99.
5. Nowak, E. J. (1981) Scientific Basis for Nuclear Waste Management, 3, 545-552.
6. Olcer, N. Y. (1964) International Journal of Heat and Mass Transfer, 7, 307-314.
7. Ogard, A., Bentley, G., Bryant, E., Duffy, C., Grisham, J., Norris, E., Orth, C., and Thomas, K. (1981) Scientific Basis for Nuclear Waste Management, 3, 331-337.

Published 1982 by Elsevier Science Publishing Co
SCIENTIFIC BASIS FOR RADIOACTIVE WASTE MANAGEMENT - V
Werner.Lutze, editor

ENGINEERED COMPONENTS FOR SPENT FUEL RADIOACTIVE WASTE ISOLATION SYSTEMS--ARE THEY TECHNICALLY JUSTIFIED?

H. C. BURKHOLDER*
Battelle Memorial Institute
Office of Nuclear Waste Isolation
Columbus, Ohio

INTRODUCTION

In response to draft radioactive waste disposal standards, R&D programs have been initiated in the United States which are aimed at developing and ultimately using radionuclide transport-delaying (e.g., long-lived waste containers) and radionuclide transport-controlling (e.g., very low release rate waste forms) engineered components as part of the isolation system.[1] Before these programs proceed significantly, it seems prudent to evaluate the technical justification for development and use of sophisticated engineered components in radioactive waste isolation.

BASIS

Consistent with concepts proposed by the regulatory authorities, the isolation system is divided into the natural subsystem and the man-made subsystem. The natural subsystem includes the far-field host rock, surrounding geologic media, and surface features. The man-made subsystem includes the waste package, the repository, and the near-field host rock. The boundary between the natural and man-made subsystems is located at those positions in the host rock where the presence of the repository and the emplaced waste have a marginal effect on host rock properties. Thus, waste and repository-caused phenomena, such as thermo-driven convection, are confined to the man-made subsystem.

The evaluation requires selection of performance measures for the overall system and each of four subsystem functions. The chosen overall system performance measure is the composite expected <u>maximum</u> total body radiation dose to an <u>average</u> future individual in a <u>local</u> population whose dietary and living habits mirror those of the present. Dose is chosen because it is the overall consequence that the isolation system is designed to control. The composite

---

* PRESENT ADDRESS: Battelle Memorial Institute
                    Pacific Northwest Laboratory
                    P.O. Box 999
                    Richland, Washington 99352

620

expected (or probability-weighted) dose is chosen so that the doses from all
phenomena that could cause the initiation of radionuclide movement away from
the emplaced waste in a given year are considered together and on a common
basis. The maximum dose (rather than the average dose) is chosen because it
is the dose quantity reported in the literature. The total body dose is
chosen so that the study results can be easily related to the natural background dose. The average individual (rather than the maximum individual) in
the population is chosen as a means of keeping in perspective situations where
a few individuals are predicted to receive relatively large doses and many
individuals are predicted to receive relatively small doses. Thus, the composite expected dose to an average member of that particular future local population which receives the largest dose is the overall system performance
measure of this study. Average individuals in the local population at other
times would receive doses less than those reported here.

The chosen subsystem performance measures are:

1) the time period to initiation of transport agent contact with the package
   (the transport agent intrusion time) for the natural subsystem delay
   function

2) The container lifetime for the man-made subsystem delay function

3) The fractional nuclide transport rate across the man-made/natural subsystem boundary (man-made subsystem transport rate) for the man-made subsystem control function

4) The groundwater travel time for the natural subsystem control function.

The last three of these are performance measures currently proposed by the
regulatory authorities. A determination of the appropriateness of these performance measures must await publication of the technical basis for the regulatory standards, and this in turn must await the completion of an overall
system performance assessment by the regulatory authorities.

The evaluation also requires selection of an acceptability standard
against which overall system performance can be judged and a population to
which the performance measure is applied. The chosen acceptability standard
for the particular performance measure used in this study is 0.05 rem per
year, and the chosen population is 100,000 persons. When the transport process involves a domestic well or an exploratory drilling operation, a group of
10 persons out of the total population is assumed to receive the radiation
dose. Thus, the exposed to total population ratio for these situations is

10^{-4}. For all other situations the value is unity. This acceptability standard is chosen based on the principle that the largest dose to an average individual in a future generation should not exceed the largest dose to an average individual in the present generation that results from the variation in natural background dose. Natural background in the United States averages 0.1 rem per year and ranges from 0.09 to 0.25 rem per year.[2] Thus, natural background in the United States varies by as much as 0.15 rem per year from the average, and there are many individuals who receive this dose involuntarily as a result of the location of their birth. Furthermore, society does not appear to use this factor in making voluntary decisions about where to live, work, or travel. For example, the background dose in Denver, Colorado is about 0.14 rem per year above the national average natural background dose, and this fact has not deterred the influx of people into the Denver area from other locations. Application of this principle suggests that all of the activities of nuclear power, considered together, should not cause the largest or maximum incremental dose to an average individual in any of the future local populations at a given site to exceed 0.15 rem per year. The fraction of this value which should be reserved for waste disposal is difficult to determine. For the first few generations following the present generation when the nuclear power economy is active, 10 percent of this value, or 0.015 rem per year seems appropriate. After the first few generations when the nuclear power economy is inactive and waste disposal is the only remaining activity, up to 100 percent of this value, or 0.15 rem per year, seems appropriate. Because the maximum expected doses from waste disposal would likely always be received by far-future generations, there is a strong rationale for adopting the latter value. However, in order to be conservative, the logarithmic average of the two values (0.05 rem per year) is adopted as an acceptability standard for the particular overall system performance measure applied in this study.

APPROACH

The evaluation combines the results of waste isolation performance assessments from the literature and focuses those integrated results on the engineered component technical justification question. The performance assessments in the literature consider two kinds of phenomena, processes and events, on isolation system performance. The processes (e.g., groundwater flow) are modeled deterministically according to known physical laws or empirical relationships. The events (e.g., drilling a borehole) are modeled probabilistically and characterized as the product of the cumulative probability that the event occurs in the year of interest, the exposed to total population ratio, and the maximum dose consequences. Catastrophic events (e.g., meteorite impacts) and intentional intrusion events (e.g., waste recovery) are not considered. It does not make much sense to calculate radiation doses from events whose occurrence would destroy the site and its local society before release of radiation occurred or from events that result from deliberate human actions taken with knowledge about the presence of the waste.

Spent fuel isolation systems in dome salt, bedded salt, basalt, and granite geologic media are considered. The spent fuel inventory is that of a typical repository as defined by a generic environmental impact statement for commercial waste management and is described elsewhere.[3,4] For each system, the effect of each phenomenon is analyzed separately for a chosen set of values for isolation system parameters, including the regulatory performance measures for the subsystem functions (container lifetime, man-made subsystem transport rate, and groundwater travel time). After the expected maximum doses are calculated for each phenomenon, they are summed to yield composite expected maximum doses for an isolation system defined by that set of parameter values. Following these dose summation calculations, a new set of values for the subsystem performance measures is chosen and the dose calculation and summation procedure is repeated while all other parameters are held constant. The procedure is repeated using many sets of values for the performance measures over reasonable ranges of values for isolation systems in each of the geologic media. Some of the calculations from this procedure are in turn repeated for sorption coefficients that are either 10 times or 1/10 of the base values for each geologic medium.

The results from all calculations are next plotted as functions of the transport initiation delay time (the transport agent intrusion time plus the container lifetime), the man-made subsystem transport rate, and the groundwater travel time by a procedure that is conservative (i.e., dose increasing), particularly for dome salt and bedded salt. This procedure is described elsewhere.[4] These graphs thus provide a basis for evaluating the general acceptability of nuclear waste isolation systems and the potential benefits from increasing the performance levels of the subsystem functions above that required for acceptability.

The composite expected maximum doses for the many possible isolation systems are next compared with the assumed acceptable dose standard. If the composite maximum expected dose for a given system is equal to or less than the acceptable dose, that system is considered acceptable; otherwise the system is considered unacceptable. If a system's performance is acceptable, the sensitivity of the dose to the performance levels of the subsystem functions is determined at the lowest levels for the appropriate performance measures (zero for container lifetime, infinity for man-made subsystem transport rate, and zero for groundwater travel time). If the dose is decreased by either an incremental increase in container lifetime, an incremental decrease in man-made subsystem transport rate or an incremental increase in groundwater travel time, an "as low as reasonably possible" (ALARP) incentive is considered to exist for better performance from that subsystem function. As any subsystem performance measure is further increased (in the case of container lifetime or groundwater travel time) or decreased (in the case of man-made subsystem transport rate), this ALARP incentive continues to exist until the slope of the dose versus subsystem performance measure curve is essentially zero. This condition defines the most sophisticated system that can possibly be technically justified because the benefit from further incremental performance improvements disappears. Thus, an ALARP incentive exists until the first zero slope region in the dose versus subsystem performance measure curve is encountered as the performance of the subsystem function is improved above that required for acceptability. Finally, an "as low as reasonably achievable" (ALARA) incentive is evaluated. Because the literature does not contain the cost information necessary for a rigorous cost-benefit analysis, the existence of an ALARA incentive is only qualitatively evaluated. The ALARA limit is by definition between the acceptability limit and the ALARP limit. Because the potential for benefit is very small even if further incremental system improvements are very inexpensive, it seems doubtful that an ALARA incentive could possibly exist for an overall system performance measure (dose) value below 1% of natural background (0.001 rem per year). Therefore, the ALARA incentive is assumed to cease existence between the acceptability and ALARP limits and at 1% of natural background in those cases when the ALARP limit is below that level.

RESULTS AND CONCLUSIONS

The literature studies consider the effect of many processes and events on isolation system performance for one or more of the four geologic media.[5-24] The processes include groundwater flow, shaft and borehole seal degradation, natural salt dissolution, natural salt diapirism, waste movement, criticality, erosion, sedimentation, tectonism, uplift and subsidence, sea-level fluctuations, diageneis, metamorphism, surficial fissuring, hydro-carbon contamination, stored energy releases, and thermal and excavation-induced fracturing. The events include salt solutioning, exploratory drill-ing, faulting, flooding, weapons testing, undetected boreholes, non-nuclear waste disposal, nuclear warfare, breccia pipes, and domestic water well opera-tion. Collectively, the literature studies consider several situations involving one or a combination of phenomena to be the most important. These situations are listed in Table 1 along with their annual probabilities of occurrence. Further descriptions are provided elsewhere.[4] For consistency, the consequence or dose estimates are taken from only four studies even though the results of many such calculations are available.[5,7,18,24]

Figures 1, 2, and 3 are examples of the overall system performance mea-sure (dose) versus subsystem performance measure plots. Figure 1 presents the dose as a function of the transport initiation delay time for bedded salt isolation systems and shows that overall isolation system performance is always acceptable and insensitive (zero slope) to the lifetime of the waste

TABLE 1.

MOST IMPORTANT DOSE-CAUSING SITUATIONS FOR NUCLEAR WASTE DISPOSAL

Situation	Probability (yr[-1])
Dome Salt	
1. solutioning of cavity which connects repository with opening at salt dome boundary followed by groundwater transport to surface water body	10^{-5}
2. solution mining followed by direct ingestion of contaminated salt	10^{-7}
3. faulting followed by groundwater transport to surface water body	10^{-7}
4. exhumation by drilling and worker exposure	10^{-5}
5. natural salt dissolution followed by groundwater transport to surface water body (slow process that will expose waste in 10^6 to 10^7 years)	process
Bedded Salt	
1. exploratory boreholes through repository which connect two aquifers followed by groundwater transport to surface water body	10^{-5}
2. same as dome salt #3	10^{-7}
3. same as dome salt #4	10^{-5}
4. same as dome salt #5	process
Basalt	
1. groundwater transport to surface water body	process
2. groundwater transport to domestic well	10^{-3}
3. same as dome salt #3	10^{-6}
4. same as dome salt #4	10^{-5}
Granite	
1. same as basalt #1	process
2. same as basalt #2	10^{-4}
3. same as dome salt #3	10^{-8}
4. same as dome salt #4	10^{-5}

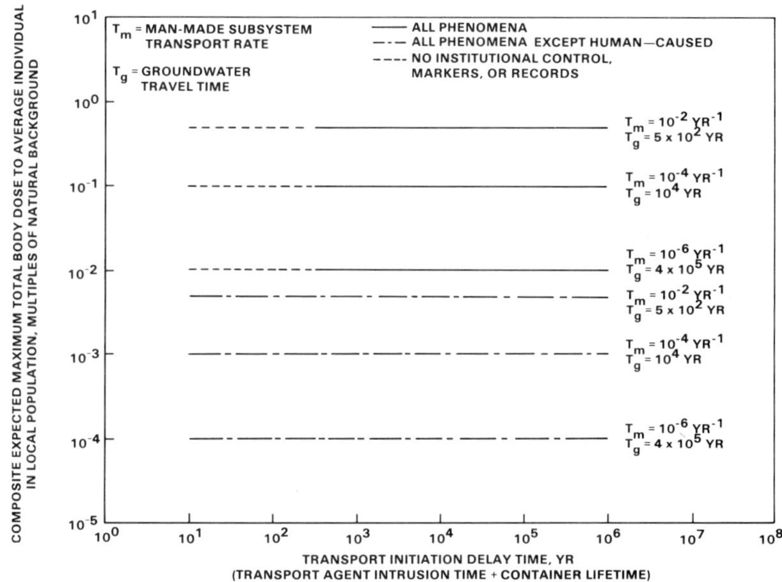

FIG. 1. The Effect of Transport Initiation Delay Time on Dose for Spent Fuel Isolation Systems in Bedded Salt

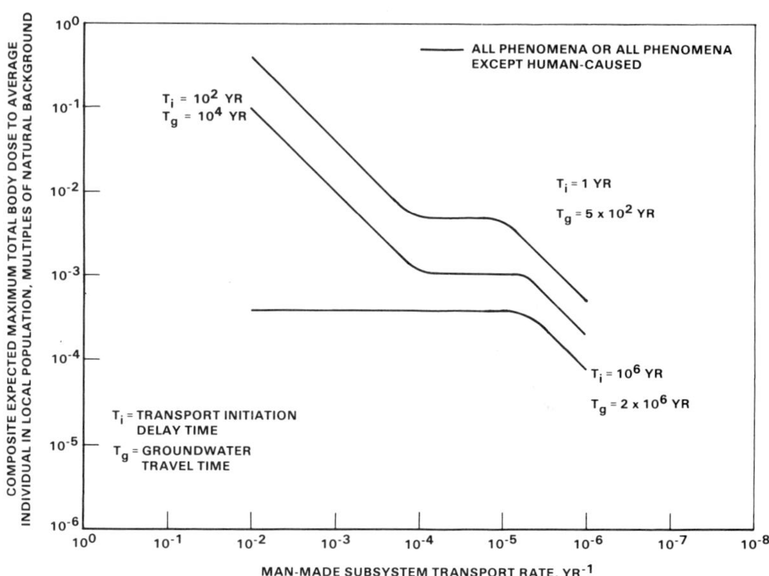

FIG. 2. The Effect of Man—Made Subsystem Transport Rate On Dose for Spent Fuel Isolation Systems in Basalt

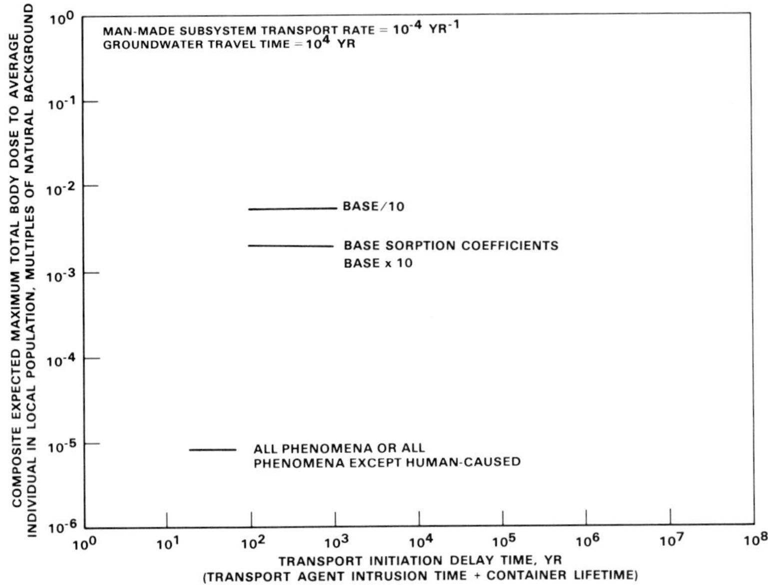

FIG. 3. The Effect of Sorption Coefficient Values on the Dose-Transport Initiation Delay Time Time Curve for Spent Fuel Isolation Systems in Granite

container even for relatively poor systems otherwise. Using the previously described evaluation approach, waste container lifetimes of zero years are necessary to meet acceptability, ALARA, and ALARP limits for the systems described in this figure. Figure 2 presents the dose as a function of the man-made subsystem transport rate for basalt isolation systems. As shown, overall system performance is sometimes sensitive to this parameter in the 10^{-2} to 10^{-4} per year range but becomes insensitive (zero slope) as the man-made subsystem transport rate gets smaller than 10^{-4} per year. For the top curve in Figure 2, man-made subsystem transport rate values of 10^{-2}, 3×10^{-4}, and 10^{-4} per year are necessary to meet acceptability, ALARA, and ALARP limits, respectively. Values of 10^{-2}, 10^{-3}, and 10^{-4} per year are necessary to meet the same limits for the middle curve, and a value of 10^{-2} per year is necessary to meet all limits for the bottom curve. Figure 3 shows the effect of varying the sorption coefficients on the dose-transport initiation delay time curves for granite isolation systems with groundwater travel of 10^4 years and man-made subsystem transport rates of 10^{-4} per year. Although the sorption coefficient values affect the absolute value of the dose, the insensitivity of the dose to container lifetime is unaffected, and waste container lifetimes of zero years are necessary to meet acceptability, ALARA, and ALARP limits for this situation.

The results from all of the evaluations are summarized in Table 2. The table shows the transport initiation delay time and the man-made subsystem transport rate values required to meet acceptability, ALARA, and ALARP limits for ranges of groundwater travel times and sorption coefficients. As shown, the transport initiation delay time values are zero for every situation investigated for all four types of repository systems and all three limits. Therefore, after the repositories have been closed, there is neither an acceptability incentive, an ALARA incentive, nor an ALARP incentive for waste container lifetimes greater than zero years even when the groundwater travel time is short (500 years) and the man-made subsystem transport rate is relatively large (10^{-2} per year). As also shown, the man-made subsystem transport rate values are 10^{-2} per year for the acceptability limit and 10^{-2} to 10^{-4} per year for the ALARA and ALARP limits. Therefore, after the repositories have been closed, there is neither an acceptability incentive, an ALARA incentive, nor an ALARP incentive for man-made subsystem transport rates less than the 10^{-2} to 10^{-4} per year range even when the groundwater travel time is short and no waste container is present. Further, for reasonable spent fuel isolation systems with groundwater travel times of 10^{4} years and man-made subsystem transport rates of 10^{-4} per year, both of these conclusions are unaffected by either a factor of 10 increase or a factor of 10 decrease in the base sorption coefficients.

The simplified approach to the evaluation reported here is not likely to be sufficient for the performance assessment of a specific isolation system because it does not explicitly consider all potentially relevant phenomena and all their potentially complex system-specific interactions. Furthermore, it treats the uncertainties in our understanding of the effects of isolation system phenomena by conservative use of models and data rather than by performing an uncertainty analysis. However, overall geologic isolation system performance is so robustly insensitive to the time that radionuclides first begin to move away from the waste form that the possibility of a more detailed, system-specific analysis reversing the conclusions regarding container lifetime seems remote. Likewise, the large conservatisms inherent in the literature studies upon which this evaluation is based and the wide range of parameters investigated make it unlikely that system-specific analyses will weaken the conclusions regarding the man-made subsystem transport

TABLE 2.

ENGINEERED COMPONENT PERFORMANCE MEASURE VALUES NECESSARY TO MEET THREE TYPES OF LIMITS FOR SPENT FUEL WASTE ISOLATION SYSTEMS.

Natural Subsystem			Acceptability Limit		ALARA Limit*		ALARP Limit	
Geology	T_g (yr)	Sorption Coeff.	T_i (yr)	T_m (yr^{-1})	T_i (yr)	T_m (yr^{-1})	T_i (yr)	T_m (yr^{-1})
Dome Salt	500	Base values	0	10^{-2}	0	10^{-2}	0	10^{-2}
	10^4	Base/10	0	10^{-2}	0	10^{-2}	0	10^{-2}
	10^4	Base	0	10^{-2}	0	10^{-2}	0	10^{-2}
	10^4	Base x 10	0	10^{-2}	0	10^{-2}	0	10^{-2}
	4 x 10^5	Base values	0	10^{-2}	0	10^{-2}	0	10^{-2}
Bedded Salt	500	Base values	0	10^{-2}	0	10^{-2}	0	10^{-2}
	10^4	Base/10	0	10^{-2}	0	10^{-2}	0	10^{-2}
	10^4	Base	0	10^{-2}	0	10^{-2}	0	10^{-2}
	10^4	Base x 10	0	10^{-2}	0	3×10^{-3}	0	3×10^{-4}
	4 x 10^5	Base values	0	10^{-2}	0	10^{-2}	0	10^{-2}
Basalt	500	Base values	0	10^{-2}	0	3×10^{-4}	0	10^{-4}
	10^4	Base/10	0	10^{-2}	0	10^{-3}	0	10^{-4}
	10^4	Base	0	10^{-2}	0	10^{-3}	0	10^{-4}
	10^4	Base x 10	0	10^{-2}	0	3×10^{-3}	0	3×10^{-4}
	2 x 10^6	Base values	0	10^{-2}	0	10^{-2}	0	10^{-2}
Granite	500	Base values	0	10^{-2}	0	3×10^{-4}	0	3×10^{-4}
	10^4	Base/10	0	10^{-2}	0	10^{-3}	0	10^{-3}
	10^4	Base	0	10^{-2}	0	10^{-3}	0	3×10^{-4}
	10^4	Base x 10	0	10^{-2}	0	10^{-2}	0	3×10^{-3}
	10^6	Base values	0	10^{-2}	0	10^{-2}	0	10^{-2}

T_g = groundwater travel time; T_i = transport initiation delay time = transport agent intrusion time + container lifetime; T_m = man-made subsystem transport rate
*ALARA limit is between acceptability and ALARP limits by definition and is conservatively estimated to have been reached when overall system performance measure (dose) is at 1% of natural background in those cases when the ALARP limit is below that level.

rate. Furthermore, the results of this study are in substantial agreement with the results of other similar analyses in the nuclear waste isolation literature.[25-27] Thus, the use of waste containers with long lifetimes and man-made subsystems with very low transport rates seems unjustified from a technical standpoint and a geologic isolation point of view. The design objective for the waste container lifetime should not be greater than that desired for retrievability (0 to 100 years), and the design objective for the man-made subsystem transport rate should not be less than the 10^{-2} to 10^{-4} per year range.

ACKNOWLEDGMENT

This report was prepared by the Battelle Project Management Division, Office of Nuclear Waste Isolation, and the Pacific Northwest Laboratory under Contract No. DE-ACO6-76-RLO1830 with the U.S. Department of Energy.

REFERENCES

1. U.S. Nuclear Regulatory Commission. (1981) 10 CFR 60, Disposal of High-Level Radioactive Wastes in Geologic Repositories. Federal Register, 46 FR 35280. Office of Federal Register, Washington, D.C.
2. Klement, A. W., et al. (1972) Estimates of Ionizing Radiation Doses in the United States 1960-2000. ORB/CSD 72-1. U.S. Environmental Protection Agency, Washington, D.C.
3. U. S. Department of Energy. (1980) Final Environmental Impact Statement--Management of Commercially Generated Radioactive Waste. DOE/EIS-0046F. Washington, D.C.
4. Burkholder, H. C. (1982) Engineered Components for Radioactive Waste Isolation Systems--Are They Technically Justified? ONWI-286. Battelle Project Management Division, Office of Nuclear Waste Isolation, Columbus, Ohio.
5. Cloninger, et al. (1980) An Analysis on the Use of Engineered Barriers for Geologic Isolation of Spent Fuel in a Reference Salt Site Repository. PNL-3356. Pacific Northwest Laboratory, Richland, Washington.
6. A. D. Little, Inc. (1980) Technical Support of Standards for High-Level Radioactive Waste Management. EPA-520/4-79-007D. U.S. Environmental Protection Agency, Washington, D.C.
7. Harwell, M. A., et al. (1981) Reference Site Initial Assessment for a Salt Dome Repository. PNL-2955. Pacific Northwest Laboratory, Richland, Washington.
8. Claiborne, H. C. and Gera, F. (1974) Potential Containment Failure Mechanisms and Their Consequences at a Radioactive Waste Repository in Bedded Salt in New Mexico. ORNL-TM-4639. Oak Ridge National Laboratory, Oak Ridge, Tennessee.
9. Girardi, F., et al. (1977) Long-Term Risk Assessment of Radioactive Waste Disposal in Geological Formations. EUR 5902.e. Commission of European Communities, Ispra, Italy.
10. Logan, S. E. and Berbano, M. C. (1978) Development and Application of a Risk Assessment Method for Radioactive Waste Management. EPA 520/6-78-005. U.S. Environmental Protection Agency, Washington, D.C.
11. Campbell, J. E., et al. (1978) Risk Methodology for Geologic Disposal of Radioactive Waste: Interim Report. SAND-78-0029. Sandia Laboratories, Albuquerque, New Mexico.
12. Berman, L. E., et al. (1978) Analysis of Some Nuclear Waste Management Options. UCRL-13917. The Analytic Sciences Corporation, Reading, Massachusetts.
13. Bingham, F. W. and Barr, G. E. (1979) Scenarios for Long-Term Release of Radionuclides from a Nuclear Waste Repository in the Los Medanos Region of New Mexico. SAND-78-1730. Sandia Laboratories, Albuquerque, New Mexico.
14. U. S. Department of Energy. (1979) Draft Environmental Impact Statement Waste Isolation Pilot Plant. DOE/EIS-0026-D. Washington, D.C.
15. Raymond, J. R., et al. (1980) Test Case Release Consequence Analyses for a Spent Fuel Repository in Bedded Salt. PNL-2782. Pacific Northwest Laboratory, Richland, Washington.

16. International Fuel Cycle Evaluation Working Group 7. (1980) Release Consequence Analysis for a Hypothetical Geologic Radioactive Waste Repository in Salt. INFCE/DEP/WG7/16. International Atomic Energy Agency, Vienna, Austria.

17. Burkholder, H. C. (1981) The Development of Release Scenarios for Geologic Nuclear Waste Repositories--Where Have We Been?--Where Should We Be Going? Radionuclide Release Scenarios for Geologic Repositories, OECD Nuclear Energy Agency, Paris, France.

18. Cloninger, M. O. and Cole, C. R. (1981) An Analysis on the Use of Engineered Barriers for Isolation of Spent Fuel in Three Reference Geologies. PNL-3530. Pacific Northwest Laboratory, Richland, Washington.

19. Dove, F. H., et al. Reference Site Initial Assessment for a Basalt Repository. PNL-3632. Pacific Northwest Laboratory, Richland, Washington.

20. Nuclear Fuel Safety Group. (1978) Handling of Spent Nuclear Fuel and Final Storage of Vitrified High-Level Reprocessing Wastes. Karn-Bransle Sakerhet, Stockholm, Sweden.

21. Nuclear Fuel Safety Group. Handling and Final Storage of Unreprocessed Spent Nuclear Fuel. Karn-Bransle-Sakerhet, Stockholm, Sweden.

22. International Fuel Cycle Evaluation Working Group 7. (1980) Release Consequence Analysis for a Hypothetical Geologic Radioactive Repository in Hard Rock. International Atomic Energy Agency, Vienna, Austria.

23. Lyon, R. B. (1980) Environmental and Safety Assessment for Nuclear Waste Disposal--The Canadian Approach. Waste Management 81 Symposium, University of Arizona, Tucson, Arizona.

24. Cloninger, M. O. and Sneider, S. C. (1982) Sensitivity of Spent Fuel Repository Performance Assessment to Variation in Radionuclide Retardation Coefficients. PNL-4296. Battelle Memorial Institute, Pacific Northwest Laboratory, Richland, Washington.

25. Burkholder, H. C., et al. (1976) "Incentives for Partitioning High-Level Waste. Nuclear Technology, 31, 202.

26. Hill, M. D. (1980) The Effect of Variations in Parameter Values on the Predicted Radiological Consequences of Geologic Disposal of High-Level Waste. Scientific Basis for Nuclear Waste Management, 2, 753.

27. Sutcliffe, W. G., et al. (1981) Uncertainties and Sensitivities in the Performance of Geologic Nuclear Waste Isolation Systems. ONWI-352. University of California, Lawrence Livermore National Laboratory, Livermore, California.

BACKFILL MATERIALS AND CANISTER CORROSION

RECENT ADVANCES IN REPOSITORY SEAL MATERIALS

D.M. ROY* AND F.L. BURNS**
*Materials Research Laboratory, The Pennsylvania State University, University
Park, PA 16802; **Formerly, Office of Nuclear Waste Isolation, Battelle,
Columbus, OH 43201, USA

INTRODUCTION

One long-term solution proposed for isolating radioactive waste from the biosphere is containment in underground geologic repositories. A requirement for
this procedure is that most penetrations into the repository region be sealed
following emplacement of the waste.   The general requirements of repository seal
materials and the seal system are that these potential pathways to the biosphere
be sealed adequately to prevent a) excess fluid flow into the repository and b)
radionuclides in the waste from reaching the biosphere in quantities in excess of
acceptable levels.[1] For boreholes, shafts and tunnels it is generally recommended that in the post-closure phase, the permeability of sealed boreholes
and shafts be approximately the same as the host rock or the aquitards in the
zones where aquitards are intersected.[2] Material properties and performance characteristics which contribute to the above requirements include the following:
They should be 1) mechanically adequate, in some cases self-supporting, and well
bonded to the host rock; they should 2) restrict severely the permeation of
fluids, particularly in the interfacial zone between seal materials and high
integrity host rock, and any disturbed zone; they should be 3) stable, chemically
durable, compatible with the surrounding rock and groundwater environment in the
long term, 4) resistant to destructive expansion and contraction, 5) resistant to
radionuclide transport; and finally, 6) emplaceable by feasible techniques.
Cementitious materials including grouts, mortars and concrete, clays,
crushed rock, earth fill and zeolitic materials, are likely components of borehole plugs and shaft and tunnel bulkheads and backfill, and storage room backfill[3,4] Candidate materials are being investigated to determine their properties
and expected longevity of performance[5] under the specific geologic repository
conditions. Cementitious materials because of high strength and relative
rigidity should serve major structural roles while more deformable clays and
zeolites have complementary roles. Here we report mostly on cementitious seal
materials, with some studies of clays and zeolites.

RESULTS OF STUDIES

General Approach. Materials are selected for further study based upon knowledge of their performance both from the literature and from short-term test data.
The first stage in a specific seal material formulation involves a thorough characterization of the starting components for factors affecting their reproducibility of performance, such as chemical composition, particle size distribution,
phase content variability within a component, and sensitivity to changes taking
place during storage. A variety of experimental and test methods are used to
accomplish this, including analytical, surface area measurements, porosity, DSC
and TG, heat of hydration, x-ray diffraction and thermal conductivity.[7] The
design of seal materials both for normal and remote placement requires knowledge
and control of the early stage rheological properties and development of structure. The effects of changes in composition, chemical additives, particle size
and other component parameters upon rheological[8] and mechanical properties are
determined for proper proportioning of mixtures and achieving satisfactory workability and placing characteristics.

Studies addressing materials longevity utilize information from a variety of sources, including tests upon experimental mixtures and formulations; and preparation and detailed characterization of experimental seal materials, including some designed for compatibility with bedded salt[7,9] and basalt lithologies. Where possible specimens from plugs that have served in the field for extended time periods are studied; evaluations of materials' thermodynamic stability, determining reaction rates and kinetic factors,[10] using selection criteria based on knowledge of natural mineral associations; studies of ancient concrete analog materials; and investigating other relevant factors, are used to determine probable compatibility of seal material with certain host rock/environment combinations.

Thermodynamic Stability. Seal materials most likely to perform for long time periods are those which are thermodynamically stable both intrinsically and under environmental effects of the geologic repository. Literature studies, calculations, and generation of needed new thermodynamic data are combined to calculate the lowest free energy (i.e., most stable) state for fixed seal material compositional conditions[11] including ground water compositions.

An example of the differences in free energy of hydration (ΔG) of Ca_3SiO_5, the most important cement component, with different reaction paths, is illustrated as follows, $\Delta F = -67$ kJ(-16.0 Kcal)/mole Ca_3SiO_5:

$$Ca_3SiO_5 + 3H_2O = \frac{1}{2}(Ca_3Si_2O_7 \cdot 3H_2O) + \frac{3}{2}Ca(OH)_2 \qquad [1]$$

"C-S-H", a subcrystalline calcium silicate hydrate and calcium hydroxide are formed as products. However, if the reaction product calcium hydroxide [$Ca(OH)_2$] is consumed by reacting with silica (SiO_2) and the product is crystalline tobermorite, the ΔG change is more negative (even more favorable): $\Delta G = 134.9$ kJ (-32.24 Kcal)/mole Ca_3SiO_5

$$Ca_3SiO_5 + \frac{13}{5}SiO_2 + \frac{16.5}{5}H_2O = \frac{3}{5}(5CaO \cdot 6SiO_2 \cdot 5.5H_2O) \qquad [2]$$

The important fact is that reaction [2] is more thermodynamically favored, and therefore the product is more stable than in [1].

Continuing studies are examining the effects of concentrations of various species in groundwaters on cementitious and other phase stabilities. Stability diagrams are widely used, for instance in studies of groundwater and in high temperature and pressure systems. In these diagrams, the difference between pH and negative log of concentration of the ion of interest, usually K^+, Na^+, or Ca^{++}, is plotted against log concentration of dissolved H_4SiO_4. Preliminary calculations show that the zeolite leonhardite (a partially dehydrated laumontite) has a stability field covering most of the area usually attributed to anorthite, restricting anorthite to the very low silica part of the diagram. Some concentrations obtained in this laboratory for solutions in contact with portland-cement-waste systems are in the leonhardite (zeolite) field.[15]

Field Plug Materials. Useful information regarding intermediate to long-term seal material stability is available from field performance data on analogous materials. Those materials investigated so far vary from 5- to 17-year-old modern cement plugs, to ancient building materials more than 2000 years old.

Duvall Potash Mine. Borehole plugging materials relatively recent on this scale ($\sim$17 years old) were first characterized.[12] The plug material used was an apparently salt-saturated high water/cement ratio grout which, while mechanically weak, had maintained its physical integrity in the salt environment. The given plug would not be satisfactory for stopping the flow of water, but mechanically is adequate for backfilling. The overall findings showed that the plug had performed satisfactorily over the 17-year time frame; but at the same time, it had a relatively high permeability (to gas), $\sim 10^{-3}$ μm^2 and would not be an appropriate material for radioactive waste repository sealing in a critical area.

BCT. For comparison, data on an experimental test plug placed more recently, and the accompanying laboratory test data showed that it is now possible to emplace much more satisfactory materials. Water permeabilities measured in the laboratory for small simulated borehole specimens prepared by casting the cementitious mixture given in Table 1, in cored anhydrite specimens were in the 10^{-6} μm^2 (microdarcy) range (see Table 2). Equivalent field permeabilities in the Bell Canyon Test (BCT) showed an approximate $\leq \sim 10^{-5}$ μm^2 permeability, which adequately stopped the flow of an aquifer which produced an ~ 12 MPa (1800 psi) initial pressure differential on the plug.[13,14]

Ancient Cements, Plasters, Mortars Concrete. The remarkable longevity of ancient cementitious mortars, plasters, and concrete which are compositionally similar to modern counterparts suggested that important information was to be gained from their detailed evaluation, regarding long-term stability of cementitious seal materials for geologic repositories.[16] The materials studied are from structures used to hold water, from buildings and port installations and incorporate concrete aggregates of several rock types pertinent to the current seal material studies.

Current studies of systematic suites of samples from three major source areas: in Greece, Cyrpus and Italy, with materials ranging back to some 2600 years age or older are in process. The results show that the materials may be classified into four major types (plus some other miscellaneous samples): a) gypsum plasters and mortars; b) hydrated limt-based mortars and plasters, c) hydraulic hydrated lime-based and hydraulic lime-based/soil plasters, and d) pozzolana/hydrated lime-based plasters, mortars and concretes. The strongest and generally most satisfactory materials which were developed most highly in Rome from about the 1st century BC to 3rd century AD appear to those with a hydrated lime-pozzolana cementitious matrix. The pozzolana (usually fine-grained volcanic ash, or finely crushed terra-cotta) reacted with lime to produce a material in some ways comparable to the cementitious matrix of modern portland cement-based materials. In fact, the performance of such analogs incorporating a large percentage of pozzolana help substantiate the durability observed over a much shorter time scale with portland cements incorporating pozzolanic or silica-rich additives.

TABLE 1.

PSU CEMENTITIOUS FORMULATIONS #79-091*

	Wt (g)
Solid	
Class H Cement (H-4)	66.6
Class C Fly Ash (B15)	22.4
Expansive Additive (A29)	8.9
Superplasticizer (A28)	0.3
Liquid	
Water	29.4

*Similar to BCT-IFF.[13]

TABLE 2. WATER PERMEABILITY OF PSU #79-091 SAMPLES CAST IN ANHYDRITE[a]

Curing Time (wk)	Elapsed Time Between Cure and Test (days)	Time of Test (hrs)	Water Pressure (MPa)	Permeability[d] (10^{-6} μm^2)
1	1	41	2.7	1.5
2	3/5	8/9	2.7	2.2/7.0
3	1/2	21/90	2.7	22.0/2.3[b]
4	12/31	29/50	4.4	<0.25/<0.16[c]

[a]Specimens prepared using modified API mixing procedures, vibrated for 60 seconds, cured at 38°C in WIPP-A brine.[13]
[b]Specimens cured at 38°C in saturated air.
[c]Essentially impermeable under experimental time and pressure conditions. However, a maximum permeability value was calculated assuming a hypothetical 1-mL flow in the given time period.
[d]$1 \mu m^2 \simeq 10^{-3}$ cm sec$^{-1} \simeq 1$ Darcy.

TABLE 3. GROUT COMPONENTS (80-081)

Dry Solids	Percent
Class H Cement	65
High-lime Fly Ash	22
Calcium Sulfate Additive	8.0
NaCl	3.9
Superplasticizer	1.1

Liquids	g per 100 g dry solids
defoaming agent	0.04
deionized water[a]	28.2
or brine (E17)[b]	40.9

[a] Used for pre-placed aggregate concrete.
[b] Used for mixed concrete; Table 5.

Salt-Compatible Seal Materials.
Cementitious Grout Properties. A salt-compatible cementitious grout was prepared in connection with the Bell Canyon Test as well as a formulation without salt, discussed previously.[13] Additional materials have been designed and studied as candidates for use in a bedded salt repository area. An intermediate composition grout for use in transition areas and mixed evaporite horizons is given in Table 3. The U.S. Army Engineer Waterways Experiment Station in Vicksburg, MS, collaborated in studies of this material.[7] Early age Young's modulus measurements made by a resonance technique of the mixture cured at 38°C (a typical temperature below the repository horizon in the WIPP site) were 20.5 and 19.1 GPa for the two laboratories, in good agreement. Figure 1 shows the effect of curing time and temperature on the dynamic modulus.

The stress is caused by the crystallization of a needle-like calcium aluminosulfate hydrate phase, ettringite. Bar measurements of restrained expansion showed a log-linear increase in expansion with time with a total 0.392% length change up to 365 days. Permeabilities of specimens cast in hollow stainless steel cylinders, 51 mm diameter x 26 mm height, capped and cured under saturated $Ca(OH)_2$ were compared with others in unrestrained glass vials, as well as others exposed fully to

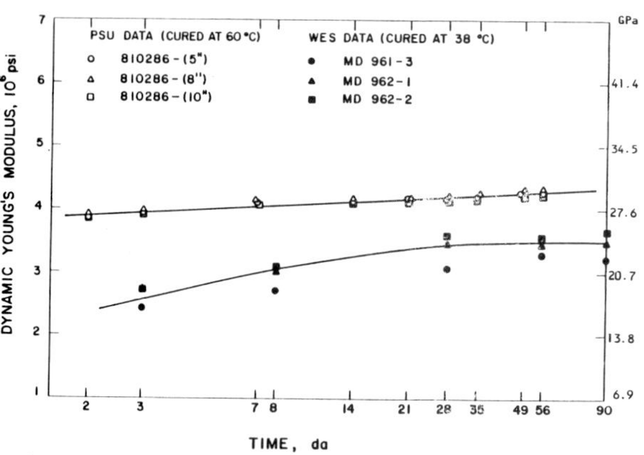

Fig. 1. Effect of curing time and temperature on the dynamic modulus of 80-081 grout components.

saturated $Ca(OH)_2$ solution. Table 4 shows the results for samples cured in restrained rings, which are essentially impermeable for all curing times between 3 and 365 days. The samples cast in glass vials also were generally impermeable, while unrestrained expansion occurred in specimens cured for 56-90 days in saturated $Ca(OH)_2$ solution, where ample $Ca(OH)_2$ was available to help form ettringite, resulting in some cracking, which was reflected in a higher measured permeability, of about 9 x 10^{-5} μm^2 at 56 days.

Effect of Ground Water. The effect of groundwater composition was studied upon the properties of unrestrained cubes and restrained cylinders of this candidate seal material, exposed at 38°C for periods of up to 6 months in contact with 4 different solutions, and a fifth control set under water-saturated vapor conditions. Environmental effects on cube strengths were apparent, but all had very high strengths (Fig. 2) including these samples which exhibited surface expansion cracks [those in $CaSO_4$, saturated $Ca(OH)_2$ and deionized water]. No cracking was observed in the restrained samples. This material appears to develop high strength and an expansive stress sufficient to maintain a good wall-rock bond under all the environmental conditions indicated. The tests are being continued for prolonged periods.

Grouted Gravels and Concretes Using Test Grout Modifications. Preliminary compatibilities of the cementitious matrix with repository rock types were investigated through studies of concrete and other test specimens prepared with very limited amounts of rock from cores, as well as a local commercial dolostone aggregate, from SE New Mexico. The brine in Table 5 was used for curing all test samples, most of which were made by pouring fluid grout into molds containing preplaced aggregate, 12.7-mm maximum diameter.[9] The permeabilities of the aggregates were as follows: dolostone, 6×10^{-8}; anhydrite, $<10^{-8}$; and sandstone, $\sim 10^{-2}$ (all μm^2). Typical permeabilities for the concretes are given in Table 6. It was possible to prepare a low permeability concrete even with mudstone, though the mechanical properties were not outstanding. Compressive strengths ranged from 25-31 MPa for the initial

TABLE 4. WATER PERMEABILITY OF SAMPLES CAST AND CURED AT 38 °C IN 50.8-mm DIAMETER STAINLESS STEEL RINGS

Time (days)	Sample (number)	Height (mm)	Pressure (MPa)	Time Interval (hr)	Permeability[a] (10^{-8} μm^2)
3	801651	25.4	0.63-2.95	94	<2.0
7	801652	25.2	0.70-2.79	163	<1.1
28	801653	25.8	0.72-2.76	172	<1.2
56	801654	27.6	0.87-2.79	71	<2.7
90	801660	26.8	0.76-1.02	73	<5.4
365	801657	26.8	0.69-3.11	72	<8.8

[a] Calculated value assuming 1 mL had passed through the sample during the total time of the test under an average of the test pressures. There was no measurable flow and hence the actual permeability value is much less (<) than the value listed.

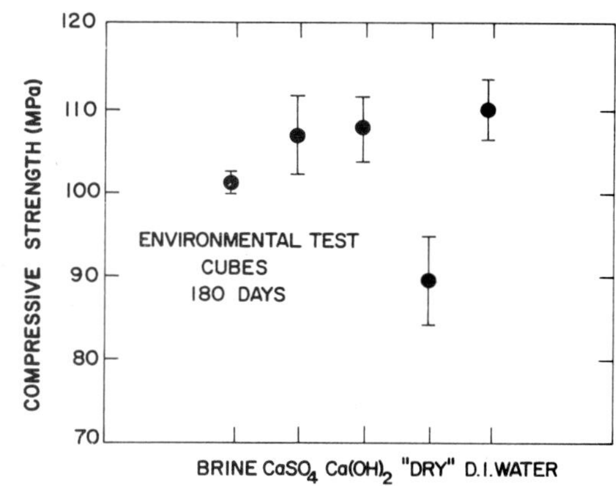

Fig. 2. Test cube strengths of mixture 80-081 exposed to different environments for 180 days.

TABLE 5. COMPOSITION OF E17 (WIPP B) BRINE TO FORM 1L

$Na_2B_4O_7 \cdot 10H_2O$	0.016 g
$SrCl_2 \cdot 6H_2O$	0.045 g
$NaHCO_3$	0.013 g
KBr	0.045 g
$MgSO_4 \cdot 7H_2O$	0.101 g
$CaCl_2$	2.44 g
NaBr	0.474 g
$Na_2SO_4 \cdot 10H_2O$	11.45 g
NaCl	292.25 g
density	1.18 kg/ L

TABLE 6. BRINE PERMEABILITIES OF SALT-COMPATIBLE CONCRETES PREPARED WITH GROUT FROM TABLE 3.

Gravel or Rock Type	Test Pressure (MPa)	Age (days)	Permeability (μm^2)
anhydrite	3.4	40	$\sim 8 \times 10^{-8}$
dolostone	4.1	62	$<10^{-8}$
mudstone	2.7	79	9×10^{-7}

grout cured 28-56 days, to 22-40 MPa for the dolostone concrete, to 19-28 MPa
for the anhydrite concrete. Even halite-concrete prepared with halite aggregate had an acceptable strength, 10-16 MPa when the mixing water was WIPP-B
brine and the concrete was mixed in the normal manner.

These studies have demonstrated that it is possible to prepare concretes using
an intermediate composition salt-containing grout, which appears reasonably compatible with several rock types from the Los Medaños bedded salt region, and produces composites which are relatively impermeable to site-derived brines. When
halite is used as aggregate, additional salts are required in the mixing water.

Basalt-Compatible Seal Materials. Seal materials for a basalt repository are
also being investigated. Some materials incorporated crushed basalt in the sand
and coarse aggregate fractions to generate greater compatibility with the host
rock (see Table 7 which gives compositions of a number of the mortars and concretes being investigated). Studies at elevated temperatures and pressures
provided accelerated test equivalent data, and also information concerning
effects of repository heating. Unusually high compressive strengths of more
than 200 MPa have been produced in some of the materials cured at temperatures
up to 175°C (mixture 81-21). Most mixtures showed a progressive strength
increase with curing temperature up to 175°C, and somewhat lower strengths again
at 250°C. The same trend was observed in Young's modulus (Fig. 3) of mixture 20,
although the lower values of the 250°C sample at 28 and 56 days are not readily
explainable. Most mixtures contained excess silica sufficient to react with
Ca_3SiO_5, as in equation [2], hence having potentially greater stability. Alternative expansion mechanisms are being investigated; expansion due to ettringite
formation, discussed earlier, is limited to below ∼100°C, depending on the
pressure.[17]

Clays, Zeolites and Derivative Seal Material Stabilities. Clay minerals and
zeolites are candidate seal components because of their sorption properties for
radionuclides and, in the case of the former, when highly compacted they can be
expected to reduce fluid flow. Certain basic characteristics of the clay minerals determine the effectiveness of each of these sealing functions, and the
type of seal required will determine those clay minerals best suited. Furthermore, these principal properties depend upon the chemistry and structural configuration of the clay minerals, and diagenetic transformation under repository
conditions can affect their behavior as sealants.

It is important to know if candidate clay minerals will be stable in the expected environment, or if not then it is necessary to determine the rates at
which these clays will alter to a more stable phase. This is particularly true
with respect to thermal stability. Heating of expansive clay minerals to moderate temperatures can reduce the exchange capacity and alter the relative re-
placeability series by dehydration of interlayer cations and increasing the
potential for cation fixation. With respect to the ability to retard radionuclide transport, inasmuch as the radionuclides will comprise only a fraction
of the total dissolved solids in a breached repository, it is not sufficient to
show a clay's selectivity for these species, but it is necessary to prove their
retention in the face of abundant competition for exchange sites.

The second problem involves the change in physical properties that often
accompany the appearance of new interlayer exchange species. Highest swelling
properties are attained in smectite clays with a Na^+ exchange cation, which is
very susceptible to exchange within a repository environment. Dehydration of
interlayer cations and cation fixation is of course, accompanied by decrease in
swelling.

Numerous studies show the high selectivity of various clays, primarily smectites, for aqueous radionuclides; however, no single clay can absorb all the
available cations. Smectites have selectivity for transuranics, but these species may not be fixed within the lattice. Smectites also selectively adsorb fission products, e.g., Cs, Sr, but their retention of these species is limited.[18]

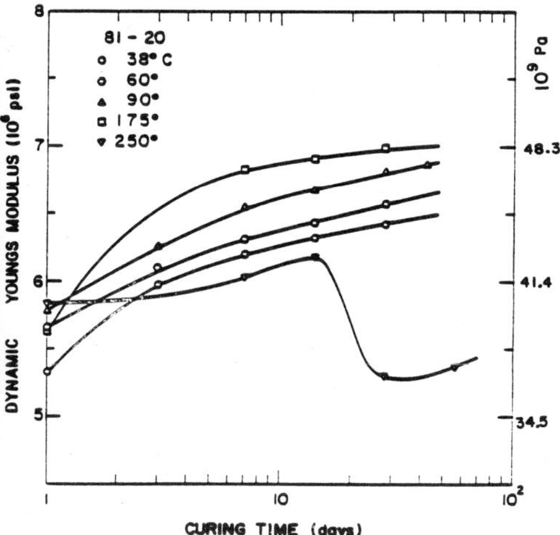

Fig. 3. Dynamic Modulus of Mixture 81-20 as a Function of Time and Curing Temperature.

The transuranics may be selectively adsorbed by smectites or other lattice-substitution dominated clays on the basis of high valences and large atomic numbers, but their fixation is limited by an equally large energy requirement for dehydration of adsorbed water. The studies in progress are addressing such questions, including the kinetics of phase (and consequent properties) transformations.[21]

For parts of the seal system, the most satisfactory materials will be those which are relatively unsusceptible to the type of alteration which would decrease adsorptive/fixation capacity for radionuclides. Therefore, the stability of zeolitic and clay backfill materials in a simulated bedded salt repository environment under elevated temperature conditions currently is being investigated. Results of treatment at 200°C under a confining pressure of 30 MPa for 4 weeks in a NaCl-rich brine and a $MgCl_2$-rich brine showed that the zeolites did not seem to alter in the $MgCl_2$ - rich brine, while montmorillonite seems to have crystallized more completely. Some of the vermiculite transformed to irregularly interstratified vermiculite plus mica by the uptake of K^+ from the $MgCl_2$ brine, causing decreased exchange capacity. Some zeolites altered to analcime, others showed no change, while montmorillonites appeared stable in a NaCl brine.[3]

Since clay minerals and zeolites specifically sorb ions such as Cs and Sr,[18] their alteration can also be detected by a change in their selective ion sorption properties. Specific sorption of Cs and Sr by the various hydrothermally altered minerals is given in Table 8. Among all the minerals which were hydrothermally treated at 200°C, only in the case of phillipsite did selective Cs sorption decrease because its open structure altered to the non-porous analcime which does not easily exchange its cations. An increase of specific sorption of some minerals can be attributed to a change in the nature of the exchangeable ions by the NaCl brine treatment, i.e., the natural cations such as K, Ca and Mg in these minerals were displaced by Na which competes less efficiently with Cs than either K, Ca or Mg. Therefore, Cs sorption increased in these minerals which were treated at 200°C. Studies are also being carried out on minerals treated

TABLE 7. FORMULATIONS OF BASALT-COMPATIBLE MORTARS (ALL MATERIAL EXPRESSED IN GRAMS EXCEPT DEFOAMER, WHICH IS IN mL).

Mix Number (81-)	Water (E1)[a]	Class H Cement (H-3)	Slag (B49)	5 μm Silica Flour (B22)	Silica Fume (B38)	Sand [ASTM C 109] (C15)	Basalt Sand (C31)	Super-plasticizer (A32)	Defoamer (A27)
19	250[b]	300	---	150	50	250	---	5	0.08
20[d]	139[c]	308	---	150	42	810	---	6	0.16
21[e]	140	300	---	150	50	250	---	6	0.08
22	180	200	200	200	--	300	---	6	0.08
23	180	200	200	200	--	---	300	6	0.08
24[de]	158	250	150	200	--	---	300	6	0.1

[a]Numbers in parentheses ( ) are PSU numbers. [b]Not tested for workability. [c]Tested for workability. [d]Concrete formulation also prepared with basalt gravel. [e]Expansive component also included.

638

TABLE 8. SPECIFIC SORPTION OF Cs AND Sr AS AFFECTED BY THE HYDROTHERMAL ALTERA-
TION (4-WEEK TREATMENT) OF BACKFILLS IN SATURATED NaCl BRINE

Sample	Cs sorption K_d^1 (mL/g)		Sr sorption K_d^1 (mL/g)	
	untreated	200°C/30 MPa	untreated	200°C/30 MPa
Montmorillonite, WY	410	400	170	200
Montmorillonite, TX	310	685	310	370
Vermiculite, SC	8900	10100	2000	980
Phillipsite, NV	11700	240	460	25
Mordenite, AZ	8300	9500	310	670
Mordenite, NV	6400	6700	170	370
Clinoptilolite, ID	3900	4800	3000	6100
Clinoptilolite, CA	5600	5600	4700	4700
Erionite, CA	9500	11700	490	590
Chabazite, AZ	6600	8900	310	400

$^a K_d$ is a distribution coefficient and is defined as the ratio of the amount of
Cs sorbed per gram to the amount of unsorbed Cs remaining per milliliter of the
solution.

at 300°C, and in $MgCl_2$-rich brines (at both 200° and 300°C). Both of these conditions give rise to considerably greater changes. Accompanying physical properties studies are in progress on smectites and mixed layer species.

Compatibilities of clay mineral types with basalt and groundwater compositions are also under investigation. Among the family of clays, clay composites
and zeolitic materials, there appear likely to be seal materials having superior
swelling properties, which will remain sorptive for radionuclides, and which will
not lose their physical properties drastically with prolonged existence in the
basalt repository.

SUMMARY

Studies of radioactive waste repository seal materials have shown the apparent
compatibility of some cementitious seal materials with the repository environments for which they were designed, both stratigraphic sequences from a bedded
salt area, and plateau basalt flows. The accompanying mechanical and physical
properties appear favorable. Accelerated test data on some clays and zeolitic
materials have placed tentative limits on conditions for their adequate performance. Low permeability has been shown to be a key factor in assuring durability; however, chemical compatibility also is important for preventing slow
changes from taking place in the seal materials, which over prolonged periods
might have cumulative harmful effects on the materials' performance. Thermodynamic data and information from old and ancient analog materials are used as
supplemental evaluatory data concerning long-term durability.

ACKNOWLEDGEMENTS

This research was supported under Subcontract E512-04200 with Battelle Project
Management Division, Office of Nuclear Waste Isolation under Contract Number
DE-AC06-76-RL01830-ONWI with the U.S. Department of Energy.

REFERENCES*

1. ONWI (1980). Repository Sealing Design Approach-1979, Report No. ONWI-55,
    D'Appolonia Consulting Engineers, Inc., 10 Duff Rd., Pittsburgh, PA.
2. Wigley, M.R., Tammenagi, H.Y., and Gnirk, P.F. (1981). Proc., 1981 NWTS
    Program Information Meeting, U.S.D.O.E./NWTS-15, November 1981, 143-148.
3. Roy, D.M. (1981). Proc., 1981 NWTS Program Information Meeting, U.S.D.O.E./
    NWTS-15, November 1981, 55-64.
4. Ellison, R.D., Kelsall, P.C., Schubert, C.E., and Stephenson, D.E. (1981).
    Proc., 1981 NWTS Program Information Meeting, U.S.D.O.E./NWTS-15, November
    1981, 51-54.
5. Roy, D.M., Grutzeck, M.W., and Licastro, P.H. (1979). Evaluation of Cement
    Borehole Plug Longevity, Report No. ONWI-30.
6. Roy, D.M. (1980). Proc., Workshop on Borehole and Shaft Plugging, OECD/DOE,
    Columbus, OH, May 7-9, 1980; OECD, 75775 Paris, Cedex 16, France, 353-368.
7. Roy, D.M., Grutzeck, M.W., Mather, K., and Buck, A.D. (1981a). PSU/WES Inter-
    laboratory Study of an Experimental Cementitious Repository Seal Material,
    ONWI-324.
8. Roy, D.M. and Asaga, K. (1980). Rheological Properties of Cement Mixes:
    V. The Effects of Time on Viscometric Properties of Mixes Containing
    Superplasticizers; Conclusions, Cem. Concr. Res. 10, 387-394.
9. Wakeley, L.D., Roy, D.M., and Grutzeck, M.W. (1981b). Experimental Studies
    of Seal Materials for Potential Use in a Los Medaños-Type Bedded Salt
    Repository, ONWI-325.
10. Goto, S., and Roy, D.M. (1981). Diffusion of Ions Through Hardened Cement
    Pastes, Cem. Concr. Res. 11, 751-757.
11. Sarkar, A.K., Barnes, M.W., and Roy, D.M. (1980). Longevity of Borehole and
    Shaft Sealing Materials: I. Thermodynamic Properties of Cements and Related
    Phases, ONWI-201, MRL-PSU.
12. Wakeley, L.D., Scheetz, B.E., Grutzeck, M.W., and Roy, D.M. (1981a), Cement
    and Concrete Research 11, 131-142.
13. Gulick, C.W., Jr., Boa, J.A., Jr., and Buck, A.D. (1980). Bell Canyon Test
    (BCT) Cement Grout Development Report, SAND 80-1928, Sandia National Labora-
    tories, Albuquerque, NM 87185 (contributions by WES and MRL-PSU).
14. Hunter, T.O. (1982). Borehole and Facility Sealing Activities for the Waste
    Isolation Pilot Plant, SAND 81-2034.
15. Roy, D.M., Scheetz, B.E., and Barnes, M.W. (1981b). Preparation of Cement
    Composite Nuclear Waste Forms and Their Chemical and Physical Properties.
    Report PSU-021, MRL-PSU, to Rockwell International (contract DE-AC09-79ET-
    41900).
16. Langton, C.A. and Roy, D.M. (1980). Longevity of Borehole and Shaft Sealing
    Materials: II. Characterization of Cement-Based Ancient Building Materials,
    ONWI-202, MRL-PSU.
17. Ogawa, K. and Roy, D.M. (1981-1982); Cement and Concrete Research 11, 741-
    750 (1981); 12, 101-110 (1982); 12, 247-256 (1982).
18. Komarneni, S. and Roy, D.M. (1980). Clays and Clay Minerals 28, 142-148.

*References with "ONWI" refer to Office of Nuclear Waste Isolation, Battelle
Memorial Institute, Columbus, OH. MRL-PSU denotes Materials Research Laboratory,
The Pennsylvania State University, University Park, PA. NWTS denotes National
Waste Terminal Storage, WES denotes, U.S. Army Engineer Waterways Expt. Station.

Published 1982 by Elsevier Science Publishing Co
SCIENTIFIC BASIS FOR RADIOACTIVE WASTE MANAGEMENT - V
Werner.Lutze, editor

DEVELOPMENT OF A BACKFILL FOR CONTAINMENT OF HIGH-LEVEL NUCLEAR WASTE[*]

FLOYD N. HODGES, JOSEPH H. WESTSIK, JR., and LANE A. BRAY. Pacific
Northwest Laboratory,[**] P. O. Box 999, Richland, Washington 99352,
U.S.A.

ABSTRACT

Sodium and calcium bentonites, pressed to densities between 1.9 and 2.2
g/cm^3, have hydraulic conductivities in the range of 10^{-11} to 10^{-13} cm/s.
Batch sorption distribution ratios (R_d) indicate that Sr, Cs, and Am are
strongly sorbed on bentonites and zeolites, that Np and U are moderately
sorbed on bentonites and zeolites, and that Am, Np, U, I, and Tc are
strongly sorbed on charcoal. Sorption results with basalt and tuff ground
waters are similar; however, iodine in tuff ground water sorbs more
strongly on bentonites. Thermal diffusivity measurements for dry,
compacted ($\rho \sim 2.1$ g/cm^3) sodium bentonite indicate that the thermal
conductivity of a high density bentonite backfill should be roughly similar
to that of silicate host rocks (basalt, granite, tuff). These results
indicate that a bentonite backfill can significantly delay the first
release of many radionuclides into the host rock and that by forming a
diffusion barrier a bentonite backfill can significantly decrease the long-
term release rate of radionuclides from the waste package.

Introduction

The inherent uncertainties involved in the predicting of long-term behavior of large-scale geologic and hydrologic systems make the use of engineered backup systems an important part of geologic nuclear waste
isolation. A backfill (buffer) material, placed between the containerized
waste (waste form plus canister and overpack) and the host rock is an
important waste package component that may perform several useful
functions. A properly designed backfill may: 1) control the migration of
water through the waste package; 2) retard the migration of radionuclides
away from the waste form; 3) retard the migration of corrosive agents and
complexants inward to the containerized waste; 4) control the Eh and pH of
water reaching the containerized waste, and 5) provide mechanical isolation
and protection for the containerized waste (1,2,3,4,5).

Initial backfill work at Pacific Northwest Laboratory (PNL), intended to
develop backfill materials for non-salt geologies (i.e., basalt, tuff,

---

[*] Work performed for the U. S. Department of Energy under Contract DE-
AC06-76RLO 1830.
[**] Operated for the U. S. Department of Energy by Battelle Memorial
Institute.

etc.) has followed the lead of the Swedish KBS program (1) and has concentrated on montmorillonite clays (bentonite) as potential backfill materials. The choice of bentonite is reasonable because of its highly desirable properties, including low permeability, high sorption ratios, plasticity, and ability to swell. The major question about the use of bentonite is its stability at the higher temperatures (T > 100°C) proposed for U.S. high-level nuclear waste repositories.

Backfill development work at PNL is part of the PNL Waste Package Program sponsored by the Office of Nuclear Waste Isolation (ONWI). The PNL Waste Package Program is an integrated research and development activity that includes development of materials for canister and overpack, determination of waste form leaching mechanisms, multicomponent testing of waste package materials under simulated repository conditions, and modeling of waste package tests, in addition to backfill development.

Water Migration

Water is involved in most, if not all, credible waste package degradation mechanisms and transport with or through an aqueous phase is the only reasonable mechanism for transport of radionuclides in a geological environment. Therefore, control of ground water migration through the waste package is the most important of the potential backfill functions.

During the early, high-temperature phase of waste package life, water will be driven away from the waste package and the backfill will be dehydrated. Later, as temperatures fall and thermal gradients lessen, water will move back into the waste package and the backfill will become water saturated. Therefore, an understanding of the ability of a bentonite backfill to control the migration of water through a waste package requires the study of two distinct processes: 1) the initial saturation of the bentonite with water, and 2) the flow of water through a saturated bentonite.

Studies carried out as part of the Swedish KBS program (6) indicate that water moves into compacted bentonites rapidly and Pusch (7) has used the time required for a sample to reach maximum swelling pressure to show that the saturation mechanism can be modeled as a diffusion process. As part of the PNL effort to understand the saturation process, we have carried out a series of short term water saturation tests. The tests were carried out in

a permeability cell (4,5), with reference basalt ground water (8). In
these tests, cylinders of Na-bentonite, 5 cm in length, were pressed to an
initial density of approximately 2.1 g/cm^3 ($\sim$ 5 wt % H$_2$O), and one face was
subjected to a water pressure of 15 MPa (2200 psi) for pre-determined
times. At the end of each test the clay cylinder was sectioned into a
series of disks, and the water content of the disks determined. The
results of these tests, for times ranging from 4-hours to 28-days are
presented in Figure 1. The form of the water concentration profiles indicates that water is moving into the clay by a diffusion process. The
operative process is probably diffusion of water vapor; however, other
possibilities, such as surface diffusion, cannot be ruled out at the
present time.

Although water (vapor?) moves into the compacted clay rapidly and reacts
with the clay causing the clay to expand, the movement of liquid water
through clay compacts is a much slower process. Water can be forced
through a thin sample rather quickly. Liquid water breakthrough requires 5
to 6 days for a bentonite sample 0.5 cm thick and only 12 to 14 days for a
bentonite sample 1 cm thick; however, a 2 cm bentonite sample has been

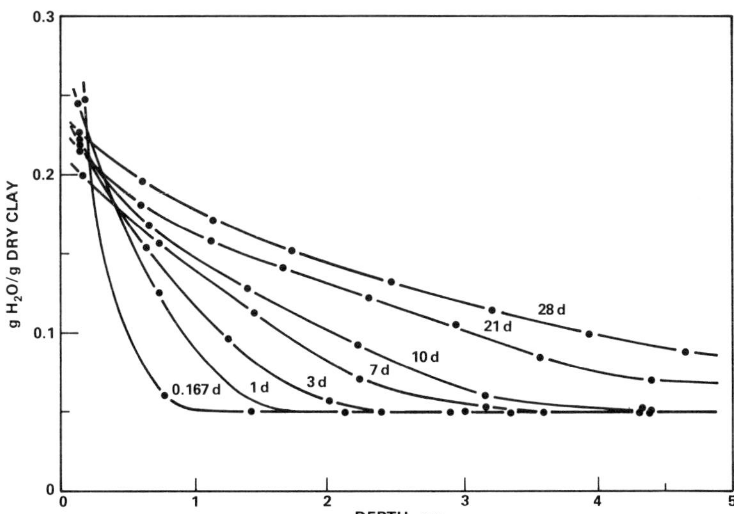

Fig. 1. Plot of measured water contents of sodium bentonite compacts
($\rho\approx$2.1gcm^3) held under water pressure (15 MPa) for times from 4 hours
(0.167 days) to 28 days. All compacts were approximately 5 cm in length.

under pressure (15 MPa) for almost one year without breakthrough of liquid
water. Negative results on a single experiment hardly form the basis for
a physical theory; however, the lack of breakthrough for the 2 cm bentonite
compact seems reasonable if we consider the low hydraulic conductivity and

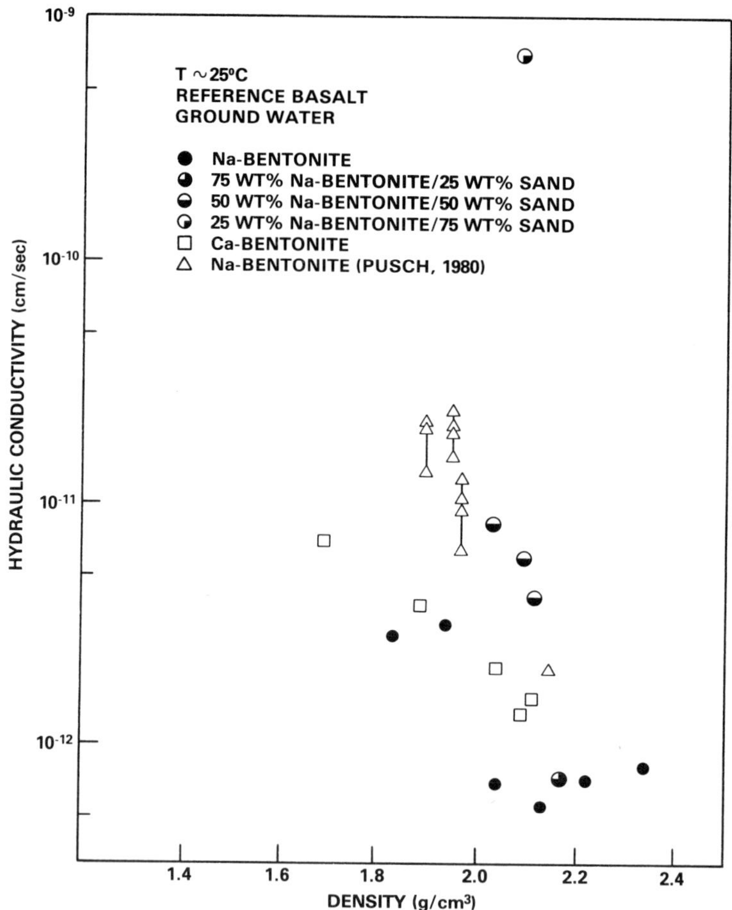

Fig. 2. Plot of hydraulic conductivity vs density for sodium and calcium
bentonites and sodium bentonite/quartz sand mixtures. All measurements,
except those of Pusch, were made with reference Hanford basalt ground
water.

high swelling pressure that will develop in the clay layer adjacent to the water source. The difficulty involved in transporting liquid water through this initial zone of saturation makes saturation of the remainder of the sample a very slow process.

Permeability determinations for compacted bentonites have been carried out with cylindrical clay compacts 1 cm in length (4,5). Hydraulic conductivities for compacted sodium bentonite, calcium bentonite, and sodium bentonite/quartz sand (80 to 120 mesh) mixtures are plotted as a function of compaction density in Figure 2. The data indicate that sodium bentonite is less permeable than calcium bentonite; however, the hydraulic conductivities of sodium and calcium bentonites are remarkably similar at high density, a result that would not be expected on the basis of their low-density swelling properties. The low hydraulic conductivities achievable with compacted bentonites, indicate that bentonite can be used to form a diffusion barrier around the waste. In this case water movement would be so slow that the dominant mechanism of radionuclide transport would be diffusion through the interstitial aqueous phase. The trend of increasing hydraulic conductivity with decreasing density is well defined and is similar to the trend shown by the KBS data for sodium bentonite (9). The difference between the two data sets is apparently mainly a result of differences in the way densities are reported. The KBS data is apparently reported in terms of a theoretical water-saturated density and the PNL data is reported in terms of an "as pressed" density ($\sim$ 5 wt % H_2O) measured prior to the test. At constant volume a water-saturated clay will have a higher density than an unsaturated clay, resulting in the apparent shift between the two data sets.

Radionuclide Sorption

The sorption of radionuclides on backfill materials will retard or slow transport of the radionuclides through the backfill and will strongly effect the release of radionuclides from the waste package (3). The retardation effect may completely prevent release of short-lived isotopes and delay the release of longer lived isotopes sufficiently to allow for significant decay.

Batch sorption distribution ratios[*] (R_d) have been determined for several commercially available, candidate backfill materials; including bentonites, zeolites, and charcoal; for a number of important nuclear waste components (Sr, Cs, Am, Np, U, I, and Tc). The results of distribution ratio determinations in reference basalt ground water (exposed to air at 25°C) have been reported previously (4,5). These results indicate that Sr, Cs, and Am are strongly sorbed (R_d = 10^2 to 10^5) by bentonites and some zeolites, that Np and U are moderately sorbed (R_d = 10 to 10^2) by bentonites and some zeolites, and that Am, Np, U, I, and Tc are strongly sorbed (R_d = 10^2 to 10^5) by charcoal. The high results for charcoal may, at least partly, indicate reduction and precipitation rather than true sorption.

A limited number of distribution ratio determinations have been carried out with tuff ground water from the Nevada Test Site (J-13 well). Both basalt and tuff ground waters are relatively dilute solutions and as expected the distribution ratios for the two ground waters are similar. One interesting difference, however, is the much lower distribution ratios for iodine in the basalt ground water (e.g., for sodium bentonite: R_d = 1.4×10^{-3} for basalt; R_d = 6.3 for tuff). This difference in distribution ratios for iodine is probably a result of the much higher concentrations of chlorine and fluorine in the basalt ground water.

Thermal Properties

The thermal conductivity of the backfill can have an important effect on the thermal history of the waste package. In this regard, it is important that the thermal conductivity of the backfill not be significantly lower than that of the host rock. If the thermal conductivity of the backfill is significantly lower than that of the host rock it will lead to appreciable increases in the maximum temperature experienced by the waste package.

The thermal conductivity of a backfill will be most important during the early, high-temperature phase of waste package life. During this period a bentonite backfill will probably be dehydrated and, as a result, its thermal conductivity will be at a minimum. Thus, the measurement of thermal conductivities for dry, compacted bentonites is very important for

---

[*] Determinations are 28-day batch contact. Rd is used rather than Kd to avoid implications of equilibrium or reversability.

waste package design. The determination of thermal conductivities for wet, compacted bentonites is a less significant but technically more difficult problem.

As a first step in determining the thermal conductivities of dry, high-density bentonite, thermal diffusivities have been measured using a laser thermal pulse technique (10). In this technique a short burst of laser light is used to heat one face of a thin (1 mm) compressed disk of bentonite and the temperature rise on the other face is measured. Determinations were made as the temperature was raised to approximately 600°C and were repeated as the sample was cooled to room temperature. The average value for fifteen thermal diffusivity measurements between 25°C and 600C° on a compressed sodium bentonite sample ($\rho\sim2.1g/cm^3$) is 3.25 $\pm$ 0.22 (1σ) x 10^{-3} cm^2/sec. Preliminary determinations of heat capacities for bentonite have not been completed; however, most silicates have Cp values in the range 0.2 to 0.3 cal/g°C over the temperature range of interest. If this range of values is assumed for bentonite, the thermal conductivities of the compacted bentonite (2.1 g/cm^3) will be in the range 0.6 to 0.9 watt/m°C, which is reasonably close to the range of values reported for various proposed silicate repository host rocks.

In addition to the thermal diffusivity measurements described above, direct determinations of thermal conductivity of dry compacted bentonites using a steady state thermal gradient technique are being initiated and preparations are underway for using a transient technique to determine the thermal conductivities of wet, high density bentonites.

<u>Summary</u>

A properly designed backfill can perform several useful waste package functions, providing both chemical and mechanical support for the containerized waste. Test results indicate that compacted bentonites have very low permeabilities and a high capacity for sorption of important radionuclide species. Therefore, a backfill of compacted bentonite can produce a large delay in the release of radionuclides from the waste package. Furthermore, if, as seems entirely possible, bentonite can be used to form a diffusion barrier between the waste and the host rock, the backfill will produce a significant decrease in the long-term release from the waste package. Determinations of thermal diffusivity for dry, compacted bentonites indicate that heat transfer through a bentonite

648

backfill should not be a serious problem for waste package design.

REFERENCES

1. R. Pusch, "Highly Compacted Sodium Bentonite for Isolating Rock-Deposited Radioactive Waste Products", Nuclear Technology 45, 153-157 (1979).
2. M. J. Smith et al., Engineered Barrier Development for a Nuclear Waste Repository in Basalt: An Integration of Current Knowledge, RHO-BWI-ST-7, Rockwell Hanford Operations, Richland, Washington (1980.)
3. E. J. Nowak, "The Backfill as an Engineered Barrier for Nuclear Waste Management", in Scientific Basis for Nuclear Waste Management, Volume 2, Plenum Press, New York, New York (1980).
4. E. J. Wheelwright, F. N. Hodges, L. A. Bray, J. H. Westsik, Jr. and D. H. Lester, Development of Backfill Materials as an Engineered Barrier in the Waste Package System -- Interim Topical Report, PNL-3873, Pacific Northwest Laboratory, Richland, Washington (1981).
5. J. H. Westsik, Jr., L. A. Bray, F. N. Hodges, and E. J. Wheelwright, "Permeability, Swelling and Radionuclide Retardation Properties of Candidate Backfill Materials", in Scientific Basis for Nuclear Waste Management, Volume 6, Plenum Press, New York, New York (in press).
6. R. Pusch, Water Uptake in a Bentonite Buffer Mass. A Model Study, KBS Teknisk Rapport 23 (1977).
7. R. Pusch, Water Uptake, Migration and Swelling Characterisitcs of Unsaturated and Saturated, Highly Compacted Bentonite, KBS Teknisk Rapport 80-11 (1980).
8. M. I. Wood, BWI Data Package for Reference Data on Ground Water Chemistry: I-Reference Grande Ronde Ground Water Composition. II-Recipe for Making Synthetic Grande Ronde Ground Water. III - Procedure for Making Synthetic Grande Ronde Ground Water. RSD-BWI-DP-007, Rockwell Hanford Operations, Richland, Washington, (July 11, 1980).
9. R. Pusch, Permeability of Highly Compacted Bentonite, KBS Teknisk Rapport 80-16 (1980).
10. J. L. Bates, High Temperature Thermal Conductivity of "Round Robin" Uranium Dioxide, BNWL-1431, Battelle Northwest Laboratory, Richland, Washington (1970).

ION/WATER MIGRATION PHENOMENA IN DENSE BENTONITES

Roland Pusch[+], Trygve Eriksen[++] and Arvid Jacobsson[+++], + +++ Div. of Soil
Mechanics, University of Luleå, 951 87 Luleå (Sweden), ++ Dept. of Nuclear
Chemistry, Royal Institute of Technology, 100 44 Stockholm (Sweden).

INTRODUCTION

The development of a suitable technique for isolating unreprocessed nuclear
reactor wastes from the biosphere has led to the Swedish multibarrier concept
KBS 2[1] with two engineered components, a thick-walled copper canister and a
clay body which confines the canister, Fig. 1. The clay consists of well fitting blocks of highly compacted Na bentonite made by "isostatic" compression of
bentonite powder. They are not water saturated when placed in the deposition
holes but take up water from the surrounding rock, swell and ultimately form a
tight contact with the rock and the canisters. When the bentonite is in physical
equilibrium with the surroundings it forms a medium with a number of valuable
properties, such as self-healing and ion exchange capacities. The healing means
that initial joints between blocks and voids, or local passages in the clay
cased by slight rock or canister displacements, will be sealed by the swelling
potential of the clay. The high ion exchange capacity retards the migration of
radionuclides through the clay barrier after corrosion of the canisters. The
most important feature is, however, the very low hydraulic conductivity and the
low ion diffusivity. These properties will be discussed in the present article.

CLAY MATERIAL AND PORE WATER

Granulometry and mineralogy One purpose of the KBS study of clay buffers is to
develop a technique for preparation and application of rather large quantities
of clay masses, which implies that only easily available and cheap soils are
considered. Thus, once the decision was made that smectites are most suitable,
the interest was focused on commercial bentonites. Two, fairly well defined
materials were selected for the study: The American Colloid Co. type MX-80
(Wyoming Na bentonite) being the main KBS reference material, and the Erbslöh
Ca-bentonite, representing natural calcium-saturated clay materials.

Both bentonites are characterized by a clay content (<2 μm) of approximately
85%, and a montmorillonite content of about 80-90 wt.% of this fraction. Silt
is the dominant remaining fraction which mainly contains quartz and feldspars
as well as some micas, sulphides and oxides.

650

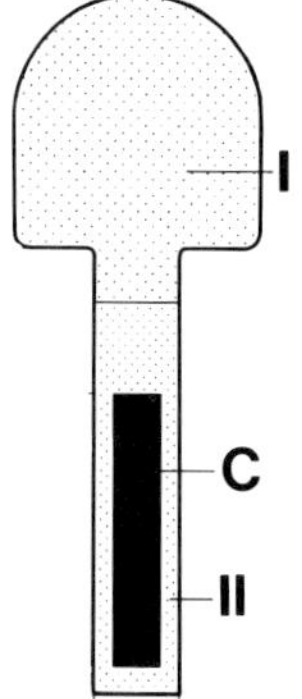

Fig. 1 Schematic view of tunnel and
deposition hole with canister
for unreprocessed reactor wastes
(C). I represents in-situ compacted sand/bentonite backfill,
II represents closely fitting
blocks of highly compacted
bentonite. The diameter and
length of the deposition hole
are 1.5 m and 7.7 m, respectively.

The two bentonites are not purely sodium or calcium saturated. Thus, spectrometric analyses have shown that the Erbslöh Ca bentonite contains 20-60 mg
Ca, 15-30 mg Mg, and 20-40 mg Na per liter of pore water, while the Wyoming
bentonite has a content of about 30 mg Ca, 15 mg Mg and 70 mg Na per liter pore
water. This relatively small difference points to fairly similar physical properties.

<u>Water</u> The KBS reference water, "Allard's" solution, which is a synthetic
ground water (91 mg of cations and 215 mg of anions per liter), was used for
water saturation and percolation in all the tests. It is considered to be
representative of the ground water at 500 m depth in Swedish crystalline rock.

<u>Microstructure, clay/water interaction</u> Heavily preconsolidated natural
bentonite clay beds, such as the Wyoming deposits, are characterized by a
marked orientation of the majority of the montmorillonite particles. The industrial processing of the mined clay material involves drying and grinding
which yields a powder with a grain size ranging from a few microns to a few
tenths of a millimeter. Each grain of the dried powder consists of a very large
number of 10 Å thick, more or less aligned montmorillonite crystal sheets, the
interlamellar spacing being about 3 Å.

When the powder is compressed to produce blocks of highly compacted bentonite,
the grains become welded together, but even at pressures of 50-100 MPa, inter-
aggregate pores with a diameter of up to about 50 microns may persist. In the

course of the water saturation after application in the deposition holes in the
rock, these voids become partly filled by the expanding grains. This expansion
leads to a two- to four-fold increase of the interlamellar spacing in the
grains and, thus, to an improved homogeneity with respect to the average interparticle distance (Figs. 2 and 3). The stronger swelling ability of Na
bentonite than of Ca bentonite means that more voids in the latter clay stay
open.

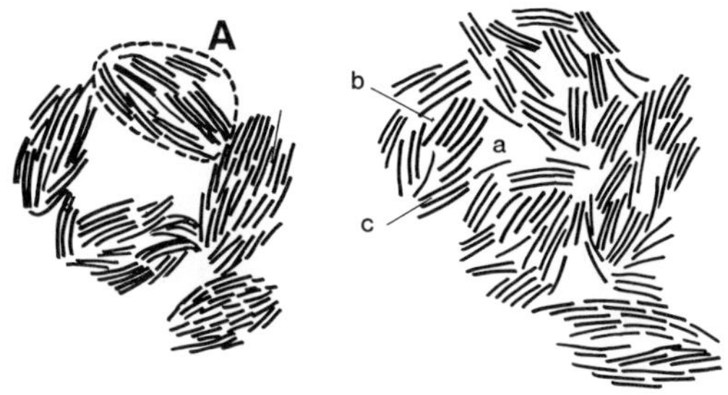

Fig. 2   Schematic particle arrangement in dense bentonites. Left picture:
         powder grains in an air-dry state. Right picture: state after water
         uptake and redistribution leading to fairly homogeneous conditions.
         A = dense particle aggregate, a = large interparticle void, b = small
         interparticle void, c = interlamellar space.

    Fig. 3, in which the fluffy character of water-exposed Na montmorillonite is
very obvious, illustrates the generally observed trend of "self-healing" on a
microscopic scale in the sense that the voids appear smaller in matured samples.
The study comprised about 100 electron micrographs which offered a fairly safe
statistical base, particularly with respect to the microstructural differences
between Na and Ca bentonites (Fig. 4). We see that the Na bentonite consists of
a fairly homogeneous montmorillonite matrix, while the equally dense Ca
bentonite is characterized by a more open, aggregated pattern. This is in per-
fect agreement with literature-reported observations[2].

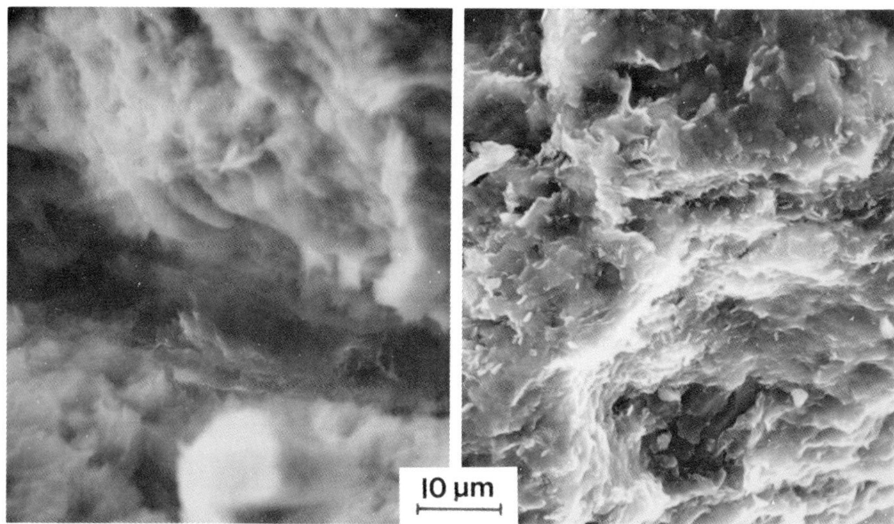

Fig. 3   Example of "maturing" of Na bentonite with a bulk density of 2.1 t/m^3.
Left SEM micrograph represents 14 days old sample with large void which
is getting occupied by expanding clay aggregates. Right micrograph
shows a sample confined in a cell for 1 year. Both were fresh-water
saturated immediately after the compression and freeze-dried soon before
the microscopy.

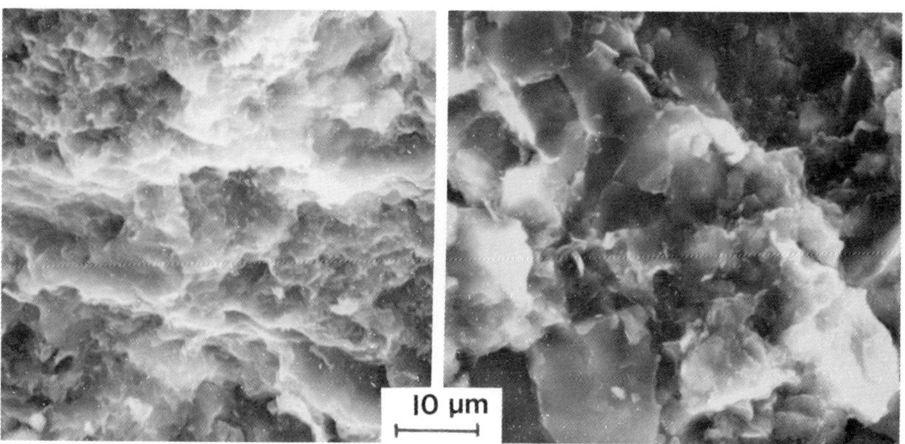

Fig. 4   SEM-pictures of Na bentonite (left) and Ca bentonite (right) both
matured for 1 year at a bulk density of 2.0 t/m^3.

EXPERIMENTAL

General   Precompaction under a pressure of 50-100 MPa yields blocks with a bulk
density of about 2.1 t/m^3 and approximately 60% water saturation of the pore
space. In the case of the KBS 2 concept, the density of confined samples increases to about 2.2 t/m^3 at complete water saturation, but in practice, a
certain amount of swelling will occur so that the final bulk density may ultimately be about 2 t/m^3. The preparation of clay samples for the present investigation yielded bulk densities ranging from 1.87-2.21 t/m^3 when water saturated.

Permeability tests   The clay powders were compacted in oedometers to various
densities. After saturation with "Allard water" to reach a high degree of water
saturation (> 97%), they were subjected to an external water pressure. In most
tests, the pressure was varied systematically in the order 500 kPa, 250 kPa,
100 kPa, and 50 kPa to find out whether the hydraulic gradient affects the
permeability. Each resulting gradient (10^4, $5 \cdot 10^3$, $2 \cdot 10^3$, and 10^3,
respectively) was held constant until stationary flow conditions were approached. The total testing time required for each sample was therefore rather
long; 3-4 weeks as an average.

654

The percolated water quantities were extremely small and the flow rate was
therefore determined by observing the rate of displacement of a water meniscus
in a calibrated capillary connected to the filter at the water exit. The
accuracy of the evaluated permeability was estimated at approximately $\pm 5 \cdot 10^{-15}$
m/s. The evaluated permeability was found to be a function of the hydraulic
gradient as illustrated in Fig. 5. This confirms the often debated idea that
the rate of water flow through very fine-grained clays often deviates from
Darcy's law[3].

The results are summarized in Fig. 6 which shows the coefficient of permeability as a function of the bulk density for the two investigated bentonites.
The MX-80 values are higher than literature-derived coefficients for Na
montmorillonite, the latter ones being represented by the area between the two
broken curves. This discrepancy may be explained by different hydraulic gradients and purities.

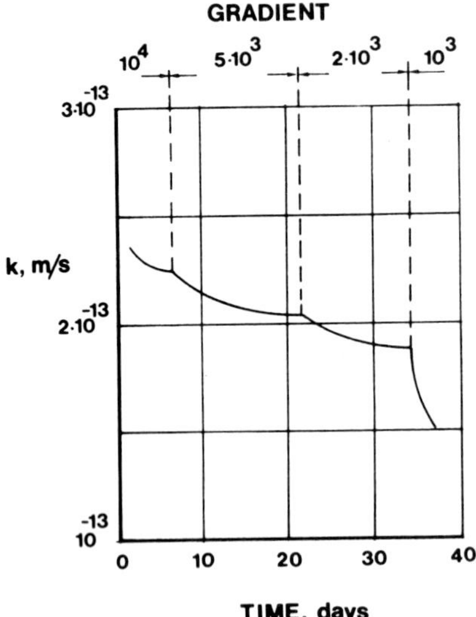

Fig. 5 Example of permeability
test with smoothed curve
for the coefficient of
permeability k versus time
after the onset of perco-
lation.

The gradient-dependence is of utmost importance, and at the low gradients
which will finally be operative in a repository, the permeability will most
probably be of the order indicated by the literature survey for Na montmorillonite (lower boundary) or less than that. For Ca-bentonite the permeability is

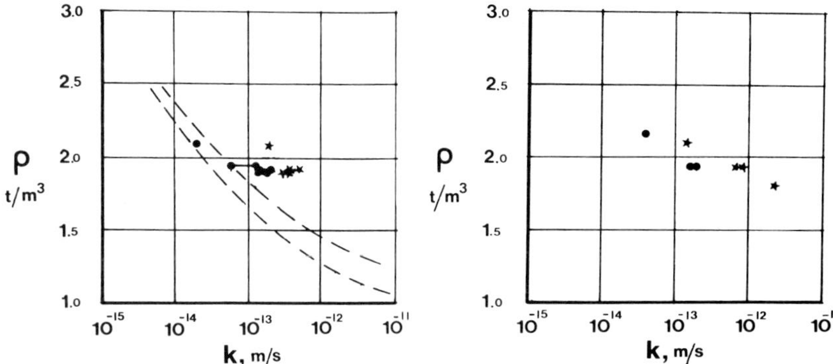

Fig. 6 The coefficient of permeability versus bulk density of Na-bentonite
(left) and Ca-bentonite (right). Where two values are plotted for
one and the same density, the left represents the gradient $i = 10^3$,
and the right $i = 10^4$. Stars represent tests run at 70 $^\circ$C.

approximately 2 to 5 times higher. In an initial period in the repository when
the temperature is expected to be of the order of 70 $^\circ$C in the bentonite, the
permeability is estimated to be approximately 5-10 times higher than at 20 $^\circ$C
for both clays. We can conclude from this that practically impervious conditions
prevail when the gradients are low. Thus, with a regional gradient of 10^{-2} and
a permeability of 10^{-13} m/s, the average flow rate will not be higher than
approximately 1 mm in 30,000 years.

The physical explanation of the obvious deviation from Darcy's law may be
variations in the mobility of the pore water. Thus, a higher viscosity of
mineral-adsorbed than of surface-distant pore water implies that the mobile
fraction of the pore water is determined by the hydraulic gradient. If so, the
evaluated permeability should also be gradient-dependent.

Substantial evidence has accumulated pointing to a strong fixation of the
first molecule layers which seems to determine interparticle and interlamellar
spacings[4]. The current study of the swelling pressure shows that tests run with
distilled water give considerably higher values than those obtained from tests
with saline water. Qualitatively, this is in agreement with electrical double-
-layer theories but quantitatively there is a large discrepancy between the
experimental data and theoretically derived curves. The most important finding,
however, is the obvious, unique relationship between swelling pressure and bulk

density when the latter exceeds 2 t/m^3. Thus, neither the salinity of the pore water nor its composition seem to affect, substantially, the swelling pressure at such high densities. The probable explanation is that electrical double-layers are not or only partly developed and that strongly adsorbed water determines the swelling pressure. This suggests that surface/water interaction is the primary cause of swelling pressures in dense montmorillonites.

Diffusion tests Two main series have been conducted. In one of them, highly compacted Na and Ca bentonites were placed in contact with aqueous solutions of ^{134}Cs$^+$, ^{85}Sr^{2+}, ^{131}I$^-$ and ^{36}Cl$^-$, and the diffusivities were calculated from the tracer concentration/distance profiles in sectioned samples 10 days after the onset of diffusion[5]. A second series involved water saturated samples with tracer-doped, saturated samples to avoid possible disturbing effects of the porous filters used in the first test series[6], cf. Fig. 7.

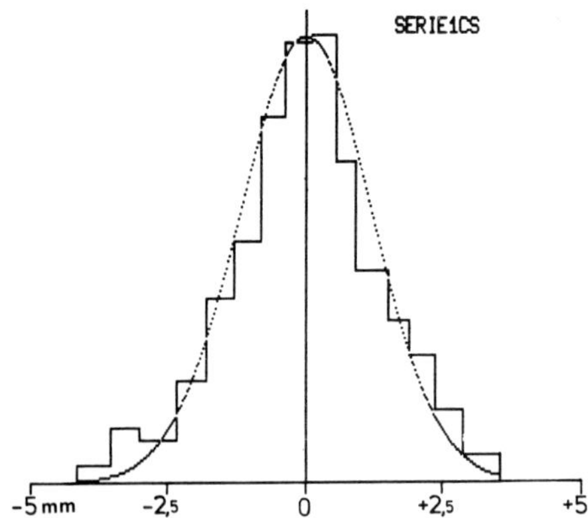

Fig. 7 Distribution of ^{134}Cs on both sides of the interface between non-doped and doped Ca bentonite samples after about 300 hours. Dotted line represents the theoretical distribution assuming instantaneous radio-active source and D = 7.5 x 10^{-13} m^2/s.

A first attempt to interpret the experimental results was based on a simple pore diffusion model, which implies that the ion migration takes place in the pore water and that the ions are retarded by sorption processes. The experiments were run as one-dimensional tests which yields:

$$\partial c/\partial t = D_w \, \partial^2 c/\partial x^2 \tag{1}$$

where c = concentration in the pore water

t = time after onset of diffusion

D_w = diffusivity (in water)

x = distance from interface between tracer solution and sample

Since there is sorption onto clay minerals we have:

$$q = K_d \cdot c \tag{2}$$

where q = concentration in solid

K_d = distribution coefficient

If K_d is taken as a constant, combined diffusion/sorption equations take the form:

$$\partial c/\partial t = D \partial^2 c/\partial x^2 - \rho_d \cdot \partial q/\partial t \tag{3}$$

where ρ_d = dry density

Eqs. (2) and (3) yield:

$$\partial c/\partial t = D/(K_d \rho_d + 1) \, \partial^2 c/\partial x^2 \tag{4}$$

$$\text{or } D_{obs} = D_w/(1 + K_d \rho_d) \tag{5}$$

Since ρ_d is somewhat less than 2 t/m^3 and K_d is known to be at least 50-100 this model suggests very low migration rates of Sr^{2+} and Cs^+ compared to those of the non-sorbed ions I^- and Cl^-. This is in marked contrast to the experimental findings, however, which gave $D_{Sr^{2+}} = 2.3 \times 10^{-11} - 4.8 \cdot 10^{-11}$ m^2/s, $D_{Cs^+} = 3.4 \times 10^{-12} - 7.5 \cdot 10^{-12}$ m^2/s, $D_{I^-} = 10^{-12} - 4 \cdot 10^{-12}$ m^2/s, $D_{Cl^-} = 6 \times 10^{-12}$ m^2/s (MX-80 only). This suggests that non-sorbing ions migrate only in the continuous but tortuous system of voids with varying width, while cations also seem to move through interlamellar space, probably according to some "ion-exchange" - type model. This fits with the observation that the cation

658

diffusion rates were practically the same in both bentonites, despite the obvious microstructural differences. The low diffusion rate of the investigated anions, as compared with the corresponding rate in bulk water, verifies that the diffusive resistance is strong.

It is also concluded that the experimental data cannot be explained by a model in which hydrated ions are free to move in an aqueous phase since their radii increase in the order; $Cs^+ < Cl^- < I^- < Sr^{2+}$. Instead, it is possible that the ions move through the clay fully or at least partly stripped of their hydration shells, which is in agreement with the concept of largely immobilized inter-lamellar water in highly compacted bentonite. The main determinant of diffusion rates may therefore be the ionic size.

REFERENCES

1. KBS (1978) Handling and Final Storage of Unreprocessed Spent Nuclear Fuel. Vol. II Technical, pp. 33-44, 127-134.

2. Smart, P. and Tovey, N.K. (1981) Electron Microscopy of Soils and Sediments: Examples. Clarendon Press, Oxford.

3. Hansbo, S. (1960) Consolidation of Clay with Special Reference to Influence of Vertical Sand Drains. Swed. Geotechn. Institute, Proc. No. 18.

4. Pusch, R. (1982) Mineral/Water Interactions and their Influence on the Physical Behaviour of Highly compacted Na Bentonite. Can. Geot. J. (in press).

5. Eriksen, T.E., Jacobsson, A. and Pusch, R. (1981) Ion Diffusion through Highly Compacted Bentonite. SKBF/KBS, Teknisk Rapport 81-06.

6. Eriksen, T.E. and Jacobsson, A. (1982) Ion Diffusion in Compacted Sodium and Calcium bentonites. SKBF/KBS Teknisk Rapport 81-12.

Published 1982 by Elsevier Science Publishing Co
SCIENTIFIC BASIS FOR RADIOACTIVE WASTE MANAGEMENT- V
Werner.Lutze, editor

TRANSPORT OF ACTINIDES THROUGH A BENTONITE BACKFILL

B. TORSTENFELT, H. KIPATSI, K. ANDERSSON, B. ALLARD AND U. OLOFSSON
Department of Nuclear Chemistry, Chalmers University of Technology,
S-412 96 Göteborg, Sweden.

INTRODUCTION

Compacted bentonite has been proposed as a suitable backfill material in the Swedish concept for underground storage of high-level waste[1]. The backfill barrier will serve both as a physical barrier, preventing convective water flow and allowing radionuclide penetration only by diffusion, and as a chemical barrier capable of chemically interacting with the radionuclides.

Previously the diffusion of fission products (Sr, Tc, and Cs) in compacted bentonite has been investigated within the present program[2]. This paper presents measurements of the diffusion and sorption of actinides in bentonite and the technique used for these studies.

EXPERIMENTAL

Radionuclides and clay-water systems

The studied systems are given in Table I. The short-lived nuclides ^{234}Th and ^{233}Pa were recovered from ^{238}U and ^{237}Np, respectively. In all experiments Wyoming Bentonite MX-80 (a Na-bentonite) was used[3].

The aqueous phase was a synthetic groundwater[4] with a total carbonate concentration of 123 mg/l, total salt concentration of 306 mg/l and pH around 8.2, preequilibrated with bentonite. In one of the U-experiments the total carbonate concentration was increased from 123 to 600 mg/l and in another one 10 mg/l of a humic acid[5] was added. Some U-experiments were also performed using a clay containing 1% $Fe_3(PO_4)_2$(s) or 0.5% Fe(s), which were thoroughly mixed with the clay.

Diffusion measurements

The clay was compacted to a density of $(1.9-2) \times 10^3$ kg/m^3, which is representative of the backfill barrier proposed in the Swedish concept for final storage of radioactive waste[3]. The compaction was performed with a KBr-pellet press used for IR-spectrometry.

To avoid concentration gradients in the compacted clay, due to incomplete equilibrium between the clay and the aqueous phase, the clay was preequilibrated with water and thereafter dried in an oven at 105°C before the pressing.

TABLE I

Nuclides and clay-water systems

Radionuclide	Solid phase	Aqueous phase
^{85}Sr	Bentonite[a]	G.W.[b]
^{134}Cs	Bentonite	G.W.
^{234}Th	Bentonite	G.W.
^{233}Pa	Bentonite	G.W.
^{233}U	Bentonite	G.W.
^{233}U	Bentonite	G.W. with 600 mg/1 HCO_3^-
^{233}U	Bentonite	G.W. with 10 mg/1 humic acid
^{233}U	Bentonite + 1% $Fe_3(PO_4)_2$	G.W.
^{233}U	Bentonite + 0.5% Fe	G.W.
^{237}Np	Bentonite	G.W.
^{239}Pu	Bentonite	G.W.
^{241}Am	Bentonite	G.W.

[a] Wyoming Bentonite MX-80[3]

[b] Synthetic groundwater[4]

A diffusion cell for studies of radionuclide diffusion in compacted clay has been constructed (See the description in ref 2).

The diffusion cylinder containing the bentonite, pressed as dry clay (water content 10%), was submerged in preequilibrated water for about 2 weeks for homogenization of the clay sample. Half of the required clay volume was pressed into an empty diffusion cylinder. A thin clay plate containing the radionuclide was placed on top of the cylinder, and wet homogenized clay was pressed on top of the radioactive layer until the whole diffusion cylinder was filled with a homogenous clay body, containing the radionuclide. The end surfaces of the cylinder were covered with metal frits and the whole diffusion cell was placed in the preequilibrated water. After a period of 2 - 3 months the cell was opened, and a cylindrical core of the clay was punched out. The core was cut into 0.1 - 0.2 mm thick slices, and the radioactivity was determined using a NaI well-type detector for the γ-emitters and with a liquid scintillation counter for the α-emitters.

The clay plate containing the radionuclide was prepared by adding a small volume (e.g. 0.05 ml) of a stock solution to a thin (0.1 mm) clay plate and gently evaporate to dryness under an IR-lamp.

The distribution coefficient (K_d, m^3/kg) between clay and water was measured by a batch technique[4]. Clay (0.5 - 1 g) and water (20 - 25 ml) containing the radionuclide (two concentrations, 10^{-7} M and 10^{-9} M) were contacted, and the concentration remaining in the aqueous phase was measured (phase separation by centrifugation at 27000 g for 1 h) as a function of time (6 h - 6 w)[6]

DETERMINATION OF DIFFUSIVITY

The diffusivity (D) is given by the diffusion equation

$$\frac{dc}{dt} = \frac{d}{dx}\left(D\frac{dc}{dx}\right) \tag{1}$$

(valid for one-dimensional diffusion), which has to be solved by applying appropriate initial and boundary conditions[7]. In the present case with a plane source containing a limited amount of substance which is diffusing in a cylinder of infinite length and assuming a concentration independent D, the solution is

$$\frac{c}{M} = \frac{1}{2(\pi Dt)^{1/2}}\, e^{-x^2/4Dt} \tag{2}$$

where c = concentration (mol/m^3), M = total amount of diffusing species added (mol/m^2), x = distance from source (m), D = diffusivity (m^2/s), and t = time (s).

The diffusivity obtained by applying eqn (2) to measured concentration profiles is an apparent diffusivity, i.e. the effect of sorption on the solid is included in D. In case of a reversible sorption on the clay, defined by a concentration independent distribution coefficient (K_d, m^3/kg), the relation between the apparent diffusivity (D_a) and the diffusivity not affected by sorption (D) is

$$D = D_a(1+K_d\rho) \tag{3}$$

where ρ = density of the solid (kg/m^3). If K_d is concentration dependent, also the apparent diffusivity will vary with the concentration, and eqn (2) will not be valid. In this case, eqn (1) must be solved by numerical methods, taking the sorption isotherm into account.

RESULTS AND DISCUSSION

The concentration profiles measured for Cs and Sr, Th, U, Np, Pu and Am, respectively, are shown in Figures 1 - 10, and the calculated diffusivities (average values $(D_a)_{av}$ and D_{av}, and maximum values $(D_a)_{max}$ and D_{max}) are given in Table II.

Almost the same apparent diffusivities were obtained for Sr and Cs as previously reported (2.6×10^{-12} m^2/s and 9.0×10^{-13} m^2/s for Sr and Cs, respectively[2]. However, the experimental technique has been modified in two significant ways. In the previous experiments the radionuclides were added to the clay mixed with agar agar in a gel plate, while they in the present studies were added absorbed in a thin clay plate. All compacted clay samples were homogenized by contact with water prior to the radionuclide addition which was not the case previously.

Recently reported measurements, using a similar technique but a different diffusion cell have given higher apparent diffusivities (1.5×10^{-11} m^2/s and 2.5×10^{-12} m^2/s for Sr and Cs, respectively[8,9].

Just as in previous measurements the diffusivity (D) of both Sr and Cs are of the order of 10^{-9} m^2/s, which would in fact be representative of a cation transport in water[10].

All of the actinides, except Pa, behaves in a fairly uniform way. The diffusivities when the sorption is subtracted (D) are similar and all very low ($< 3 \times 10^{-10}$ m^2/s).

The high distribution coefficients and corresponding low apparent diffusivities for Th and Am seem to be representative of the behaviour of actinides in their lower oxidation state (highly hydrolyzed species with a low mobility)[6]. Evidently, Pu exists predominantly in one of the lower oxidation states, most likely in the tetravalent state[11]. Indications of a reduction of U(VI) to U(IV) in the presence of Fe(s) are evident, while the presence of $Fe_3(PO_4)_2(s)$ has little effect during the short contact time in the present experiments.

For all of the systems, the measured concentration profiles do not entirely agree with theoretical curves that can be obtained from eqn (2). The calculated apparent diffusivities are significantly lower close to the radionuclide source than at some distance from it. Non-linear sorption isotherms could lead to such results[12]. However, both for Sr[4] and the actinides in their lower oxidation state[6] the observed sorption isotherms are almost linear in the concentration range 10^{-4} - 10^{-8} M (for Sr) and 10^{-7} - 10^{-9} M (for the actinides). However, for the actinides in the tri- and tetravalent state (Th, Am, and Pu in the present experiment) it is likely that the solubility products for the trivalent carbonate and tetravalent oxide/hydroxide, respectively, have been exceeded. Under the present conditions the maximum solubilities of trivalent and tetravalent actinides would be of the order of $10^{-6.5}$ M and $10^{-8.5}$ M, respectively[11]. This would correspond to c/c_0 = 1.9×10^{-3} and 2×10^{-5} respectively, considering the amount of radionuclide added and the available pore water in the clay (c.f. Figure 1-10). Thus, only the maximum diffusivities, representative of species that have travelled far from the source, would correspond to the migration of truly soluble species. For the penta- and hexavalent states (Pa, Np and U, except possibly for the U-Fe-system), the solubility limit would not be exceeded.

TABLE II

Measured diffusivities and distribution coefficients

Element	$\log C_i$ [a]	Time	$(D_a)_{av} \times 10^{14}$	$(D_a)_{max} \times 10^{14}$	K_d [g]	$D_{av} \times 10^{10}$	$D_{max} \times 10^{10}$
		d	m^2/s	m^2/s	m^3/kg	m^2/s	m^2/s
Sr	-11	50	180	1200	2.9	100	670
Cs	-10	53	140	200	1.4	39	57
Th	-14	64	0.46	0.85	≥ 6	≤ 0.55	≤ 1.0
Pa	-13	76	≥ 57	≥ 60	5.0	≥ 55	≥ 60
U	-8	62	58	82	0.093	1.1	1.5
U[b]	-8	54	19	34			
U[c]	-8	62	57	94			
U[d]	-8	63	20	38			
U[e]	-8	53	4.5	11			
Np	-9	69	22	37	0.12	0.52	0.88
Pu	-9	53	0.69	3.0	3.5	0.48	2.1
Am	-9	77	0.40	1.4	6.6	0.53	1.8

[a] Number of moles initially added
[b] 600 mg/1 HCO_3^- in aqueous phase
[c] 10 mg/1 humic acid in the aqueous phase
[d] 1% $Fe_3(PO_4)_2(s)$ in the clay
[e] 0.5% Fe(s) in the clay
[f] Preliminary values
[g] For total nuclide concentrations of $\leq 10^{-9}$M

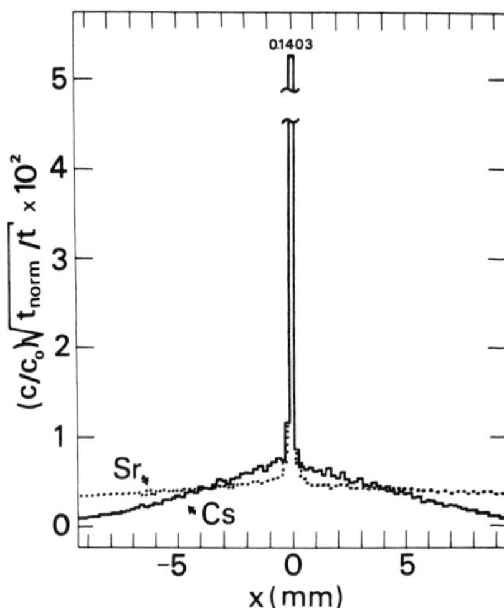

Fig. 1. Concentration profiles for Cs and Sr.

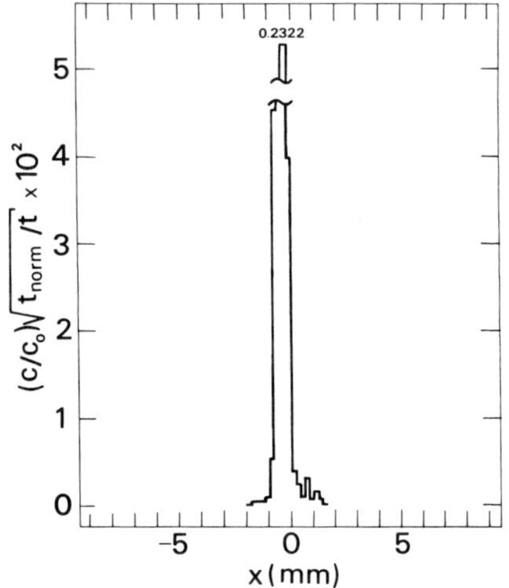

Fig. 2. Concentration profile for Th.

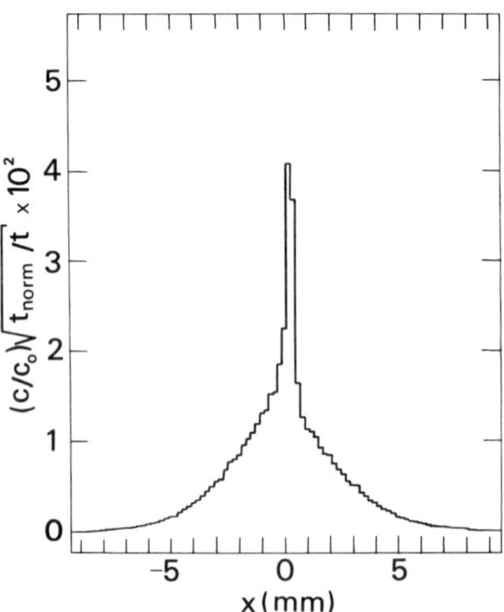

Fig. 3. Concentration profile for U.

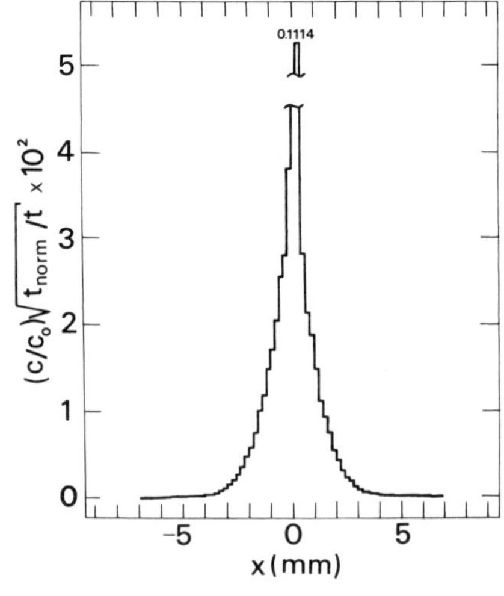

Fig. 4. Concentration profile for U with 600 mg/l HCO_3^- in the aqueous phase.

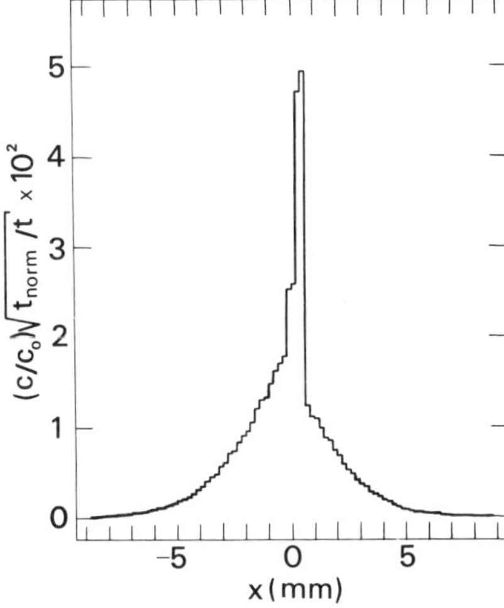

Fig. 5. Concentration profile for U with 10 mg/l humic acid in the aqueous phase.

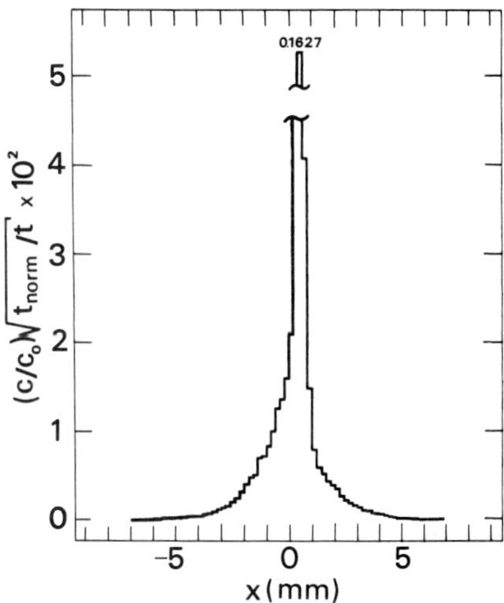

Fig. 6. Concentration profile for U with 1% $Fe_3(PO_4)_2(s)$ in the clay.

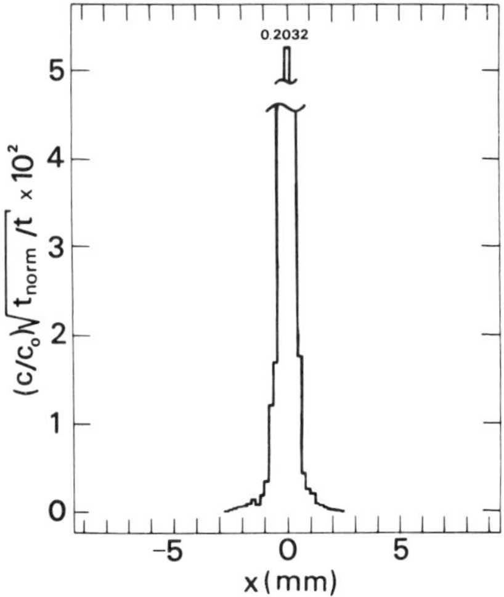

Fig. 7. Concentration profile for U with 0.5% Fe(s) in the clay.

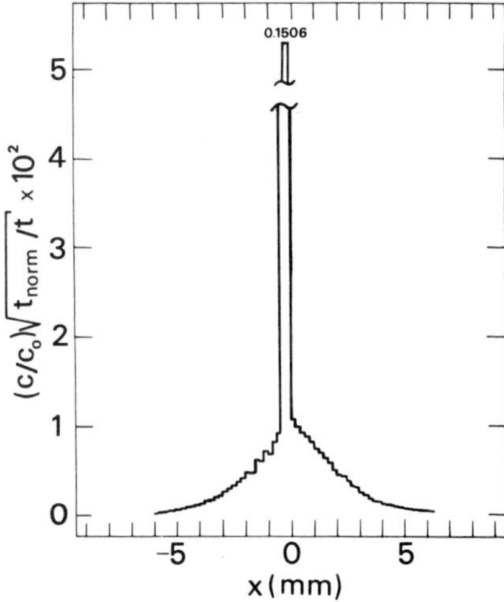

Fig. 8. Concentration profile for Np.

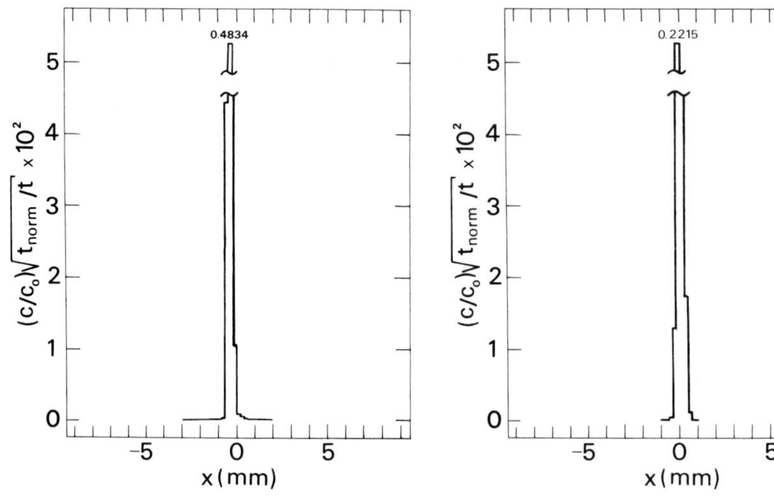

Fig. 9. Concentration profile
for Pu.

Fig. 10. Concentration profile
for Am.

RETENTION IN A BACKFILL BARRIER

For a diffusion process, the retention time t would be given by

$$t = 0.1 \, z^2/D_a \tag{4}$$

where z = barrier thickness (m)[13]. Breakthrough is here assumed to occur when
the concentration outside the clay barrier reaches 5% of the concentration at the source.

In Table III the retention time t according to eqn (4) as well as the
remaining fraction for z = 0.5 and z = 1 are given for some of the long-lived
radionuclides in spent fuel, using the measured apparent diffusivities in
Table II. The only one of these nuclides that is not significantly retarded
is ^{237}Np. Even after a reduction to Np(IV) and increasing the barrier
thickness to 1 m about 90% of the nuclide would remain after breakthough. In
order to reduce ^{237}Np to less than 10^{-3} a barrier thickness of ca 4.5 m has
to be introduced, assuming reducing conditions. These results are in fair
agreement with the general suggestions in ref. 13, as well as with the
predictions previously given[14].

TABLE III

Retention of long-lived radionuclides in 0.5 m compacted bentonite

Nuclide	t (y)	$t/t_{\frac{1}{2}}$ [a]	Remaining fraction after t	$4\ t$ [d]
^{90}Sr	68	2.4	0.9	1.3×10^{-3}
^{137}Cs	3.9×10^{2}	13	1.2×10^{-4}	<[c]
^{229}Th	9.3×10^{4}	13	1.5×10^{-4}	<
^{237}Np	2.2×10^{3}	1.0×10^{-3}	1.0	1.0
^{237}Np [b]	9.3×10^{4}	4.4×10^{-2}	1.0	0.9
^{239}Pu	2.6×10^{4}	1.1	0.5	5.2×10^{-2}
^{240}Pu	2.6×10^{4}	4.0	6.3×10^{-2}	1.6×10^{-5}
^{241}Am	5.8×10^{4}	135	<[c]	<
^{243}Am	5.8×10^{4}	7.9	4.2×10^{-3}	<

[a] $t_{\frac{1}{2}}$ = half-life, y; t according to eqn (4) for $z=0.5$
[b] Assuming reduction to Np(IV); values for Th(IV) used
[c] $<10^{-9}$
[d] For $z=1$

CONCLUSIONS

It has been confirmed that Sr and to some extent Cs migrates faster than expected, considering measured sorption data. A fraction of these elements seems to migrate rapidly.

The actinides U and Np in their higher oxidation states (V, VI) exhibit a mobility similar to but slightly lower than what is observed for Cs, considering only the apparent diffusivities. For Th, Pu and Am which are in the lower oxidation states (III, IV), the apparent diffusivities are 1-3 orders of magnitude lower. For all of the actinides (with the exception of Pa) similar diffusivities (without sorption) are obtained. These are significantly lower than for Sr and Cs.

The dominating oxidation state for Pu is probably IV, while U(VI) seems to be reduced to U(IV) by Fe in the clay.

The presence of a high carbonate concentration or humic acid has little effect on the U(VI)-diffusivity.

All of the nuclides ^{137}Cs, ^{229}Th, ^{241}Am and ^{234}Am would decay to insignificant concentration levels before breakthrough, assuming a 1 m thick clay barrier. For ^{90}Sr, ^{239}Pu and ^{240}Pu a significant reduction of the concentration would be expected, while the concentration of ^{237}Np would be very little reduced.

ACKNOWLEDGEMENT

This work has been funded by the Nuclear Fuel Safety Project (KBS). The skilful experimental help by Ms. M. Bengtsson, Ms. L. Eliasson, Mr. T. Eliasson and Ms. W. Johansson is gratefully acknowledged. The purified and characterized humic acid was provided by Prof. G. Choppin, Florida State Univ.

REFERENCES

1. "Handling of Spent Nuclear Fuel and Final Storage of Vitrified High-Level Reprocessing Waste" and "Handling and Final Storage of Unreprocessed Spent Nuclear Fuel", Kärnbränslesäkerhet (KBS), Stockholm 1977 and 1978.
2. B. Torstenfelt, K. Andersson, H. Kipatsi, B. Allard, U. Olofsson, "Diffusion Measurements in Compacted Bentonite" in S. Topp (Ed.), Scientific Basis for Nuclear Waste Management, Vol.6 , Elsevier, New York 1982, in press.
3. A. Jacobsson, R. Pusch, "Egenskaper hos bentonitbaserat buffertmaterial" (Properties of bentonite based buffer substances), KBS-TR-32, Kärnbränslesäkerhet, Stockholm 1978 (in Swedish).
4. B. Torstenfelt, K. Andersson, B. Allard, "Sorption of Sr and Cs on Rocks and Minerals. Part I: Sorption in Groundwater", Report Prav 4.29, National Council for Radioactive Waste, Stockholm 1981.
5. G. R. Choppin, L. Kullberg, "Protonation Thermodynamics of Humic Acid", J. Inorg. Nucl. Chem. 40 651 (1978).
6. B. Allard, U. Olofsson, B. Torstenfelt, H. Kipatsi, K. Andersson, "Sorption of Actinides in Well-defined Oxidation states on Geologic Media", Proc. Mat. Res. Soc. Ann. Meeting, Berlin, June 7-10 1982.
7. J. Crank, The Mathematics of Diffusion, (Oxford University Press, London 1956).
8. T. Eriksen, A. Jacobsson, "Ion Diffusion Through Highly Compacted Bentonite", KBS-TR-81-06, Kärnbränslesäkerhet, Stockholm 1981.
9. T. Eriksen, A. Jacobsson "Ion Diffusion in Compacted Sodium and Calcium Bentonites", KBS-TR-81-12, Kärnbränslesäkerhet, Stockholm 1982.
10. R. C. Reid, J. M. Prausnitz, T. K. Sherwood, The Properties of Gases and Liquids, 3rd ed. (Mc Graw-Hill, New York 1977).
11. B. Allard, "Solubilities of Actinides in Neutral or Basic Solutions", in N. Edelstein, Proc. of the Actinides-81 Conf., Pergamon Press, Oxford 1982, in press.
12. C. B. Amphlett, Inorganic Ion Exchangers, (Elsevier, Amsterdam 1964).
13. I. Neretnieks, "Transport of Oxidants and Radionuclides Through a Clay Barrier", KBS-TR-79, Kärnbränslesäkerhet, Stockholm 1978.
14. B. Allard, G. W. Beall, "Actinide-retaining Backfill Materials in Underground Repositories for Alpha-Contaminated Wastes", in Management of Alpha-Contaminated Wastes. (IAEA-SM-246/13, Vienna 1981, p. 667).

ANALYSIS OF THE CORROSION PRODUCTS FORMED ON Ti AND A Ti-Pd ALLOY DURING EXPOSURE IN HOT WATER

INGEMAR OLEFJORD AND HÅKAN MATTSSON

Department of Engineering Metals, Chalmers University of Technology,
S-412 96 Göteborg, Sweden

ABSTRACT

This is a preliminary report dealing with the surface analysis of reaction products formed on Ti and a Ti-Pd alloy during their exposure in hot water. The compositions of the aqueous media were varied with respect to the dissolved oxygen and the content of chloride ions. The temperature was $60^{o}C$ and the exposure times were 10 min. and 6 months. Work is in progress in which samples are exposed at $80^{o}C$ and $95^{o}C$ in the aqueous solutions. Surface analysis was also performed on a sample which had been exposed in water-saturated bentonite.

It appears from the ESCA spectra that the oxide products formed on the surface consist of TiO_2. The results also indicate that the thickness of the film formed at $60^{o}C$ in water is in the range 50 Å to 100 Å. This is somewhat more than that obtained after exposure in water at room temperature. Exposure for 6 months increases the thickness of the oxide two to three times compared to that obtained during the short exposure at $60^{o}C$. The analyses of the samples that had been embedded in bentonite indicate that the surface reaction products are thinner than those found on the surface after exposure in an open vessel.

INTRODUCTION

This investigation is one part in a research program dealing with the possibility of using Ti or a Ti 0.2 weight % Pd alloy as a material for the encapsulation of radioactive waste. The canister will be placed in deposition holes in bedrock at approximately 500 m depth. These holes will then be refilled with a sand-bentonite mixture. The design of the capsule system, details of the refilling and the composition of the environments are described elsewhere (1).

The aim of this project is to study by surface sensitive techniques (ESCA) the reaction products formed on the surfaces of Ti and a Ti-Pd alloy during exposure in an environment similar to that existing in bedrock at a depth of 500 m. The oxygen content at this depth is very low and the reducing components in the bentonite will consume oxygen so that the capsule environment will be almost oxygen free. The temperature is assumed to attain a maximum of 95°C during the first period after the waste is deposited. The real situation is simulated on the laboratory scale. The metals are exposed to water-saturated bentonite clay at a realistic pressure. Exposure times of up to a maximum of three years are planned.

In parallel with the water-saturated bentonite exposure, a second experimental series has been set up. The aim of this work is to make more fundamental studies of the reactivity of the metals with hot water containing various amounts of oxygen and sodium chloride. In this case the maximum exposure time is planned to be about half a year.

THE EQUIPMENT

Bentonite exposures

The real deposition milieu was simulated on the laboratory scale by embedding the samples in blocks of water-saturated dense bentonite. To achieve long term exposure while maintaining high pressure and high temperature the samples have to be located in thermostated pressure cells. For this purpose eight cells and one insulated vessel were made. The box marked A in figure 1a is a double-walled thermostated bath in which the pressure cells are located.

Figure 1b shows details of the pressure cell. It consists of a cylinder in which two pistons are fitted. Two water tubes are welded to one of the pistons. The sample is positioned in a block of bentonite. To prevent a change in the volume of the bentonite block due to swelling, two yokes are fixed to the two pistons by six bolts. Transportation of the bentonite into the water tubes is prevented by a porous filter of stainless steel located on the top of the bentonite clay.

The sample was pretreated by polishing on emery paper down to 600 mesh and then positioned in the center of a loosely packed volume of bentonite powder and compressed between two pistons in a hydraulic press giving a density of $2.15 \, \text{g/cm}^3$. Then, one of the pistons was removed and the cell was assembled. The bentonite was saturated with water having a composition (2) corresponding to the ground water in the actual bedrock. Table 1 shows the chemical composition of the water.

Table 1. Chemical composition of artificial ground water

Species	HCO_3^-	SiO_2	SO_4^{2-}	Cl^-	Ca^{2+}	Mg^{2+}	K^+	Na^+
Composition (ppm)	123	12	9.6	10	18	4.3	3.9	6.5

Fig. 1. A: Overview of the experimental set up. B: Pressure cell for the
bentonite exposure. C: Water exposure cell.

Water exposure

The aim of this part of the project was to study the influence of exposure
time, oxygen- and NaCl-content on the reaction products formed on the surface
of Ti and Ti-Pd alloy at 60°C. The exposure media and conditions were as

follows: low oxygen-containing water and oxygen saturated water, chloride free
water and a 1% NaCl solution. The exposure times were planned to be in the
range of a few minutes to half a year.

For adequately controlled long term exposure specially designed exposure
cells were built. Figure 1a shows the exterior of the three baths (marked B).
On a bench above the vessels, four water reservoirs made of Pyrex glass are
placed. Each of these contains one of the four different solutions. Teflon tubes
connect the reservoirs to the cells located in the baths. Figure 1c shows a
schematic view of the bath and the cell arrangement. The exposure water enters
the cell via a glass tube at the bottom. The flow rate - 0.3 1 per day - is low
enough to allow the water to achieve the desired temperature. The flow of water
through the cell ensures that the contents of oxygen and sodium chloride are
constant throughout the experiment.

RESULTS

Calibration

Analyses of cleaned Ti metal and TiO_2 were performed to obtain photoelectron
yield constants, Y, of Ti and oxygen and also to obtain the shapes and the positions of the individual signals. The ESCA signal representing the metallic state
of Ti was recorded from a sample of pure Ti, cleaned by ion etching (P_{Ar} = $5 \cdot 10^{-4}$
torr, acceleration voltage 3 kV, emission current 30 mA, etching time 10 min.).
The TiO_2 was formed on an ion-cleaned Ti sample by oxidizing it at $400^{\circ}C$ in
oxygen at 50 torr for 10 min. The oxidation was performed in the furnace
attached to the preparation system of the ESCA (3).

The ESCA spectra recorded from Ti and TiO_2 are shown in figure 2. The solid
curves are the as-recorded signals while the dotted peaks represent the single
core electron levels and the backgrounds caused by inelastic scattering of the
photoelectrons. The signals at 453.9 eV and 460.0 eV in row A represent the $2p_{3/2}$
and the $2p_{1/2}$ levels of the metallic state of Ti respectively. The small peak
at 454.7 eV belongs to the metallic state and is a shake-up satellite while the
component at 455.5 eV represents the divalent state. Row B in the figure shows
the spectrum recorded from the oxidized sample. The positions of the $2p_{3/2}$ and
the $2p_{1/2}$ peaks of Ti are at 459.0 eV and 465.0 eV respectively. This spectrum
corresponds to the four valence state of titanium, Ti^{4+}. The O 1s peak is positioned at 530.5 eV. The binding energies are very close to the values given in
the literature (4,5).

The thickness of a uniform surface oxide layer can be estimated from the
intensities of the ESCA signals representing the oxide and the metallic states

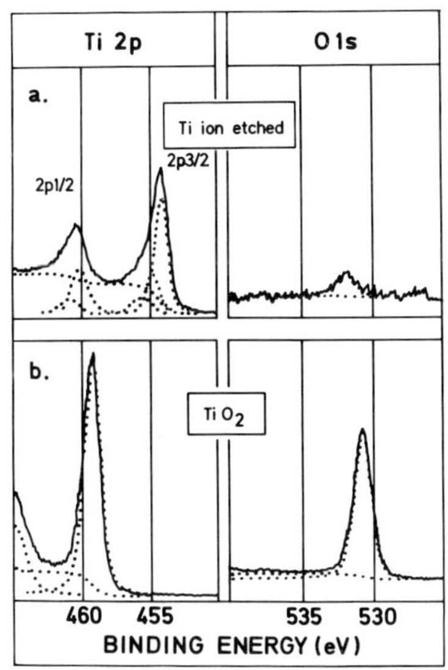

Fig. 2. ESCA spectra recorded from the metal and the oxide phases.

if its thickness is of the same order of magnitude as the attenuation length of the photoelectrons. The expression for the thickness, a, is:

$$a = \overline{\lambda}_{Ti} \cos 51.5^{\circ} \ln \left[1 + \frac{D_{Ti}^{met} I_{Ti}^{ox}}{D_{Ti}^{ox} I_{Ti}^{met} \beta} \right] \qquad (1)$$

where $\overline{\lambda}_{Ti}$ = 15 Å is the average value of the attenuation length of the Ti 2p electrons in the metal and the oxide phases; the angle 51.5° is the angle between the normal of the surface and the spectrometer axis; D is the density of Ti in the two phases; I is the measured intensity (Gaussian peak area); β is a factor which takes into account the relative probability of inelastic scattering of the photoelectrons. Details of conditions are given elsewhere (3,6).

Water exposure

The surfaces were pretreated by polishing on emery paper down to no. 600. The exposure times were 10 min. and 6 months at 60°C. The samples for the shortest exposure were prepared by polishing on emery paper in water at 60°C. The intention was that an oxide-free surface is created and instantaneously exposed to the oxygenated water. The long-term exposed samples were polished on emery paper at $25^{\circ}C$ and then assembled in the exposure cells. After exposure the samples were rinsed in methanol, dried and moved to the analyser.

Figure 3a shows signals recorded from samples exposed for short time (solid curves) and for long time (dotted curves) in oxygen saturated water. The measurements were performed at the etch levels marked in the figure. It appears from the spectra that the positions and the shapes of the 2p and the O 1s signals are the same as were obtained from the calibration studies of TiO_2. This indicates that at least the composition of the outermost region of the oxide product corresponds to TiO_2. The thickness of the oxide must be more than 40 Å since no contribution from the metal phase below the oxide is detected. After ion etching to a depth of 40 Å a peak representing the metal phase is recorded for the short-term sample. Further etching to 100 Å gives a signal originating almost entirely from metal. This shows that the thickness of the oxide film is in the range of 50 to 100 Å.

After exposure for 6 months the intensity of the recorded Ti 2p signals from the as-exposed surface is noticeably lower than that obtained from the short exposure time. However, the intensity of the oxygen signal is at least as high as before. The reason for the low ratio between the Ti and O signals is contamination by Si, Sn, Pb and Cu present as oxides. Analysis by atomic adsorption spectroscopy of the test water before and after it had passed the cell showed clearly that the source of Si, at least, was the pyrex glass wall. The contents of the other impurities were too low to be detected. From the ESCA and from the atomic adsorption spectroscopy measurements, it appeared that the dissolution of the glass was more pronounced when Cl^- ions were present in the solution. The surface has to be ion etched down to 100 Å before the impurity elements are removed.

It was pointed out above that ion etching of TiO_2 creates extra peaks in the ESCA spectrum. The effect is demonstrated in figure 3b. The Ti spectra obtained at 35 Å and 190 Å levels are shown deconvoluted. The positions of the peaks marked 2+ and 3+ in the left part of the figure (35 Å etching) agree within the experimental accuracy with the values given by Sayers and Armstrong (7). The origin of the peak marked E+ (extra peak) is at the moment not known. This peak has not been mentioned in the literature.

The deconvoluted spectrum representing the state after ion bombardment to the 190 Å level shows beside the peaks discussed above also two additional signals. One is the metallic state of Ti. Its occurrence shows that the surface is partly oxide free. The other signal marked "Ti-H" is positioned at the same binding

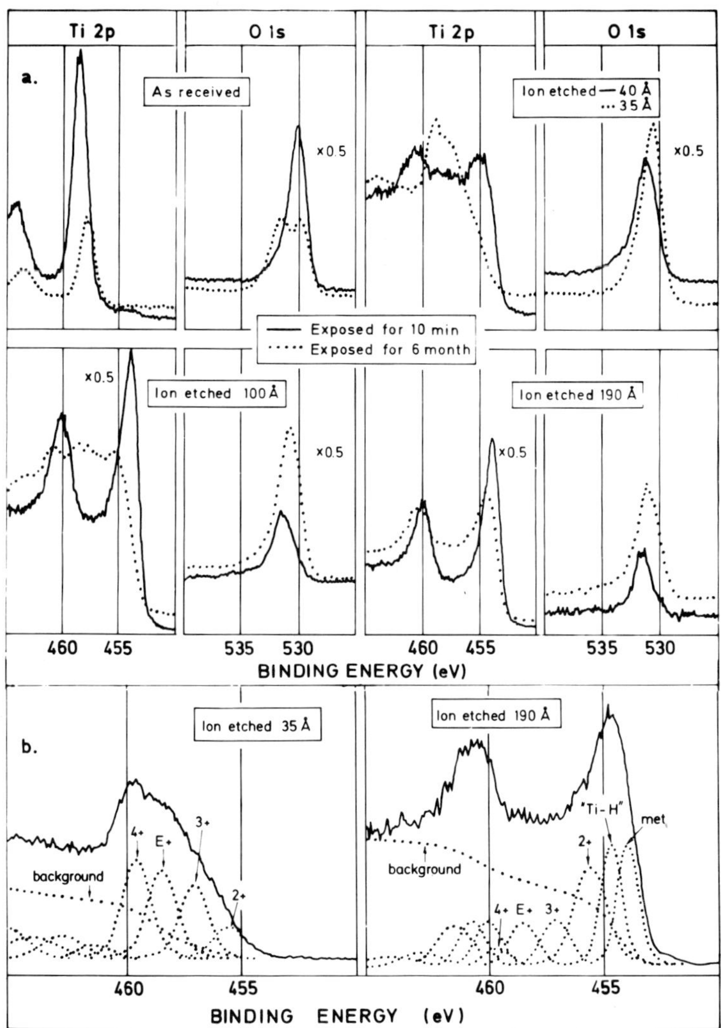

Fig. 3a. ESCA spectra recorded after ion etching of a sample exposed for 10 min. (solid lines) and for 6 months (dotted lines) at 60°C in oxygen saturated water.
   b. Deconvoluted Ti-spectrum recorded from the long-term exposed sample ion etched to 35 Å and 190 Å.

energy as the Ti 2p signal in TiH_2 (7). Its position is the same as the position of the shake up satellite of the metallic state, but its intensity is much higher. Further investigations have to be made to confirm whether this peak stems from TiH_2 present in the metal/oxide interface or is an artifact created by ion etching.

Etching has been performed on an oxidized sample covered with thick TiO_2 (rutile). By doing so, oxides of lower valency appeared (2+, 3+, E+ and 4+) and a stationary state was obtained between 35 Å and 100 Å where the peak heights were of the same magnitude. The oxide formed in water at prolonged exposure shows the same feature when etched to 100 Å (not shown in figure). Further etching to 190 Å (fig. 3b) will however give an increase in the divalency state and a decrease in the four valency state which indicates that the oxide is not TiO_2 at this depth.

The occurrence of both the metallic and the oxide states at 190 Å and lower etching levels shows that the surface is partly freed from the surface oxide. The surface coverage of the oxide, A, can be estimated from the intensities of the signals if it is assumed that the oxide island is much thicker than the attenuation length of the photoelectrons. Details of the technique are described elsewhere (3). The expressions used are:

$$I_{Ti}^{met} = (1 - A)\ D_{Ti}^{met}\ \lambda_{Ti}^{met}\ Y_{Ti}\ \cos 51.5^{\circ} \tag{2}$$

$$I_{Ti}^{ox} = A\lambda_{Ti}^{ox}\beta Y_{Ti}\cos 51.5^{\circ}\left[\overline{D}^{rox}\left[1-\exp\left(-\frac{a'}{\lambda_{Ti}^{ox}\cdot\cos 51.5^{\circ}}\right)\right] + D_{Ti}^{TiO_2}\cdot\exp\left(-\frac{a'}{\lambda_{Ti}^{ox}\cdot\cos 51.5^{\circ}}\right)\right] \tag{3}$$

where A is the surface coverage, $\overline{D}^{rox}$ is the mean density of oxides formed by ion sputtering, $D_{Ti}^{TiO_2}$ is the density of rutile, λ_{Ti}^{met} and λ_{Ti}^{ox} are the attenuation lengths in the metal and the oxide respectively. Their values have been estimated to 13 Å (λ_{Ti}^{met}) and 17 Å (λ_{Ti}^{ox}). a' is the thickness of the oxide layer formed by ion sputtering; by using an equation similar to eqn. 1, a' is found to be 17 Å. By dividing the two equations and solving for A, the coverage can be calculated.

Figure 4 shows the surface coverage versus the etch depths after long term exposure of pure Ti in all four exposure media. It appears from the figure that the surfaces of all samples are completely covered with oxide after etching to 100 Å. After etching to 190 Å, the coverage differs between the four samples. The sample exposed in the salt-free oxygen saturated water shows the lowest degree of coverage. At the etching depth of 450 Å this sample is almost oxide free, while the samples exposed in salt water are covered to about 15 %. The average thickness of the oxide can be defined as the etch depth where the cove-

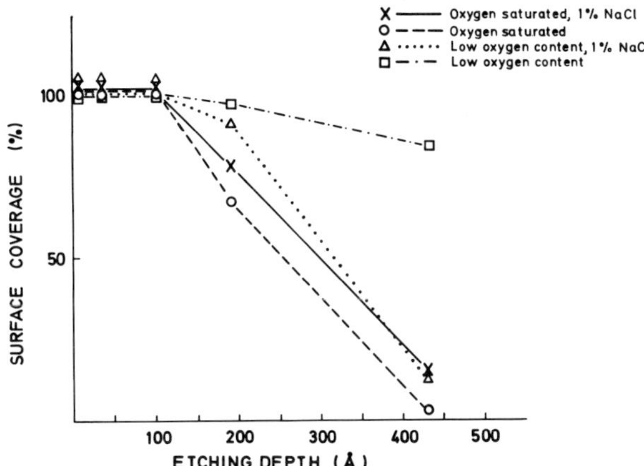

Fig. 4. The surface coverage of oxides vs. etching depth for the samples
exposed in aqueous solutions at 60°C.

rage is 0%. This gives an average oxide thickness of about 250 Å after exposure
in salt free solution and somewhat higher after exposure in the salt containing
solution.

Figure 5 shows the surface coverage for pure Ti and Ti-Pd alloy after exposure for four months in water-saturated bentonite clay at 95°C. After ending
the exposure, the samples were ultrasonically cleaned in water, dried and moved
to the spectrometer. It is very interesting to note that the thickness of the

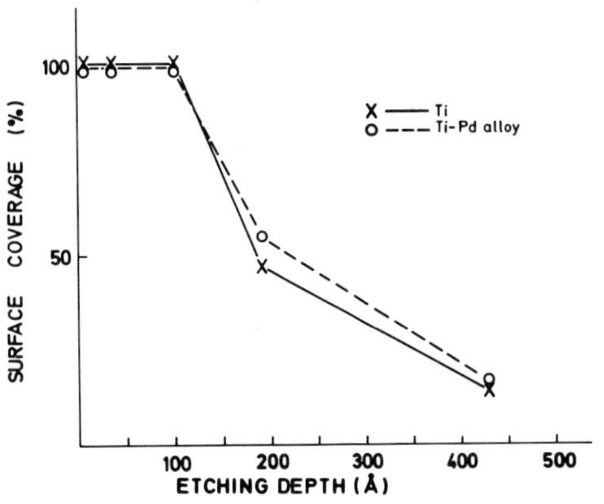

Fig. 5. The surface
coverage of oxide vs.
etching depth for Ti
and TiPd alloy exposed
in water saturated
bentonite clay.

reaction products seems to be lower after exposure in the bentonite than in the water. Also noteworthy is the very good reproducibility between the two materials. The agreement indicates that the Pd alloying element has no influence on the formation of reaction products.

DISCUSSION

Short-term exposure at $60^{\circ}C$ gives TiO_2 on the surface. The thickness of the oxide is in the range 50 to 100 Å. This is somewhat more than is obtained after exposure at room temperature. It has not been possible to determine whether any other oxide phase is present in the layer. The analysis of the long-term exposed samples gives an average thickness of the oxide of about 250 Å. The measurement indicates that the thinnest layer is obtained in oxygen-saturated salt-free water. However, the increase in thickness of the oxide formed in salt-containing solution is limited to 20% which is close to the accuracy of the technique used. Due to reduction of titanium from its four valency state to products in lower valency states during ion etching it has not been possible to make any statement about the composition and the structure in the underneath layers of the oxide. Nevertheless, by comparing the spectra from ion-etched TiO_2 with the oxide product formed during exposure in water we obtain some indications that at least one lower valency oxide is present at the metal/oxide boundary.

The reaction product is mainly oxide. Only in the outermost layer might hydroxide be present. The normal behaviour for most oxides is to form hydroxide in the outermost atomic planes (8,9). It has not been possible in any case to detect chloride ions within the film formed in the hot water.

ACKNOWLEDGEMENTS

The Swedish Nuclear Fuel Supply Company is gratefully acknowledged for the financial support.

REFERENCES

1. E. Mattsson et al., KBS Teknisk Rapport, Kärnbränslesäkerhet, Fack, 102 40 Stockholm.
2. B. Allard and G.W. Ball, J. Environ. Sci. Health, 14 (1979) 511.
3. I. Olefjord et al., Applications of Surface Science, 6 (1980) 241.
4. L. Ramqvist et al., J. Phys. Chem. Solids, 30 (1969) 1835.
5. D. Simen et al., Ed. H. Kimura and O. Izumi, Proc. of the Fourth International Conference on Titanium, Titanium 80, Kyoto, Japan, AIME, p. 2859.
6. I. Olefjord, Met. Sci. and Engineering, 42 (1980) 161.
7. C.M. Sayers and N.R. Armstrong, Surface Science, 77 (1978) 301.
8. I. Olefjord and H. Fischmeister, Corrosion Science, 15 (1975) 697.
9. P. Marcus and I. Olefjord, Surface and Interface Analysis, 4 (1982) 29.

EVALUATION OF SOLUBILITY AND SPECIATION OF ACTINIDES IN NATURAL GROUNDWATERS

Marc R. Schweingruber

Swiss Federal Institute for Reactor Research, CH-5303 Würenlingen, Switzerland

INTRODUCTION

Groundwater circulation is in the long term probably the most important mechanism for the transport of radionuclides from deep underground repositories to the biosphere. The fate of leached long-lived nuclides, such as the actinides, is of particular interest to repository safety analysis. Both experimental and theoretical studies should lead to a comprehensive description of their behaviour in the geosphere.

First quantitative prognoses on concentrations of such elements in aqueous solution and the pattern of their complexation may be obtained by a theoretical speciation analysis based on the premises of chemical equilibrium. Results of this kind can form a basis to justify boundary conditions and parameter choices in geospheric transport models. The application of speciation models to nuclide migration scenarios is worldwide in progress [1-5].

This paper summarizes first modeling efforts undertaken to evaluate the speciation and solubility of U, Th, Np and Pu in two selected Swiss groundwaters. Starting from a rather general box model, two simple case studies will be defined. The first could be visualized as pertaining to the far-field, the second to the near-field of a repository. For both cases the required model parameters are specified. Some emphasis is laid upon the effect of kinetically hindered solid phase precipitation.

MODEL

As a simplification, the geosphere can be divided into a series of boxes with characteristic, uniform internal properties. The boxes' dimensions will, of course, strongly depend both on the processes to be modelled and on the scale of the whole system under consideration. Figure 1 depicts the structure of such a box chosen to model actinide speciation and solubility in groundwater. A well-mixed, mobile aqueous phase contacts a set of non-aqueous phases, which may interact with the solution. If the system is not in overall chemical equilibrium, some reactions might change the composition and/or the speciation of the aqueous phase. Asymptotically, partial equilibria will be reached by reversible reactions which are fast in comparison to water residence time. The leaching of an actinide source embedded in the box (e.g. uranium ore body, waste repository) is clearly induced by a non-equilibrium situation and may lead to partial or complete dissolution of phases initially present as well as to the formation of secondary solids.

In principle, the following parameters must be defined to model the consequences of actinide leaching with this approach: (1) temperature and pressure; (2) water residence time; (3) elemental composition of the inflowing solution; (4) chemical composition and masses of non-aqueous phases (with respect to solution volume); (5) thermodynamic and kinetic characterization of both the formation of dissolved species and the precipitation/dissolution of non-aqueous phases; (6) composition, thermodynamic stability and leaching behaviour of the actinide source.

Hence, the information required for a strict analysis is rather complex, extensive and frequently - e.g. for kinetic aspects - not at hand. If we restrict ourselves to systems which are assumed to be at least in partial chemical equilibrium, we can apply existing speciation programs. Among these, the code MINEQL[6] has been adapted and extended to match the model under the condition of overall chemical equilibrium except for actinide dissolution. The leaching of the actinide source is described by a solid/solution titration concept.

PARAMETER SELECTION

In Switzerland, today's plans foresee a high-level waste repository in the granitic basement of the northern part of the country and medium level waste deposits in sedimentary layers. Therefore, two representative Swiss groundwaters have been selected as leaching agents. Their characteristics are summarized in Table 1. The thermal spring of Zurzach originates in the uppermost part of the granitic basement. The sedimentary layer at 300 m below Beznau consists dominantly of clay, limestone and anhydrite. The redox potential of both waters is unknown at present. Therefore Eh will be used as an independent variable.

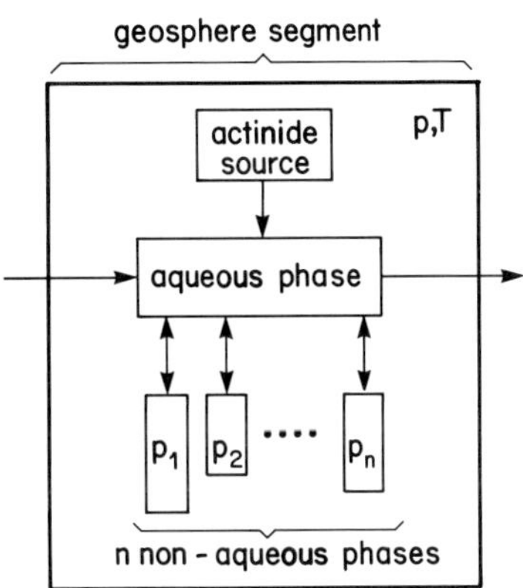

Fig. 1: Model concept

Table 1 : Characterization[*] of Zurzach mineral water[15] (Z) and water sample 7904/5 from a bore hole at Beznau[16], depth 300 m (B).

(a) analytical composition

component	Z	B
H^+	5.14[=]	8.90[=]
Li^+	0.174	0.40
Na^+	12.6	50.89
K^+	0.202	3.30
Mg^{2+}	0.058	7.77
Ca^{2+}	0.38	15.47
ΣFe	0.001	0.388
ΣU	$3 \cdot 10^{-6}$[†]	$2.1 \cdot 10^{-6}$[†]
F^-	0.545	0.084
Cl^-	3.75	36.67
Br^-	0.0078	0.002
SO_4^{2-}	2.65	29.35
CO_3^{2-}	4.29[=]	7.19[=]
$H_2SiO_4^{2-}$	0.384	0.182
$B(OH)_4^-$	0.039	0.33
PO_4^{3-}	0.0005	0.0002

(b) other parameters

parameter	Z	B
$T_{in\ situ}$	39.9	22.8
pH	8.0	7.05
Alkalinity	4.24	6.25
Ionic str.[§]	.0165	.1202

*) units: millimole/l for concentrations, mval/l for alkalinity, celsius for T.

†) from Ref. 17

=) calculated from measured alkalinity and pH.

§) calculated from speciation analysis, assuming precipitate-free solutions (Z is slightly oversaturated with respect to CaF_2, B with respect to calcite and siderite).

The MINEQL library of equilibrium constants, covering complex formation in solution as well as redox and solid/solution equilibria, was revised (e.g. based on Ref. 7) and extended. Additional actinide data were taken from recent compilations[1, 4, 8-11]. Conversion of equilibrium constants from 25° C to other temperatures in the interval 10 to 100° C is based on the approximation of temperature independent standard reaction enthalpies[12]. Necessary ΔH° values at 25° C and 1 atm are extracted as far as possible from the sources indicated above. A complete survey of used thermodynamic data is given elsewhere[13]. It should be stressed that the thermodynamic data available for a single species often differ markedly. These uncertainties may have considerable effects within the model limits.

The activity coefficients of charged soluble species are estimated with the Davies approximation[12]; for all other species these coefficients are set to unity.

APPLICATIONS: PROBLEM SPECIFICATION, RESULTS AND CONCLUSIONS

First we estimate the solubility and speciation of leached actinides in the far field of a repository where the leachate can be expected to be highly di-luted by uncontaminated groundwater. Formally, this situation may be simulated by a titration of the original groundwater with a combination of actinide compounds of appropriate valence states with respect to both the prevailing redox potential and the analytical water composition. The nature of the counter-ions in these compounds can be neglected as long as the titrant addi-tion is small compared to the amounts of important components already present in solution. We may then presume that the speciation of the non-actinide com-ponents will not be altered during the titration, and calculate the possible maximum actinide concentration (referred to as maximum solubility) present in solution without precipitation of any actinide solid phase. The results are plotted as a function of Eh in Figs. 2 and 3.

First note that the maximum solubility of Th, Pu and Np is limited by their tetravalent oxide under all conditions considered. For uranium, these phases are coffinite ($USiO_4$) and $CaUO_4$ under reducing and oxidizing regimes, re-spectively. The formation of $USiO_4$ by precipitation may be kinetically hin-dered[8, 14]. If we exclude this phase from computation, UO_2 and U_4O_9 limit the solubility at low Eh at a somewhat higher level.

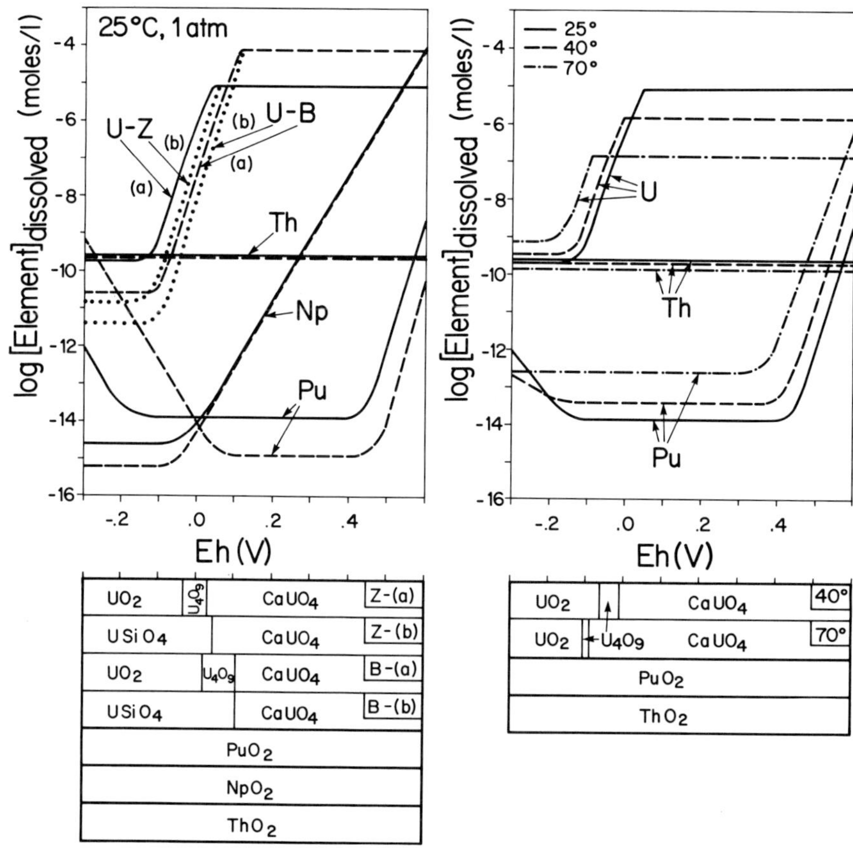

Fig. 2: Maximum soluble concentrations of some actinides in Zurzach mineral water (Z, full line) and Beznau water sample 7904/5 (B, dashed) at 25° C and 1 atm in function of Eh. Indication of solubility limiting phases.
U: (a) exclusive, (b, dotted) inclusive $USiO_4$ formation.

Fig. 3: Maximum soluble concentrations of some actinides in Zurzach mineral water at 25, 40, 70° C and 1 atm as a function of Eh. Indication of solubility limiting phases.
U: exclusive $USiO_4$ formation.
Dominant dissolved species with increasing Eh : $U(OH)_5^-$ changing to $UO_2(CO_3)_2^{2-}$ / $UO_2(CO_3)_3^{4-}$; $Th(OH)_4^0$; $PuCO_3^+$ through $Pu(OH)_5^-$ to $PuO_2(OH)CO_3^-$; $Np(OH)_5^-$ changing to NpO_2^+ / $NpO_2HCO_3^0$.

The curves for maximum actinide concentrations show plateaus whenever the valency of the dominating dissolved species coincides with that of the critical solid phase. Otherwise they exhibit a more or less pronounced redox potential dependency. The most important valency states in solution are +IV for Th, +IV and +VI for U, +III, +IV and +V for Np, +III, +IV and +VI for Pu, dominantly as hydrolysis products and carbonato complexes[a].

Within the expected Eh range of deep groundwaters (-300 to 0 mV at pH between 7 and 8, see e.g. Ref.1), the calculated maximum concentrations are very low except for uranium. The solubility controlling reaction for thorium, $ThO_{2(s)} + 2H_2O = Th(OH)_{4(aq)}^{o}$, does neither include any ligand listed in Table 1 nor charged species, leading to a single, Eh-independent curve for both waters. Differences between the other curves for the two waters (Fig. 2) are mainly determined by their pH difference. The trend of temperature effect upon uranium solubility (Fig. 3) is in accordance with Goodwin[3]. Note that the uranium concentrations found in Zurzach and Beznau water ($10^{-8.5}$ and $10^{-8.7}$ moles/1) are compatible with the model if the in situ Eh is not too low.

At higher maximum solubilities, the model may yield incorrect results, because the charge compensating ions have not been defined explicitly. Further, note that Figs. 2 and 3 do not represent conservative estimates. Future modelling efforts should not be based on thermodynamic data alone, but also be supported by kinetic evidences. Indeed, the precipitation of secondary solids from oversaturated solutions may be controlled rather by kinetic favour than thermodynamic stability in some cases, resulting in the formation of metastable phases. The transformation of such solids into more stable compounds might proceed very slowly at low temperatures, and the solution could maintain apparent oversaturation over long times. This aspect clearly deserves additional investigation.

If the box of Fig. 1 is confined to the near field zone of a waste repository, possible changes of groundwater chemistry - e.g. due to increased temperature, radiolysis, dissolution of waste backfill and canister material - can not necessarily be neglected. This rather complex situation has been highly simplified in a first attempt to describe progressing waste dissolution.

---

a) A detailed discussion including speciation patterns and intercomparison
   with similar studies is given in Ref.13 .

In terms of Fig. 1, the actinide source was taken to be UO_2 only [b], and any host rock/solution interaction was excluded [c]. Thus, the curves of Fig. 4 account for the evolution of dissolved uranium concentration, pH and Eh after an equilibration of increasing UO_2 amounts with Zurzach mineral water at 25° C. The consequences of UO_2 dissolution are conveniently stated in function of the solid amount dissolved per unit water volume [5]; this scale may be transformed to absolute time once the dissolution rate law has been established.

Three initial Eh conditions are treated in Fig. 4. An oxygenated solution was chosen to stand for an aggressive lixiviant, whereas potentials around 0 mV may be appropriate for undisturbed deep groundwaters. Again, the effect of $USiO_4$ formation was investigated. Note that the redox behaviour of the solution has been described only by O_2/H_2O, Fe(II)/Fe(III), and U(IV)/U(V)/U(VI) equilibria. Up to 10^{-5} moles UO_2/l can be dissolved in the oxygenated solution (A-curves) before $CaUO_4$ starts to precipitate, eliminating part of the added uranium from the solution. Until the oxygen reservoir is exhausted, the pH is continually lowered by this reaction as well as by formation of the important U(VI)-tricarbonato complex. A sudden Eh drop occurs when all oxygen has been consumed. The curves are stabilized afterwards due to formation of additional secondary solids (U_3O_8, $USiO_4$ (if included), U_4O_9). After redissolution of U_3O_8 and $CaUO_4$, a saturation with UO_2 is reached. At this point we are left with a reduced water with $10^{-6.7}$ moles U/l and a pH comparable to that of the initial solution. These statements are, of course, again subject to the reservations about secondary solid formation expressed above. In Fig. 4, B- and C-curves differ markedly from the A-curves. Instead of oxygen, Fe(III)-hydroxide [d] is used here to oxidize UO_2. Only a small amount of titrant is necessary to saturate the solution, provided that coffinite is not formed (open circles). Near and far field model should yield identical results in these cases. The corresponding dissolved uranium concentrations and Eh values at the titration endpoints indeed fit to the relationship expressed by curve U-Z-(a) in Fig. 2.

---

b) UO_2 may be a model substance for unreprocessed fuel without canister or backfill material.

c) This restriction is by no means necessary for the program; rock/solution equilibria can easily be included with MINEQL.

d) Based on total Fe concentration (Table 1), this phase should be present above an Eh of about -80 mV.

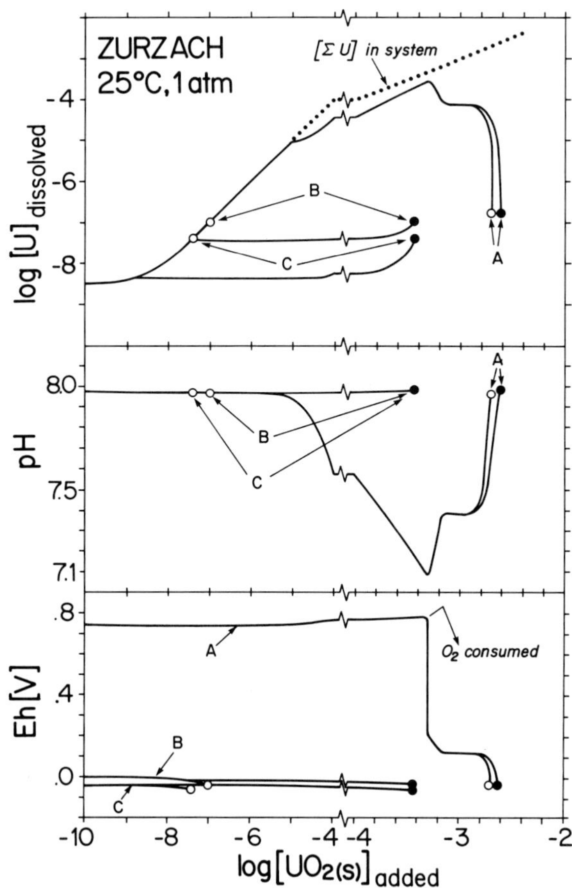

Fig. 4: Titration of Zurzach mineral water with solid UO_2 at $25°$ C and 1 atm. Initial redox potentials defined by : (A) oxygen saturation (8 mg O_2/l); (B) Eh = O mV; (C) Eh = -45 mV. Concentration unit : mole/l. Circles denote titration endpoints (chemical equilibrium between solution and UO_2) : open if $USiO_4$ formation excluded, full if included. Note scale change.

To reach saturation under inclusion of coffinite precipitation (full circles), an additional amount of UO_2, essentially equivalent to the total silicate present, is required in all cases, attributable to the reaction $UO_{2(s)}$ + $H_4SiO_4 \rightarrow USiO_{4(s)}$ + $2H_2O$. Although silicate is highly depleted after UO_2 saturation, the final characteristics of the corresponding A-, B- and C - curves are very similar, because this ligand has a very minor influence on uranium speciation compared to other ligands in Zurzach mineral water. A major conclusion to be drawn is that the uranium concentration maximum in solution is not necessarily appearing after equilibration of waste and solution. Higher concentration levels may be found during the dissolution process under special conditions. This is, however, not the case at typical deep groundwater conditions (B- and C - curves).

ACKNOWLEDGEMENT

The author would like to thank J. Hadermann, R. Grauer, J. Westall for constant support, discussions and valuable criticisms, and H. Flury for his interest.

Partial financial support by NAGRA is acknowledged.

REFERENCES

1.Skytte Jensen, B. (1980) The Geochemistry of Radionuclides with long Half-Lives - Their Expected Migration Behaviour. Report of Risø Nat. Lab. to ELSAM/ELKRAFT.
2.Allard, B., Kipatsi, H., Liljenzin, J.O. (1980) J. Inorg. Nucl., 42, 1015.
3.Goodwin, B.W. (1980) Maximum Uranium Solubility Under Conditions Expected in a Nuclear Waste Vault. Atomic Energy of Canada Report TR-29.
4.Rai, D., Serne, J. (1978) Solid Phases and Solution Species of Different Elements in Geologic Environments. PNL-2651/UC-70, Pacific Northwest Lab.
5.Wolery, T.J. (1980) Chemical Modeling of Geologic Disposal of Nuclear Waste: Progress Report and a Perspective. UCRL-52748, Lawrence Livermore Lab.
6.Westall, J.C., Zachary, J.L., Morel, F.M.M. (1976) MINEQL - A Computer Program for the Calculation of Chemical Equilibrium Compositions of Aqueous Systems. MIT Technical Note 18.
7.Smith, R.M., Martell, A.E. (1976) Critical Stability Constants. Plenum Press New York.
8.Langmuir, D. (1978) Geochim. Cosmochimica Acta, 42, 547.
9.Lemire, R.J., Tremaine, P.R. (1980) J. Chem. Eng. Data, 25, 361.
10.Langmuir, D., Herman, J.S. (1980) Geochem. Cosmochimica Acta, 44, 1753.
11.Allard, B., Kipatsi, H., Rydberg, J. (1977) Sorption av länglivade radionuklider i lera och berg. KBS Teknisk Rapport 55, Stockholm.

12.Stumm, W., Morgan, J.J. (1970) Aquatic Chemistry. Wiley-Interscience,
   New York.
13.Schweingruber, M. (1981) Löslichkeits- und Speziationsberechnungen für U,
   Pu, Np und Th in natürlichen Grundwässern. EIR-Report No. 449/NAGRA Technical Report NTB-81-13, Würenlingen/Baden (Switzerland).
14.Keller, C. (1963) Nukleonik, 5, 41.
15.Högl, O. (1980) Die Mineral- und Heilquellen der Schweiz. Paul Haupt, Bern.
16.Baertschi, P. (1981) personal communication.
17.Baertschi, P., Keil, R. (1980) EIR internal report TM-44-80-12, Würenlingen,
   Switzerland.

STUDY OF RADIONUCLIDE MIGRATION FROM DEEP-LYING REPOSITORY SITES WITH OVER-LYING SEDIMENTARY LAYERS

JOERG HADERMANN[+], FRITZ ROESEL[+], and CHARLES MCCOMBIE[++]

+ Swiss Federal Institute for Reactor Research, CH-5303 Würenlingen;
++ National Cooperative for the Storage of Radioactive Waste, CH-5401 Baden, Switzerland

INTRODUCTION

In Switzerland the host rock formations currently considered for disposal of high-level wastes lie in the northern part of the country. They form the crystalline basement and are covered by substantial layers of sedimentary rocks.

First predictions of regional groundwater flow fields for this hydrologic system have become available recently. These define preferential migration paths to the biosphere and groundwater velocities to be used as input to radionuclide transport model calculations. We consider a one-dimensional transport along these preferential paths from a source described by a leaching model. The source is located at notional repository depths of 1500 and 2400 meters, respectively. In each case the depth in granitic rock is 1000 meters. As a source inventory we have chosen the ^{245}Cm chain because it includes the nuclide ^{237}Np which appears to be crucial in long-term safety analyses. As far as possible field and laboratory data are used in the calculations. However, at present most of these data have very large uncertainties, so that we consider the influence of their variation on the resulting nuclide concentration at the outlet of the geosphere.

MODELS

The leaching model RNRM[1] is based on a mass balance equation for the radio-nuclide inventory of a waste matrix. It takes into account time-dependent leach rates for the individual nuclides and a constant corrosion rate for the glass matrix. Various representations can be chosen for the matrix geometry. The resulting radionuclide injection rates are transformed to a concentration boundary at the geosphere inlet with the aid of the code CONZRA[2]. This will allow for a check whether solubility limits are surpassed at the source. We realize that representing the source term for the far-field migration simply by a leach model constitutes a gross simplification.

Radionuclide transport through the geosphere is calculated by the RANCH code[3]. This models one-dimensional transport through layered media with an arbitrarily time-dependent concentration boundary taking into account dispersion, advection and linear sorption equilibrium characteristics.

PARAMETERS

The parameters used in the present calculations are discussed in detail elsewhere[4]. We restrict ourselves to a few remarks.

Hydraulic parameters, i. e. migration distances from repository to the biosphere and water velocities, are taken from ref. 5 (tables 1 and 2) in which a two-dimensional porous flow model was employed with broad estimates for various geological data which are not yet available. This is especially true for hydraulic conductivities. This is not currently of great importance since our principal objective was to test the coupled transport models in preparation for future work using results from hydrologic fracture flow models, based upon data from the extensive field tests just starting. Some potential advantages at granite sites with overlaying sediments can, however, be noted (see also ref. 6) and some general conclusions readily drawn.

Table 1 : Hydrologic data for the trajectory from the deeper repository site (- 2400 m). Data of columns 1 to 4 are from ref. 5.

Layer	Conductivity (m/s)	Flow velocity (m/s)	Migration length (m)	Water velocity (m/d)	Water transit time (yr)
Intermediate granite	1.-11	3.-14	4900	5.-5	2.6+5
Upper granite	1.- 7	2.5-10	500	4.-3	320
Lower and intermediate Muschelkalk	1.-11	5.-12	70	9.-5	2.2+3
Uppermost layers	2.-6	6.-9	1320	1.-1	30

Table 2 : Hydraulic data for the trajectory from the upper repository site
(- 1500 m). Data of columns 1 to 4 are from ref. 5.

Layer	Conductivity (m/s)	Flow velocity (m/s)	Migration length (m)	Water velocity (m/d)	Water transit time (yr)
Upper granite	1.-7	3.1-10	10000	5.-3	5100
Lower and intermediate Muschelkalk	1.-11	5.1-12	70	9.-5	2200
Uppermost layers	2.-6	1.2- 8	1120	2.-1	17

First, in granite water flow velocities are extremely small and consequently
transit times large. This could change dramatically if flow is primarily through
extended fracture systems. Second, the geologic barrier has two distinct effects
on radionuclide migration. Granite yields the large delay times and, transit
times through the upper sedimentary layers of 500 and 1400 meters thickness,
respectively, can be neglected. However, in these layers the small amounts of
contaminated water emerging from the crystalline basement can be strongly diluted. The water yield at the geosphere exit is taken to be . 1 m^3/min, a typical
number for a thermal spring. Third, a one-dimensional transport model seems
not to be applicable to the lower limestone layer. In addition it is doubtful
whether a porous model approach is adequate at all for this layer consisting
of limestone, clays, anhydrite, gypsum and salt. We have not taken into account
this layer in our transport calculations. For granite we have chosen a dispersion length of 100 meters.

The inventory considered in the nuclear waste repository consists of the
nuclides ^{245}Cm, ^{241}Am, ^{237}Np, ^{233}U and is given in table 3. The quantities of
the nuclides correspond to a single 150 l glass canister 1000 years after discharge from the reactor. For the dissolution of the radioactive material from
the solid waste matrix into the groundwater the glass is assumed to be decomposed into spheres of 0.05 m radius. Thus the release rate $A_i(t)$ for the nuclide i passing from the solid into the water is given by

$$A_i(t) = L_i(t)\frac{O(t)}{M(t)} N_i(t) \qquad (1)$$

Table 3 : Nuclide dependent data. Inventory is given for time zero corresponding to 1000 years after discharge from reactor. Retention factors are calculated from surface based equilibrium constants and pertain to the upper granite. For the intermediate granite they are twice the values given.

Nuclide	Inventory at time zero (Ci)	Retention factors reducing	oxidizing conditions
^{245}Cm	0.4	1.+5	15
^{241}Am	300	2.+5	34
^{237}Np	0.58	1.3+4	30
^{233}U	0.0021	1.3+4	28

where $N_i(t)$, given by a set of coupled differential equations, denotes the amount of the nuclide i within the waste matrix at time t. $O(t)$ and $M(t)$ are the surface and mass of the waste matrix. The leachability $L_i(t)$ may be a function of time. Short-time experimental results indicate power laws between t^o and t^{-1}. A time dependence of the form

$$L_i(t) = L_i \qquad \text{for } t < T$$
$$L_i(t) = a_o + \frac{b_i}{t} \qquad \text{for } t > T \qquad (2)$$

may be assumed for a representation of a general case. To investigate the influence of the leaching behaviour to the nuclide concentration at the outlet of the geosphere, three different calculations have been performed. In the first case, the leaching rates $L_i(t)$ are assumed to be constant for all nuclides and the waste matrix, i. e. $L_i = L_o = 10^{-7}$ (g/cm^2· day). The second case is characterized by nuclide dependent rates, where the following values are chosen[4] : $L_{Cm} = 10^{-6}$, $L_{Am} = 10^{-7}$, $L_{Np} = 10^{-5}$, $L_U = 10^{-5}$ and $L_{matrix} = 10^{-7}$ (g/cm^2· day), respectively. Finally, for the time dependent case, the constants a_o and b_i are adjusted in such a way that for t < T=2 years the function $L_i(t)$ is given by the same set of parameters as in case 2 and as time goes to infinity $L_i(t)$ approaches the value of 10^{-7} for all nuclides.

The release rates $A_i(t)$ for the first two cases may be calculated by a modified Bateman equation whereas for case three the coupled differential equations for $N_i(t)$ have to be solved numerically.

In fig. 1 the release rates $A_i(t)$ are shown as a function of time, whereas

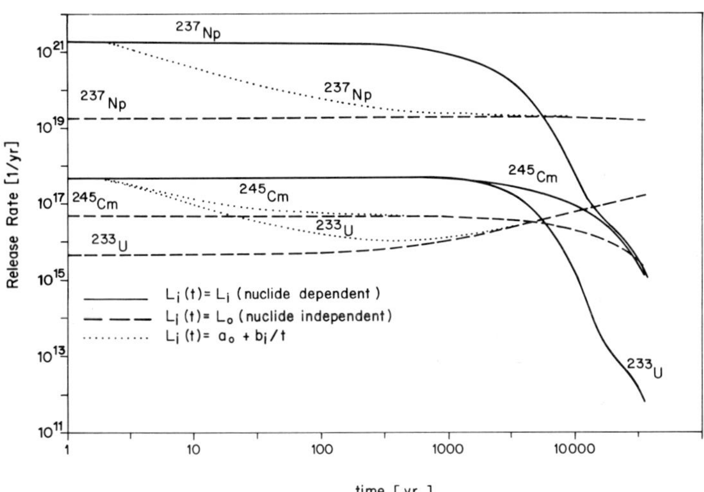

Fig. 1 : Release rate as a function of time after start of leaching. Three different assumption for the leaching rate L_i (t) are taken into account. The release rate of the nuclide 241-Am is not shown explicitly.

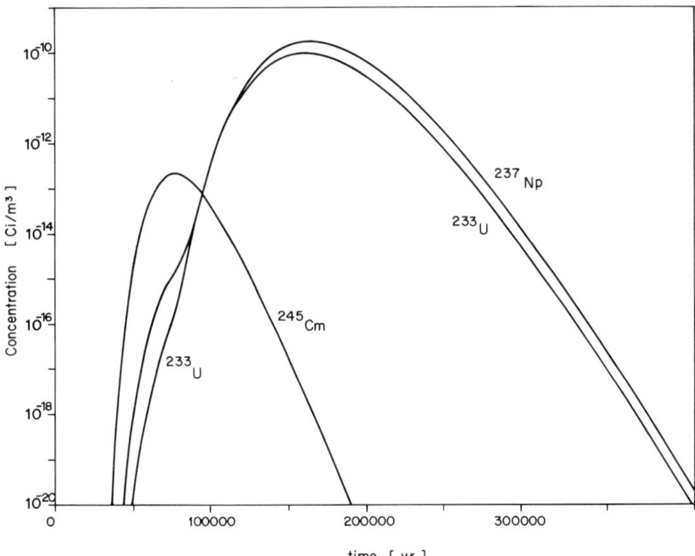

Fig. 2 : Concentrations of nuclides at the outlet of the geosphere as a function of time for the upper repository site (- 1500 m) and the nuclide dependent leaching rates. The nuclide 241-Am is in secular equilibrium with its precursor and not shown explicitly.

only the nuclides ^{245}Cm, ^{237}Np and ^{233}U entering in our transport model calculation are given. The different behaviour of the nuclide dependent and the nuclide independent case is easily understood from the Bateman equations and the
chosen forms of the leach rates. However, the interesting feature is the fact
that the time dependent case approaches the nuclide independent values within
a short time interval. Thus with an appreciable dispersion it is expected that
the resulting concentration at the outlet of the geosphere is the same for both
cases. In our calculations we assume that nuclide injection is up to 10^5 years
after time zero.

Two sets of retention factors for granite have been used, table 3. The larger
numbers correspond to reducing conditions, the smaller numbers to oxidizing
conditions. In absence of in-situ redox potential measurements we believe to
consider thus a reasonable band width.

RESULTS AND CONCLUSIONS

A quick estimate from the tables shows readily that for the large retention
factors, concentrations at the geosphere exit are reduced to total insignifi-
cance (less than $10^{-20} Ci/m^3$). This shows the well-known importance of determining the redox potential in great depths. Thus, results from calculations
with the low retention factors are discussed, only.

The calculation with nuclide dependent leach rates do not differ too much
from those with nuclide independent leach rates, figs. 2 and 3. The first one
gives rise to somewhat higher values since most of the inventory is leached
during the leaching period of 10^5 years. Calculations with the time-dependent
leach rates coincide with those resulting from constant rates since the effect
of decreasing injection rate (for the first 500 years) is smeared out by
dispersion.

One notices that for the upper repository site, figs. 2 and 3, neptunium
and uranium retardation times are short compared to half-lives and essentially
the full inventory exits the geosphere. Here, dilution by the groundwater
bearing overlaying sediments is very important. In contrast, for the lower repository site, fig. 4, retardation leads to a strong decay of these nuclides.

Within the hydraulics model of flow through porous media our results are
certainly very conservative for the following reasons. First, we have considered one-dimensional transport, only. With such long water transit times,

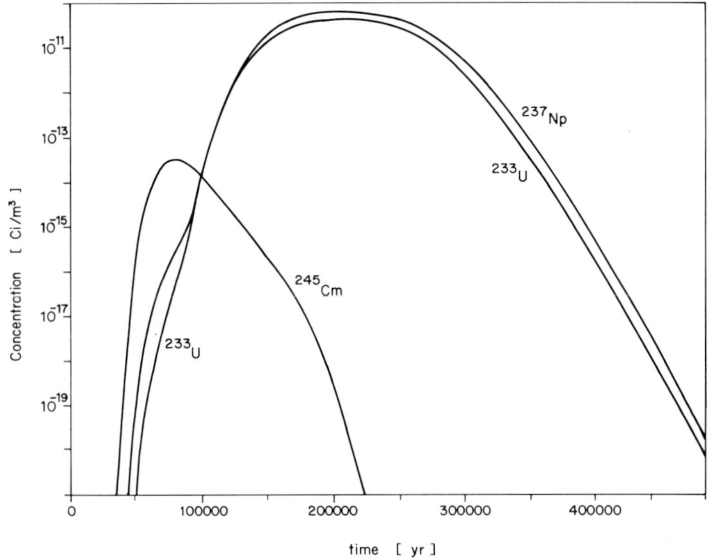

Fig. 3 : The same as fig. 2, but for nuclide independent leaching rates.

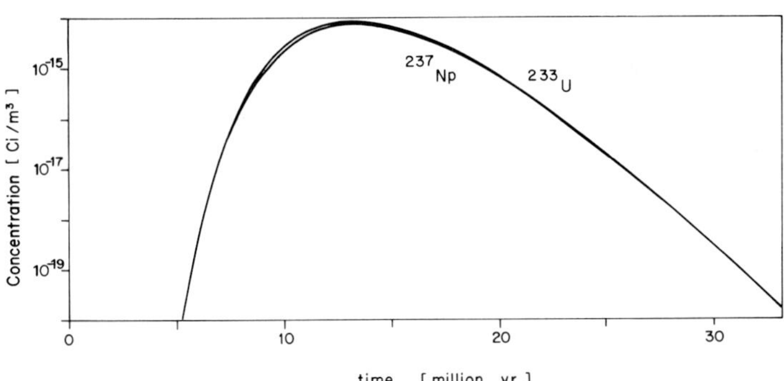

Fig. 4 : Concentrations of nuclides at the outlet of the geosphere as a function of time for the lower repository site (− 2400 m) and the nuclide independent leaching rates.

transversal dispersion will contribute appreciably to a reduction of nuclide concentrations. Second, dilution might be two orders of magnitude larger, if the contaminated water does not exit in a localized spring, but is distributed over a larger geographical area. Third, the leach rates assumed seem to be unrealistically high in view of the low water supply at the repository. At the geosphere inlet corresponding to fig. 2 we calculate neptunium concentrations of 10^{-6} mol/liter. This is far above expected[7] solubility limits. Taken all-together a more refined consideration based on experimental data not available at present could lead to concentrations at the geosphere outlet which might be many orders of magnitude lower than those presented here.

ACKNOWLEDGEMENTS

We would like to thank F. v. Dorp and M. Schweingruber for helpful discussions, B. Housley for computer assistance and H.P. Alder and H. Flury for their interest in this work.

REFERENCES

1. F. Rösel and J. Hadermann, Mathematical formulation of nuclide release models : A computer program, EIR-internal report TM-45-81-48, Würenlingen 1981
2. U. Schmocker, J. Patry, and J. Hadermann, On the Connection Between Injection Rate and Repository Boundary Concentration in Geospheric Transport Models, EIR-Bericht Nr. 426, Würenlingen 1981
3. J. Hadermann and J. Patry, Nucl. Technol. <u>54</u> (1981) 266
4. J. Hadermann, Atomkernenergie/Kerntechnik, to appear (May/June 1982)
5. B. Saugy et al., Modèle hydrogéologique en rapport avec le transfert des radionuclides, IENER Lausanne 1980.
6. J.D. Bredehoeft and T. Maini, Science <u>213</u> (1981) 293
7. M. Schweingruber, contribution to this conference.

Published 1982 by Elsevier Science Publishing Co
SCIENTIFIC BASIS FOR RADIOACTIVE WASTE MANAGEMENT - V
Werner.Lutze, editor

A COMPARISON OF IN-SITU RADIONUCLIDE MIGRATION STUDIES IN THE STUDSVIK AREA AND LABORATORY MEASUREMENTS

O. LANDSTRÖM[+], C.E. KLOCKARS[++], O. PERSSON[++], K. ANDERSSON[+++],
B.TORSTENFELT[+++], B.ALLARD[+++], S.Å. LARSSON[++] AND E.L. TULLBORG[++]
[+]Studsvik Energiteknik AB, S-611 82 NYKÖPING, Sweden; [++]Geological Survey
of Sweden (SGU), Box 670, S-75 128 UPPSALA, Sweden; [+++]Department of Nuclear
Chemistry, Chalmers University of Technology (CTH), S-412 96 GÖTEBORG, Sweden

ABSTRACT

A series of in-situ radionuclide migration tests are in progress in the
Studsvik area on the Swedish east coast. Three well defined flowpaths have
been located and characterized using non-sorbing tracers (I-131 and H-3), and
one of these pathways have been used for a study of the migration of sorbing
elements (Sr-85).

Laboratory sorption studies with Sr-85 have been performed on materials
(rock-water) from the same location as the field tests and with variation of
parameters not easily varied in-situ (e.g. pH and nuclide concentration).

A fairly good correlation was obtained between retention in natural fractures (field experiment) and laboratory column studies, although the latter
were performed on crushed samples of whole rock (retention factors 15 - 30).

INTRODUCTION

In-situ migration studies were suggested by O. Brotzen in 1976 as part of
the Prav and KBS program and a series of experiments were carried out in
fractured crystalline rocks at the Studsvik site during 1977 and 1978[1]. In
another part of the Studsvik site a new test area has recently been prepared
and the first series of migration experiments have been carried out and are
reported in this paper. This work was initiated and supported by Prav but is
now carried out on commission from KBS.

The main objectives with the present work are a comparison between migration parameters measured in field and laboratory experiments, and a critical
examination of the methods and materials used. This is achieved by a coordination between different KBS-supported projects, such as sorption studies at
CTH, fracture mineral investigations at SGU and field migration experiments
at Studsvik.

FIELD MEASUREMENTS

Preparation of the test site

The Studsvik test site is located about 90 km south of Stockholm (Cf
Figure 1). The experimental area is situated approximately 400 metres from the

sea, adjacent to an outcrop formation at about 30 metres above sea level. Based on the results of geological and tectonic mapping of the bedrock and geophysical studies from the ground surface eight hammer-drilled (Ø 115 mm) and one diamond-drilled (K1N, Ø 56 mm) holes were made.

Borehole measurements (partly to obtain background values) were carried out and fractures and their orientation were identified by TV-logging.

In order to determine hydraulic conductivities around and between the boreholes and to find suitable flow paths, water injection tests in single holes and interference tests were carried out. Based on the results of these hydraulic tests, the boreholes B6N (pumphole), B1N, B5N and B8N (injection holes) were selected for the migration studies (cf Figure 1). These boreholes are all drilled in the direction S70°W, with a slope of 80° and to a length of about 100 metres. The main hydraulic contacts (flow paths) and estimated flow path lengths are summarized in Table 1. Double packers enclosed the injection zones and single packers were placed at 30 metres level in B5N, B6N and B8N. The pump was placed at the 60 metres level.

Details and results of the preparatory work and investigations are reported elsewhere[2,3].

Geology and fracture minerals

The bedrock in the region belongs to the Svecokarelian belt of Sweden (1 800 - 2 000 Ma). The rock type present at the test site is a migmatite gneiss of sedimentary origin, which in parts is strongly migmatised and

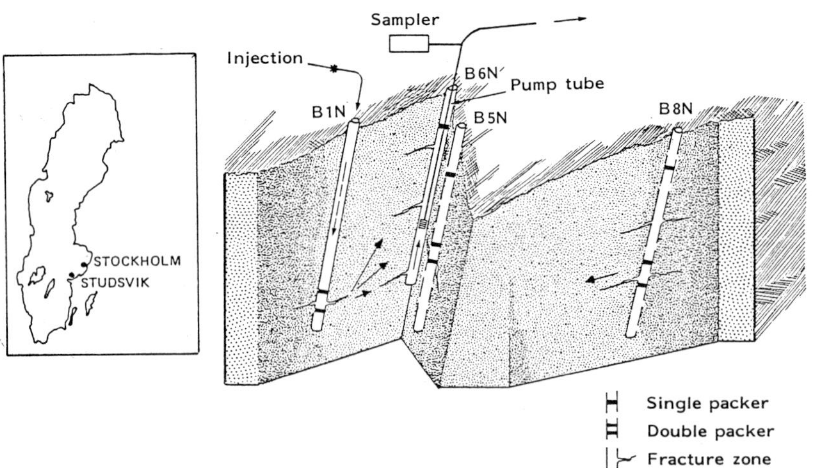

Fig. 1 Cutaway view of the in situ test site

TABLE 1

GEOMETRIC AND HYDRAULIC DATA FOR THE MAIN HYDRAULIC CONTACTS (FLOW PATHS),
UTILIZED IN THE MIGRATION STUDIES

Flow paths Sections of injection holes	Sections of pumphole B6N	Distance between sections
B1N: 91.0 - 92.3 m	94 - 102 m	11.8 m
B5N: 78.8 - 80.1 "	64 - 66 "	14.6 "
B8N: 76.0 - 77.3 "	64 - 66 "	22.6 "

veined. The core of the diamond-drilled hole K1N (drilled with a slope of 60°
and of 200 metres length), was mapped for rock types, fractures and fracture
fillings and the results are summarized below.

Three rock types have been distinguished. The most common type is a veined
gneiss which by Lundström[4] is characterized as a very heterogeneous rock with
biotite-foliated paleosomes. Another rock type is a metabasite-metadiorite,
dominated by plagioclase, hornblende and biotite and the third rock type distinguished is a granite-pegmatite.

The most frequent fracture fillings are calcite, chlorite and smectite.
Other filling materials observed are feldspar, pyrite, palygorskite and some
mica. The thickness of the filling material is usually less than 1 mm. Most of
the fractures observed in the core are coated and open. However, many of them
have probably been opened during the drilling operation and it is difficult
to decide whether a fracture originally was open or sealed. However, feldspar
filling is probably more frequent within sealed fractures than within open
fractures. In contrast palygorskite has only been observed within open fractures. The frequency of fracture filling minerals is shown in Figure 2.

From the TV-logs, totally 310 fractures were oriented. Of these, 63 % dip
towards the north and most (132) within the interval 30° - 60°. Of the fractures that dip towards the north, most (71 %) have a strike of between N and
N40W.

Groundwater chemistry

Chemical analysis of pumped water from borehole B6N has been made at regular intervals throughout the experiment in order to check possible changes
in the chemical environment. In Table 2 are shown typical values of the water
composition during the non-sorbing tracer tests. The analysis "80-12-10" shows
the effect of a packer accident which caused the inflow of surface water (with

700

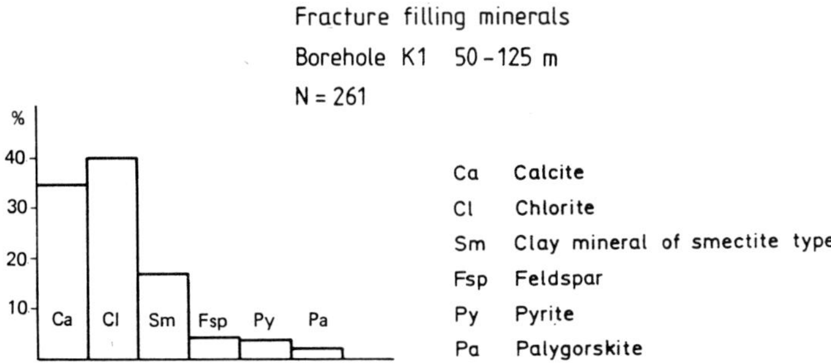

Fig. 2. Frequency of fracture filling minerals in the K1N drill core

a lower mineral content) to the flow paths. As is clear from the analyses 80-12-29 and 81-12-24 the previous equilibrium values are fairly soon recovered.

<u>Hydraulic properties of the flow paths</u>

The hydraulic properties of the flow paths and water transit times were determined using a two-well pulse method with I-131 as non-sorbing tracer. The pumping rate in B6N was 1.2 l min^{-1}. About 370 MBq of I-131 in 0.5 liters of groundwater was injected during 1 hour. Prior to and after the tracer injection, groundwater was pumped down into the injection zone with a rate of 0.01 l min^{-1} in order to develop a steady state flow into the fractures. In both the pumped groundwater and in the tracer solution NaI was added (0.08 M) as a carrier.

Water samples for monitoring tracer concentration were taken from pumphole B6N and in Figure 3 is shown the experimental breakthrough curve for I-131 in the flow path B1N - B6N. By fitting breakthrough curves according to the method described by Zuber[5] and Lenda and Zuber[6] the mean transit time t_o and the dispersion parameter D/v is obtained. Values of these parameters are marked in Figure 3. The TV-log showed 4 fractures in the injection section 91.0 - 92.3 of borehole B1N and the three partial curves in Figure 3 are interpreted as representing different flow paths.

For the fracture zone of B1N - B6N a mean value of $1.4 \cdot 10^{-7}$ m s^{-1} was obtained for the hydraulic conductivity of rock mass (injection tests of 2 metres section in B1N and B6N, respectively). From data of the I-131 test the value $6.6 \cdot 10^{-5}$ m s^{-1} for the hydraulic conductivity of the flow path B1N - B6N was calculated. Details and results of the non-sorbing tracer tests are given elsewhere[8].

TABLE 2

ANALYSES OF WATER PUMPED FROM BOREHOLE B6N

		80-06-16	80-07-09	80-09-17	80-10-28	80-12-10	80-12-29	81-02-24
Conductance, $\mu S/cm$		300	300	330	350	245	360	380
pH		7.2	6.6	6.5	6.7	6.4	6.4	7.0
$KMnO_4$,	mg/l	20	23	26	27	32	33	28
Ca,	"	31	43	30	34	26	37	44
Mg,	"	9	7	12	12	8	12	13
Na,	"	39	40	43	43	26	45	39
K,	"	2.8	2.9	2.9	3.7	2.4	2.8	2.9
Cl,	"	12	11	12	12	10	12	12
SO_4,	"	29	29	29	30	34	29	26
HCO_3,	"	195	200	218	230	133	227	256
NH_4,	"	0.04	0.04	0.04	0.05	0.01	0.01	0.02
NO_2,	"	<0.01	<0.01	<0.01	<0.01	<0.01	<0.01	<0.01
NO_3,	"	0.03	0.04	0.03	0.03	0.39	0.01	<0.01
PO_4,	"	0.02	0.01	0.01	0.01	0.01	0.01	0.01
Fe (II)	"	0.07	0.04	0.04	0.04	0.05	0.09	0.05
Fe (tot)	"	0.07	0.06	0.07	0.07	0.18	0.11	0.15
Mn,	"	0.28	0.29	0.30	0.33	0.10	0.36	0.41
F,	"	0.30	0.26	0.22	0.23	0.12	0.24	0.27
SiO_2,	"	11	11	12	11	10	11	11
C,	"	4.6	4.7	4.8	5.4	14.9	14.3	6.6

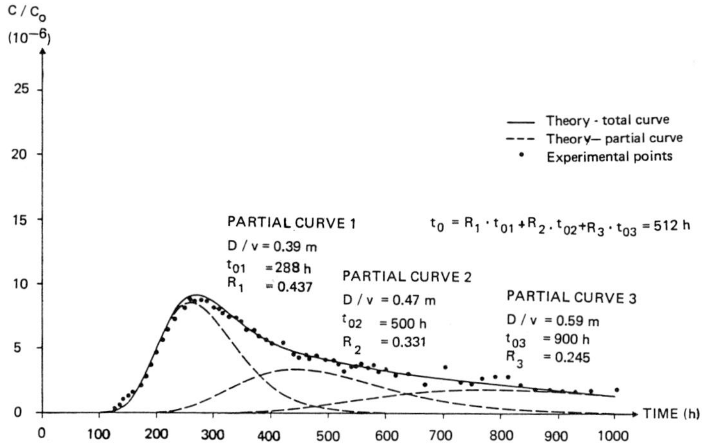

Fig. 3. **Experimental breakthrough curve for I-131 in B1N - B6N and fit with theoretical curves**

Sr-85 migration test

The migration experiments with sorbing tracers started with Sr-85 which was injected in B1N together with the non-sorbing tracers I-131 and H-3, which were compared with one another and used as reference for the water flow. Tracer data are summarized in Table 3.

TABLE 3

DATA ON THE Sr-85 MIGRATION TEST

Tracer injected	Chemical form	Total radioactivity	Concentrations (carriers included)
I-131	NaI	352 MBq	0.8 M
H-3	Tritiated water	357 "	
Sr-85	SrCl	707 "	10^{-4} "

The injection was performed in the same way as in the flow path mapping. Water samples from the pump hole were analyzed for Sr-85 and I-131 by gamma spectrometry and H-3 by a liquid scintillation technique.

The results are shown in Figure 4, where the H-3 and Sr-85 breakthrough curves are compared. The tracer concentration C is compared to the concentration C_o in the injected 0.5 l tracer solution. The I-131 curve was almost identical with that of H-3. The retention factor is about 15, i.e. Sr is transported 15 times slower than the water. A minor amount of Sr, however, seems to migrate with the water.

LABORATORY MEASUREMENTS

Laboratory measurements are performed on three types of solids:

1. Rock samples, obtained by sampling drill cuttings from borehole B8N (cf ref 3). Dry samples were obtained down to 76 m.

2. Fracture surface samples, obtained from the core of K1N. (Sorption measurements are in progress and will not be reported here).

3. Pure minerals (not from the test site), chosen considering the mineralogy of the rocks and fracture fillings in the test area.

The results of sorption measurements on these three types of solids would allow the prediction of probable intervals of retention factors in a field experiment.

Experimental

Sorption measurements have been performed using a batch method[9] as well as column experiments[10]. Some measurements on natural fracture filling surfaces from other sites[11] have also been carried out. The following measurements have been performed for Sr-85 (10^{-8} M) (water data acc. to Table 4): Distribution coefficient (K_d) for unweathered as well as artificially weathered (autoclave 150°C, 1 week) rock in synthetic groundwater S. K_d for rock in natural water from borehole B6N, as well as for pure minerals in synthetic groundwater A. Sorption isotherm (10^{-8} - 10^{-4} M) for rock in synthetic groundwater S, using Na-22 and I-131 as non-sorbing tracers.

TABLE 4

WATER COMPOSITIONS

Specie [mg/l]	Natural groundwater Borehole B6N (800609)	Synthetic groundwater S	Synthetic groundwater A[11]
Na^+	37	93	65
K^+	3.6	4.0	3.9
Ca^{2+}	36	14	18
Mg^{2+}	8.7	9.0	4.3
Cl^-	7	28	70
SO_4^{2-}	35	34	9.6
HCO_3^-	204	200	123
Si_{tot}	5.3	5.5	5.3
pH	8.1	7.8 - 8.0	8.0 - 8.2

Results

The breakthrough curve from the column experiment is shown in Figure 5. I-131 seems to be a "better" non-sorbing tracer than Na-22. A retention factor of 30 - 35 for Sr-85 is obtained. This is equal to a distribution coefficient of 0.006 - 0.008 m^3/kg as is indicated in Figure 6, where the batch measured distribution coefficients are given.

The sorption isotherm was found to be almost linear, - the best fit to a Freundlich isotherm was $q = 6 \cdot 10^{-3} c^{0.89}$ at [Sr] = 10^{-5}-10^{-8} M (q = mol/kg solid, c = mol/m^3 liquid). At 10^{-4} M there was a deviation indicating a limited sorption capacity.

The pH-dependent sorption is typical of an ion-exchange process[13].

The difference between natural and synthetic water disappeared at longer contact times.

Among the pure minerals, those from fracture fillings had higher K_d values (cf ref 12).

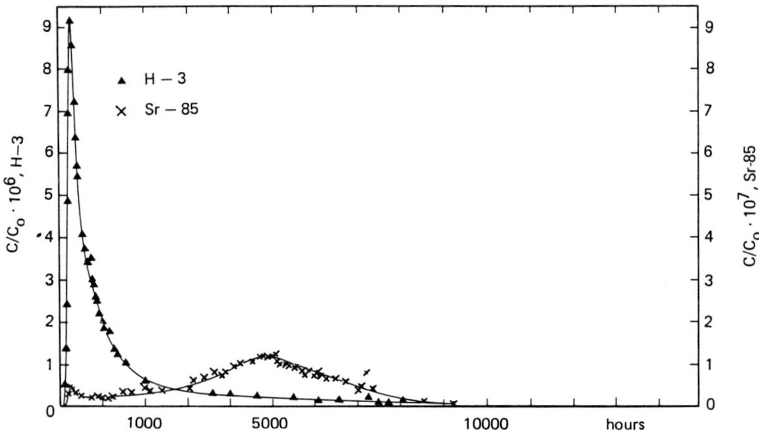

Fig. 4. Field experiments H-3 and Sr-85 breakthrough curves
(flowpath B1N - B6N).

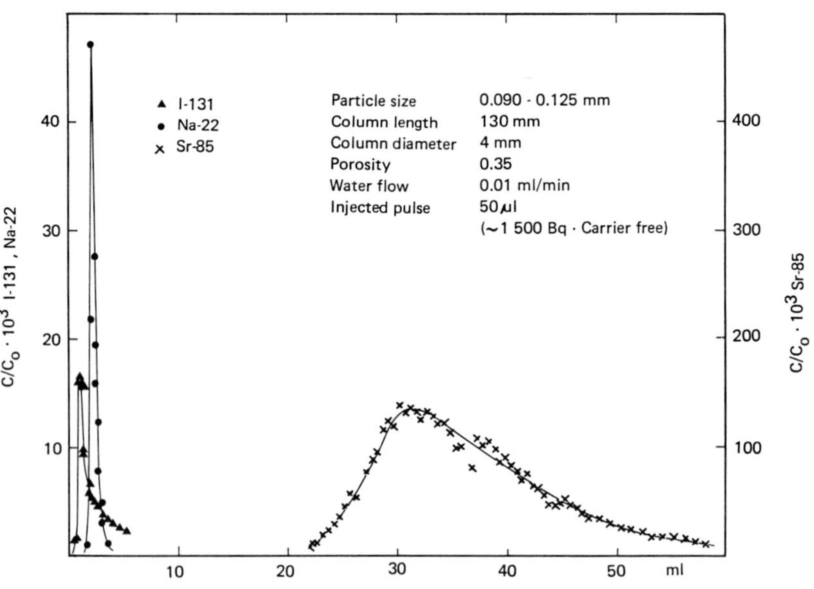

Fig. 5. Laboratory experiment I-131, Na-22 breakthrough curves

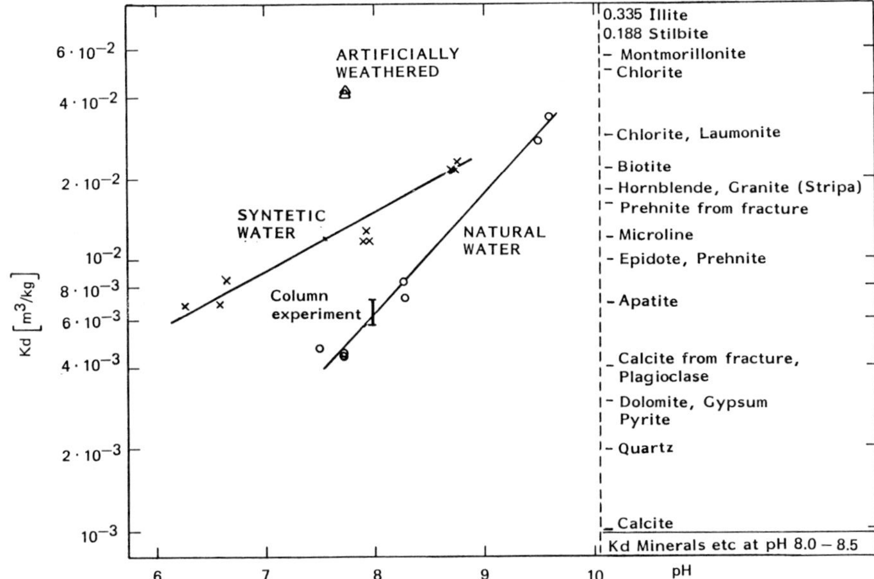

Fig. 6. Distribution coefficients for Sr-85. 1 week contact time, ~ 10⁻⁸ M [Sr]

DISCUSSION

There is a considerable retention/sorption of Sr in both the laboratory and field experiments. The retention factor is of the same order of magnitude (15 - 17 in field and 30 - 35 at lab). The difference may be explained by the fact that the column study was performed on drill cuttings from borehole B8N (representative for the average rock composition) whereas in the field test the radionuclides interact with the fracture surfaces, which may have other sorption properties. The results of the laboratory experiments on fracture fillings from the test area (now in progress) are thus important for the interpretation of the field data.

The distribution coefficients for Sr do not differ by more than one order of magnitude for most rocks and minerals studies. Expected differences between field and laboratory retention values due to different mineral composition are thus of the same magnitide as that measured. For nuclides with larger differences in distribution coefficient between minerals, the knowledge of the mineral composition of the reacting solids (fracture fillings and rocks) is more important.

The early pulse of Sr in the field study may be due to some transport of Sr sorbed on particles in the water (i.e. on FeOOH(s) which may precipitate when the water is aerated during the injection). An attempt to obtain this fast transport of Sr in the column by adding a clay saturated water when injecting the nuclide did not succeed although the effect has been observed for hydrolyzed Am in other column experiments[14].

CONCLUSION

Retention data for Sr-85 from field migration experiments and laboratory sorption/retention measurements show quite a good correlation although only the laboratory measurements on rock samples were considered.

REFERENCES

1. Landström, O. et al. (1978), KBS-TR-10. Kärnbränslesäkerhet, Stockholm, Sweden.
2. Klockars, C.-E. et al. (1980), Report Prav 4.17.
3. Landström, O. (1980), Report Prav 4.18.
4. Lundström, I. (1976), Sveriges Geologiska Undersökning, Serie Af Nr. 114. English summary.
5. Zuber, A. (1974), Isotope Techniques in Groundwater Hydrology. (Proc. Symp. Vienna, 1974), IAEA, Vienna.
6. Lenda, A., Ruber, A. (1970), Proceedings of a symposium arranged by IAEA, Vienna, 1970.
7. Gustafsson, E., Klockars, C.-E. (1981), KBS-TR-81-07.
8. Klockars, C.-E. et al. To be published in KBS Teknisk rapport series 1982.
9. Allard, B. To be published.
10. Allard, B. (1981), Scientific Basis for Nuclear Waste Management, Vol. 3, Plenum Publ. Corp., N.Y.
11. Torstenfelt, B. et al. (1981), Prav Report 4.29, Stockholm, Sweden.
12. Allard, B. et al. (1981), Proc. OECD/NEA Workshop on Near Field Phenomena in Geologic Repositories for Radioactive Waste, Seattle, 1981.
13. Torstenfelt, B. et al., Journ. of Chemical Geology. In press.
14. Olofsson, U. et al. (1982), Scientific Basis for Nuclear Waste Management, Vol.6 , Elsevier N.Y. In press.

RADIONUCLIDE RETARDATION AND RELEASE RATES FOR OCEANIC SEDIMENTS AND CLAY[*]

F. SCHREINER, S. FRIED AND A. FRIEDMAN
Chemistry Division, Argonne National Laboratory, 9700 South Cass Avenue,
Argonne, Illinois 60439 USA

INTRODUCTION

The concept of emplacing nuclear waste in the argillaceous sediments covering
extensive regions of the ocean floor is being given serious consideration by a
group of interested nations.[1] For such a subseabed repository to be acceptable
the effectiveness in isolating the radioactive material must be demonstrated.
Since the waste form cannot be relied upon to retain its integrity in the sub-
oceanic environment, the subseabed repository relies principally on the containment qualities of the sediment that forms the geologic barrier impeding the
spread of radionuclides.

For the past several years our laboratory has been engaged in the measurement of the parameters that characterize the mobility of actinide element ions
and other radioactive species in marine sediments and in terrestrial clay. As
a result of this work we are able to present convincing evidence showing the
sediment to represent an adequate barrier for the isolation of nuclear wastes.

DIFFUSION EXPERIMENTS

In order to obtain direct information on the mobility behavior of radionuclides, laboratory tests were used to determine the effective coefficient of
diffusion. The substrates for the diffusion experiments included three different oceanic sediments and two clays. The sediments were two red clays that had
been collected at depths near 6000 m, and a reducing sediment from a shallower
location in the Pacific ocean.[2] The clays were a bentonite and a hectorite,
both considered as backfill materials in the overpack of land-based repositories.[3]

For the measurement of ion migration the sediments and clay were shaped into
sample cylinders of 12.25 mm height and 14.4 mm diameter. This was accomplished
by packing the sediment into plastic disposable syringe tubes whose tips were
cut off. The plungers of the syringes served to control the quantities of material transferred. First a 1 cm^3 plug of sediment to which the desired radionuclide tracer had been added previously was packed into a tube and squared off

[*]This work was performed under the auspices of the Sandia National Laboratory
 Program Contract Numbers 13-9940 and 74-1160.

carefully at the open end. Next, another plug of equal volume that did not contain any radioactive tracer was butted up to the face of the first plug to be in intimate contact across the entire cross section. The finished sample contained exactly 2 cm^3 of material and was stored in the plastic tube after closing the top with a rubber stopper and the bottom by attaching a closed end tube containing a small amount of interstitial liquid. The small volume of liquid prevented the sample from drying out during the diffusion period.

At the end of the diffusion period, which varied with the type of the sample, 2mm segments of the sediment were rubbed off on strips of filter paper moistened with an aqueous solution of an organic binder. These strips were dried and cut into smaller pieces ($4cm^2$) for assay of the radioactivity with solid-state gamma-ray counting equipment. The use of radioisotopes emitting gamma or X-rays made it unnecessary to isolate the radionuclide chemically prior to the assay.

The activity measured at various distances of the paper strips was correlated to the original height level in the sample core and a distribution curve of relative specific activity as a function of height level was generated. This distribution curve was analyzed in terms of simple linear diffusion in a bounded region. The boundary condition was taken from the distribution of radioactivity at the beginning of the diffusion process, i.e. at time t_o. It was represented by a constant 100% level of activity in the upper half of the sample, and zero activity for the lower half.

The curves obtained from the experiments were analyzed by a graphical interpolation procedure that used curves derived from the analytical solution to the linear diffusion problem. These curves had been calculated for increments of the parameter $D{\cdot}t$, i.e. the product of diffusion coefficients and time. The best fit was obtained by minimizing the variance of the data points. It is estimated that the probable error of the diffusion coefficients calculated in this manner is $\pm15\%$. For the purpose of the present study this precision was judged sufficient, since variations of the nature and particularly the moisture content of the substrate for different samples are expected to limit a more precise determination of D.

Graphical representations of the measured activity levels for several cases are shown in Figs. 1 through 5. The ordinate represents the height level in the sample cylinder measured from the bottom plane, while the abscissa shows the tracer activity on a relative scale. Relative values of the activity for a given sample at different height levels were calculated by assigning a suitable average value of the activity at the high end. This latter average was near 100% in cases in which no diffusion was discernible and less in cases that showed extensive ion migration.

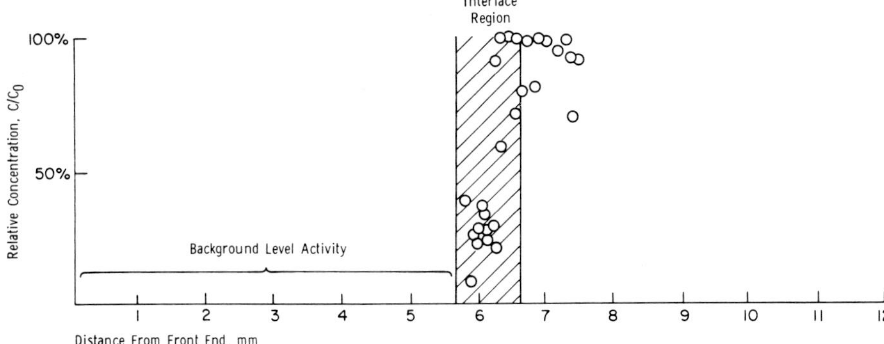

Fig. 1: Concentration profile of plutonium in smectite-rich oceanic sediment after 28.2 days.

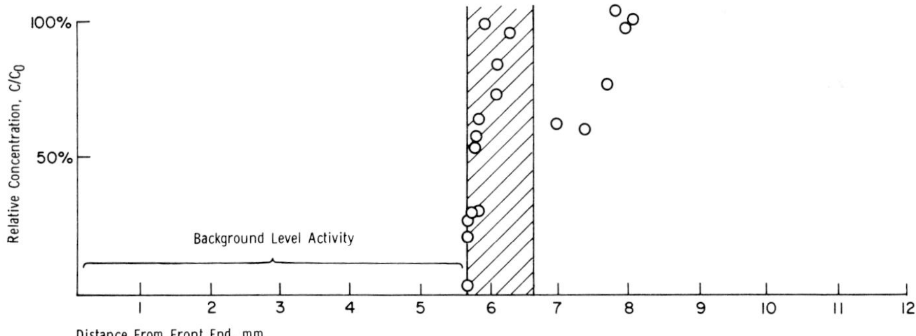

Fig. 2: Concentration profile of americium in smectite-rich oceanic sediment after 35.9 days.

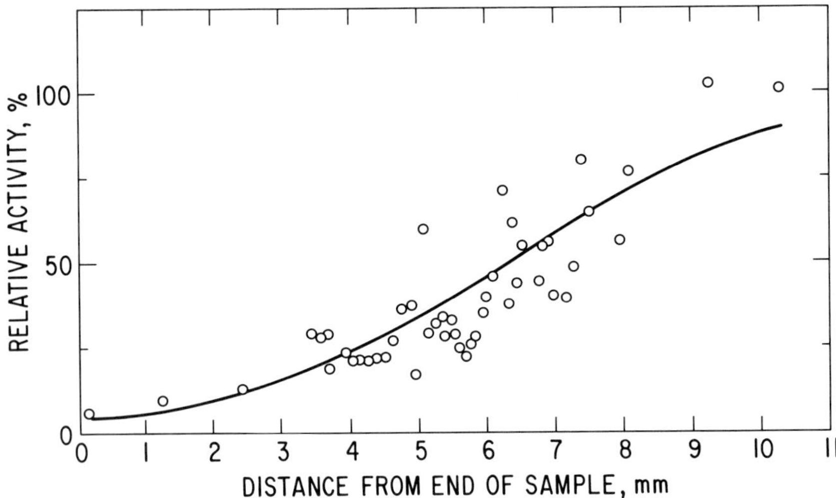

Fig. 3: Concentration profile of pertechnetate ion in smectite-rich oceanic sediment after 4.7 hours. Solid curve: calculated profile for a diffusion coefficient $D_{eff}= 3 \cdot 10^{-10} m^2 s^{-1}$.

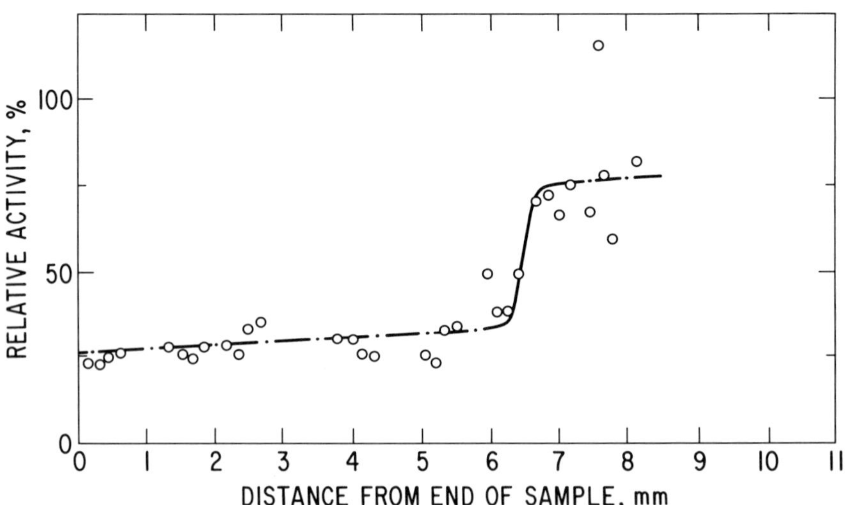

Fig. 4: Concentration profile of pertechnetate ione in reducing sediment after 1.12 days.

DISCUSSION

The data displayed in Figs. 1 through 5 are drawn from a large number of observations and exemplify the diffusion of patterns of the actinide species Pu, Am, and Np, and the fission product Tc. Table 1 lists the various combinations of substrate, tracer nuclide, and interstitial solution that have been studied. The storage times for the samples range from several hours for the fairly mobile pertechnetate species to several hundred days for samples containing plutonium and americium.

The results of the diffusion tests may be summarized as follows. For all samples made up with plutonium and americium as tracers the spread of activity beyond the contact plane of the two sample halves did not exceed the limits of resolution of our method, even for samples that had been stored for as long as 311 days. We can therefore only assign an upper limiting value of 10^{-14} m^2s^{-1} to the effective coefficient of diffusion.

Neptunium, which was originally added to the sediment and clay as neptunyl(V) chloride, NpO_2Cl, showed complex behavior. No diffusion was detected in a sample analyzed after 5.1 hours. However, after 46.2 days the pattern of Fig. 5 was obtained. This profile does not follow a simple diffusion curve. It is obvious that part of the neptunium had the capability of penetrating to the inactive end of the sample while a significant fraction appeared to have stayed stationary. A tentative interpretation can be given in terms of immobilization

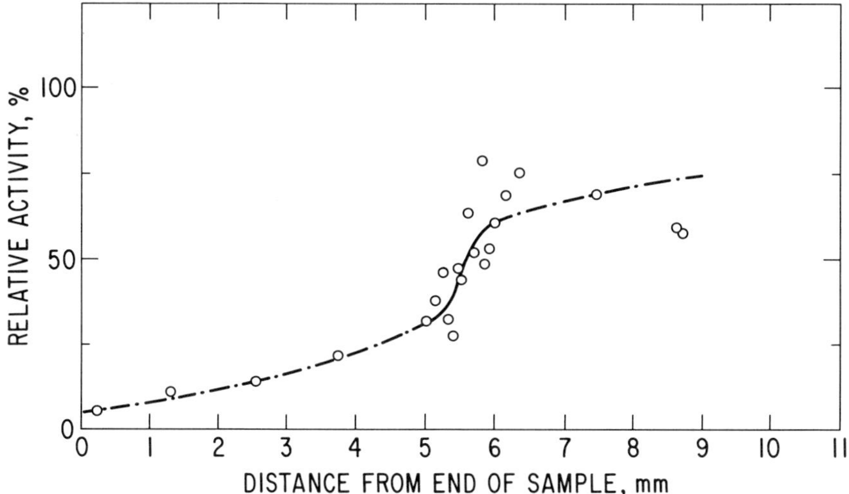

Fig. 5. Concentration profile of neptunyl ion in smectite-rich oceanic sediment after 46.2 days.

TABLE 1

EFFECTIVE DIFFUSION COEFFICIENTS FOR SELECTED RADIONUCLIDES IN OCEAN FLOOR SEDIMENT AND CLAY

Substrate	Nuclide			
	Pu	Am	Np	Tc
Ocean floor Sediment Illite from MPG III		$D_{eff} < 10^{-14}$ $m^2 \cdot s^{-1}$ $-\cdot-$ 2 samples: Am^{+3} diff. time: 134&240 d		
Ocean floor Sediment Smectite from MPG I	$D_{eff} < 10^{-14}$ $m^2 \cdot s^{-1}$ $-\cdot-$ 3 samples, Pu^{+4}&Pu^{+6} diff. time: 28-311 d	$D_{eff} < 10^{-14}$ $m^2 \cdot s^{-1}$ $-\cdot-$ diff. time: 35.9 d	D_{eff}: est. 1.5×10^{-12} $m^2 \cdot s^{-1}$ $-\cdot-$ 2 samples; Np^{+5} diff. time: 0.2-46.2 d	$D_{eff} = 3 \times 10^{-10}$ $m^2 \cdot s^{-1}$ $-\cdot-$ 2 samples; Tc^{+7} (TcO_4^-) diff. time: 0.19, 17.2 d
Reducing Ocean floor Sediment			D_{eff}: est. 2×10^{-12} $m^2 \cdot s^{-1}$ $-\cdot-$ diff. time: 50.2 d	D_{eff} undertermined because of immobilization 2 samples; Tc^{+7} (TcO_4^-) diff. time: 1.1, 65.1 d
Bentonite clay w. WIPP A&B Brine	$D_{eff} < 10^{-13}$ $m^2 \cdot s^{-1}$ $-\cdot-$ 5 samples, Pu^{+4}&Pu^{+6} diff. time: 51.3-153 d	$D_{eff} < 10^{-13}$ $m^2 \cdot s^{-1}$ $-\cdot-$ 3 samples; Am^{+3} diff. time: 21.9-93.9 d		
Hectorite clay w. WIPP A&B Brine	$D_{eff} < 10^{-13}$ $m^2 \cdot s^{-1}$ $-\cdot-$ diff. time: 118.1 d	$D_{eff} < 10^{-13}$ $m^2 \cdot s^{-1}$ $-\cdot-$ 4 samples; Am^{+3} diff. time: 81-217 d		

d = days

of the neptunium species by reduction from the pentavalent to the tetravalent
state and precipitation as insoluble product. In order to account for the observations this reaction must take place over a period of several days so that
some of the neptunium has a chance to migrate over a distance of 6 millimeters.
Further experiments aiming at the resolution of this matter are in progress.

Technetium, which was applied in anionic form as ammonium pertechnetate,
shows different behavior in non-reducing and reducing sediment. The data plotted in Fig. 3 were obtained from a sample of non-reducing smectite-rich sediment
to which Na_4TcO_4 had been added. The pattern shown in the figure was established after 4.7 hours diffusion time. Consistent with this high mobility, a
sample analyzed after 17.2 days showed a uniform technetium concentration
throughout. In a reducing sediment, however, there is evidence for immobilization of the technetium. Figure 4 shows the distribution time of 1.12 days.
Had the pertechnetate diffused with a similar high mobility as in the non-
reducing sediment, a uniform level of activity should be observed. It is obvious, however, that only a fraction of the technetium has left the upper half
of the sample. This observation suggests that the reducing sediment has the
capability of immobilizing the technetium by reducing the pertechnetate to a
tetravalent species, which is insoluble and therefore does not diffuse. As in
the case of the neptunium, further experiments to clarify the chemistry are in
progress.

Knowledge of the diffusion behavior of radionuclides in the sea floor sediment
is essential for the evaluation of the effectiveness of the geologic barrier
surrounding a subseabed repository. This effectiveness is expressed in terms of
release rates for the radioactive elements into the ocean. Release rates have
been calculated with effective diffusion coefficients based on retention coefficients K_d. This was done by Coplik and Ensminger, who utilized the ion
transport model TRION as part of a comprehensive safety analysis of the subseabed
disposal concept.[5] The calculated release rates show the sediment to be an
effective barrier for the isolation of nuclear wastes since it permits only interstitial diffusion as a mechanism for the spread of radionuclides.

The diffusion data listed in Table 1 lead to the same conclusion. Since no
release criteria are available for subseabed repositories, only the criteria set
by the U.S. Environmental Protection Agency for land-based repositories can be
used as a measure for the adequacy of radionuclide isolation. These criteria
set levels of cumulative radionuclide releases over a period of 10000 years.
For ^{239}Pu, ^{237}Np, and ^{99}Tc the leakage of radioactivity from a repository must
not exceed 100, 20, and 2000 curies, respectively, over a period of 10000 years.

714

If it is assumed that the subseabed repository consists of 50000 waste canisters
emplaced at a depth of 30 m in the ocean floor sediment and the radionuclides
are free to diffuse upward through the sediment the effective diffusion coeffi-
cients of ^{239}Pu and of ^{237}Np must be smaller than 10^{-10} $m^2 \cdot s^{-1}$ in order to meet
the release criterion. Table 1 shows this to be the case with a very large mar-
gin for plutonium. Of the relatively fast moving technetium, on the other hand,
about 4000 curies, or about twice the EPA level could reach the ocean water in
10000 years. It would therefore help to maintain a reducing environment in which
the technetium would be immobilized to the extent of meeting the EPA standard.

CONCLUSIONS

The diffusion of plutonium, americium, neptunium, and technetium was studied
in a variety of substrates including commercial clays and reducing as well as
non-reducing marine sediments. No direct diffusion was observed for the actinide elements plutonium and americium, placing an upper limit on the effective
diffusion coefficient of 10^{-14} $m^2 \cdot s^{-1}$. The pertechnetate ion being an anionic
species diffuses rapidly by comparison with an effective diffusion coefficient
in the range of 10^{-10} $m^2 \cdot s^{-1}$. The complex pattern of the neptunium behavior is
not yet understood. But even the mobile fraction of this element has diffuses
slowly in comparison to an anion. Its diffusion coefficient is approximately
10^{-14} $m^2 \cdot s^{-1}$.

As long as molecular diffusion is the only mechanism that permits the migration of radionuclides from the waste towards the sediment-ocean interface no
significant releases of the most hazardous transuranium elements will occur from
a subseabed repository. As was shown above, even the comparatively mobile anionic species can lead to cumulative releases that may be considered acceptable
in an oceanic environment.

REFERENCES

1. Proceedings of the Sixth Annual NEA-Seabed Working Group Meeting, Anderson,
   D. R., editor, Sandia National Laboratories report no. SAND81-0427,
   Albuquerque, New Mexico, 1981.
2. Schreiner, F., Fried, S., and Friedman, A., Subseabed Disposal Program
   Annual Report 1980, Hinga, K. R., editor, Sandia National Laboratories
   report no. SAND81-1095/II, Albuquerque, New Mexico, 1981, volume II, p. 267.
3. Fried, S., Friedman, A., and Schreiner, F., (1981) "A Study of the Mobility
   of Plutonium and Americium in Clay", Sandia National Laboratories report no.
   SAND81-7114, Albuquerque, New Mexico, 1981.
4. Schreiner, F., Fried, S., and Friedman, A., Nuclear Technology, in press.
5. Koplik, C. M., and Ensminger, D. A., (1982) Proceedings of the 1981 Subseabed Disposal Program Annual Workshop, Sandia National Laboratories report
   no. SAND81-2498, Talbert, D. M., Workshop Chairman, Albuquerque, New Mexico,
   p. 208.

SCIENTIFIC BASIS FOR RADIOACTIVE WASTE MANAGEMENT - V
Werner.Lutze, editor

LIQUID CHROMATOGRAPHY IN MIGRATION STUDIES

LARS CARLSEN and WALTHER BATSBERG

Chemistry Department, Risø National Laboratory,
DK-4000 Roskilde, Denmark

INTRODUCTION

A detailed knowledge of the geochemical environment of a site for the disposal of radioactive waste is of fundamental importance. To evaluate the migration
behaviour of radionuclides in geological media a series of data are needed,
amongst others a number of physico-chemical properties of the media, such as
permeability, porosity, dispersion-, diffusion-, and sorption characteristics.
In this connection liquid chromatography appears to be advantageous as a facile
experimental technique to obtain relevant data for these physico-chemical properties.

The capabilities of the liquid chromatography technique in connection with
migration studies are in the following illustrated by examples from recent studies on the cretaceous formation overlying the Erslev salt dome (North Jutland,
Denmark).[1,2]

BASIC PRINCIPLES OF THE LIQUID CHROMATOGRAPHY TECHNIQUE

Liquid chromatography is a column technique normally used for the separation
of different organic as well as inorganic components in a solution.

A typical experimental set-up for a liquid chromatographic separation is
shown in figure 1. The pump delivers a solvent (eluent) flow at a preset, constant flow rate. Samples of known volume are injected into the column via the
injection port and the eluate from the column is continuously monitored by a detector.

For a given column material, e.g. chalk, the separation mechanism will be
controlled by the physical and chemical nature of the interaction between the
molecules present in the injected sample and the column material. Adsorption-,
partition-, substitution-, and ion exchange processes, separately or in combination, can occur in the columns, being responsible for the actual shape of the
chromatogram. Figure 2 depicts a typical chromatogram used for the determination
of the different physico-chemical characteristics of the columns.

The shape of the peak corresponding to an unretarded solute characterizes the
flow dispersion in the column (vide infra), whereas the elution volume (or time)

716

as well as the shape for a retarded peak characterizes sorption phenomena, related to the solute under investigation, on the column packing material. Symmetrical peak shapes are normally obtained in cases where a single mechanism, with a
concentration independent distribution coefficient, can describe the sorption
phenomena. Cases where several mechanisms are operating simultaneously will generally lead to skewed peaks.

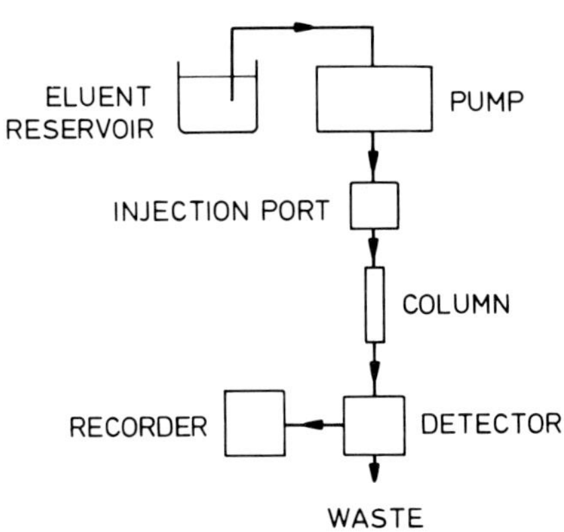

Fig. 1. Experimental Liquid Chromatography set-up

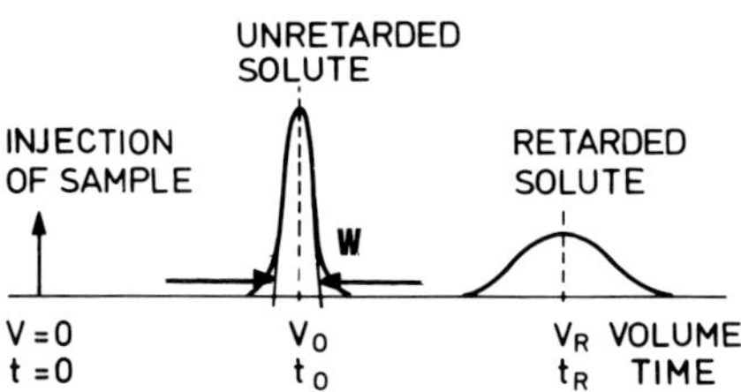

Fig. 2. Schematic Chromatogram defining elution
volume and time.

Permeability. The permeability, K, for a given column is determined by the
volume flow rate, Q (cm^3/sec), through the column, the length of the column, L
(m), the cross sectional area of the column, A (cm^2), and the pressure drop over
the column, h (metre of water).

$$K = Q \cdot L/A \cdot h \qquad (cm/sec)$$ (1)

The present experimental set-up (vide infra) allows studies of columns with
permeabilities down to approximately $5 \cdot 10^{-9}$ cm/sec, however, a further lowering

of this limit seems achievable by minor modifications of the column and/or pumping system.

Volume Porosity. The volume porosity, ε, of the column is determined by injection of a solute, which will not be retarded by the column packing material:

$$\varepsilon = V_o/A \cdot L \tag{2}$$

where V_o is the so-called dead volume (cf. Fig. 2). It should be remembered that
the dead volume has to be corrected, owing to contributions from the liquid
chromatography system, i.e. dead volumes in the different fittings.

In principle no limitations of the magnitude of the volume porosity exist.
However, very low volume porosities in general means correspondingly low permeabilities, which accordingly will lead to certain limitations as mentioned above.

Flow Dispersion. The flow dispersion, σ, is determined from the width of the
peak in the chromatogram for an unretarded solute, w, measured at the base line
(cf. Fig. 2). For a Gausian peak shape the flow dispersion is given by Eqn. 3.[3]

$$\sigma = w/4 \tag{3}$$

If the flow dispersion observed is due to diffusion phenomena only, the diffusion coefficient, D, can be calculated according to the Einstein-Smolukowski
equation (Eqn. 4),[4] the term $\sigma_c \cdot L$ being equal to the mean free path of the molecules undergoing diffusion

$$D = \tfrac{1}{2}(\sigma_c \cdot L)^2/t \qquad (m^2/sec) \tag{4}$$

where σ_c is the flow dispersion for an unretarded solute in the column,[5] L is
the length of the column, and t is the rentention time (corrected for contributions due to the dead volume of the chromatographic system). It should, however,
in this connection be noted that in general this relation will be valid only in
cases of no flow or very low flow rates through the column, otherwise convection
will give rise to considerable contributions.

Sorption. The retention factors for the single solutes are easily derived
from the chromatogram according to Eqn. 5.

$$R_f = V_o/V_R = t_o/t_R \tag{5}$$

where V_R is the elution volume for the retarded species (cf. Fig. 2).

The retention factors reflects the movement of a solute, M, relative to the
solvent front, and is related to the distribution coefficient, K_D, by the Eqn. 6.

$$K_D = ((1-R_f)/R_f)\varepsilon/(1-\varepsilon)\rho \tag{6}$$

where ε and ρ are the volume porosity of the column and the bulk density of the
solid phase, respectively.

Although sorption phenomena preferentially are studied on columns of undis-

718

turbed material (in the present case chalk rods, cf. experimental section), it shall be emphasized that retention volumes, V_R, higher than 10-15 times the corresponding dead-volume, V_o, i.e. retention factors less than ca. 0.07, in general will give rise to unsatisfactory broad peak, which at the limit may escape detection. For studies of sorption phenomena, resulting in very low retention factors, so-callled 'diluted' columns may advantageously be applied. These columns consist of a certain known amount of crushed material diluted with a non-sorbing material, which affords a decrease in the apparent distribution coefficients, K_D^{obs} proportional to the dilution

$$K_D^{obs} = K_D \cdot a^{-1} \tag{7}$$

where K_D^{obs} is the distribution coefficient in the diluted material and a is the dilution factor. From the column experiments the corresponding apparent retention factors, R_f^{obs}, for the diluted columns are obtained, the true distribution coefficient for the undiluted material being obtained according to Eqn. 8.

$$K_D = a \cdot (\varepsilon'/\rho'(1-\varepsilon')) (1-R_f^{obs})/R_f^{obs} \tag{8}$$

where ε' and ρ' are porosity and bulk density corresponding to the 'diluted' column.

It is important to note that the distribution coefficients calculated by Eqn. 8 corresponds to crushed material, and may differ from the values true for the corresponding undisturbed columns.

EXPERIMENTAL

Liquid Chromatography equipment. High pressure pumps (Knauer model Fr-30 or Kontron model LC-410) were used together with a Rheodyne injection port model 7120, a compression module RCM-100 from Waters Ass, retrofitted with an accurate manometer, a refrative index detector (Waters model R-401), and a radioactivity monitor (Dr. Berthold model LB-503), the latter being based on a liquid scintillation principle, which allows detection of α, β, and γ emitters.

Chalk Columns. Homogeneous, cylindrical rods of chalk from the Erslev chalk formation[2] were provided with end fittings and encapsulated in a heat shrinkable tubing (polyethylene). The total length and diameter of the column were adjusted to dimensions compatible with the compression module, i.e. 100 mm x 10 mm $\emptyset$.

The hydrostatic pressure in the compression module was adjusted to be higher than the pressure drop over the column at the actual flow rate. Before injection of the solute the column was equilibrated until a stable base line of the detector signal was obtained, typically corresponding to approximately 10 pore volumes.

Determination of Apparent Diffusion Coefficients. To determine apparent diffusion coefficients of a solute within the column material, the column in the liquid chromatography system was substituted with a 'diffusion unit'. The cylindrical column, saturated with the solution containing the solute under investigation, encapsulated in shrinkable polyethylene tubing, is placed with an open end in a 6 mL reservoir filled with eluent (the same solution as in the column, however, without the solute). The upper end of the column is closed in order to prevent an unwanted flow up through the column. Variation of the concentration of the solute in the reservoir are followed by slow passage of eluent through the latter and consecutively through the detector. Uniform concentration and insignificant concentration polarization in the diffusion unit is ensured by continuous stirring while the eluent flows through the unit. The actual magnitude of the apparent diffusion coefficient is obtained by computer simulation (vide infra).

APPLICATIONS

The applicability of the liquid chromatography technique to determination of permeabilities, porosities, dispersion-, diffusion-, and sorption characteristics is illustrated in the following by examples mainly taken from recent investigations on the Erslev chalk formation (North Jutland, Denmark).[2]

Permeabilities are determined according to Eqn. 1. In Fig. 3 the permeabilities for a series of chalk samples are visualized as a function of depth. The pronounced decrease in permeability observed around -350 m is associated with

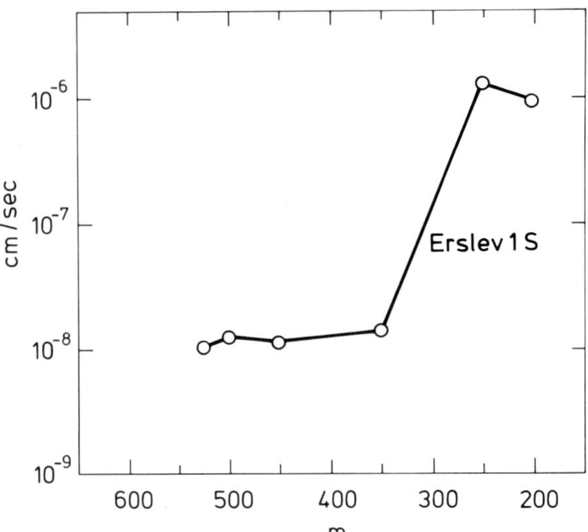

Fig. 3. Permeabilities as function of depth for chalk samples from the Erslev 1s well.

the Maastricht chalk - Campanien chalk transition.[1]

It was found that the permeability was nearly independent of the pressure drop over the column over a wide range of flow rates. However, at very low pressure drops, i.e. at very low flow rates, which most probably corresponds to the actual conditions within the formation, distinct deviations from D'Arcy's law are observed,[6] strongly suggesting a further lowering of the water flow.

Finally, it should be noted that if very high hydrostatic pressures on the columns are applied, the permeabilities appear, not unexpected to decrease slightly. A similar trend is observed in the values for the pore volume.

Porosities are obtained as the elution volume for non-retarded species. In the investigations on the Erslev chalk formation Cl^--36 or D_2O was used, the elution volume being estimated by means of a radioactivity monitor or refractive index detector, respectively. Especially in cases of low permeability, the application of the radioactivity monitor was advantageous, owing to the lack of long term stability of the RI-detector.

Good agreement between values derived from the liquid chromatography data and those obtained by density measurements was obtained. In the latter type of experiments the density of non-porous $CaCO_3$ was assumed to be 2.7 g/cm^3; the densities of the actual salt solution in the pores were established using the sodium chloride molarity in the pore water.[7]

The formation was found to be rather porous, with a distinct variation of porosity as function of depth. Porosities between 0.45 (around -200 m) and 0.25 (around -500 m) were determined.

Dispersion phenomena in the columns may be investigated, as mentioned above, by examinations of the actual shape of the single peaks in the chromatogram, the flow dispersion being equal to the standard deviation, σ, of the peak. Typically the dispersion in the Eslev chalk was estimated to be 0.05 - 0.06 (in units of pore volume) for a non-retarded species. Consistent data were obtained using D_2O/RI-detector and Cl^--36/radioactivity monitor.

It should be noted that the flow rates applied in the liquid chromatography experiments in general far exceed those operating under in-situ conditions. In cases where only moderate water flows occur, the dispersion of the solutes may be dominated by contributions from diffusion.

In porous media the apparent diffusion coefficient, D, measured relative to the open ends of a pore, is less than the intrinsic diffusion coefficient in the pore fluid by a factor equal to the square of the tortuosity of the pore.[8] Additionally a distinct decreasing effect of the apparent diffusion coefficient with decreasing pore size has been observed.[9] The actual magnitude of the decrease is dependent of the nature of the porous media and the diffusing material.

Substitution of the column in the liquid chromatography system with the "diffusion unit" (vide supra) enables us to determine apparent diffusion coefficients of geological material, e.g. diffusion of sodium ions in chalk. The columns were, prior to the diffusion experiment saturated with 1M sodium chloride containing Na^+-22, by application of the liquid chromatography system. During the diffusion experiment the diffusion unit was eluted with 1M sodium chloride, the variations in Na^+-22 concentration in the reservoir being followed by a radioactivity monitor. The magnitude of the apparent diffusion coefficient were obtained by simulation of the concentration variation by application of the computer program DIFMIG,[10] the latter being designed to handle diffusive migration of radionuclides through column systems (Fig. 4).

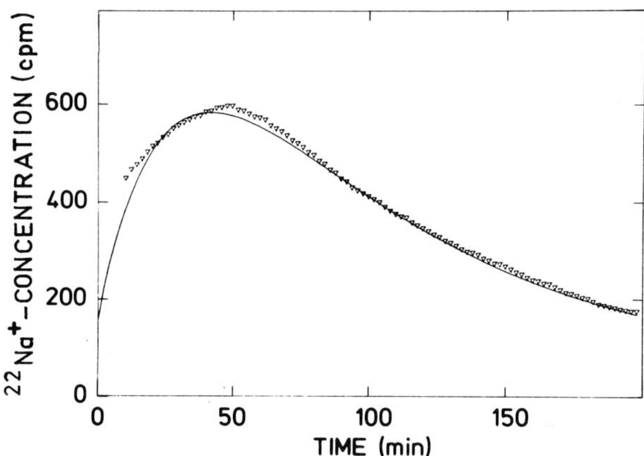

Fig. 4. Diffusive leaching of Na^+-22 from a 10 cm sodium chloride saturated chalk column, (∇: experiment, -: theory).

In the Erslev chalk the apparent diffusion coefficient for Na^+-22 was found to be $2\text{-}5\cdot10^{-8}$ cm^2/s, which is approximately three orders of magnitude lower than the self diffusion of sodium ions in pure water, in good agreement with the rather low permeability of the material, i.e. high degree of very small pores. It should be noted that somewhat higher diffusion coefficients, however still considerably lower than the self diffusion coefficient, have to be considered in the near surface region, apparently owing to imperfections in the surface in contact with the element reservoir of the 'diffusion unit'.

Sorption phenomena, expressed as retention factors, are readily available from the chromatograms (vide supra). The liquid chromatography technique may advantageously be applied in studied, where the influence of ground water composition on radionuclide migration is investigated. In the table the retention factors for Cs^+-134 and Sr^{2+}-85 in Erslev chalk, derived from a 10 mm column, as a function of the bulk sodium chloride concentration in the eluent are given. The values exhibit a slight tendency towards higher retention by decreasing salt content, in agreement with previous investigations, reporting very high retention of cesium in chalk formations.[11]

TABLE
RETENTION FACTORS FOR Cs^+-134 AND Sr^{2+}-85 AS FUNCTION OF BULK SODIUM CHLORIDE CONCENTRATION (10 mm Chalk sample from Erslev)

[NaCl]	0.5M	1.0M	2.0M	4.0M
R_f(Cs)	0.12	0.26	0.38	0.51
R_f(Sr)	0.15	0.21	0.28	0.28

Similarly the influence of e.g. complexing agents as EDTA, can be studied. In Fig. 5 the eluted Sr^{2+}-85, as function of time and EDTA concentration in the eluent from a 10 cm test column is visualized. The retention factors for Sr^{2+}-85 are calculated to be 0.59, 0.61, and 0.82 for EDTA concentrations 10^{-5}, 10^{-4}, and 10^{-3} mol/L, respectively.

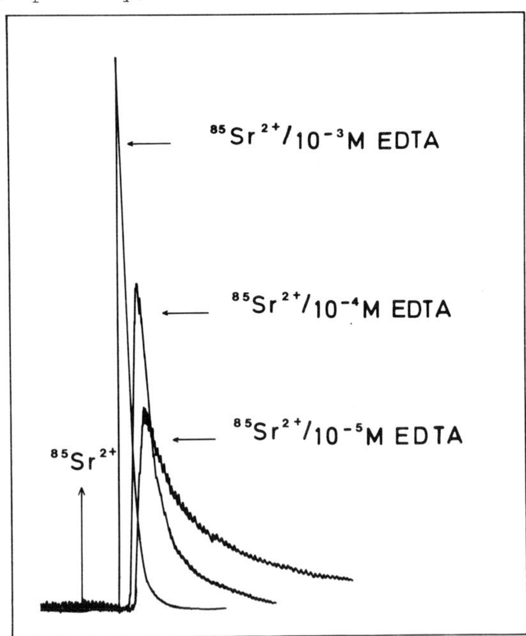

Fig. 5. Retention of Sr^{2+}-85 in a 10 cm test column as function of eluent EDTA content.

Apart from the variation in retention factors, a pronounced change in the actual peak shape as function of EDTA concentration is noted, i.e. distinct peak broadening and tailing at low EDTA concentrations, strongly suggesting the involvment of several comcurrent retention mechanisms.

CONCLUSION

Liquid chromatography has been suggested as a facile experimental technique to determine important physico-chemical properties, as permeability, porosity, dispersion-, diffusion-, and sorption characteristics for geological material as chalk samples. The feasibility of the technique as a rapid method to evaluate the possible influence of changes in ground water composition on the migration behaviour of radionuclides has been demonstrated.

ACKNOWLEDGMENT

The authors are indepted to Drs. Bror Skytte Jensen and Peter Bo for valuable discussions and to Ms. Lisbeth Halby and Ms. Jeanette Meldgaard for technical assistance.

REFERENCES AND NOTES

1. Disposal of High-Level Waste from Nuclear Power Plants in Denmark. ELSAM/ELKRAFT, 1981, volume 2, Geology.
2. Carlsen, L. et al. (1981) Permeability, Porosity, Dispersion-, Diffusion-, and Sorption Characteristics of Chalk Samples from Erslev, Mors, Denmark, Risø-R-451, Risø National Laboratory.
3. Yau, W.W. et al. (1979) Modern Size-Exclusion Liquid Chromatography, Wiley, New York, p. 59.
4. Atkins, P.W. (1982) Physical Chemistry (2nd ed.), Oxford University Press, Oxford, chapter 26.
5. The flow dispersion in the column, σ_c, will differ from the experimentally obtained flow dispersion, σ, owing to dispersion phenomena in the liquid chromatography system, σ_s (fittings, detector), the latter being corrected accordingly ($\sigma^2 = \sigma_c^2 + \sigma_s^2$).
6. Carlsen, L. et al. to be published.
7. Solgård, P. and Skytte Jensen, B. (1981) Chemical Analysis of Pore Water in Chalk samples from Erslev, Mors, Denmark, Risø National Laboratory, Chemistry Department.
8. Meares, P. (1968) in Diffusion in Polymers (Crank, J. and Park, G.S., ed.), Academic, London, chapter 10.
9. Beck, R.E. and Schultz, J. (1970) Science 170, 1302-5.
10. Bo, P. and Carlsen, L. (1981) DIFMIG - A Computer Program for Calculation of Diffusive Migration of Radionuclides through Multibarrier Systems, Risø-M-2262, Risø National Laboratory.
11. Seitz, M.G. et al. (1979) Nucl. Technol. 44, 284-96.

SOME ASPECTS OF THE INFLUENCE OF SURFACE AND GROUND WATER CHEMISTRY ON THE
MOBILITY OF THORIUM IN THE "MORRO DO FERRO" - ENVIRONMENT.

NORBERT MIEKELEY[+], MARIA G.R. VALE[+], TEREZINHA M. TAVARES[+] AND WAYNE LEI[++]
[+]Catholic University of Rio de Janeiro (PUC), R. Marquês de S. Vicente 225,
Rio de Janeiro (Brazil); [++]New York University Medical Center, New York (USA)

INTRODUCTION

Information concerning the migration of thorium in natural environments is of
interest in studies related to impact evaluations of mining and milling operations, to geochemical prospecting and at some point, to predict the long-term
behavior of critical waste nuclides.

As reported earlier[1], it has been proposed that the thorium-rare-earth deposit "Morro do Ferro" in Minas Gerais, Brazil, could be used as a natural analog
to study interactions of long lived actinide waste nuclides with the environment.
It is assumed that thorium and some of the rare-earth elements behave, under
certain environmental conditions, like Pu(IV) and trivalent actinides[1,2]. Since
the more general aspects of this project are reviewed elsewhere in this Symposium[3], this paper focuses especially on the behavior of "dissolved" thorium
in the "Morro do Ferro" environment.

The geochemical retention and, especially, the uptake of a radionuclide by
plants are strongly influenced by the radionuclide's ability to form stable
solution species and therefore this study may provide some insight into the
ultimate biological availability of thorium and its transuranium analog
plutonium.

During one year the elemental composition and some relevant physico-chemical parameters of stream water, which receives drainage water from the radioactive deposit, have been measured. In addition, ground water samples from bore
holes, as well as stagnant waters from trenches, and percolation waters from an
old mine gallery have been analyzed. Although the data are still incomplete,
we felt encouraged to publish preliminary results, noticing that information
concerning thorium concentration in natural waters, especially ground waters,
are very scarce[4].

MATERIALS AND METHODS

Water samples of 10-20 l were collected at the sampling points indicated in
Fig. 1 and filtered (0.45μm) as soon as possible. The filtrate, considered to
contain the operationally defined "soluble" fraction of the elements, was
processed as described below.

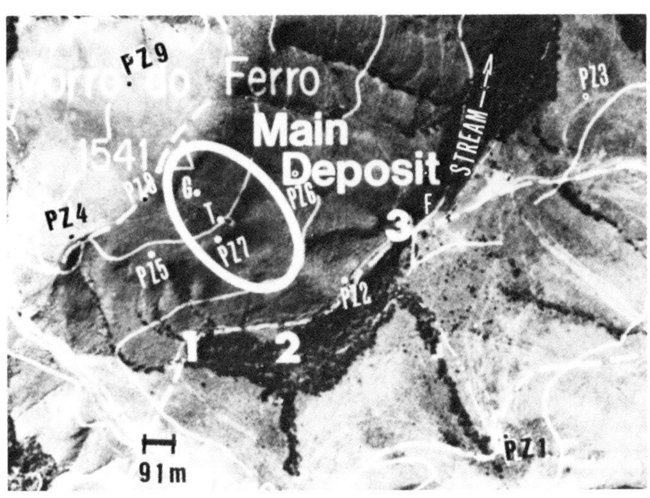

Fig. 1. Aerial photograph of the region with indication of the approximate
sampling points. (PZ = bore hole locations; 1,2,3 = surface sampling sites).

Determination of Th-isotopes. After acidification with HNO_3 the filtrate
was spiked with Th-234 or Th-229 tracer, evaporated nearly to dryness and then
transferred to a Teflon beaker. Silicate was removed by repeated evaporations
with HNO_3/HF and organic material was decomposed with HNO_3/H_2O. Thorium was
separated by extraction with TOPO, reextracted with H_2SO_4 and co-precipitated
with 50μg of LaF_3. The coprecipitate was collected directly on a membrane filter (0.1μm), where 100μg of La^{3+} as fluoride were previously deposited. This
filter served as a counting sample[5]. The limit of detection of the method is
0.09μg of Th per filter (counting time 40Ksec), which corresponds to 0.005 ppb
if a 20 l water sample is initially used. Energy resolution of the Th-232-peak
(E_α=4.01 MeV), determined with a surface barrier detector (ORTEC BR-024-450-
500), is 70-80 KeV typically.

<u>Preliminary Experiments to Identify Associations of Thorium with Humic Compounds.</u> 18.4 l of percolation water from the mine gallery, where humic compounds were suspected to be present because of the brown colour of the samples, were filtered (0.45μm) and spiked with Th-234. After equilibrating for 4 days the filtrate, having a natural pH-6.55, was passed through a first column (XAD-7) and then acidified with HCl to pH-2. The precipitated humic acid fraction (HA) was filtered off, dissolved with 0.1N NaOH and analyzed by spectrophotometry in the 300-800 nm range. The alkaline solution was then acidified with HNO_3, evaporated and, after destruction of the organic material, Th was separated and determined in the usual manner. The HCl-solution, containing possible fulvic acids (FA), was passed through a second XAD-7 column and then treated similarly. Humic compounds retained in the first column were eluted with NaOH but, because of their low concentrations, could not be identified by spectrophotometry, therefore, only Th was determined. The FA-fraction of the second column was eluted in the same way and then analyzed for FA by spectrophotometry, and for Th by α-spectroscopy. In an aliquot of the percolation water, total dissolved Th was determinded by the same technique.

<u>Leaching Experiments.</u> Leaching experiments were performed with two types of material: a composite sample of high grade ore and a composite soil sample from the upper horizon. Ten g of these materials were leached under batch conditions, with deionized water containig the complex forming agents to be studied (F^-, PO_4^{3-} and HA-compounds). After a given leaching time the liquid phase was separated by filtration (0.45μm), and Th-isotopes determined as already described.

RESULTS AND DISCUSSIONS

<u>Th-232-concentrations.</u> Table 1 summarizes Th-isotopic data for surface and ground waters collected at the indicated sampling points and times. As can be seen, the concentration of dissolved Th-232 in the South Stream, which is believed to receive a considerable fraction of the drainage water from the radioactive mineralized zone, is relatively low and constant ($\bar{C}_{Th}$=0.07±0.01 ppb). This is comparable with the concentrations found outside the "Morro do Ferro" environment (Opposite Ridge) and is also similar to data for non-radioactive areas (e.g. Miyake et al.[6] found 0.0087-0.045ppb in ten Japanese river waters). This is surprising, since the radioactive deposit, which is very close to the stream, is believed to contain about 20,000 MT of thorium in a highly decomposed matrix. Chemical data of an ore sample, confirming this last statement, are given in Table 2, which also shows that the ore contains fluoride

and phosphate, reported as efficient complexing agents for thorium[3].

TABLE 1

CONCENTRATIONS OF DISSOLVED THORIUM IN WATERS FROM THE MORRO DO FERRO ENVIRONMENT

DATE	LOCATION	Th-228 / Th-232	Th-232 pCi/l	Th-232 ppb	Th-230 pCi/l	Th-228 pCi/l	
21/08/81	T	–	2.46	22.55	–	–	Dry Season
19/10/81	T	–	1.89	17.34	–	–	Dry Season
19/10/81	PZ-6	0.8	0.26	2.38	0.06	0.21	Dry Season
17/11/81	T	–	1.04	9.50	–	–	Transition Time
17/11/81	PZ-2	1.4	0.37	3.37	0.05	0.38	Transition Time
17/11/81	PZ-4	1.7	0.02	0.14	0.01	0.03	Transition Time
17/11/81	PZ-5	6.0	0.06	0.58	0.01	0.39	Transition Time
17/11/81	PZ-6	4.0	0.22	2.01	0.03	0.89	Transition Time
17/11/81	G	42	0.56	5.13	0.03	23.72	Transition Time
04/01/82	PZ-1	9.7	0.01	0.06	0.01	0.07	Season
05/01/82	PZ-3	50.3	0.02	0.19	ND	1.06	Season
06/01/82	PZ-4	27.3	0.01	0.08	ND	0.25	Season
04/01/82	PZ-5	124.2	0.10	0.89	ND	12.05	Season
05/01/82	PZ-6	5.4	0.02	0.20	ND	0.12	Season
02/04/82	T	13.9	0.25	2.27	ND	3.44	Season
02/04/82	G	32.7	0.83	7.58	ND	0.11	Rainy
02/04/82	PZ-5	–	ND	ND	ND	2.74	Rainy
02/04/82	PZ-6	2.8	0.04	0.36	ND	0.11	Rainy
02/04/82	PZ-7	45.6	0.01	0.06	ND	0.27	Rainy
02/04/82	PZ-9	196.7	0.004	0.04	0.01	0.79	Rainy
21/08 – 17/11/81	SOUTH STREAM-F (*)	10.5± 4.7	0.007± 0.001	0.07± 0.01	0.005± 0.001	0.07± 0.03	
19/10/81	OPP.RID- GE (*)	2.1± 0.4	0.006± 0.001	0.05± 0.001	0.004± 0.001	0.01± 0.001	

(*) mean value $\bar{x} \pm 1\sigma$ G = Gallery   T = Trench

TABLE 2

TYPICAL COMPOSITION OF A HIGH-GRADE THORIUM-RARE-EARTH ORE AS DETERMINED BY XRFA (values in wt%)

SiO_2	Al_2O_3	Fe_2O_3	MgO	CaO	TiO_2	S	F	FeO	P_2O_5
18.9	26.6	22.4	0.21	0.11	2.0	0.03	0.62	<0.07	0.64

ThO_2	Na_2O	K_2O	MnO	CO_2	La_2O_3	CeO_2	Y_2O_3	Pr_6O_{11}	Eu_2O_3
1.3	0.06	2.14	0.76	0.20	2.35	3.6	0.22	0.56	0.06

Nd_2O_3	Gd_2O_3	Dy_2O_3	Sm_2O_3	Yb_2O_3	loss on ignition
1.4	0.12	0.16	0.27	0.008	15.5

As shown in Table 1, significantly higher concentrations of dissolved Th were found in ground waters from bore holes localized in or near the radioactive mineralized zone. The highest values (8-23 ppb), however, were found in percolation waters from an old mine gallery and from a trench, situated nearby.

To understand the differences in concentration levels of dissolved Th, an attempt was made to explain the Th-232 concentrations by the chemical characteristics of the waters. Since the more general aspects of water chemistry are still under study, only some preliminary conclusions concerning the participation of some relevant complex forming agents, can be presented here.

Typical, pH, Eh and conductivity values of surface and ground waters are 5.5-6.5, 420-630mV and 7-20μmhos/cm, respectively. Phosphate concentrations are below 50 ppb and fluoride varies between 70-170 ppb in the South Stream and between 600-800 ppb in the bore hole, trench and percolation waters[7]. Different characteristics are observed for the wells PZ-6 and PZ-9. While the bore hole water from PZ-9 showed high pH (9-11) and conductivity values (400-1000μmhos/cm) and low Eh (350-450mV) during the sampling period from October 1981 to April 1982, the pH of bore hole water from PZ-6 increased from 6 to 9 and the Eh decreased from 540 to 400 mV. With the change in pH the Th-concentration decreased from 0.26 to 0.02-0.04 ppb. This might indicate that solid-water interactions, increasing the pH of the medium, act as an efficient chemical barrier against migration. As shown by Langmuir and Herman[4] at pH-values 8 the insoluble species, $Th(OH)_4^o$, predominates.

Fluoride concentrations in bore hole waters from PZ-6 and PZ-9 were 1.6 and 2.4 ppm, respectively. Although fluoride concentrations were in most cases higher (except for PZ-6 and PZ-9) in those samples with higher Th-concentrations, a clear correlation was not apparent.

The observation that the highest Th-values were found in the percolation and stagnant waters from the surface region, suggested that organic complex formers, like humic compounds, could be responsible for these higher concentrations. To check this hypothesis, 18.4 l of percolation water from the gallery (7.6 ppb of Th) were passed through a XAD-7 column as already described, and the distribution of the Th between the different fractions determined by a α-spectroscopy.

The most important findings were:

1. The presence of HA and FA was confirmed. The accurate concentration levels have not been determined yet, but seem to fall in the 1-10 mg/l range.

2. Th-232 from the water with natural pH (6.55) is only about 10% retained in the column. This indicates that thorium is predominantly in ionic form (organic or inorganic).

3. During acidification of the water with HCl to pH 2, humic acids precipitate and carry about 64% of the Th, complexed or adsorbed, as indicated by Kerndorf and Schnitzer[8].

4. The FA-fraction retained in the column, after passing the acidified water through it, contained about 16% of the thorium. About 10% of the total Th was not retained under these conditions. This could indicate that this fraction is not associated with humic compounds and/or was released from these compounds at pH 2.

<u>Leaching Studies.</u> To obtain a first approximation as to the ability of F^-, PO_4^{3-} and HA to mobilize thorium, preliminary leaching experiments were performed (Table 3). Under the same experimental conditions, fluoride is more efficient in solubilizing thorium from ores and soils than phosphate, which is even less effective than pure water. Higher solution concentrations and fractional releases where obtained by leaching soils with humic acid solutions.

TABLE 3

EFFECT OF SOME IONS ON THE LEACHING BEHAVIOR OF THORIUM (leaching conditions: $T=30^{\circ}C$, t=24h, 10g/1; *0.5g/10ml, t=2h, $T=100^{\circ}C$)

SAMPLE	LEACHING CONDITIONS			Th-232 IN SOLUTION (ppb)	FRACTIONAL RELEASE (%)	ISOTOPIC RATIO Th-228/Th-232
	agent	C (mg/1)	pH			
ORE (1.13% Th)	H_2O	–	4	0.5	0.0004	33
	PO_4^{3-}	10	4	0.4	0.0003	45
	F^-	10	4	2.5	0.002	9
	HCl*	10%	<0	–	2.8	1.1
SOIL (0.102% Th)	H_2O	–	4	0.40	0.004	3.4
	PO_4^{3-}	10	4	0.2	0.002	1.3
	F^-	10	4	2.1	0.02	1.4
	HA	10	8	8.9	0.09	1.6
	HA	1	8	1.8	0.02	2.3
	H_2O	–	8	0.6	0.005	3.0
	HCl*	10%	<0	–	40	1.0
MONAZITE (4.63% Th)	HCl*	10%	<0	–	0.03	1.1

Although one cannot expect that these leaching conditions – which in this first set of experiments were chosen to reinforce the effect of these ions on Th-dissolution – reflect the natural conditions, at least some trends can be observed. Fractional release of thorium from the soil seems to be more pronounced than from the ore sample. This tendency could reflect differences in adsorption/desorption behavior of the two materials, and/or in the chemical binding of the thorium.

Although the identification of radioactive minerals in the ore is not yet concluded, there are indications that the predominant host phases for thorium are monazite and cheralite, minerals which are known to have very low solubility in water and even with HCl are only slowly attacked[9]. Table 3 contains leaching data for monazite, soil and ore with HCl. As expected, monazite is only slightly attacked. Thorium release from the ore is much higher, indicating that part of the Th is bound to less stable materials. Up to 40% of the Th is released from soil samples. These data suggest that different portions of Th are bound to mineral phases which are easily dissolved in dilute acids. Probably, these are hydrous iron and aluminium oxides known to be present in significant quantities. In soil samples clay minerals could play this role additionally. The strong desorption effect obtained with humic acids could be explained by the association of Th with these materials, as shown by Bondietti[10].

Isotopic and nuclide ratios. It is interesting to compare the isotopic ratios of Th-228/Th-232 of the leaching solutions between each other and with those of the natural water samples. Strong leaching with diluted hydrochloric acid removes the two isotopes in nearly proportional activities.

Leaching under weaker conditions favours the dissolution of Th-228 and it seems that there is an inverse correlation between fractional release of Th-232 and isotopic fractionation. Alpha-recoil could be partially responsible for this effect, but its magnitude can be better explained, assuming that an excess of Th-228 was formed recently by disintegration of Ra-228 on surface layers (limonites, clays,etc.), or by direct adsorption of Th-228 from solution. Th-228 could be readily desorbed, from these positions, by complexing solutions. Excess Th-228 was also found in sediment samples from the South Stream[1], giving still more support to this model.

Although there are only few data available at the moment, it seems that the isotopic ratios of Th-228/Th-232, in ground waters, increase from the dry to the rainy season. One of the reasons could be that during the dry season, water in the bore hole vicinity, is only slightly mixed with surface waters, allowing disintegration of excess Th-228. During rainy seasons, ground water becomes readily mixed - because of the high permeability of the overlying layers - with recent surface waters, which are enriched in Th-228 for reasons already discussed.

Figure 2 is a typical α-spectrum of a radioactive ore sample, showing a small Th-230 peak derived from the U-238 decay series. The much lower activity of this isotope in comparison with that of Th-232 confirms the fact that mineralization in the "Morro do Ferro" environment caused a preferential enrichment of thorium (and rare-earth elements). Th-230 is also present in the waters in varying activities (Table 1).

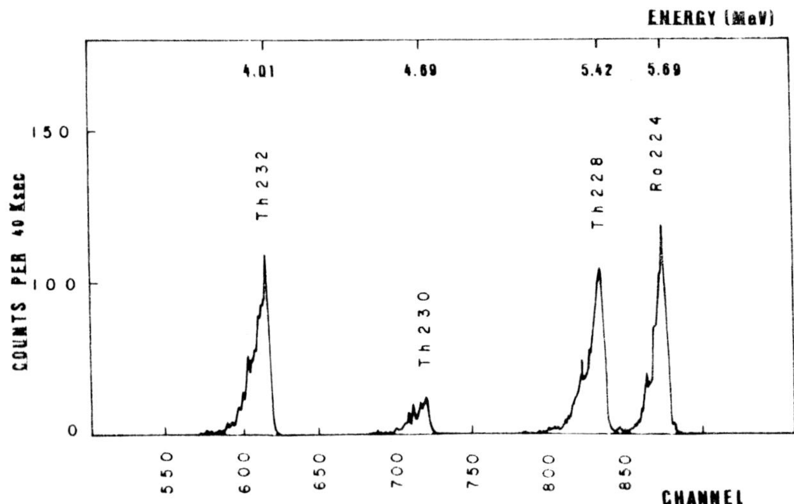

Fig. 2. Typical α-spectrum of the radioactive ore (4,603 ppm Th; 16 ppm U=5.4 pCi/g U-238; 112 pCi/g Th-230; U-238/Th-230=0.05)

Under most environmental conditions uranium is more soluble than thorium. This effect promotes radioactive disequilibrium between U-238 and Th-230, which can be used to make rough estimates of the mobilized uranium.

Fluorimetric analysis of the ore sample gave 16 µg/g of uranium, corresponding to 5.38 pCi/g of Th-230 if radioactive equilibrium is considered. Under the assumption that this element is more or less uniformly associated with the thorium ore - estimated to contain about 20,000 MT of thorium - this would indicate that the deposit contains about 70 MT of uranium, presently. This is in reasonable agreement with estimates published by Frayha[11]. Calculating "equivalent uranium" based on the measured Th-230 activity of the ore sample, a value of 333 µg/g of uranium is obtained, which would correspond to 1,448 MT of this element in the ore body.

If the assumptions are realistic, this shows that more than 1,300 MT (90%) of uranium have already been leached from the ore. The amount of mobilized uranium could be even higher, considering that a small part of Th-230 is also leached (Table 1).

This example illustrates well the differences in the migration behavior between thorium and uranium in the "Morro do Ferro" - environment. At the same time it shows some limitations in utilizing thorium as a chemical analog for plutonium, having in mind that this element can form ions (PuO_2^+ and PuO_2^{2+})

with chemical properties similar to UO_2^{2+}. Although it seems reasonable to assume that Pu (IV) would be the most stable species in the "Morro do Ferro" environment[1], higher oxidation states may be possible. Recent data published by Wahlgren and Orlandini[12] showed that oxidation state ratios of fallout plutonium (V+VI to III+IV) in lakes ranged from as high as 7 to 0.05, but even so, thorium correlated well with plutonium.

ACKNOWLEDGMENTS

The present work was supported financially by CNEN, FINEP, CAPES and CNPq from Brazil and DOE from the USA. The authors are indebted to C. Pires Ferreira (CNEN),M. Eisenbud (NYU), E. Penna Franca (UFRJ) and T. L. Cullen (PUC-RJ) for their interest in the work and to Ruy Frayha, M. R. L. Nascimento (Nuclebras/Pocos de Caldas), A. S. Mangerich (UFRJ), C. A. da Silva and D. J. Santos(PUC-RJ) for their technical assistance. The collaboration of J. A. Medeiros and F. R. Pivetta is also gratefully acknowledged.

REFERENCES

1. Eisenbud, M. et al. (1981) in "International Symposium on Migration in the Terrestrial Environment of Long-Lived Radionuclides from the Nuclear Fuel Cycle", Knoxville, Tennessee.
2. Weimer, J. C. et al. (1980) in "Contaminants and Sediments", ed., Ann Arbor Science Publishers, Inc., Ann Arbor, MI, Vol. 2, pp. 465-484.
3. Eisenbud, M. et al. (1982) "Studies of the Mobilization of Thorium from the Morro do Ferro". MRS Symposia Proceedings, Vol. 11, Scientific Basis for Nuclear Waste Management,Lutze, W. ed.
4. Langmuir, D. and Herman, J. S. (1980) Geochim. Cosmochim. Acta __44__, (11) pp. 1753-1766.
5. Publication of the analytical procedure is in preparation.
6. Miyake, Y. et al. (1964) in "The Natural Radiation Environment", eds. J. A. S. Adams and W. M. Lowder, Univ. Chicago Press, pp. 219-225.
7. Data furnished by Medeiros, J. A. and Pivetta, F. R. - PUC-RJ.
8. Kerndorf, H. and Schitzer, M. (1980) Geochim. Cosmochim. Acta __44__, (11) pp. 1701-1708.
9. Klockmanns Lehrbuch der Mineralogie (1978), 16. Edition, revised by P. Ramdohr and H. Strunz, ed. Ferdinand Enke Verlag, Stuttgart, p. 626.
10. Bondietti, E. A. (1974) in "Adsorption of U (+4) and Th (+4) by Soil Colloids". Agronomy Abstracts.
11. Frayha, R. (1962) "Uranio e Tório no Planalto de Pocos de Caldas". Brazilian National Department of Mineral Production, Bull. 116.
12. Wahlgren, M. A. and Orlandini, K. A. (1981) in "International Symposium on Migration in the Terrestrial Environment of Long-Lived Radionuclides from the Nuclear Fuel Cycle", Knoxville, Tennessee.

Published 1982 by Elsevier Science Publishing Co
SCIENTIFIC BASIS FOR RADIOACTIVE WASTE MANAGEMENT - V
Werner.Lutze, editor

STUDIES OF THE MOBILIZATION OF THORIUM FROM THE MORRO DO FERRO

M. EISENBUD,+ W. LEI,+ R. BALLAD,+ E. PENNA FRANCA,++ N. MIEKELEY,+++ T. CULLEN,+++ AND K. KRAUSKOPF++++

+New York University Medical Center, New York, New York, USA; ++Federal University of Rio de Janeiro, Rio de Janeiro, Brasil; +++Catholic Pontifical University, Rio de Janeiro, Brasil; ++++Stanford University, Stanford, California, USA.

The Morro do Ferro[1,2,3] is a hill on the Pocos de Caldas plateau in the state of of Minas Gerais, Brazil which, except for a few monazite beaches, may have the highest levels of natural radioactivity of any place on the surface of the earth (1-3 mR/hr). The radioactivity originates from an ore body located on the upper slopes of the hill (Fig. 1), which rises about 140 m above its surroundings to a maximum altitude of 1540 m. The ore body is estimated to contain about 20,000 metric tons of Th and a somewhat greater quantity of rare earths. The Morro do Ferro has been the site of a number of radiobiological studies conducted during the past 20 years.[4,5,6,7]

Our studies have three objectives: 1) to predict by analogy with the thorium deposit, the environmental implications of an ancient residue of plutonium in a nuclear waste repository that has been eroded to the surface or has been subject to groundwater intrusion; 2) to develop models for transport of Th, U, Ra-226, Ra-228, rare earths, and possibly other elements; and 3) to evaluate the dosimetric implications to humans living near the site. This paper is a progress report that will be limited to our findings with respect to the first objective, the use of this plutonium analogue to predict the environmental implications of a nuclear waste repository that has been breached. The rationale for the assumption that thorium is a valid analogue for Pu, by which one can predict the mobilization of Pu from an underground repository, was discussed in a previous paper.[1]

The physical circumstances for our study are favorable: there is an average of approximately 170 cm of rain per year, most of which falls during four months. There have not been major disturbances of the hill or its drainage basin by human activity, and the ore body exists in an advanced stage of weathering, relatively close to the surface. The drainage basin is self-contained to a distance of several kilometers, and a stream (South Stream) that receives most of the Morro do Ferro drainage originates at the base of the hill (Fig. 1). A second stream, the North Stream, which is of secondary

importance, drains the north face of the hill. One farm family has lived for
several years years beside the South Stream about 2 km from the deposit.

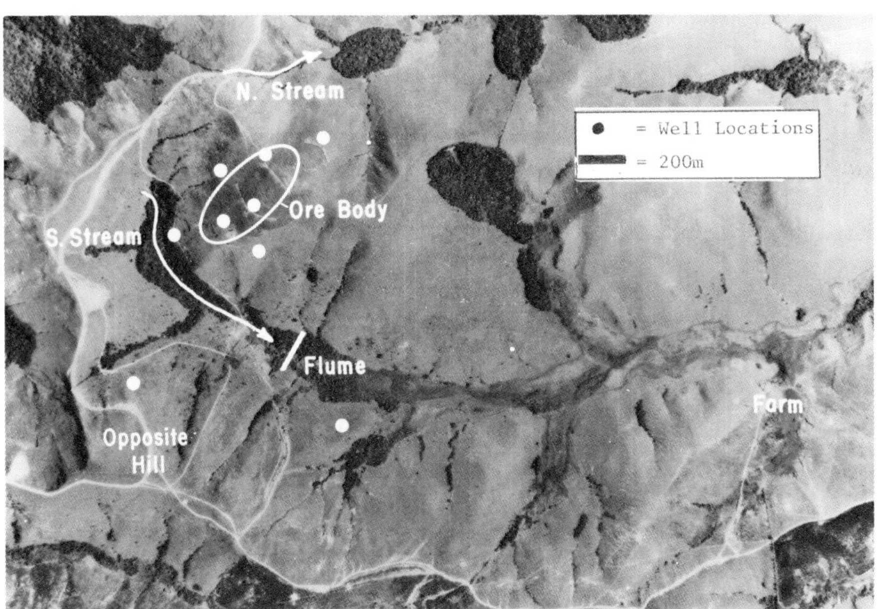

Fig. 1. Aerial photograph of the Morro do Ferro.

METHODS OF STUDY

The mobilization rate of Th is estimated from knowledge of the mass of Th
in the deposit and estimates of the annual Th flux in the streams that drain

the hill. Twenty liter water samples are collected from an instrumented flume
constructed for the purpose, and are filtered routinely through a 0.45 μm Millipore. Selected samples have been passed through a sequence of progressively
finer filters. The filtrates and filtered solids are analyzed separately,
using radiochemical (alpha spectrometric) methods. In addition, routine measurements are made of suspended solids, Eh, pH, SO_4^{2-}, F^-, and other ions of
interest.

In addition to the routine sampling program, more detailed studies have
been undertaken during rainstorms. A recording rain gauge provides information about rainfall rates. Information about ground water level, ground water
chemistry, direction of ground water flow, and permeability are obtained from
nine wells that have been drilled in and around the ore body.

An unfavorable feature of the Morro do Ferro is that it is located in an
area that is generally thoriferous and the South Stream upstream from the
flume drains not only the Morro do Ferro, but also the hills that rise from
the right bank of the stream (opposite hills). The concentration of Th in
these hills is far less than in the Morro do Ferro, but sufficient to partially mask the thorium mobilized from the ore body. An attempt to correct
for this contribution has been made in the following way.

It is assumed that the amount of thorium mobilized in particulate form is
proportional to the concentration of thorium in the surface soil and the area
presented to falling rain. We take the average soil concentration over the ore
body to be 1,000 μg/g and in the remaining portion of the drainage basin to be
70 μg/g. The area of the ore body is estimated to be 48,000 m^2. The area of
the remaining portion of the drainage basin above the flume is estimated to be
5.0 x 10^5 m^2. With these assumptions, we calculate that the surface of the
ore body contributes 60% of the total mobilized particulate thorium.

The relative contributions of the ore body and surrounding areas to the
soluble fraction of the thorium flux were similarly estimated, assuming that
the contributions would be proportional to the volume through which the
groundwater passes, as well as the concentration of Th in the geomedia. By
this method we estimate that the ore body contributes 78% of the measured filtrate Th concentration. We appreciate that this is at best a coarse estimate:
for example, the groundwater could be channeled in passing through the magnetite stockwork associated with the deposit, or the concentration of thorium
could be solubility-limited. Moreover, thorium is undoubtedly dissolved in

surface runoff. However, at this stage of the investigation, we do not have sufficient information to permit more than this first approximation.

GEOLOGY AND HYDROLOGY OF THE MORRO DO FERRO

The Pocos de Caldas plateau is believed to be a deeply eroded caldera, meaning that the underlying rocks were originally a molten mass that punched its way upward into the surrounding ancient gneisses and that the center of the mass then collapsed. Age determinations[8] show that intrusion took place over a long period: phonolites and tinguaites give ages between 75 and 87 million years, and foyaites 63 to 64. Thus the body of igneous rocks was built up by piecemeal additions over some 20 million years toward the end of the Cretaceous period.

The only rock outcrops at the Morro are magnetite, which occurs as a set of subparallel dikes up to a few meters in thickness on the south face of the hill. Much of the magnetite is so massive and free of cracks that it has resisted weathering except for surface films of limonite.

Th and rare earths are widespread in the surface material and are especially concentrated near some of the dike contacts. However, the Th-rich material is apparently not directly related to the magnetite. The depth of Th mineralization remains uncertain, but examination of cores shows concentrations of radioactivity to depths of nearly 200 m.

Study of thin sections of some of the deeper cores indicates the presence of zircon, monazite and cheralite, a rare monazite-like mineral. The zircon is present as crystals on the order of 100 μm that appear to be a primary phase in the rock. ThO_2 is present in the range of 1-10wt% in the zircon and monazite, but occurs up to 30 in the cheralite. The monazite and cheralite are both present as crystals of 1-5 μm dimensions that sometimes occur as aggregates 10-50 μm in diameter.

About 83% of the rainfall which averages 170 cm/yr occurs during the wet season, from October through March. The groundwater level is a subdued replica of the surface topography with recharge from precipitation at relatively higher elevations and discharge into a network of seepages at or near stream level. The highest point of the water table during the 1981 rainy season was about 65 m below the surface of the ridge or about 75 m above the base of the hill. During the 1981-82 rainy season, the groundwater level in the vicinity of the ore body rose by about 2 m.

Weathering has produced a zone of high permeability in the upper 150 m of
the ore deposit, which permits uniform ground water flow. Locally, however,
in the vicinity of magnetite dikes, which are relatively resistant to weathering, flow may be controlled primarily by fractures. Borehole spoils indicate
the predominance of a material consisting largely of argillaceous silt.
Laboratory examination of surface soils ($<$ 0.3 mm particle diameter) show that
up to 50% of the volume in a compacted column can be saturated with water.
The composition of the borehole spoils suggests that the zone of higher permeability exists throughout much of the Morro do Ferro and the South Stream
drainage basin. Permeability measurements made primarily in the region of the
ore body give values generally between 10^{-4} to 10^{-5} cm/sec. The average
groundwater velocity from the highest point of the water table in the direction of the South Stream has been estimated to be about 1 cm/day.[9]

It is important to note that the area of the watershed that drains through
the flume is only 0.5 km^2, and that there are no impoundments. Surface runoff
is thus rapid during a rainfall, and is often accompanied by intense scouring.
In a typical shower, the flow in the South Stream can increase in a matter of
minutes from a year-round dry weather base flow of 0.7 m^3/min (stage height 3
cm) to flows as high as 66 m^3/min (stage height 50 cm). In the absence of
additional rain, the flow then returns to its baseline value in less than one
day. During periods of base flow, water reaches the stream by seeping from the
ground into small tributaries and thus contains thorium that has been mobilized chemically as well as in suspended solids eroded during the brief period
of surface flow. During and immediately following a rainfall, the stream
water is largely the accumulation of surface runoff.

RESULTS OF WATER ANALYSIS

A continuous proportional sampler has been installed in the flume, and grab
samples are collected elsewhere as needed. Because of the low concentrations
of thorium, 20 L samples are required. Eh and pH measurements are made in the
field. The samples are filtered, acidified by addition of 80 mL of concentrated HNO_3, and are then evaporated to about 250 mL on hotplates before being
sent to the laboratory for analysis. Estimates of the sediment load are made
by weighing previously tared filters.

Routine analyses are now being undertaken for thorium, uranium, rare earths, and both radium-228 and radium-226. However, only the thorium data will be discussed here.

The analytical procedure has been modified from that of Sill,[10] and has been described elsewhere.[1]

The lower limit of detection at the 95% confidence level for Th-232 is about 0.12 µg of Th-232, or about 0.006 µg/L for a 20 L water sample.

Table 1 presents the concentrations of thorium in base flow water samples collected from the South and North Streams, as well as from tributaries flowing into the South Stream from hills opposite the Morro do Ferro. For purposes of comparison, samples were also collected from Sao Joao do Boa Vista, which is located outside of the Pocos de Caldas plateau and from the Hudson River in New York.[11] Even during periods of base flow, 90% of the thorium in South Stream water is present in suspended solids.

TABLE 1

CONCENTRATIONS OF THORIUM IN BASEFLOW WATER SAMPLES

Locations	No. Samples	Filtrate	Suspended Solids As Reported	Normalized*	Total
South Stream	59	0.053 ± 0.013	0.54 ± 0.11	0.12	0.55 ± 0.11
North Stream	14	0.027 ± 0.006	0.26 ± 0.03	0.11	0.29 ± 0.04
Opposite Hills	14	0.051 ± 0.012	0.87 ± 0.32	0.056	0.92 ± 0.33
Sao Joao do Boa Vista	6	0.022 ± 0.012	0.79 ± 0.28	0.015	0.81 ± 0.28
Hudson River	3	0.006	0.17 ± 0.10	0.009	0.18 ± 0.10

Mean Th Concentrations $(\mu g\ L^{-1} \pm SE)$

* See text for normalization procedure.

The concentration of thorium in the water samples is thus highly dependent on the concentration of suspended solids. For purposes of comparison, the data in Table 1 are normalized as µg Th/L per milligram of suspended solids per liter. When this is done, the turbid waters of the stream at Sao Joao do Boa Vista and the Hudson River, as expected, have very much lower concentrations of Th than suspended solids from samples collected within the Pocos de Caldas plateau. The mean concentration of thorium in the Brazilian filtrate samples vary only from 0.022 at Sao Joao do Boa Vista to 0.053 in the South Stream. It is surprising that the Th concentrations in water from the opposite hills and South Stream are similar. This suggests that the thorium in surface water samples is solubility-limited.

Three of the water samples were passed sequentially through membrane filters of 1.2, 0.45 and 0.22 μm and for one sample through a 0.10 μm filter. We estimate, on the basis of the data obtained, that 97% of the particulate fraction is retained on the 0.45 μm filter.

As reported earlier,[1] the thorium-228:thorium-232 ratios for the filtered solids are slightly greater than unity and for the filtrates are considerably greater. The isotopic ratios for the filtrates are highly variable. We interpret the slight thorium-228 enrichment of the suspended solids to be due to ingrowth of this nuclide from radium-228 adsorbed on particulate surfaces. The ratio in South Stream suspended solids is 1.6 ± 0.8, which is not very different from the ratios obtained at other locations for which we have data.

The filtrate Th-228:Th-232 ratios for the 59 South Stream samples range from 1.3 to 56.5 and average 14.2 ± 15.5. The South Stream values are comparable with the range in North Stream and Sao Joao do Boa Vista samples but, for some reason, water from the opposite hills has an average ratio of only 3.1. Thorium-228 enrichment in filtrates would be expected as a result of either enhanced leaching of Ra-228 and/or nuclei recoil mechanisms similar to those hypothesized to explain U-234/U-238 disequilibrium.[12] We have not as yet examined the reason for the variability we are finding.

The results of Th speciation studies will be reported elsewhere in these proceedings.[13] It is of particular interest that as much as 40% of the Th appears to be complexed with humic acids. No significant associations have been found with F^-, SO_4^{2-}, or HOP_4^{2-}.

Suspended Solids in the South Stream during Periods of Rainfall

During the rainy season 1981-82, a total of 45 samples of South Stream water were collected during 10 periods of rainfall. Because the filtrates have not as yet been analyzed, only the thorium content of the suspended solids will be reported here.

As would be expected, the thorium concentration is highly variable during periods of rainfall, and can be shown to be closely dependent on both stream flow and concentration of suspended solids. The thorium concentrations were more than 700 times the average concentrations found in suspended solids during periods of base flow.

The relationship between Th flux and flow rate was found to be:

$$\phi_{Th} = 4.5 \ Q^{2.4} \qquad n = 45 \qquad (r^2 = 0.82) \qquad [1]$$

where: ϕ_{Th} = mg Th/min
$Q = m^3/min$
n = number of samples
r^2 = coefficient of determination

It was found, not surprisingly, that ϕ_{Th} could be determined with considerable precision from its relationship to the flux of suspended solids as follows:

$$\phi_{Th} = 45 + 200 \ \phi_{ss} \qquad n = 45 \qquad (r^2 = 0.98) \qquad [2]$$

where ϕ_{ss} = Kg suspended solids per minute

CALCULATION OF MOBILIZATION RATE

Our estimates of Th mobilization have been based on South Stream measurements primarily. However, it has been recognized from the start that Th from the ore body could be mobilized via the stream (North Stream) that drains the north face of the mountain. Based on limited measurements, the flow rate in this stream is estimated to be about 400 m^3/day under conditions of base flow. The mean concentration is 0.29 µg/L, which gives an annual Th flux of 43 g/yr. This is about 25% of the South Stream flow flux. In view of these findings, our estimates of the Th mobilization rate are adjusted by a factor of 1.25 to account for the estimated loss of Th via the North face of the mountain. We estimate that 98% of all particulates are mobilized during periods of rainfall, and that the annual runoff of thorium in particulate form is about 11.9 kg, which is more than an order of magnitude greater than our 1981 estimate of 0.8 kg, based on analysis of samples collected under base flow conditions and not including the North Stream contribution.[1]

The mobilization rate due to surface erosion is estimated to be 5.9 x 10^{-7}/yr, equivalent to an average life of 1.7 x 10^6 years. The contribution of the filtrates can be neglected in this estimate.

Estimate of the Mobilization by Groundwater Solubilization Mechanisms

The mobilization rate due to chemical factors can be calculated from the assumption that the mean filtrate concentration (0.05 µg/L) during conditions of base flow in the South Stream results from solubilization of Th by groundwater. To differentiate the contribution of the ore body from that from other sources upstream from the flume, we have made the additional assumptions that the flux of Th in solution is dependent on 1) the volume of groundwater in contact with Th-bearing solids, and 2) the surface area of the contacted geomedia. If we further assume that the particle size of the material is similar both in and out of the ore body, and that the Th is present in similar chemical form, then leaching would be dependent only upon the Th concentration in the ore body and elsewhere in the basin upstream of the flume. If the concentration in the ore body is taken to be 1,000 µg/g and the concentration elsewhere upstream of the flume to be 70 µg/g, then it is estimated that the ore body can contribute up to 78% of the Th that is present in soluble form. The mobilization rate of Th in the ore body is in this way estimated to be 7.5 x 10^{-10} per year, and the average life of the Th deposit would be 1.3 x 10^9 yr.

From the point of the view of the primary purpose of this investigation, mobilization by ground water is more relevant than the effect of erosion, since it is unreasonable to assume that a deep geological repository will erode to the surface in the lifetime of plutonium. A more probable assumption is that mobilization would take place by intrusion of ground water into the repository so that the plutonium would be exposed only to the reducing, neutral to slightly alkaline conditions generally encountered in deep groundwater environments. As reported earlier, our Eh and pH measurements, both in surface and ground water, lead us to believe that the behavior of thorium at the Morro do Ferro should be a reliable guide to the behavior of plutonium in a repository invaded by groundwater at depth.[1]

ACKNOWLEDGEMENTS

This research was undertaken with the support of the Department of Energy (Contract No. DE-AC97-79ET46606), the Office of Waste Isolation, Battelle Memorial Institute (Contract No. E512-07900), and Comissao Nacional de Energia

Nuclear. The authors are especially grateful for the assistance rendered by Dr. Carlos Pires Ferreira of the latter organization. This research was also supported by core grant Nos. ES 00260, National Institute of Environmental Health Sciences, and CA 13343, National Cancer Institute.

REFERENCES

1. Eisenbud, M., Lei, W., Ballad, R., Krauskopf, K., Penna Franca, E., Cullen, T. and Freeborn, P. (1982 in press) in Proc. Int. Symp. on Migration in the Terrestrial Environment of Long-Lived Radionuclides from the Nuclear Fuel Cycle, SM-257/49, Int. Atomic Energy Agency, Vienna.

2. Wedow, H. (1967) The Morro do Ferro Thorium and Rare-Earth Deposit, Pocos de Caldas District, Brazil. U.S. Geological Survey Bull. 1185-D.

3. Frayha, R. (1962) Uranio de Torio no Planalta de Pocos de Caldas. Brazilian National Dept. of Mineral Production, Bull. 116.

4. Roser, F.X. and Cullen, T.J. (1964) in The Natural Radiation Environment. Univ. of Chicago Press, Chicago IL, p. 825.

5. Eisenbud, M., Petrow, H., Drew, R.T., Roser, F.X., Kegel, G. and Cullen, T.L. (1964) in The Natural Radiation Environment. Univ. of Chicago Press, Chicago IL, p. 387.

6. Cullen, T.L. and Paschoa, A.S. (1976) in Proc. 10th Midyear Topical Symp., Health Physics Soc., Saratoga Springs, N.Y., p. 301.

7. Drew, R.T. and Eisenbud, M. (1966) Health Physics, 12, 1267.

8. Bushee, J.M. (1971) Geochronological and Petrographic Studies of Alkaline Rocks from Southern Brazil. Ph.D. Diss., Univ. of Calif., Berkeley CA.

9. Instituto de Pesquisas Tecnologicas do Estado de Sao Paulo (1982) Estudos para a caracterizacao do comportamento de lencol freatico na area do Morro do Ferro--Pocos de Caldas, M.G. Rept. 16, 423, Sao Paulo, Brazil.

10. Sill, C.W. and Williams, R.L. (1981) Anal. Chem., 53, 412.

11. Linsalata, P. (1980) New York University Medical Center, New York N.Y. (Personal communication).

12. Fleischer, R.L. and Raabe, O.G. (1978) Geochim. et Cosmochim. Acta, 42, 973-978.

13. Miekeley, N., Vale, M.G.R., Tavares, T.M. and Lei, W. (1982) Some Aspects of the Influence of Surface and Groundwater Chemistry on the Mobility of Thorium in the "Morro do Ferro" Environment. Scientific Basis of Nuclear Waste Management, Vol.11, Lutze, ed.

Published 1982 by Elsevier Science Publishing Co
SCIENTIFIC BASIS FOR RADIOACTIVE WASTE MANAGEMENT - V
Werner.Lutze, editor

POTENTIAL FOR THE RAPID TRANSPORT OF PLUTONIUM IN GROUNDWATER AS DEMONSTRATED BY CORE COLUMN STUDIES

D.R. CHAMP, W.F. MERRITT, and J.L. YOUNG
The Chalk River Nuclear Laboratories
Atomic Energy of Canada Limited
Chalk River, Ontario
Canada, K0J 1J0

INTRODUCTION

The mobility of radionuclides in groundwater flow systems can be significantly altered by processes such as complexation, sorption on particulates, or hydrolysis, precipitation and the formation of colloids. Such processes provide potentially significant pathways for transport of radionuclides in the biosphere. Previous soil column studies[1] demonstrated that radiocesium could be transported by particulates and that the process was likely mediated by micro-organisms. That observation provided a plausible mechanism to explain the anomalously rapid transport of small amounts of cesium-137 released from glass blocks buried below the water table at the Chalk River Nuclear Laboratories[2].

The behaviour of plutonium in soils is apparently covered largely by processes which influence the chemistry of the Pu (IV) ion. Wildung et al[3] maintained that higher oxidation states are reduced to Pu (IV) which can then undergo extensive hydrolysis to produce insoluble products. Such reactions should effectively inhibit Pu transport in the biosphere. However, complexation with organic and inorganic ligands can stabilize Pu (IV) against hydrolysis and increase its solubility. Analysis of surface soils that contain Pu has shown that a small percentage of the Pu remains in solution predominantly as colloidal particles[4] and therefore is available for transport. This study has examined the potential for the transport of Pu in a groundwater flow system through the use of laboratory columns prepared from "undisturbed" horizontal cores taken from saturated zone soils adjacent to the glass block site. The columns were spiked with ^{239}Pu, continuously eluted with groundwater from the glass block site, and the activity and physical-chemical characteristics of the ^{239}Pu in the effluent determined following various manipulations of the column elution conditions.

MATERIALS AND METHODS

Pu Solution - The Pu solution contained 76.8, 19.7, 2.9, 0.5 and 0.1 atom % ^{239}Pu, ^{240}Pu, ^{241}Pu, ^{242}Pu, and ^{238}Pu respectively. In the text the predominant isotope, ^{239}Pu, is used when referring to this mix of isotopes.

Soil Columns - Soil columns were prepared from undisturbed horizontal cores of uncontaminated saturated zone sand taken upgradient of the glass block site. The columns (4.5 cm x 11 cm, void volume ~54 mL) were equilibrated for 1 to 3 months with groundwater, taken from a well adjacent to the site, with an upward flow of approximately 45 cm·d^{-1} (approximately 3 times the groundwater flow rate at the site). Columns were loaded with 80 μg (10 μCi or 3.1x10^5 Bq) of ^{239}Pu, obtained by a 100-fold dilution of a stock ^{239}Pu solution (8 M HNO$_3$), and eluted with groundwater. The column effluent was collected for analysis.

Effluent Analysis - ^{239}Pu in the column effluent was determined as either total Pu or as particulate Pu. Total ^{239}Pu was determined by evaporating 10 mL of effluent directly onto a stainless steel planchet and α-counting. Particulate ^{239}Pu was determined by filtering an aliquot of the column effluent through 0.45 μm Millipore filters and then counting the filtrate and filter. The reported ^{239}Pu activities have a statistical accuracy of ±12% or less with 95% confidence limits.

Geochemistry - The geochemical characteristics of the groundwater, before and after equilibration in the laboratory, and of the column effluent were measured. Redox and pH measurements were made in line using a flow cell. Dissolved anions in the groundwater were measured by ion chromatography using a Dionex system while cations were measured by atomic absorption spectroscopy. Dissolved oxygen was measured on a YS1 Model 57 dissolved oxygen meter. Alkalinity was determined by potentiometric titration[5] of the sample following 0.45 μm filtration.

Temperature Cycling - Columns were routinely eluted at 22±2°C. For low temperature conditions (~0°C) the column and the eluant were kept in an ice bath during the experiment.

Antibiotic Treatment - The columns were eluted in the normal manner at ~22° with groundwater containing 50 mg·L^{-1} streptomycin sulfate, 50 mg·L^{-1} penicillin G, plus 100 mg·L^{-1} chloramphemical. The antibiotic solution was shielded from light and changed after 2 days. Recovery was achieved by re-eluting with untreated groundwater.

Ultrafiltration - The standard protocol for ultrafiltration was as follows; Amicon ultrafiltration membranes (UM05 and YM 10, apparent exclusion limits of 500 and 10,000 molecular weight, respectively) were presoaked overnight in double-distilled deionized water, the membranes were placed in the ultrafiltration cell and washed with 1 cell volume of double-distilled deionized water, the cell was then rinsed with an aliquot of the sample which was then discarded, the sample (prefiltered through a 0.05 μm Millipore filter) was then loaded into the cell and concentrated to ~10% of the cell volume (the first ~6 ml of ultrafiltrate containing some distilled water from the membrane rinse was discarded). The filtrate and retentate were accurately measured for the mass and activity balance calculations. UM05 membranes and YM 10 membranes were run at N_2 pressures of 60 and 20 psi (414 and 138 kPa) respectively.

Chemical Speciation - The oxidation states of the Pu in the column effluents were determined using modifications of the methods reported by Foti and Freiling[6].

NdF Coprecipitation - The NdF_3 coprecipitation was conducted using a [236]Pu (IV) tracer which was prepared by adding 100 mg of $NaNO_2$ to 400 μL of [236]Pu tracer in 8 M HNO_3, evaporating to dryness, and then resuspending in 5 mL of distilled H_2O. The coprecipitation was carried out by adding 5 mL of tracer solution to 10 mL of column effluent containing [239]Pu. This solution was adjusted to 1 M in HNO_3, 100 μg of Nd (III) carrier in 1 M HNO_3 (1 mL) was added and dispersed, and then 2 mL of 49% HF was added to precipitate NdF_3. A NdF_3 bed was prepared by mixing 50 μg Nd (III), 5 mL of 1 M HNO_3, and 1 mL of 49% HF and filtering through a 0.1 μm Tuffryn filter. The sample was similarly filtered, washed, and then mounted on a stainless steel planchet for determination Pu (III), (IV) or (V) by α-spectrometry. The filtrate which contained any Pu (VI) present was then reduced with $NH_2OH \cdot HCl$ plus Fe and any reduced Pu (VI) was coprecipitated on CeF_3 and determined by α-counting. The tracer yield in the above procedure was 85 to 90%.

TTA Extraction - A 10 mL sample of column effluent was extracted sequentially with 2-thenoyltrifluoracetone (TTA) in toluene at pH's 0.4 and 4.3. 10 mL of column effluent was adjusted to pH 0.4 with hydrochloric acid, extracted with 10 mL of 0.4 M TTA for 20 minutes, and the phases separated. The aqueous phase was then adjusted to pH 4.3 with ammonium acetate, extracted for 20 minutes with 0.4 M TTA, and the phases were again separated. The organic phases were each washed twice with 10 mL of distilled water for 10 minutes and the Pu was then stripped out by extracting twice with 6 M hydrochloric acid. These acid extracts were pooled, evaporated just to dryness, resuspended in nitric acid and dried twice, and finally resuspended in nitric acid, heated, and then transferred to stainless steel planchets for evaporation and α-counting.

RESULTS AND DISCUSSION

Low levels of plutonium were continuously released from the core columns

748

into the groundwater used for column elution. Figure 1 shows the ^{239}Pu release
rates observed on a column over a period of ~8 months during which time ~800
void volumes of solution were passed through the column; low levels of Pu were
detected in the effluent within 1 day of loading. The transport of Pu associated with particulates is shown in the lower curve of Figure 1 which represents the activity of plutonium in solution that was retained on 0.45 μm
filters. The amount of Pu associated with particulates increased gradually
with time and, at steady state approximately 50% of the plutonium released was
associated with particulates.

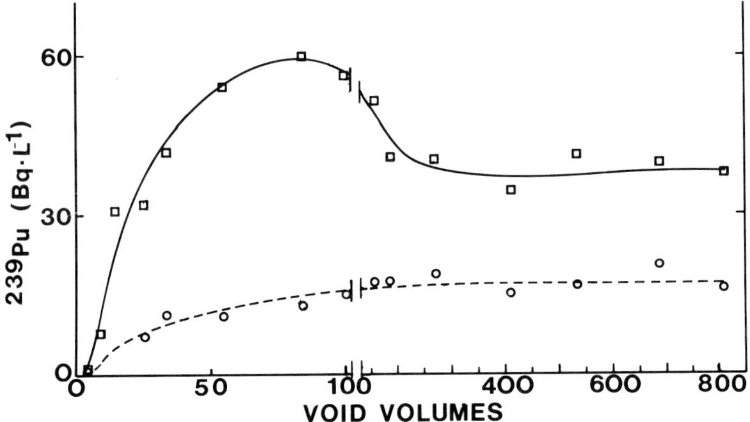

Figure 1 Release of plutonium from Soil Column. The column was eluted at
220 mL·d^{-1} and the total, □————□ , and particulate, o--------o, ^{239}Pu
activity in the effluent determined as described in Methods.

Steady-state releases from individual columns have varied from approximately
40 Bq·L^{-1} for the column shown in Figure 1 (release of the total Pu inventory
would require ~115 years) to a maximum steady state release rate of approximately 130 Bq·L^{-1}. In the latter case the initial peak was only 140 Bq·L^{-1}.
In spite of the 3 fold higher release rate for total Pu from the latter column
the percentage associated with particulates was similar; 60±10% for the latter
versus 50±10% for the former. The different release rates from the different
columns appear to be a function of either the mineralogy and/or the associated
microbiology of the sediments since all columns were equilibrated and eluted
with groundwater from the same well.

TABLE 1

GEOCHEMICAL DATA

Water Source	pH	E_H	DO	HCO_3^-	NO_3^-	SO_4^{2-}	Ca^{2+}	Mg^{2+}	K^+	Na^+	Si^+	Al^+
Groundwater at Well Head	5.3	445	2.5	8.7	0.2	18.8	5.5	1.6	0.9	1.9	5.0	0.2
Groundwater in Laboratory	5.6	490	8.5	8.2	0.2	17.4	5.1	1.6	0.9	2.0	5.0	0.2
Column Effluent	6.7	520	n·a·[b]	9.5	0.2	18.5	6.1	1.6	0.9	2.0	4.0	0.5

a   All data in $mg \cdot L^{-1}$, except E_H (mV)
b   n·a = not analyzed

Significant alterations in the various geochemical parameters characterizing
the well water, Table 1, were not noted during the course of these experiments.
Fe and Mn concentrations were both at the limits of detection (0.05 $mg \cdot L^{-1}$).
The well water is relatively oxidizing as it contains appreciable dissolved
oxygen and its oxidizing potential was only marginally elevated by laboratory
equilibration and passage through the column.

Size Distribution of Plutonium Species - The size characteristics of the Pu
species released from the column have been examined by filtration and ultrafiltration techniques. For the purposes of this study particulate species are
those excluded by a 0.45 μm (450 nm) filter, the remaining species are considered "dissolved". The dissolved species are further subdivided into colloidal and soluble components. Colloidal material is that excluded by a YM 10
(10,000 molecular weight exclusion limit, effective pore diameter of 2.8 nm)
ultrafiltration membrane as proposed by Schindler and Alberts[7]. Soluble components are those passing through a YM 10 ultrafiltration membrane. Two size
classes of colloids were determined since ultrafiltration studies were always
performed on samples which had been prefiltered through 0.05 μm filters. The
soluble components were similarly divided into two size classes, one comprising
intermediate molecular weight (500-10,000) species and the other (<500
molecular weight) including ionic and small organic or inorganic complexes.

We have determined the size distribution for the Pu species in the effluent
from the column illustrated in Figure 1. Table 2 contains results for samples
taken prior to the attainment of steady state conditions (85 void volumes),
column A of Table 2, and for samples taken at steady state (250 and 350 void
volumes), column B of Table 2. The size distribution at both steady state
sampling points were very similar and have been averaged. Under conditions of
steady state release (column B, Table 2) 46% of the Pu was associated with

particulates, a further 23% was colloidal and 31% was soluble. Two-thirds of the colloidal species were 2.8 to 50 nm in size. Approximately 87% of the soluble species had apparent molecular weights of 500 to 10,000, the remaining 13% were present likely as small complexes with apparent molecular weights of less than 500. The data in column A (Table 2) indicates that during the approach to steady state there has been a shift in the size distribution of the Pu species, most notably in the particulate versus large colloidal fraction. The significance of this observation is as yet unclear.

TABLE 2

SIZE DISTRIBUTION OF PLUTONIUM IN COLUMN EFFLUENTS

Size Category	Percent of Pu in Column Effluent	
	A	B
Particulate (>450 nm)	21	46
'Dissolved' (<450 nm)		
Colloidal (50-450 nm)	23	8
(2.8-50 nm)		15
Soluble[b] (500-10,000 molecular weight)	48[a]	27
(<500 molecular weight)	8	4

[a]Ultrafiltration with a UM05 membrane only was performed.

[b]Ultrafiltration percentages have been corrected assuming 100% recovery. Average recoveries were 87 and 92% for UM05 and YM10 ultrafiltration respectively.

Avogadro et al[8] have also reported on Pu soil column experiments where a continuous input of Pu to the column was derived from leaching of actinide bearing glass. Under these conditions ~80% of the input Pu was not retained by the soil columns and of this only ~20% was colloidal or particulate as assayed in the column effluent, substantially less than we have observed. The generation of particulate or colloidal Pu species on the column, as opposed to that which may have been present in the column input, could not be assessed on the basis of the information presented. Our observation of a large fraction of colloidal plus particulate material is in general agreement with observations for Pu in other systems, such as various surface soils[2]. The colloidal material in the column effluent appears to be relatively stable since no statistically significant difference in the percentage retained by a 0.05 μm filter was observed following storage at ~22°C for 4 weeks.

Distribution of Oxidation States - The oxidation states of ^{239}Pu in the column effluents were determined by a combination of NdF_3 coprecipitation and 2-thenoyltrifluoracetone (TTA) extraction. From the NdF_3 coprecipitation results, shown in Table 3, we conclude that most if not all ^{239}Pu in the column effluent was present in the lower oxidation states.

TABLE 3

NdF_3 COPRECIPITATION OF ^{239}Pu IN COLUMN EFFLUENTS

Pu Species	Percent of Activity[a]	
	^{236}Pu Tracer[b]	^{239}Pu
Pu(III), (IV) and (V)	86	81
Pu(VI)	n.d.[c]	n.d.

[a] Values are means of triplicate determinations. Bulk effluent activity was 37 $Bq \cdot L^{-1}$ ^{239}Pu

[b] ^{236}Pu(IV) was added as a tracer

[c] Not detectable.

From the results of the TTA extraction experiments (Table 4) approximately 80% of the ^{239}Pu was extracted from the column effluents by TTA at pH 0.4 and was therefore likely present as Pu (IV). The remaining 20% of the activity extracted at pH 4.3 must have been present at Pu (III) since only Pu (VI) would have coextracted with Pu (III) and we have previously shown the absence of Pu (VI). Extraction experiments on filtered and unfiltered samples yielded very similar results from which we conclude that the Pu present as or associated with particulate species and the dissolved species are equally well extracted by the procedure used. The results also show that the distribution of oxidation states is very similar for both fractions.

TABLE 4

TTA EXTRACTION OF ^{239}Pu IN COLUMN EFFLUENT

Source	pH	Percent of ^{239}Pu Recovered in Extract[a]
Bulk Column Effluent	0.4	80
	4.5	20
0.45 µm Filtered Column Effluent	0.4	86
	4.5	14

[a] Values are means of triplicate determinations assuming 100% recovery. The mean recovery was 82%.

The overall chemical speciation results have shown that the Pu in the column effluents was present as Pu (III) and Pu (IV) in a ratio of approximately 1 to 4. The ^{239}Pu stock solution in 8 M HNO$_3$ likely contained only Pu (IV)[9], however upon dilution the pH change may have resulted in a change to Pu (III) and Pu (VI)[9]. The complexity of these reactions makes it very difficult to decide whether the effluent speciation is what one would expect on the basis of the input conditions. However since Pu (III) and Pu (IV) were observed in the effluent then the Pu must have been either sorbed as Pu (III) and Pu (IV) or if Pu (VI) was also present it must have undergone reduction to the lower oxidation states during residence on, or transfer through, the column.

Mechanisms of Pu Release - The mechanisms of release of the Pu have been studied by temperature cycling and treatment with antibiotics.

Figure 2 illustrates the effect of temperature on release rates for 2 columns whose steady state release rates differed by a factor of 3. After cooling to 0°C there was an approximately 4-fold decrease in the release rate of total Pu within approximately 48 hours. However the release rate of particulate Pu decreased by approximately 8-fold suggesting different mechanisms for the release of particulate versus other fractions. Within 20 days of shifting back to ~22°C both columns had re-established the original release rates for both total Pu and particulate Pu species. At present we have no explanation for the unexpectedly long recovery time.

The influence of a suite of antibacterial agents on Pu release was also tested and the results for 2 columns are presented in Figure 3.

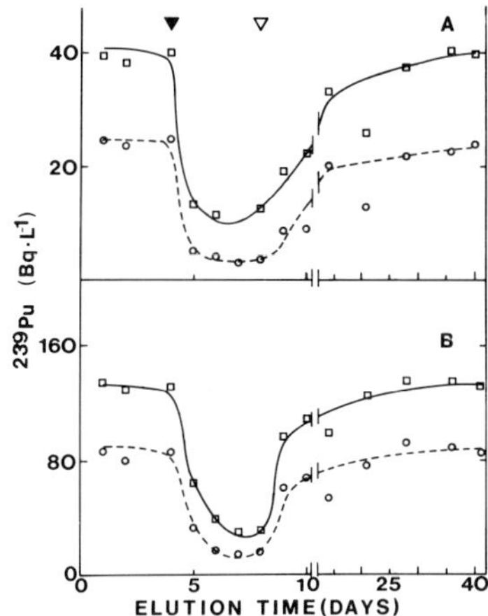

Figure 2. Temperature Effects on ^{239}Pu Release for 2 Different Columns.▼, shift to ~0°C; ▽, shift to ~22°C.

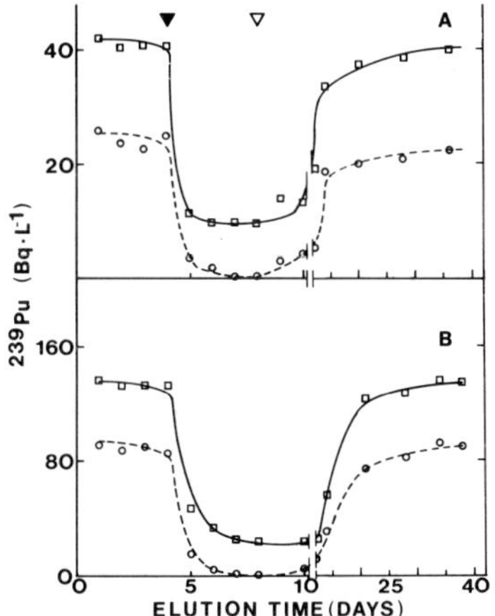

Figure 3. Effect of Antibiotics on ^{239}Pu Release for the same Columns as in Figure 2.▼, beginning of elution with antibiotic ▽, cessation of antibiotic solution.

These results provided further support for differing mechanisms for the release of colloidal or particulate material versus soluble material. Following 3 days of continuous exposure to the antibiotic solution there was no longer any Pu in the effluent that could be removed by filtration through 0.45 μm filters, from which we conclude that the release of Pu in association with particulates was totally inhibited by the antibiotic treatment. The total Pu activity in the effluent was only 20 to 25% of the activity levels before antibiotic treatment and following recovery. If we compare these residual activity levels with the size determination data of Table 2 we see that the residual activity levels are equivalent to the percentage of the Pu in the column effluent that was present as soluble species. This comparison suggests that the antibiotic treatment inhibited the release of colloidal species as well as particulate species. This prediction will be tested in future studies. If indeed the antibiotic effects were due to the inhibition of bacterial metabolism then the slow but total recovery of Pu release rates would suggest that either the

754

bacterial species responsible for the release were not completely removed by the treatment and repopulated the column or that bacteria with similar metabolic properties were present in the groundwater eluent and repopulated the column.

The introduction of antibiotics at relatively high concentrations may have had effects other than removing bacteria but these effects if any have not been defined. However the authors[1] have also observed the transport of Cs in association with particulates. In that instance not only antibiotic treatment but also irradiation of the column and eluent interfered with the release of particulates, which lends further support to bacterial involvement. The nature of the particulates and of the bacterial involvement in their release is still undergoing investigation.

SUMMARY

Through the use of core columns we have demonstrated the potential for the release of low-levels of Pu from contaminated saturated zone soils. Transport of the reduced species of Pu occurs as both dissolved (soluble plus colloidal) species and as particulates. The colloidal plus particulate species account for 50 to 70 % of the total Pu released under our oxidizing conditions. The mechanisms for release of soluble versus colloidal plus particulate species appear to be different with bacterial involvement in the release of the latter species being highly probable.

REFERENCES

1. Champ, D.R. and Merritt, W.F. (1981) in Proceedings of the Second Annual Conference of the Canadian Nuclear Society, McDonnell, F.N. ed., pp 66-69.
2. Walton, F.B. and Merritt, W.F. (1980) in The Scientific Basis for Nuclear Waste Management, Vol. II, Plenum Press, pp 155-166.
3. Wildung, R.E., Garland, T.R., and Cataldo D.A. (1979) in Biological Implications of Radionuclides Released from Nuclear Industries, IAEA, Vienna, Vol. II, pp 319-334.
4. Garland, T.R., and Wildung, R.E. (1977) in Biological Implications of Metals in the Environment, CONF-750929, National Technical Information Service, U.S. Dept. of Commerce, pp 254-263.
5. Barnes, I. (1964) U.S. Geol. Surv. Water-Supply Paper 1535-H.
6. Foti, S.C., and Freiling, E.C. (1964) Talanta, II, 385.
7. Schindler, J.E., and Alberts, J.J. (1974) Arch. Hydrobiol. 74, 4, 429.
8. Avogadro, A., Murray, C.N., and De Plano, A. (1979) in The Migration of Long-Lived Radionuclides in the Geosphere, OECD, pp 137-146.
9. Allard, B., Kipatsi, H., and Liljenzin, J.O. (1980) J. Inorg. Nucl. Chem., 42, 1015.

Published 1982 by Elsevier Science Publishing Co
SCIENTIFIC BASIS FOR RADIOACTIVE WASTE MANAGEMENT - V
Werner.Lutze, editor

PROPERTIES AND MOBILITIES OF ACTINIDE COLLOIDS IN GEOLOGIC SYSTEMS

U. OLOFSSON, B. ALLARD, B. TORSTENFELT AND K. ANDERSSON
Department of Nuclear Chemistry, Chalmers University of Technology,
S-412 96 Göteborg, Sweden

INTRODUCTION

The interaction between radionuclides in solution and exposed geologic media, e.g. in connection with underground storage of radioactive waste, is largely determined by the chemical properties of the system. The mobility of radionuclides in contact with low-capacity minerals would depend on the chemical state of the element (e.g. degree of hydrolysis, formation of organic and inorganic complexes, etc.). Particularly for the actinides in their lower oxidation states (III, IV), however, would a formation of colloid particles be feasible in the groundwater environment. These colloids could be either true radiocolloids, i.e. aggregates of the radionuclide itself, or pseudo-colloids, i.e. colloid material already present in the groundwater, onto which the radionuclide has sorbed [1,2]. These colloidal particles may be very poorly sorbed on water exposed geologic media in comparison with radionuclides in true solution. Studies on the formation and properties of some actinide colloids (Am, Np, Pu) are discussed in this paper.

EXPERIMENTAL

Chemicals

All salt solutions were prepared from p.a. $NaClO_4$, filtered through Nuclepore filters with a pore size of 400 nm and centrifuged at about 27000 g for one hour in order to remove all impurities prior to the experiments.

Radionuclides

The radionuclides used were ^{241}Am, ^{237}Pu with ^{239}Pu as carrier and ^{235}Np with ^{237}Np as carrier. All stock solutions were filtered through Nuclepore filters with a pore size of 400 nm.

Centrifugation experiments

Solutions of $NaClO_4$ at various ionic strengths (20 ml) were poured into 50 ml centrifuge tubes and the radionuclide was added to desired concentration. (See below). The pH-value was adjusted with $HClO_4$ or NaOH.

The solutions were centrifuged at about 27000 g for half an hour, and 0.5-1.0 ml samples were withdrawn and measured in a well-type scintillation counter. Samples were also taken from agitated solutions. This separating and sampling procedure, as well as measurements of pH, but without any pH-adjustments, were repeated after various storage times.

The following parameters were varied:
- pH-value of the solution (2-12)
- initial nuclide concentration (10^{-7} M, 10^{-9} M)
- ionic strength of the aqueous phase ($NaClO_4$; 0.01 M, 1.0 M)
- storage time (6 h, 27 h, 1 w, 4 w, 6 w).

All experiments were carried out in polypropylene tubes in order to avoid any particle release from the tube walls, which could serve as sources of pseudocolloids.

The experiments with plutonium and neptunium were performed in a glove-box with N_2-atmosphere in order to avoid any formation of carbonate complexes at high pH due to the contact with the air containing CO_2.

Diffusion experiments

Two 25 ml polyethylene vessels were put together and a Nuclepore filter with pore sizes 400 nm was placed between them, Figure 1.

Solutions of Am and Pu in 0.01 M $NaClO_4$ of various pH were prepared and stored in polypropylene vessels for about one week. In some of these solutions a formation of colloidal particles would be expected, according to the centrifugation experiments. After the storage the solution was poured into one of the diffusion vessels. A solution of the same composition, but without the radionuclide was poured into the other vessel. The solutions were stirred with magnetic stirrers, and samples were withdrawn from the vessels after various times.

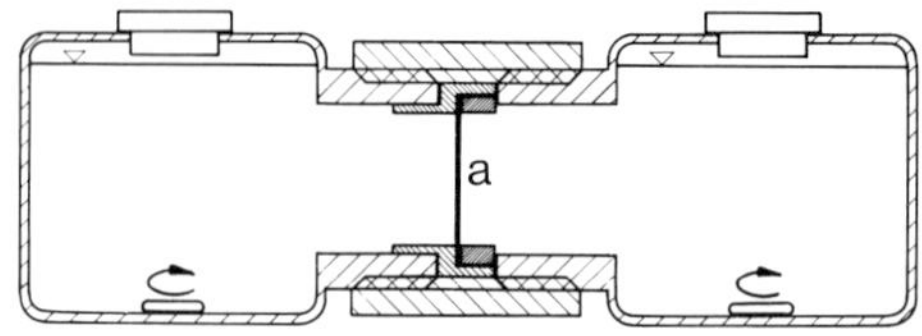

Fig. 1. Cell for diffusion measurements. a. Membrane.

RESULTS AND DISCUSSION

Centrifugation experiments

Since colloid formation of e.g. actinides is strongly related to hydrolysis and hydroxide precipitation in most cases [3-5], the pH-value of the solution has been selected as an independent chemical parameter in all the present experiments.

Ionic strength and storage time. At the lower ionic strength (0.01 M) the removal of Am from the solution starts already at pH about 3, and in the pH-range 6-8 more than 50% is removed after 6 h [6]. For Pu this removal is even greater than for Am (Figure 2) and in the pH-range 6-10 more than 60% is removed. For Np at the lower ionic strength there is no removal at all (Figure 2) not even after 4 weeks. Both for Am and Pu the removal increases with the storage time with a maximum at pH 6-7.5.

At the higher ionic strength (1 M $NaClO_4$) the removal of Am and Pu is more than 80% in the pH-range 6-11 already after 6 h. However, there is a decrease in the removal at a pH-value about 12. For Np there is a removal of 10% after 6 h at pH above 6 and this fraction is increasing with the storage time.

The total remaining nuclide concentration for Am and Pu in the pH-range 8-10 is about 10^{-11} M and for Np 10^{-9} M. These concentrations are much lower than the calculated maximum solubilities for $Am(OH)_3$, $Am_2(CO_3)_3$, PuO_2, NpO_2 and NpO_2OH [7].

The centrifugable fraction. The removal of the actinides from the solution can be caused either by sorption on the tube walls or by formation of colloid particles which can be centrifuged out of the solution. To determine the particle fraction, samples were taken both after centrifuging at about 27000 g for half an hour, (only particles with a radius less than 50 nm remain in solution) and after agitating the solutions. The difference between these samples is denoted the centrifugable fraction. Figure 3 indicates that for both Am, Pu and Np most of the removal is due to sorption on the tube walls.

For Am the centrifugable fraction has two maxima, at pH about 5 and at pH above 12. However, at the high ionic strength there is no maximum at pH 12. The centrifugable fraction at pH around 5 can probably be ascribed to impurities in the system (not in the salt solution, but in the ^{241}Am stock solution [6]).

For Pu there is no centrifugable fraction at pH below 8. At pH above 8-9, however, this fraction is increasing with the pH-value, and at pH about 12 most of the removed Pu-fraction seems to be in particulate form which can be centrifuged out of the solution.

758

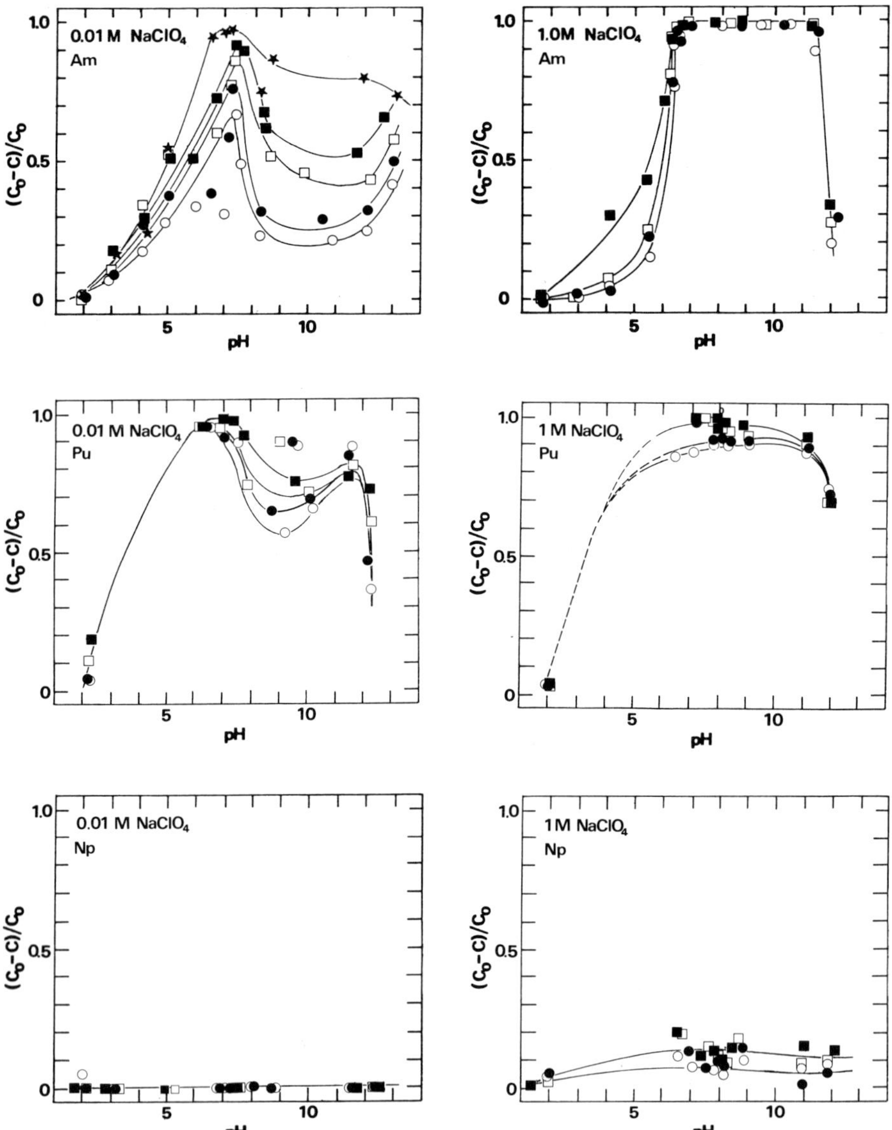

Fig. 2. Fraction of Am, Pu and Np removed from solution vs pH after various storage times and at various ionic strengths. [Am] = 2.3×10^{-9} M, [Pu] = 6.0×10^{-10} M, [Np] = 2.1×10^{-9} M; centrifuged samples (27000 g), no sorbent added. ○ 6h, ● 27h, □ 1w, ■ 4w (Pu, Np) or 6w (Am), ✶ 6m.

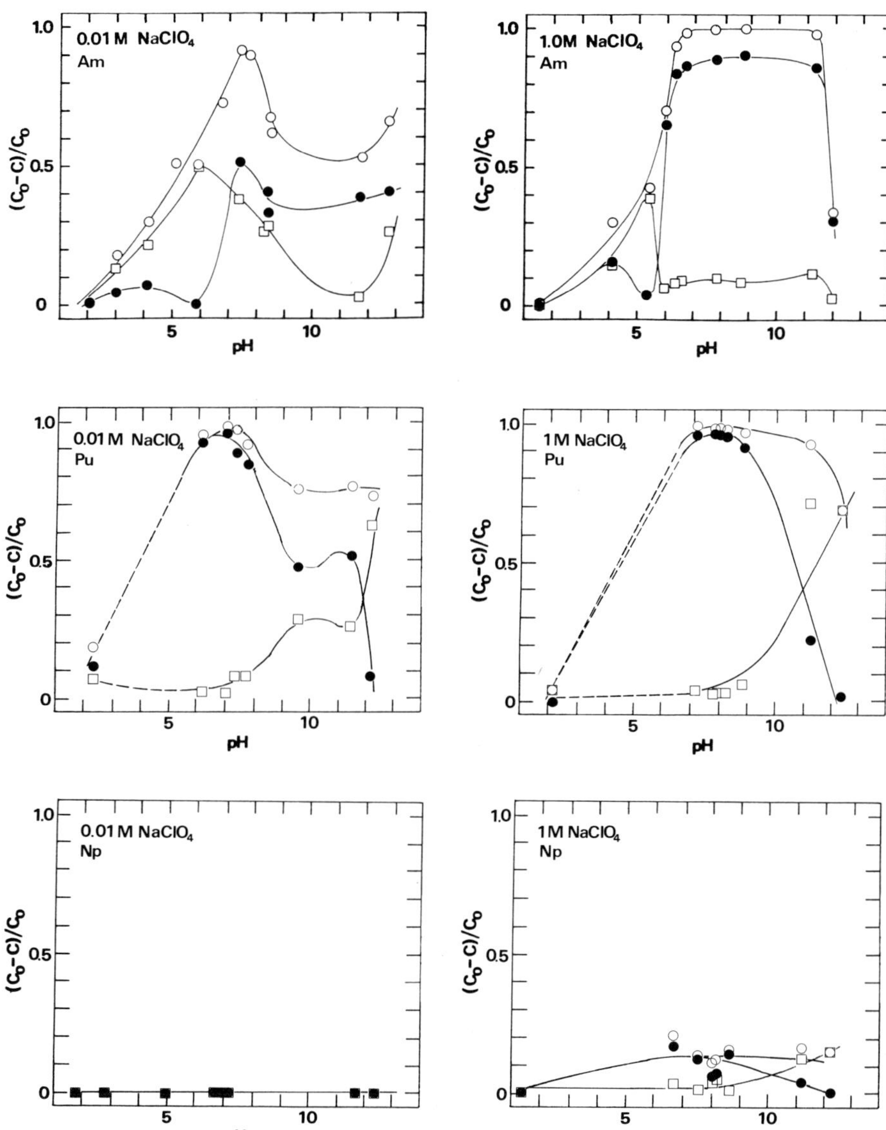

Fig. 3. Centrifugable fraction of Am, Pu and Np vs pH at various ionic strengths. Concentrations as in Fig. 2, 6w storage time, no sorbent added. ○ centrifuged samples (27000 g), ● agitated samples, □ particle fraction (=centrifuged-agitated).

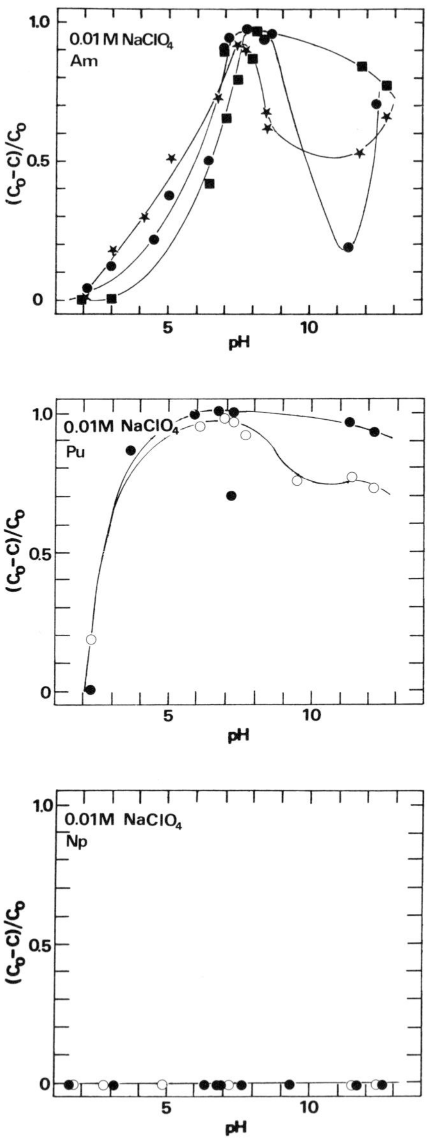

Fig. 4. Fractions of Am, Pu and Np removed from solution vs pH for various initial concentrations in 0.01 M $NaClO_4$; 6w storage time; centrifuged samples (27000 g), no sorbent added. Am: ● 2.9×10^{-7} M, ✶ 2.3×10^{-9} M, ■ 1.8×10^{-11} M; Pu: ● 6.0×10^{-8} M, ○ 6.0×10^{-10} M; Np: ● 2.1×10^{-7} M, ○ 2.1×10^{-9} M.

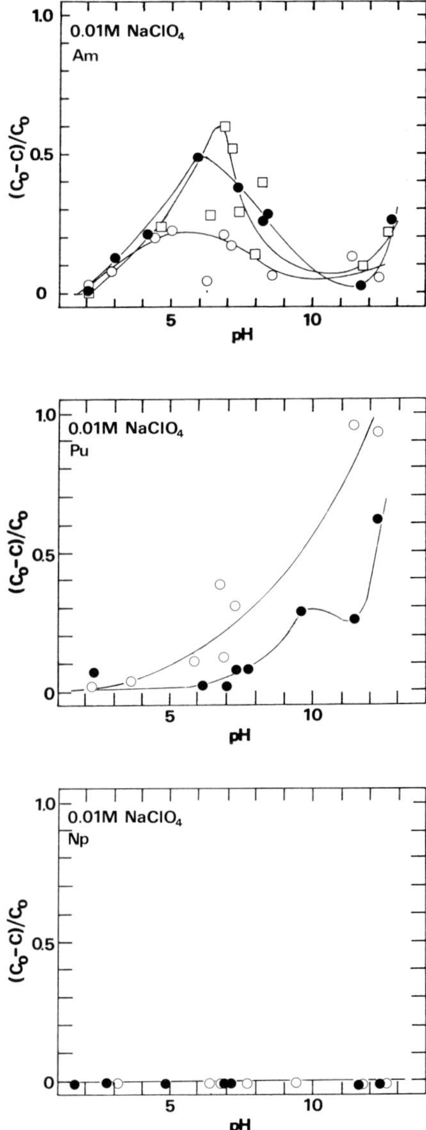

Fig. 5. Centrifugable fraction (particle fraction) of Am, Pu and Np vs pH for various initial concentrations in 0.01 M $NaClO_4$, 6w storage time, no sorbent added. Am: ○ 2.9×10^{-7} M, ● 2.3×10^{-9} M, □ 1.8×10^{-11} M; Pu: ○ 6.0×10^{-8} M, ● 6.0×10^{-10} M; Np: ○ 2.1×10^{-7} M, ● $2.\overline{1} \times 10^{-9}$ M.

For Np there are no indications of centrifugable fractions at lower ionic strength. At higher ionic strength and at pH above 10, however, there seems to be a small centrifugable particulate fraction.

The nuclide concentration

The removal of both Am and Pu appears at the same pH-values independent of the nuclide concentration. It starts at pH about 3 and reaches a maximum at pH about 7 for Am. For Pu it seems to start at somewhat lower pH-values and reach a maximum already at pH 5 (Figure 4 and 5). For Np there is no removal at all at the lower ionic strength. At the higher ionic strength and at pH-values above 6, however, there is a small removal, similar for both nuclide concentrations. For Am the formation of a centrifugable fraction at pH about 5 cannot be explained by precipitation of $Am(OH)_3(s)$ or $Am_2(CO_3)_3(s)$, when the initial Am-concentration is as low as 10^{-11} M. At pH above 12, however, a precipitation of $Am(OH)_3(s)$ is possible [6]. There is a formation of a centrifugable fraction for Pu, starting at pH about 8 and increasing with the pH. This particulate fraction may be a true colloid consisting of $Pu(OH)_4$.

For Np there is a small centrifugable fraction at pH above 10 for the higher ionic strength. Possibly this is a true colloid consisting of NpO_2OH.

Diffusion experiments

Diffusion through a membrane can be used as a method for separating ions and molecules from colloid particles in solutions. The colloid particles have a much slower diffusion rate than ions and molecules [8].

Figure 6 shows that at pH 2 there seems to be no colloid particles in the solutions, neither for Am nor Pu. For Am there seems to be colloids at the higher of the studied pH-values. At pH 4.3 there seems to be no particle fraction for Pu, while there are indications of colloid particles at pH above 10. However, it is impossible to determine, if the observed particle behaviour would be due to the formation of true colloids or pseudocolloids, because filters with only one pore size have been used so far in the experiments. Further measurements using filters with other pore sizes as well as calculations on particle size distributions are in progress.

CONCLUSIONS

The loss of Am in the pH-range 7-11 is mostly due to sorption on the tube walls. The centrifugable Am-fraction between pH 3 and 8 could be a pseudocolloid, where Am is sorbed on impurities in the solution. The particle fraction at pH above 12 may be a true colloid of $Am(OH)_3$.

Most of the Pu-fraction removed from the solution in the pH-range 5-9 is sorbed on the tube walls. There is a centrifugable Pu-fraction at pH above 8, which can possibly be a true colloid of $Pu(OH)_4$.

Neptunium is very poorly sorbed on the tube walls. However, there is a small fraction sorbed at pH between 6 and 9. A small centrifugable fraction is formed at pH above 10. This fraction may be a true colloid of NpO_2OH.

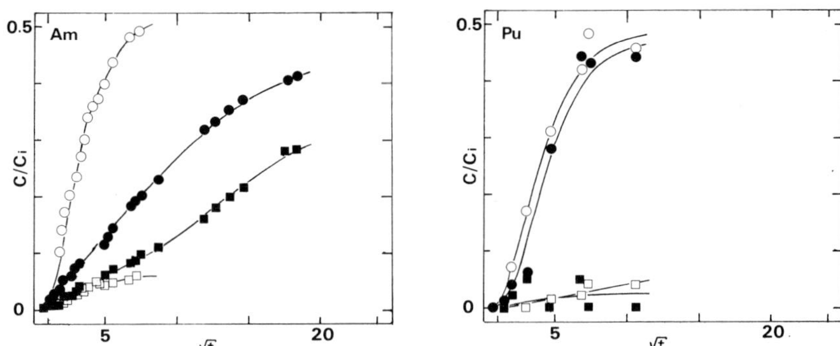

Fig. 6. Diffusion of Am and Pu vs time. C = Am- or Pu-concentration in the vessel with inactive start solution. C_i = Initial Am- or Pu-concentration – Am or Pu sorbed on the vessel walls and the filters. Am: ○ = pH 2, ● = pH 6.5, □ = pH 9.8, ■ = pH 12.2; Pu: ○ = pH 1.8, ● = pH 4.3, □ = pH 10.6, ■ = pH 11.8. Initial concentrations, [Am] = 2.9×10^{-9} M, [Pu] = 6.0×10^{-8} M.

ACKNOWLEDGEMENT

This work was funded by the Nuclear Fuel Safety Project (KBS). The skilful laboratory work of Ms. M. Bengtsson is gratefully acknowledged.

REFERENCES

1. F. Kepák, "Adsorption and Colloidal Properties of Radioactive Elements in Trace Concentrations", Chemical Reviews, 71 357 (1971).
2. P. Benes and V. Majer, Trace Chemistry of Aqueous Solutions, (Elsevier Scientific Publishing Company, Amsterdam-Oxford-New York 1980).
3. I. E. Starik and F. L. Ginzburg, "the Colloidal Behavior of Americium", Radiokhimiya, 6 685 (1961).
4. E. L. King, "Radiocolloidal behavior of Neptunium and Plutonium", in The Transuranium Elements, (McGraw-Hill Book Company, Inc, 1949, pp. 434-444).
5. L. D. Sheidina and E. N. Kovarskaya, "Colloidal State of Pu(IV) in Aqueous Solutions", Radiokhimiya, 12 253 (1970).

6. U. Olofsson, B. Allard, K. Andersson and B. Torstenfelt, "Formation and Properties of Americium Colloids in Aqueous Systems", in S. Topp. (Ed.) <u>Scientific Basis for Nuclear Waste Management</u>, Vol.**6**, (Elsevier, New York 1982), in press.
7. B. Allard, "The Solubilities of Actinides in Neutral or Basic Solutions", in N. Edelstein, Proc. of the Actinides-81 Conf., (Pergamon Press, Oxford 1982), in press.
8. P. Podhajecký, A. Gosman and P. Benes, "Radiotracer Analysis of the Physico-Chemical State of Trace Elements in Aqueous Solutions", J. Radioanal. Chem. <u>57</u> 253 (1980).

SCIENTIFIC BASIS FOR RADIOACTIVE WASTE MANAGEMENT - V
Werner.Lutze, editor

THE MIGRATION OF RADIONUCLIDES WITH GROUND WATER.

A DISCUSSION OF THE RELEVANCE OF THE INPUT PARAMETERS USED IN MODEL CALCULATIONS

BROR SKYTTE JENSEN

Department of Chemistry, Risø National Laboratory, DK-4000, Roskilde, Denmark

The plans for the disposal of radioactive waste leave very little time for testing long term performance of a repository so the evaluation of the hazards involved in the operation relies heavily on model calculations. It is therefore of utmost importance that these model calculations take all important parameters into account and are based on a thorough understanding of the possible physical and chemical processes in which the migrating species take part.

Although I am personally convinced that geological disposal can be done without any risk for mankind I am also fully aware of our lack of deep insight into many of the geophysical and geochemical processes in which migrating species can take part.

In the following I will discuss some of these processes and pinpoint some which I - at present - find the more important.

THE MIGRATION EQUATION

The differential equation governing the transport of a solute with flowing ground water[1] is given by

$$\mathrm{div}(D\mathrm{grad}\ C) - \mathrm{div}(Cv) + F = \frac{\partial C}{\partial t} \qquad (1)$$

stating that the rate of increase in concentration of solute in a given volume element can be expressed as a sum of three terms.

The term $(-\mathrm{div}(Cv))$ takes care of concentration changes due to passive transport with flowing ground water, and the term $\mathrm{div}(D\mathrm{grad}\ C)$ collects the effects of diverse mixing processes including diffusion. The term, F, takes care of sources and sinks changing the concentration of solute species in ground water. Chemical reactions like complex formation and adsorption/desorption are typical examples of such mechanisms.

CONVECTIVE TRANSPORT, HYDROLOGY

Laboratory experiments have clearly demonstrated an almost linear variation

766

of the dispersion coefficient, D, with the porewater velocity, v; v also varies
with porosity which again varies with space coordinates. In equation (1) the
dispersion coefficient is therefore assumed dependent on space coordinates. Slow
adsorption kinetics may also give rise to dispersion terms in F, whose magnitude depends on the porewater velocity. Before realistic model calculations can
be made it is therefore important to obtain data concerning the distribution of
porewater velocities and directions in and around the geological formation under
consideration. Such data are derived by hydrologists from the measurements of
hydraulic heads in boreholes taking the knowledge of the structure of the formation into account. Subsequent analysis is based on the application of d'Arcy's
law

$$\bar{v} = -Kgrad\ h \tag{2}$$

and the assumption of continuity in an incompressible medium,

$$div\ \bar{v} = 0 \tag{3}$$

where $\bar{v}$ is the seepage velocity of water and K the permeability constant for
the volume of the formation considered and h the piezometric head.

It is obvious that in practice it is impossible to obtain a very detailed
picture of water flows in and around a repository, but main flows will beyond
doubt be identified and characterised, and this seems to be adequate for the
present purpose because by far the largest and fastest transport of radionuclides will occur with the main flows.

Recent experiments in Japan[2] and Denmark[3] have indicated deviations from
d'Arcy's law at low water velocities in microporous media, which means that at
small hydraulic heads some types of microporous media may in practice be impervious. The experimental data obtained at Risø are shown in figure 1.

The experiments at Konan University were made on packed columns and at Risø
on pieces of unconsolidated chalk. Both investigations indicate a fall in permeabilities at the low water velocities most often found in deep ground waters.
At Risø additional experiments are being performed which hopefully will describe
the transition region in more detail.

If the observations are verified, convective transport in microporous media
like clays, silts, chalk etc. may be smaller than usually assumed, or even zero.
This fact will probably not simplify model calculations, but may make the construction of artificial barriers easier.

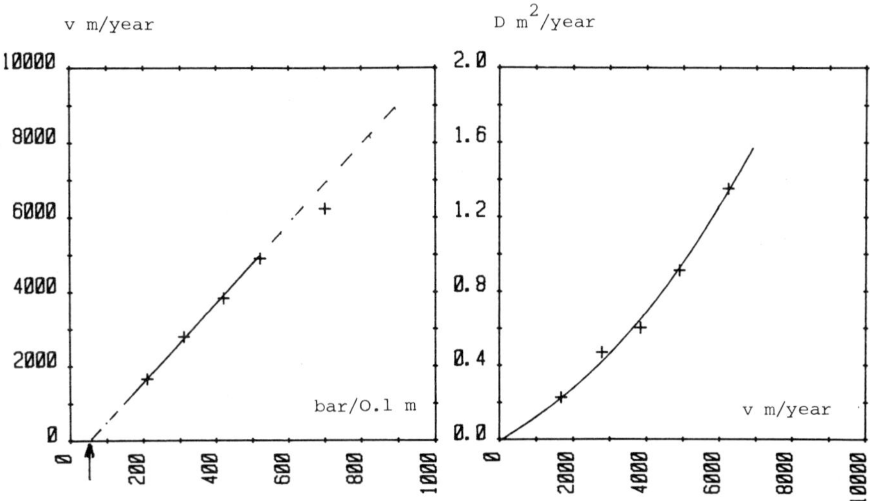

Fig. 1. An experimental velocity/
gradient curve indicating a deviation from d'Arcy's law.

Fig. 2. The variation of the dispersion coefficient with velocity.
Dispersivity $\simeq$ 2 mm.

THE DISPERSION PROCESS

The first term in equation (1) describes the effect of mixing neighbouring
volume elements of ground water, a simple physical process caused by concentration dependent diffusion, characterised by a diffusion coefficient, D_o, and diverse mechanisms induced by the flow of water in the pores. The overall mixing
is characterised by the dispersion coefficient, D, which experimentally has been
found to be related to D_o by the following approximate equation

$$D = D_o + \alpha v \tag{4}$$

where α is called the dispersivity and is measured in units of length. For
$v = 0$, D is identical with D_o, the diffusion coefficient in the medium.

The effect of a large dispersion coefficient will be that the solute is rapidly diluted and at the same time it is spread out in space. Experiments in the
laboratory have demonstrated that this effect has to be described by a dispersion coefficient, D_L, valid in the direction of flow and of transverse dispersion
coefficients, D_T, valid in directions perpendicular to the flow. The latter are
most often smaller by a factor of 5-20 than D_L. In figure 2 are shown experimen-

tal data from Risø of dispersion coefficients, D_L, measured on chalk samples. The dispersivity has been determined to $\simeq 2$ mm, which corresponds well with laboratory determinations made elsewhere.[2,6]

When field measurements are compared to the result of the model calculations it has been found that the best fit between experiments and the model is found when dispersivities of the order of $\alpha_L \simeq 100$ m and $\alpha_T \simeq 5$ m, i.e. several orders of magnitude larger than the laboratory data, are used.[6] Why is it so? One explanation may be that the large dispersivities take care of the effect of unknown inhomogeneities in the formation. When dispersivities of the order of 100 m are used, concentration variations over smaller distances are purely hypothetical and computational lattices of the same or larger size are used.

On this scale the distortion of concentration profiles caused by flow around inhomogeneities (figure 3) and the effect of fingering (figure 4) due to permeability variations may well be described as a dispersion effect, and may, if many such irregularities occur, be characterised by a dispersion coefficient. The effect of unknown irregularities in and around a possible site for geological disposal will probably be determined best from a study of the distribution of some natural tracer in the ground water. Salt seems to be an obvious tracer for mapping the flow around an isolated salt dome.

Another factor to be taken into account is the variation of the diffusion coefficient with coordinates. It is well known that D_o varies with the porosity of a medium – or more correctly with its poresize distribution. An approximate expression relating the effective diffusion coefficient to that of pure water is[4]:

$$\bar{D}_o = D_o \cdot g(\varepsilon/(2 - \varepsilon))^2 \tag{5}$$

where ε is the porosity of the medium and g a factor which is less than 1, and which reflects other characteristics of the medium such as poresize distribution.

The diffusion coefficient for the sodium ion in a chalk sample has at Risø been determined to $5 \cdot 10^{-8}$ cm^2/sec, and given a porosity of 40%, the g-factor has been calculated to the rather small value of 0.062.[5]

This discussion demonstrates clearly that there is every reason to show reservation towards the results of first attempts to model-calculations based on too simple ideas about the physical processes of migration phenomena; needless to say the same is also true when chemical interactions are considered.

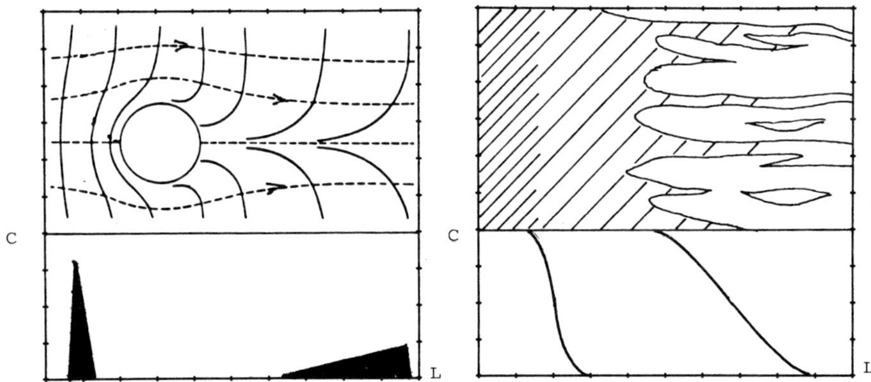

Fig. 3. The dispersion of a solute
front bypassing an obstacle.

Fig. 4. The dispersion effect of
fingering in an inhomogeneous medium.

CHEMICAL INTERACTIONS

The term F in equation 1 collects different physical and chemical interactions and may even include the source term, i.e. the leaching provided it is within the space coordinates. The leaching will not, however, be discussed here, but only the far-field migration phenomena of chemical nature.

Mass balance considerations make valid a substitution of F by $-\frac{\partial q}{\partial t}$, indicating that an increase in the aqueous concentration of the solute, C, is accompagnied by a simultaneous decrease in the concentration, q, of the species adsorbed on the solid. If it is furthermore assumed that there exists chemical equilibrium between these concentrations at any time and any place, then $q = K_D fC$, where f is a factor which converts q and C to the same dimensions. When C is measured in mol/L and q in mol/kg, f is $((1 - \varepsilon)/\varepsilon)\sigma$, where σ is the density of the solid phase.

An almost incalculable number of K_D determinations has been made throughout the years, which suffices for classifying elements into groups - as - not, slightly, medium or strongly adsorbed, but very few investigations have gone so much in detail, that a clear understanding of the adsorption mechanism has been obtained. It must also be admitted that even in the case such attempts have been

770

made, the interpretation of results is not always straightforward. One obvious
reason is that soil- or rock-samples are always mixtures of components, which
may show widely different adsorption behaviour. Classical ion-exchange has been
demonstrated to occur with clay minerals, at least with the simpler ions (figure
5). An anomalous curve in figure 6 may be interpreted as due to only 1% impurity
of a smectic clay in kaolinite, but may also be interpreted as due to different
adsorption sites on kaolinite crystallites. The author prefers the first interpretation.[1]

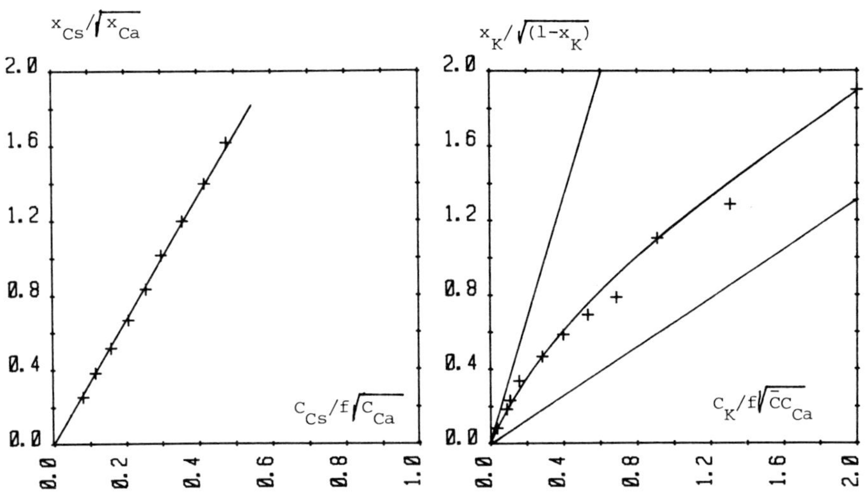

Fig. 5. An ion-exchange case according to theory.

Fig. 6. An anomalous ion-exchange
curve (impurity?).

Another complication arises because minerals as such are by no means unreac-
tive systems. They dissolve quite rapidly when being out of equilibrium with the
aqueous phase, albeit only to a minor degree. Most minerals dissolve incongruently, wherefore new phases of unidentified composition tend to be formed on
their surfaces probably exerting strong adsorptive properties. Experimental K_D
determinations may therefore to a smaller or lesser degree reflect the properties of these weathering products rather than these of the parent mineral surface.

All mineral surfaces aquire a charge when immersed into an aqueous solution.
The sign of the charge depends on the composition of the liquid phase, but is

negative in most of the cases encountered in nature. A charge distribution is built up in the solution layer close to the surface including the excess of positive charge needed for neutralisation. The solvated ions in the neutralising layer distribute themselves according to the same rules which govern ion-exchange, most often with preferential adsorption of the heavier elements. This adsorption is directly proportional to the available surface, wherefore finely divided or microporous matter may show a quite large adsorption capacity per weight.

If ions penetrate the double layer, they may be incorporated into the mineral surface or bound by strong chemical bonds to specific centers on the surface. These chemical forces are expected to be rather specific, limited by the same factors which regulate f.inst. the mutual solubility of compounds. The capacity for adsorption will probably be rather low, but corresponding affinities may be very large.

The processes mentioned above are expected to occur with almost every mineral phase imagined, whereas the more pronounced ion-exchange reactions occur with zeolites and some clay minerals, which have a built-in negative charge in their main lattice and have replaceable ions more loosely bound.

The ion-exchange properties of clays have been demonstrated beyond any doubt with simple ions like Na, K, Ca etc., but for ions of higher charge it is extremely difficult to obtain reproducible results and to interpret the adsorption mechanism. Both field observations and laboratory investigations have shown that some of the expandable clays collapse to clays of the illitic type under the influence of moderate heat in the presence of larger amounts of less hydrated cations, like potassium and cesium. The mechanism is believed to be due to a charge redistribution within the clay minerals favouring strong interactions between layers. Similar reactions may well be envisaged to occur pointwise where multicharged ions are adsorbed, whereby they may be caught mouse-trap-wise within clay minerals. Further transport may occur as particle transport or by dissolution reactions. That such mechanisms may actually occur is indicated by the close correspondence between the migration pattern of plutonium and cesium in topsoils exposed to infiltration, where one would expect different behaviour when considering their chemistry.

This last retardation mechanism will probably be characterised by slow kinetics without well defined equilibrium data. Other mechanisms may reach equilibria, but very slowly. This will affect the migration behaviour of a species very much if chemical rate constants are comparable to velocity parameters.

In figure 7 is shown the result of model calculations of a migrating species

772

assuming slow adsorption kinetics. It can be seen that a solute will experience
a kinetic dispersion, becoming highly diluted with passage through the formation. Slow kinetics will give rise to concentration distributions on the solid
phase which are displaced, relative to the concentration distribution in the
aqueous phase. Slow adsorption kinetics have to my knowledge not been investigated yet, but may well be an important case which deserves to be studied and to
be taken into account in modelling work.

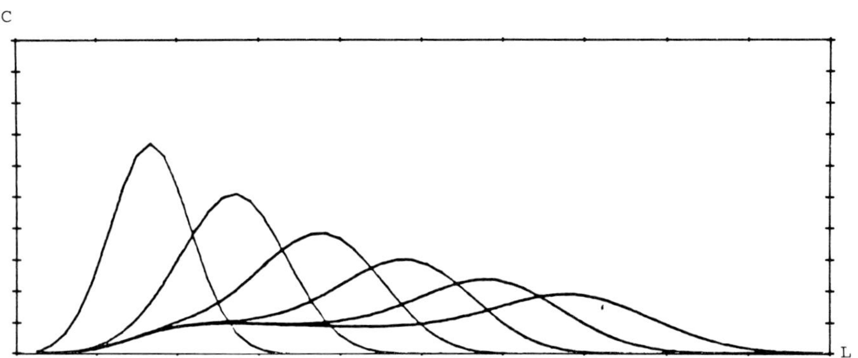

Fig. 7. Model calculations showing the dispersion effect of slow adsorption kinetics.

COMPLEX FORMATION

Experience shows that positively charged ions invariably are adsorbed onto
soil components, however to a varying degree, whereas negatively charged species
mostly remain unadsorbed in solution. Exceptions are found with anions like
phosphate which are known to form insoluble compounds and therefore may be adsorbed to surfaces with chemical bonds. Complex formation of positively charged
cations with negatively charged ligands, thereby forming complexes with lower
and even negative charge, will therefore tend to reduce adsorption. The experimental equilibrium distribution coefficient, K_D, for a z-valent cation will in
the case of complex formation be expressed by

$$K_D = \frac{((M) + \sum_i (MA_i))_s}{([M] + \sum_i [MA_i])_{aq}} = \frac{K_{D(0)} + \sum K_{D(i)} \beta_i [A]^i}{1 + \sum_i \beta_i [A]^i} \tag{6}$$

where the β's are the formation products of the complexes in the aqueous solution and the $K_{D(i)}$'s are the distribution coefficients for the individual species. For $i > z$, the distribution coefficients will, according to the above discussion, tend towards zero. Consequently, with increasing ligand concentration, the experimental distribution coefficient will approach zero, which has often been observed. If the stepwise formation of complexes is slow, a column experiment may separate the different complexes, which will be eluted according to the magnitude of their own distribution coefficient, but if their formation is fast, the elution will be described by the average distribution coefficient (equation (6)).

With other reactions similar arguments are valid. If redox equilibria are slow, individual species may show up in column experiments, and if fast elution according to an average value will be experienced.

REDOX POTENTIALS

Another factor which influences the speciation and thereby the adsorption and migration of at least some radionuclides is the redox potential, E_h.

The redox potential is a measure of the potential of the electron at equilibrium conditions between a redox pair, i.e.

$$E_h = E_o + \frac{RT}{nF} \ln(Ox/Red) \tag{7}$$

Although the definition of the redox potential and its numerical manipulation is straightforward, its experimental determination is often uncertain or difficult. Just like pH determinations in pure water may present difficulties, the determination of redox potentials in solutions without any buffer capacity for electron transfer may be wildly misleading. A further difficulty is due to the fact that many redox reactions are slow to very slow both in solution and at the electrode surface, a complication which normally does not interfere with pH determinations.

It is probable that the redox potentials of ground water are determined by the relatively fast electron exchange between ferrous and ferric species in solution and not by slow reactions like those between sulphur species. The readiness with which the ferrous/ferric ions takes part in reactions with other species strengthens the importance of the ferrous/ferric couple in groundwater systems.

Minerals containing ferrous and ferric ions are ubiquitous in amounts more than sufficient to supply the soluble species needed to determine the redox potentials. The extreme insolubility of ferric hydroxide (oxide) at pH > 7 will

774

secure reducing conditions in ground water contacting ferrous-containing minerals. A semiquantitative estimate indicates a probable range from 0 to -0.3 volts for the redox potential, which secures that the actinides will be found in the almost immobile tetravalent or trivalent states under deep soil conditions.

CONCLUSIONS

It is probably obvious to all, that establishing the scientific basis of geological waste disposal by going deeper and deeper in detail, may fill out the working hours of hundreds of scientists for hundreds of years. Such an endeavour is, however, impossible to attain, and we are forced to define some criteria telling us and others when knowledge and insight is sufficient.

In the present case of geological disposal one need to be able to predict migration behaviour of a series of radionuclides under diverse conditions to ascertain that unacceptable transfer to the biosphere never occurs. We have already collected a huge amount of data concerning migration phenomena, some very useful, other less so, but we still need investigations departing from the simple ideal concepts, which most often have provided modellers with input data to their calculations.

I therefore advocate that basic research is pursued to the point where it is possible to put limits on the effect of the lesser known factors on the migration behaviour of radionuclides. When such limits have been established, it will be possible to make calculations on "the worst cases", which may also occur. Although I personally believe, that these extra investigations will prove additional safety in geological disposal, this fact will convince nobody, only experimental facts will do.

REFERENCES

1. Jensen Skytte, B. (1982, in press) Critical Review of Available Information on Migration Phenomena of Radionuclides into the Geosphere.
2. Kaizuma, H., Ogawa, K. and Yamakita, I. (1977) LA-TR-77-12.
3. Jensen Skytte, B. (1982, in press) Proceedings of the ELSAM/ELKRAFT meeting in Copenhagen.
4. Mears, P. (1968) in: Diffusion in Polymers, Crank, J. and Park, G.S., ed., Acad. Press, London.
5. Carlsen, L., Batsberg, W., Jensen Skytte, B. and Bo, P. (1981) Risø-R-451.
6. Freeze, R.A. and Cherry, J.A. (1979) "Ground Water". Prentice Hall Inc.

SORPTION OF ACTINIDES IN WELL-DEFINED OXIDATION STATES ON GEOLOGIC MEDIA

B. ALLARD, U. OLOFSSON, B. TORSTENFELT, H. KIPATSI AND K. ANDERSSON
Department of Nuclear Chemistry, Chalmers University of Technology,
S-412 96 Göteborg, Sweden

INTRODUCTION

The long-lived actinides and their daughter products largely dominate the biological hazards from spent nuclear fuel already from some 300 years after the discharge from the reactor and onwards [1]. Therefore it is essential to make reliable assessments of the geochemistry of these elements in any concept for long-term storage of spent fuel or reprocessing waste, etc.

It is well known that the interaction between e.g. actinides in groundwater and exposed geologic media, and the retention of them in a geologic system due to this interaction, are largely dependent on the chemical state of the element in question (e.g. oxidation state, existence of complexes etc.). The effects of the chemical conditions is of particular importance for the actinides, considering

- redox properties; various oxidation states are possible in the environmental potential-pH-range (U(IV)-U(VI), Np(IV)-Np(V), Pu(III)-Pu(IV)-Pu(V)-Pu(VI), Am(III)).

- complexation, especially with hydroxide (hydrolysis) and carbonate (always present at a total concentration usually in the mM-range).

Much of the confusion concerning the interpretation of experimental actinide distribution data can most likely be related to a poor characterization and control of important chemical parameters such as

- pH

- redox potential

- concentrations of complexing anions (including carbonate from the water-air equilibrium).

This is particularly the case for plutonium.

The purpose of the present study is to get information on the mechanisms of actinide sorption on geologic material under well-defined conditions and using actinides in discrete oxidation states For americium, the only oxidation state, that would be obtained under oxic conditions and in the absence of strong complexing agents, would be the trivalent state [5]. For thorium, only the tetravalent state would exist in aqueous solution. Under oxic conditions and in the pH-range of environmental interest neptunium and uranium would exist entirely as penta- and hexavalent species, respectively, in solution [5] . By performing experiments under oxic conditions Am(III), Th(IV), Np(V) and U(VI) would be obtained purely in one single oxidation state without any rigorous control of the redox potential. Thus, they may serve as reference systems. Hereby, it is possible to illustrate the importance of characterizing the redox properties of e.g. a potential geologic repository site and the effect of a sudden change of the pH or redox potential on the retention of a specific actinide.

EXPERIMENTAL

Radionuclides

Radionuclides according to Table I are being used in this study. For those of the elements where long-lived as well as short-lived isotopes were available, the sorption studies were performed at two different nuclide concentrations. All nuclides were stored in stock solutions of 1-2 M HCl.

The short-lived isotopes ^{235}Np and ^{237}Pu were obtained from Harwell, UK, and ^{234}Th and ^{233}Pa recovered from ^{238}U and ^{237}Np, respectively, by the use of a sorption procedure [2] .

The concentrations of ^{233}U was measured by liquid scintillation technique (alpha-activity) while all the other nuclide concentrations were determined from measurements of the gamma-activity.

Distribution measurements

The distribution coefficient (K_d; mol/kg solid per mol/m^3 solution), was measured by a batch technique [3] under the conditions given in Table II.

TABLE I

Radionuclides used in the sorption studies

Radionuclides	Initial concentrations [a]	
	I, Mx10^7	II, Mx10^9
^{232}Th + ^{234}Th	3.0	2.5
^{233}Pa	–	0.004
^{233}U	2.1	–
^{235}Np + ^{237}Np	1.9	1.9
^{237}Pu + ^{239}Pu	0.6	0.6
^{241}Am	2.9	2.3

[a] Concentration in the solution at the start.

Thoroughly sieved and washed and well characterized solid sorbents and water were contacted and the acidic radionuclide stock solution was added (typically 0.1 ml per batch of 20 ml). After an initial pH-adjustment (with NaOH-solution, 8-12 simultaneous experiments at various pH) the distribution coefficients were determined as a function of contact time. No further adjustments of pH were made. Only samples from the water phase were taken, and the distribution coefficient was calculated from

$$K_d = [(c_o - c)/m]/[c/V] \qquad (1)$$

where c_o = initial element concentration in solution (mol/l), c = element concentration after certain contact time (mol/l), m = mass of sorbent (kg) and V = volume of the solution phase (m^3). No correction for sorption on the vessel walls was made. This should not be required, when considering that the surface areas of the exposed fine grained solids are several orders of magnitude larger than the surface area of the vessel.

All of the experiments reported in this paper were performed in air except two series of measurements on Np and Pu, which were performed in N_2-atmosphere.

TABLE II

Conditions for the distribution measurements.

Solid sorbents:	Al_2O_3, SiO_2; particle size: 0.090-0.125 mm, montmorillonite [a]
Solutions:	0.01 M $NaClO_4$ (for Al_2O_3 and SiO_2) Synthetic groundwater (for montmorillonite) (total salt concentration: 306 mg/l, total carbonates: 123 mg/l, pH: 8.2), variation of pH (2-12)
Radionuclides:	See Table I
Experimental conditions:	Temperature: $25 \pm 1\,^{o}C$ Solid/liquid: 0.2 g/20 ml Contact time: 6h, 1d, 6d, 6w [b] Phase separation: Centrifugation; 27000 g for 1 h. Equipment: Polypropylene

[a] Wyoming Bentonite MX-80 [4]

[b] Only data for 6 d given in this paper

RESULTS AND DISCUSSION

The measured distribution coefficients vs pH for Th, Pa, U, Np, Pu and Am are given in Figure 1.

The correlation between sorption (given as $(c_o-c)/c_o$) and the degree of hydrolysis (1%, 50% and 99% of the total concentration as hydrolyzed species) is illustrated in Figure 2. The degree of hydrolysis was calculated using constants from ref. 5. For Am two sets of constants were used. In Figure 2 the dotted lines correspond to the set of higher constants.

The extrapolated distribution coefficients for pH 8.5 (from Figure 1) are given in Table III together with the measured values for bentonite/groundwater.

The observed sorption behaviour can be summarized qualitatively in the following way:

1. For all systems the sorption is drastically increased when hydrolysis starts.
2. Sorption maxima are obtained when neutral hydroxy complexes would dominate in solution (>99% hydrolysis).
3. The decrease in sorption at high pH seems to coincide with the formation of anionic species (e.g. $Am(OH)_4^-$ or $Am(CO_3)_2^-$ and $UO_2(CO_3)_3^{4-}$) [5].
4. There is very little difference between the sorption on bentonite (with high ion exchange capacity) in groundwater and on Al_2O_3 (with low ion exchange capacity) in 0.01 M $NaClO_4$.
5. The sorption is fairly independent of the nuclide concentration for the lower oxidation state systems (III and IV).

In general three basic kinds of sorption can be distinguished [6]. Due to non-specific forces of attraction between the sorbent and the solute, a physical adsorption would occur. This sorption results in the binding of the solute to particles or solid surfaces in several consecutive layers. This would be a reversible and rapid process.

Due to the action of chemical forces between solute and sorbent a chemisorption process may occur. This process would be specific, concentration dependent, and possibly slow and partly irreversible.

The most frequent type of interaction would be electrostatic adsorption (ion exchange), due to the action of attractive coulomb forces between charged particles in solution and solid surfaces. Also this process would be concentration dependent.

Evidently, for the actinides in the present type of experiments, the sorption process can largely be looked upon as a physical adsorption process. Ion exchange processes, as defined above, would be less significant, except for non-complexed and non-hydrolyzed systems. However, there is an apparent contribution from electrostatic interaction, indicated by the decrease of the sorption, when anionic species would be expected to form. (e.g. for U(VI) in equilibrium with air-CO_2). Similar conclusions have previously been suggested for the sorption of actinides [7,8].

No efforts have been made to correlate the sorption of Pa(V) with its hydrolysis. However, the sorption is very pronounced at pH above 5-6.

The data for Pu indicate, that the oxidation state at pH below 4-5 is not likely to be V, and not purely IV or III, but possibly a mixture of all of these oxidation states. The higher oxidation states do not appear to dominate at any pH, since predominantly Pu(V) would lead to a significantly lower sorption than observed for pH below 9-10, and Pu(VI) would lead to a significant reduction of the sorption at pH above 8-9 in the aerated system.

TABLE III

Actinide distribution coefficients for bentonite in groundwater (I = 0.008) at pH 8.5 and for Al_2O_3 in 0.01 M $NaClO_4$-solution at the same pH (from Figure 1).

Element	Initial concentration	$\log K_d$, m^3/kg Bentonite [a]	Al_2O_3 [b]
Th(IV)	I	> 0.8	
	II	> 0.8	1.0
Pa(V)	II	0.70	1.2
U(VI)	I	− 1.03	−0.8
Np(V)	I	− 1.31	
	II	− 0.94	−2.2
Pu	I	0.55	
	II	0.54	0.5
Am(III)	I	0.79	
	II	0.82	0.8

[a] Cation exchange capacity 750 − 800 meq/kg
[b] Cation exchange capacity <1 meq/kg

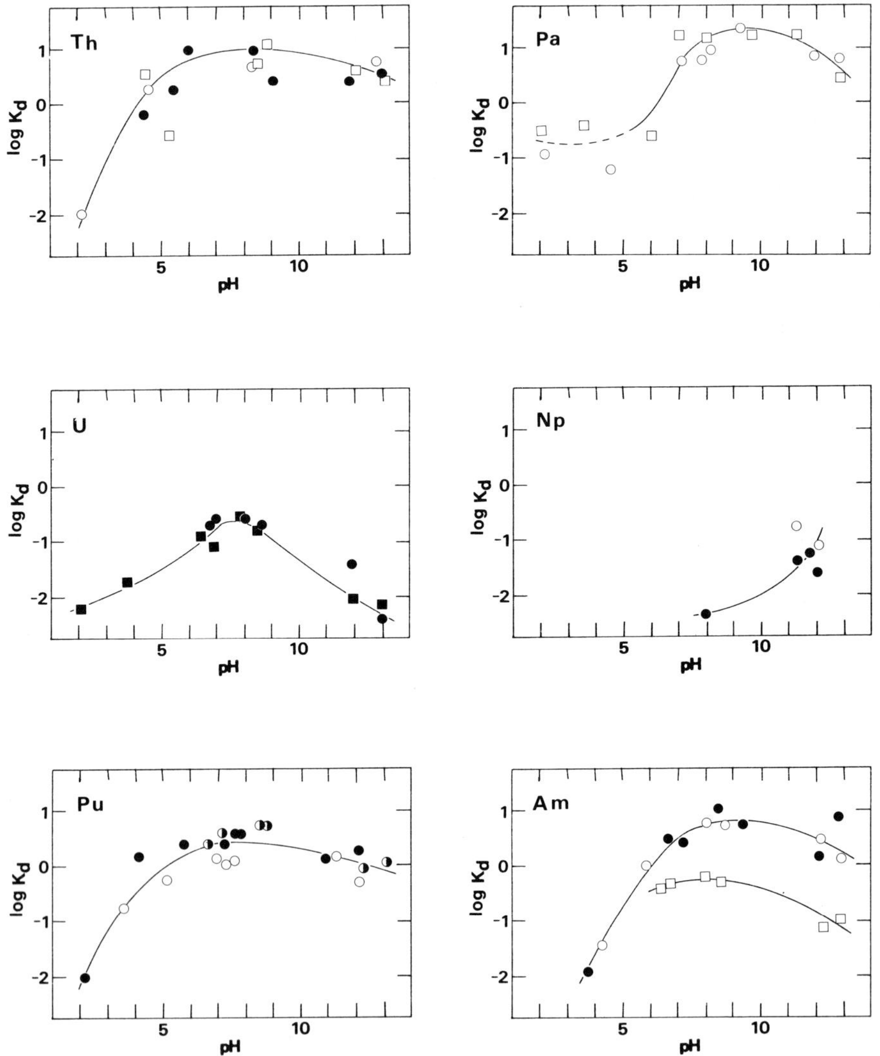

Fig. 1. Distribution coefficients (K_d, m^3/kg) for Th, Pa, U, Np, Pu and Am in 0.01 M NaClO$_4$ (c.f. Table I and II).

● Al$_2$O$_3$-I, ○ Al$_2$O$_3$-II, □ SiO$_2$-II, ■ SiO$_2$-I, ◐ Al$_2$O$_3$-I measured in air (for Pu).

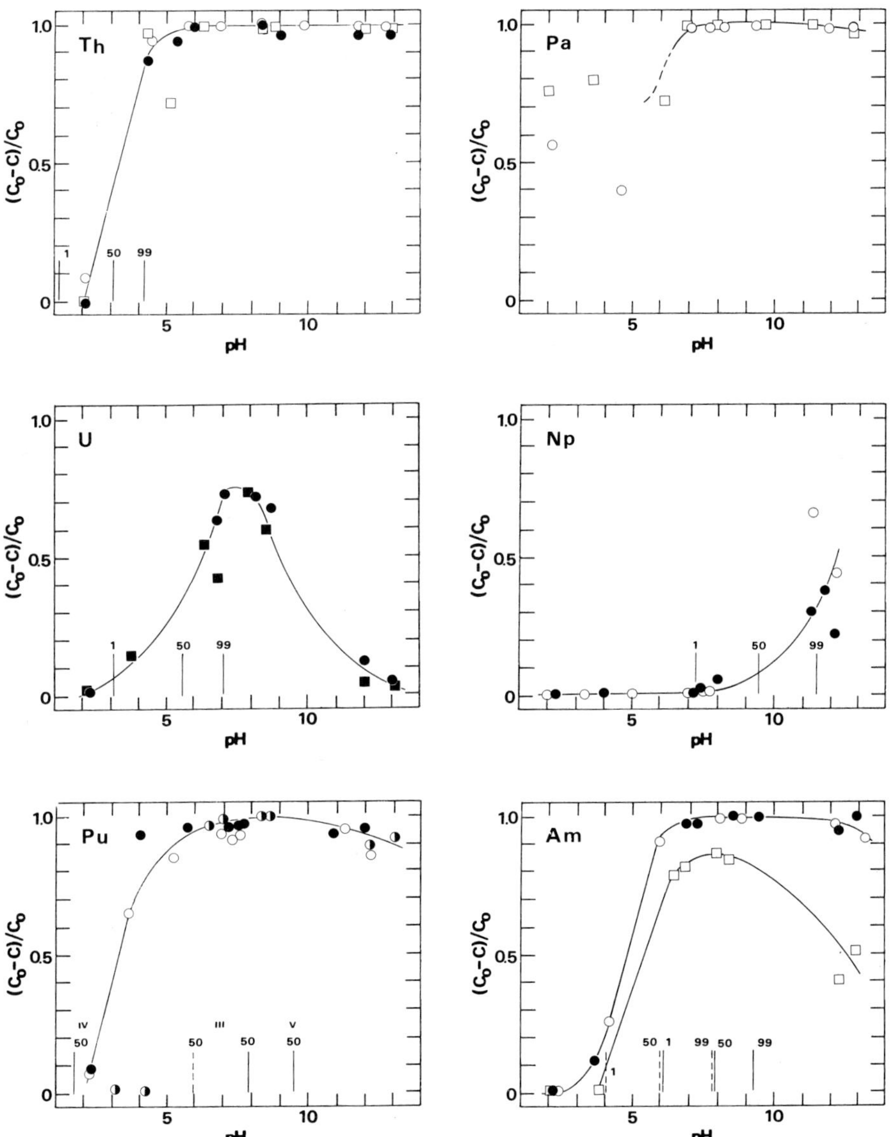

Fig. 2. Sorbed fraction vs degree of hydrolysis (1%, 50% and 99%). Symbols as in Fig. 1.

The sorption behaviour of Am does indicate that the set of higher hydrolysis constants is more correct than the set of lower constants, which previously has been suggested [5].

Further discussions of the sorption mechanisms as well as presentation of additional results from the present project will be given elsewhere [9].

CONCLUSIONS

The observed actinide sorption behaviour in the present experiments indicate a predominantly physical adsorption mechanism, largely related to the degree of hydrolysis. The formation of anionic species would reduce the sorption.

Under the present conditions Pu seems to exist largely as a mixture of various oxidation states, possibly III-V-IV, as has been suggested previously [5]. The reduction of actinides from the penta- or hexavalent state (e.g. for Np(V) or U(VI)) to the tetravalent state would give a higher distribution coefficent (by 2-3 orders of magnitude) in the environmental pH-range.

ACKNOWLEDGEMENTS

This work has been financed by the Nuclear Fuel Safety Project (KBS).

The skilful experimental help by Ms. M. Bengtsson, Ms. W. Johansson and Ms. L. Eliasson is gratefully acknowledged.

REFERENCES

1. "Handling of Spent Nuclear Fuel and Final Storage of Vitrified High-Level Reprocessing Waste" and "Handling and Final Storage of Unreprocessed Spent Nuclear Fuel", Kärnbränslesäkerhet (KBS), Stockholm 1977 and 1978.
2. B. Allard, U. Olofsson and Y. Albinson, to be published.
3. B. Torstenfelt, K. Andersson, B. Allard, "Sorption of Sr and Cs on Rocks and Minerals. Part I: Sorption in Groundwater", Report Prav 4.29, National Council for Radioactive Waste, Stockholm 1981.
4. A. Jacobsson, R. Pusch, "Egenskaper hos bentonitbaserat buffertmaterial" (Properties of bentonite based buffer substances), KBS-TR-32, Kärnbränslesäkerhet, Stockholm 1978 (in Swedish).
5. B. Allard, "Solubilities of Actinides in Neutral or Basic Solutions", in N. Edelstein, Proc. of the Actinides-81 Conf., Pergamon Press, Oxford 1982, in press.
6. P. Benes, V. Majer, Trace Chemistry of Aqueous Solutions, (Elsevier Scientific Publ. Comp., Amsterdam-Oxford-New York 1980).
7. B. Allard, G. W. Beall, T. Krajewski, "The Sorption of Actinides in Igneous Rocks", Nuclear Techn. 49 474 (1980).
8. G. W. Beall, B. Allard, "Sorption of Actinides from Aqueous Solutions under Environmental Conditions", in P. H. Tewari (Ed.) Adsorption From Aqueous Solutions, (Plenum, New York 1981).
9. B. Allard, U. Olofsson, B. Torstenfelt, work in progress.

LABORATORY TESTS ON THE MIGRATION BEHAVIOUR OF SELECTED FISSION PRODUCTS IN AQUIFER MATERIALS FROM A POTENTIAL DISPOSAL SITE IN NORTHERN GERMANY

H. BEHRENS, D. KLOTZ, H. LANG, H. MOSER
Institut für Radiohydrometrie der Gesellschaft für Strahlen- und Umweltfor-
schung mbH München, D-8042 Neuherberg, Fed. Rep. Germany
G. BARKE, H. BRÜHL, S. GEHLER AND U. MÜHLENWEG
Institut für Angewandte Geologie, Freie Universität Berlin, D-1000 Berlin 33,
Fed. Rep. Germany

ABSTRACT

The migration behaviour with respect to retardation by sorption of radioactive Ce, Cs, I, Nb, Ru, Sm, Sr, Tc and Zr in fluviatile and aeolian sand aquifers has been examined in laboratory tests. In these tests it was attempted to simulate the natural conditions of the field. The tests were performed as batch and column experiments. The results of both types of experimental procedures agree within the error limits and show

- small sorption for I^- and TcO_4^-
- medium sorption for Sr^{2+}
- high sorption for Cs^+, Ce^{3+}, Sm^{3+}, Ru^{4+}, Zr^{4+} and Nb^{5+}

The investigations will be continued with other geologic materials from the same site.

INTRODUCTION

Within the frame work of a safety investigation program with regard to the final disposal of radioactive waste, work is being done to simulate the migration behaviour of selected fission products by a mathematical model. As a contribution it was necessary to perform experiments for the determination of distribution coefficients and retardation factors for the radionuclides. The material under investigation was sampled in the surroundings of a salt dome, near Gorleben in the North West German Basin, which was chosen as a possible repository for radioactive waste in the Federal Republic of Germany.

MATERIALS

Clayey sand, sandy and gravelly material of fluviatile and aeolian origin were sampled from cores, which were drilled between surface and 50 m depth, in amounts of several kilograms. Three samples of medium grain size sand of the same origin were sampled in larger amounts from a depth of 1.5 m below surface. The latter three samples were taken in disturbed form for batch and column experiments as well as in undisturbed form for some special column tests. The grainsize distribution of the samples showed high uniformity ($1.5 \leqslant U < 7$), the porosity was in the range $0.2 < n < 0.4$, and the permeability

784

was between 10^{-7} m/s and 10^{-3} m/s. The longitudinal dispersivities of these materials were measured with conservative tracers in column arrangements and were in the range of $0.2 < L_D < 5.1$ cm for water saturated flow. The clayey fractions of the samples were separated for identifying their mineral components by x-ray diffraction; chlorite, illite, kaolinite and small amounts of montmorillonite were found.

Because of the inavailability of water taken from the site in sufficient amounts for the experiments, water was prepared artificially according to the average analytical composition of groundwater and rainwater in the test area. The ground water had a total salt concentration of 4.3×10^{-3} val/l, with an ionic strength of 6.2×10^{-3} molal. The rain water had a total salt concentration of 2.4×10^{-4} val/l, with an ion strength of 3.8×10^{-4} molal.

The tested radionuclides in their initial chemical forms are $^{144}Ce^{3+}$, $^{134}Cs^{+}$, $^{125/131}I^{-}$, $^{95}Nb^{5+}$, $^{106}Ru^{4+}$, $^{153}Sm^{3+}$, $^{85}Sr^{2+}$, $^{95m/99}TcO_4^{-}$ and $^{95}Zr^{4+}$. Nuclide preparations with the lowest possible carrier concentrations were selected. The typical chemical concentrations of the nuclides together with the carrier material was smaller than 7×10^{-8} moles/l. In the case of acidic spike solutions, pH-values were adjusted according to the requirements of the experiments.

METHODS

The aim of all tests was to simulate the natural conditions found in the field. Batch and column experiments were performed. The determination of <u>distribution coefficients</u> (K_d) by batch experiments was done in three ways:
- by shaking on a vibrating table
- by shaking in an over head rotating machinary
- by processing in a continuously recirculating device

Great care was taken to assure that the measured K_d values were not disturbed by the experimental set up, e.g. sorption on walls or wear by grinding. The experiments were carried out using a 1 to 2.5 ratio of g sand to ml water. Usually the experiments were started with the radionuclides in the water, i. e. adsorption distribution coefficients were determined. In most cases the experiments were carried out as multitracer tests.

Identical samples as in the batch experiments were used for the determination of _retardation factors_ (R_D) by column experiments. These tests were done with columns of different dimensions. The diameters of columns containing the disturbed sample materials had diameters in the range of 50 mm to 295 mm, with length in the range of 50 mm to 2000 mm. Those containing undisturbed sample materials were in the range of 58 mm to 84 mm with length of 69 mm to 250 mm. Some of the column experiments were carried out under the condition of unsaturated water flow, simulating continuous or interrupted rain events of 800 mm/a using artificial rain water. The retardation factors of the radionuclides were determined by spiking with the tracer solutions as a thin layer on the top of the column and measuring the tracer passage curves at the column outlet, or by slicing the column packings and measuring the tracer distribution over the different layers. Tritium in form of $^1H^3HO$ was used as a reference tracer for determining the water movement [1,2].

RESULTS

Tab. 1 shows the ranges of the K_d values measured with artificial groundwater and up to 36 sand samples.

Radionuclide	Range of K_d-values		
$^{144}Ce^{3+}$	15	–	12000
$^{134}Cs^{+}$	30	–	2100
$^{125/131}I^{-}$	0.01	–	5.2
$^{95}Nb^{5+}$	10	–	3600
$^{106}Ru^{4+}$	4	–	1900
$^{85}Sr^{2+}$	1	–	22
$^{95m/99}TcO_4^{-}$	0.4	–	3.5
$^{95}Zr^{4+}$	6	–	2800

Tab. 1: Distribution coefficients K_d of selected radionuclides in up to 36 sand samples, obtained by means of batch procedures

The pH-values in the solutions were in the natural range from pH 4 to pH 7. The results were obtained at room temperature and under aerobic conditions. Thus the solutions showed E_H values between 350 and 400 mV.

786

Using batch procedures, influences of various parameters on the K_d values
were examined in detail e.g. the influences of different electrolytes and of
nuclide carrier substances in various concentrations. It could be shown that,
for radionuclides of the elements Sr, Cs and Ce, the salt content of the
ground water exerts in general only a small influence on the results if the
concentration of an added electrolyte is small in comparison to the total ion
content of the ground water. If not, than the obtained K_d values of Cs and
Ce depend in a strong way on added nuclide carriers.

In Fig. 1. an example of the influence of added electrolytes is shown. In
these experiments the distribution coefficients of Sr and Cs were determined
between an aeolian sand and artificial groundwater with an added salt
(Na_2CO_3, $CaCl_2$, KCl) and a carrier ($SrCl_2$ or CsCl) in concentrations
from 10^{-8} moles/l to 10^{-2} moles/l. In case of the Cs the added $CaCl_2$ and
Na_2CO_3 exerts no influence if the added amounts are smaller than the total
ion concentration of 4.3×10^{-3} val/l; however the added carrier CsCl and
KCl deminuite the K_d-Cs in a significant manner.

In the case of Sr, the added electrolytes show no selective influence on the
measured distribution coefficients. Only the total salt contents of the
solution determines the position of the K_d-Sr.

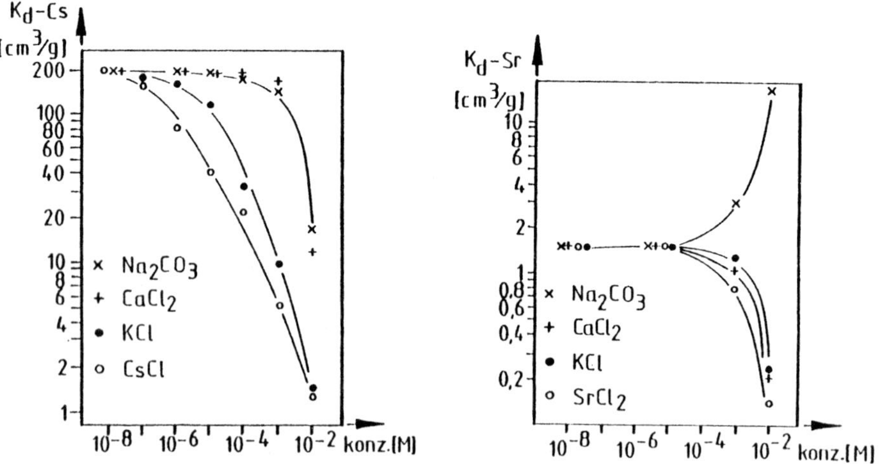

Fig. 1: Change of distribution coefficients K_d-Cs and K_d-Sr between an
aeolian sand sample and artificial groundwater with added electrolytes
in different concentrations M=moles/l.

Tab. 2 shows the ranges of <u>retardation factors</u> (R_D) measured in column arrangements under saturated flow conditions. The results were obtained at room temperature and under aerobic conditions. The filter velocity in these experiments was about 5×10^{-7} m/s. Fast migrating nuclides could be measured by means of the tracer passage curves. Some radionuclides with high sorption coefficients showed during the observation time of three years only a migration distance of some mm. In such cases the retardation factors have been determined by the radioactivity distribution within the columns.

Radionuclide	Range of R_D-values
$^{144}Ce^{3+}$	10000 — 12000
$^{134}Cs^{+}$	530 — 1300
$^{125/131}I^{-}$	1.0 — 1.5
$^{95}Nb^{5+}$	390 — 9000
$^{106}Ru^{4+}$	390 — 10000
$^{95}Sr^{2+}$	5.5 — 12
$^{95m/99}TcO_4^{-}$	1 — 1400
$^{95}Zr^{4+}$	490 — 10000

Tab. 2: Retardation factors R_D of selected radionuclides in up to 36 sand samples, obtained by means of column arrangements.

The retardation factors measured for unsaturated water flow are on the same order of magnitude as the values in Tab. 2 excepting for Ce with R_D values in the range from 1000 to 3000 [3].

DISCUSSION

The measured distribution coefficients show characteristic differences according to the different physicochemical properties of the used nuclide ions and those of the tested sands. We found

- small sorption for I^{-} and TcO_4^{-}
- medium sorption for Sr^{2+}
- high sorption for Cs^{+}, Ce^{3+}, Sm^{3+}, Ru^{5+}, Zr^{4+} and Nb^{5+}.

The ion content in the ground water does not influence the sorption constants selectively; only the total ion strength of the ground water affects these constants.

The sorption constants were determined in four different ways (vibration table, rotation machinary, continuous recirculation column and single pass column). The results agree well within the error limits. The results can thus be regarded as free of systematic errors. Fig. 2 shows R_D values calculated from K_d values obtained by the batch method, in comparison with those obtained by column arrangements.

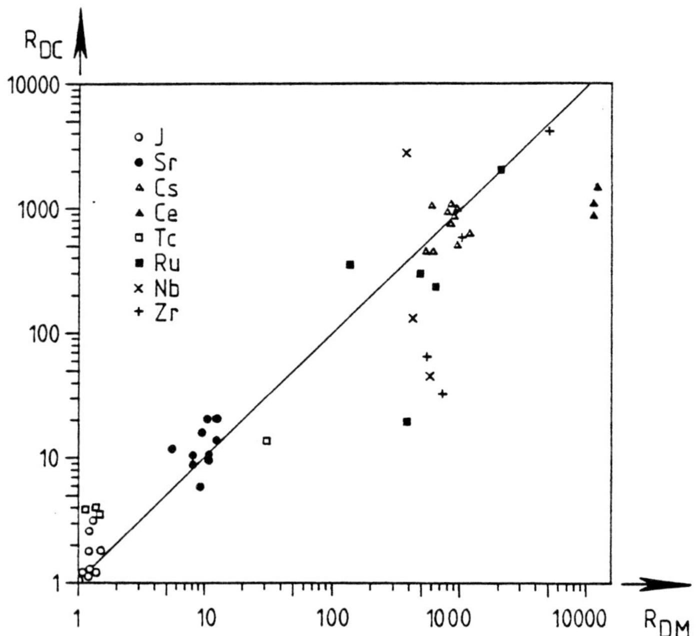

Fig. 2: Comparision of retardation factors directly measured by column experiments (R_{DM}) with retardation factors calculated from batch K_d-values (R_{DC}) by means of the equation $R_D = \varrho/n \cdot K_d + 1$ [4]. (ϱ =bulk density, n=porosity).

The sorption properties of the tested sands is in the expected range known from the literature for other mineralogical materials. As an example in Fig. 3 K_d values measured in our experiments are compared with those measured by BURKHOLDER et al.[5] on single mineral components.

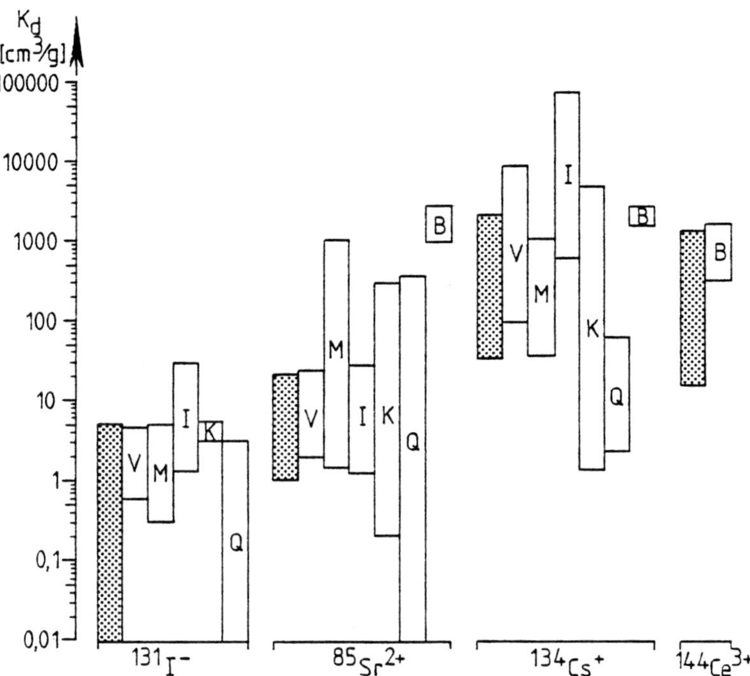

Fig. 3: Measured K_d-values ▒▒▒▒ of selected radionuclides in comparison with K_d-values of the same radionuclides determined from pure minerals BURGHOLDER et al. [5]:V = Vermiculite, M = Montmorillonite, I = Illite, K = Kaolinite, Q = Quartz,.B = Bentonite.

CONCLUSIONS

The investigations are still in progress for materials from deeper geological formations at the same site. In addition radionuclides of Co, Pd, Se, Sn and Pm will be included in the experiments. The work is now performed with original groundwater and if necessary under nonaerobic conditions for a still better approach to the natural conditions.

ACKNOWLEDGEMENT

The Federal German Ministry of Research and Technology provided financial support for these studies within the project "Sicherheitsstudie Entsorgung ". The authors are grateful to P. Gügel, C. Lutz, M. Baumann, F. Oliv for

assistance in the experimental work.The Tritium measurements were performed by
W. Rauert and H. Rast.

REFERENCES

1. BEHRENS, H., KLOTZ, D., LANG, H., MOSER, H., (1979)   Ausbreitung von
   Radionukliden (Sr, J, Cs, Ce) im oberflächennahen Boden und in
   Lockergesteinen. Report PSE-Nr. 80/15, Berlin

2. WOLTER, R., MÜHLENWEG, U., GEHLER, S., BARKE, G., SAMMLER, H., BRÜHL, H.
   (1979) Ausbreitung von Radionukliden (Zr, Nb, Tc, Ru, J) im
   oberflächennahen Boden und in Lockergesteinen. Report PSE-Nr. 80/16, Berlin

3. KLOTZ, D., LANG, H., MOSER, H., BEHRENS, H. (1982) Ausbreitung von
   Radionukliden der Elemente J, Sr, Cs und Ce in oberflächennahen
   Lockergesteinen. GSF-Bericht R 289, Neuherberg

4. MAYER, S.W., TOMPKINS, E.R. (1947) Ion Exchange as a separation method. A
   theoretical analysis of column separations process. J.Amer. Chem. Soc. 69:
   2866 - 2874

5. BURKHOLDER, H.C., GREENBORG, J., STOTTLEMEYRE, J.A., BRADLEY,
   D.J., RAYMOND, J.R., SERNE, R.J. (1979) Waste Isolation Safety Assessement
   Program. Technical Progress Report for FY-77 PNL-2642, UC-70

Published 1982 by Elsevier Science Publishing Co
SCIENTIFIC BASIS FOR RADIOACTIVE WASTE MANAGEMENT- V
Werner.Lutze, editor

MODELLING OF THE MIGRATION OF LANTHANOIDS AND ACTINOIDS IN GROUND WATER; THE MEDIUM DEPENDENCE OF EQUILIBRIUM CONSTANTS.

G. BIEDERMANN, J. BRUNO, D. FERRI, I. GRENTHE, F. SALVATORE AND K. SPAHIU
Department of Inorganic Chemistry, Royal Institute of Technology,
S-10044 Stockholm, Sweden

INTRODUCTION

Radioactive products may be released from a nuclear waste repository
to the biosphere through dissolution in, and transport by ground-water.
The fundamental chemical phenomena involved are solubility, complex formation, redox and sorption (e.g. ion-exchange) processes. A first estimate
of the behaviour of the waste-repository-ground water system is usually
obtained by using a chemical equilibrium model in order to calculate
properties such as chemical speciation and solubilities of important radionuclides. For practical reasons the modelling is often made in two different regions, the near field and the far field. In the latter, the main
components of the aqueous phase are essentially those of undisturbed groundwater, while in the former substantial changes in composition may occur.
The total electrolyte concentration is usually small, at least in the far-
field region.

Thermodynamic models are no better than the quality of the data base
used, some aspects of the selection of thermodynamic data and their validity in different ground-water media will be discussed in this communication.

Thermodynamic data always refer to a chosen standard state, for solutes
this standard state is often the infinite dilute aqueous solution.
However, data for many equilibria cannot be determined accurately, or at
all, in dilute solution. This is invariably true for equilibria involving
ions of high charge, e.g. most actinoid species. Precise thermodynamic
information for these systems can only be obtained in the presence of an
inert electrolyte of fairly high concentration (0.5 - 4 M) in order to
ensure that activity factors are reasonably constant or estimable.
This is the only possibility to unravel a complicated chemical equilibrium
system. We are then faced with the problem of converting these data to
other media. Two common procedures are:
-   calculation of the equilibrium constant as a function of ionic strength
    by using the "empirical" Davies equation.

792

- calculation of activity factors as a function of the ionic composition by
  using semi-theoretical electrolyte models, as used e.g. by Baes and
  Mesmer[1].

THE ESTIMATION OF ACTIVITY COEFFICIENTS

The Debye-Hückel term, which is the dominant term in the expression
for the activity coefficients in dilute solution, accounts for long-
range electrostatic interactions. At higher concentration also short-
range non-electrostatic interactions have to be taken into account. This
is usually made by adding ionic strength dependent terms to the Debye-
Hückel expression. The methodology was first outlined by Bronsted[2] and
later elaborated by Guggenheim[3] and Scatchard[4]. The basic assumptions in
the so called specific ion interaction theory (referred to as S.I.theory) are:

- the activity factor, γ_j, of an ion of charge z_j in a solution of the
  ionic strength I may be described by the equation

$$\log \gamma_j = -z_j^2 D + \sum_k \epsilon(j,k,I) m_k \tag{1}$$

  where

  $D = 0.5107\sqrt{I}/(1 + 1.5\sqrt{I})$, is the Debye-Hückel term at $25^{\circ}C$. The summation
  extends over all ions, k, present in solution with the molality m_k.

- the interaction coefficients ϵ are zero for ions of the same charge.

By using (1) one can make fairly accurate estimates of ionic activity
coefficients in mixtures of electrolytes, provided the interaction coefficients are known. Interaction coefficients involving simple ions can usually be determined from tabulated data of mean activity coefficients of
electrolytes or from the corresponding osmotic coefficients. Interaction
coefficients for complexes must be either estimated from the charge and
size of the ion or determined experimentally. This point will be discussed in the next section.

Strictly, the interaction coefficients are not constants. Their concentration dependence varies with the charge type and is small for 1:1, 1:2
and 2:1 electrolytes at total molalities less than 3.5 m, and can often be
neglected. This point has been emphazised by Guggenheim, who has presented
a large experimental material supporting this approximation.[3] The concentration dependence is larger for electrolytes of higher charge type and in
order to reproduce the activity coefficient data for these accurately, one

must use concentration dependent interaction coefficients c.f. Pitzer and
Brewer[4] or Baes and Mesmer[1].

By using a more elaborate virial expansion Pitzer has managed to describe measured activity coefficient of a large number of pure electrolytes
with a high precision over a large concentration range. Pitzers model contains three parameters as compared to one in the S.I.-theory and seems to
be difficult to use in the complicated nuclear repository-ground water
systems.

In the following section we will demonstrate the applicability of the
S.I. theory on some examples of interest for nuclear waste management.

APPLICATIONS OF THE SPECIFIC INTERACTION THEORY

The first example concerns the formation of carbonate complexes in the
Y^{3+}- CO_2(g) - H_2O system as studied by emf and solubility methods. An
outline of the experimental procedure is given in ref. 5. The calculation
methods will be described in a forthcoming publication[6]. The measurements
were performed in aqueous media at three constant perchlorate concentrations.
For each perchlorate level several series of measurements were made, in
which the total metal-ion concentration was kept constant, whereas the
acidity was varied. The experimental matrix is given below.

TABLE 1

EXPERIMENTAL MATRIX FOR THE Y(III)-CO_2(g)-H_2O SYSTEM

Perchlorate concentration C (M)	Total metal-ion concentration B (M)
3	1, 0.75, 0.5, 0.3, 0.1
1.5	0.5, 0.3, 0.2, 0.1
0.9	0.3, 0.2, 0.1

From the table it is obvious that the composition of the ionic medium
varies strongly at each perchlorate concentration e.g. at the 3 M level
from pure $Y(ClO_4)_3$ to 0.1 M $Y(ClO_4)_3$ + 2.7 M $NaClO_4$. Such large changes
in the ionic medium are expected to affect the activity coefficients.

794

The equilibrium constant for the reaction
$$pY^{3+} + qH_2O + rCO_2(g) \rightleftharpoons Y_p(OH)_q(CO_2)_r^{3p-q} + qH^+$$
is denoted β_{pqr} and β_{pqr}^o, respectively at the reference states C M $NaClO_4$ and pure water. Note that $(OH)CO_2^- \equiv HCO_3^-$ and $(OH)_2CO_2^{2-} \equiv CO_3^{2-}$.

The chemical model and the equilibrium constants are deduced from experimental measurements of the free hydrogen ion concentration, h, in solutions of known and constant total concentration of yttrium, B, and known analytical excess of H^+, H c.f. ref. 5, p. 404.

From each experimental series one can calculate two constants K_1 and K_2 where
$$K_1 = \beta_{011} + \beta_{111} \cdot B + \beta_{221}B^2 + .. \tag{2 a}$$
$$K_2/B = \beta_{121} + \beta_{221} \cdot B + \beta_{32}B^2 + .. \tag{2 b}$$

Plots of K_1 and K_2/B vs B are given in Figure 1. The curves are linear at low values of B, but bend towards the abscissa at larger B. This type of deviation cannot be due to incompleteness in the chemical model, but rather to changes in activity factors. The full-drawn curve is calculated by using the S.I. theory.

The analysis of the experimental data will be performed in two steps.

First, we will calculate equilibrium constants in $NaClO_4$ solvents of concentrations 3, 1.5 and 0.9 M at trace yttrium concentration. This means that we must calculate activity coefficient changes when part of the $NaClO_4$ solvent is replaced by $Y(ClO_4)_3$.

Second, we will calculate the activity factor changes when the solvent is changed from $NaClO_4$ - H_2O to pure water. In this way we obtain thermodynamic data corresponding to pure water

<u>S.I. interaction theory in $Y(ClO_4)_3$-$NaClO_4$ mixtures</u>. From the S.I. theory we obtain for the molar activity coefficient f_i, where $f_i \rightarrow 1$ in the pure solvent (3 M $NaClO_4$).
$$\log f_i = -z_i^2(D(I)-D(3)) + \varepsilon(i,Na^+)([Na^+] -3)+\varepsilon(i,Y^{3+})[Y^{3+}] \tag{3}$$

The terms involving interaction coefficients $\varepsilon(i, ClO_4^-)$ are zero because the perchlorate concentration is constant, 3 M.

Introduction of activity factors in (2) gives

$$K_1 = \beta_{011} \cdot f_{H^+}^{-1} \cdot f_{HCO_3^-}^{-1} + \beta_{111} \cdot \frac{f_{Y^{3+}}}{f_{YHCO_3^{2+}} \cdot f_{H^+}} [Y^{3+}] \tag{4}$$

Activity changes of water have also been estimated by the S.I. theory.
Introduction of eqn (3) in eqn. (4), and rearrangment
gives

$$F = \{K_1 - \beta_{011} \; antilog[2(D(I)-D(3))-\varepsilon(Na^+,HCO_3^-)([Na^+]-3)-\varepsilon(Y^{3+},HCO_3^-)$$

$$[Y^{3+}]]\} \cdot antilog\{4(D(I)-D(3)\} = \beta_{111} \cdot [Y^{3+}] \tag{5}$$

The interaction coefficient $\varepsilon(Na^+, HCO_3^-) = 0.02$ was calculated from
the data of Frydman et al[7], of Güntelberg and Schiödt [8] and
of Harned and Davies[9] who studied the carbonic acid protolysis in several
ionic media. No experimental value is available for $\varepsilon(Y^{3+}, HCO_3^-)$, instead
we assume $\varepsilon(Y^{3+}, HCO_3^-) = \varepsilon(Y^{3+}, Cl^-) = 0.26$ as calculated from the data
of Mason[10]. This analogy is based on the comparison of several $\varepsilon(Me^+, HCO_3^-)$
with $\varepsilon(Me^+, Cl^-)$ values. ε values in molar scale were calculated by using
the measured densities.

A plot of F vs B (Figure 2) is very near linear and the deviations of
the experimental data are thus well explained by activity factor variations.
By repeating the procedure outlined above at the other perchlorate concentrations, we obtain the following set of constants valid in $NaClO_4-H_2O$
solvents.

TABLE 2

STABILITY CONSTANTS AT TRACE LEVELS OF Y(III) IN DIFFERENT $NaClO_4$ MEDIA.
β_{111}, β_{121} AND β_{221} REFER TO THE COMPLEXES $YHCO_3^{2+}$, YCO_3^+ AND $Y_2CO_3^{4+}$,
RESPECTIVELY.

C_{NaClO_4}	$-log\beta_{111}$	$-log\beta_{121}$	$-log\beta_{221}$
3.0	6.73±0.06	11.48±0.1	10.62±0.08
1.5	6.46±0.04	11.33±0.1	10.37±0.05
0.9	6.30±0.02	11.21±0.1	10.24±0.05

<u>S.I. theory for extrapolation of constants at zero ionic strength.</u>
The set of β_{pqr} constants given in Table 2 may now be used in the second
step of the application of S.I. theory where we calculate equilibrium
constants referring to the pure water standard state.

We have

$$\beta_{pqr} = \beta^o_{pqr} \cdot \gamma^p_{Y^{3+}} \cdot \gamma^{-q}_{H^+} \cdot \gamma^{-1}_{pqr} \tag{6}$$

where β_{pqr} on the molal scale were calculated from the values in Table 2
and published density data for $NaClO_4$.[11]

Substituting γ values calculated with (1) in (6) and rearranging one
obtains

$$\log\beta_{111}+4D=\log\beta^o_{111}+[\varepsilon(Y^{3+},ClO_4^-)-\varepsilon(H^+,ClO_4^-)-\varepsilon(YHCO_3^{2+},ClO_4^-)]I \tag{7}$$

$$\log\beta_{121}+6D=\log\beta^o_{121}+[\varepsilon(Y^{3+},ClO_4^-)-2\varepsilon(H^+,ClO_4^-)-\varepsilon(YCO_3^+,ClO_4^-)]I \tag{8}$$

$$\log\beta_{221}=\log\beta^o_{221}+[2\varepsilon(Y^{3+},ClO_4^-)-2\varepsilon(H^+,ClO_4^-)-\varepsilon(Y_2CO_3^{4+},ClO_4^-)]\ I \tag{9}$$

As seen from fig (3) these equations represent straight lines from
which β^o_{pqr} values are determined as the intercept while the interaction
coefficients for the complexes can be evaluated from the slopes. The
values of β^o_{pqr} are given in Table 3 together with the values of β^o_{pqr} obtained by applying the Davies equation to values of β_{pqr} at 3, 1.5 and
0.9 M ClO_4^-.

TABLE 3

VALUES OF LOG β^o_{pqr} CALCULATED FROM THE S.I. THEORY AND FROM THE DAVIES
EQUATION.

pqr	log β^o_{pqr} from Davies eqn			log β^o_{pqr}
	3 M	1.5 M	0.9 M	S.I. theory
111	−7.47	−6.27	−5.85	−5.40
121	−12.56	−11.03	−10.54	−10.00
221	−10.55	−10.34	−10.22	−10.11

The superiority of the S.I. approach is clearly seen by the consistency
of the log β^o_{pqr} values deduced from the various perchlorate media. Davies
equation, however, gives rise to considerable systematic deviations,
when applied to data obtained in solutions of high ionic strength.

ON THE MAGNITUDE OF INTERACTION COEFFICIENTS

Interaction coefficients can be calculated[12] if mean activity coefficient data of the corresponding electrolytes are available.

It is possible to calculate interaction coefficients for complexes from equilibrium constant measurements at different ionic strengths, as shown in the previous section. From the slopes of the lines in fig (2), using $\varepsilon(H^+, ClO_4^-) = 0.14$ as evaluated from mean activity coefficient data of $HClO_4$ solutions[13] and $\varepsilon(Y^{3+}, ClO_4^-) = \varepsilon(Ho^{3+}, ClO_4^-) = 0.49$ as evaluated from data of Spedding and coworkers[14] we found:

$$\varepsilon(YHCO_3^{2+}, ClO_4^-) = 0.40, \quad \varepsilon(YCO_3^+, ClO_4^-) = 0.17, \quad \varepsilon(Y_2CO_3^{4+}, ClO_4^-) = 0.80$$

Since this procedure in general is not practical because of the very large experimental efforts involved, comparison of interaction coefficients of ions of the same charge type makes it possible to make reasonable estimations as shown from the following data

$$\varepsilon(HgCl^+, ClO_4^-) = 0.20, \quad \varepsilon(H^+, ClO_4^-) = 0.14 \quad \varepsilon(YCO_3^+, ClO_4^-) = 0.17$$

$$\varepsilon(Cd^{2+}, ClO_4^-) = 0.36, \quad \varepsilon(Hg^{2+}, ClO_4^-) = 0.30 \quad \varepsilon(YHCO_3^{2+}, ClO_4^-) = 0.40$$

$$\varepsilon[Fe_2(OH)_2^{4+}, ClO_4^-] = 0.75, \quad \varepsilon(Y_2CO_3^{4+}, ClO_4^-) = 0.80$$

The equilibrium constants do not depend too strongly on errors in the interaction coefficients. An error of ±0.1 in the value of interaction coefficients of the complexes in the yttrium case would cause an error ±0.1 I in β_{pqr}^o, where I represents the ionic strength at which the constant is determined. These errors are in general much smaller than errors introduced by using "empirical" equations of the Davies type.

The ionic strength dependence of the interaction coefficients increases with the charge of ions and some experimental efforts must be made to get an idea of the interaction coefficients of ions such as $U(CO_3)_5^{6-}$. In the following section we will describe the results of S.I. calculations on another system of interest for modelling of radionuclide ground-water systems.

THE MEDIUM EFFECTS ON THE SOLUBILITY OF $UO_2(s)$ IN CARBONATE SOLUTIONS

The precise estimation of the medium effect on the standard potential (η_o) of the reaction

$$UO_2(s) + 3CO_3^{2-} \rightleftarrows UO_2(CO_3)_3^{4-} + 2e \qquad (10)$$

requires measurements at minimum two medium levels because the equation

$$\eta^o = e^o + 29.58 \log \frac{\gamma_{4-}^{3}}{\gamma_{CO_3^{2-}}^3 \cdot \gamma_{H^+}^2} = e^o \quad +$$

$$+ 29.58\{-2D +[\varepsilon(Na^+,UO_2(CO_3)_3^{4-})-3\varepsilon(Na^+,CO_3^{2-})-2\varepsilon(H^+,ClO_4^-)]I\} \qquad (11)$$

contains two unknowns e^o and $\varepsilon(Na^+,UO_2(CO_3)_3^{4-})$. The other interaction coefficients in eqn. (11) are available from literature data.

The experimental determination of η^o for the reaction (10) is rather laborious, therefore, a single determination in 3.5 m $NaClO_4$ is available. Under these circumstances we have relied on an approximate treatment where $\varepsilon(Na^+,UO_2(CO_3)_3^{4-}) \simeq \varepsilon(K^+,W(CN)_8^{4-}) = -0.1$.

In general interaction coefficients of electrolytes of the same charge type show a simple dependence on the ionic size. For 1:4 electrolytes we have $\varepsilon(K^+,Fe(CN)_6^{4-}) = -0.2$, $\varepsilon(K^+,W(CN)_8^{4-}) = -0.1$, $\varepsilon((CH_3)_4N^+,Mo(CN)_8^{4-} = -0.07$. The $W(CN)_8^{4-}$ ion was chosen as a model for $UO_2(CO_3)_3^{4-}$ because of its size. By using these interaction coefficients we estimate a difference $e^o - \eta^o = -21$ mV. This is the expected medium effect when going from the 3.5 m $NaClO_4$ to the pure water solvent. This corresponds to a change of solubility of less than one order of magnitude.

CONCLUSIONS

The examples given in this communication indicate that it is possible to obtain a good estimate of the medium dependence of equilibrium constants by using the S.I. theory. The theory is applicable both when extrapolating equilibrium constants to zero ionic strength and for the estimation of activity coefficients in mixtures of electrolytes.

Many interaction coefficients are available in the literature, or can be calculated from published mean activity coefficient or isopiestic data.

The magnitude of interaction coefficients can often be correlated with the charge and size of ions. This offers a possibility to estimate the coefficients for complexes, for which direct experimental information is difficult to get.

The S.I. theory is superior to the empirical equations of the Davies type. There is sufficient experimental information on interaction coefficients to warrant the implementation of the S.I. approach in existing speciation codes.

ACKNOWLEDGEMENTS

This work has been financially supported by the Swedish KBS-project and by the Swedish Natural Science Research Council.

REFERENCES

1. Baes, C.F. and Mesmer, R.E. (1976) The Hydrolysis of Cations. John Wiley and Sons, New York.
2. Brønsted, J.N. (1922) J. Am. Chem. Soc. 44, 938 and 877
3. Guggenheim, E.A. (1966) Applications of Statistical Mechanics. Oxford, Clarendon Press.
4. Scatchard, G. (1936) Chem. Revs. 19, 309.
5. Ciavatta et al. (1981) Acta Chem. Scand. A35, 403.
6. Spahiu, K. et al. (1983) Acta Chem. Scand. To be published.
7. Frydman, M. et al. (1958) Acta Chem. Scand. 12, 878-884.
8. Güntelberg, C.E. and Schiödt, E. (1928) Z. Ph. Chem. 135, 393.
9. Harned, H.S. and Davies, R. Jr. (1943) J. Am. Chem. Soc. 65, 20.
10. Mason, C.M. (1938) J. Am. Chem. Soc. 60, 1638.
11. Wirth, H.E. and Collier F.N. (1950) J. Am. Chem. Soc. 72, 5292.
12. Biedermann, G. (1975) in Goldberg, E.D. Ed., On the nature of sea water, Dahlem Konferenzen, Berlin.
13. Robinson, R.A. and Stokes, R.M. (1965) "Electrolyte Solutions", 2nd ed., revised, Butterworths, London.
14. Rard, A.J. et al. (1977) J. Chem. Eng. Data 22, 187.

800

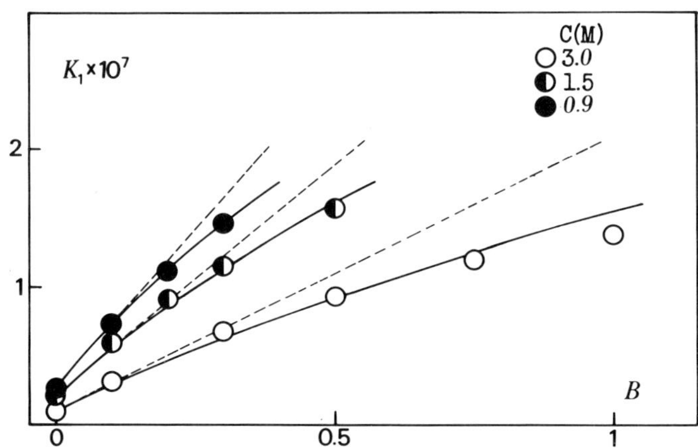

Fig. 1. K_1 as a function of B(mol/l). The dashed lines represent the expected functional behaviour in absence of medium effects due to replacement of Na^+ by Y^{3+}. The full drawn curves are calculated taking the change of activity factors into account

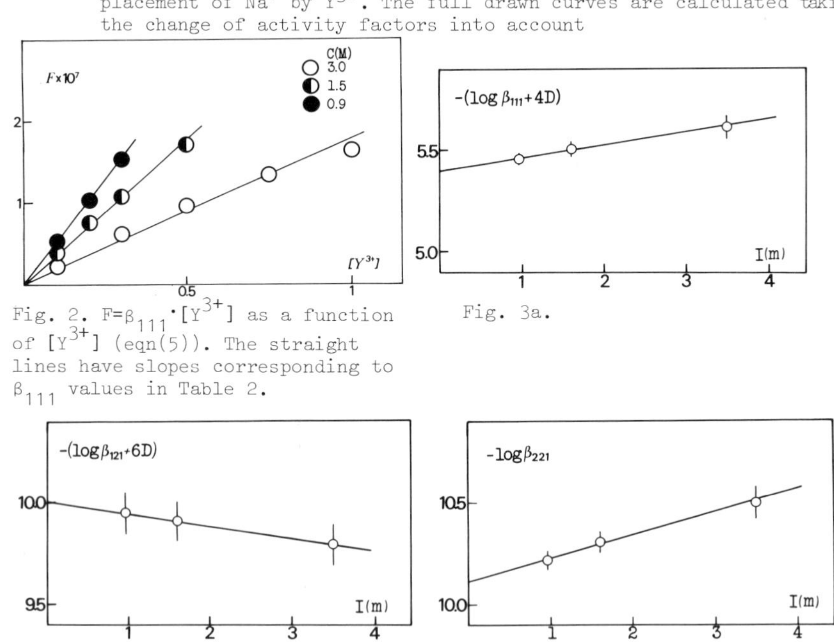

Fig. 2. $F=\beta_{111}\cdot[Y^{3+}]$ as a function of $[Y^{3+}]$ (eqn(5)). The straight lines have slopes corresponding to β_{111} values in Table 2.

Fig. 3a.

Fig. 3a,b,c. Variation of the equilibrium constants corresponding to the formation of $YHCO_3^{2+}$, YCO_3^+ and $Y_2CO_3^{4+}$, respectively with ionic strength. The straight lines represent equations 7, 8 and 9, respectively.

Published 1982 by Elsevier Science Publishing Co
SCIENTIFIC BASIS FOR RADIOACTIVE WASTE MANAGEMENT - V
Werner.Lutze, editor

EVALUATION OF RADIONUCLIDE TRANSPORT: EFFECT OF RADIONUCLIDE SORPTION AND SOLUBILITY

P. F. SALTER AND G. K. JACOBS
Basalt Waste Isolation Project, Rockwell Hanford Operations, P. O. Box 800, Richland, Washington 99352

INTRODUCTION

The current strategy for the permanent isolation of nuclear wastes in the United States involves the storage of these wastes within repositories mined in deep geologic formations. In this disposal strategy, the isolation of nuclear wastes relies on a series of natural and engineered barriers to prevent the unacceptable release of radionuclides to the accessible environment. An integral part of the development of a qualified subsurface nuclear waste repository, therefore, is the assessment of the ability of these barriers to adequately prevent or retard the migration of radionuclides to the accessible environment. The Basalt Waste Isolation Project (BWIP) under guidance from the Department of Energy (DOE) is investigating the feasibility of storing nuclear w_stes in the basalts beneath the Hanford Site.

The most probable mechanism for radionuclide release from a repository in the basalt geohydrologic system at the Hanford Site is dissolution and subsequent transport in groundwater. Therefore, the most important elements to be considered in evaluating radionuclide release and transport are: (1) the hydrologic properties of the system, (2) the physicochemical environment of the system, and (3) radionuclide sorption and solubility. The physicochemical parameters are particularly important because they control radionuclide sorption and solubility, the two dominant mechanisms for retardation of radionuclide transport in the basalt-groundwater system. An understanding of the sorption and solubility behavior of key radionuclides under relevant physicochemical conditions is necessary, therefore, to complete a safety assessment of the permanent storage of nuclear waste in basalt.

In this paper, the sorption and solubility behavior of several potentially hazardous (key) radionuclides in the Columbia River Basalt geohydrologic system is discussed. Utilizing the transport characteristics of the site and the available sorption and solubility data, a preliminary site performance analysis then is performed through comparison of the resultant predicted radionuclide transport behavior with proposed U.S. environmental regulations.

COLUMBIA RIVER BASALT ENVIRONMENT

   Geohydrology. The Columbia Plateau basalts in Southeastern Washington are tholeitic, continental, flood basalts. The basalt beneath the Hanford Site consists of a thick sequence of basalt flows of the Columbia River Basalt Group. Three formations within this group are of interest in the assessment of site performance. Sequentially up the stratigraphic column, they are the Grande Ronde ($\sim$50 flows), the Wanapum ($\sim$14 flows), and the Saddle Mountain ($\sim$10 flows, shallowest) basalts. The Umtanum basalt flow within the Grande Ronde formation has been designated the reference repository horizon. A typical basalt flow is several tens of meters thick and consists of an upper vesicular flowtop and an interior region composed of vertical to subvertical massive basalt columns.[1] Fractures (cooling joints) between the columns are usually filled or lined with secondary minerals (clays, silica, zeolites).[2] As a result, the more porous, vesicular flowtop zones of each flow typically have higher permeability than the flow interiors. In addition, many of the flows within the Umtanum and Saddle Mountain basalts are separated by thin sedimentary layers (interbeds) which also exhibit higher permeabilities than the basalt flow interiors. Consequently, these physical characteristics of the Columbia River basalt system exert a strong influence on groundwater flow. Although groundwater does move vertically through the cooling joints and tectonic fractures in basalt, hydrologic modeling studies for the reference repository location indicate that the dominant groundwater flow is lateral and exhibits an upward stair step pattern that is controlled by the basalt flowtops and interbeds.[3,4] The site-specific geologic and hydrologic characteristics of the Columbia River Basalts are discussed in detail elsewhere.[1,5]

   Geochemistry. The geology of the Columbia River basalts also strongly influences the chemistry of the geohydrologic system. The typical flow in the Grande Ronde basalts consists principally of pyroxene, plagioclase, magnetite, olivine and a silica rich ($\sim$70%) glassy mesostasis.[1,6] Important secondary minerals found associated with the basalts include: pyrite, smectite clays, zeolites and various silica phases.[2,6] The Eh of the system is moderately reducing and is probably controlled by interactions in the groundwater/basalt/secondary mineral system.[7] The reactive silica rich, glassy mesostasis of the basalt strongly influences the basalt groundwater chemistry; the groundwater is a low ionic strength, high pH, sodium-chloride solution that is saturated with silica. A summary of the expected physicochemical parameters for a repository in basalt is presented in Table 1.

TABLE 1

SUMMARY OF THE BASALT GEOCHEMICAL ENVIRONMENT AT THE REFERENCE REPOSITORY LEVEL[5,7,8]

Parameter	Ambient	Expected Maximum Repository Range
Temperature	59°C	59°C - 300°C
Pressure	114 bars	1 - 114 bars
pH	9.6	6.0 - 9.6
Eh	-0.48 V	+0.5 V to -0.61 V
Groundwater Composition (Ionic Strength $\sim$650 mg/l)	$Na^+ >> K^+ \sim Ca^{2+} \sim Mg^{2+} > Fe^{2+}$ $Cl^- \sim SO_4^{2-} > CO_3^{2-} > F^-$ SiO_2 saturated	——

RADIONUCLIDE SORPTION BEHAVIOR

One of the major mechanisms for removing or retarding the movement of radionuclides in groundwater is the sorption of radionuclides onto basalts, secondary minerals, interbed materials or engineered barrier materials from groundwater in the Columbia River basalt geohydrologic system. A commonly used measure of the ability of geologic materials to sorb radionuclides is the radionuclide distribution coefficient (Kd). Assuming a reversible, equilibrium sorption reaction and a linear sorption isotherm, the distribution coefficient can be used to calculate the retardation factor used in radionuclide transport modeling. However, if these conditions are not valid for a system, a more complex sorption isotherm function, usually describing radionuclide sorption behavior as a function of radionuclide concentration, is necessary to accurately calculate the retardation factor used in transport modeling. Recent investigations comparing the effects of different sorption isotherms on predicted contaminant transport indicate that the linear (Kd) isotherm does not necessarily give the most conservative predictions.[9] The indiscriminate use of Kd sorption values, therefore, could introduce a significant error in radionuclide transport calculations for a specific geohydrologic system.

Radionuclide distribution coefficients and isotherms have been determined for the Columbia River basalt geohydrologic system using an equilibrium batch technique.[10,11] Because radionuclide sorption behavior is strongly influenced by many environmental parameters, radionuclide Kd values and isotherms have been determined under conditions which simulate actual repository conditions as closely as possible. A synthetic groundwater simulating actual groundwater composition and pH was used. Reducing conditions ($\sim$ -0.4 V) were established experimentally by adding a reducing agent (hydrazine, 0.1M) to the system. Initial results of these investigations are presented in Tables 2 and 3.

The sorption behavior of those radionuclides (U, Pu, Np, Tc and Se) which can exhibit more than one stable oxidation state in the environment will be significantly influenced by the ambient Eh conditions. Under the oxidizing conditions expected during the operational period of the repository, these radionuclides are poorly sorbed by the basalts. This probably reflects the fact that these radionuclides are present dominantly as anions. Most geologic media have a poor affinity for anions. However, after repository closure, reducing conditions will be re-established by the basalt geohydrologic system. The basalts exhibit a stronger affinity for these radionuclides under the ambient reducing conditions. The increased Kd values probably reflect changing radionuclide speciation (increase in cationic species) with changing oxidation states. Initial radionuclide concentrations are below expected solubilities for the experimental Eh-pH conditions, eliminating precipitation as a possible cause for the increased Kd values. The higher Kd values observed for these radionuclides by the Mabton Interbed materials reflects the shift in the ambient groundwater pH from $\sim$9.4 for the basalt to $\sim$8.0 and the greater percentage of highly sorptive clays and zeolites in the interbed material. Americium, Ra, Cs and Sr are strongly sorbed by both the basalt and the Mabton Interbed material. This reflects the fact that these radionuclides are present as readily sorbed cations in the groundwaters. Sorption isotherms, describing the dependence of radionuclide sorption on radionuclide concentration in the groundwater are presented in Table 3. These equations can be used to calculate a retardation factor when radionuclide concentrations are not at the trace concentrations (e.g., non-linear isotherm) where a simple Kd value can often be used. A more detailed discussion of the

TABLE 2

CURRENT CONSERVATIVE ESTIMATES FOR RADIONUCLIDE Kd VALUES AND SOLUBILITIES IN THE COLUMBIA RIVER BASALT GEOHYDROLOGIC SYSTEM

| Radionuclide | Kd^a, ml/g | | | | Solubility Estimates[b] (Reducing Conditions) | |
| | Umtanum Basalt | | Mabton Interbed Materials | | | |
	Oxidizing Conditions	Reducing Conditions	Oxidizing Conditions	Reducing Conditions	Solid Phase	Solubility M/l
U	0	40	25	100	UO_2, $USiO_4$	1E-8
Pu	20	20	300	500	PuO_2	1E-9
Np	15	200	40	500	NpO_2	1E-15
Am	340	NA	>10,000[c]	>10,000[c]	$Am(OH)_3$	1E-10
Ra	200	---	1,500	---	---[d]	---
Tc	0	29	0	70	TcO_2	1E-12
Se	0	8	0	5	Se^0	1E-15
Cs	500	---	1,000	---	---[d]	---
Sr	170	---	400	---	---[d]	---

[a]Data from Barney (1981)[11], Salter, et al. (1981)[10]; Reducing conditions $\sim$ -0.4V, established by addition of hydrazine to the system; Error associated with values generally less than ±25%.
[b]Solubilities estimated from available thermodynamic and laboratory data.
[c]Greater than 98% removal.
[d]Not considered to be a solubility controlled radionuclide.
NA - No data available.

sorption behavior of these radionuclides and sorption isotherms is presented in Salter, et al. (1981)[10] and Barney (1981)[11].

TABLE 3

SORPTION ISOTHERMS FOR SELECTED RADIONUCLIDES ON UMTANUM BASALT

Radionuclide	Eh[a]	Isotherm Type[b]	Equation
^{233}U	R	F	$S = 0.06C^{0.99}$
	O	DR	$S = (5.0E-6) \exp[-0.042(RT \ln(1 + 1/C))^2]$
^{237}Pu	O	F	$S = 0.98C^{1.12}$
^{226}Ra	O	F	$S = 356C^{1.29}$
^{137}Cs	O	DR	$S = (1.8E-5) \exp[-0.0031(RT \ln(1 + 1/C))^2]$
^{85}Sr	O	F	$S = 0.17C^{0.97}$

[a] R = Reducing Conditions; O = Oxidizing Conditions.
[b] F = Freundlich[12], $S = KC^N$; DR = Dubinin-Radushkevich[13],
$S = X_m \exp B(RT \ln(1 + 1/C)]^2$ where:

S = equilibrium concentration of sorbed ion per gram of solid, mol/g
C = equilibrium concentration of sorbing ion in bulk solution, mol/l
K = constant related to affinity of ion for substrate
B,N = constants related to energy of sorption
R = 8.314 E-3 kJdeg^{-1} mol^{-1}
T = temperature, K (333K for basalt)
X_m = maximum solute loading, mol/g

RADIONUCLIDE SOLUBILITY BEHAVIOR

A second major mechanism for controlling the release of radionuclides from a repository in basalt is the precipitation or incorporation of radionuclides into new mineral phases, e.g., solubility constraints. Assuming steady-state conditions, radionuclide solubilities can be used to calculate the maximum possible concentration, and therefore release, of a radionuclide in a specific system. Radionuclide solubility, like sorption is highly dependent on environmental conditions such as Eh, pH and groundwater composition. At this time, there are few solubility data available for conditions relevant to a repository in basalt. Based on available thermodynamic and laboratory data[14,15,16,17], solubilities have been estimated for the basalt geohydrologic system. These solubility estimates are presented in Table 2.

EFFECT OF RADIONUCLIDE SORPTION AND SOLUBILITY ON RADIONUCLIDE TRANSPORT

A simple analysis of the effect of sorption and solubility on radionuclide transport in the Columbia River basalt geohydrologic system can provide a preliminary assessment of the ability of the site to control the release of radionuclides to the accessible environment. The effect of radionuclide sorption on radionuclide release for both a non-disruptive and a disruptive scenario is evaluated through the determination of radionuclide travel times to the accessible environment (Table 4). The two scenarios and assumptions used in the analyses are as follows:

Case I: Nondisruptive transport through a basalt flow top to the accessible environment (10km)
- Porosity = 0.05; Water velocity = 0.06 m/yr; Density = 2.7 g/cc
- Zero travel time to the flow top from the repository

Case II: Disruptive scenarios with resultant transport through a confined aquifer (interbed zone) to the accessible environment (10 km)
- Porosity = 0.1; Water velocity = 3.2 m/yr; Density = 2.6 g/cc
- Zero travel time to the interbed zone from the repository

By comparing radionuclide travel times with their respective half-lives, an evaluation of potential problem radionuclides for the repository can be made. For the radionuclides considered for Case I, only the U isotopes and ^{99}Tc under oxidizing conditions and only ^{235}U and ^{238}U under reducing conditions will reach the accessible environment before radioactive decay reduces their concentrations to insignificant levels. For Case II, the disruptive case, ^{235}U, ^{236}U, ^{238}U, ^{237}Np, ^{99}Tc, and ^{79}Se under oxidizing conditions and only ^{235}U, ^{236}U and ^{238}U under reducing conditions will reach the accessible environment at significant activity levels.

Radionuclide solubilities represent the maximum amount of a given radionuclide that can be in solution at any one time. As such, radionuclide solubilities can be used to calculate a maximum possible radionuclide release from the repository. Solubility limited release rates for U, Np, Up, Am, Tc and Se for a repository in basalt have been determined for Case I and Case II (Table 5). When these solubility constraints on radionuclide release are considered for these potential problem radio-

TABLE 4

COMPARISON OF RADIONUCLIDE TRAVEL TIMES TO THE ACCESSIBLE ENVIRONMENT[a] WITH THEIR RESPECTIVE HALF-LIVES

| Radionuclide | Half Life (yr) | Radionuclide Travel Times[b] | | | |
| | | Case I | | Case II | |
		Oxidizing Conditions (yrs)	Reducing Conditions (yrs)	Oxidizing Conditions (yrs)	Reducing Conditions (yrs)
^{233}U	1.62E+5				
^{234}U	2.47E+5				
^{235}U	7.1E+8	1.6E+5	3.4E+8	2.1E+6	8.2E+6
^{236}U	2.39E+7				
^{238}U	4.51E+9				
^{237}Np	2.14E+6	1.3E+8	1.7E+9	3.3E+6	4.1E+7
^{238}Pu	8.64E+1				
^{239}Pu	2.439E+4				
^{240}Pu	6.58E+3	1.7E+8	3.4E+8	2.5E+7	4.1E+7
^{241}Pu	1.32E+1				
^{242}Pu	3.79E+5				
^{241}Am	4.58E+2				
^{242m}Am	1.52E+2	2.9E+9	N.A.	>8.2E+8	>8.2E+8
^{243}Am	7,95E+3				
^{226}Ra	1.6E+3	1.7E+9	---	1.6E+7	---
^{99}Tc	2.1E+5	1.6E+5	2.5E+8	3.2E+3	5.8E+6
^{79}Se	6.5E+4	1.6E+5	6.9E+7	3.2E+3	4.2E+5
^{135}Cs	3.0E+6	4.3E+9	---	8.2E+7	---
^{90}Sr	2.8E+1	1.6E+5	---	3.3E+7	---
Groundwater Travel Time		1.6E+5	1.6E+5	3.2E+3	3.2E+3

[a]As defined by EPA, 1981.[18]

[b]Travel Time = $\frac{XR}{V}$ where X = distance, m; R = 1 + (ρ/ϕ) Kd, and V = groundwater velocity. Kd is the distribution coefficient, ml/g, ρ = density, and ϕ = porosity.

N.A. No data available.

nuclides, in addition to sorption constraints, it becomes apparent that none of the radionuclides considered can be released from the repository at rates that will exceed the current regulatory criteria for a nuclear waste repository. The basalt site, therefore, appears to be capable of adequately controlling the release of U, Np, Pu, Am, Ra, Tc, Se, Cs and Sr to the accessible environment through sorption and solubility constraints.

TABLE 5

COMPARISON OF ESTIMATED SOLUBILITY LIMITED RADIONUCLIDE RELEASE RATES* FROM A REPOSITORY IN BASALT WITH RELEASE RATES REQUIRED TO MEET CURRENT EPA LIMITS

Radionuclide	Solubility (M/1)	Release Rate Required to Meet EPA Limits (Year^{-1})	Case I Release Rate (Year^{-1})	Case II Release Rate (Year^{-1})
^{233}U		9E-3	5E-12	5E-10
^{234}U		2E-6	5#-12	5E-10
^{235}U	1E-8	1E-5	5E-12	5E-10
^{236}U		8E-6	5E-12	5E-10
^{238}U		6E-6	5E-12	5E-10
^{237}Np	1E-15	3E-6	3E-16	3E-14
^{239}Pu		7E-8	1E-10	8E-9
^{240}Pu	1E-9	5E-9	8E-11	8E-9
^{242}Pu		1E-5	1E-10	8E-9
^{243}Am	1E-10	6E-8	1E-10	1E-8
^{99}Tc	1E-12	1E-5	1E-13	1E-11
^{79}Se	1E-15	1E-4	2E-14	2E-12

*Based on a repository inventory of 50,000 MTHM (1/2 spent fuel and 1/2 commercial high level waste), 1,000 year containment of the waste by the waste package and water flow rate of 56,700 l/yr for Case I and 5,670,000 l/yr for Case II through the repository.

REFERENCES

1. Myers, C. W. and Price, S. M., ed. (1981) Subsurface Geology of the Cold Creek Syncline, RHO-BWI-ST-14, Rockwell Hanford Operations, Richland, Washington.

2. Benson, L. V. and Teague, L. S. (1979) A Study of Rock-Water-Nuclear Waste Interactions in the Pasco Basin, Washington, LBL-9677, Lawrence Berkeley Laboratory, Berkeley, California.

810

3. Arnett, R. C., Mudd, R. N., Baca, R. G., Martin, M. D., Norton, W. R., and McLaughlin, D. B. (1981) Pasco Basin Hydrologic Modeling and Far-Field Radionuclide Migration Potential, RHO-BWI-LD-44, Rockwell Hanford Operations, Richland, Washington.

4. Deju, R. A. (1981) A Summary of the Groundwater Hydrology of the Hanford Site as it Affects a Potential Nuclear Waste Repository in Basalt, RHO-BWI-LD-53, Rockwell Hanford Operations, Richland, Washington.

5. Gephart, R. E., Arnett, R. C., Baca, R. G., Leonhardt, L. S., and Spane, R. A. (1979) Hydrologic Studies Within the Columbia Plateau, Washington: An Integration of Current Knowledge, RHO-BWI-ST-5, Rockwell Hanford Operations, Richland, Washington.

6. Noonan, A. F., Frederickson, C. K. and Nelen, J. (1980) Phase Chemistry of the Umtanum Basalt: A Reference Repository Host in the Columbia Plateau, RHO-BWI-SA-77, Rockwell Hanford Operations, Richland, Washington.

7. Jacobs, G. K. and Apted, M. J. (1981) "Eh-pH Conditions for Groundwater at the Hanford Site, Washington: Implications for Radionuclide Solubility in a Nuclear Waste Repository Located in Basalt," EOS, Trans. Amer. Geophys. U. 62, 1065.

8. Smith, M. J. ed. (1980) Engineered Barrier Development for a Nuclear Waste Repository in Basalt, An Integration of Current Knowledge, RHO-BWI-ST-7, Rockwell Hanford Operations, Richland, Washington.

9. Goodwin, B. W., O'Connor, P. A. and Vandergraaf, T. T. (1982) Newsletter, Radionuclide Migration in the Geosphere, #6, OECD Nuclear Energy Agency 5.

10. Salter, P. F., Ames, L. L. and McGarrah, J. E. (1981) Sorption Behavior of Selected Key Radionuclides in Columbia River Basalts, RHO-BWI-LD-48, Rockwell Hanford Operations, Richland, Washington.

11. Barney, G. S. (1981) Radionuclide Interactions with Groundwater and Basalts from Columbia River Basalt Formations, RHO-SA-217, Rockwell Hanford Operations, Richland, Washington.

12. Hayward, D. O. and Trapnell, B. M. W. (1964) Chemisorption, Butterworth and Company, Ltd., London, England.

13. Dalal, R. C. (1979) "Applications of the Dubinin-Radushkevich Adsorption Isotherm for Phosphorus Sorption by Soils," Soil Sci. 128, 65-69.

14. Cleveland, J. M. (1979) "Critical Review of Pu Equilibria of Environmental Concern" in Chemical Modeling of Aqueous Systems, Jenne, E. A., ed., ACS Sym. Ser. 93.

15. Rai, D. and Seine, R. J. (1978) Solid Phases and Solution Species of Different Elements in Geologic Environments, PNL-2651, Pacific Northwest Laboratory, Richland, Washington.

16. Rai, D., Strichert, R. G. and Moore, P. A. (1981) Influences of Solid Phases on Am Concentrations in Solutions, PNL-SA-9106, Pacific Northwest Laboratory, Richland, Washington.

17. Langmuir, D. (1978) "Uranium Solution-Mineral Equilibria at Low Temperature with Applications to Sedimentary Ore Deposits," Geochim. Cosmochim. Acta, 42, 547-569.

18. EPA (1981) Radiation Protection Programs, subpart B--Environmental Standards for Disposal, 40 CFR 191 (draft), Code of Federal Regulations, U.S. Environmental Protection Agency, Washington DC.

RADIONUCLIDE RETARDATION DURING TRANSPORT THROUGH FRACTURED GRANITE

Ian G.McKinley, Julia M.West
Environmental Protection Unit, Institute of Geological Scinces, Building 151,
AERE Harwell, Oxfordshire. OX11 ORA, United Kingdom.

INTRODUCTION

In several countries low permeability crystalline rocks (e.g. granites) are
under consideration as potential hosts for radioactive waste repositories. In
such formations groundwater flow occurs predominantly in specific fractures
rather than being a general porous flow through the entire rock matrix. By
considering fractures to be simple parallel plates various authors[1,2,3] have
demonstrated the potential importance of diffusion into dead-end pores and the
rock matrix itself ('matrix diffusion') as a mechanism for the retention of
migrating radionuclides. Complementing these theoretical studies, several in-
situ migration experiments are planned in single fissures in crysalline rocks
in Sweden, the U.K. and the U.S.A. The 'parallel plate' approximation to a
single fissure is, however, acknowledged to be a gross simplification of any
real case where "flowing" fractures are expected to be either filled or coated
with secondary minerals, formed by hydrothermal alteration of fracture surfaces.
In the evaluation of net radio-nuclide retardation, therefore, the effect of
sorption onto such secondary minerals must be carefully considered.

In this paper the results of laboratory sorption studies of particular
radionuclides (isotopes of Cs, Sr, Co and Ce) onto naturally weathered granite
fracture infill are reported. Complementary mechanistic studies investigating
the effect of reaction direction, temperature and competing ions on sorption
isotherms are also briefly summarised. By use of simple computer models of
groundwater flow, the consequences of observed 'isotherm non-linearity' on
resultant nuclide migration are illustrated and compared with retardation
calculated assuming 'matrix-diffusion' effects.

MATERIALS AND METHODS

Sorption studies are reported on weathered granite fracture infill from a
proposed fissure migration site in a near surface mine at Camborne, Cornwall.[4]

The fracture infill consisted of granite residuum with a fine grained
component including:- quartz, smectite, mica, chlorite and kaolinite. For this
weathered rock assemblage sampling and pre-treatment were designed to

812

minimise pertubation (e.g. loss of natural moisture). It must be emphasised
that mineralogical analysis of such material gives an incomplete picture of
composition as many poorly ordered or amorphous intermediates in the process of
weathering from component granite minerals to thermodynamically stable clays
are present. Poorly formed surfaces present will be very complex chemically
but are likely to be important in the sorption of trace concentration species.

A standard batch reaction technique was utilised in which rock samples are
equilibrated with groundwater spiked with γ-emitting radionuclides under
conditions of controlled temperature, atmosphere, and intensity of agitation.
In kinetics studies, little change in solution chemistry was observed after
~5-10 days but, in these experiments, ~30-60 days were allowed for equilibration. Radionuclide partitioning between solid and liquid phases is
evaluated, after centrifugation, by measurement of the change in activity
of the solution phase as evaluated by NaI(Tl) or Ge(Li) γ-spectrometry.
Subsequent to sorption experiments, the aqueous phase concentration was
decreased and the resultant re-equilibration (desorption) measured. Backgrounds,
blanks and tube sorption effects are taken into account during spectrum
analysis and subsequent calculation to equilibrium elemental concentrations in
solid (C_R) and aqueous (C_W) phases. Methodology has been described in detail
elsewhere.[6,7]

RESULTS AND DISCUSSION

A general finding in this work is that the partitioning of radionuclides
between rock and aqueous phases is often dependent on elemental concentration
and thus the concept of a concentration independent distribution coefficient
or 'Kd' is not applicable. To prevent confusion with more rigorously defined
Kd terminology, the empirical rock/water partition coefficient (= C_R/C_W) will
be denoted by K (in ml/g) with subscripts if referring to particular reaction
directions (K_{sorp} or K_{desorp}).
Variation of K with concentration has been observed even when aqueous
concentrations may be regarded as at 'trace' levels (10^{-10} M), solid phase
concentrations are far below cation exchange capacities and perturbing
reactions (e.g. precipitation) are unlikely.[4,5,6,7] When such variation exists,
plots of log (C_R) vs log (C_W) are often linear and thus observed sorption may
be described empirically by a Freundlich isotherm of form:-

$$C_R = aC_W^b$$

Despite the 'linearising' effects of such log-log transform, it must be noted
that occasional data sets are only poorly described by Freundlich isotherms.
Nevertheless, this procedure provides a much better empirical measure of

concentration dependent radionuclide sorption than a single $'K_d'$ value.

Sorption isotherms for Cs, Sr and Co on the weathered fissure infill at 25° and 5°C are listed on Table 1. From these data it may be noted that:-

1. Cs and Sr sorption data at both temperatures are quite well described by the Freundlich equation and have gradients slightly greater than 1. Co sorption is poorly described by the Freundlich equation but any forced empirical linear fit to the data yields a gradient much less than one.

2. In all cases the gradient of Freundlich curve (b in the equation above) is little affected by a temperature change of 20°C but the intercept (a in the equation) is increased by a factor of ~2 as temperature increases from $5-25^\circ$C.

Complementary studies of sorption on a silt/clay grade component of surficial weathered granite show similar data trends (Table 1) while, additionally, sorption/desorption hysteresis has been demonstrated on this substrate and, in multiple spiking experiments, competition between nuclides has been observed in binary (Cs/Co) and ternary (Cs/Co/Ce) systems.[7]

Isotherm gradients = 1 correspond to 'ideal' concentration independent sorption (e.g. ion exchange) and gradients less than 1 may be attributed to the heterogeneity of available rock surfaces as saturation of limited numbers of high affinity sites would result in the observed decrease in sorption with increasing concentration. Gradients greater than 1 have, however, rarely been reported in inorganic systems and the mechanistic basis for such behaviour is currently under investigation. Contributions from colloidal or organically complexed species in production of anomalous isotherms cannot be excluded although filtering experiments set an upper size limit for aqueous phase components of 0.45 microns.

MIGRATION MODELLING

From the sorption data presented above it is apparent that retardation of migrating nuclides by weathered granite facture infill is likely to be a complex process depending on the concentrations of the nuclides involved and of competing ions, temperature and sorption/desorption hysteresis. By use of a simple box model of groundwater flow, however, the consequences of empirically derived trends in sorption data may be assessed in terms of resultant radionuclide retardation. As a particular example, retardation due to concentration dependent sorption onto fracture infill in a simple fissure tracer experiment will be analysed and compared with that predicted due to simple $('K_d')$ concentration independent sorption and matrix diffusion processes.

In the calculation of retardation due to matrix diffusion, nomenclature

TABLE 1

FIT TO FREUNDLICH EQUATION $C_R = a\, C_W^b$: C_R µg/g, C_W µg/ml

ROCK	NUCLIDE	TEMPERATURE	REACTION DIRECTION	a	b	CORRELATION COEFFICIENT	C_W RANGE	(MOLARITY)
Fracture Infill	^{137}Cs	25°C	S	794	1.2	0.95	$10^{-5} - 10^{-1}$	$(10^{-10} - 10^{-6})$
"	"	5	S	501	1.1	0.98	$10^{-5} - 10^{-1}$	$(10^{-10} - 10^{-6})$
"	^{85}Sr	25	S	380	1.08	0.96	$10^{-4} - 10^{-1}$	$(10^{-9} - 10^{-6})$
"	"	5	S	195	1.05	0.99	$10^{-4} - 10^{-1}$	$(10^{-9} - 10^{-6})$
"	^{60}Co	25	S	0.77	0.33	0.75	$10^{-4} - 10^{-1}$	$(10^{-9} - 10^{-6})$
"	"	5	S	0.42	0.32	0.66	$10^{-4} - 10^{-1}$	$(10^{-9} - 10^{-6})$
Surficial Weathered granite	^{137}Cs	25	S	776	1.03	0.99	$10^{-6} - 10^{-4}$	$(10^{-11} - 10^{-1})$
"	"	25	D	2818	1.17	0.94	"	"
"	"	5	S	37	0.51	0.999	"	"
"	"	5	D	62	0.58	0.98	"	"
"	^{60}Co	25	S	0.06	0.33	0.57	$10^{-7} - 10^{-2}$	$(10^{-12} - 10^{-7})$
"	"	25	D	0.09	0.22	0.37	"	"
"	"	5	S	0.13	0.42	0.78	"	"
"	"	5	D	2.63	0.69	0.73	"	"
"	^{139}Ce	25/5	S/D	0.07	0.18	0.32	$10^{-7} - 10^{-2}$	$(10^{-12} - 10^{-7})$

and background hydrogeological data were selected from the work of
Neretnieks.[1] An isolated fracture of thickness 1.3 x 10^{-4}m is assumed to be
surrounded by rock with 'apparent diffusivity' (Da) of 10^{-10}m^2/sec for non-
sorbed species. This will be compared to an infilled fissure of width 6.5 x
10^{-6}m and porosity 20% (i.e. of identical water content) from which there is no
diffusion into the surrounding rock matrix. To allow sorption equilibrium to be
assumed on the timescale of a 5m in-situ experiment, a water flow rate of 5 x
10^{-6}m/sec (~0.4m/day) is assumed.

An extremely simple box model is utilised in which a 5m flow path is
regarded as being composed of 10 boxes in each of which rock and water phase
concentrations are homogenous and in equilibrium. At each time step of 10^5 sec
(equivalent to the water flow time through a box) the re-distribution of radionuclide between solid and liquid phases is calculated for each box. Because of
the time stepping process, the output from this model is rather crude but, due
to the mathematical simplicity of the calculations involved at each time step,
the model may be easily evaluated using a benchtop microcomputer.

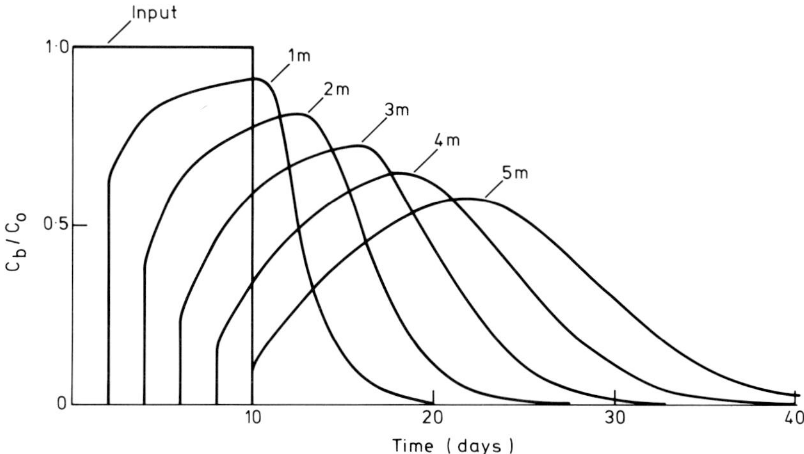

FIG. 1. Output profiles for tracer retarded by matrix diffusion only.

816

An example of the output from this model is shown on Fig 1 which presents breakthrough curves expressed as the ratio of water content in each box (C_b) to input concentration (C_o) at varying distances from the source of a step-pulse input. In this calculation the retardation and dispersion are caused by matrix-diffusion effects only and would reflect the type of breakthrough curve expected for a non-sorbed tracer. The tracer penetration thickness $(\bar{\eta})$ at each time step (t) is calculated by the equation of Neretnieks[1] $\bar{\eta} = 1.13 \sqrt{D_a\ t}$. The curves at 1m and 5m are also presented on log (C_b/C_o) vs time plots on Figs 2 and 3 respectively (curve b) and may be compared with the output curves at these distances resulting from ideal plug flow (a).

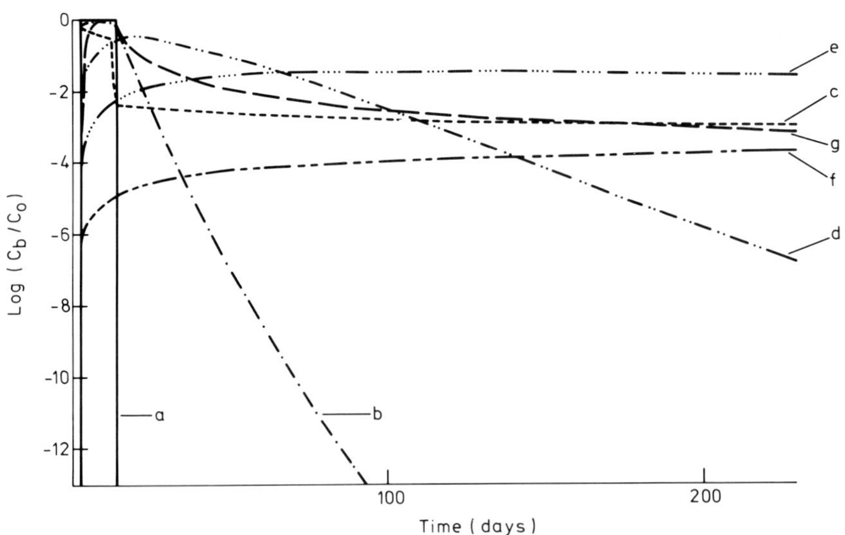

FIG. 2. Output profiles at 1m (see text)

The effect of coupled matrix diffusion and sorption has been considered by Neretnieks and can be readily calculated for any value of Kd. The exact shape of the diffusion/sorption curve is, to some extent, dependent on cal-

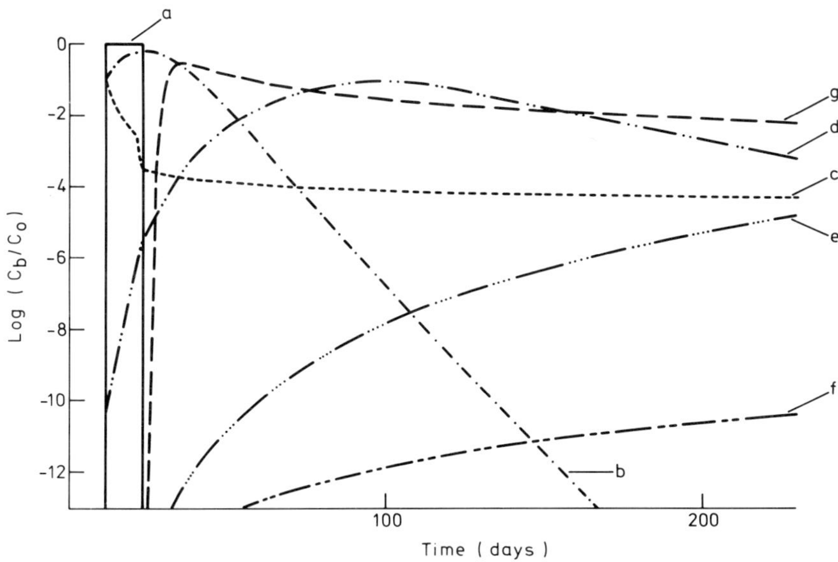

FIG. 3. Output profiles at 5m (see text).

culational methods, but the dominant effects of including a sorption term are
reduction of the peak output concentration (especially evident at 5m) and
formation of a long low-concentration tail to the output curve. For a
relatively poorly sorbed species (Kd = 1 ml/g = .001 m^3/kg) the calculated
output profiles are plotted as curve (c) on Figs 2 and 3. Increasing the
'distribution coefficient' of the rock matrix has a slight effect on the peak
output but a larger effect on the concentration of the peak tail - an increase
of Kd from 0.001 to 0.01 m^3/kg.causes a corresponding factor of ten decrease
in tail concentration.

By considering the alternative model of an infilled fracture, the effects
of concentration independent sorption of infill can be readily calculated and
is shown on Figs 2 and 3 for Kd values of 0.001 and 0.01 m^3/kg (curves d and e
respectively). For low values of Kd (~0.001 m^3/kg) the resultant curves at
both 1 and 5m are similar in shape to the matrix diffusion curves (b) with
slightly greater retardation and spreading of the output peak and a more

818

gradual tail-off of concentration. As the value of Kd increases the output peak
is greatly retarded and spread over such a long time that output is apparently
peakless.

     For all cases considered so far presentation of output as a fraction of
input concentration has been quite sufficient but, for concentration dependent
sorption, the numerical value of input concentration is extremely important.
For the infilled fracture, empirical sorption isotherms may be used to calculate resultant nuclide migration. The inventory of radionuclide in each box
(I_b) is given by $I_b = rC_R + wC_W$

$$\text{where} \quad r = \text{rock mass}$$
$$w = \text{water volume.}$$

For a general Freundlich equation $C_R = a\, C_W^b$, the inventory becomes $I_b =
r\, a\, C_W^b + wC_W$ which may be solved iteratively by Newton's method. From these
calculations output curves for Co and Cs are presented in Fig 4 based on the
sorption isotherms at 25°C on weathered granite infill presented earlier
(Table 1). It is immediately apparent that the breakthrough curves are very
sensitive to input concentration (in μg/ml $\equiv$ mg/l) and that the shape of output
curve varies dramatically with the form of the sorption isotherm. For Co the
isotherm $C_R = 0.77\, C_W^{0.33}$ reflects increased sorption as concentration
decreases resulting in self-sharpening of the migration plume front. In the
case of Cs, however, the equivalent isotherm $C_R = 794\, C_W^{1.2}$ indicates much
greater retardation at unit concentration (1 μg/ml) but a decrease in sorption
as C_W decreases resulting in considerable retardation of a high concentration
input pulse but rapid transport of a low concentration front. Because of these
differences it is noted that the relative retardation for Co increases as input
concentration decrease while for Cs retardation decreases with decreasing input.

     To allow direct comparison, output curves for Cs and Co resulting from an
input pulse of a reasonable 'trace' concentration in this context (10 μg/ml $\equiv$
0.01 kg/m^3) are plotted on Figs 2 and 3 (curves f and g respectively). The
main points from this figure are that the low concentration breakthrough front
from Cs (f) is similar in form to that resulting from high 'Kd' sorption (e)
while the Co output curve (g) is, at short distances (Fig 2) quite similar
in shape to the matrix diffusion/sorption curve (c).

     Finally it must be emphasised that this model (and all others) considerably
oversimplifies physical and chemical processes, in particular by
1) excluding complicating flow parameters such as dispersion, flow path variability and fissure tortuosity
2) modelling retardation processes in isolation whereas, for example, matrix

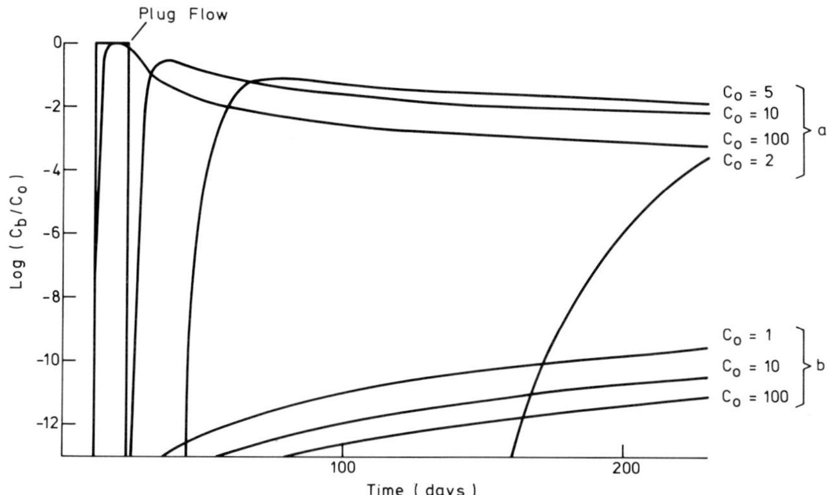

FIG. 4. Cs and Co output profiles at 5m:- variation with input concentration
a) Co   b) Cs

diffusion, 'linear' sorption on the rock matrix and 'non-linear' sorption on
infill could be present together in a real system
3) excluding sorption/desorption hysteresis, ion-competition and the mechanistic
analysis of sorption data. For example, if isotherm gradients greater than 1
were due to colloids, a more complex model describing transport of both solutes
and colloids and their different retardation processes would be required.
It is nevertheless apparent that almost any form of output curve produced in an
in-situ tracer experiment could be explained by several different mechanistic
models but by experiment design the applicability of such models can be tested.
For example, as shown above, models assuming retardation by concentration dependent sorption can be tested against alternatives incorporating 'Kd' or matrix diffusion by observing the effects of varying tracer input concentration.

CONCLUSIONS

From this work it is apparent that sorption of radionuclides onto
weathered fracture infill is likely to be a complex process although data

820

trends may be presented as empirical relationships (e.g. Freundlich isotherms).
Detailed studies of reaction mechanisms are in progress and indicate that, for
even simple chemical species, sorption can be a complex function of reaction
time, temperature, elemental concentration, concentration of competing ions and
reaction direction. By use of simple models of groundwater flow, sorption data
can be placed in a realistic geological context and results indicate that
output curves resulting from fissure tracer experiments could be explained in
terms of totally different retardation mechanisms. While various mechanisms
might give similar retardation at short path lengths, more detailed information
on mechanisms is required to justify extrapolation of data to the long path-
lengths and timescales required in safety assessment and it is thus apparent
that laboratory and in-situ studies are both essential and complementary in
this field.

ACKNOWLEDGEMENTS

   This work was carried out as part of the Department of the Environment research programme on radioactive waste management, and is published with the permission of the Director of the Institute of Geological Sciences. Useful comments
on colloidal transport from B. Skytte Jensen are gratefully ackowledged.

REFERENCES

1. Neretnieks, I., 1979. KBS 79-19.
2. Glueckauf, E., 1980. AERE R 9823.
3. Wadden, M.M., Katsube, T.J., 1981. Geol. Survey. Canade Rep. 8182-GP4.
4. McKinley, I.G., West, J.M. 1982. Rep. Inst. Geol. Sci. (in press).
5. McKinley, I.G., Greenwood, P.B. 1980. Rep. Inst. Geol. Sci. ENPU 80-7.
6. McKinley, I.G., West, J.M. 1981. a. Rep. Inst. Geol. Sci. ENPU 81-6.
7. McKinley, I.G., West, J.M. 1981. b. Rep. Inst. Geol. Sci. ENPU 81-14.

A COMPARATIVE ANALYSIS OF FRACTURED AND POROUS MEDIUM RADIONUCLIDE TRANSPORT

B. J. TRAVIS* AND H. E. NUTTALL**

*Earth and Space Sciences Division, Los Alamos National Laboratory, Los Alamos, New Mexico 87545, **Department of Chemical and Nuclear Engineering, Univerity of New Mexico, Albuquerque, New Mexico 87131

INTRODUCTION

Modeling of radionuclide transport in geological formations has until recently been limited to the treatment of nonfractured porous media. Recently researchers in Sweden, Canada, France, and the USA have become interested in the problems of nuclide transport in geological formations which may contain fractures. Since many conventional porous-medium ion-transport codes are available, the question arises as to when fracture widths and fracture frequency dictate the use of these existing codes.

This paper provides a comparative study of porous and fractured medium nuclide transport by modeling the far field (1 km) problem for three different types of geomedia, i.e., a homogeneous porous medium, a single fracture in the porous medium, and a multi-fracture medium. Analytical models were used to study the homogeneous and single fracture geometries while an advanced numerical code was used to model the multi-fractured medium.

Analytical Models

An analytical model for the homogeneous porous medium case with an instantaneous point source, radioactive decay, and retardation was presented by Nuttall and Ray[1]. Several papers[2,3,4] have presented essentially identical models for nuclide transport in a single fracture with diffusion into the surrounding porous medium. The paper by Nuttall and Ray[1] presents a more complex model permitting flow and nuclide migration in both the porous medium and the fracture.

Multi-Fracture Model

For the study of multi-fractured medium, a numerical model and code recently developed by Travis[5] was used. The TRACR3D code is a modified version of a code used to study tracer migration and breakthrough in large rubblized oil shale retorts. The code solves the three-dimensional, time-dependent mass conservation equations and modified momenta equations for single- or two-phase flow with transport of a tracer in either phase. The code can handle single phase (air or water) flow or two-phase (air and water) flow in steady state or transient conditions with (or without) tracer transport. Transport mechanisms include advection, diffusion and velocity dependent dispersion[6]. This is not simply a dual porosity model but a detailed treatment of both the porous and fracture regions.

In addition the code models radioactive decay, and has linear, Langmuir and Freundlich sorption models. The medium can have spatially varying properties. Discrete, finite width, fractures can be modeled as well but fractures must be orthogonal. Each fluid phase (air, water) has its own velocity field which can either be specified as an input (in the case of steady-state flow) or can be calculated by the code. If calculated, the code solves the coupled mass conservation equation, equation of state, and reduced momenta equations (Forchheimer flow in porous medium, Fanning equation in fractures) for the pressure and saturation fields at each numerical time step. At low Reynolds number flow, the momenta equations reduce to Darcy's equation. This is the usual situation. Once the pressure and saturation fields are known, velocities can be calculated. Then transport can be computed. This computational procedure is repeated for each time step.

Any computational zone can be specified as a source or sink of mass and tracer species. Time dependent sources/sinks and boundary conditions are allowed. Boundary condition types include: 1) no flow, 2) ambient conditions for pressure, saturation, tracer concentration, 3) specified history for pressure, saturation, tracer concentration, and mass flux.

Zones that represent the intersection of cracks are treated strictly as mixing zones; outflow is apportioned according to velocities. Numerical integration is accomplished with an integrated finite difference scheme[7] with special treatment of convective terms to minimize numerical diffusion. At crack-porous medium interfaces, pressure is made continuous as is mass flux.

MODELING STUDIES

A series of modeling calculations are presented to illustrate the fact that fractures increase nuclide transport rates. A two-dimensional (x-y) geometry with an impulse source was used to investigate the problem of far field nuclide migration in complex geological formations. A case study approach was used and the effect of fracture width was investigated. Formation properties (see Table 1) were chosen to represent a typical low permeability medium such as tuff. In all cases a constant pressure gradient in the "x" direction was assumed. Three different types of geo-medium are investigated, i.e., (1) nonfractured, (2) multi-fractured, and (3) single fracture. The system geometry and types of geo-medium are illustrated in Fig. 1. Since the problem is symmetric about the source, only the upper region was treated in the study. To compare the effects of fractures, concentration contours are presented as well as the downstream nuclide discharge rate versus time curves for a location 1 km from the nuclide source. Retardation was studied using an adsorption coefficient of "2".

The nonfractured porous medium was investigated first because it provides the greater resistance to both nuclide migration and fluid flow and thus forms a lower limit to which the subsequent fractured medium studies can be compared. The homogeneous porous medium model present by Nuttall and Ray[1] was applied using the study parameters given in Table 1. The calculated concentration contours at 7500 years are presented in Fig. 2. Comparing the nonfractured and fracture medium contours illustrates how a fracture accelerates and spreads the migrating nuclides. Further, since the porous medium retards nuclide migration by both adsorbing the species and reducing the fluid flow, the rate of migration in the nonfractured medium case is a limiting lower bound and the presence of fractures, as illustrated, tends to enhance nuclide migration.

To further illustrate the effects of fractures, the discharge rates for nuclides passing through an imaginary plane drawn normal to the flow and downstream of the source are presented in Fig. 3. The discharge rate versus time curves were evaluated at a distance 1 km downstream of the source. The resulting discharge rate curves are shown for three cases, i.e. (1) nonfractured medium, (2) single fracture case (a), and (3) single fracture case (b). The small 50 μm fracture discharge curve is similar to the nonfracture case producing only a slight acceleration while the case (b) 200 μm fracture shows a very early breakthrough curve.

Finally the effects of multiple fractures was studied using the TRACR3D numerical code.   The nuclide transport results were intermediate to those for the single fracture and the nonfracture cases. In general, the effect of the multiple fractures is to widen the lateral dispersion and provide more nuclide retardation because of the increased surface area when compared to the single fracture case. The increased lateral dispersion is caused by fluid flow in the "y" direction as illustrated by the "x–y" velocity field shown in Fig. 4.   A summary of findings are discussed in the following section.

CONCLUSIONS
A modeling study was used to show the effects of fractures on the rate and behavior of radionuclide migration as it may relate to problems of geological storage of nuclear waste. The following list summarizes the conclusions of this paper.

A. A set of models were developed and used to investigate the effects of fractures on nuclide migration rates in several types of geological media.

B. A single uninterrupted fracture is in a sense the worst case for nuclide migration. (It provides early breakthrough of the nuclides.)

C. The nonfractured and single fracture cases set lower and upper limits on the rate of nuclide migration with the multi–fracture cases giving intermediate results.

D. A single fracture or multiple fractures lower the level of the nuclide concentration in the porous medium by spreading the material over a greater area.

E. Fractures tend to accelerate nuclide transport and overall migration rates.

TABLE 1

STUDY PARAMETERS

Fracture half width, 2.5×10^{-5} m or 25 μm (case a)
100 μm (case b)

Width of line source and fracture, 1.0 m

Effective diffusivity coefficient in both the fracture and porous medium, 1×10^{-9} m^2/sec

Fluid velocity in the porous medium, 1.59×10^{-9} m/s

Fluid velocity in the fracture, 625 m/yr (case a)
10,000 m/yr (case b)

Retardation factor, 16

Corresponding adsorption coefficients, 2

Permeability of the porous medium, 4.9×10^{-15} m^2

Void fraction, 0.3

Reference concentration, 0.001 kg/m^3

Mass of contaminant released, 0.001 kg

Bulk density of the geomedium, 2300 kg/m^3

Pressure gradient, 95 N/m^3

Fluid viscosity, 1×10^{-3} kg/m sec

(For the analytical modeling, flow velocity for the porous medium was calculated from Darcy's equation and the fluid velocity in the fracture was determined from the following equation[1].)

$$\frac{u_c}{u_p} = \left(\frac{\varepsilon}{k}\right)\left(\frac{d^2}{3}\right)$$

where

u_c = Fracture fluid velocity

u_p = Porous fluid velocity

ε = void fraction

k = Permeability

d = Fracture half width

REFERENCES

1. Nuttall, H. E. and A. R. Ray (1981) "A Combined Fracture/Porous Media Model for Contaminant Transport," Scientific Basis for Nuclear Waste Management, Vol. 3, Plenum Press, 1981.

2. Neretnieks, Ivars (1980) "Diffusion in the Rock Matrix: An Important Factor in Radionuclide Retardation?", Journal of Geophysical Research, Vol. 85, No. B8, August 10, 1980, pp. 4379-4397.

3. Grisak, G. E. and J. F. Pickens (1981) "An Analytical Solution for Solute Transport Through Fractured Media with Matrix Diffusion", Journal of Hydrology, 52, pp. 47-57.

4. Erickson, K. L. and D. R. Fortney (1981), "Preliminary Transport Analyses for Design of the Tuff Radionuclide Migration Field Experiment," SAND81-1253, Sandia National Laboratory, September 1981.

5. Explosively Produced Fracture of Oil Shale, LA-8956-PR, Progress Report, Los Alamos National Laboratory, January-March 1981, pp. 73-80.

6. Bear, Jacob (1979) <u>Hydraulics of Groundwater</u> , McGraw-Hill, Inc., New York.

7. Roache, Patrick J. (1972) <u>Computational Fluid Dynamics</u>, Hermosa Publishers, Albuquerque, New Mexico.

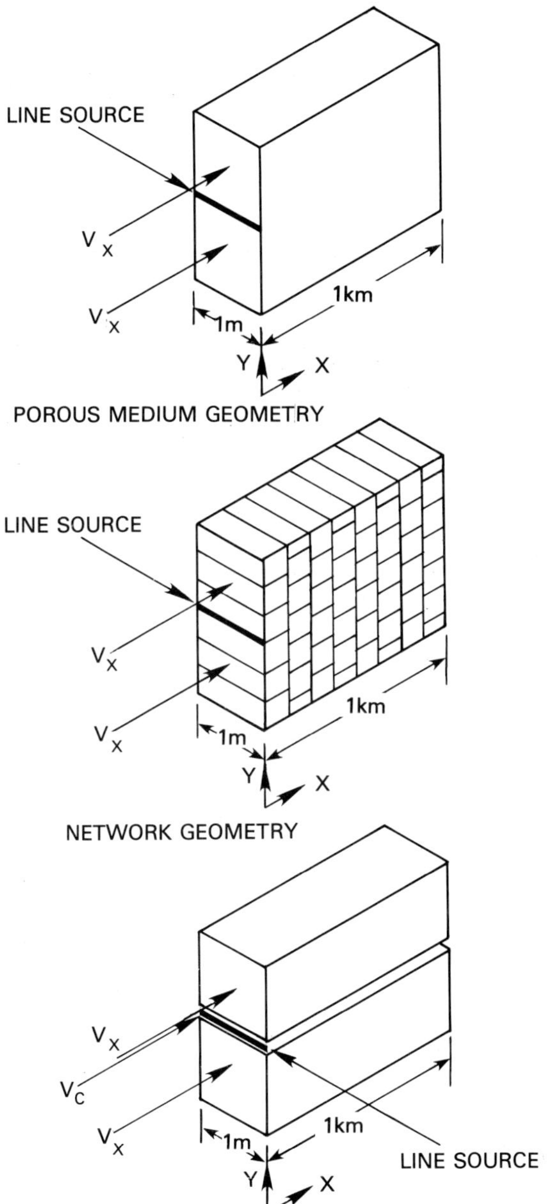

Fig. 1 STUDY GEOMETRY

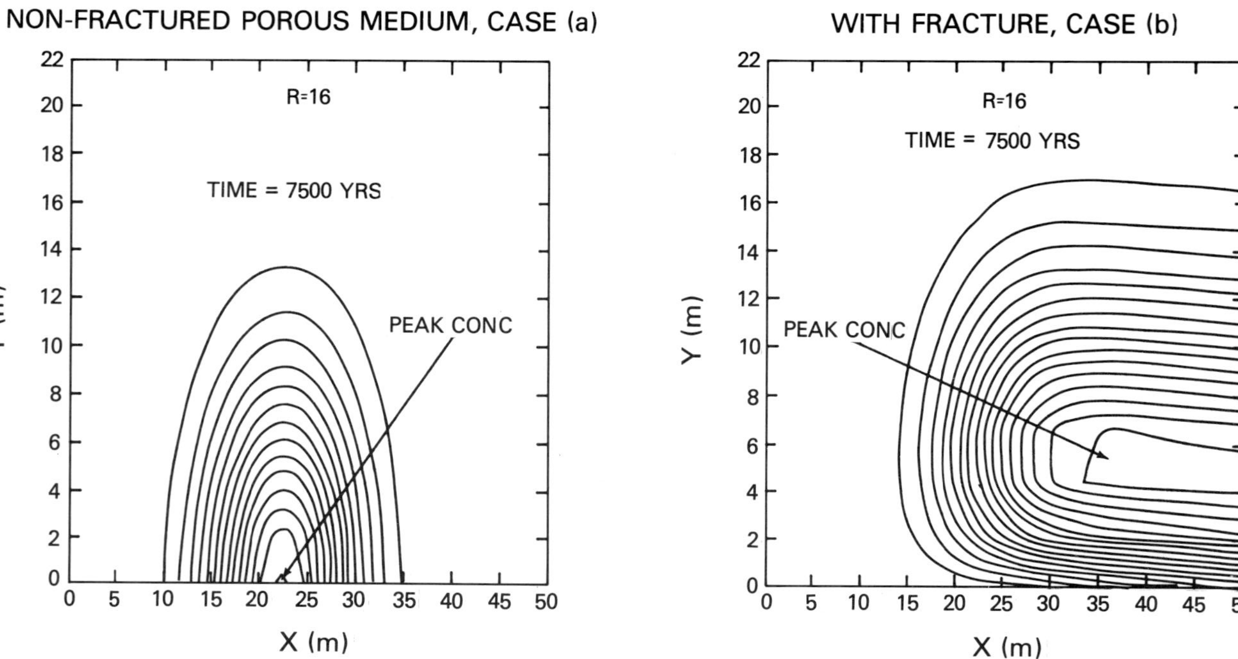

Fig. 2.   Concentration contours (7500 yrs)
2a.   Nonfractured porous medium (Peak concentration ~1 x 10^{-10} kg/m^3)
2b.   Single fracture – case (b) (Peak concentration ~1 x 10^{-13} kg/m^3).

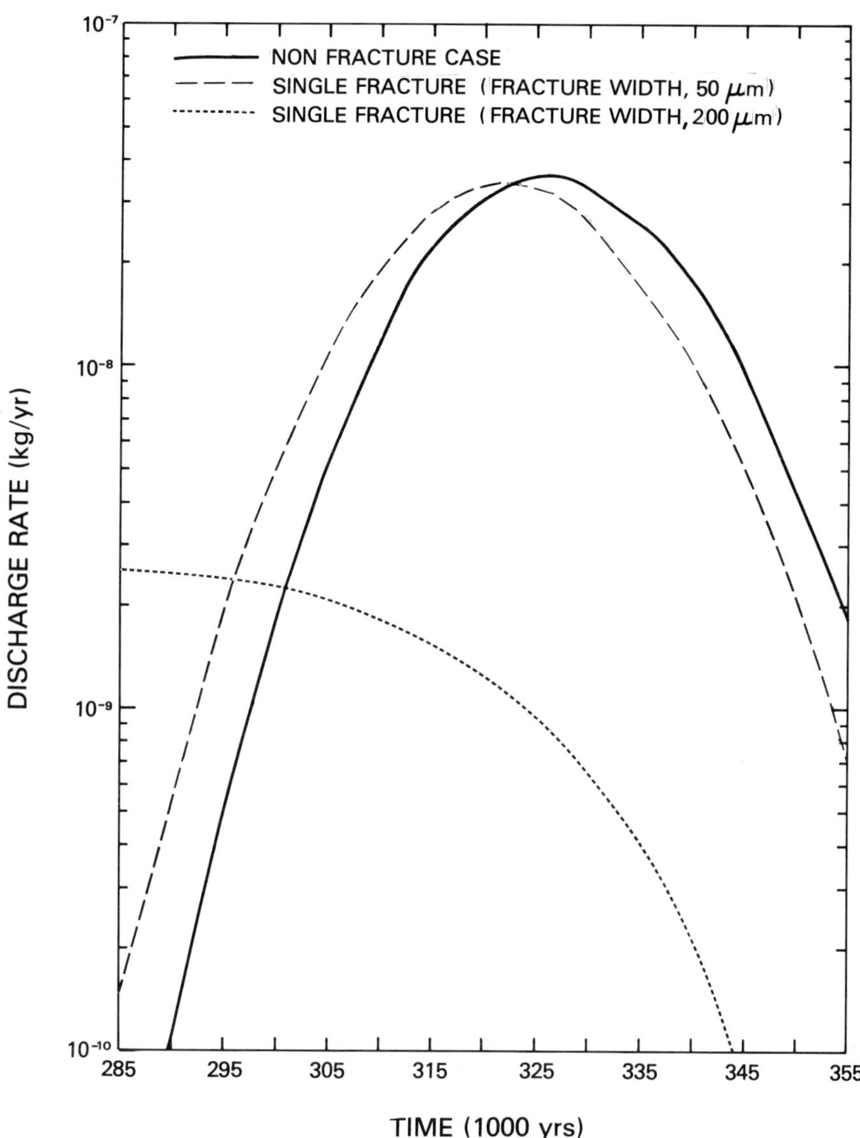

Fig. 3.   Discharge rate versus time at 1 km.

830

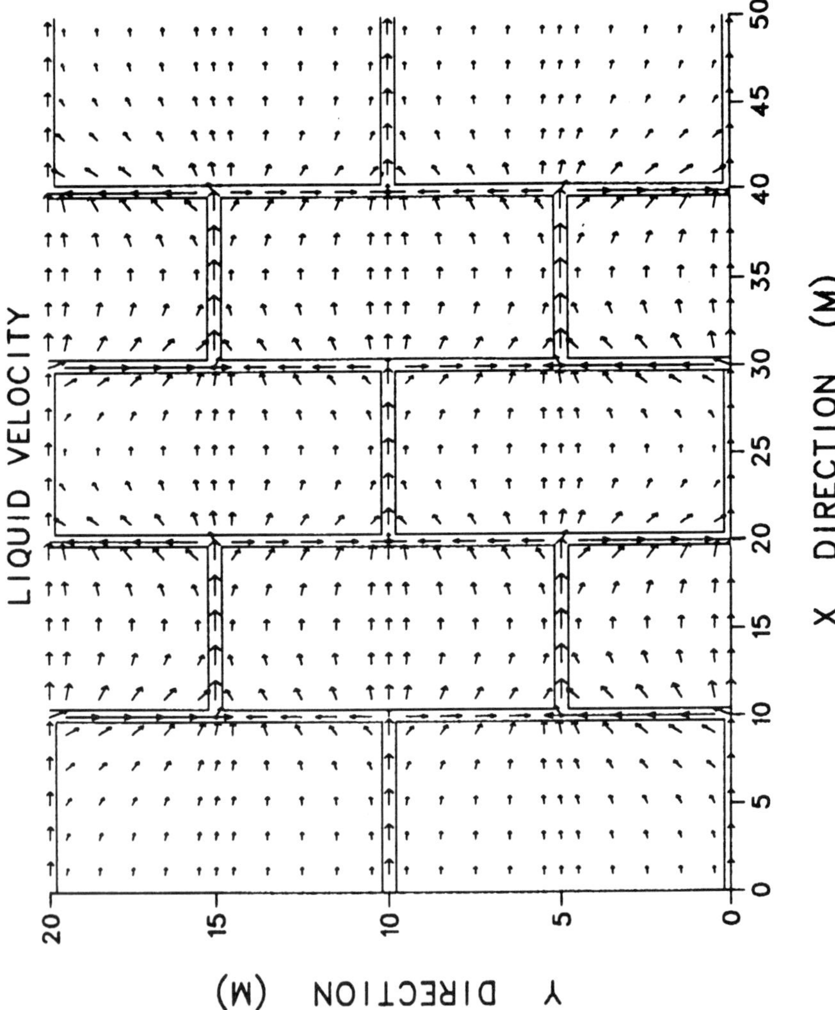

Fig. 4.   Two-dimensional velocity field for the multi-fracture case.
Fracture width of 200 μm has been exaggerated in the figure.
Velocities in the porous matrix have been multiplied by 3×10^5.
Maximum velocity in fractures is $\approx 9.7 \times 10^3$ m/yr.

THE EFFECT OF MICROBIAL ACTIVITY ON THE CONTAINMENT OF RADIOACTIVE WASTE IN A DEEP GEOLOGICAL REPOSITORY

JULIA M. WEST, IAN G. McKINLEY AND NEIL A. CHAPMAN
Environmental Protection Unit, Institute of Geological Sciences, Building 151, Harwell Laboratory, Oxfordshire. OX11 ORA, United Kingdom.

INTRODUCTION

Much effort is currently centred on the construction of elaborate computer programs to model the release to the biosphere of radionuclides from proposed high level radioactive waste repositories in deep geological formations. Although the lack or poor quality of background data is often emphasised, it is generally considered that by examining a wide range of possible values of the important parameters involved, likely processes are assessed or their omission and possible relevence is acknowledged. However, one factor that has been almost totally ignored is the presence of micro-organisms in deep geological formations. The common assumption is that the biosphere is limited to the earth's surface, and soil to a depth of a few metres. Recent research has shown, however, that viable micro-organisms can inhabit deep groundwaters and that the biosphere can extend to depths of at least 5 km. Examples of organisms tolerant to extreme environments are given in Table 1 although ranges are likely to be conservative due to the difficulties of culturing these organisms, which tend to have exotic nutritional requirements, and sampling them in extremely hostile environments.

Limited studies of the microbial populations in mine waters and oil wells show that micro-organisms thrive at depth and influence industrial processes (by the production of acid mine waters and the corrosion of well casings and by aiding the secondary recovery of oil from wells).

A repository built at a depth of <5 km (in fact maximum proposed depths vary only between 200-1000 m) may be sited in an area containing a significant microbial population and would also be contaminated by microbes introduced by the process of excavation and waste emplacement. Thus the possible significance and consequences of microbial activity on the performance of a waste disposal system need to be examined closley. For example, microbes alter the stability of materials and a literature survey[1] has identified species which could attack both waste forms and materials used in repository construction; concrete, metals, glass, bitumen and organic materials such as ion-exchange

TABLE 1

TOLERANCE OF MICROBES TO EXTREME ENVIORNMENTS (after West et al)

ENVIRONMENTAL CONDITIONS		EXAMPLES OF RESISTANT ORGANISMS	LIMIT OF GROWTH
High Temperature	(Thermophiles)	Bacillus stearothermophilus	$>\sim$ $80^{\circ}C$
		Sulphate reducing bacteria	$\gtrsim$ $140^{\circ}C$ (at 10^3 atm)
Low Temperature	(Psychrophiles)	Sporotrichum carnis	$<$ $-20^{\circ}C$
High pH	(Basophiles	Nitrobacter sp, Nitrosomanas sp	$>$ 13
Low pH	(Acidophiles)	Thiobacillus ferrooxidans	$\sim$ 0
High Salinity	(Halophiles)	Halobacterium halobium	$\gtrsim$ 50% salt by wgt
Low Salinity	(Non-Halophiles)	Salmonella oranienburg	$\lesssim$ 70 ppb dissolved salt
High Pressure	(Barophiles)	Vibro desulphuricans	$\gtrsim$ 180 MPa
		(Desulfovibrio desulfruricans)	
High Radiation		Micrococcus radiodurans	Single dose 5 x 10^5 rad
		Saccharomyces cerevisiae	Single dose 10^5 rad

NB: The resistant organisms listed are those characterised for a particular resistance and do not necessarily show the extreme tolerance shown for a particular environmental condition.

resins. Radionuclide migration rates could also be influenced by a microbial
presence. Mobile microbes may transport, either internally or externally,
'sorbed' ionic or particle bound radionuclides while micro-organisms may cover
the surfaces of rocks thus decreasing their retardation of groundwater transported species. In addition, the presence of micro-organisms can affect
groundwater chemistry by altering pH and E_h, catalysing specific redox reactions
chemically altering mineral surfaces and by introducing labile organic by-
products. Such changes in groundwater chemistry would alter radionuclide
solubility, speciation and sorption in a complex manner.

This paper describes preliminary experiments designed to study one of these
factors namely the effect of ambient microbial populations on radionuclide
migration.

METHODS AND MATERIALS

A standard batch reaction technique developed to study radionuclide sorption
on naturally disaggregated rock material has been modified for these experiments.

In the standard technique rock samples are equilibrated with groundwater
spiked with γ-emitting radionuclides for a set time under conditions of
controlled temperature, atmosphere and intensity of agitation. Radionuclide
partitioning between solid and liquid phases is evaluated, after centrifugation,
by measurement of the change in activity of the solution phase as evaluated by
NaI (Tl) and Ge (Li) γ-spectrometry[2,3,4].

In the modified technique rock, micro-organisms or a mixture of rock and
micro-organisms were equilibrated with spiked groundwater or spiked distilled
water in aerobic conditions and at three temperatures, 5°C, 25°C and 40°C.
Radionuclides used were ^{139}Ce and ^{137}Cs. An inherent difficulty in such work
is the variation in microbial numbers during each experiment and the fact that
sorption could occur onto viable, non-viable and dead micro-organisms. Solid
and liquid phases were separated by filtration in this study as it was found
that micro-organisms (whether living or dead) readily stayed in suspension even
after extensive centrifugation. Biomass at the end of each experiment was
simply measured by weighing dried filters before and after filtration and this
includes the mass of all viable and non-viable organisms plus any associated
organic or inorganic material.

In addition culture experiments, using serial dilutions of groundwater mixed
with nutrient agar, were incubated at the same temperatures and for the same
length of time.

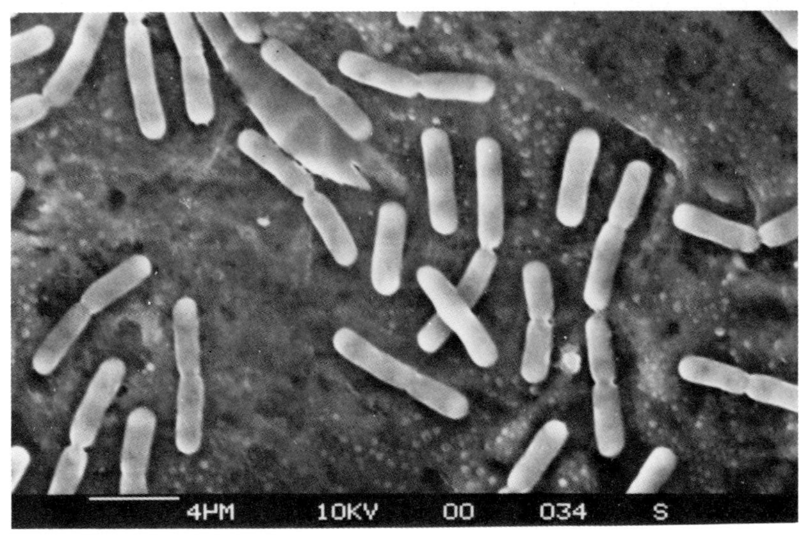

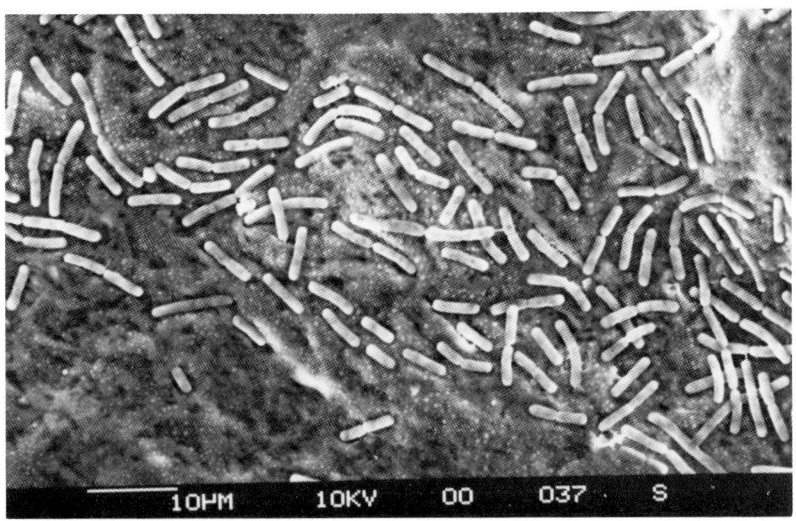

FIGS. 1. and 2. 'Rod' shaped bacteria cultured from Uffington groundwater at 25°C in aerobic conditions. (Scanning Electron Microscope (SEM) photographs)

The rock phase used in these experiments was a Lower Greensand from a lysimetry site at Uffington, Oxfordshire, U.K., which has been characterised in detail elsewhere.[5] Groundwater used was streamwater from the same area from which the micro-organisms and organic material were isolated. The micro-organisms involved included desmids, unicellular and filamentous species from the Cyanophyta and bacteria (Figs. 1 and 2).

RESULTS AND DISCUSSIONS

Results of the experiments are expressed as partition coefficients which are the ratio of the nuclide concentration on the solid phase (rock and/or microbe) to that in the aqueous filtrate. Data from the sorption experiments are shown in Table 2 and show some interesting trends. First, it is clear that microbes do indeed act as quite strongly sorbing media. Sorption of Ce and Cs onto organisms alone (Table 2a) shows an increase in partition coefficients with decreasing (initial) concentrations of nuclide in solution suggesting that saturation effects occur. At low concentrations the radionuclide may be sorbed with high efficiency onto (or into) a limited number of sites on the micro-organisms which saturate at higher concentrations when sorption is dominated by large numbers of relatively low affinity sites. Processes occuring appear to be temperature dependent, the highest partition coefficients for a given nuclide concentration at 40°C for both nuclides. This is probably due to increasing population of viable microbes as culturing experiments revealed that the heterotrophs present in the water grew best at this temperature (5 colonies/ml at 5°C, 120 colonies/ml at 25°C and 900 colonies/ml at 40°C). It must be emphasised, however, that the measured weight of biomass did not change significantly during these experiments and thus the viable component must be a small percentage of the total biomass. If viable organisms are, however, responsible for much of the total sorption onto biomass, their increase would cause a net increase in the total effective partition coefficient. Alternatively, the increase in partition coefficient could be due to increase in reaction rate, as kinetic studies have not been completed, but the con-centration dependence of this increase tends to make this explanation less likely.

Sorption onto rock only (Table 2b) using distilled water plus spike instead of groundwater again shows the partition coefficients to increase with de-creasing initial nuclide concentration in solution which again is probably due to saturation effects. The lower partition coefficients onto rock relative to microbes indicate that per unit weight less sites for radionuclide uptake are

TABLE 2

PARTITION COEFFICIENTS (C_R/C_W) FOR [139]Ce AND ^{137}Cs FOR THE THREE
EXPERIMENTAL SYSTEMS

INITIAL CONCENTRATION IN SOLUTION (M)	TEMPERATURE		
	5°C	25°C	40°C
a. Sorption onto micro-organisms			
[139]Ce			
1.423×10^{-5}	4.82×10^{3}	9.19×10^{3}	4.9×10^{3}
1.423×10^{-6}	2.8×10^{5}	7.05×10^{5}	5.01×10^{5}
1.423×10^{-7}	3.95×10^{6}	5.5×10^{6}	9.02×10^{6}
[137]Cs			
1.504×10^{-5}	1.03×10^{3}	1.59×10^{3}	1.64×10^{3}
1.504×10^{-6}	5.42×10^{4}	5.38×10^{4}	5.38×10^{4}
1.504×10^{-7}	5.98×10^{5}	5.26×10^{5}	7.03×10^{5}
b. Sorption onto rock			
[139]Ce			
1.423×10^{-5}	6.2	$1.43 \times 10_{2}$	$1.7 \times 10_{2}$
1.423×10^{-6}	$9.71 \times 10_{3}$	1.09×10^{2}	$1.48 \times 10_{3}$
1.423×10^{-7}	1.45×10^{3}	9.93×10^{2}	1.06×10^{3}
[137]Cs			
1.504×10^{-5}	3.37	7.44	$1.81 \times 10_{2}$
1.504×10^{-6}	1.94×10^{2}	3.87×10^{2}	$6.66 \times 10_{3}$
1.504×10^{-7}	3.39×10^{3}	1.43×10^{3}	5.34×10^{3}
c. Sorption onto rock and organisms			
[139]Ce			
1.423×10^{-5}	$3.13 \times 10_{2}$	$3.53 \times 10_{2}$	$6.81 \times 10_{2}$
1.423×10^{-6}	8.18×10^{2}	5.15×10^{2}	$5.79 \times 10_{3}$
1.423×10^{-7}	4.44×10^{3}	2.51×10^{3}	4.36×10^{3}
[137]Cs			
1.504×10^{-5}	4.62	2.29×10	$2.0 \times 10_{2}$
1.504×10^{-6}	1.41×10^{2}	$8.73 \times 10_{3}$	$8.73 \times 10_{3}$
1.504×10^{-7}	1.26×10^{4}	3.91×10^{3}	4.46×10^{3}

available on the rock surface than are present on the surfaces of micro-organisms. This is probably due to a reduction in the weight to surface area ratio. Sorption of Ce appears to show little temperature dependency at any concentration, all results for the varying concentrations remaining stable throughout the temperature ranges. Cs sorption, however, is considerably increased with increasing temperature, the relative increase in partition coefficient from 5 to 40°C being even greater than was observed for the sorption onto micro-organisms.

Comparing partition coefficients from the rock sorption experiment with those from the combined rock plus organism experiment (Table 2b and c) it is apparent that sorption of Ce increases at all concentrations and temperatures in the presence of micro-organisms. The increase is approximately four fold independent of concentration or temperature. This would suggest that a physical sorption process onto the micro-organisms is occuring which is relatively insensitive to changes in these parameters. However, it also suggests that the microbial population remains static throughout the experiment. which contradicts the culturing results and microbial sorption experiments. Results for Cs are more complicated. The partition coefficient varies in the presence of microbes from a four fold increase at a concentration of 1.5×10^{-7} M at 5°C to no significant increase at 1.5×10^{-6} M at 40°C. A complex process is occuring , which may be wholly or partially metabolic with temperature and concentration playing an important role. However, the major drawback of this explanation is that the metabolic rate should be at its peak at 40°C (maximum growth) and hence the increase in partition coefficients should be at its highest, but this is not observed. In fact at 40°C the overall increases are the lowest of all the temperatures and concentrations measured.

CONCLUSIONS

The sorption experiments described above were intended as part of a pilot study and are consequently of a preliminary nature and difficult to interpret. However, they have shown very clearly that micro-organisms play a significant quantitative role in the uptake and transport of radionuclides and that the topic requires serious consideration.

Several features have emerged that point to further studies and technique development. For example, sorption data are presented here in terms of the total 'biomass' in solution. Techniques are required for the separate quantification of each part of the biomass (viable, non-viable and dead organisms and organic and associated inorganic material) in both solution or on the rock phase. A realistic explanation of sorption processes where micro-organisms are involved is lacking at present and is likely to be very complex. Eventually partition coefficients could be calculated for each individual biogenic component. More data is required on the identification and quantification of microbial populations in relevant deep formations. Problems arise in such analysis due to the unusual metabolisms of such organisms and very slow growth in culture (especially in the case of autotrophs). Information on the bioenergetics of these organisms and possible sources and inputs of energy into deep geological environments may allow the constraints of the population to be calculated and assessed. With such data the in-situ effect of micro-organisms on the containment of radioactive waste in a deep geological repository could be examined realistically.

ACKNOWLEDGEMENTS

This study was funded by the Natural Environment Research Council as part of a special project on radionuclides in the environment and is published by permission of the Director of the Institute of Geological Sciences (N.E.R.C.).

REFERENCES
1. West, J.M., McKinley, I.G., and Chapman, N.A. (1982) Radioactive Waste Management, (in press).
2. McKinley, I.G. and Greenwood, P.B. (1980) Rep. Inst. Geol. Sci. ENPU 80-7.
3. McKinley, I.G., and West, J.M. (1981) Rep. Inst. Geol. Sci. ENPU 81-6.
4. McKinley, I.G., and West, J.M. (1981) Rep. Inst. Geol. Sci. ENPU 81-14 (in press).
5. Ross, C.A.M. (1980) Q.J. Eng. Geol. London. $\underline{13}$, 177-187.

Published 1982 by Elsevier Science Publishing Co
SCIENTIFIC BASIS FOR RADIOACTIVE WASTE MANAGEMENT - V
Werner.Lutze, editor

INTRACOIN - AN INTERNATIONAL NUCLIDE TRANSPORT CODE INTERCOMPARISON STUDY

Secretariate : K. Andersson, Swedish Nuclear Power Inspectorate, Box 27 106, S-102 52 Stockholm, B. Grundfelt, Kemakta Konsult AB, Luntmakargatan 94, S-113 51 Stockholm, and J. Hadermann, Swiss Federal Institute for Reactor Research, CH-5303 Würenlingen.

INTRODUCTION

The importance of safe final disposal of high-level wastes has led to extensive research and development activities in many countries. In addition to the need of adequate data on the characteristics of repository sites, appropriate tools for the evaluation of safety of the disposal system are necessary. Mathematical models are used in describing the mechanisms involved in the nuclide transport from the repository to the biosphere as an essential part of safety assessments.

To improve the understanding of various strategies for geospheric radionuclide transport modelling an international cooperation project with ten participating organisations has been set up. In the project - INTRACOIN (International Nuclide TRAnsport COde INtercomparison) study - a comparison between models is made on three levels with increasing complexity. Calculations are performed with some 20 computer codes representing different modelling strategies.

The INTRACOIN study is limited in its scope to far-field radionuclide geosphere transport models. Thus, near-field models, groundwater hydraulics and models for transport in the biosphere are not included. However, since the geological barrier is an important part of the safety system the study should contribute to the understanding of important phenomena related to long-term safety assessments. In addition it provides a forum for discussion and for direct exchange of experience between experts in the field.

MODELLING STRATEGIES

The various mathematical models of geospheric nuclide migration differ mainly in two aspects. They may model different phenomena through differing equations and they may use various algorithms solving these equations. Both respects have to be considered in model comparisons.

Among the growing number of transport models now becoming available there

840

is a clear tendency for the models to become more complex by including more
and more physico-chemical phenomena. The first generation models include such
phenomena as chain decay, hydrodynamic dispersion and retardation given by a
linear sorption isotherm describing equilibrium between the liquid and the
solid phase. These models are typically one-dimensional and based on analytical solutions of the governing set of differential equations assuming transport in a homogeneous geologic medium.

To overcome these constraints, computer codes based on various numerical
algorithms have been developed. These second generation models have a greater
flexibility with respect to boundary conditions and parameter variability both
in space and time domains. In addition to the natural extension to two or three
spacial dimensions, developments concentrate on description of diffusion from
flowing water into stagnant regions (matrix diffusion), non-linear sorption
equilibrium, sorption kinetics and multispeciation of radionuclide in liquid
and solid phase. Furthermore, there are integrated models which combine contaminant and heat transport with hydraulics calculations. Here, in the INTRACOIN
study, only the nuclide transport part is to be compared between the models.

TECHNICAL REALISATION OF THE STUDY

In order to guarantee an optimal foundation when comparing the model strategies, comparisons are performed on three levels aimed at examining

    1. the numerical accuracy of the codes compared,

    2. the capabilities of models and codes to describe in-situ measurements,

    3. the quantitative impact of choosing either modelling strategy on the nu-
       clide transport calculations in a typical repository scenario assessment.

At the _first level_ the numerical accuracies of the codes should be established to provide a basis for understanding deviations at comparison levels
two and three. This is achieved by comparing computations with results from
analytical solutions in such cases where this is possible, as well as by inter-
comparing results from codes using different numerical methods to solve the
same problem. To this end seven benchmark cases with additional parameter variations have been defined. A more detailed description is given in section 4.

At _level two_ a careful choice of in-situ tracer tests is made to serve as
a basis for the comparisons. Two types of field experiments have been chosen,
one sample case for transport in a fractured medium and one in a porous medium.

For the latter a further highly idealised two-dimensional case is defined as an intermediate step. This takes care of the fact that the interests of the participating countries are divided between these two types of media and that some of the codes are especially designed for particular situations. The computations are to be performed in two steps. In a first one hydraulics and physico-chemical processes are predetermined. In a second step such conditions are relaxed and information is obtained on possible descriptions concerning water flow fields and reaction mechanisms.

Level three of the comparison is intended to demonstrate the influence of the various physico-chemical effects included in the different models on the results from a repository simulation. A working group has been formed to evaluate candidate sites for such an intercomparison on large scales. As for level two, one representative case for fractured and one case for porous media are foreseen.

LEVEL 1 BENCHMARK CASES

As has been stated in section 3, the level 1 benchmark cases have been defined with the aim to compare code accuracies and solution algorithms. The cases are divided into seven groups corresponding to the different model assumptions :

1. One-dimensional advection-dispersion with chain decay, constant migration parameters (ground water velocity, retention factors and dispersivity) and constant leach rate.
2. One-dimensional advection-dispersion in a layered medium (piece-wise constant migration parameters).
3. One-dimensional advection-dispersion with continuously varying migration parameters.
4. Two-dimensional advection-dispersion with constant retention factors and dispersivity.
   a) Parallel velocity field and radial dispersion.
   b) Two-dimensional velocity field and radial dispersion.
5. One-dimensional advection-dispersion with diffusion into the rock matrix.
6. Two-dimensional advection-dispersion with matrix diffusion.
   a) Parallel velocity field and radial dispersion.
   b) Two-dimensional velocity field and radial dispersion.

7. One-dimensional advection-dispersion with linear mass transfer kinetics,
   chain decay and constant migration parameters.

Case 1 was defined in order to have a common basis of comparison for all the
codes. This case includes several parameter variations, such as two different
radionuclide inventories (three member chains), leach durations, sets of
retention factors, migration distances and three different dispersivity values.
This benchmark case is calculated by all participants.

As an example we show in figures 1 and 2 a comparison of selected preliminary results for level 1 case 1 benchmarks. The pertinent parameters are given
in table 1. The figures show the outlet concentration as a function of time for
four of the participating codes. Beyond that, comparisons in tabular form of
peak concentrations, times of peak occurence and times to reach half the peak
concentration will be made. Computer requirements in terms of CPU seconds
relative to computer speed at the installations used will also be compared.

Table 1.   Parameter values for sample cases.

Leach duration : 10^5 yr

Inflow boundary condition :

    For the models MMT1D, RANCH and COLUMN : fixed concentration at the inflow

    For the model SWIFT : fixed injection rate

Migration length :   500 m

Interstitial water velocity : 1 m/yr

Peclet number :   10

Retention factors :

Element	R
Cm	5 000
Np	700
U	300
Th	20 000
Ra	10 000

Inventory at t=0 :

    Long-lived parent nuclide (Fig. 1)

Nuclide	$t_{1/2}$ (yr)	I
U-234	$2.445 \cdot 10^5$	1.000
Th-230	$7.7 \cdot 10^4$	0.010
Ra-226	$1.6 \cdot 10^3$	0.004

Short-lived parent nuclide (Fig. 2)

Nuclide	$t_{1/2}$ (yr)	I
Cm-245	$8.500 \cdot 10^{3}$	0.700
Np-237	$2.14 \cdot 10^{6}$	1.000
U-233	$1.592 \cdot 10^{5}$	0.004

PROJECT ORGANISATION AND TIME SCHEDULE

The organisation of the INTRACOIN study is regulated in an agreement between the participating organisations (Parties). According to the agreement the study is directed by a coordinating group consisting of one representative from each Party. The Swedish Nuclear Power Inspectorate (SKI) acts as managing participant. Each Party organises one or several project teams which performs the calculations defined by the coordinating group and formulated in detail by the project secretariate.

The project secretariate has been set up by SKI in cooperation with the Swiss organisation NAGRA. The Swedish consulting company KEMAKTA is acting as principal investigator for the study and performs the technical work within the project secretariate in cooperation with the Swiss Federal Institute for Reactor Research (EIR).

Each Party covers the costs for its participation in the study and is re-ponsible for organising the budget for its own project team or teams. The administrative effort for the project secretariate is covered by SKI while the cost for the technical and scientific work within the secretariate is covered by SKI and NAGRA according to a special agreement.

Table 3 gives a list of parties and project teams participating in the INTRACOIN study. Table 4 shows a list of codes used by the project teams. A short characterisation of the codes is given in the INTRACOIN Progress Report No 2 (November 1981 to February 1982).

The INTRACOIN study is planned to have a duration of approximately two years. Progress reports are issued which describe the advancement of the study. After conclusion of each level results of the calculations are compiled and evaluated by the project secretariate in cooperation with the project teams. A final report at the end of the study will summarise these intermediate results and present the conclusions.

844

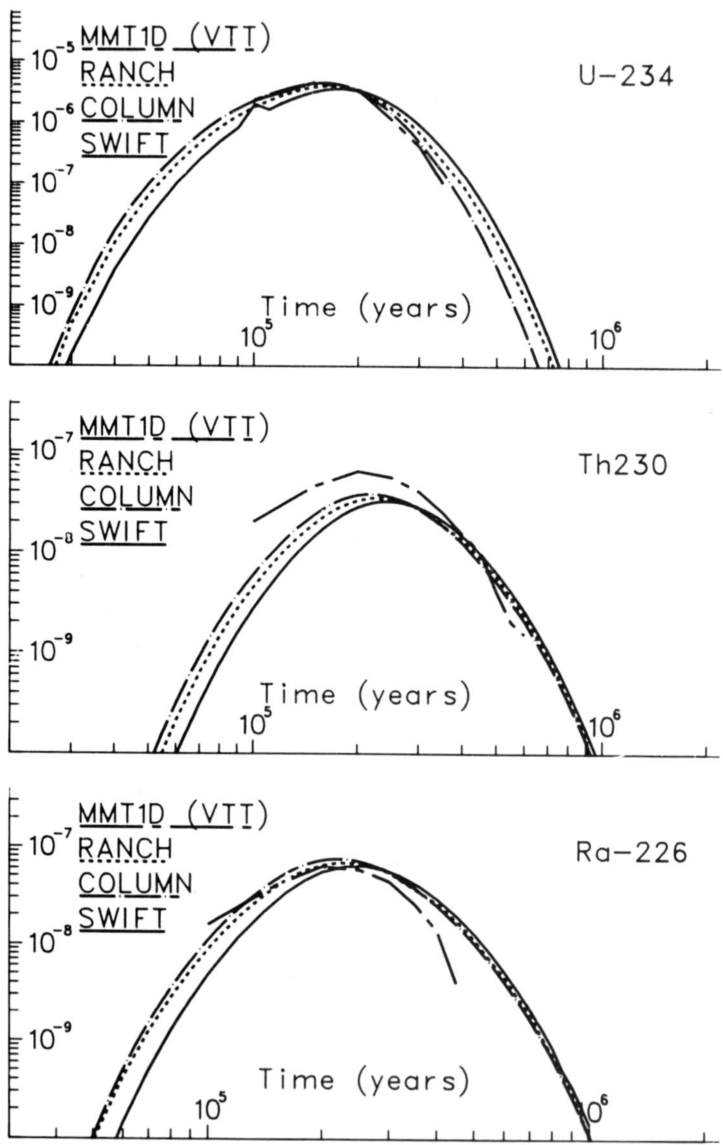

Fig. 1 : Comparison of preliminary results from the codes indicated for level 1 case 1 benchmarks. Plotted are arbitrary activity concentrations as a function of time for the U-234 chain.

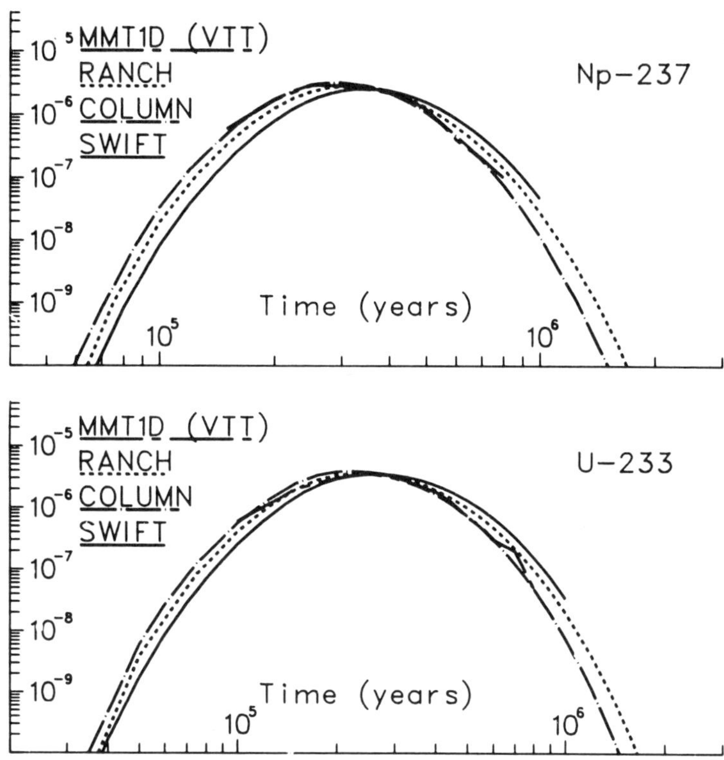

Fig. 2 : Comparison of preliminary results from the codes indicated for level 1 case 1 benchmarks. Plotted are arbitrary activity concentrations as a function of time for the Cm-245 chain.

Party	Project teams
Atomic Energy of Canada Ltd (AECL)	Whiteshell Nuclear Research Establishment (WNRE)
Commissariat à l'Engergie Atomique/ Institut de Protection et de Sûreté Nucléaire (CEA/IPSN)	Ecole Nationale Supérieure des Mines de Paris (ENSM)
Nationale Genossenschaft für die Lagerung radioaktiver Abfälle (NAGRA)	Swiss Federal Institut for Reactor Research (EIR) Polydynamics Ltd.
National Radiological Protection Board (NRPB)	NRPB
Project Sicherheitsstudien Entsorgung (PSE)	Technische Universität Berlin (TUB)
Swedish Nuclear Fuel Supply Co (SKBF/KBS)	Royal Institute of Technology (KTH)
Swedish Nuclear Power Inspectorate (SKI)	Kemakta Konsult AB
Technical Research Centre of Finland (VTT)	VTT
U.K. Atomic Energy Authority/Atomic Energy Research Establishment (UKAEA/AERE)	AERE
U.S. Department of Energy (U.S. DOE)	Office of Nuclear Waste Isolation (ONWI) Battelle Pacific Northwest Laboratory (PNL) Intera Environmental Consultans Inc. University of California, Berkeley (UCB)

1)

1) It was decided at the first coordinating group meeting to leave a place open for participation of U.S. Nuclear Regulatory Commission (NRC).

Tab. 3 : Parties and project teams participating in the INTRACOIN study.

Party	Code	Code Developer
AECL	DRAMA	WNRE, Pinawa
	GARD2S	WNRE, Pinawa
	TRANSAT	National Hydrological Research Institute, Ottawa
CEA/IPSN	METIS	ENSM, Paris
NAGRA	CONZRA/RANCH	EIR, Würenlingen
	TROUGH	Polydynamics, Zürich
NRPB	GEOS	NRPB, Harwell
PSE	SWIFT	Intera, Houston
SKBF/KBS	TRUCHN	KTH, Stockholm
SKI	COLUMN	Risø Laboratories, Roskilde
VTT	GETOUT	PNL, Richland
	MMT1D	PNL, Richland
UKAEA/AERE	DIFFUSE	AERE, Harwell
	NAMTAR	AERE, Harwell
USDOE	FRANS	Intera, Houston
	GETOUT	PNL, Richland
	MMT	PNL, Richland
	SWENT	Intera, Houston
	UCB-NE-X[1]	UCB, Berkeley

1) UCB-NE-X denotes a number of analytical codes developed at UCB.

Tab. 4 : Codes used in INTRACOIN study

REFERENCES

INTRACOIN Progress Report No 1 (June - October 1981), Stockholm, 1981.
INTRACOIN Progress Report No 2 (November 1981 - January 1982), Stockholm, 1982.

IN-SITU SOLIDIFICATION OF LOW -
AND MEDIUM-LEVEL RADIOACTIVE WASTE

IN-SITU SOLIDIFCATION OF LOW AND MEDIUM LEVEL WASTES

R. KRAEMER and R. KROEBEL
Kernforschungszentrum Karlsruhe GmbH., Projekt Wiederaufarbeitung und
Abfallbehandlung, Postfach 3640, D-7500 Karlsruhe, Federal Republic of Germany.

1. INTRODUCTION

Since 1976, a German R & D project has been carried out to find an alternative concept for the treatment and disposal of MLW and LLW arisings generated
mainly in the planned German reprocessing plant and other nuclear facilities
(LWR, fuel fabrication, R & D establishments).

The main feature of this concept is an in-situ solidification of precondi-
tioned waste granules in large salt caverns located in the deep geological
underground, thus avoiding such non-radioactive ballast as lost concrete
shielding and container material.

The necessary R & D work is sponsored by the Federal Ministry for Research
and Technology and performed in three phases, which may be characterized as
follows

Phase 1: 1976 - 1978 Feasibility study and definition of a reference system /1/.

Phase 2: 1978 - 1981 Experimental demonstration of relevant components on a
semi-industrial, non-radioactive scale.

Phase 3: 1981 - 1984 Implementation of a large scale test involving the production and direct filling of 1000 m^3 reference waste
product from above ground into the 10,000 m^3 prototype
cavern located 1000 m below ground in the ASSE salt mine.

This survey paper and accompanying papers /2, 3, 4, 5/ describe recent
R & D results obtained through Phase 2, which caused the sponsor to grant
further support to the project in the third and final phase.

2. DESCRIPTION OF THE IN-SITU CONCEPT

The in-situ concept as developed was selected among ten different alternative proposals on the different possibilities of waste forms, transportation
techniques and filling procedures. The chief selection criteria were
simplicity of the concept as well as operational safety, health and safety,
and economic benefit.

850

The main process steps of the in-situ concept are summarized as follows and illustrated in Figure 1.

(1) Prefabrication of granules at the waste producer's site from MLW and LLW arisings and cement. The granules (grain size: 0.3 - 5 mm) are kept in an interim storage facility for a setting time of approx. one week.

(2) Transportation of the granules in special shielded containers from the waste producer's site to the final repository.

(3) Unloading of granules at the final repository site and mixing of the granules (60 vol.%) with cement grout (40 vol.%), using tritiated waste water from the reprocessing plant, if the occasion arises (i.e., H3-waste water transport is licensed).

(4) Vertical transport by gravity of the reference mixture from above ground into a salt cavern of 75,000 m^3 located at 900 - 1000 m depth through a relatively thin pipeline, which keeps the flow speed below 0.5 m/s to reduce erosion damage.

(5) In-situ solidification of the concrete-like waste form in layers, each of which represents one filling campaign, The open cavern space is finally completely back-filled by a large monolith consisting of granules embedded in solidified grout.

The underground cavern system is illustrated in Figure 2, which shows three caverns in different states, i.e., construction, operation and the final re-filled and sealed position.

The heat of hydration released during in-situ solidification of the cementitious waste product was identified as a substantial problem, which led to major consequences with respect to the selected waste form.

In order to avoid excessive product temperature rise, i.e., above the boiling point, it was decided to release a considerable part of the heat of hydration already above ground by mixing prefabricated granules in the grout. The first of the papers related to the same subject /2/ summarizes the computations of the temperature profiles in the waste product as well as in the surrounding rock salt.

On the other hand, this pretreatment of liquid wastes (MLW and LLW) in granules offers a number of advantages, such as:

- Transport of liquid radioactive wastes are avoided where the waste producer's site is some distance away from the final disposal site.

- Reworking batches is possible in case unknown chemicals in the waste solutions prevent the cement granules from hardening.

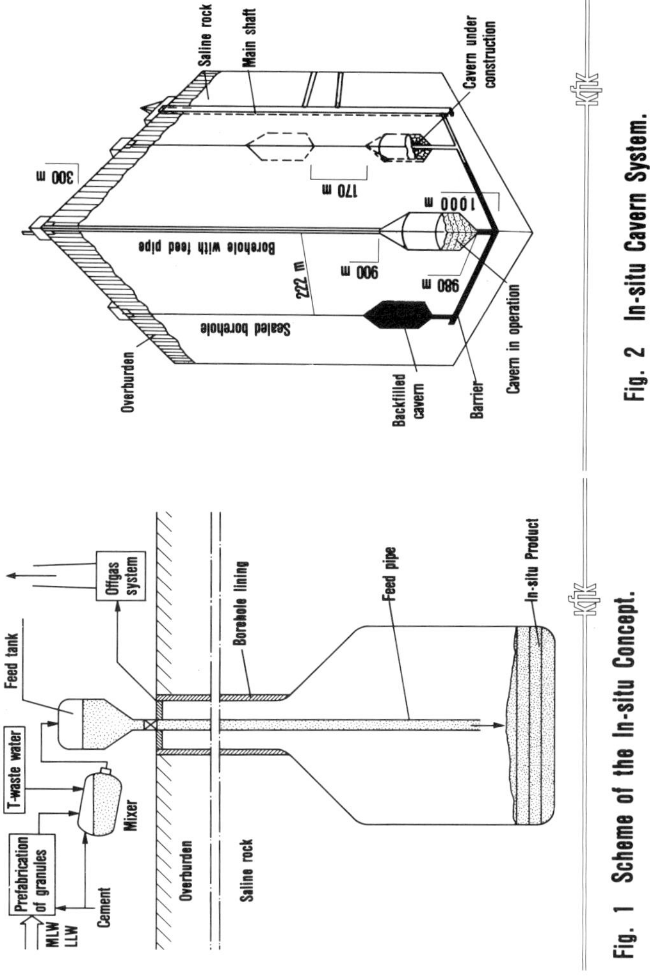

Fig. 2 In-situ Cavern System.

Fig. 1 Scheme of the In-situ Concept.

- Quality control of the waste-bearing solids (the granules) is fairly easy, in contrast to a grout system made up of MLW solution and cement, which sets only underground.

3. WASTE SPECIFICATIONS

The waste arisings which can be disposed of by the in-situ concept cover a great variety with respect to physical state, origin and activity concentration. The main criteria are as follows:

(i)   The waste must be compatible with a cement matrix.

(ii) The activity level is restricted due to the evolution of radiolytic gases.

(iii) The waste must be able to be transformed into granules, thus forming with the grout a medium that is pourable through a relatively thin pipeline (50 - 80 mm diameter). The granule size is to be between 0.3 mm and 5 mm.

The following waste categories, generated mainly in a reprocessing plant and in nuclear power plants, have been investigated:

Medium and Low Level Aqueous Waste Concentrates

This category represents process and decontamination waste solutions with high salt contents (350 g $NaNO_3$/l) and specific activity levels of about 300 - 700 Ci/m^3, which occurs in amounts of up to 5 m^3 per ton heavy metal reprocessed. The main R & D activities during Phase 2 of the in-situ project were focused on the granulation technique for these wastes.

Ion Exchange Resins (grain and powder)

This waste is generated mainly in nuclear power plants and already represents a waste form, which can be mixed directly with grout and disposed of through the in-situ concept.

Ashes from Waste Incineration

The ashes appear primarily as slags, which can be shredded to the proper dimensions of less than 5 mm grain size, which then are compatible with the in-situ concept.

Tritiated Waste Water

The in-situ concept represents an important alternative disposal concept for H 3 waste water which is generated at reprocessing plants in relatively large specific volumes of about 2 m^3 per ton heavy metal reprocessed. This waste water is used in preparing the filler suspension in the reference waste form. As the waste receiving cavern is not man-operated, contamination of the cavern atmosphere can be tolerated even to high activity levels.

Iodine Absorber Material

Loaded iodine absorbers in the form of silver nitrate impregnated silica grain, as used in nuclear power plants and reprocessing plants, are also compatible with the in-situ concept and may be disposed of in the grout without any pre-treatment.

4. PRECONDITIONING WASTES BY GRANULATION

Developing the granulation technique, which converts liquid raw waste arising into granules as specified, was a major task in the second project phase. The second accompanying paper /3/ outlines in greater detail the technique and the properties of the granules.

The specifications of the granules were defined to meet the following criteria:
- The grain size was to be polydispersive in order to arrive at an optimal volume filling ratio in the cavern.
- The maximum grain size was to be less than 5 mm, which is about 10% of the pipe diameter, in order to guarantee safe operation during vertical transport through the pipeline.
- As the granules represent the primary matrix material for the activity inventory to be disposed of (except H 3 waste water) the leaching rate especially for Cs and Sr, and the radiolytic gas production were to be minimized.
- The granules must be stable enough to withstand specified handling procedures, such as loading and unloading, silo storage and transportation, without losing integrity.

As a result of the relatively low water/cement ratio of 0.15 achievable in the technical granulation process, liquid waste solutions must be concentrated as much as possible to produce salt concentrations in the granules, which are comparable with those in conventional homogeneous cement products. Therefore, the nitrate content in the simulated waste solutions was increased to 500 g/l.

The technological development of the granulation process has been carried up to a non-radioactive pilot plant scale with a daily throughput of 1.6 tons. About 100 t. of specified granules have been produced to specifications and used for various investigations of vertical transport as well as of in-situ solidification.

5. VERTICAL TRANSPORT INTO THE UNDERGROUND CAVERN OF THE REFERENCE MIXTURE

Vertical transport of the concrete-like reference product through a relatively thin pipeline (50 - 60 mm) only by gravity, without any additional pressure, can be considered a unique concept for the disposal of radioactive wastes from above ground in prefabricated salt caverns in the deep geological underground.

Considerable efforts were made to find the proper rheological data, because no parallel application in conventional industry was available. Professor Weber et al. in the third accompanying paper /4/ report on the main results of the experiments performed with an 8.5 m test pipeline at the University of Karlsruhe.

The rheological behavior of the granules/grout mixture provides the major control on product specifications with respect to grain size spectrum, granule fraction, type of cement and additives and water/cement ratio in the grout.

The final scientific goal is a comprehensive rheological theory for this type of pouring medium.

The technological goal of the project is the experimental determination of rheological data necessary for the layout of a reference facility with above-ground installations for mixing and feeding the reference mixture into the vertical pipeline, thus defining the conditions for safely operating a system with an underground vertical flow of 900 - 1000 m. The controlled constant speed of the reference mixture is achieved by counterbalancing gravity and friction along the walls.

The main characteristics determined to date for the in-situ reference waste form, suitable for the treatment of specified MLW/LLW arisings from nuclear fuel cycle facilities with a reference reprocessing capacity of 1400 MTU/a, are summarized in Table 1.

With one vertical pipeline of 64 mm diameter one would need less than 4000 hours per annum to dispose of the reference waste form specified in Table 1. At 24 h/day operating time, one would spend 160 net days of operation per annum, which will be broken down into three filling campaigns.

TABLE 1

REFERENCE IN-SITU WASTE FORM

Reference capacity 15,000 m^3/a		Granules for MLW/LLW	Filler suspension for H3 waste water
Volume ratio	%	60	40
Net volume	m^3/a	9000	6000
Density	t/m^3	2,6	1,6
Mass	t/a	23400	9600
Mass ratio	%	71	29
Additives	t/a	1170 (bentonite)	32 (retarder)
Water/cement ratio	1	0.15	0.5
Wastes	t/a	4000	3200
Salt/cement ratio	%	6 - 9	-

6. IN-SITU SOLIDIFICATION AND WASTE PRODUCT PROPERTIES

The in-situ concept employs a ventilation system for the cavern atmosphere, which does not produce an air exchange rate, but only a relatively low discharge rate equivalent to the inlet flow rate of the reference mixture ($4 - 5$ m^{3}h^{-1}). In-situ solidification can be considered to take place in a quasi-closed system with only very limited exchange with the open air above ground. This design keeps the release of tritiated waste water during setting at a minimum. Calculations have shown, that less than 100 Ci/a HTO will be released from the offgas system resulting in a dose rate of 2.8 E-7 Sv/a which is equivalent to 10^{-3} of the maximum permissible dose rate according to German standards.

The in-situ solidification process and the saline rock/waste interaction has been closely investigated. The radionuclides embedded in the granules can interact with the saline rock and other leaching media only if they are able to penetrate the filler suspension during and after setting.

Tracer experiments have shown that $3 - 5$ % of the Cs and Sr activity may be leached by the filler suspension during transport and setting of the product. No diffusion into the saline rock was to be observed. At the waste and rock salt interface a cement layer of about 5 mm thickness could be identified, which was enriched in chlorides.

In order to evaluate the consequences of accidental brine intrusion during the operational period extensive leaching experiments of reference products have been performed with different kinds of leaching media, i.e., saturated NaCl

brine and quinary brine, as well as demineralized water for reference
purposes. Nuclide specific leaching rates, averaged over a leaching time of
200 days according to ISO draft specifications, are summarized in Table 2.

TABLE 2

MEAN LEACHING RATES OF THE IN-SITU REFERENCE WASTE FORM AFTER 200 DAYS OF
LEACHING TIME

Leaching media	Mean leaching rates (g cm^{-2} d^{-1})		
	Cs-137	Sr-85	H-3
Demineralized water	3.2 E-5	1.1 E-4	2.4 E-3
Saturated NaCl brine	4.8 E-5	1.1 E-4	–
Quinary brine	2.3 E-5	3.1 E-5	–

The results show leaching rates significantly lower (factor 3 - 10) than
those of homogeneous cement products. In addition, the surface-to-volume ratio
of the waste monolith may be expected to be smaller by a factor of 100 than that
of corresponding waste forms in drums.

7. CONSTRUCTION AND STABILITY OF LARGE SALT CAVERS

The stability of a cavern in a salt formation at 1000 m depth and under atmospheric pressure conditions was the subject of close geomechanical investigations. From the view point of safe operation and economical benefit, very large
volumes are aimed at, the only constraints being stability reasons.

In the fourth accompanying paper /5/, W. Raab et al. report on the most
suitable geometries applicable to a system of three caverns with each providing
an open cavern volume of 75,000 m^3 (see Fig. 2). This volume represents a maximum with respect to the stability requirements under the specified conditions.

The conditioned waste arisings from five years of operation of the reference
fuel cycle can be disposed of in one cavern volume. This means that after five
years of operation the open cavern space is completely backfilled with a stable
waste monolith with mechanical properties similar to those of the surrounding
saline rock.

With respect to the most suitable mining techniques to be applied in constructing the caverns, two alternative techniques are discussed, i.e., solution
mining and conventional mining by means of drilling and blasting. Since both
methods have their pros and cons, a decision in favor of either technique cannot
be made until the actual site has been identified.

8. CONCLUSIONS

The in-situ concept as developed represents an alternative disposal concept to the currently demonstrated drum-disposal technique. Only significant advantages with respect to safe operation, health and safety, and economic benefit can justify the R&D efforts already spent since 1976 and to be spent in the third forthcoming project phase.

The drum disposal technique requires much more open cavern volume (up to a factor of 6), which must be stable for a period of operation of at least fifty years. The stability requirements for such a large man-operated mine are considered to be much more stringent than for an anticipated in-situ cavern volume of 75,000 m^3, which will be completely backfilled with a stable waste monolith after only five years. A backfilled in-situ cavern restabilizes the geological strata and stops convergence movements. On the other hand, convergence is a very positive safety factor, for it closes all gaps between the monolith and the salt formation so that no liquid can move along this interface and potentially carry activity. A filled and sealed cavern needs no more surveillance.

Even during the time of operation, water intrusions do not pose major problems. Calculations of the activity releases into groundwater tables with contaminated brine solutions from the flooded in-situ cavern did not show groundwater contamination above MPC levels beyond a distance of 300 m from the borehole.

As a further remedial action it is feasible, after a water inrush, to overcast the waste product in the cavern with a layer of insoluble matter, such as bentonite, which will result in a very long term diffusion process for the underlying activity. This layer further reduces groundwater contamination by several orders of magnitude in case of this very improbable accident.

A similar approach in a man-operated disposal mine is, at a minimum much more complicated, if not impossible. Application of the drum disposal technique always leaves some free space between and above the drum layers, which cannot be completely backfilled. Hence, to arrive at a safety level equivalent to that of the in-situ cavern, the underground disposal chambers must be sealed by dams of great dimensions, which is only possible in the post-operational phase for the respective chamber.

Furthermore, some nuclides that have the potential to contaminate the mine ventilation air such as tritium, carbon-14 and Kr-85 bearing wastes, can be disposed of. Ventilation of the cavern during or after operation is not necessary and is even to be avoided. As the in-situ solidification technique does not require any person to work underground, a considerable reduction of the dose to

858

personnel can be attained.

In addition of all these favorable safety features of this new concept, calculations show that the economics are very attractive. Though a detailed comparison is not yet possible, space requirements are much lower for in-situ solidification and operational and transport costs are also much lower, so that a
price reduction for disposal by factors rather than percentages can be predicted.

ACKNOWLEDGMENTS

The R&D work for this project is sponsored by the Federal Ministry for
Research and Technology and is executed jointly by the following institutions:

- Amtliche Materialprüfanstalt für Steine und Erden, Clausthal-Zellerfeld
- F.J. Gattys, Verfahrenstechnik GmbH, Neu-Isenburg
- Gesellschaft für Strahlen- und Umweltforschung mbH, Institut für Tieflagerung,
  Clausthal-Zellerfeld
- Kernforschungszentrum Karlsruhe GmbH
- NUKEM GmbH, Hanau.

The authors, who are members of the Project Management would like to express
their thanks to numerous scientists at the institutions mentioned above, whose
contributions are summarized in this paper.

REFERENCES

1. R. KÖSTER, R. KRAEMER, R. KROEBEL, Disposal and Fixation of Medium- and Low-
   Level Liquid Wastes in Salt-Caverns, IAEA-SM 243/16, Otaniemi near Helsinki,
   2-6 July, 1979.

2. W. HAUSER, E. SMAILOS, R. KÖSTER (1982), Computations of Temperature Profiles
   in the Medium/Low Level Waste Cement and the Rock Salt Surrounding the Disposal
   Cavern. Proceedings of this symposium.

3. Z. BOGUSLAWSKI, E. DUMONT, W. GUBER (1982), Granules, a new Waste Form for
   In-situ Solidification of MLLW. Proceedings of this symposium.

4. M. WEBER, Y. DEDEGIL, H.H. HOMANN (1982), Transport of Waste into Caverns by
   Vertical Flow of Concrete Suspensions, Proceedings of this symposium.

5. C. FROHN, W. FISCHLE, W. RAAB, M.W. SCHMIDT (1982), Stability of Salt Caverns.
   Proceedings of this symposium.

SCIENTIFIC BASIS FOR RADIOACTIVE WASTE MANAGEMENT - V
Werner.Lutze, editor

COMPUTATIONS OF TEMPERATURE PROFILES IN THE MEDIUM/LOW-LEVEL WASTE CEMENT AND IN THE ROCK SALT SURROUNDING THE DISPOSAL CAVERN

W. HAUSER, E. SMAILOS, R. KÖSTER
Kernforschungszentrum Karlsruhe GmbH., Institut für Nukleare Entsorgungstechnik, Postfach 3640, D-7500 Karlsruhe, Federal Republic
of Germany

1. INTRODUCTION

Prior to disposal of cemented medium level waste and low level
waste in large caverns 1000 m below ground it is important to
know what temperatures will be caused by heat generation in the
waste product. During the disposal period the heat produced by hydration of the cement used as the binder as well as by decaying radionuclides in the waste may give rise to inadmissibly high temperatures in the waste product resulting in a deterioration of the
product qualities. To avoid such undesirable processes, a maximum
permissible product temperature of $90^{\circ}C$ was specified for the investigations, i.e. a temperature that is significantly below the
boiling point of the water contained in the cemented product.

Earlier investigations have shown that temperatures occur in
the salt cavern during disposal and fixation of a cemented waste
suspension which are well above this maximum acceptable temperature. The fixation of the liquid wastes in cement granules which
release their heat of hydration already above ground, and the disposal of a granule/grout product results in temperatures during
setting in the cavern which are significantly below 90° C /1/.
This has been shown by one-dimensional calculations, based on literature data, for the hydration heat of the cements used.

In the investigations discussed in this paper, a three-dimensional computer program was employed using experimentally determined
data on the heat generation of the granule/grout product. In addition, this made it possible to calculate temperature profiles in
the waste product as well as in the surrounding rock salt.

2. DESCRIPTION OF THE MODEL OF CALCULATION
2.1 HEAT SOURCES IN THE IN-SITU WASTE PRODUCT

For calculating the temperatures in the cavern and in the surrounding salt rock which undergo variations with time, the time dependent heat power of the waste product must be known. The heat source to be taken into account in the computation model is, besides the decay heat of radionuclides contained in the waste, the heat of hydration resulting from cement setting. This variable was determined by calorimetric measurements made on an in-situ reference product.

The investigations have revealed that in the process of setting the development with time of the hydration heat $\dot{Q}(t)$ can be described mathematically in good approximation by the parameters characteristic of radioactive decay such as the decay constant λ and the initial heat power $\dot{Q}_o$. Accordingly the following equation applies:

$$\dot{Q}(t) = \dot{Q}_o \exp (-\lambda \cdot t)$$

Fig. 1 shows the development with time of the heat power of the in-situ waste product /2/ made up of the contributions of the indi-

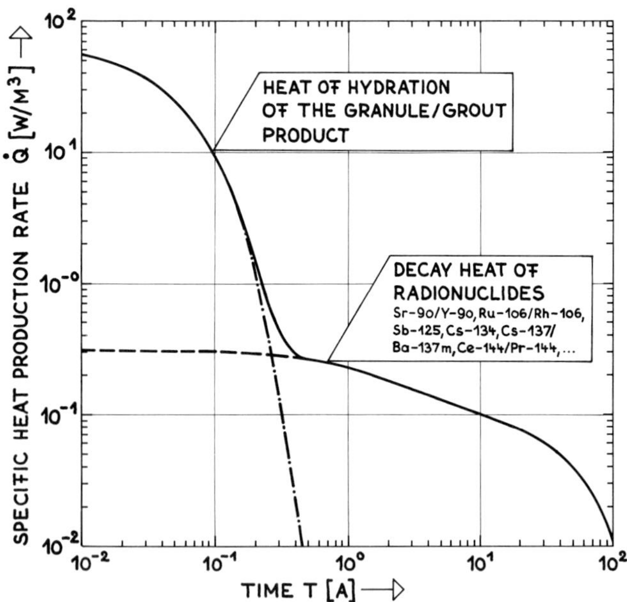

Fig. 1. Specific heat production rates of the in-situ waste product

vidual heat sources. It is evident from this figure that for periods shorter than 0.2 years after production of the final product the hydration heat governs the temperature whilst for periods longer than 0.2 years the decay heat will be the dominant factor.

2.2 SIMULATION OF THE PROCESS OF FILLING INTO THE SALT CAVERN

For calculation of the temperature evolution in the product and in the surrounding salt rock a three-dimensional Fortran computer code was used /3/. The geometry of the in-situ cavern in the salt dome was approximated by a cuboid model body consisting of the cavern and the surrounding salt rock (Fig. 2).

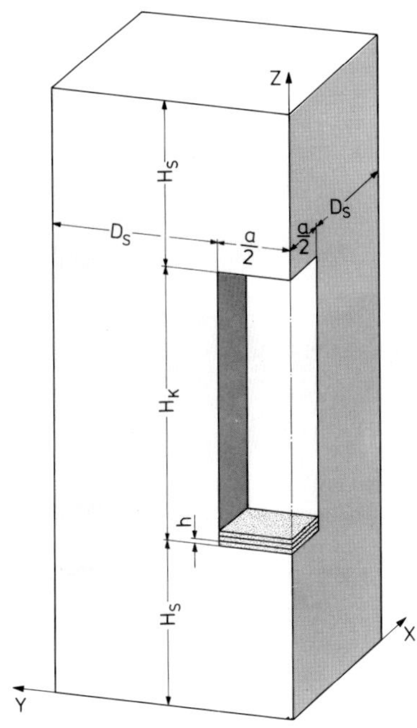

Fig. 2. Geometry of the model body
H_S= 20 m
D_S= 15 m
H_K= 82 m
$\frac{a}{2}$ = 15 m
h = 5.5 m

Data applicable to the in-situ reference waste product (60 vol.% granules, 40 vol.% grout):

specific weight: $\rho = 2.2$ g/cm^3

specific heat: $c_p = 0.96$ J/g·K

heat conductivity: $\lambda^* = 1.0$ W/m·K

heat transition coefficient of air: $\alpha = 2.0$ W/m^2·K

product temperature at the time of disposal: 30°C

Data applicable to the cuboid model cavern:

cavern volume: $V = 75\ 000$ m^3

edge length of the surface: $a = 30$ m

cavern height: $H_k = 82$ m

rate of disposal: $15\ 000$ m^3/a

period of filling: 5 a

If the final waste volume to be disposed of every year is divided over three disposal campaigns, a thickness of the successive disposal layers of h = 5.5 m is obtained. In the computation model used the disposed layers are considered as filled instantaneously. This is a conservative consideration since, in practice, the cavern is filled continuously /2/ which means a better heat removal in the z-direction (Fig. 2).

Data applicable to the surrounding salt rock:

For the temperature calculations an initial temperature of the surrounding rock of 40° C was assumed, corresponding to a depth of 1000 m. On the outer faces of the salt cube (Fig. 2) the temperature is at a constant level of 40° C (boundary condition). For rock salt the temperature dependent values for $\rho \cdot c_p$ (T) and λ^* (T), respectively, are used which have been proposed by Ploumen /3/.

3. CALCULATED TEMPERATURES IN THE SETTING WASTE PRODUCT AND IN THE SURROUNDING ROCK SALT

3.1 TIME DEPENDENCE OF TEMPERATURES

The calculations have shown that over the period of waste product filling the temperatures in the cavern are mainly governed by the temporary heat of hydration of the setting cements. In contrast at the end of the storage period the temperatures occurring are caused by the decay heat of the radionuclides present in the waste.

The calculated development with time of the product temperature

in the intervals between disposal of the first three layers and
after disposal of the last layer, respectively, in the cavern has
been plotted in Fig. 3a and b. It appears from the calculations
that already after 0.13 years (47 days) after introduction of the
first layer, 95 % of the temperature maximum is attained in the
cemented waste. Subsequently, the temperature in this layer falls
again. In the following disposal layers introduced at a 0.3 year
time interval the temperature attains its maximum always in the
layer filled last. It is evident from Fig. 3b that the maximum product temperature is measured after the last disposal campaign
(15th layer after five years). In the case discussed here it
attains 80° C and is thus 10° C below the specified maximum permissible product temperature.

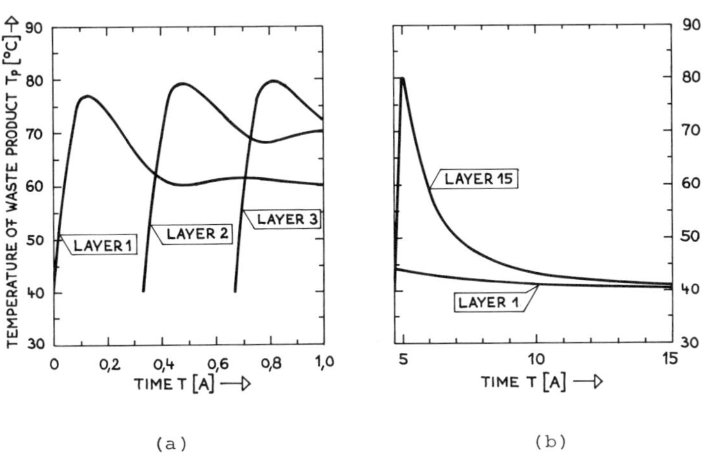

(a) (b)

Fig. 3. Calculated temperatures for different filling campaigns
starting for the first three layers (a) up to total cavern
filling, respectively (b)

At the end of the filling period the temperature in the cavern
decreases again and is determined over a significantly longer period by the diminishing decay heat of the radionuclides. Five
years after the end of filling the product temperature in the layer introduced last exceeds the original rock temperature by only
about 4° C.

3.2 TEMPERATURE PROFILES

Fig. 4 shows the temperature development plotted in the z-direction of the salt cube after termination of the third disposal campaign at the time of temperature maximum. The maximum lies at approximately 80° C and occurs in the center of the topmost layer. Temperature gradients appear in this layer, especially towards the air space of the cavern, the order of magnitude being about 15° C/m. The temperatures in the layers previously filled have already dropped considerably.

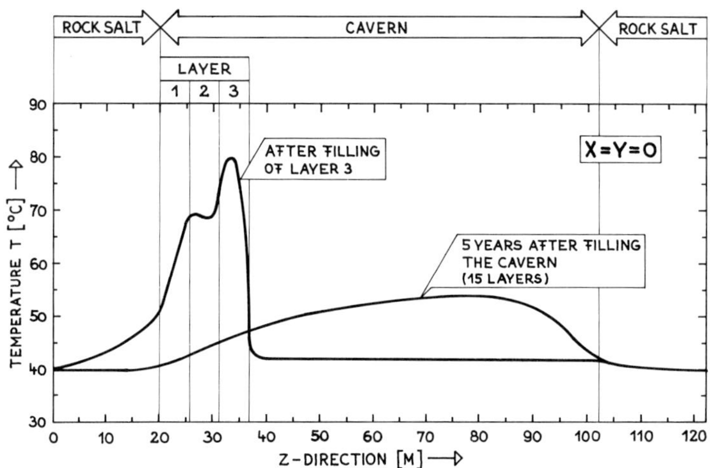

Fig. 4. Temperature profiles in the z-direction of the model body (plane X=Y=O) after the filling of the 3rd layer and 5 years after total cavern filling, respectively.

The calculations showed that a maximum salt temperature of 52°C is attained on the floor of the cavern after filling of layer 3. Neither is this salt temperature much surpassed on the lateral faces of the cavern. But on account of the better heat conduction of the salt rock, the temperature gradients in the salt are much lower than in the waste product and attain about $1\text{-}2^{\circ}$ C/m, which means that within about 15 m distance from the surface of the cavern the undisturbed rock temperature of 40° C prevails.

Five years after filling the cavern a nearly symmetric temperature profile is obtained in the z-direction of the model body

(Fig. 4). The maximum temperature in the cavern now mainly controlled by the decay heat is 54° C.

4. CONCLUSIONS

It became evident in the course of in-situ project implementation that particular attention must be paid to the heat which is generated in the waste product and which is caused above all by the setting of the cement. By a multitude of temperature calculations performed for different compositions of the waste product decisive incentives have been given to optimizing the method.

The computation model employed for this purpose and described in this paper is characterized by the following features:

- The geometries prevailing in the cavern of the salt dome are reflected by the choice of a three dimensional model body.
- The total heat output, composed of hydration and decay heat, which decreases with the time is taken into account. The hydration heat was determined by calorimetric measurements performed on an in-situ reference product.
- The time dependence of the hydration heat output can be mathematically described in terms of the decrease of the heat output of the radionuclides. For this purpose, the characteristic parameters of radioactive decay are used such as decay constant and initial heat output.
- The disposal by layers of the waste product in the scheduled disposal campaigns is taken into account.

Calculations based on the available input data have shown that the maximum product temperature occurring (temperature in the cavern) does not exceed 80° C at any moment. It is mainly determined by the temporary heat of hydration of the setting cements. Thus, the temperature remains clearly below the release temperature of the water contained in hydrated cement. The maximum temperature in the surrounding salt rock reaches about 52° C and hence surpasses the rock temperature by 12° C.

Thus it can be demonstrated by the temperature calculations that the concept of in-situ solidification and storage of a mixture consisting of the waste product hardened above ground and grout filling can be realized without problems resulting from the temperature.

ACKNOWLEDGEMENT

The heat of hydration data have been measured and supplied for the computations by R. Struth, AMPA/Clausthal.

REFERENCES

1. Hauser, W., Smailos, E., Köster, R. (September 1979)Berechnungen zur Wärmeentwicklung bei der Lagerung und Verfestigung von MAW/LAW in großräumigen Kavernen (Referenzsysteme A und B des "in-situ Projektes").KfK 2857

2. Kraemer, R.H., Kroebel, R.H., In-situ solidification of low and medium level wastes. Proceedings of this symposium.

3. Ploumen, P., Strickmann, G., Winske, P. (1979) Untersuchungen zur Temperaturentwicklung bei der Endlagerung hochradioaktiver Abfälle. Atomwirtschaft, Jahrg. 24, Nr. 2, S. 85-91.

Published 1982 by Elsevier Science Publishing Co
SCIENTIFIC BASIS FOR RADIOACTIVE WASTE MANAGEMENT - V
Werner.Lutze, editor

GRANULES, A NEW WASTE FORM FOR IN-SITU SOLIDIFICATION OF LOW• AND MEDIUM-LEVEL WASTE

Z. BOGUSLAWSKI, E. DUMONT and W. GUBER

F.J. Gattys Verfahrenstechnik GmbH, D 6078 Neu-Isenburg, Federal Republic of Germany.

1. INTRODUCTION

As already mentioned in the overview paper /1/ the in-situ solidification process requires a preconditioning of MLW/LLW arisings by granulation. The main purpose thereof is the dissipation of the heat of hydration already above ground and to replace in the final concrete like waste form a considerable part (i.e. 60 vol.%) with inert granules.

In conventional industry granulation is a well known process but the application in radioactive waste management in compliance with a number of specifications required considerable R & D efforts which are described in this paper.

Two different systems which are most appropriate for this process i.e. the rotating dish as used by R. Köster and coworkers /2/ and the ploughshare mixer, have been investigated till the extend that a comparison between the two processes and their products could be performed.

In contrast to the rotating dish the ploughshare mixer represents a closed system which can be well adapted to treat radioactive wastes. Fabrication of specified polydispersive granules is possible which show a significant higher density of 2.5 - 2.7 gcm^{-3} than pellet products from a rotating dish thus indicating higher mechanical strength and better leaching resistance.

These facts were the basis for the decision to use the ploughshare mixer for the granulation process in the in-situ solidification concept and to develop this process till an inactive pilot plant scale.

2. DESCRIPTION OF THE GRANULATION PROCESS

Fig. 1 shows the pilot plant which was constructed and operated during the second project phase. It was designed for a capacity of one m^3 or 1.6 t granules in one shift. The total equipment covers storage tanks for simulated MLW concentrates, silos for cement and bentonite, weighing and dosage equipment for solid and liquid materials, ploughshare mixer, conveyer belt for the discharged granules, sieving device, transportation equipment and two storage silos for the granules. During one year operation about 100 t specified granules have been produced.

The granulation process is operated batchwise and represents a sequence of different operational steps as follows:

- charging the mixer with cement (PZ 35 F) and bentonite,
- injecting MLLW-concentrates for 10 minutes,
- 30 minutes granulation process under permanent rotation of ploughshares in combination with a rotating cutter (see Fig. 2) which keeps the grainsize below 5 mm,
- discharging of the granules through the bottom door onto a slowly moving conveyer belt where the "green" granules set for about 25 - 30 minutes in a relatively thin layer till they have enough stability for further handling. This time of "careful" interim storage can be reduced by heating the product to about 40°C,
- removing over- and undersized grain by sieving and transportation of specified granules through pneumatically driven pipelines into special silos where they must set for about 10 days to dissipiate their heat of hydration and to arrive at their final stability.

The different operations can be superimposed properly so that the effective batchtime reduces to about one hour.

Meeting a specified granule spectrum is very important for the fabrication of the reference waste form, i.e. the granules/grout mixture.

The tolerance of the granule spectrum is within 0,3 mm and 5,0 mm with a maximum at 1,85 mm.

The over- and undersized grain fraction of one production charge (about 3,6 % wt and 9,0 % wt respectively) are separated by a sieving process step and are pneumatically recycled to the ploughshare mixer where they are smashed by the rotating cutter within the next production batch.

The ploughshare mixer should be operated at a temperature of $35^{\circ} \pm 5^{\circ}$C. A higher temperature leads to a prematured hardening of the mix whereas a lower temperature favors the formation of bigger granules than specified.

Another important process parameter is the water/cement ratio which was optimised at 0,15. A lower ratio leads to prematured hardening without proper formation of granules whereas a water/cement ratio up to 0,2 decreases the stability of the "green" granules up to desintegration.

Fig. 3 shows a random sample of a granulation batch spread on draft paper in millimeter squares which illustrates the grain spectrum of the polydispersive granules. Their main characteristics can be extracted from table 1.

TABLE 1

MAIN CHARACTERISTICS OF GRANULES

Composition	76 % cement
	5 % bentonite
	19 % MLLW-concentrate
Product density	2,6 t m^{-3}
Bulk density	1,56 t m^{-3}
Open pores	15.0 Vol.%
Pore size	$10^{-5} - 10^{-3}$ mm
Grain spectrum (after sieving)	0.3 - 5.0 mm
Impact strength /3/	10^{-3} m^2 J^{-1}

3.   PRODUCT INVESTIGATIONS

Investigations on the structure of granules using scanning electron microscope and X-ray microprobe show a total encapsulation of the waste particles
within the cement matrix.

Figures 4, 5   show secondary electron pictures of a granule slice and
the element distribution in the granules. Sodium was used as guide element for
the waste salt, and calcium for the cement matrix. The large smooth particle is
a waste salt mixture whereas the fine crystals represent the cement-bentonite-
matrix.

Other scanning electron microscope pictures show the presence of open pores
and micro cracks. These cracks have a maximum width of about 10 µm. The number
of the open pores and their dimensional spectrum were determined by means of a
mercury pressure porosity meter.

The pore size spectrum shows four peaks at about $7 \cdot 10^{-3}$ µm, $2 \cdot 10^{-2}$ µm,
$4,5 \cdot 10^{-1}$ µm and $3,7 \cdot 10^{1}$ µm. Most of the pores are in the dimension of $2 \cdot 10^{-2}$ µm.
The fraction of the open pores in the granules is less then 15 Vol.% which is
in agreement with the relatively high density of the granules.

The mechanical strength of the granules was measured with the help of the
impact-test according to Wallace and Kelley /3/. A certain quantity of granules
with known grain size is smashed by a determined mechanical shock, and the increase of the grain surface is related to the applied kinetic energy.

4. FUTURE R & D WORK

Due to the relatively low water/cement ratio the final salt content in the granules is limited to maximum 7 % wt if waste salt solutions are used. Dewatering of the MLLW up to dry salt and granulation thereof may increase the final salt content up to 25 % wt which results in a considerable volume reduction of the in-situ waste form. Furthermore increasing salt content and a low water/cement ratio decrease the formation of radiolytic hydrogen.

Also other waste categories than MLW salt solutions will be investigated with the final goal to use the in-situ concept for special non radioactive wastes too which require similar equipment due to their potential hazard.

REFERENCES

1. KRAEMER, R., KROEBEL, R. (1982) In-situ solidification of low and medium level waste. Proceedings of this symposium.

2. AKBAR, A.H., KÖSTER, R. and RUDOLPH, G. (1980) Solidification of radioactive waste solutions by pelletizing technique. KfK-Report 2941.

3. WALLACE, R.M., KELLEY, J.A. (1976) An impact test for solid waste forms. Report DP-1400.

Fig. 1. General view of the pilot plant.

Fig. 2. Ploughshare mixer with open side door.

Fig. 3. Polydispersive granules.

Fig. 4. Secondary electron photo of a granule slice (750x).

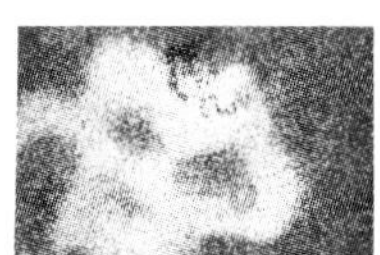

Fig. 5. Sodium distribution (bright) and calcium in the dark areas from the same detail.

TRANSPORT OF WASTE INTO CAVERNS BY VERTICAL FLOW OF CONCRETE SUSPENSIONS

M. WEBER[+], Y. DEDEGIL[+] AND H. H. HOMANN[++]
[+] Universität Karlsruhe, 7500 Karlsruhe, Germany
[++] Nukem GmbH, 6450 Hanau, Germany

CONSIDERATIONS TO THE TRANSPORT SYSTEM

In the process described in the first paper of this session for container-less final storage of radioactive wastes, a granule-cement suspension (waste grout) is transported by means of a vertical pipe from above ground into the underground cavern.

In the mining industry gravity-feed pipes are usual for downward conveying mainly of backfilling material. Other materials like gun powder and liquid concrete are also transported in this way whereby the solids are dry or are part of a suspension. In these cases high velocities are reached since the material is dropped nearly free falling. It occupies only partly the cross-section of the pipe and falls just slightly hindered by the pipe wall. Although these gravity-feed pipes are very simple, great disadvantages exist with re-spect to operational safety and availability (segregation, blocking, wear, attrition). Therefore, the downward transport of radioactive waste was designed in a new way as a simple gravity-feed plant without pumps, however, a small diameter pipe was used whose cross-section is completely filled over its whole length of about 900 m. This configuration leads to a slow, hindered fall velo-city avoiding the above-mentioned disadvantages as far as possible.

The mixture to be conveyed consists of granules, cement, water and possibly additives and flows through this vertical pipe from above ground into the cavern. The process is inherently safe and operates under atmospheric pressure. The fact that mechanical conveyor mechanisms, for ex. pumps, are not required, results in reducing maintenance and repair work, while simultaneously increa-sing the system availability.

In order to minimize abrasion of the suspension and the system equipment, the average speed of the suspension in the pipe is selected much lower than the corresponding rate of free fall. This lower speed is achieved by dimensioning the diameter of the vertical pipe as a function of the quantity of suspension transported and its rheological characteristics in such a way that the opposing forces of wall friction and gravity in the pipe counterbalance each other.

874

EXPERIMENTAL OBJECTIVES AND RESULTS

In order to determine the rheological characteristics of the suspension and the correct diameter required for design of the industrial-scale system, an experimental program was carried out with the following objectives and results.

Optimizing the product composition with respect to maximum granule loading and adequate transportability

A parameter of crucial importance for the degree of utilization of the cavern is the proportion of granules, the actual source of activity, in the granule-filler suspension. Earlier studies of the temperature development showed the necessity for an above-ground preconditioning of the MAW/LAW waste concentrates because otherwise -- if pure cement suspensions are fed into the cavern -- the temperature within the product rises to values higher than 150°C due to the release of hydration heat. Since improved filling of the cavern results with polydispersed granules in comparison with monodispersed, a material with a polydispersed grain spectrum, as shown in Fig. 1, was used for the studies. Furthermore such a grain spectrum favours non-Newtonian flow behaviour which in turn helps to keep the particles in suspension. A blast-furnace cement, Type HOZ 35 L NW/HS was used for production of the filler suspension. As additives, a liquifier and a retarder, both produced by Halliburton, were used.

Adequate transportability was attained using a product containing the materials mentioned above and with the composition listed in Table 1; this product was specified as the reference mixture for all further experiments.

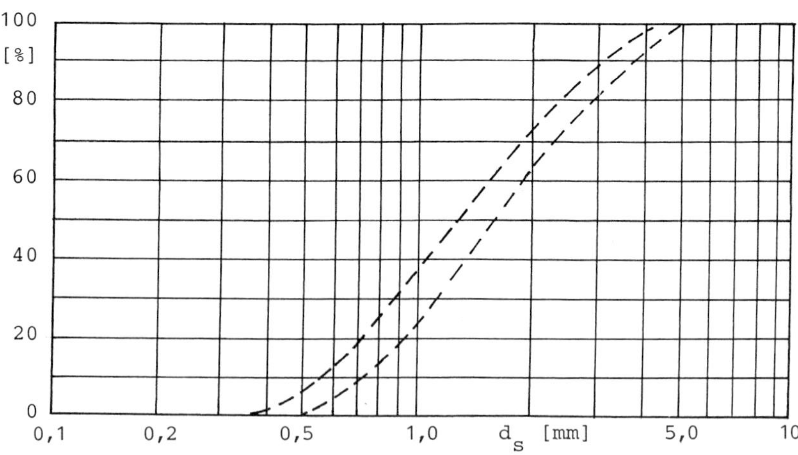

Fig. 1: Grain spectrum of the granules used

TABLE 1

COMPOSITION OF THE REFERENCE MIXTURE AT T = 17°C

Material	Mass %	Vol. %	Wtr.-cem. ratio	Theor. density g/cm^3
Granules	71.6	66.2		
Cement	19.2	13.6	0.45	2.20
Water	8.6	19.0		
Retarder	0.3	0.6		
Liquifier	0.3	0.6		
Total	100.0	100 %		

Determination of throughput as a function of the pipe diameter

Using the reference mixture specified above, the pipe characteristics for various pipe diameters were determined in a test stand. The results for the diameters 52, 64 and 82 mm are compiled in Table 2.

TABLE 2

TRANSPORT PIPE CHARACTERISTICS AS A FUNCTION OF THE VERTICAL PIPE DIAMETER

Pipe diameter	Density g/cm^3	Mass flow rate kg/s	Volume flow rate l/s	m^3/h	velocity m/s
52 mm	1.676	0.870	0.513	1.85	0.24
64 mm	1.745	2.105	1.206	4.34	0.375
82 mm	1.745	3.86	2.21	7.86	0.42

When carrying out these experiments using a system with a transport length of about 8.5 m, it was found that for start-up operation a "lubricant-mixture" composed of a cement suspension without granules and with a water-cement ratio of at least 0.45 must first be passed through the system, because otherwise the tip of the mixture stream loses water as a result of wetting the inside wall of the pipe and thus the pipe becomes clogged.

Determining the spreading behaviour, angle of repose and separation of the mixture into its components

An essential parameter for the degree of filling of a cavern is the angle of repose that develops during the filling operation. In addition, knowledge of the degree of separation of the product into its components as a result of radial spreading in the cavern is essential to ensure the quality of the stored product following solidification.

As early as in the transport experiments mentioned above, the reference mixture considered showed good spreading characteristics with a free-flow cone of 3 - 4°. Even in experiments with free-fall heights up to 20 m, there was no change in the angle of repose behaviour. The study of the concentration distribution of the granules as a function of the distance from the center of the "cake" revealed no significant deviations from the original values, so separation processes, due to radial spreading, can be regarded as negligible, if they are present at all.

In order to evaluate questions that are still open such as testing special components, determining operational needs and requirements, testing a suitable mixing and dosing subsystem, etc., an experimental system with a transport length of 50 m and a mixing and transport capacity of about 2 m^3/h is currently being constructed and operated in conjunction with a supplementary testing program at the experimental final storage facility ASSE II. It is assumed that, when these experiments have been concluded, the data required for the construction and operation of a prototype system will be available.

Analysis of the rheological model of the suspension

Since an extrapolation of the measured data, for throughput V towards zero, shown in Table 2, seemed too uncertain, additional tests were performed to determine the pipe diameter for which the mixture is not flowing out anymore.

The tests yielded a diameter of $D_o \cong 12$ mm corresponding to a shear stress (yield stress) τ_0 of about 46 N/m^2, which cannot be overcome by the weight (Fig. 2).

These empirically determined relationships lead to the conclusion that the flow of suspension behaves as a Bingham fluid and can be represented by the generalised Bingham model:

$$\tau = \tau_0 - k\left(\frac{dv}{dr}\right)^{1/n} \qquad (1)$$

τ being the shear stress   k   being consitency coefficient
τ_0 being the yield stress   n   being power law constant
dv/dr being the shear rate

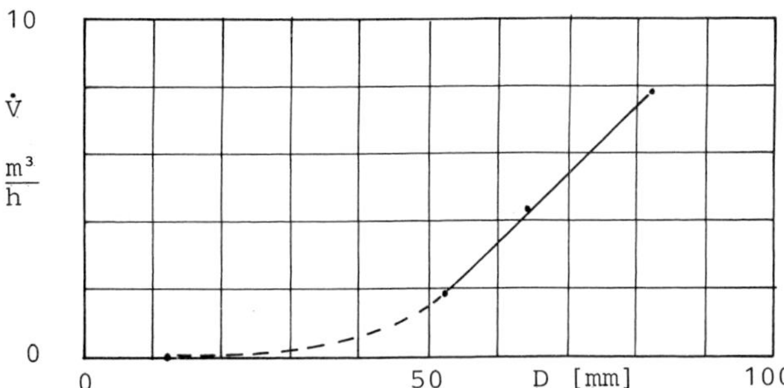

Fig. 2: Throughput of the suspension versus pipe diameter

In order to set up equation (1) exactly as rheological model for scale up calculations further experiments have to be carried out, particularly when the falling-pipe has horizontal or inclined parts, too. For vertical pipes Fig. 2 can be used to scale up.

CONCLUSIONS

The results of these studies can be comprised as follows:
The vertical transport of a waste-concrete suspension by gravity without pumps is feasible if a suitable composition of the mixture and a convenient pipe diameter are provided. Then the flow velocity of the suspension in the pipe adjusts corresponding to the pipe diameter, it is much lower than the free-fall velocity.

A suitable composition of the mixture must ensure sufficient freedom of movement of the solids, which means the portion of coarser granules (> 400 μm) should not exceed about 50%. This figure should be met in order to allow for the remainder - the fine portion of granules, the cement, water and additives - at least 50%.

STABILITY AND CONSTRUCTION OF CAVERNS FOR THE DISPOSAL OF CEMENTED LOW-AND MEDIUM-LEVEL WASTES

W. RAAB, C. FROHN, M.W. SCHMIDT
Gesellschaft für Strahlen- und Umweltforschung mbH München, Institut für Tief-
lagerung, Wissenschaftliche Abteilung, Theodor-Heuss-Str. 4, 3300 Braunschweig,
FRG

ABSTRACT

The geomechanical and mining-technological aspects of the construction of salt caverns as disposal chambers have been investigated during project phase 2, completed by mid 1981. With a view towards the stability analysis of such a cavern, FEM-estimates have been carried out and evaluated. From these it can be derived that

- a rotational ellipsoid would be the most suitable shape
- its dimensions should be 82 m (vertical axis) and 42 m (horizontal axis)
- the distance (safety pillar) between the neighbouring caverns should be 170 m (vertical) and 180 m (horizontal).

For practical engineering purposes the rotational ellipsoid can be modified into a cylinder with conic bottom and top. The numerical model simulated the short term as well as the long term characteristics of the surrounding salt rocks. The short term characteristics were assessed by an elastic approach, the long term characteristics by a rheological model. The input parameters have been determined by means of laboratory tests on ASSE rock salt.

In a second step the characteristics of partially and completely filled caverns were simulated. It was shown clearly that deformation of the salt rock comes to a halt when counteracted by the filling.

Based upon the results of the stability analysis, investigations were made to find out a suitable mining technique for the construction of the cavern. Solution mining and conventional development by means of drilling and blasting have been studied alternatively. Since both methods have their advantages and disadvantages a decision in favour of the one or the other cannot be made until the actual site has been defined.

INTRODUCTION

A bulk disposal of low and medium level waste and a subsequent in situ-solidification requires the preparation of stable boreholes for access to the caverns which are to serve as disposal cavities for the concrete monolith.

During project phase 1 (1977 - 1978) a disposal system has been developed in accordance with the anticipated waste volume, the resultant operating time of

the system, and the heat-release from within the product due to both the hydra-
tization of the cement and the decay of the radioactive contents. The system is
based upon caverns with some 75.000 m^3 effective volume employing rockmechanical
considerations that proceed from experience with comparable underground openings.

During project phase 2 (1979 - 1981) this system had to be evaluated by GSF
with a view towards its stability and construction technology.

GEOLOGICAL POSITION

The caverns should be constructed in a virgin part of the salt dome. Depending on the excavation technique to be applied, these caverns either can be constructed completely isolated from the accessible section of the disposal mine
by solution, or can be connected to it by drifts (conventional mining).

In the first stage of planning, the "Anhydrite-Region" of the Staßfurt
Halite (Na2β) has been chosen as the stratigraphic horizon to accomodate the
cavern field. A sufficiently large safety distance to the carnallite and to an-
hydritic and clayey rocks is a basic requirement.

The overlying rocks are assumed to consist of some 300 m of unconsolidated
Quarternary and Tertiary layers underlain partially by Mesozoic sediments. The
thickness of the cap rock can be on the order of 50 m. Generally a geological
situation has been assumed that can be derived from information generated at
the German final repository site presently under investigation.

  0 - 250 m: unconsolidated to partly diagenetically consolidated rocks of the
             Quarternary and Tertiary, eventually underlain with discordant
             Mesozoic rocks of small thickness.
250 - 300 m: cap rock in the transition zone between the overburden and the
             salt dome.
from 300 m: salt dome with partially inverted layers. Anhydrite (A3 and A4)
             and clay layers (T3 and T4) of varying thicknesses can occur.
             Furthermore, a complex folding of the carnallite must be assumed.

GEOMECHANICAL CRITERIA FOR THE DESIGN OF THE CAVERN FIELD

The cavern field has been designed on the basis of a rock mechanical stability analysis, the essential feature of which is a continual summarization and
review of the state of geotechnical knowledge (ALBRECHT, H. et al.[1], 1978).

The geotechnical assessment is founded upon geological and geophysical investigations of the rocks and on a determination of the rock mechanical parameters.

The actual stability analysis and its continual reworking is being carried out in three parts, each of which equally complements the other:

1. Evaluation of experience in the construction and operation of comparable underground openings.
2. Mathematical estimation with numerical modelling of rock stresses and time-strain behaviour, e. g. by means of FEM.
3. Verification of the numerically evaluated data by in situ measurements on suitable structural elements.

Experience. A great deal of know how in constructing large underground-openings in salt rock does exist. Most of this experience, however, is derived from the excavation of oil- and gas-storage caverns. Since these caverns are being operated under interior pressure, they are comparable only up to a point. In a way, the interior pressure acts to support the cavern walls thereby providing a more advantageous rock mechanical state than exists in a cavern under atmospheric conditions. However, experience with large glory-hole cuts in North German salt mines shows that openings of the required size - even with unfavourable geometry - can be safely constructed and operated over tens of years. Similar experience has been obtained in the Asse salt mine. Large chambers of some 80.000 m³ prove that such openings in salt rock can be constructed without support, and maintained over periods greater than 20 years. Intolerable large scale deformations in the surrounding rocks have nowhere been observed.

Calculations. On behalf of GSF, the Technical University Braunschweig (Institute for Statics, Prof. Dudek) has investigated the deformation of the cavern walls. The model applied uses the linear-elastic approach (Hook's Law) for the instantaneous strain and a rheological model (Newton-Voigt/Kelvin-Body) to simulate the time dependent characteristics, the input parameters being determined in the GSF-laboratories on rock-salt samples from the Asse mine.

The short term (elastic) characteristics have been evaluated for two cavern shapes:

Variant 1: The rotational ellipsoid (axes length: 82 m vertical, 42 m horizontal) offers the optimal solution from a rockmechanical point of view. However, it requires a great deal of mining effort (small advances per round, continual measurement, manual refinishing of the walls).

882

Variant 2: A <u>circular cylinder with conic top and bottom</u> creates a less advan-
          tageous secondary stress field. However, significantly construction
          efforts are some 30 % less and cavern dimensions are much easier to
          guarantee.

Variant 1 was used to study the long-term (creep) characteristics with varying input parameters. An example is given in figures 1a and 1b which show the
resulting deformations of an elliptic cavern at different construction stages.
Fig. 1c shows the convergence of an empty cavern after two years; the loss of
volume amounts to 2.000 m³ which is equivalent to 3 % of the total opening.

A half-filled cavern after two years shows a state of deformation as represented in fig. 1 d. The counter-pressure of the filling material results in an
evident reduction of convergence in the lower half of the cavern. The deformation parameters of the pellet-containing concrete are based upon GSF-laboratory
tests.

Due to enhanced convergence, the lifetime of the open, i. e. unfilled cavern
during the construction phase is a decisive parameter in optimizing the cavern
shape. This parameter favours variant 2, because of its shorter construction
time. If one also includes the constructional expenditures into the optimization, variant 2 is at least equal to variant 1, despite its more stable shape.

To completely assess the stability of the system under consideration, the
stress-strain behaviour of the bottom drifts, which are necessary in the case
of conventional mining, had to be estimated.

A two-stage-mining of the drifts has been simulated by FEM. They show clearly
that the drifts can be mined with sufficient stability. Local stress concentrations will be reduced rapidly because of the plastic reaction of the rock salt.
Before performing the deformation calculations, stability of the total cavern
field was considered. Results indicate that at a horizontal distance 60 m from
the edge of the cavern the secondary stress is merely 20 % higher than the
assumed primary stress. This value is considered to be tolerable. By doubling
the distance (for two neighbouring caverns), and with a safety allowance of
50 %, the horizontal safety pillar becomes 180 m. A similar procedure leads to
a vertical safety pillar of 170 m. Experience from the mining practice verifies
these theoretical findings.

<u>Verification.</u> The R&D-Programme established by GSF for the prototype cavern
in the Asse salt mine (SCHMIDT, M.W. et al.[2], STAUPENDAHL, G. et al.[3]) provides
a means of gathering valuable information for the system. Geotechnical measurements carried out for more than five years in the cavern, the surrounding rock,
and the adjacent shaft permit the following generalized statements:

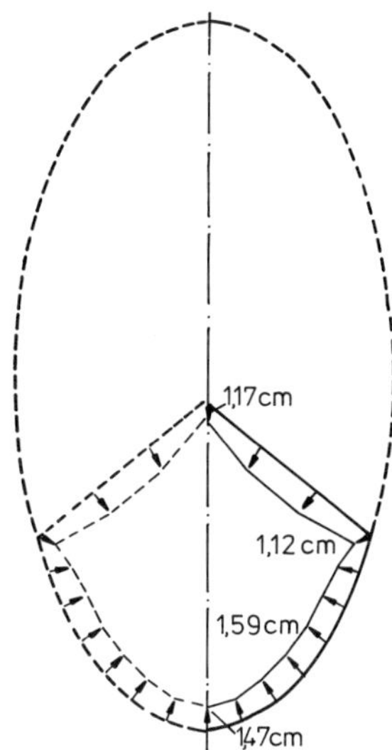

Fig. 1a: Convergence of cavern walls after development of 30 % of the cavity volume. Assumptions: instantaneous excavation with elastic response, only.

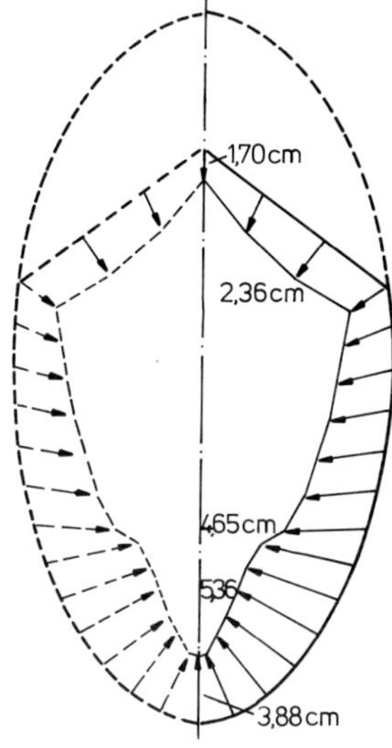

Fig. 1b: Convergence of cavern walls 107 days after the state described in Fig. 1a with plastic deformation.

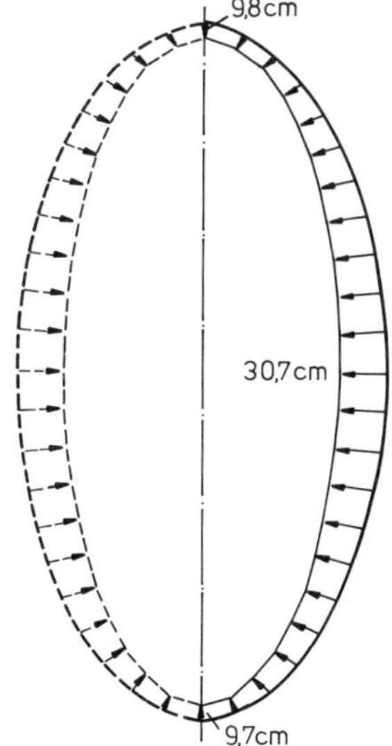

Fig. 1c: 700 days after development. Assumptions: elastic response followed by plastic (creep) deformation.

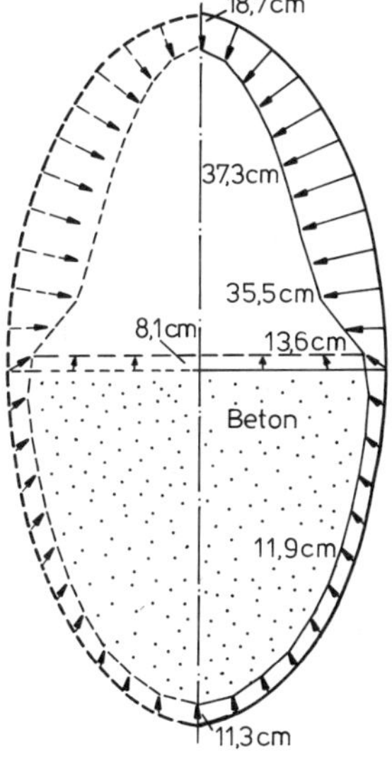

Fig. 1d: 1400 days after development Assumptions: as in fig. 1c, but with 50 % filling after 68 days.

1. Deformation around the cavern proceeds at a decreasing rate with time.
2. The stability of the cavern is ensured as defined above.
3. The deformation state, as derived from extensometer measurements, is largely homogeneous and can be modelled by FEM.

MINING TECHNIQUES

Preconditions. Two alternative mining techniques were compared and evaluated by various criteria:
1. Conventional mining (drilling and blasting) of the caverns and drifts.
2. Solution mining of the caverns (drifts not necessary).

The caverns had to be arranged in a way that they could be filled directly from the surface by means of of a vertical conveyor pipe. Perforation of the overburden has to be minimized.

In the case of conventional mining, it must be ensured, that the man-operated part of the repository mine does not become endangered by the connecting drifts. Both sections, the cavern and the mine, must be sealed from one another in such a way that, even after a water inrush into one section, an operation of the unaffected section remains possible. Basically it is required that the permeability of the sealing dam for gas and liquids is not greater than that of the surrounding salt rock.

The boreholes which connect the caverns to the surface facilities have to be sunk, cased, and their tightness tested prior to the mining of the cavern. The boreholes should be vertical to a high degree of accuracy (maximal plumb line deflection = 1,5 degrees), their effective diameter being dependent upon the method of mining the cavern (between 300 and 600 mm). According to investigations carried out by the Technical University Clausthal (Institute for Deep Well Drilling, Prof. Marx) a 4-fold casing has to be provided in order to protect the borehole from a water inflow.

Conventional Mining. The concept for mining a cavern conventionally has been prepared by Thyssen Schachtbau, Mühlheim (West Germany) on behalf of GSF.

In accordance with the above mentioned preconditions, access to the cavern has to be provided by a drift at the bottom, thereby introducing the first requirement for a gas and watertight seal with respect to the man-operated repository. Consequently, only two cavern excavation methods are possible:
1. bench stoping
2. overhead stoping

Due to the rock mechanical and economic reasons the second variant is favored: mining the rock from the bottom to the top and subsequently hauling the broken material from the top to the bottom reduces the open lifetime of the cavern and permits the broken material to exert support pressure onto the rock as long as possible.

Conventional mining of the cavern by one of the above mentioned methods is a proven technique and does not require new development.

<u>Solution Mining.</u> On behalf of GSF, the Kavernen Bau- und Betriebs-GmbH, Hannover (West Germany) has studied solution mining of the cavern.

Some fundamental advantages make the solution technique an interesting alternative to the conventional method:
- separation from construction and operation of a man-operated repository,
- gentle treatment of the rock since drilling and blasting can be abolished,
- no drifts to be re-sealed,
- economic provision of cavity volume.

The solution-mining concept envisions three phases:
1. After completion of a borehole, a sump cavern has to be prepared in order to accomodate the insoluble rock remnants (anhydrite and clay).
2. Subsequently the cavern proper is mined in a process called "indirect solution" (Fig. 3).
3. The required top shape is developed in the last phase at which the "blanket" level, consisting of liquified gas (LPG) which prevents solution attack of higher levels in the cavern, is lifted in several steps until the given top level is reached.

Experience shows, however, that the designed cavern shape can never be achieved, even in absolutely homogeneous rock salt. Thus an adequate safety allowance must be provided.

For rock mechanical reasons, the cavern should be maintained under a supporting pressure (brine and/or gas) during the time preceeding filling and be emptied shortly before the start of active operation. The displacement of the brine by gas can serve simultaneously to provide a pressure test (about 140 bar) of the cavern.

FINAL REMARKS

Drifts, boreholes, and caverns, as they are required for the in situ disposal and solidification of LLW and MLW, can be constructed with sufficient stability and be operated safely.

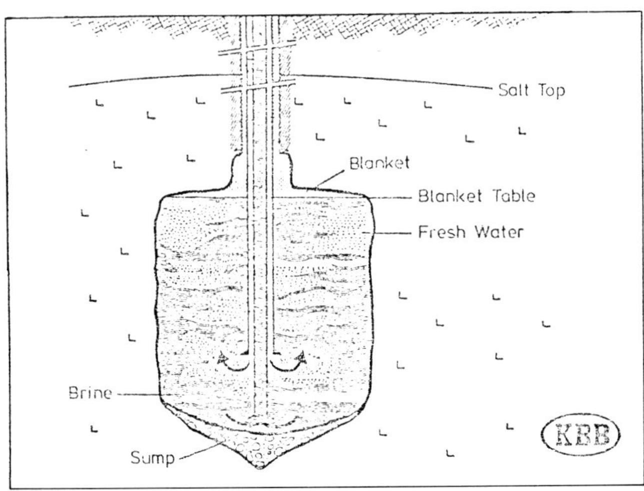

Fig. 2: Development of disposal cavern by "indirect solution".

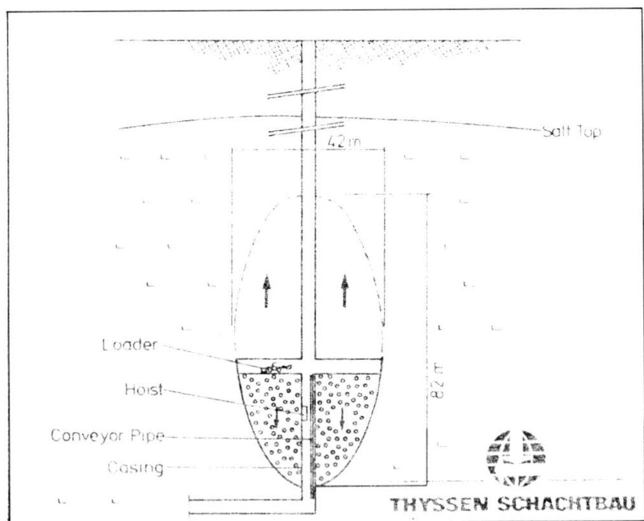

Fig. 3: Development of disposal cavern by "overhead stoping".

CRITERIA	SOLUTION MINING	CONVENTIONAL MINING
1. Geological Site Investigation		
- Site Selection	Flexible, since connection with repository-mine is not necessary. But restricted to salt.	Shaft and headings are necessary. Others than salt rock are possible.
- Max. Depth	About 1500 m. Limitations given by max. allowable temperature during hydratization.	To about 1100 m, due to mining methods, ventilation, ergonomy.
- Required Gross Rock Volume	Relatively large, due to uncertainty in geological predictability and geometrical tolerances.	Smaller, since the accessibility permits more thorough investigation. Shape of the cavern can be better controlled.
- Effectiveness of Geological Barrier	High. Only one borehole must be sealed. Sufficiently large safety pillars required.	Less. Larger borehole diameter and connecting drifts to be sealed.
2. Rockmechanics		
- Alternative Cavern Geometry	Only cylinders with irregular walls. Tops tapering in steps are possible.	All reasonable shapes are possible.
- Preservation of Rock Quality	Very good.	Fissuring, due to drilling and blasting.
- Geomechanical Monitoring	Possible only by remote sensing.	Possible by installation of instruments.
- Open Lifetime	Convergence can be counteracted by pressure until filling starts.	Temporary support by loose material only.
3. Technology		
- Know How	Various. But only under support pressure.	Two examples in German salt mines (ASSE, RIEDEL).
- Underground Operations	No.	Yes.
- Ventilation	Not necessary.	Necessary and expensive.
- Disposal of Mined Material	Only in liquid state (brine): - pipeline to the sea - injection to the ground - dumping to the sea - flooding of a shut down mine - selling (if there is a market)	Only solid material: - backfill of a mine (repository mine) - dumping - selling (if there is a market)
- Vertical Arrangement of Caverns (necklace)	Difficult. Sealing problems between the caverns.	Possible. However, long open life time of the upper caverns.
- Construction Time	About 2 years.	About 3 years. Difficult timing, due to probable linkage to an operating mine.
4. Costs	55 ... 67 DM/m³	120 ... 150 DM/m³

Tab. 1: Comparison of the two processes for construction of disposal caverns.

Two alternative mining techniques, i. e. conventional mining and solution mining, have been investigated. In discussing the advantages and disadvantages of both methods the following criteria should be referred to:

- rock formation (evaporite rocks or others),

- geological suitability of the site,

- stress/strain behaviour,

- costs.

It is the opinion of the authors that a final evaluation should not be made before a site for application of the concept has been selected and a decision to apply the concept has been made.

REFERENCES

1. Albrecht, H., Meister, D., Stork, G.H. and Wallner, M. (1978) Zur Frage des Standsicherheitsnachweises von Hohlräumen in Salzgesteinen. Proc. V. Int. Salzsymposium, Hamburg 1978.
2. Schmidt, M.W., Kolditz, H., Staupendahl, G., Thielemann, K. (1979) Bau einer Prototyp-Kavernenanlage im ehemaligen Steinsalzbergwerk ASSE zur Durchführung von Forschungs- und Entwicklungsaufgaben auf dem Gebiet der Endlagerung radioaktiver Abfallstoffe. Rock Mechanics, suppl. 8, 1979.
3. Staupendahl, G., Schmidt, M.W., Meister, D., Wallner M. (1979) Geotechnische Untersuchungen an der Prototyp-Kaverne in der Schachtanlage ASSE. 4. Internationaler Kongreß für Felsmechanik, Montreux 1979.

RESEARCH PRIORITIES FOR NUCLEAR WASTE ISOLATION

Published 1982 by Elsevier Science Publishing Co
SCIENTIFIC BASIS FOR RADIOACTIVE WASTE MANAGEMENT - V
Werner.Lutze, editor

RESEARCH PRIORITIES FOR NUCLEAR WASTE ISOLATION

Summary prepared by
T.H. Pigford, (panel chairman), University of California, Berkeley

Brief presentations made by panel members are as follows:

D. E. GORDON, E. I. du Pont de Nemours, Savannah River Laboratory, USA

Mr. Gordon's comments are specific to defense waste only. The research priorities center around:

1. Improvements in the technology for immobilizing high-level waste. For waste vitrification, these improvements include lower corrosiveness, less crystal formation, lower volatility, reduction in glass cracking, increased life of glass-melter, nondestructive testing methods, and remote handling techniques.

2. Better understanding of the performance of the waste package components in the repository environment. This requires additional research in the areas of repository characterization, the mechanisms of waste form corrosion, radionuclide migration in geologic media, predictive models for projecting waste system performance, and synergistic effects of waste system components.

3. Confirmation of waste disposal risk assessments and their establishment as the technical basis for regulatory criteria.

4. Research priorities for transuranic (TRU) waste isolation are focused on the development of processing technologies for immobilization and of improved instrumentation for classification and nondestructive examination.

H. C. BURKHOLDER, Battelle PNL, Richland, WA, USA

1. We need technically-based performance requirements for both waste products and treatment processes to prevent an endless sequence of product and process development activities.

2. We need to demonstrate the operability and maintainability of the liquid-fed ceramic melter for vitrifying nuclear wastes under pilot-scale, radioactive conditions as the final step to full-scale application.

3. We need to bring transuranic waste treatment technology up to the same level of development as high-level waste treatment technology.

4. We need to test waste products under the dynamic conditions of transportation, storage and disposal and develop models which describe both the internal and

external processes by which radionuclides are transported from waste forms.

5. We need to develop techniques to assure waste product quality without destructive inspection.

6. Finally, we do not need to perform research and development on either long-lived waste containers or on waste forms with low release rates. Waste container need not last longer than length of time desired for retrievability (0 to 100 years). Fractional release rates below 10^{-2} to 10^{-4} per yr are not required.

F. FEATES, Dept. of Environment, U.K.

1. The priority rests with satisfying the public on the acceptability of a disposal method. We must beware that peripheral research is not justified by the disposal programme, thereby suggesting to the public that problems are much further from a solution than is actually the case.

2. We must now demonstrate to the public the capability of geologic disposal.

3. The research programme should be organised to support well defined needs within an overall systems approach. The contribution of the research towards solving problems should be spelt out.

4. An acceptable scenario must be developed, it must be technically possible, and it must be acceptable to the public.

5. In the first instance cost should not be a dominant factor. It will clearly be important when we come to build a repository, but first we have to develop a technically acceptable solution.

6. We need work on specific repositories. Practical experience is needed. Research will follow the development of this acceptable scenario.

7. We must begin by considering time scales and potential hazards.

8. The work must be broadened to include the lower-level waste.

E. R. MERZ, KFA Jülich, Germany

1. In developing overall protection from nuclear waste isolation the multi-barrier system should be emphasized, because it is the most convincing.

2. The waste and the repository should have matching properties.

3. The waste canister should last 10 - 200 years.

4. We need leach experients in the more complicated, realistic situations, with real radionuclides.

5. Additional areas of study are protective layers, helium formation, and microbe effects.

6. Quality assurance, i.e., characterization of the waste form, needs more emphasis.
7. The canister material is important, and its corrosion mechanisms require further study.
8. More study is needed on radiolysis and other such long-term radiation effects.
9. Backfill materials are important, and their chemical, mechanical, and thermal effects need study.
10. Safety analysis should include starting with a specified dose goal and calculating backward to see what waste-isolation barriers will be sufficient.
11. We should not always feel obliged to use too sophisticated a science.

T. PAPP, Swedish Nuclear Fuel Supply, Sweden

1. Sweden believes that the feasibility of a safe final storage and disposal of high-level waste is proven.
2. The next goal is to obtain such an understanding of the various components of the waste isolation system that a suitable balance can be created between the barriers in the total system.
3. The planning is to have the final repository ready for operation in 2020.
4. The most complex part of system is the near field where there is a very strong interaction between the various components. Presently the concept is based on a near field temperature below 100° C.
5. To model and evaluate these interactions, a high priority has been given to the task of getting better, data, especially for the actinides, of the thermodynamics and reactions in the actual environment.
6. Another priority area is the need for better models to describe the alpha-radiolysis and its consequences for the redox-interactions in the near field.
7. After a site selection the groundwater chemistry of the repository is defined. An early survey of a number of possible sites is necessary to define the variability of possible sites for a deep geologic repository.

A. E. RINGWOOD, Australian National University, Australia

1. The plan for waste isolation should have the greatest safety at reasonable cost, and it must in principal be explainable to the layman, to gain public acceptance.
2. SYNROC satisfies both of these requirements because of its natural analogues and better performance than glass at elevated temperatures. Glass requires too many additional considerations, e.g., a good canister.

3. Glass waste cannot be deposited very deep because of the increase of ambient temperature with depth. The public would not accept centralized, shallow repositories.SYNROC can survive the elevated temperatures in geologies that have lower permeabilities.
4. Disposal in deep drill holes is more understandable to the layman.
5. Research priority should be aimed at advanced waste forms that (a) posses capacity for public acceptance, (b) are superior to borosilicate glass, particulary with regard to lower leachability at high temperature and (c) provide maximum flexibility for subsequent strategies of final geologic disposal.

Y. SOUSSELIER, CEA,Fontenay-aux-Roses, France

1. The present situation is paradoxical. Many processes are already implemented on an industrical scale and repositories for disposal are almost already realized, yet there is a simultaneous increase in the programs of research and development.
2. This situation arises in part because of the early development of waste management without a precise knowledge of the waste objectives. There are a large number of parameters, creating an ambitious and large program, with the necessity for following the iterative process.
3. Existing techniques allow for an acceptable solution.
4. An objective should be defined in terms of the ICRP principles. This requires a better knowledge of the present situation, which is possible only by a scientific approach.
5. There must be a recheck of the results already obtained.
6. A scientific approach is needed to (a) predict the waste package performance, (b) to convince the public, and (c) to justify our choices.
7. There must be international cooperation and final judgment of acceptability on an international level.

Th. H. PIGFORD, University of California, Berkeley, Ca., USA

In addition to research priorities mentioned by some of my colleagues, I would like to add a selected few:
1. There needs to be a critical evaluation of the applicability of data from laboratory leaching experiments to the prediction of performance of a waste form in the repository. There is no clear meaning to the concept of some finite volume of water in a repository to be associated with the dissolving

waste form, such that the resulting surface-to-volume ratio could be used to select laboratory experiments carried out at the same ratio.

2. In current laboratory leaching experiments, the concentration profiles of dissolved constituents in the liquid adjacent to the waste sample, whether static or flowing experiments, are not the same as will occur in a repository. For constituents that are solubility limited in their dissolution rate, the resulting dissolution rates in laboratory experiments will not be the same as in the repository. Experiments need to be designed to verify the predictions of solubility-limited dissolution rates under repository conditions, wherein the dissolution rate is controlled by the concentration boundary layer adjacent to the dissolving waste. The time required to establish these steady-state concentration profiles should be considered in designing the experiments, and should also be considered in predicting repository performance.

3. Attention should be given to the prediction of the dissolution rate, under repository conditions, of waste constituents such as cesium that are not solubility limited and which may not necessarily undergo congruent dissolution.

4. The temperature effects of the dissolution rate of individual waste constituents under repository conditions needs further study.

DISCUSSION

Pigford: In opening the discussion, I would like to pose a question to my fellow panelists.

In the U. S. there is a research need to establish a clear and firm overall performance goal for a geologic repository. Whether such a performance goal or safety standard is in the form of some maximum dose rate to the maximally exposed individual, as normally used in licensing nuclear fuel cycle facilities, or in the form of an integrated population dose, as now under consideration by the U. S. Environmental Protection Agency, the form of the performance goal and the magnitude of the specified maximum dose or dose rate have a strong impact upon the research needs. Some waste forms and waste packages are quite adequate for one performance goal, but are inadequate for another. When statements have been made that some waste form is or is not adequate, it has meaning only if the criterion for adequacy is specified. In the U. S. the lack of an overall performance goal has been a major problem, it is difficult to define necessary research priorities without having such a performance goal.

Burkholder: It is most logical to define the overall performance goal in terms of some fraction of natural background radiation. It is such a definition that

894

has been adopted in the studies of overall performance, reported in ONWI-286.

Feates: Plutonium has such a high retention in the geologic isolation system that it makes little difference which goal is adopted.

Gordon: The estimated release from a repository results in doses of the magnitude of background radiation or less.

Pigford: The U. S. Nuclear Regulatory Commission's proposed 10 CFR 60 regulations for geologic repositories purport to implement the yet undefined radiation standard of the U. S. Environmental Protection Agency for geologic repositories. From public presentations by representatives of the EPA, we know that the EPA thinking is to establish a limit on the integrated radioactivity release, calculated over a specified time of 10,000 years. For some proposed repository locations, the equivalent individual dose rate from such a limit will be a very small fraction of background radiation, far less than has been assumed as a reference goal in the published studies of the long-term release from conceptual repositories. The technology that satisfies one possible overall radiation standard, does not necessarily satisfy the other.Research objectives must be framed in terms of accomplishing some technology to meet a specified performance goal, and that is now missing.

Eisenbud (U.S.): An important research need is to learn more about radionuclide release and transport from the natural analogues that are available, such as the transport of thorium and its decay daughters, as well as neodymium, at the Morro do Ferro site in Brazil, and the transport of fission products and actinides at the OKLO site in Africa. Data from these locations show that natural barriers to the transport of actinides are insurmountable. There should be more effort to study other natural analogues and to use these results in verfying the expected performance of repositories.

Questioner: : It is important to consider the transuranic wastes because of the long-term effect of their actinides. Is there need for research on the disposal of fuel-cladding hulls as waste from fuel reprocessing?

Burkholder: It is important to address these transuranic wastes.

Pigford: Before asking for comment from other panel members, please consider also other wastes that may be destined for the geologic repository, such as carbon-14 and iodine-129 that may be recovered separately in fuel reprocessing.

Sousselier: There is a need for better knowledge of nuclides in the fuel cladding. Some reports indicate low actinide condent.

Ringwood: The SYNROC process is good for handling fuel cladding.

Papp: We do not have adequate information on what the transuranic wastes contain.

Merz: We must consider the costs of technologies, particularly the costs of those proposed by Ringwood.

Feates: The cladding hulls must be decontaminated, if at all possible.

Kühn (Germany): What are the specific scenarios considered by Mr. Feates?

Feates: Storing the wastes above ground for several decades is an important first step to a technical solution.

Baetsle (Belgium): The Belgium authorities do not accept scenarios of waste disposal if there is no technical proof of the scenarios. With regard to the cladding hulls, Belgium experience indicates that hulls may contain as much as 5% of the actinides in the high-level waste. Belgium and Switzerland have sea disposal for low-level wastes, but they cannot accept sea disposal for intermediate level wastes or of high-level waste, so the issue of cladding hulls is important.

Campbell (U.S.): Is there a need to reduce costs?

Sousselier: Yes, if we look at the overall waste treatment system, one should not only try to reduce the amount of radioactive waste but also the potential radiation doses to personl and future generations as well as cost.

Merz: It is important to consider costs.

Pigford: It is not clear to what extent costs are actually being considered. In program decisions for example, in the U. S., the assumption that there should be a corrosion-resistant overpack, i.e., an additional metallic envelope around the waste form, evidently adds about \$20,000 to \$40,000 per canister, yet the functional requirement and technical justification for the overpack is far from clear.

Feates: The first priority should be given to public acceptance, and then to cost.

Hamstra (Netherlands): The entire system and its performance requirements must be considered. If the purpose for this long-lived canister is to aid retrievability, then one should question whether retrievability is desirable or necessary.

Burkholder: Retrievability has a negative effect on long-term disposal and is more expensive.

Pigford: Evidently the overpack is thougth to be needed for more than retrievability. The proposed regulations of the U. S. N. R. C. would require retreivability capability for as long as 140 years for high-level waste packages. However, the corrosion resistant overpack is now thought to require a 1000-year life. Its purpose evidently is to avoid what are said to be uncertainties in predicting the performance of a repository during the "thermal period". The uncertainties are thought to arise from the effect of

896

temperature on the leach behavior of borosilicate glass, or of alternate waste
forms, and from the affect of temperature on the velocity of groundwater in the
repository, which can further affect the dissolution rate of radionuclides. We
know from other calculations that waiting 1,000 years before exposing the waste
to groundwater accomplishes little in reducing the later radiation doses from
groundwater transport of radionuclides to the biosphere, unless it avoids
appreciably higher concentrations of dissolved radionuclides in water during
the thermal period. The extent to which greater radionuclide concentrations in
the biosphere could result if waste were exposed to groundwater during the
thermal period should be determined by experiment and analysis. This has not
yet been done also, the "thermal period" is not sufficiently defined. Depending
upon the temperature level of concern, depending upon the concern about thermally
induced flow, and depending upon whether the waste is unreprocessed spent fuel
of high-level waste form fuel reprocessing, the thermal period could be found
to extend from as little as a few hundred years to many thousands of years.
Finally, such a proposed "technical fix" as the corrosion-resistant overpack
does not avoid the question of waste-package performance and waste-form
performance during the thermal period, because some fraction, however, small,
of the waste packages are likely to be faulty and will still expose the contained
radioactive solid to leaching and radionuclide transport during the thermal
period. Therefore, analyzing the waste-package performance during the thermal
period is important and must be done.

Northrup (U.S.A.): Is there some complete system study, beginning with the
fabrication of the waste solid, that looks at cost and overall performance?

Gordon: Yes, this has been done as part of the environmental review. But,
this is an area that requires further work.

Papp: In Sweden it is a requirement that a cost analysis be provided to the
government each year. The first will be delivered in June this year.

Feates: There are a number of system analyses of exposure and costs.

Pigford: Radiation effects is a research area that many of us had thought
had been put to rest years ago. What are the new research needs and why?

Merz: We must have additional investigation to come to a final conclusion.

Neretnieks (Sweden): Alpha radiolysis does have an effect on the liquid
adjacent to the waste package, particulary in that radiolysis can change
conditions from reducing to oxidizing, depending upon the possibility of escape
of hydrogen, leaving behind peroxide in the solution. This can have an important
effect upon the solubility of critical radionuclides, such as plutonium and
technetium, and it can affect their long-term release to the biosphere. The

radiation effects on the waste form solid are not so important.

Pigford: Even without regard to radiolysis, the slow transition from oxidizing to reducing conditions during the reflooding period after repository closure, and the effect upon solubilities and long-term radionuclide transport needs further study.

Macedo (U. S.): There are several important questions concerning the possible effects of radiolysis upon the chemical environment adjacent to the waste package and the effect on dissolution rates. Sea bed may be considered, because of the low temperatures of waste packages stored therein.

Merz: The geological barriers can save the waste forms.

Ringwood: There is concern about the stress-corrosion leaching of glass caused by the crystalline phases that will be present.

Author Index

Subject Index

A

B

C

D

E

F

G

H

904

I

In situ experiment, 135, 153, 477,
  519, 529, 579, 597, 697, 849
In situ solidification, 867
INTRACOIN, 839
Iodine, 509, 569, 641, 697, 783
Ion implantation, 339, 349, 357,
  379
Isotopic fractionation, 399

L

Leachability, 1, 15, 25, 37, 45, 57,
  71, 83, 93, 103, 113, 125, 135,
  145, 153, 163, 173, 181, 203, 211,
  219, 273, 319, 329, 339, 349, 357,
  389, 399, 559, 725, 849
  compositional effects, 1, 273
  pH-effects, 83, 93, 113, 125, 181,
    203, 273
  pressure effects, 163, 173
  radiation effects, 163
  SA/V-effects, 15, 83
  saturation effects, 93, 203
  selective ion leaching, 83, 103,
    113, 145, 181
  temperature dependence, 173
  time dependence, 173
Leaching mechanism, 15, 25, 45, 57,
  93, 103, 173
Low-level waste, 849, 859, 867, 873,
  879

M

Medium-level waste, 849, 859, 867,
  873, 879
Metal corrosion, 669
Metamictization, 369, 379, 389, 409
Microbial activity, 745, 831
Microstructure, 309
Modeling, 15, 45, 57, 145, 497, 539,
  549, 559, 597, 611, 619, 679, 689,
  765, 791, 811, 821, 839, 859, 879
  temperature effect, 859
Molybdate phase, 279, 289
Monazite, 399, 725
Morro do Ferro, 725, 735
MORS, 419, 429

N

Natural analogues, 37, 329, 379, 389,
  399, 409, 725, 735
Natural deposits, 725, 735
Neptunium, 83, 113, 559, 641, 659,
  679, 689, 707, 755, 775, 801

O

Overpack, 135, 153
Oxidation state, 279, 775

P

Palagonite, 37
Permeability, 649
Perovskite, 299, 309, 329
Phase separation, 279, 289
Plutonium, 83, 103, 113, 339, 559,
  659, 679, 707, 745, 755, 775, 801
Processing, 289, 299, 867
Pyrochlore, 369

Q

Quality assurance, 289

R

Radiation effects, 369, 379, 389, 399,
  409
Radiolysis, 339, 467
Radionuclide migration, 529, 539, 549,
  559, 569, 587, 597, 611, 619, 659,
  679, 689, 697, 707, 715, 725, 735,
  745, 755, 765, 775, 783, 791, 801,
  811, 821, 831, 839
  thermal effects, 587, 597

S

Safety assessment, 611, 619
Safety study, 497
Salinity profile, 429
Salt brine, 25, 173, 181, 203, 449,
  487, 491, 509